ADVANCES IN CHEMICAL PHYSICS

VOLUME XLII

ADVANCES IN CHEMICAL PHYSICS—VOLUME XLII

I. Prigogine and Stuart A. Rice—Editors

POTENTIAL ENERGY SURFACES

Edited by

K. P. LAWLEY

Department of Chemistry
University of Edinburgh

AN INTERSCIENCE® PUBLICATION

JOHN WILEY & SONS

CHICHESTER · NEW YORK · BRISBANE · TORONTO

British Library Cataloguing in Publication Data:

Advances in chemical physics.
Vol. 42: Potential energy surfaces
1. Chemistry, Physical and theoretical
I. Prigogine, Ilya II. Rice, Stuart Alan
III. Lawley, Kenneth Patrick IV. Potential energy surfaces
541′.3 QD453.2

ISBN 0 471 27633 2

Photosetting by Thomson Press (India) Limited, New Delhi
Printed in Great Britain at The Pitman Press, Bath, Avon.

INTRODUCTION

Few of us can any longer keep up with the flood of scientific literature, even in specialized subfields. Any attempt to do more, and be broadly educated with respect to a large domain of science, has the appearance of tilting at windmills. Yet the synthesis of ideas drawn from different subjects into new, powerful, general concepts is as valuable as ever, and the desire to remain educated persists in all scientists. This series, *Advances in Chemical Physics*, is devoted to helping the reader obtain general information about a wide variety of topics in chemical physics, which field we interpret very broadly. Our intent is to have experts present comprehensive analyses of subjects of interest and to encourage the expression of individual points of view. We hope that this approach to the presentation of an overview of a subject will both stimulate new research and serve as a personalized learning text for beginners in a field.

Ilya Prigogine

Stuart A. Rice

CONTRIBUTORS TO VOLUME XLII

C. Bottcher, Oak Ridge National Laboratory, Oak Ridge, Tennessee, USA

J. Scott Carley, Department of Chemistry, University of Waterloo, Waterloo Ontario, Canada

A. E. DePristo, Department of Chemistry, Princeton University, Princeton, New Jersey, USA

Dennis Diestler, Department of Chemistry, Purdue University, West Lafayette, Indiana, USA

Karl Freed, The James Franck Institute, University of Chicago, Chicago, Illinois, USA

K. Lacmann, Hahn Meitner Institut für Kernforschung, Berlin, West Germany

R. J. Le Roy, Department of Chemistry, University of Waterloo, Waterloo, Ontario, Canada

M. C. Lin, Chemistry Division, Naval Research Laboratories, Washington DC, USA

H. J. Loesch, Fakultät für Physik, Universität Bielefeld, Bielefeld, West Germany

Michael Menzinger, Lash Miller Chemical Laboratories, University of Toronto, Toronto, Canada

Herschel Rabitz, Department of Chemistry, Princeton University, Princeton, New Jersey, USA

John C. Tully, Bell Laboratories, Murray Hill, New Jersey, USA

CONTENTS

Potential Energy Surfaces
Edited by K. P. Lawley

ELECTRONIC CHEMILUMINESCENCE IN GASES

MICHAEL MENZINGER

Department of Chemistry, University of Toronto, Toronto, M5S 1A1, Canada

CONTENTS

I. INTRODUCTION

Electronic chemiluminescence (CL) in gases may arise through a variety of processes like recombination of atoms and radicals or its reverse, dissociation, and from chemical exchange (metathesis) reactions proper. I shall confine myself in this review to the discussion of exchange processes for two reasons: first, a great number of them have been discovered in recent years, primarily in the course of the quest for electronic transition laser systems, without having been comprehensively reviewed, and secondly, atomic and molecular recombination processes have already been adequately covered by previous reviewers of the CL field (Thrush, 1968; Carrington and Garvin, 1969; Carrington and Polanyi, 1972; Carrington, 1973; Golde and Thrush, 1975).

CL reactions are studied for a variety of reasons. As indicated, the possibilities of practical uses, such as chemical lasers in the visible (Wilson *et al.*, 1977) or efficient chemical light sources in general (Cormier *et. al.*, 1973), as well as analytical applications (Mendenhall, 1977), have been dominant sources of impetus. Most appealing to the fundamentalist are the facts that they provide the simplest examples of reactions governed by more than one potential energy surface PES, and that the high information content of the CL spectra can be studied with relative ease. Much has become known about the dynamics of single PES reactions since the mid-fifties, but our understanding of the much more complicated non-adiabatic reactions is still in its infancy. It is an interesting historical fact that the chemiluminescent alkali–halogen reactions provided Michael Polanyi and his school (Polanyi, 1932) with the first paradigms for formulating many of the ideas of modern reaction dynamics which would three decades later (i.e. $\geq$ 1956, the beginning of the 'modern alkali age') bear fruit in the development of modern reaction dynamics. A timely monograph is being published by Davidovits and McFadden (1978) devoted to alkali reactions, in which McFadden (1978) reviews also the CL aspects relevant in the present context.

The orientation towards applications has resulted in a rich inventory of new CL reactions which are summarized in Tables I–XI of the Appendix. A few of them have been extensively investigated but less can lay claim to reasonably well-understood kinetics and dynamics, and even fewer have found the helping attentions of theorists.

In their recent articles, Carrington (1973) and Golde and Thrush (1975) have taken a deductive approach towards classifying CL processes, based on a hierarchy of theoretical concepts and models. Proceeding from the simplest and best understood case, the recombination of atoms along a single potential energy curve to recombination *via* curve crossing (the inverse of predissociation) and the equivalent recombination processes in three atomic systems, they showed that exchange reactions in triatomic, and to a much higher degree in polyatomic systems, were the most complex in terms of theoretical description as well as experimental accessibility. While adiabatic atom recombination can be analysed experimentally and theoretically with nearly ultimate rigour, only qualitative and general arguments can be advanced in the interpretation of multi-PES exchange reactions.

The central concept in CL reactions are ground- and excited-state potential energy surfaces (PES's, the subject of the present volume), whose symmetry, and with it their mutual strength of interaction, may change with internuclear geometry as described by group theory. The topologies of PES intersections and avoided intersections, mostly in triatomic systems, have been reviewed by Carrington (1974). The dynamical problem governed by three-atom PES's may in simple cases be amenable to (nearly) rigorous computational solutions by trajectory surface hopping and other semiclassical techniques (Tully, 1976, 1977). Polyatomic systems, in contrast, are characterized by much higher vibronic state densities and, in general, by the absence of symmetry elements. Only in the presence of high barriers will symmetry considerations still be important since then the reaction will be confined to conformations of high symmetry, as Metiu *et al.* (1974) have shown. These circumstances cause the vibronic states to interact strongly over extended ranges of the multidimensional configuration space for which realistic and quantitative dynamical calculations become far too complex.

In contrast to the more recent of the above mentioned reviews, a more inductive point of view is taken here by organizing the chapters in terms of a hierarchy of questions that might arise in the course of typical CL studies. No attempts will be made to discuss a reaction exhaustively at any given point but different experimental facts of one system may be used to illustrate points of the systematic discussion. The first steps in characterizing a CL reaction, before questions of dynamic nature can be addressed, concern the chemiluminescence light yield, the spectroscopic emitter analysis, and the reaction mechanism, as well as molecular dissociation energies, an important by-product of the latter points. These are discussed in Section II. Section III deals with dynamic interpretations of measurements. The nuclear and electronic rearrangements occurring in the course of a reaction are loosely classified into a nuclear and an electronic problem. The former concerns mainly the energy partitioning among nuclear degrees of freedom, while the latter deals with the coupling mechanism between reactant and product electronic states. Depending on the representation, reactions may be symmetry allowed, dynamically induced, or caused by spin–orbit interaction. In all three cases one can theoretically and experimentally distinguish electronic from nuclear or Franck–Condon-like components of the coupling terms. The experiments discussed under these headings include energy consumption and disposal among nuclear and electronic degrees of freedom, angular momentum consumption and disposal, specifically anisotropy effects of the reactant and product rotational angular momentum distribution, and angular distributions of CL products. Section IIIC on theory is necessarily brief: methods for obtaining ground- and excited-state PES's on the one hand, and the formal dynamical theory of multi-PES reactions and a variety of semiclassical models on the other, have existed for some years, practical difficulties however have prevented them from being applied to the point of a cooperation of theory and experiment. Statistical and information theory analyses represent the majority of efforts. The epilogue closes this review with some personal views on desirable objectives for future CL research.

The tables in the Appendix contain a compilation of electronically chemiluminescent exchange reactions in the gas phase, primarily covering the last decade. The selection is personal and no claim of completeness is made. I apologize to those authors whose work was not included either by oversight or by practical limitations. The literature cited includes papers up to the summer of 1978 as well as papers made available to me in preprint form. I thank the authors of those references especially for their help.

II. A HIERARCHY OF PROBLEMS AND RESULTS

The aim of this chapter is to discuss, from an experimental point of view, a hierarchy of problems one might encounter in the course of a typical CL study, and to illustrate them where appropriate. Consider the flow diagram Fig. 1. Independently of the *experimental* apparatus, i.e. regardless of whether the CL phenomenon is observed as the bulk or in molecular beams, the natural first questions concern the *quantum yield* Φ, i.e. what fraction of reactive events produces CL emitters, and secondly the *identification and analysis of the emitter*. One cannot embark on the dynamic interpretation of the spectrum without having clarified the *reaction mechanism*. If the process is bimolecular and vibrational structure of the emission spectrum is well resolved, energy balance arguments can be used to obtain limits to the *dissociation energy* D_0^0 of product or reactant molecules. The product internal state distribution may then be obtained as the primary dynamic signature of the system. The *dynamical aspects* are considered in Section III.

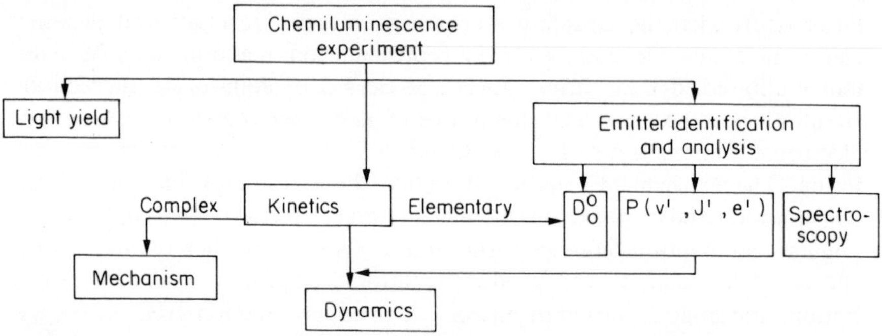

Fig. 1.

A. Experimental

CL experiments combine well-established *spectroscopic* techniques for detection, with *bulk-flow* (Polanyi, 1932; Felder and Fontijn, 1975, 1978; West *et al.*, 1975), *molecular beam* (Fluendy and Lawley, 1973; Manos and Parson, 1975, 1978; Redpath *et al.*, 1978), or *ion beam* techniques (Ottinger, 1976) for inducing the reaction and controlling the reaction parameters.

These methods, their applications, potentials, and limitations are sufficiently well known for this section to be brief and sketchy.

Our inventory of CL reactions has been dramatically expanded in recent years through the use of atomic and molecular reactants, neutral and charged, that are unstable under normal conditions. Radicals are obtainable by thermal dissociation or in discharges, and an arsenal of techniques common to high-temperature chemistry (Gole, 1976) is in use for evaporating even highly refractive metals. Traditional methods of resistance and electron bombardment heating have been supplemented by the shock-tube evaporation of metal powders (Johnson *et al.*, 1974) report evaporation of refractives such as B, C, Ti, W, etc.) and by laser-vaporization of metal films (Utterback *et al.*, 1976; Tang *et al.*, 1976; Wicke *et al.*, 1978). Exceptionally high metal vapour densities (typically $\sim 10^{15}$–10^{20} cm^{-3}) have been obtained in the heat pipe oven (Hessel *et al.*, 1975; Luria *et al.*, 1976a, b).

Reactants have been excited to metastable states by electron bombardment in a discharge (Brinkmann and Telle, 1977; Dagdigian, 1978; Kowalski and Heldt, 1978). Reactant state distributions have been varied thermally *via* the source temperature and by taking into consideration the internal state relaxation during the formation of supersonic beams (Redpath *et al.*, 1978; Wren and Menzinger, 1975, 1979a). Lasers were used to state select reactants (Estler and Zare, 1978a), and to diagnose products (Pruett and Torres-Filho, 1977). We conclude by mentioning the direct photographic (or image-intensifier and optical multichannel analyser) observation of the CL angular distribution in crossed beam experiments (Siegel and Schultz, 1978).

B. Light Yields Φ_{CL}

Foremost in the mind of the applications-oriented researcher (the laser builder and developer of chemical light sources) is the question of the CL quantum yield Φ_{CL}. It is defined as

$$\Phi_{CL} = \frac{r_{CL}}{r_T} \tag{1a}$$

the ratio of CL photon production rate to the rate of (reactive) consumption of reactants. In the special cases where both factors obey the same rate law equation (1a) reduces to

$$\Phi_{CL} = k_{CL}/k_T \tag{1b}$$

the ratio of the corresponding rate constants. This is necessarily the case at sufficiently low pressures (the 'single-collision regime') for both terms to be bimolecular. The CL yield is to be distinguished from $\Phi^* = r^*/r_T \simeq k^*/k_T$ the yield (or branching ratio) of electronically excited products regardless of whether they decay radiatively or not. In cases where 'reservoir states' are produced with high efficiency Φ^*, e.g. $Ba + N_2O$, $Sm + N_2O$, $Sm + F_2$, etc., the CL yield will be low under single-collision conditions but will increase substantially at higher pressures due to collisional intramolecular energy transfer to radiating states which entails a higher-than-second-order rate law for the rate of CL production r_{CL}.

Actual Φ_{CL} measurements may follow two methods based on equations (1a) and (1b) respectively. In *bulk-flow experiments* the rates of photon production (flux) and reactant consumption required in equation (1a) are measured directly. By working under pseudo-unimolecular conditions with one reactant in excess, the rate of reactant consumption is automatically given by the flux of the minor reactant, and the principal task remains the determination of absolute photon fluxes with a properly calibrated detection system (Black *et al.*, 1974; Edelstein *et al.*, 1974; Eckstrom *et al.*, 1974, 1975; Field *et al.*, 1974; Jones and Broida, 1973, 1974). A novel method for producing a calibrated light source has been reported by Lee *et al.* (1976) and Woolsey *et al.* (1977).

Equation (1b) forms the basis of quantum yield measurements in beam/gas experiments. Both factors in (1b) are obtainable from the dependence of the CL photon flux I_{CL} on target gas density n_{BC}:

$$I_{CL}(n_{BC}) = k_{CL}\, n_A^p\, n_{BC}^q \tag{2}$$

where p and q are the partial orders of the CL reaction. The density of beam particles in the plane of the detector obeys the (bimolecular) Lambert–Beer law

$$n_A(l) = n_A(0)\exp(-\,n_{BC}\, l\sigma_{att}) \tag{3}$$

Here l is the (effective) scattering path length and σ_{att} the attenuation cross section. The dependence $I_{CL}(n_{BC})$ at low reactant densities gives the reaction orders p, q, and the initial slope (or curvature, as the case may be) yields k_{CL} values. An upper limit of k_T is given by $k_T \geq k_{att} \simeq \bar{v}\sigma_{att}$, the rate constant for (reactive + non-reactive) attenuation of the (atomic) beam A, where $\bar{v}$ is the average relative collision velocity. According to equation (3), σ_{att} is obtainable either from $I_{CL}(n_{BC})$ at fixed l or from $I_{CL}(l)$ at fixed n_{BC} (Dickson *et al.*, 1977).

To circumvent tedious detector calibrations relative Φ_{CL}^{rel} values are frequently normalized by a standard reaction (Yokozeki and Menzinger, 1976). Dickson *et al.* (1977) propose the $Sm + N_2O$ reaction as a standard (see Fig. 2) for which they measure a single-collision CL yield of $\Phi_{CL} = 0.0039$. The energy dependence of $\Phi_{CL}(Sm + N_2O)$ has been reported by Yokozeki and Menzinger (1977a). An ingenious method yielding absolute Φ_{CL} of reactions of metastable metal atoms M* that eliminates detector calibrations has been reported by Dagdigian (1978). For the $Ca^*(^3P) + N_2O$ reaction it yields $\Phi_{CL} = 0.10 \pm 0.03$ and $\sigma_{CL} = 6.2 \pm 1.7\,\text{Å}^2$.

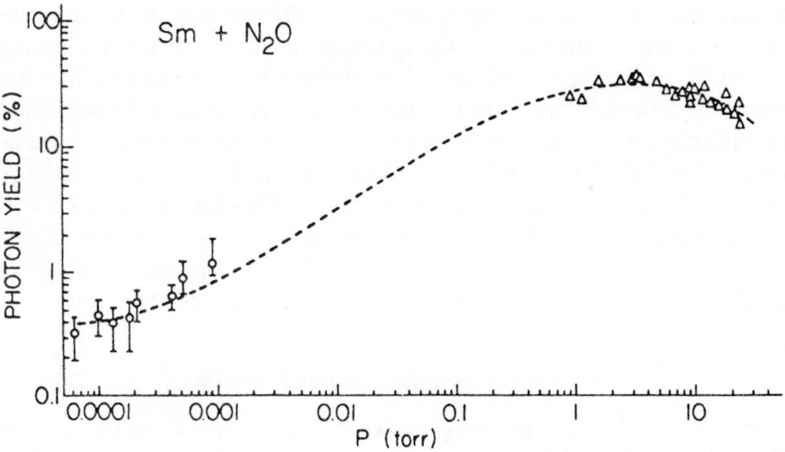

Fig. 2. Absolute photon yields as a function of pressure for the Sm + N_2O reaction. From Dickson *et al.* (1977). Reproduced by permission of American Institute of Physics.

The quantum yields of known CL reactions extend from the detection limit ($\Phi_{CL} \geq 10^{-8}$, e.g. for the ubiquitous autoxidation of organic material by the atmosphere (Mendenhall (1977) to essentially unity. The majority of the gas-phase processes studied so far have disappointingly low $\Phi_{CL} \lesssim 10^{-}2$. It is well to keep in mind that one deals in these cases with *minor reaction channels* (the proverbial 'tip of the iceberg'), which are of minor significance on a global scale. Nevertheless such processes inform us in a direct way about the system's excited PES's and they frequently display unusual kinetic and dynamic peculiarities.

Single-collision light yields of the order of unity are rare: a noteworthy exception is the reaction of alkali dimers with halogen atoms, e.g.

$$Na_2 + Cl \rightarrow NaCl + Na^{*,**} \tag{4}$$

(Struve *et al.*, 1975). In alkali-rich diffusion flames the overall quantum yield reaches $\Phi_{CL} \sim 0.4$ (Bogdandy and Polanyi, 1928). As the authors have pointed out, this requires a near unit quantum yield for the elementary processes which will be discussed later.

The reaction of group IVa atoms with oxygen donors, for instance $Sn(^3P) + N_2O \rightarrow SnO^* + N_2$ are likewise thought to produce predominantly excited triplet products $SnO(a^3\Sigma^+, b^3\Pi)$—based on adiabatic correlations (Donovan and Husain, 1970) and on experimental evidence (Wiesenfeld and Yuen, 1976; Felder and Fontijn, 1975, 1978).

The explanation for high CL quantum yields at thermal energies is usually found in the existence of adiabatic correlations to the emitter state, while diabatic transitions that are often required to reach the emitter state act as the dynamical bottlenecks responsible for low Φ^* and Φ_{CL}. A striking exception to this rule of thumb—the CL decomposition of tetramethyldioxetane into one S_0 ground state and one T_1 triplet acetone (Wilson, 1976)—has a plausible dynamical explanation which will be discussed in Section III.B.3. As mentioned earlier, consecutive reaction mechanisms may entail drastically pressure-dependent Φ_{CL}. We merely point here to Fig. 2 and to the Sm + $F_2 \rightarrow SmF^*$ reaction (Eckstrom *et al.*, 1975; Dickson *et al.*, 1977). The reactions between metal atoms (Ba, Sm, Al) and NF_3 (Eckstrom *et al.*,1975; Rosenwaks, 1976; Rosenwaks and Broida, 1976) are noteworthy for their high Φ_{CL} through a multistep mechanism involving metastable N_2^*.

C. Emitter Identification and Analysis

The celebrated state resolving power of spectroscopic methods *ideally* endows CL studies with a maximum of product state information. This is represented by detailed rate constants k_{fi} or cross sections σ_{fi} for forming (e.g. diatomic) products in individual $|v', J', e'\rangle$ states and distributions over m_J, states, all of which we collectively label $|f\rangle$. The notation i,f stands for reactant and product states, primes and double primes for upper and lower states of radiative transitions.

The relation

$$I_{f'' f'} = A_{f'' f'} N_{f'} \tag{5}$$

between the observed intensity $I_{f'' f'}$ of an (ideally rotationally resolved) spectral line and the steady-state emitter concentration $N_{f'}$ and spectral transition probability $A_{f'' f'}$ has been documented in detail by Nicholls and Stewart (1962), Albritton *et al.* (1980), and Manos (1976). This is done by writing the Einstein A-factor as

$$A_{f'' f'} = \alpha g_{f'}^{-1} \nu_{f'' f'}^3 R_e^2(\bar{R}_{v'' v'}) q_{v'' v'} S(J'\Lambda', J''\Lambda'') \tag{6}$$

where $\alpha = 64\pi^4/3h$ and $g_{f'} = 2J' + 1$ is the degeneracy of the emitter state. $\nu_{f'' f'}$ is the frequency of the spectral transition, $R_e^2(\bar{R}_{v'' v'})$ the electronic transition moment evaluated at the R-centroid $\bar{R}_{v'' v'}$, $q_{v'' v'}$ the Franck–Condon factor, and $S(J'\Lambda', J''\Lambda'')$ is the (analytically given) Hönl–London factor (Herzberg, 1950) for the $J'\Lambda' \rightarrow J''\Lambda''$ transition where Λ is the projection of the electronic angular momentum on the internuclear axis. Franck–Condon factors are calculated using standard computer programs (Zare, 1963,1964; Jarmain and McCallum, 1970; Albritton *et al.*, 1980) from the potential curves $V(R)$ derived from spectroscopic constants or from numerical RKR potentials. The necessity of extrapolating $V(R)$ to high quantum numbers beyond the range of validity of the input data may introduce substantial errors. Electronic transition moments are normally not known.

The possible errors associated with the usual assumption of an $\bar{R}_{v''v'}$-independent R_e ought to be kept in mind.

The next step in relating an observed spectrum or equivalently the steady-state emitter population $N_{f'}$, to the desired microscopic rate constant k_{fi} is to consider the kinetics of emitter formation and decay, including all competing loss channels. The latter include radiative and non-radiative intramolecular decay, drift out of the viewing range of the detector during the radiative lifetime, and collisional relaxation and quenching. This yields a relation between phenomenological and microscopic rate constants for CL radiation and an interpretation of the steady-state emitter concentration $N_{f'}$ in terms of the rate constants of all contributing processes (Carrington, 1961; Polanyi and Woodall, 1972; Redpath *et al.*, 1978). The simplest case corresponds to a short-lived emitter (say $\tau_R < 10^{-6}$s) being formed at low enough pressures for single-collision conditions to prevail.

So far the CL emission has been assumed to correspond to well-known electronic band systems. Frequently however, previously unknown electronic states may be populated, or known electronic states may be formed with higher vibrotational excitation than under conventional excitation methods. Such spectra may represent new sources of spectroscopic information (Oldenburg *et al.*, 1975; Dickson and Zare, 1978) or frustration, as the case may be.

D. Dissociation Energies

The energy balance of bimolecular CL reactions may provide convenient estimates of molecular dissociation energies (Zare, 1974; Gole, 1976). The method applies also to laser-induced fluorescence and complements D_0^0(AB) determinations based on the study of chemical equilibria by mass sepctrometric (Drowart and Goldfinger, 1967) and flame photometric methods (Gaydon, 1968). For a reactive event of the $(A + BC)_i \rightarrow (AB^* + C)_f$ type, energy conservation requires

$$E_T + \sum_i E_i + D_0^0(\mathrm{AB}) - D_0^0(\mathrm{BC}) = E'_T + \sum_f E'_f \tag{7}$$

In accord with common usage in this context, primed quantities denote products, unprimed ones reactants. The available energy, i.e. the sum of reactant translational and internal energies and the energy released, given by the difference of the product and reactant bond energies, equals the product internal plus translational excitation. The D_0^0 value of either AB or BC could be determined exactly from (7) if all other factors were known. *In praxi* this is not the case and D_0^0 estimates can be obtained, based on a number of *approximations and caveats* which, although obvious, have in a number of cases bedevilled the application of this method: 1. The reaction has to be bimolecular, a condition that is by no means guaranteed even in beam-gas (Jonah and Zare, 1971; Gole *et al.*, 1977) experiments. 2. Since the product translation E'_T is generally unknown, this term is neglected and the highest product quantum state $(\sum_f E'_f)_{max}$ observable in the CL spectrum is used to derive an obvious inequality from (7) which provides lower and upper limit estimates for D_0^0(product) and D_0^0(reactant), respectively. The CL spectrum ought to be analysable for the highest quantum state, or else additional assumptions must be made. The quality of this approximation depends clearly on the *a priori* unknown dynamics: if in a detectable fraction of emitters the *total* available energy is deposited as internal energy then the equality (7) is maintained in (9). In general, (9) gives only lower- (and upper-) limit estimates for D_0^0(AB) (or for D_0^0(BC)). 3. Matters are complicated since one is usually faced with a transvibronic *distribution* of reactants, and since the energy consumption, i.e. the distribution of states that actually do react, is frequently unknown. The consequence of translational and/or internal energy barriers to reaction has been pointed out by Thrush (1973), Preuss and Gole (1977a, b, c). One replaces the sharp values in (7) by the average translational energy $\langle E_{T,I}\rangle^*$ at which state $|i\rangle$ reacts and by the average internal energy $\langle \sum_i E_i\rangle^*$ at which reaction *actually* occurs, and noting that these are related to $\langle E_T\rangle$ and $\langle \sum_i E_i\rangle$, the unstarred averaged energies of all (i.e. reactive and non-reactive) collisions by (Menzinger and Wolfgang, 1969; Redpath *et al.*, 1978; Preuss and Gole, 1977a,b):

$$\langle E_{T,i}\rangle^* + \left\langle \sum_i E_i \right\rangle^* = \langle E_T\rangle + \left\langle \sum_i E_i \right\rangle + E_a \tag{8}$$

where E_a is the (Arrhenius) activation energy, one finally obtains the desired inequality

$$D_0^0(\mathrm{AB}) \geq D_0^0(\mathrm{BC}) + \left(\sum_f E'_f\right)_{max} - \left\langle \sum_i E_i \right\rangle - \langle E_T\rangle - E_a \tag{9}$$

for D_0^0.

Applications of the method have been reviewed by Gole (1976), and I refer to that work for a data compilation. Discrepancies between CL and mass spectrometric D_0^0(AB) values (the latter lie usually lower) could in some cases be resolved by establishing the *involvement of metastable reactant states.* A good example is an ingenious experiment, reported by Estler and Zare (1978b), in which a pulsed metastable Ba beam was used to record time-resolved TOF-CL spectra in order to observe directly the E_T^*-distribution of reactive events, and to demonstrate the previously unrecognized involvement of $Ba^*(^3P)$ metastables. The earlier value of $D_0^0(\mathrm{BaI}) \geq 102 \pm 1$ kcal/mol (Dickson *et al.*, 1976) is now corrected to $D_0^0(\mathrm{BaI}) \geq 72.9 \pm 2$ kcal/mol, which agrees well with mass spectrometric values (Hildenbrand, 1977, 1978). Dynamic interpretations of many experiments have likewise suffered from unrecognized metastable contaminations, although clues about their involvement have been recognized in some studies (Dubois and Gole, 1977; Gole and Chalek, 1976; Dickson and Zare, 1978; Rosenwaks and Broida, 1976). Small concentrations of metastables reacting with large cross sections (Brinkmann and Telle, 1977; Kowalski and Heldt, 1978; Pasternack and Dagdigian, 1978) may cause the small phenomenological σ_{CL} of the overall system, which are traditionally interpreted as arising from ground-state reactants.

E. Reaction Mechanisms

Since radiative lifetimes for allowed electronic transitions lie in the $\tau_R \sim 10^{-9}$–10^{-6} s range one expected originally that secondary collisions would not compete noticeably with radiation from a newly formed CL emitter at pressures below, say 10^{-2}–10^{-1} Torr. While this is usually the case for short-lived species, metastable intermediates have been found to play a crucial role in many systems. The additional fact that their internal excitation may endow them with much higher cross sections for secondary collisions than those which one is used to expect from their ground-state counterparts, is the reason for higher-than-second-order kinetics that are occasionally observed even in beam experiments at pressures well below 10^{-3} Torr, which were traditionally believed to define the single-collision regime (Wren and Menzinger, 1974; Yokozeki and Menzinger, 1976; Dickson *et al.*, 1977; Brown and Menzinger, 1978; Kusunoki and Ottinger, 1978a). Bulk-flow experiments—typically in the 10^{-2}–10 Torr range—are frequently affected by secondary processes, and dynamic interpretations of experiments of this type have to be taken with great care.

A great variety of kinetic schemes is possible, depending on the *emitter formation* via metathesis, energy transfer or recombination/dissociation, and on the *first-order loss rates* (radiation, intramolecular radiationless processes, diffusive loss, etc.) and *collisional rates* (quenching, energy transfer, etc.). Typical mechanisms have been considered by Gole *et al.* (1977). The three examples presented below are meant to illustrate some of these aspects.

1. *The Alkali–Halogen Systems*

These deserve mention not only for their venerable history but also for the central role which reaction mechanisms play in the overall CL production. The discussion is kept brief since McFadden (1978) and the book edited by Davidovits and McFadden (1978) review this subject in depth. In skeletal form, the mechanism of CL production from the reagents $(M, M_2) + X_2$, present at thermal equilibrium, is

$M + X_2 \rightarrow MX + X$	primary reaction	(10)
$X + M_2 \rightarrow MX + M^{*,**}$	direct electronic excitation	(11)
$\rightarrow MX^{\ddagger} + M$	vibrational energy release	(12)
$MX^{\ddagger} + M' \rightarrow MX + M'^{*}$	indirect (non-reactive) excitation	(13)
$\rightarrow M'X + M^{*}$	indirect (reactive) excitation	(14)
$M_2 + X_2 \rightarrow MX + M^{*} + X$	atomic excitation (4 centre)	(15)
$\rightarrow MX^{*} + MX$	molecular excitation (4 centre)	
$M + X + S \rightarrow MX^{*} + S$	molecular excitation (atom recombination)	(16)

From their well-reviewed (Polanyi, 1932; Laidler, 1955; Kondratiev, 1964; McFadden, 1978) 'highly dilute flame studies'—particularly from the fact that N_2 buffer gas quenched the $Na^{*}(^2P)$ chemiluminescence faster than it quenched resonance fluorescence—Polanyi and coworkers were led to infer that the 'indirect excitation' pathway (10) + (12) + (13,14) involving the fast $V \rightarrow E$ transfer (13, 14) dominated over the direct mechanism (10) + (11) (Beutler and Polanyi, 1928; Polanyi and Schay, 1928). This requires predominantly vibrational energy release in (10) and it is the earliest evidence for such dynamically biased energy partitioning. Moulton and Hershbach (1966) substantiated the indirect mechanism by a triple-beam experiment, and showed that reactive $V \rightarrow E$ transfer (14) was substantially ($\sim 100\times$) faster than the non-reactive mode (13). On a statistical basis one would have expected equal efficiency for both channels. Hershbach sees in this dynamical bias evidence for the importance of electron transfer in reactive electronic excitation. The complementary $T \rightarrow E$ transfer was found to be much slower than $V \rightarrow E$ transfer (13 + 14) by Neoh and Hershbach (1975). Based on his qualitative PES's obtained from correlation considerations, Magee (1940) criticized the above-mentioned conclusion of the Polanyi school, and showed that the direct mechanism should even dominate the indirect one. More recent semiempirical M_2X surfaces of Struve (1973) confirm some of Magee's early work. The crossed $M_2 + X$ beam experiments of Struve *et al.* (1971, 1975) agree beautifully with this early inference by giving $Na_2 + Cl \rightarrow Na^{*}$ cross sections of the order of 10–100 Å^2. No self-consistent kinetic analysis has been given to date which would reconcile the early quenching results with the later evidence for the importance of both mechanisms (10)–(15) in flame emission.

The early observation of highly excited alkali states M^{**} (Beutler *et al.*, 1926) was originally also ascribed to vibrational-to-electronic energy transfer (Schay, 1931) but again the beam experiments of Struve *et al.* (1975) exposed the direct excitation (11) as an important competitor, particularly since the (statistical) M** distributions from flow tube and beam experiments were in essential agreement.

The molecular MX* emission observed by Beutler *et al.* (1926), and later analysed and interpreted by Levi (1934) in terms of two- and three-body atom recombination (16), was eventually found to occur by a four-centre exchange step (15) (Oldenburg *et al.*, 1974; Kaufmann *et al.*, 1974a, b; Struve *et al.*, 1975). The latter authors have also discussed a mechanism involving the rearrangement of a $M–M^+X^-–X$ complex in which the central M^+X^- plays the role of the chaperone in the recombination of the end-standing atoms to yield electronically excited MX*.

2. *The Group IIa Metal–Halogen Reactions*

These reactions proceed by a number of competing channels:

$$M + X_2 \rightarrow MX^+ + X^- \quad \text{chemiionization} \quad (17a)$$

$$\rightarrow MX^*(X, A, B, C \ldots) + X^- \quad \text{monohalide formation} \quad (17b)$$

$$\rightarrow MX_2^* \quad \text{dihalide formation} \quad (17c)$$

With M = Ca, Sr, Ba; X = F, Cl, Br, I, this family of reactions differs from the homologous alkali systems primarily through the presence of an extra valence electron. In MX monohalides it occupies a non-bonding MO outside an alkali-halide-like core centered on the M^+ ion. Its excitation gives rise to the alkali-like Rydberg series of low-lying product states $A^2\Pi$, $B^2\Sigma^+$, $C^2\Pi$, etc., and to the low MX ionization potential, slightly less than that of the free metal atom M. This becomes manifest through the energetically accessible bimolecular CL and CI (chemiionization) channels (17b) and (17a) (Jonah and Zare, 1971; Menzinger, 1974a, 1974b; Diebold *et al.*, 1977; Wren and Menzinger, 1979b).

The dihalides MX_2 also possess low-lying excited states. They may be visualized as originating from charge transfer from the doubly ionic ground state $X^-M^{++}X^-$ to the singly ionic excited states $X^{\delta-}M^+X^{\delta-}$. In MO language the lowest energy MX_2^* excitations correspond to transitions from the highest Π_g (b_2, a_2) orbitals to the lowest unoccupied (antibonding) $\bar{\sigma}_g$ (a_1) orbital (Hayes, 1966; Gole, 1973; Gole *et al.*, 1973; Yarkony *et al.*, 1973). Although MX_2^* can principally be formed *via* radiative two-body recombination (known to be rather ineffective), most of the MX_2^* emission which dominates the CL spectra (except for X = F) at pressures round 10^{-3} Torr (Jonah and Zare, 1971; Menzinger, 1974a) obeys a third-order rate law, surprisingly even at pressures as low as $\sim 10^{-5}$ Torr (Wren and Menzinger, 1974, 1979c). This was originally explained by radiative three-body recombination involving the rapid stabilization of a long-lived $MX_2^{*\ddagger}$ collision complex. A more recent finding (Wren and Menzinger, 1979c), namely the fact that the $BaCl_2^*$ intensity proves independent of N_2 pressure when N_2 is substituted as a stabilizing gas for Cl_2 in a beam/mixed gas ($Cl_2 + N_2$) experiment, is evidence against this route. A more likely mechanism involves the fast 'avoided harpooning' reaction of vibrationally highly excited ground state $MX^\ddagger$ ($X^2\Sigma^+, v \gg 0$):

$$M + X_2 \rightarrow MX^\ddagger + X \tag{18}$$

$$MX^\ddagger + X_2 \rightarrow MX_2^* + X \tag{19}$$

Both mechanisms are consistent with the angular CL distribution, observed by Mims and Brophy (1977) to resemble *roughly* the M–X_2 centroid distribution.

Even the MX* channels, which were labelled bimolecular a moment ago when we had well-controlled beam experiments in mind, may develop mechanistic complexities at higher pressures (Eckstrom *et al.*, 1975; Yokozeki and Menzinger, 1976), or when beam experiments suffer from a contamination by M* metastables. For instance the BaCl* ($C^2\Pi$) state has been observed in beam/gas (Jonah and Zare, 1971) as well as in flow tube experiments (Menzinger and Wren, 1973; Bradford *et al.*, 1975), and thermochemical as well as dynamical interpretations have been attempted, although it is now recognized that for energetic reasons the $C^2\Pi$ cannot arise from ground-state reactants as written in (17b) but it must involve high-energy reactants such as Ba* metastables. Brinkmann and Telle (1977) and Kowalski and Heldt (1978) used metastable $M^*(^3P)$ beams to show that their (symmetry-allowed) CL reaction with Cl_2 and HCl have near gas kinetic cross sections and that they produce MCl* states that are energetically precluded for thermalized ground-state reactants. This illustrates once again how minor contaminations by highly reactive species may confuse the issues by mimicking a slow reaction of the dominant component.

The simple electronic structure and the multiplicity of reaction paths (17a–c) which can be studied with relative ease make this reaction family the *prototype CL systems* which share many characteristics with others of greater complexity. Calculations of ground and excited-state PES's are well within present-day capabilities, as the initial. CaF_2 study of Yarkony *et al.* (1973) and the paper by Tully in this volume show. It appears highly desirable to pursue this matter since the prospect of narrowing the gap between experiment and theory is most promising in the present case.

3. *The* $Ba + N_2O \rightarrow BaO^* + N_2$ *Reaction*

This reaction is the best-studied member of a family of group IIa and IIIa metal atom reactions that have attracted attention since Jones and Broida

(1973, 1974) discovered the dramatic increase of the photon yield $\Phi_{CL}(p)$ with pressure of an inert buffer gas—the principal indication for the involvement of non-radiating '*reservoir states*'. The $\Phi(p)$ dependence is very similar to that of the $Sm + N_2O$ system, shown in Fig. 2 (Eckstrom *et al.*, 1975; Dickson *et al.*, 1977). Other systems with similar kinetic signatures have been reported (Eckstrom *et al.*, 1974, 1975, 1977; Hsu *et al.*, 1974a; Palmer *et al.*, 1975). In his interesting 'case study', Field (1976) has concerned himself primarily with the kinetic and spectroscopic problems arising from the presence of long-lived precursors. He and Jones and Broida (1974) have elaborated on the pronounced spectral changes accompanying the increase of Φ_{CL} with pressure: the highly structured, many-line low-pressure spectrum which has so far defied a complete analysis, changes into the well-known $(A^1\Sigma^+ - X^1\Sigma^+)$ band system as the buffer gas pressure is raised to $\sim 0.1.–1$ Torr. Vibrational and rotational population anomalies in the $(A–X)$ spectrum below 1 Torr coincide precisely with those rotational levels of the $A^1\Sigma^+$ state which are most strongly perturbed by the nearly $a^3\Pi$ state (Field, 1974). This establishes the $a^3\Pi$ state as a precursor, but cannot exclude possible precursors to $a^3\Pi$ (Field *et al.*, 1974). The $a^3\Pi$ and $A'\ ^1\Pi$ states have virtually the same spectroscopic constants and are strongly mixed by spin–orbit coupling (Field, 1974). For this reason any discussion of $a^3\Pi$ as a precursor to $A^1\Sigma^+$ must also imply the same role for $A'\ ^1\Pi$ (Hsu *et al.*, 1974b). Kinetic modelling which includes as the crucial step the collisional transfer from a reservoir to the $A^1\Sigma^+$ emitter reproduces the measured $\Phi_{CL}(p)$ very well but cannot distinguish between possible precursors (Pruett and Torres-Filho, 1977). Whatever their nature, they must be formed at yields approximating unity to account for the maximum CL yield $\Phi_{CL}^{max} \sim 0.35$ at $p(Ar) \sim 10$ Torr (Jones and Broida, 1973, 1974). Numerous kinetically equivalent pathways can arise from the interaction of at least five states of BaO (high v states of $x^1\Sigma^+$; $^3\Sigma^+$, $a^3\Pi$, $A'\ ^1\Pi$ and $A^1\Sigma^+$) since it is known that intramolecular energy transfer between mutually perturbing states is fast and may proceed at a rate exceeding that of rotational relaxation (Radford and Broida, 1963; Pratt and Broida, 1969; Gelbart and Freed, 1973). Much remains to be learnt about these aspects.

Wicke *et al.* (1975) have considered the possible involvement of metastable $Ba^*(^3D)$ in the $Ba + N_2O$ system but found it insignificant under normal conditions. The suggestion had been made that Ba^* would provide a symmetry-allowed pathway towards $a^3\Pi$ precursors. This would support the earlier view that the oxygen atom in N_2O behaves much like $O(^3P)$ (Field *et al.*, 1974; Tully, 1974a). An alternative pathway towards $a^3\Pi$, proceeding through an ion-pair complex will be discussed in section III.B.2 in connection with the excitation function measurements. Very efficient production of metastable $M^*(^3P)$ occurs in ternary $((Mg, Ca) + N_2O + CO)$ flames (Benard *et al.*, 1976, 1977a, b; Benard and Slafer, 1977; Eckstrom *et al.*,

1977) by the following mechanisms:

$$Mg(^1S) + N_2O \rightarrow MgO^{\ddagger} + N_2 \tag{20}$$

$$MgO^{\ddagger} + CO \rightarrow Mg^*(^3P) + CO_2 \tag{21}$$

Whether $MgO^{\ddagger}$ is in the ground state or whether it is electronically excited is not clear. In either case addition of CO to $(M + N_2O)$ flames dramatically increases the CL yield through the reactions

$$Mg^*(^3P) + N_2O \rightarrow MgO^*(A^1\Sigma^+, B^1\Sigma^+, a^3\Pi, 3\Delta) \tag{22}$$

Here, $a^3\Pi$ and $^3\Delta$ are produced adiabatically, while $B^1\Sigma^+$ requires diabatic transitions. Process (22) for $Ca^*(^3P)$ has been studied by Dagdigian (1978) in a beam-gas experiment.

In conclusion we note that the gross features of the mechanism involving precursors are understood, although detailed knowledge of all precursors and their mutual interconversions must await their spectroscopic analysis under typical reaction conditions, this is being attempted in different laboratories.

III. DYNAMICS

The high product state resolution of chemiluminescence can be combined with the control over reactant conditions achievable by molecular beam and selective reactant excitation techniques to obtain, in principle, the ultimate state-to-state dynamical information symbolized by the transition probability P_{fi}. Different experiments concern themselves with different subsets of, or averages over, the quantum numbers which are symbolized by the collective labels i, f as will be exemplified below.

Product and reactant attributes can be classified into scalar and vector quantities: *energy disposal* (product state distributions or specific rate constants) indigenous to the CL method and *energy consumption* (translational and internal energy dependences) are the most frequently studied aspects. More recently, *directional information* has begun to become available. This comprises *angular distributions* of CL products (Mims and Brophy, 1977; Siegel and Schultz, 1978) and *angular momentum consumption* (orientation dependence of reaction probability) and *disposal* (product alignment) (Estler and Zare, 1978a).

Before illustrating some of these points with examples from the literature, it appears useful to place them in a systematic scheme, together with the basic problems and theoretical tools that are currently available. This is done in Fig. 3.

The complete dynamical information displayed by the partitioning of the collisional invariants (energy, linear and angular momenta) is loosely and somewhat arbitrarily divided into a *nuclear problem* which deals with

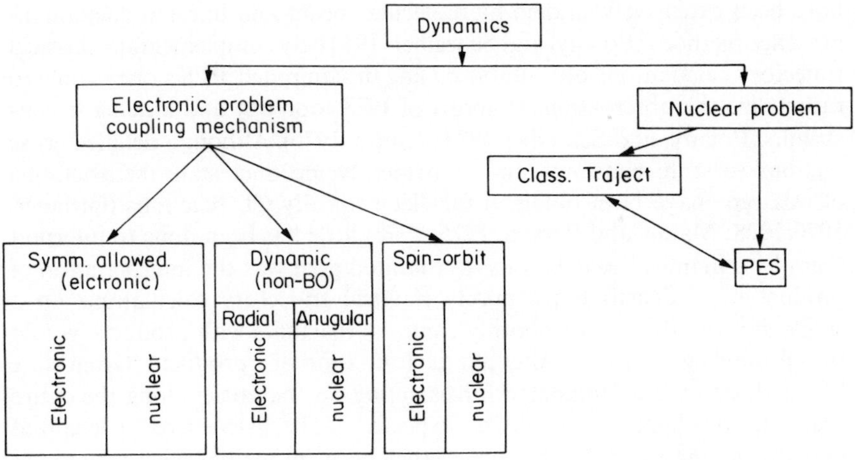

Fig. 3.

rotation–vibration–translation aspects already familiar from single PES reactions, and into the *electronic problem* ('how does the system proceed from the reactant to product potential energy surface'?) which is more specifically related to CL. Interesting aspects arising from the interplay of nuclear and electronic motion (failure of the Born–Oppenheimer separation) are incorporated into the electronic problem.

A. The Nuclear Problem

This deals with the partitioning of the energy, linear and angular momenta describing the dynamic states of reactant and product nuclear motion. The total dynamic information is contained in total and differential detailed cross sections (or the equivalent rate constants) for populating states of nuclear motion in a given electronic level $|e'\rangle$. Total state-to-state cross sections ideally given by $\sigma(v', J', m_J' | E_{T'} v, J, m_J)_{e'}$ contain *in praxi* a certain degree of reactant and product state averaging. Orientational effects, expressed by the reactant and product magnetic quantum numbers m_J, m_J' are included here.

Examples will be discussed in this section and in the following one: differential CL cross sections have been measured for total CL emission (not spectrally resolved) although differential state-to-state cross-section measurements are in principle possible for long-lived emitters. The preceding lines merely summarize the observables of chemical reaction dynamics, where the specifically CL-oriented aspect deals with the electronic states $|e\rangle$ and $|e'\rangle$.

For reactions on a single PES, partial averages over the above functions have been extensively studied by molecular beam and infrared chemiluminescence methods (Polanyi and Schreiber, 1974). By complementary classical trajectory calculations on suitable *ad hoc* or computed PES's one is able to rationalize the observations in terms of PES topology and kinematic constraints, Polanyi and Schreiber 1974; Kuntz, 1976). Although detailed cross sections (or rate constants and equivalently, product state distributions) of this type have been obtained for electronically CL reactions (Ottinger, 1976, 1978; Manos and Parson, 1975, 1978) little has been done to interpret them by dynamical calculations. If a limited goal was the interpretation of product v', J' distributions, purely classical trajectory calculations on a single *diabatic* PES that smoothly connects reactants with products, would be a promising approach. After all, the formation of a product internal state $|f\rangle$ is due to the accumulated forces acting on the nuclei along the entire reaction coordinate. It would be expected to be insensitive to localized diabatic interactions of PES's, unless they occur in regions where the forces on the nuclei (gradients of the PES's) are large.

The *kinetic energy dependence of the CL cross section* (total $\sigma_{CL}(E_T)$ or electronic-state-resolved $\sigma_{CL}(E_T)_{e'}$) is relatively easy to obtain from the CL intensity measured at a fixed laboratory angle, compared to single PES reactions which require laborious normalizations and integrations of product fluxes over scattering angles. This simplification is due to the near-isotropic angular distribution of CL light, equivalent to the low degree of CL polarization (Jonah *et al.*, 1972).

Product alignment or an anisotropic distribution of the rotational angular momentum vector manifests itself through anisotropic CL polarization. This may arise in beam experiments through the anisotropy of the initial collisional angular momentum **L** with respect to the collision axis, and through its propagation into product rotation $\mathbf{J}'$. Maximal $\mathbf{J}'$ alignment is expected in the presence of kinematic constraints, e.g. in a system of the type $HL + H \rightarrow HH + L$, where the exit collisional angular momentum is constrained by the low product reduced mass (H, L = heavy and light groups). In any case, averaging due to product rotation (Zare, 1966) reduces the degree of CL polarization P to a low value. Jonah *et al.* (1972) have shown this to be maximally $P \leq 0.3$ for a simple dynamical model and an idealized beam-gas experiment. This has been verified for $Ba + NO_2$ and for (Tl, In) + I_2^* by Jonah *et al.* (1972) and by Estler and Zare (1978a) respectively.

Quantitative information about *product orientation* (the m'_J, distribution) is obtainable from polarization measurements. Rotationally resolved spectra are required since the degree of polarization depends on the type—P, O, or R branch—of the rotational transition. While this is very promising in principle, the high spectral resolutions required are seldom achieved and the method has consequently not yet been applied. Case *et al.* (1978) have

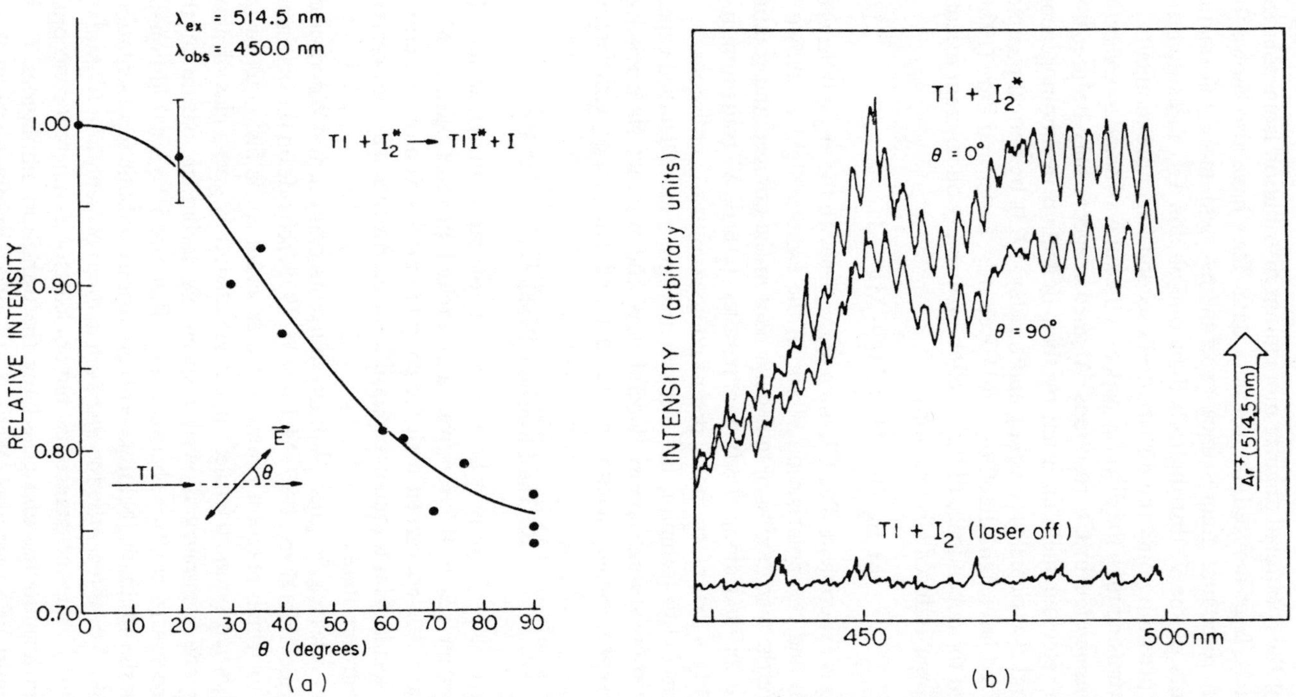

Fig. 4. (a) Laser-induced chemiluminescent spectra for $Tl + I_2^* \rightarrow TlI^*$ as a function of the angle θ between the metal beam axis and the polarization vector of the linearly polarized exciting light. (b) Variation of the chemiluminescence signal with angle θ for $Tl + I_2^*$. The data demonstrate a preference of collinear over perpendicular collisions. From Estler and Zare (1978). Reprinted, with permission, from *J. Ann. Chem. Soc.*, 100, 1323. *Copyright by the American Chemical Society.*

given a theoretical analysis of the more general problem of determining the moments of the rotational angular momentum distribution from polarized resonance (e.g. laser-induced) fluorescence data. They have also shown that two-photon resonance fluorescence experiments yield more information (12 moments of the J′ distribution) than one-photon CL measurements, capable of providing only the distribution's second Legendre moment.

More promising than the disposal aspect is the study of *angular momentum* (i.e. m_J) *consumption* in CL reactions. Aligned reagents may be prepared, apart from selecting them as usual by their deflection in inhomogeneous magnetic and electric fields (Fluendy and Lawley, 1973), by optical pumping with polarized laser light (Feofilov, 1961). Recently, Estler and Zare (1978a) have studied the dependence of the reaction probability on reactant orientation in the laser-induced CL reactions

$$\mathrm{M\,(In, Tl)} + \mathrm{I}_2^*(B^3\,\Pi) \rightarrow \mathrm{MI}^* + \mathrm{I} \qquad (23)$$

in a beam-gas experiment. The CL intensity varies with the angle θ between metal beam and the polarization vector of the exciting light as shown in Fig. 4. Since the distribution of molecular axes in the pumped and rotating molecules is peaked along $\theta = 0°$ for 'parallel' $I_2(B \leftarrow X)$ transition, it is clear from Fig. 4 that reaction (23) prefers collinear over C_{2v} collisions.

Extensions of this technique to other systems and to more detailed experiments (e.g. translational energy dependences and product state resolved experiments) promise new insight into the detailed structure of steric factors.

B. The Electronic Problem

The electronic problem may be described by the question: 'How does the reacting system reach the electronically excited product state?'. More fundamentally it is concerned with the coupling mechanism or the term in the total molecular Hamiltonian responsible for the transition to the observed product electronic state e'.

Symmetry rules (e.g. Wigner–Witmer, spin conservation, and Woodward–Hoffmann rules (Herzberg, 1967; Metiu *et al.*, 1978) provide on the one hand, a guide as to which reaction channels are expected to be fast, and on the other hand, when symmetry rules appear violated, they give a classification scheme for the symmetry-allowed respectively forbidden nature of the observed process. Interactions between more than one PES are traditionally described in the adiabatic (Nikitin, 1974) or various diabatic representations (Smith, 1969). More recently, the *quasi-adiabatic* representation (O'Malley, 1971) has found some acceptance, last but not least for its inherent conceptual simplicity in formulating and classifying the transition amplitudes T_{fi} in *concerted* multi-PES reactions (Metiu *et al.*, 1978). Briefly, quasi-adiabatic PES's (QAPES) are designed so as to exclude specifically the possibility of nuclear rearrangement on any one of them, i.e. there are reactant and product QAPES's which intersect in the interaction region. Reaction is considered as an electronic transition between reactant and product QAPES's, and symmetry rules can be derived in analogy to optical selection rules. Metiu *et al.* (1978) have shown that, given the approximate molecular Hamiltonian

$$H = H_{\mathrm{el}} + H_{\mathrm{N}} + H_{\mathrm{SO}} \tag{24}$$

whose components are the total electronic Hamiltonian H_{el}, the nuclear kinetic energy H_{N}, and the electron spin–orbit interaction H_{SO}, the total transition amplitude, whose square modulus is the transition probability from reactant $|i\rangle$ to product state $|f\rangle$, may be approximately expressed as

$$T_{fi} = T_{fi}^{\mathrm{el}} + T_{fi}^{\mathrm{N}} + T_{fi}^{\mathrm{SO}} \tag{25}$$

Each component is the matrix element of a term in the Hamiltonian operator (24) between quasi-adiabatic reactant and product vibronic states, e.g. $T_{fi}^{\mathrm{el}} = \langle f | H_{\mathrm{el}} | i \rangle$.

By making Franck–Condon and subsidiary approximations these component T matrix elements can be further decomposed into the products of a nuclear (Franck–Condon) q_{fi} and an electronic factor R_{fi}

$$T_{fi}^{\mathrm{OP}} = q_{fi} R_{fi}^{\mathrm{OP}} \tag{26}$$

The very simple factorization based on equations (25) and (26), is summarized in Fig. 3. The usual *symmetry rules* (e.g. of the adiabatic state correlation, or Woodward–Hoffmann type) serve only to predict whether the electronic R_{fi}^{el} term is Large or not, regardless of the corresponding Franck–Condon factor. For a concerted reaction to be fast both the electronic transition moment and the Franck–Condon factor must be large. An example for the controlling influence of the nuclear factor is given below.

Dynamically induced reactions (zero T_{fi}^{el}, non-zero T_{fi}^{N}) arising from the breakdown of the Born–Oppenheimer approximation are further separated into radially and rotationally induced (Russek, 1971):

$$T_{fi}^{\mathrm{N}} = T_{fi}^{\mathrm{N,rad}} + T_{fi}^{\mathrm{N,rot}} \tag{27}$$

Traditionally radial coupling is taken to be the dominant term in low-energy scattering, since in the adiabatic representation the rotational matrix element, coupling states of different electronic angular momentum $\Lambda' = \Lambda \pm 1$ is weighted by vb/R^2 where v, b, R are radial velocity, impact parameter, and interparticle distance (Barat, 1973). It remains to be seen whether rotational coupling in reactive collisions is really as unimportant as is usually assumed to be the case.

Whenever the total electronic spin is not conserved in a reaction, or when fine structure components exhibit different reactivities or are formed in non-statistical amounts, then spin–orbit coupling T_{fi}^{SO} plays a special role.

1. *Symmetry-allowed* CL *Reactions*

A good example for a symmetry-allowed CL reaction (large T^{el}_{fi}) involving orbitally degenerate reactants is the previously discussed dialkali–halogen process (equation 2.11). The high CL yield at thermal energies is mainly due to the adiabatic correlation of the one reactant component with excited ^2P products (Magee, 1940). The actual situation is much richer, since among the many observed excited M** states (Struve *et al.*, 1973, 1975) only the lowest ^2P, which also contributes heaviest to the light yield, correlates adiabatically with reactants. The higher energy and minor channels do require diabatic transitions.

A related process, in which all reactant components correlate adiabatically with excited products (Wiesenfeld and Yuen, 1976) is the

$$\mathrm{Sn}(^3P_J) + \mathrm{N_2O} \rightarrow \mathrm{SnO}^*(a^3\Sigma^+, b^3\Pi) + \mathrm{N_2} \tag{28}$$

reaction (as well as its Pb analogue, to be discussed below). The high CL quantum yield $\Phi \lesssim 0.5$ (Felder and Fontijn, 1975, 1978) and its pressure dependence are consistent with (28) as the dominant channels. The CL yield does not reach unity since reaction is slow (Wiesenfeld and Yuen (1976) have measured $k = 5 \times 10^{-13} \exp - 4000/RT$ cm^3 mol^{-1} s^{-1}), the radiative lifetimes long, and product quenching fast. Although the vibrational energy dependence of (28) has not been studied, one may expect in analogy with the $\mathrm{BaO} + \mathrm{N_2O}$ reaction (see below) that the $\mathrm{N_2O}$ bending vibration may play an important role. This would be an example for the influence of Franck–Condon overlap (26) on the rate of a symmetry-allowed reaction.

Reactions which are symmetry forbidden and slow can be greatly accelerated by electronic excitation to states that correlate adiabatically with the emitting products (Donovan and Husain, 1970). Examples are the CL reactions of metastable ^3P alkaline earth metals:

$$\mathrm{M}^*(^3P) + (\mathrm{HCl}, \mathrm{Cl_2}) \rightarrow \mathrm{MCl}^* + (\mathrm{H}, \mathrm{Cl}) \tag{29}$$

studied by Brinkmann and Telle (1977) and Kowalski and Heldt (1978) and $\mathrm{Ca}^*(^3P) + (\mathrm{O_2}, \mathrm{CO_2}) \rightarrow \mathrm{CaO}^*$ studied by Pasternack and Dagdigian (1978). The reaction

$$\mathrm{I}_2^*(B^3\Pi) + \mathrm{F_2} \rightarrow \mathrm{IF}^*(B^3\Pi(\mathrm{O}^+)) + \mathrm{IF} \tag{30}$$

however is symmetry forbidden both for ground state $\mathrm{I_2}(X)$ (Valentini *et al.*, 1977) and for the $\mathrm{I}_2^*(B)$ state. It is somewhat of an exception since it occurs readily as written (Engelke *et al.*, 1977).

2. *Dynamically Induced* CL *Reactions*

Numerous cases of dynamically induced, diabatic CL reactions are known. A particularly illustrative example for present purposes is $(\mathrm{Ba} + \mathrm{N_2O})$.

From direct photographic observation of the angular CL distribution, Siegel and Schultz (1978) have inferred the $BaO^*(A'^1\Pi)$ state, which incidently does not correlate adiabatically with reactants, to be the primary emitter. The excitation function measured at two N_2O vibrational temperatures is shown in Fig. 5 (Wren and Menzinger, 1975, 1979a). The pro-

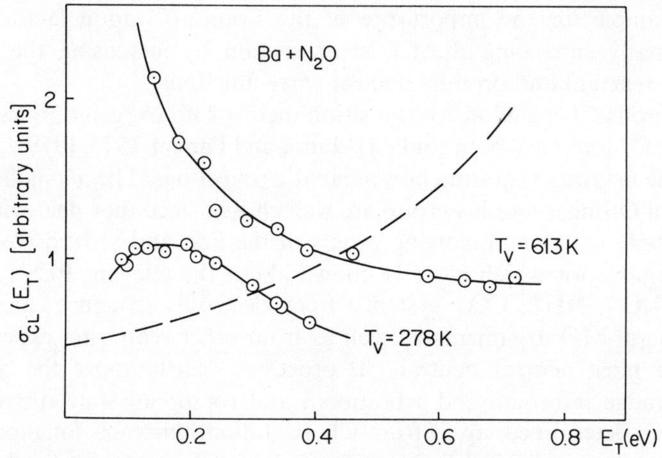

Fig. 5. Chemiluminescence excitation functions for $Ba + N_2O \rightarrow BaO^*$, measured at two N_2O vibrational temperatures (assumed to be equal to the nozzle temperature). The dashed curve represents the BaO^* ($A'^1\Pi$) product state density. From Wren and Menzinger (1975, 1979a).

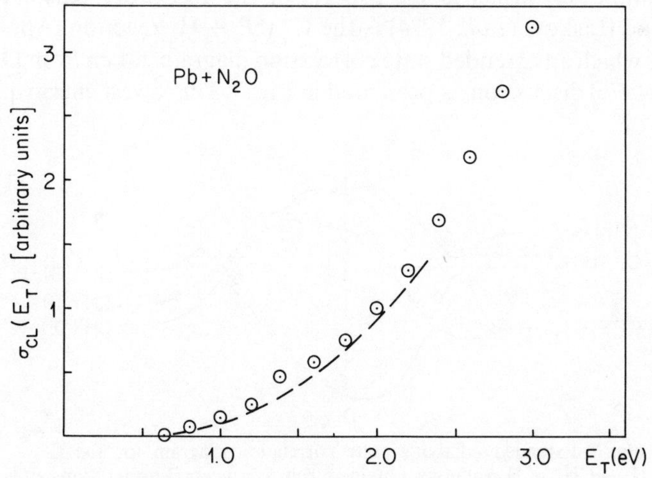

Fig. 6. Chemiluminescence excitation functions for the endoergic $Pb + N_2O \rightarrow PbO^*(B, v = 0)$ reaction. The dashed curve is the product state density as drawn by the authors. From Wicke *et al.* (1978). Reproduced by permission of the North Holland Publishing Corp.

nounced CL enhancement by vibrational excitation is believed to be due principally to the ν_2 bending rather than the ν_3 asymmetric stretch vibration. The strong increase of electron affinity upon N_2O bending (linear ground state) combined with the data, Fig. 5, suggests that formation of the ion pair $Ba^+N_2O^-$ initiates the reaction, followed by rearrangement to the observed products. Whatever the promoting mode and the emitter, we have here a clear example for the importance of the Franck–Condon factor in (26) vibrationally enhancing the CL cross section by increasing the overlap between reactant and product nuclear wave functions.

Numerous CL reactions of transition metals with oxygen donors, and of O^+, N^+, C^+ ions have been studied (Manos and Parson, 1975, 1978; Ottinger, 1978), the majority requiring non-adiabatic transitions. The *ion-molecule CL studies* of Ottinger and his group are well chosen since they deal with relatively simple reactions involving atoms of the first and second row of the periodic table for which more is known about the relevant PES's (e.g. the CH_2^+, N_2O^+, NH_2^+, CO_2^+ systems) from electronic structure calculations and general MO arguments as well as from other scattering experiments, than for most neutral–neutral CL processes. Furthermore the available energy range is broad, and vibrational and rotational state distributions have been determined, apart from CL excitation functions, for most of the reactions given in Table XI. Out of these, large CL cross sections exceeding prior expectations ($\sigma_{CL} > 1\,Å^2$) were found only for the N^+ + hydrocarbon reactions. The reviews by Ottinger (1976, 1978) of these studies, particularly (1976), contain more detail than the individual papers.

An example for which some aspects of the PES's are reasonably well understood (Liskow *et al.*, 1974) is the $C^+(^2P) + H_2$ reaction (Appell *et al.*, 1975), for which an extended state correlation diagram, taken from Ottinger's detailed (1976) discussion, is presented in Fig. 7. The lowest energy pathways

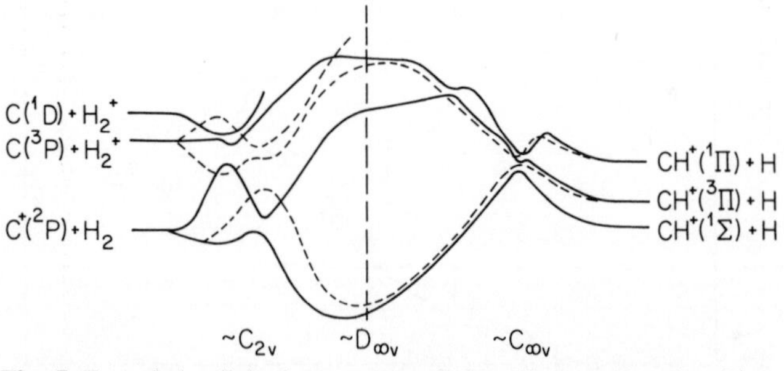

Fig. 7. Extended adiabatic state correlation diagram for the $C^+ + H_2 \rightarrow CH^+ + H$ reactions, obtained from symmetry considerations and electronic structure calculations. The state energies are drawn for complexes of slightly distorted (C_s) complexes of approximate ($C_{2v}, D_{\infty h}, C_{\infty v}$) symmetries. Solid and dashed lines correspond to states of A' and A'' symmetry, respectively. Reproduced, with permission from Ottinger (1976).

to $CH^+(A^1\Pi)$ are seen to involve the lowest $A' + A''$ surfaces, from which adiabatic transitions to the product state occur during the exit phase of the collision. It is evident that only the A' pathway will be significant at low collision energies.

Interesting product state distributions shown in Fig. 8 were found for

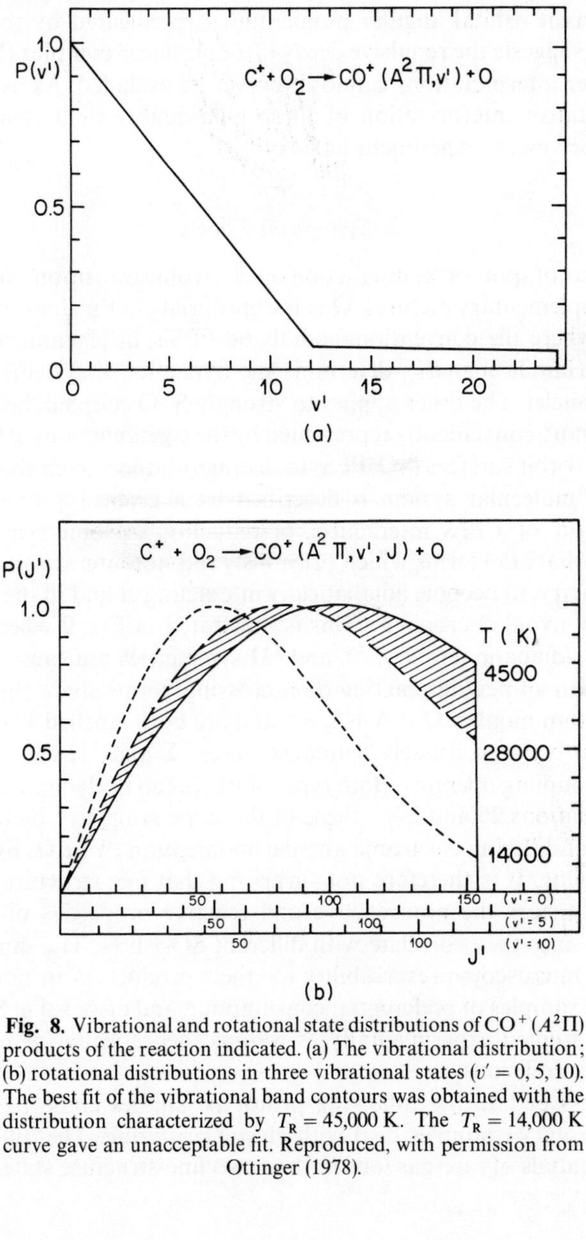

Fig. 8. Vibrational and rotational state distributions of $CO^+(A^2\Pi)$ products of the reaction indicated. (a) The vibrational distribution; (b) rotational distributions in three vibrational states ($v' = 0, 5, 10$). The best fit of the vibrational band contours was obtained with the distribution characterized by $T_R = 45{,}000$ K. The $T_R = 14{,}000$ K curve gave an unacceptable fit. Reproduced, with permission from Ottinger (1978).

the $C^+ + O_2 \rightarrow CO^+(A^2\Pi)$ reaction (Ottinger and Simonis, 1975; Ottinger, 1978). The monotonically decreasing but non-statistical $P(v')$ seems to be evidence for short-lived, bound CO_2^+ intermediates. The rotational distributions $P(J')_{v'}$ in different v' are of the Boltzmann type, truncated for angular momentum restrictions, and characterized by extremely high rotational temperatures $T_R \simeq 45{,}000$ K (average $\bar{J}' \sim 80$)! This corresponds to $\sim 50\%$ of the reactant orbital angular momentum **L** predicated by the Langevin model and suggests the repulsive decay of a non-linear complex (Miller *et al.*, 1967). Other interpretations cannot however be excluded. As usual, a more than qualitative interpretation of these high-quality data is not possible unless theory meets experiment halfway.

3. *Spin–orbit Effects*

The effects of spin–orbit interaction on electronic transitions are discussed in two complementary pictures. One is appropriate in the limit of weak S–O coupling, where the conventional adiabatic PES's, or eigenfunctions of the electronic Hamiltonian H_{el} determine the dynamics of systems containing only light nuclei. The other applies to strongly S–O coupled, heavy systems which are more conveniently represented by the eigenfunctions of $(H_{el} + H_{SO})$ called spin–orbit surfaces (SO-PES) to distinguish them from the former.

When a molecular system is described by a given approximation, the 'switching on' of a new interaction, or reduction of some symmetry, may cause two (diabatic) terms which previously did not interact and crossed at some geometry, to become adiabatically interacting at and in the neighbourhood of an 'avoided crossing'. This is illustrated in Fig. 9 where two non-interacting adiabatic terms of $^3\Sigma$ and $^1\Pi$ symmetries are caused by H_{SO} to mix. The pair of new interacting (non-crossing) terms share the remaining good quantum number $\Omega = \Lambda + \Sigma = 1$ and are both labelled 1. In this sense a transition between Russell–Saunders states $^3\Sigma$ and $^1\Pi$ can be 'induced' by the LS coupling operator. Both types of PES's can be dynamically coupled by T_{fi}^{N} (equations 25 and 27)—those of the same symmetry by radial $T_{fi}^{N,rad}$ and those of differing electronic angular momentum (Λ or Ω) by rotational $T_{fi}^{N,rkt}$ coupling. It is therefore not surprising that fine-structure states may dynamically steer the non-reactive and reactive properties of atoms and molecules, since they correlate with different SO-PES's. The same holds, of course, by microscopic reversability for their production in non-statistical amounts. Examples of preferential consumption and disposal of S–O energy have been studied experimentally as well as theoretically for non-reactive and reactive collisions.

Regardless of CL, the following prototype studies come to mind: The most elementary examples deal with elastic scattering. The different interaction potentials of rare-gas ions in their two fine-structure states $R^+(^2P_{1/2},$

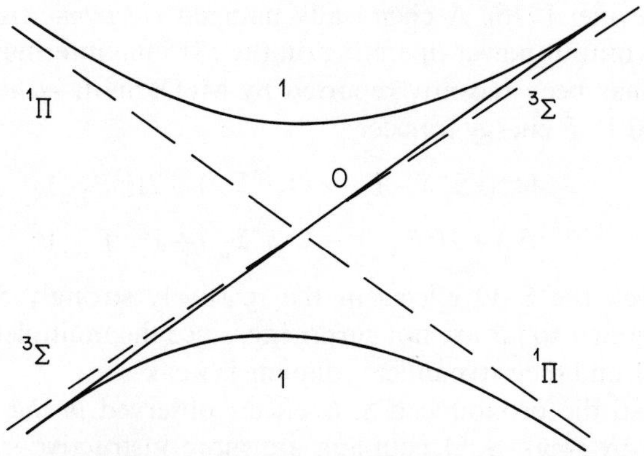

Fig. 9. Schematic diagram illustrating how two intersecting adiabatic B–O surfaces (dashed lines) of different symmetry $^1\Pi$ and $^3\Sigma$ are coupled by the S–O operator in the S–O representation, giving rise to a new set of non-intersecting SO–PES's (solid curves) labelled 0 and 1 respectively by the new quantum number $\Omega = \Lambda + \Sigma$. Note that the $^3\Sigma$ B–O state splits into a 1 state which does and a 0 state which does not interact with the 1 state originating from $^1\Pi$.

$^2P_{3/2}$) with their parent atoms endows them with unequal transport properties, as drift tube measurements of their mobilities have shown (Helm, 1976). PES's for $F(^2P_{3/2,\,1/2}) + H_2$ have been computed and studied by Tully (1974b) and Jaffe *et al.* (1975). In the related system $Br(^2P_{3/2}, {}^2P_{1/2}) + HI \rightarrow HBr + I$, Bergmann *et al.* (1975) have observed that ground-state $Br(^2P_{3/2})$ reacts rapidly while electronic excitation to $Br(^2P_{1/2})$ even inhibits the reaction! Similar selectivity and specificity has been observed in the following CL processes. The reactions of metastable $Hg^*(6^3P_{0,2})$

$$Hg^*(6^3P_{0,2}) + X_2 \rightarrow HgX^*(B^2\Sigma^+) + X \tag{31}$$

have been found by Krause *et al.* (1975) and Hayashi *et al.* (1978) to proceed primarily *via* the $Hg^*(6^3P_2)$ component. The reactive cross section of $Hg^*(6^3P_0)$ is roughly a tenth of the former, presumably due to the existence of a barrier in the corresponding SO-PES.

Of particular interest to the chemical laser community are processes (photodissociation, energy transfer, exchange reactions) that generate halogen (especially iodine) atoms with inverted fine-structure populations. As an example we mention the reaction

$$F + HI \rightarrow HF + I(^2P_{3/2,1/2}) \tag{32}$$

for which Burak and Eyal (1978) find the production rates for $I^*(^2P_{1/2})$ and $I(^2P_{3/2})$ to be non-statistical and approximately equal (Dinur *et al.*, 1975;

Dinur and Levine, 1975). A chemically pumped *electronic transition laser* with > 4 mW output power operating on the 1315 nm intermultiplet transition of $I(^2P)$ has been recently reported by McDermott *et al.* (1978). It is based on rapid E–E energy transfer

$$O_2^*(^1\Sigma_g^+) + I_2 \rightarrow O_2(^3\Sigma_g^-) + 2I(^2P_{3/2}) \tag{33}$$

$$O^*(^1\Delta_g) + I(^2P_{3/2}) \rightarrow O_2(^3\Sigma_g^-) + I^*(^2P_{1/2}) \tag{34}$$

In retrospect, the S–O effects in the relatively strongly S–O coupled systems mentioned so far are not surprising, since the multiplet surfaces are well separated, and their dynamical coupling is weak.

By constrast, the pronounced *S–O* effects observed *in the light systems* characterized by weak S–O coupling are more instructive since they are much less obvious.

A reaction of great complexity (5 atoms) whose numerous kinetic and energy partitioning aspects have been studied in some detail, is

$$NO(2\Pi_{3/2}; v) + O_3(v_1 v_2 v_3) \xrightarrow{E_T} NO_2^* + O_2(^3\Sigma_g^-) \tag{35a}$$

$$NO(^2\Pi_{1/2}; v) + O_3(v_1 v_2 v_3) \longrightarrow NO_2^\ddagger + O_2(^3\Sigma_{g'}^- \, ^1\Delta_{g'} \, ^1\Sigma_g^+) \tag{35b}$$

As written in terms of two separate reaction channels, equations (35a, b) already anticipate results to be discussed below. Thermal rate constants are known for both channels (Clough and Thrush, 1967) and their dependence on reactant vibration has been studied by laser excitation. As one would expect, NO vibration which does not lie along the reaction coordinate, is relatively ineffective in promoting reaction (Stephenson and Freund, 1976). Excitation of the $O_3(001)$ asymmetric stretch mode, followed to a certain, unspecified, degree of intermode relaxation (Hui *et al.*, 1975), enhances both channels (Braun *et al.*, 1974; Kurylo *et al.*, 1974, 1975; Gordon and Lin, 1976) by reducing their activation energies (Moy *et al.*, 1977). This acceleration by $O_3^\ddagger$ vibration exceeds prior expectations, but translational energy is even more effective.

The excitation functions have been measured in a beam-gas experiment by Redpath and Menzinger (1971, 1975) and by Redpath *et al.* (1978). Their most relevant finding in the present context was the (indirect) observation (Redpath and Menzinger, 1975) that the two relatively closely spaced ($\Delta E = 121$ cm^{-1}) fine-structure components of $NO(^2\Pi_{1/2, 3/2})$ are capable of channelling the reaction as written in (35), i.e. of causing the upper $^2\Pi_{3/2}$ state to give predominantly electronically excited NO_2^*, and ground-state $^2\Pi_{1/2}$ to give $NO_2^\ddagger$. The fact is unexpected that at collision energies ~ 10 times in excess of the fine-structure splitting dynamic coupling of SO-PES's appears negligible. A direct test of the f.s. reactivities by fine-structure selecting the NO beam in a deflection experiment is within reach and is being attempted by Stolte (1978). The CL excitation function for the

$NO(^2\Pi_{3/2})$ state was found to be well approximated by $\sigma_{3/2}(E_T) = C(E_T/E_0 - 1)^n$ in the threshold region $3.0 > E_T > 6$ kcal/mol, where the threshold is $E_0 = 3.0 \pm 0.3$ kcal, $C = 0.163$ Å^2 and $n = 2.4 \pm 0.15$. At higher E_T, σ_{CL} rises faster ($n > 2.4$) than in the threshold region. A puzzling feature of reaction (35) is the failure to observe excited $O_2(^1\Delta_g$ and $^1\Sigma_g^+)$ which are both energetically accessible and favoured by adiabatic correlations (Gauthier and Snelling, 1973). Implications of this fact have been discussed in the light of general features of the O_3NO PES's (Redpath *et al.*, 1978).

A second example for pronounced S–O effects despite weak coupling is the thermal *cleavage of tetramethyldioxetane* into predominantly S_0 ground-state

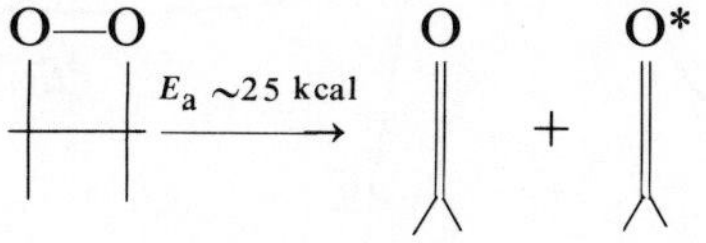

(36)

$\sim 50\%\ S_0$ $\quad \sim 49\%\ T_1$
$\sim 1\%\ S_1$

acetone and T_1 triplet acetone. It is a prime example for highly efficient production of excited states in a blatantly spin-non-conserving reaction (Turro *et al.*, 1974; Wilson, 1976). Process (36) is isomorphous with the (cyclobutane → 2 ethylene) reaction in that formation of excited products is enforced by orbital symmetry (Woodward–Hoffmann rules). This is illustrated by the schematic diabatic correlation diagram Fig. 10 showing the S_0 ground-state PES with the symmetry-enforced high barrier towards decomposition into $2A$ on the $S_{0,P}$ product surface and the intersecting T_1 triplet and S_1 singlet PES's. Configuration interaction will strongly mix the singlet surfaces as indicated by the large circles, but S–O coupling of the $S_0 \times T_1$ and $S_{0,P} \times T_1$ surfaces is very weak as symbolized by the small circles. The high triplet yield can be understood in terms of a model developed Tully (1974a, 1975) and Zahr *et al.* (1975), to explain the fast quenching of $O(^1D)$ by N_2. Applying it to the present case, the isolated S_0 molecule, vibrationally excited above the threshold E_0^T is imagined to pass many times through the $S_0 \times T_1$ intersection and has a small probability, say $P_{ST} \sim 10^{-3}$, at each passage to make a transition to T_1. After $P_{ST}^{-1} \sim 10^3$ such passages, the transition will have occurred with unit probability. The same model is believed to explain, by a long-lived $BaSO_2$ complex (Behrens *et al.*, 1976), the rapid, spin-non-conserving (but not chemiluminescent) process

$$Ba(^1S) + SO_2(^1A_1) \rightarrow BaO(^1\Sigma^+) + SO(^3\Sigma^-) \tag{37}$$

that has been first reported by Smith and Zare (1975). Dioxetanes have been activated (a) thermally (Wilson, 1976), (b) collisionally by $T \rightarrow V$ transfer

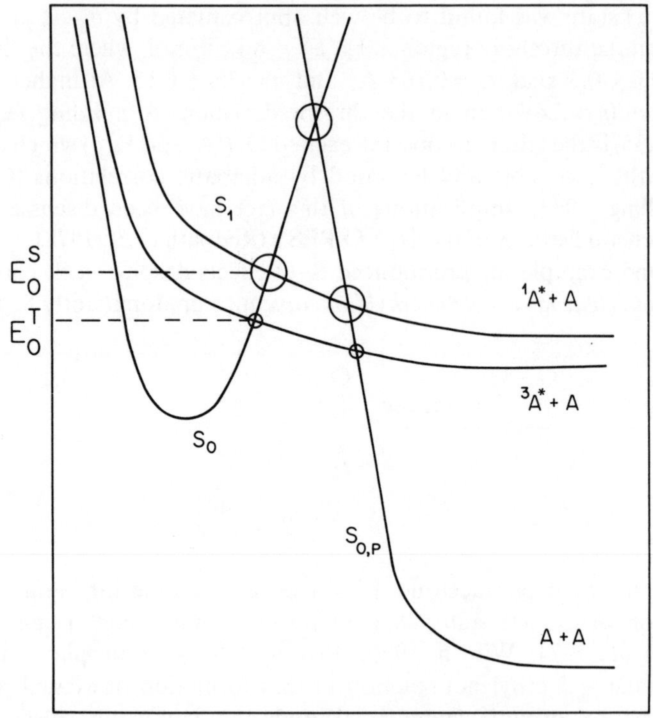

Fig. 10. Schematic diabatic correlation diagram for the decomposition of tetramethyl-1, 2-dioxetane into ground-state and electronically excited (triplet^3 A^* and singlet^1 A^*) acetone. The small circles represent the weak spin–orbit interaction of the singlet S_O, $S_{O,\mathrm{P}}$ with triplet T_4 surfaces. The big circles symbolize the strong configuration interaction of the singlet states S_0, S_1 and $S_{\mathrm{O,P}}$, which causes the transitions between diabatic surfaces to be facile. Thresholds for the slow $S_0 \rightarrow T_1$ transition and the fast $S_0 \rightarrow S_1$ transition are labelled E_{O}^T, E_{O}^S respectively, Redrawn from Brown and Menzinger (1978).

(Brown and Menzinger, 1978), (c) chemically by transiently forming a dioxetane from $O_2(^1\Delta_g)$ + alkene (in the bulk: Bogan *et al.* (1975, 1976), and in crossed beam: Alben *et al.* (1978)), and (d) photochemically by infrared $S_0 \rightarrow S_0$ (Haas and Yahav, 1977, 1978) or by direct $S_0 \rightarrow S_1$ absorption (Lechtken and Turro, 1973; Smith *et al.*, 1977).

The collision-induced dissociation (b) of tetramethyldioxetane by a variable energy xenon beam (Brown and Menzinger, 1978) shows a threshold for CL production in nice agreement with thermal activation energy and, more interestingly, a marked increase in photon yield as the collision energy is raised and the amount of vibrational energy deposited in the dioxetane molecule increases. With reference to Fig. 10 this signals the opening of a new and direct reaction channel giving singlet acetone $^1A^*$ at the second E_0^S threshold.

We have seen that common symmetry rules based on the conservation of only a subset of the collisional invariants (George and Ross, 1971) may fail through the influence of a higher-order coupling term. Other *counter-intuitive results*, which shed an interesting light on the rules being broken, are violations of the principle of least motion Clyne and Coxon, 1966; (Pearson, 1976; Engelke *et al.*, 1976), and the allocation of reaction energy in an old rather than a newly formed bond (Engelke and Zare, 1977; Kiefer *et al.*, 1978). The first case is exemplified by the reactions

$$M(Mg, Ca, Sr, Ba) + OClO \rightarrow MCl^* + O_2. \tag{38}$$

Observed emitters are $MO^*(A'^1\Pi)$ and MCl^*. The latter requires attack of the central atom accompanied by extensive nuclear and electronic rearrangement, which the authors (Engelke *et al.*, 1976) were able to rationalize by a simple MO argument. A similar mechanism is implied by the reaction $Br + OClO \rightarrow BrCl^* + O_2$, reported by Clyne and Coxon (1966).

The second case is illustrated by the processes

$$M(Ca, Sr, Ba) + S_2Cl_2 \rightarrow S_2^*(B^3\Sigma_u^-) + MCl_2 \tag{39}$$

(Engelke and Zare, 1977) and

$$Ba + CCl_4 \rightarrow CCl_2^*(\tilde{A}) + BaCl_2 \tag{40}$$

(Kiefer *et al.*, 1978). Molecular orbital arguments and the isotropic angular CL distribution respectively led both sets of authors to infer that they were dealing with long-lived collision complexes of a definite structure.

C. Interaction with Theory

Theoretical descriptions range from the *solution of the scattering problem* by classical and quantum mechanical methods and various linear combinations thereof on previously calculated *potential energy surfaces* on the upper end of the complexity scale, to *statistical theories* on the other end. Several excellent reviews describing and evaluating these methods have been written in recent years, to which the reader is referred. Tully (1976, 1977) has discussed currently promising methods for solving the scattering problem arising from the interaction of several PES's, and in this volume describes how these PES's can be obtained. Thus *ab initio* or semiempirical PES's combined either with a solution of the (close-coupled) set of quantum dynamical equations or with variants of the classical path equations (trajectory surface hopping, path integral methods, analytical continuation of classical trajectories into the complex plane, multiple crossing models, etc.) provide potentially powerful tools towards understanding at least the simplest CL systems and some of the effects discussed above on an *ab initio* basis. However these methods have with few exceptions, not yet been applied to realistic CL systems, a condition which according to Carrington (1973) is unlikely to change greatly in the near future.

The M_2(alkali) + X(halogen) PES's (Struve, 1973; Balint-Kurti and Karplus, 1974) and the $Ca + F_2$ surface calculation of Yarkony *et al.* (1973) come to mind regarding neutral–neutral CL systems. A close-coupling, one-dimensional quantum calculation on *ad hoc* PES's representative of $Ba + N_2O$ has been performed by Bowman *et al.* (1976).

Statistical theories (phase space theory, information theory) on the other hand, are easy to apply and have found correspondingly widespread applications in prediction (e.g. Pechukas *et al.*, 1966; Dinur *et al.*, 1975; Dinur and Levine, 1975; Levine and Ben-Shaul, 1977) and in separating the dynamical from statistical content of experimental data (Alexander and Dagdigian, 1978). The low quantum yields observed in so many CL reactions attest the importance of statistics, or, in microscopically meaningful language, of diabatic correlations.

The observed distributions of excited alkali atoms (e.g. K*(4*S* to 9*S*) for X = Cl) in $M_2 + X \rightarrow MX + M^*$ reactions (Struve *et al.*, 1975; Krenos *et al.*, 1975) have been shown to agree essentially with phase space predictions (Krenos and Tully, 1975; Krenos, 1978), and information theoretical analysis yields a non-surprising result (Faist and Levine, 1977). This is taken as evidence for an unusually facile electronic energy exchange among many energetically accessible states of the collision complex and represents a counter-example against the importance of adiabatic correlations. The existence of closely spaced PES's that mutually interact over a broad range of configuration space provides a plausible reason for strong coupling in these and other systems, and may be the dominating factor whenever electronic state densities become high.

A group of reactions forming electronically excited molecular products MO*, in which both the partitioning of energy among product vibration, rotation, and electronic excitation and the consumption of reactant translation (i.e. excitation functions) are found to be in essential agreement with statistical predictions, are the M(Sc, Y, La) + O_2 reactions, thoroughly studied by Manos and Parson (1975, 1978). The high density of electronic states in the intermediate complexes as well as their long life may be responsible for this behaviour. A typical result illustrating this statement is the excitation function for the

$$La + O_2 \rightarrow LaO^*(A, B, C) + O(^3P) \tag{41}$$

reactions, given in Fig. 11. They are seen to conform closely to prior expectations given by the solid lines, as do the branching ratios.

While it is gratifying to find statistical models frequently so successful in

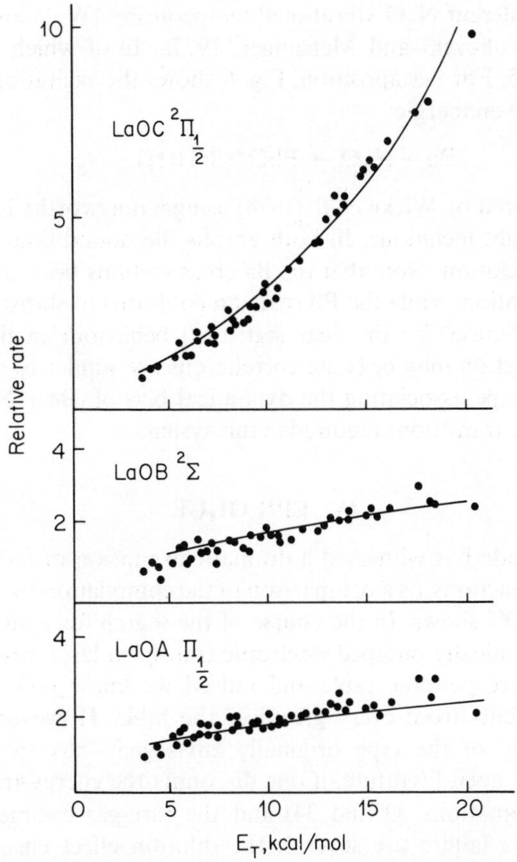

Fig. 11. Relative rate constants for production of three electronic product states in the $La + O_2 \rightarrow LaO^* + O$ reaction. The prior statistical factors, given by the solid lines, are seen to agree with the measured points very well. (Reproduced by permission of the American Institute of Physics from Manos and Parson, 1978.)

describing reality, and at the same time dispensing the theorist from performing tedious or wellnigh impossible dynamical calculations, much remains to be learnt about the different possible causes for statistical behaviour and about the kinds of systems and conditions to which it applies. Again the correlation between electronic structure and dynamic bias is of particular interest. A case in point is provided by the *reactions of N_2O with different metal atoms.* Manos (1978) has found significant dynamical bias in the excitation functions of (Sc, Y, La) + N_2O compared with their O_2 counterparts, exposing the special role played by the N_2O molecule (Tully, 1974a). This is most pronounced in the (Ba, Sm) + N_2O excitation functions measured at different N_2O vibrational temperatures (Wren and Menzinger, 1974, 1979a; Yokozeki and Menzinger, 1977a, b) of which the former is shown in Fig. 5. For juxtaposition, Fig. 6 shows the excitation function for the 14 kcal/mol endoergic

$$Pb + N_2O \rightarrow PbO^*(B^3\Pi(1)) + N_2 \tag{42}$$

reaction measured by Wicke *et al.* (1978) using a noteworthy laser vaporization/time-of-flight technique. In both graphs the dotted lines represent the statistical expectation. Note that the Ba cross sections bear no resemblance to prior expectation, while the Pb reaction conforms to statistics quite well. Whatever the 'cause' for the near statistical behaviour in the symmetry-allowed Pb reaction may be (state correlations are similar to equation (28)) one cannot escape associating the dynamical bias of ($Ba + N_2O$) with the specific diabatic transitions required in this system.

IV. EPILOGUE

The last decade has witnessed a dramatic expansion of our inventory of CL exchange reactions, as a comparison of the compilation by Thrush (1968) with Tables I–XI shows. In the course of the search for systems amenable to a purely chemically pumped electronic transition laser, researchers have grazed the entire periodic table and indeed we know now CL reactions involving elements from every group of the table. However the primary goal—a 'device' of the type originally envisaged—has so far not been reported in the open literature, if one discounts the energy-transfer atomic iodine laser (equations 33 and 34) and the rare-gas excimer lasers. The reasons for this failure are several: the dilution effect ensuing from the formation of molecular emitters in numerous v', J' states causes the inversion densities necessary for losing action to be prohibitively high, a fact that would favour atomic transition lasers. High inversion densities require (a) fast emitter production, (b) rapid depopulation of the lower level (favouring bound-free transitions), (c) a relatively long ($\tau_R > 10^{-6}$ s) radiative lifetime, and (d) last not least negligible emitter quenching (Sutton and Suchard, 1975; Herbelin and Cohen, 1975). To find a system that obeys these constraints will certainly not be easy.

As a compensation, much has been learnt about specific kinetic and dynamic aspects of CL systems. As so often happens in mission-oriented research the enquiry did not follow the course of steepest intellectual ascent and there remains much ground for further basic research in the field of multi-PES reactions. What seems to be needed is a *consensus about the important and presently soluble problems*, and a closer cooperation between theory and experiment. First of all very little is known about the relevant PES's beyond the level of minimal state correlation diagrams for any systems contained in Tables II–X. The only exceptions are the alkali–halogen (Table I) and some of the ion-molecule CL reactions in Table XI. Among the numerous methods for calculating PES's reviewed in this volume by Tully, semiempirical methods of the DIM and pseudopotential type hold promise for first systematic studies of the correlations between the electronic structure of homologous series and the PES topologies. This should be done before embarking on much more demanding *ab initio* calculations. The alkaline earth–halogen systems (prototype: $Ca + F_2$) seem to be the best starting points for such an enterprise, based on their relatively simple electronic structure, the existing experimental data base, and the relative ease of extending and refining the experiments.

We may look forward to a deepened understanding of multi-PES reactions.

Acknowledgements

This review was written while the author was on sabbatical leave at the Chemistry Department, Harvard University, the Hahn-Meitner Institut für Kernforschung in Berlin, and the Facolta di Scienze, Universitá Libera di Trento. He wishes to express his gratitude to these institutions, and in particular to Professor D.R. Herschbach, Dr. A. Ding, and Professor G. Scoles for their hospitality.

V. Tables

TABLE I. Alkali metal + halogen Reactions[a]

Reactands	Products observed	Method → results	References
Na + (Cl_2, Br_2, I_2, O_2)	Na*, molecular	$F^1 \rightarrow kin^8$	Haber and Zisch (1922)
(Na, K) + (Cl_2, Br_2, $HgCl_2$)	$M^*(n^2S, {}^2P, {}^2D)$	$F^1 \rightarrow SP^5$	Beutler and Polanyi (1925) Beutler *et al.* (1926)
Na + (Cl_2, metal halides)	molecular MX*	$F \rightarrow kin^8, \Phi^7, dyn^{10}$	Beutler and Polanyi (1928)
Na + (Cl_2, Br_2, $HgCl_2$)	Na*	$F \rightarrow kin, \Phi^7$	Bogdandy and Polanyi (1928)
Na + Cl_2		$F \rightarrow kin, \Phi, dyn$	Polanyi and Schay (1928)
(Li, Na, K) + (Cl_2, Br, I)	MX*	$F \rightarrow spy^6$	Levi (1934)
(Li, Na, Cs) + (F_2, NF_3)	M*	$FHP^2 \rightarrow sp, \Phi, kin$	Luria *et al* (1976c)
	Modern Akali Age		
$KBr^{\neq}$ + Na	$K^*(4^2P)$	$B/B^3 \rightarrow (V \rightarrow E)$ energy transf.	Moulton and Hershbach (1966)
$NaBr^{\neq}$ + K	$K^*(4^2P)$		
M + NX			
(M = K, Rb; N = Li, Na, K, Rb, Cs; X = Br, I)	N*	$B/B \rightarrow (T \rightarrow E)$ energy transf.	Neoh and Hershbach (1975).
Na + (F_2, Cl_2)	Na*, NaX_2^*	$F \rightarrow sp, kin$	Ham and Chang (1974)
Li_2 + H	Li*, Li**, LiH*	$B/B \rightarrow sp$	Crooks *et al.* (1978)
M_2 + X systems			
(M = Na, K, Rb; X = Cl, Br, I)	M*, M**	$B/B \rightarrow sp, \sigma_{CL}^9$, dyn	Struve *et al.* (1971, 1973, 1975)
	M*, M**	$F \rightarrow D_0^{0\ 14}$	Ham (1973, 1974)

$M_2 + X_2(XY)$ *systems*			
$(K_2, Rb_2, Cs_2) + (Cl_2, Br_2, I_2,$ ICl, IBr, ClF)	MX*	$B/G^4 \rightarrow$ spy, dyn,	Oldenburg *et al.* (1974) Kaufmann *et al.* (1974a, b)
$(K_2, Rb_2, Cs_2) + (Cl_2, Br_2, I_2,$ ICl)	M*, MX*	$B/B \rightarrow$ sp, σ_{CL}, dyn,	Struve *et al.* (1973, 1975)

[a] For complete bibliography see McFadden (1979). Also compiled by McFadden are reactions with HX Halide salts, polyhalides and alkyl halides.

1. F = bulk-flow experiment
2. FHP = heat pipe oven experiment
3. B/B = crossed beam experiment
4. B/G = beam-gas experiment
5. sp = CL spectra reported
6. spy = spectroscopic analysis
7. Φ = quantum yield
8. kin = kinetics, mechanism
9. σ_{CL} = CL cross section
10. dyn = dynamics
11. $P(J', v', e')$ = rot-vib-el distribution
12. $P(v'|E_T)_{e'} = E_T$-dependent $P(V')_{e'}$
13. Q = quenching
14. D_0^0 = dissociation energy
15. $d\sigma/d\Omega$ = ang. distribution
16. CI = chemiionization
17. $\sigma_{CI}(E_T)$ = CI excitation function
18. $\sigma_{CL}(E_T)$ = CL excitation function
19. TI = theory, interpretation
20. corr. diag = correlation
21. $k(T)$ = rate constant
22. pol = CL polarization

TABLE II. Group IIa metal + halogen reactions

Reactands	Products observed	Method → results	References
(Sr, Ba) + Cl_2	MCl*, MCl_2^*	B/G[4] → kin[8], sp[5]	Jonah and Zare (1971)
Ba + Cl_2	BaCl*(*C*)	F[1] → $P(v')$[11]	Menzinger and Wren (1973)
	$BaCl_2^*$	F → Q[13]	Wren and Menzinger (1973)
Ca + (F_2, Cl_2, Br_2)	CaX*, CaX_2^*	F → sp, $P(v')$	Menzinger (1974a)
(Ca, Sr, Ba) + (F_2, Cl_2, Br_2)	MX*	F → sp, D_0^0(MX)[14]	Menzinger (1974b)
(Mg, Ca, Sr, Ba) + (F_2, Cl_2, NOCl)	MX*	F → Φ	Eckstrom *et al.* (1974)
Ba + (F_2, Cl_2, Br_2, I_2, SF_6)	BaX*	F → $P(v', e')$[11]	Bradford *et al.* (1975)
(Ca, Ba) + NOCl	MCl*	F → $P(e')$, sp, Φ	Obenauf *et al.* (1973)
(Ca, Sr) + (F_2, Cl_2, Br_2)	MX*	F → $P(e')$, sp, Φ	Capelle *et al.* (1975a, b)
Ba + (Cl_2, Br_2, I_2)	BaX_2^*	B/G → kin	Wren and Menzinger (1974)
Ba + I_2	BaI*(*C*)	B/G → D_0^0(BaI)	Dickson *et al.* (1976)
			Estler and Zare (1978b)
Ba + Cl_2	$BaCl_2$	B/B → $d\sigma/d\Omega$[15]	Mims and Brophy (1977)
Ba + CCl_4	BaCl*, CCl_2^*	B/B → sp, $d\sigma/d\Omega$	Siegel and Schultz (1978)
(Mg, Ca, Sr) + (Cl_2, HCl)	MCl*	B/G → sp, dyn corr. diag[20]	Brinkmann and Telle (1977)
			Kowalski and Heldt (1978)
(Ca, Sr, Ba) + (F_2, Cl_2, Br_2, I_2, ICl)	$MX^+ + X^-$	B/G → σ_{CI}[16]	Diebold *et al.* (1977)
(Ca, Sr, Ba) + (F_2, Cl_2, Br_2)	$MX^+ + X^-$	B/B → $\sigma_{CL}(E_T)$[17]	Wren and Menzinger (1979b, c)
(Ca, Sr, Ba) + S_2Cl	MX_2^*, S_2^*	B/G → sp, $P(v')$, D_0^0	Engelke and Zare (1977)
(Ca, Sr, Ba) + ClO_2	MCl*, MO*	B/G → sp, $P(v', J', e')$, D_0^0	Engelke *et al.* (1977)

TABLE III. Group IIa metal + oxygen donors (O_3, NO_2, N_2O; O_2, ClO_2)

Reactands	Products observed	Method → results	References
(Ba, Sr) + (NO_2, N_2O)	MO*	B/G → sp, kin, $P(v')$, D_0^0	Ottinger and Zare (1970)
(Ca, Sr, Ba) + (NO_2, N_2O)	MO*	B/G → sp, kin, $P(v')$, D_0^0, pol[2]	Jonah *et al.* (1972)
Ba + (NO_2, N_2O)	MO*	F → Φ	Obenauf *et al.* (1972, 1973)
Ca + (NO_2, N_2O, O_3, O_2)	CaO*	F → sp, spy, Φ	Capelle *et al.* (1975a)
Sr + (NO_2, N_2O, O_3)	SrO*	F → Φ, sp, spy,	Capelle *et al.* (1975b)
Ba + N_2O	BaO*	F → Φ(p), kin	Jones and Broida (1973)
Ba + N_2O	BaO*		Hsu *et al.* (1974a, b)
Ba + N_2O	BaO*		Eckstrom *et al.* (1974, 1975)
Ba + N_2O	BaO*		Palmer *et al.* (1975)
Ba + N_2O	BaO*	F → sp, Φ, kin, $P(v')$	Jones and Broida (1974)
Ba + N_2O	BaO*	spy	Field (1974)
Ba + N_2O	BaO*	F → sp, $P(v')$, corr. diag	Field *et al.* (1974)
Ba + N_2O	BaO*	F → kin	Wicke *et al.* (1975)
Ba + N_2O	BaO*	B/B → $d\sigma_{CL}/d\Omega$	Siegel and Schultz (1978)
Ba + N_2O	BaO	B/B → $d\sigma/d\Omega$, dyn	Parr *et al.* (1977)
Ba + N_2O	BaO*	B/B → $\sigma_{CL}(E_T, T_V)$, dyn	Wren and Menzinger (1975)
Ba + (N_2O, NO_2)	BaO*	B/B → $\sigma_{CL}(E_T, T_V)$, dyn	Wren and Menzinger (1979a)
Ca*(^3P) + (O_2, N_2O)	CaO*	B/B → Φ	Pasternack and Dagdigian (1978)
Ba + N_2O	MO*	TI[19]	Husain and Wiesenfeld (1975)
	MO*	TI	Field *et al.* (1975)
	MO*	TI	Bowman *et al.* (1976)
M + N_2O + CO	MO*	F → kin, sp	Benard *et al.* (1976)
(M = Mg, Ca, Sr, Ba)	MO*	F → kin, sp	Benard *et al.* (1977a, b)
(M = Mg, Ca, Sr, Ba)	MO*	F → kin, sp	Benard and Slafer (1977)
(M = Mg, Ca, Sr, Ba)	MO*	F → kin, sp	Eckstrom *et al.* (1977)
Ba + O_2	BaO*	F → sp, kin	Edelstein *et al.* (1977)
(Ca, Sr, Ba) + ClO_2	MO*, MCl	B/G → sp, $P(v', J', e')$, D_0^0	Engelke *et al.* (1977)

TABLE IV. Group IIIa atom reactions

Reactands	Products observed	Method → results	References
(B, Al) + (N_2O, O_2, Cl_2, F_2)	MO*, MX*	superson. mix. → sp	Johnson *et al.* (1974)
B + N_2O	BO*	B/B → $\sigma_{CL}(E_T)$	Tang *et al.* (1976)
B + MF(M = Na, K, Rb, Cs)	M*(^2P)	F → sp, $P(e')$, kin, Φ	Sridharan *et al.* (1977)
Al + O_2	AlO*(?)	TI	Kolb *et al.* (1975)
Al + (N_2O, O_3, NO_2, O_2, CO_2)	AlO*	F → Φ, sp,	Rosenwaks *et al.* (1975)
Al + (F_2, NF_3, SF_6)	AlX*($a^3\Pi$)	F → $P(v')$, sp, spy	Rosenwaks *et al.* (1976)
Al + NF_3	AlF*, Al*	F → Φ, $P(e')$, sp, spy	Rosenwaks and Broida (1976)
Al + (F_2, Cl_2, Br_2, I_2, NF_3, SF_6)	AlX*, NF*, N_2^*	F → Φ, kin, sp	Rosenwaks (1976)
(Al, Ca) + $H_2C{=}O$	MH*	F	Sakurai *et al.* (1976)
Al + O_3	AlO*	B/G → D_0^0(AlO)	Gole and Zare (1972)
Al + O_3	AlO*	F → sp, $P(v')$, kin	Lindsay and Gole (1977)
Al + O_3	AlO*	B/G → sp, $P(v')$	Sayers and Gole (1977)
Tl + F_2	TlF*, Tl*	F → sp, kin	Maya and Nordine (1976)
(Tl, In) + $I_2^*(B^3)$	MI*	B/G → laser-induced CL react. orientation, pol.	Estler and Zare (1978a)

TABLE V. Group IVa atom reactions

Reactands	Products observed	Method → results	References
$C_2(^3\Pi) + NO(^2\Pi)$	$CN^*(B)$	B/B → sp, $P(v')$	Krause (1978)
$Si + (O_2, N_2O, F_2)$	SiO^*	F → sp	Johnson *et al.* (1974)
$(Si, Ge) + N_2O$	SiO^*, GeO^*	F → sp	Hager *et al.* (1974)
$Si + F_2$	SiF^*	F → sp	Armstrong and Davis (1977)
$Ge + (O_2, NO, NO_2, N_2O)$	GeO^*	F → spy, Φ	Capelle and Brom (1975)
$Ge + N_2O$	GeO^*	F → kin	Swearengen *et al.* (1977)
$Ge + (F_2, Cl_2, Br_2)$	GeF^*	F → spy, Φ	Capelle and Brom (1975)
$Sn + (O_2, NO_2, N_2O)$	SnO^*	F → sp, Φ, spy	Capelle and Linton (1976)
$Sn + F_2$	SnF^*	F → sp, Φ	Capelle and Linton (1976)
$Sn + N_2O$	SnO^*	F → kin, $k(T)^{21}$, corr. diag	Wiesenfeld and Yuen (1976)
		F → sp, kin, $k(T)$, Φ	Felder and Fontijn (1975, 1978)
$Pb + F_2$	PbF^*	B/G → spy, $P(v')$	Dickson and Zare (1978)
$Pb + O_3$	PbO^*	B/G → spy, kin, D_0^0	Oldenburg *et al.* (1975)
$Pb + O_3$	PbO^*	F → kin, laser excitn.	Kurylo *et al.* (1976)
$Pb + (O, O_2, O_3, N_2O)$	PbO^*	F → spy	Linton and Broida (1976)
$Pb + N_2O$	$PbO^*(B)$	B/G → σ_{CL} (E_T)	Wicke *et al.* (1978)

TABLE VI. Group IIIb + Halogen—containing molecules

Reactands	Products observed	Method → results	References
$Sc + F_2$	ScF*	F → sp, kin	Fischell and Cool (1977)
(Sc, Y) + (F_2, Cl_2, ClF)	MX*	B/G → kin, sp	Gole *et al.* (1977)
(Eu, Sm) + (F_2, NF_3)	MX*	F → sp, $\Phi(p)$	Eckstrom *et al.* (1975)
(Eu, Sm) + F_2	MX*	B/G → sp, kin, D_0^0, Φ^{rel}	Dickson and Zare (1975)
(Sm, Yb) + (F_2, Cl_2)	MX*	B/G → sp, kin, D_0^0, Φ^{rel}	Yokozeki and Menzinger (1976)
$Sm + F_2$	MX*	B/G → $\Phi(p)$	Dickson *et al.* (1977)
(Sc, Y, La) + (F_2, FCl, Cl_2)	MX*	B/G sp, spy, Φ, kin	Gole (1977)

TABLE VII. Group IIIb + Oxygen donors

Reactands	Products observed	Method → results	References
(Sc, Y, La) + O_2	MO*	B/B → sp, $P(v' \mid E_T)^{1,2}_{e'}$, σ_{CL} (E_T)	Manos and Parson (1975, 1978)
(Sc + Y) + O_2	ScO*	B/G → spy, D_0^0	Chalek and Gole (1976)
La + OCS, LaO*, LaS*	LaS*, LaO*	B/G → sp, D_0^0 (LaS)	Jones and Gole (1977)
La + (O_2, NO_2, N_2O, O_3)	LaO*	B/G → sp, D_0^0 (LaO), kin	Gole and Chalek (1976)
(Sc, Y, La) + (O_2, NO_2, N_2O)	MO*	B/G → kin	Gole and Preuss (1977)
La + O_2	LaO*	B/G → kin	Gole *et al.* (1977)
(Sm, Eu) + N_2O	MO*	F → $\Phi(p)$,	Edelstein *et al.* (1974)
(Sm, Eu, Ba) + (O_2, N_2O, O_3)	MO*	F → $\Phi(p)$, sp,	Eckstrom *et al.* (1975)
(Eu, Sm) + (NO_2, N_2O, O_3)	MO*	B/G → sp, kin, D_0^0, Φ^{rel}	Dickson and Zare (1975)
(Sm, Yb) + (O_3, N_2O)	MO*	B/G → sp, kin, D_0^0, Φ^{rel}	Yokozeki and Menzinger (1976)
Sm + N_2O	SmO*	B/B → σ_{CL} (E_T)	Yokozeki and Menzinger (1977a, b)
Ho + N_2O	HoO*	B/G → σ_{CL} (E_T)	Tang *et al.* (1976)
Sm + N_2O	SmO*	B/G → $\Phi(p)$ (abs)	Dickson *et al.* (1977)

TABLE VIII. Groups IVb–VIIIb reactions

Reactands	Method → results	References
(Ti, V, W, Zr, Cu) + (F_2, Cl_2)	F → sp	Johnson *et al.* (1974)
Cu + F_2	B/B → sp, $P(v' \mid E_T)_{e'}^{12}$, σ_{CL} (E_T)	Parson (1978)
$TiCl_4$ + K + N_2O	F → spy	Palmer and Hsu (1972)
(Ti, V, W, Zr, Cu) + (N_2O, O, O_2)	F → sp	Johnson *et al.* (1974)
Ti + (NO, NO_2, N_2O)	F → sp, $P(e')$	Linton and Broida (1977)
Ti + (O_2, NO_2, N_2O)	B/G → kin, sp, D_0^0	Dubois and Gole (1977)
(Ti, V) + (O_2, SO_2, CO_2, N_2O)	B/B → sp, $P(v' \mid E_T)_{e'}$, σ_{CL} (E_T)	Parson and Conway (1977)
		Conway and Parson (1978)
Fe + (O_2^*, NO_2, N_2O, O_3)	F → sp, Φ, spy	West and Broida (1975)

TABLE IXa. Reactions of NO and SO with O_3

Reactands	Products observed	Method → results	References
$NO + O_3$	NO_2^*	F → kin	Clough and Thrush (1967)
	NO_2^*	F → no $O_2^*(^1\Delta_g, ^1\Sigma_g^+)$	Gauthier and Snelling (1973)
	$NO_2^{\neq}$	F → vib. emission, Q	Golde and Kaufmann (1974)
$NO^{\neq}(V = 1) + O_3$	NO_2^*	F(CO laser excitation) → kin	Stephenson and Freund (1976)
$NO + O_3^{\neq}$	NO_2^*	F(CO_2 laser excitation) → kin	Gordon and Lin (1973, 1976)
			Braun *et al.* (1974)
			Kurylo *et al.* (1974, 1975)
			Moy *et al.* (1977)
			Bar-Ziv *et al.* (1977, 1978)
$NO(^2\Pi_{1/2,3/2}) + O_3$	NO_2^*	B/G → σ_{CL} (E_T)	Redpath and Menzinger (1971, 1975)
$NO(^2\Pi_{1/2,3/2}) + O_3$	NO_2^*, $NO_2^{\neq}$	B/G → σ_{CL} (E_T), sp, dyn	Redpath *et al.* (1978)
$SO + O_3$	SO_2^*	F → kin, sp	Halstead and Thrush (1964)
$SO + O_3^{\neq}$	SO_2^*	F(CO_2 laser excitation) → kin	Kaldor *et al.* (1974)

TABLE IXb. Other reactions of groups Va-VIIa

Reactands and products	Method → results	References
$N_3 + (Cl, Br, O, N) \rightarrow NX^*$	F → sp, spy, kin	Clark and Clyne (1970)
$NF_2 + (O, N, H) \rightarrow NF^*(a, b)$	F → kin,	Clyne and White (1970)
$NF_2 + (H, D, CH_3) \rightarrow NF^*(a, b)$	F → kin	Herbelin and Cohen (1973)
		Kwok *et al.* (1976)
		Herbelin *et al.* (1977)
$F_2 + I_2^* \rightarrow IF^*(B)$	B/G → sp,	Engelke *et al.* (1977)

TABLE X. 1,2-Dioxetanes

		References
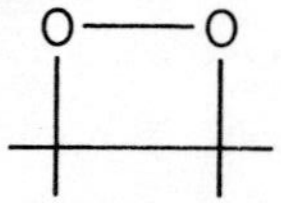	Review	Turro *et al.* (1974)
Dioxetanes general	Review	Wilson (1976)
$O_2(^1\Delta_g)$ + alkenes	F → sp, kin	Bogan *et al.* (1975, 1976)
$O_2(^1\Delta_g)$ + alkenes	B/B → σ_{CL} (E_T, T_v)	Alben *et al.* (1978)
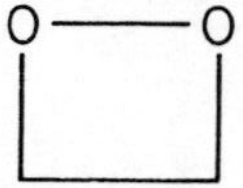	Ti → PES (+ review)	Harding and Goddard (1977)
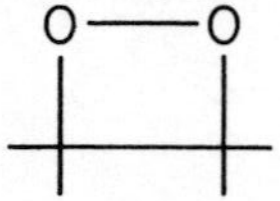	CO_2-laser i.r. absorption	Farneth *et al.* (1976)
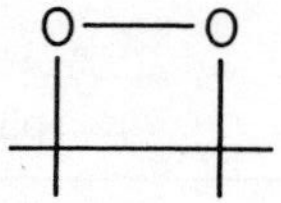	Multiphoton i.r. absorption	Haas and Yahav (1977, 1978)
	Picosecond dye laser	Smith *et al.* (1977)
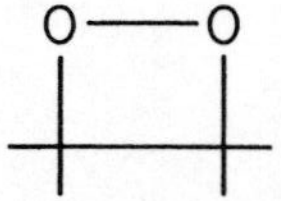	B/G → coll. ind. dissoc., σ_{CL} (E_T)	Brown and Menzinger (1978)

TABLE XI. Ion–Molecule Reactions

Reactands	Products observed	Method → results	References
$C^+(^2P) + H_2, D_2$	$CH^{+*}(A^1\Pi, b^3\Sigma^-, B^1\Delta)$	B/G → sp, $\sigma_{CL}(E_T)$, Theory	Appell *et al.* (1975)
			Ottinger (1976, 1978)
			Harris *et al.* (1975)
$O^+ + H_2$	$OH^{+*}(A^3\Pi)$, $OH^*(A^2\Sigma^+)$	B/G → sp, $\sigma_{CL}(E_T)$	Harris and Leventhal (1976)
$C^+(^2P) + O_2$	$CO^{+*}(A^2\Pi)$, $O_2^{+*}(A^2\Pi, b^4\Sigma_g^-)$	B/G → sp, $\sigma_{CL}(E_T)$, $P(v', J')$	Ottinger and Simonis (1975)
			Ottinger (1976, 1978)
$N^+(^3P) + NO$	$N_2^{+*}(B^2\Sigma_u^+)$, $NO^*(A^2\Sigma^+, B^2\Pi)$	B/G → sp, $\sigma_{CL}(E_T)$, $P(v', J')$, kin	Brandt and Ottinger (1973)
			Brandt *et al.* (1973)
			Ottinger (1976, 1978)
$C^+ + NO$	$CN^*(B^2\Sigma^+)$, NO^*	B/G → sp, $\sigma_{CL}(E_T)$, kin, $P(v')$	Brandt and Ottinger (1973)
$N^+ + RH(RH = H_2, CH_4, C_2H_4, C_2H_6, C_3H_8)$	$NH^*(A^3\Pi)$	B/G → sp, $\sigma_{CL}(E_T)$, $P(v', J')$	Kusunoki *et al.* (1976)
			Ottinger (1978)
			Kusunoki and Ottinger (1978a, b)
$O^+ + RH$	$OH^{+*}(A^3\Pi)$	B/G → sp	Ottinger (1978)

References

Alben, K. T., Auerbach, A., Ollison, W. M. Weiner, J., and Cross R. J. (1978). Molecular beam study of the activation energy requirements for the dioxetane reaction, *J. Amer. Chem. Soc.*, **100**, 3274.

Albritton, D. L., Schmeltekopf, A. L., and Zare, R. N., (1980). *Diatomic Intensity Factors*, to be published by Wiley, New York.

Alexander, M. H. and Dagdigian, P. J. (1978). Statistical and dynamical influences on electronic branching in reactions of ground and excited state alkaline earth atoms with molecular oxidants, *Chem. Phys.*, preprint.

Appell, J., Brandt, D., and Ottinger, Ch. (1975). Chemiluminescent ion–molecule reactions in the $C^+ + H_2$ system, *Chem. Phys. Lett.*, **33**, 131.

Armstrong, R. and Davis, S. (1977). *Chemiluminescent Studies of the $A^2\Sigma^+$ and $a^4\Sigma^-$ States of SiF, Electronic Transitions Lasers II* (Eds. L. E. Wikson, S. Suchard, and J. I. Steinfeld), MIT Press, p. 133.

Balint-Kurti, G. G. and Karplus, M. (1974). Atoms in molecules, in *Orbital Theories of Atoms and Molecules* (Ed. N. H. March), Clarendon Press, Oxford, p. 250.

Barat, M. (1973). Excitation and ionization in ion–atom collisions, in *The Physics of Electronic and Atomic Collisions* Invited Lectures, VIII ICPEAC, Beograd 1973 (Eds. B. C. Cobic and M. V. Kurepa), Institute of Physics, Beograd.

Bar-Ziv, E., Moy, J., and Gordon, R. J. (1977). Temperature dependence of the laser-enhanced reaction $NO + O_3$ (001). II. Contributions from reactive and non-reactive channels, *J. Chem. Phys.*, **68**, 1013.

Bar-Ziv, E., Moy, J., and Gordon, R. J. (1978). Temperature dependence of the laser-enhanced reaction $NO + O_3$ (001). II. Contributions from reactive and non-reactive channels, *J. Chem. Phys.*, **68**, 1013.

Behrens, R., Freedman, A., Herm, R. R., and Parr, T. P. (1976). Crossed beams chemistry: Ba $(^1S) + SO_2(^1A) \rightarrow BaO(^1\Sigma^+) + SO(^3\Sigma^-)$, *J. Amer. Chem. Soc.*, **98**, 294.

Benard, D. J. and Slafer W. D. (1977). Mechanism of chemiluminescent chain reactions in Mg catalyzed $N_2O + CO$ flames, *J. Chem. Phys.*, **66**, 1017.

Benard, D. J., Slafer, W. D., and Hecht J. (1977a). Chain reaction chemiluminescence of alkaline earth catalyzed N_2O–CO flames, *J. Chem. Phys.*, **66**, 1012.

Benard, D. J., Slafer, W. D., and Lee P. H. (1976). Efficient chemical production of metastable alkaline earth atoms, *Chem. Phys. Lett.*, **43**, 69.

Benard, D. J., Slafer, W. D., Love, P. J., and Lee P. H. (1977b). Modulated transmission spectroscopy of gaseous chemi-excited Ca and Sr monoxides, *Appl. Opt.*, **16**, 2108.

Bergmann, K. Leone, S. R., and Moore C. B. (1975). Effect of reagent electronic excitation on the chemical reaction Br $(^2P_{1/2,3/2}) + HI$, *J. Chem. Phys.*, **63**, 4161.

Beutler, H., Bogdandy, St.v., and Polanyi, M. (1926). Uber Lumineszenz hochverdünnter Flammen, *Naturwissenschaften*, **14**, 164.

Beutler, H. and Polanyi, M. (1925). Reaktionsleuchten und Reaktions-geschwindigkeit, Naturwissensch. **13**, 711.

Beutler, H. and Polanyi, M. (1928). Ueber hochverduennte Flammen I Flammen im einfachen Rohr. Vorläufige Analyse des Reaktionsmechanismus. Reaktionsgeschwindigkeit, Leuchtvorgang, *Z. Phys. Chem.*, **B1**, 3.

Black, G., Luria, M., Eckstrom, D. J., Edelstein, S. A., and Benson S. W. (1974). Chemiluminescence photon yields for some encapsulated metal system flames, *J. Chem. Phys.*, **60**, 3709.

Bogan, D. J., Sheinson, R. S., and Williams, F. W. (1976). Gas phase dioxetane chemistry. Formaldehyde $(A \rightarrow X)$ chemiluminescence from the reaction of $O_2(^1\Delta_g)$ with ethylene, *J. Amer. Chem. Soc.*, **98**, 1034.

Bogan, D. J., Sheinson, R. S., Williams, F. W., and Gann R. G. (1975). Formaldehyde ($A^1A_2 \rightarrow X^1A_1$) chemiluminescence in the gas phase reaction of $O_2(a^1\Delta_g)$ plus ethyl vinyl ether, *J. Amer. Chem. Soc.*, **97**, 2560.

Bogandy, St.v. and Polanyi, M. (1928). Ueber hoch verduennte Flammen II Anstieg der Lichtausbeute beiwachseudien Partialdruck des Natriunmolampfes, *Z. Phys. Chem.*, **B1**, 21.

Bowman, J. M., Leasure, S. C., and Kuppermann, A. (1976). Large quantum effects in a model electronically nonadiabatic reaction: $Ba + N_2O \rightarrow BaO^* + N_2$, *Chem. Phys. Lett.*, **43**, 374.

Bradford, R. S., Jones, C. R., Southall, L. A., and Broida, H. P. (1975). Production efficiences of electronically excited states of BaX*, *J. Chem. Phys.*, **62**, 2060.

Brandt, D. and Ottinger, Ch. (1973). $N_2^+(B^2\Sigma_u^+)$ and $CN(B^2\Sigma^+)$ emission from chemiluminescent ion–molecule collisions, *Chem. Phys. Lett.*, **23**, 257.

Brandt, D., Ottinger, Ch., and Simonis, J. (1973). Visible and UV luminescence on ion-beam-neutral collisions, *Ber. Bunsen. Ges.*, **77**, 648.

Braun, W., Kurylo, M. J., Kaldor, A., and Wayne R. P. (1974). Infrared laser enhanced reactions: Spectral distributions of the NO_2 chemiluminescence produced in the reaction of vibrationally excited O_3 with NO, *J. Chem. Phys.*, **61**, 461.

Brinkmann, U. and Telle H. (1977). Luminescent reactive collisions between excited Ca atoms and HCl, Cl_2, *J. Phys., B.*, **10**, 133.

Brown, J. C. and Menzinger, M. (1978). Molecular beam chemiluminescence: Collisional dissociation of tetramethyl-dioxetane by a fast xenon beam, *Chem. Phys. Lett.*, **54**, 235.

Burak, I. and Eyal, M. (1978). ${}^2P_{1/2}$ excited iodine formation in the F + HI reaction, (preprint).

Capelle, G. A. and Brom, J. M. (1975). Reactions of germanium vapor with oxidizers: Photon yields and a new GeO band system, *J. Chem. Phys.*, **63**, 5168.

Capelle, G. A., Jones, C. R., Zorskie, J., and Broida H. P. (1975a). Photon yields and spectra resulting from reactions of Ca with oxidants, *J. Chem. Phys.*, **61**, 4777.

Capelle, G. A., Broida, H. P., and Field R. W. (1975b). Photon yields of several reactions producing diatomic strontium oxide and halides, and $SrO(A'^1\Pi - X^1\Sigma)$: a new band system, *J. Chem. Phys.*, **62**, 3131.

Capelle, G. A. and Linton C. (1976). Chemiluminescence spectra and photon yields for several Sn—oxidizer reactions, *J. Chem. Phys.*, **65**, 5361.

Carrington, T. (1961). Transition probabilities in multilevel system: Calculation from impulsive and steady state experiments, *J. Chem. Phys.*, **35**, 807.

Carrington, T. (1973). Chemiluminescence in gases, in *Chemiluminescence and Bioluminescence* (Eds. M. H. Cormier, D. M. Hercules, and J. Lee), Plenum Press, pp. 7.

Carrington, T. (1974). The geometry of intersecting potential surfaces, *Accts. Chem. Res.* **7**, 20.

Carrington, T. and Garvin D. (1969). in *Comprehensive Chemical Kinetics* (Eds. C. H. Bamford and D. F. H. Tipper), Vol. 3, p. 107.

Carrington, T. and Polanyi, J. C. (1972). In *MTP Int. Rev. Sci.*, **9**, 135.

Case, D. A., McCelland, G. M., and Hershbach, D. R. (1978). Angular momentum polarization in molecular collisions: Classical and quantum theory for measurements using resonance fluorescence, *Mol. Phys.*, **35**, 541.

Chalek, C. L. and Gole, J. L. (1976). Chemiluminescence spectra of ScO and YO: Observation and analysis of the $A^2\Delta - X^2\Sigma^+$band system, *J. Chem. Phys.*, **65**, 2845.

Clark T. C. and Clyne M. A. A. (1970). Kinetics of chemiluminescent reactions of the gaseous azide radical, *Trans. Faraday Soc.*, **66**, 877.

Clough, P. N. and Thrush, B. A. (1967). Mechanism of the chemiluminescent reaction between nitric oxide and ozone, *Trans. Faraday Soc.*, **63**, 915.
Clyne, M. A. A. and Coxon J. A. (1966). A novel chemiluminescent reaction: Detection of electronic emission from BrCl, *Chem. Commun., (Chem. Soc.)*, 285.
Clyne, M. A. A. and White I. F. (1970). Electronic energy transfer processes in fluorine-containing radicals: Singlet NF, *Chem. Phys. Lett.*, **6**, 465.
Conway, T. J. and Parson J. M. (1978). Private communication.
Cormiër, M. J., Hercules, D. M., and Lee, J., (Eds.) (1973). *Chemiluminescence and Bioluminescence*, Plenum Press, New York.
Crooks, J. B., Way, K. R., Yang, S. C., Wu, C. Y. R., and Stwalley W. C. (1978). Photon and positive ion production from collisions of superthermal hydrogen atoms with lithium atoms and molecules, *J. Chem. Phys.*, **69**, 490.
Dagdigian, P. J. (1978). Determination of the absolute chemiluminescence cross section and photon yield for the $Ca^*(4_s4_p3_p) + N_2O$ reaction, *Chem. Phys. Lett.* **55**, 239.
Davidovits, P. and McFadden, D. L. (1978). *Alkali Halide Vapors: Structure, Spectra and Reaction Dynamics*, Academic Press.
Dickson, C. R., George, S. M., and Zare R. N. (1977). Determination of absolute photon yields under single collision conditions, *J. Chem. Phys.*, **67**, 1024.
Dickson, C. R., Kinney, J. B., and Zare, R. N. (1976). Determination of D_0^0(BaI) from the chemiluminescent reaction $Ba + I_2$, *Chem. Phys.*, **15**, 243.
Dickson, C. R. and Zare R. N. (1975). Beam-gas chemiluminescent reactions of Eü and Sm with (O_3, N_2O, NO_2, F_2), *Chem. Phys.*, **7**, 361.
Dickson, C. R. and Zare, R. N. (1978). Spectroscopic study of $Pb + F_2$ chemiluminescence, *Optical Pura y Aplicada*, accepted.
Diebold, G. J., Engelke, F., Lee, H. U., Whitehead, J. C., and Zare R. N. (1977). Chemi-ionization reactions of Ca, Sr, Ba and Yb atoms with the halogen and interhalogen molecules, *Chem. Phys.*, **30**, 265.
Dinur, U., Kosloff, R., Levine, R. D., and Berry M. J. (1975). Electronically non-adiabatic reactions: Information theoretic approach, *Chem. Phys. Lett.*, **34**, 199.
Dinur, U. and Levine, R. D. (1975). Does $H + ICl \rightarrow HCl + I^*(^2P_{1/2})$ occur? *Chem. Phys. Lett.* **31**, 410.
Donovan, R. J. and Husain, D. (1970). Recent advances in the chemistry of electrically excited atoms, *Chem. Revs.*, **70**, 489.
Drowart, J. and Goldfinger P. (1967). High temperature inorganic mass spectrometry, *Angew. Chem. (Int. Ed.)*, **6**, 581.
Dubois, L. H. and Gole, J. L. (1977). Bimolecular, single collision reaction of ground metastable excited states of titanium with O_2, NO_2 and N_2O: Confirmation of D_0^0 (TiO), *J. Chem. Phys.*, **66**, 779.
Eckstrom, D. J., Barker, J. R., Hawley, J. G., and Reilly J. P. (1977). Intracavity dye laser spectroscopy studies of the $Ba + N_2O$, $Ca + N_2O + CO$ and $Sr + N_2O + CO$ reactions, *Appl. Opt.*, **16**, 2101.
Eckstrom, D. J., Edelstein, S. A. and Benson, S. W. (1974). Chemiluminescence photon yields for several alkaline earth metal–halogen/oxygen reactions, *J. Chem. Phys.*, **60**, 2930.
Eckstrom, D. J., Edelstein, S. A., Huestis, D. L., Perry, B. E., and Benson, S. W. (1975). Chemiluminescence studies. IV. Pressure-dependent photon yields for (Ba, Sm, Eu) + (N_2O, O_3, O_2, F_2, NF_3) reactions, *J. Chem. Phys.*, **63**, 3828.
Edelstein, S. A., Eckstrom, D. J., Perry, B. E., and Benson, S. W. (1974). Chemiluminescence studies. III. Pressure dependent photon yields for several metal M(Ba, Sm, Eu, Tl, Pb, P_4) + N_2O reactions, *J. Chem. Phys.*, **61**, 4932.

Edelstein, S. A., Perry, B. E., Eckstrom, D. J., and Gallagher, T. F. (1977). Chemiluminescence from three body reactions in the Ba + O_2 flame, *Chem. Phys. Lett.*, **49**, 293.

Engelke, F., Sander, R. K., and Zare, R. N. (1976). Crossed-beam chemiluminescent studies of alkaline earth atoms with ClO_2, *J. Chem. Phys.*, **65**, 1146.

Engelke, F., Whitehead, J. C., and Zare R. N. (1977). The four centre reaction $I_2^* + F_2$ studied by laser-induced chemiluminescence in moleculear beams, *Faraday Disc. Chem. Soc.*, **62**, 222.

Engelke, F. and Zare R. N. (1977). Crossed beam chemiluminescence: the alkaline rearrangement reaction $M + S_2Cl_2 \rightarrow S_2^* + MCl_2$, *Chem. Phys.*, **19**, 327.

Estler, R. C. and Zare, R. N. (1978a). Laser induced chemiluminescence. Variation of reaction rates with reactant approach geometry, *J. Amer. Chem. Soc.*, **100**, 1323.

Estler, C. R. and Zare, R. N. (1978b). Determination of bond energies by time-of-flight single-collision chemiluminescence, *Chem. Phys.*, **28**, 253.

Faist, M. B. and Levine R. D. (1977). On the product electronic state distribution in reactions of alkali dimers with halogen atoms, *Chem. Phys. Lett.*, **47**, 5.

Farneth, W. E., Flynn, G., Slater, R., and Turro, N. J. (1976). Time resolved infrared laser photochemistry and spectroscopy: The methyl fluoride sensitized decomposition of tetramethyl-1, 2-dioxetane. An example of infrared laser induced electronic excitation, J. *Amer. Chem. Soc.*, **98**, 7877.

Felder, W. and Fontijn, A. (1975). High-temperature fast-flow reactor study of Sn/N_2O chemiluminescence, *Chem. Phys. Lett.*, **34**, 398.

Felder, W. and Fontijn, A. (1978). HTFFR kinetics studies of Sn/N_2O. A highly efficient chemiluminescent reaction, *J. Chem. Phys.*, **69**, 1112.

Feofilov, P. P. (1961). *The Physical Basis of Polarized Emission*, Consultants Bureau Enterprises, Inc., New York.

Field, R. W. (1974). Assignment of the lowest $^3\Pi$ and $^1\Pi$ states of CaO, SrO, BaO, *J. Chem. Phys.*, **60**, 2400.

Field, R. W. (1976). Long-lived, energetic products of chemical reactions: Ba + N_2O, A case study, in *Molecular Spectroscopy: Modern Research*, Vol. II, Academic Press, New York.

Field, R. W., Jones, C. R., and Broida, H. P. (1974). Gas-phase reaction of Ba with N_2O. II. Mechanism of reaction, *J. Chem. Phys.*, **60**, 4377.

Field, R. W., Jones, C. R., and Broida, H. P. (1975). Reply to comment by D. Husian and J. R. Wiesenfeld, *J. Chem. Phys.*, **62**, 2012.

Fischell, D. R. and Cool T. A. (1977). Spontaneous emission from SCF in a supersonic mixing flame, *electronic Transition Lasers*, **II**, 166.

Fluendy, M. A. D. and Lawley, K. P. (1973). *Chemical Applications of Molecular Beam Scattering*, Chapman and Hall, London.

Gauthier, M. and Snelling D. R. (1973). Possible production of $O_2(^1\Delta_g)$ and $O_2(^1\Sigma_g^+)$ in the reaction ($NO + O_3$), *Chem. Phys. Lett.*, **20**, 178.

Gaydon, A. G. (1968). *Dissociation Energies and Spectra of Diatomic Molecules*, 3rd edn., Chapman and Hall, London.

Gelbart, W. M. and Freed, K. F. (1973). Intramolecular perturbations and quenching of luminescence in small molecules, *Chem. Phys. Lett.*, **18**, 470.

George, T. F. and Ross J. (1971). Analysis of symmetry in chemical reactions, *J. Chem. Phys.*, **55**, 3851.

Golde, M. F. and Kaufmann, F. (1974). Vibrational emission of NO_2 from the reaction of NO with O_3, *Chem. Phys. Lett.*, **29**, 480.

Golde, M. F. and Thrush, B. A. (1975). Chemiluminescence in gases, *Adv. Atom. Mol. Phys.*, **11**, 361.

Gole, J. L. (1973). Nonempirical LCAO–MO–SCF studies of the low lying states of BeF_2^*, *J. Chem. Phys.*, **58**, 869.

Gole, J. L. (1976). High temperature chemistry: Modern research and new frontiers, *Ann. Rev. Phys. Chem.*, **27**, 525.

Gole, J. L. (1977). Development of visible chemical lasers from reactions yielding visible chemiluminescence, *Electronic Transition Lasers*, **II**, 136.

Gole, J. L. and Chalek, C. L. (1976). Characterization of the ground and excited states of lanthanum oxide through biomolecular oxidation of La metal with O_2, NO_2, N_2O and O_2, *J. Chem. Phys.*, **65**, 4384.

Gole, J. L. and Preuss, D. R. (1977). The temperature dependence of single collision bimolecular beam-gas chemiluminescent reactions. II. Experimental studies, *J. Chem. Phys.*, **66**, 3000.

Gole, J. L., Preuss, D. R. and Chalek, C. L. (1977). Kinetics of metastable excited state products in a beam-gas chemiluminescent reaction, *J. Chem. Phys.*, **66**, 548.

Gole, J. L., Siu, A. K. O., and Hayes, E. J. (1973). Nonempirical LCAO–MO–SCF studies of the group IIa dihalides BeF_2, MgF_2 and CaF_2, *J. Chem. Phys.*, **58**, 857.

Gole, J. L. and Zare, R. N. (1972). Determination of D_0^0(AlO) from crossed beam chemiluminescence of $Al + O_3$, *J. Chem. Phys.*, **57**, 5331.

Gordon, R. J. and Lin, M. C. (1973). Laser enhancement of the reaction $NO + O_3 \rightarrow NO_2^* + O_2$, *Chem. Phys. Lett.* **22**, 262.

Gordon, R. J. and Lin M. C. (1976). The reaction of nitric oxide with vibrationally excited ozone. *J. Chem. Phys.*, **64**, 1058.

Hass, Y. and Yahav, G. (1977). Gas phase unimolecular decomposition and chemiluminescence of tetramethyldioxetane by a TEA CO_2 laser, *Chem. Phys. Lett.*, **48**, 63.

Haas, Y. and Yahav, G. (1978). Infrared multiphoton dissociation of tetramethyldioxetane: Disect detection of triplet acetone, *J. Amer. Chem. Soc.*, **100**, 4885.

Haber, F. and Zisch, W. (1922). Anregung von Gasspektren durch Chemische Reaktionen, *Z. Physik*, **9**, 302.

Hager, G., Wilson, L. E., and Hadley, S. G. (1974). Reactions of (Si, $Ge(^3P) + N_2O \rightarrow$ SIO*, GeO*, *Chem. Phys. Lett.*, **27**, 439.

Halstead, C. J. and Thrush, B. A. (1964). The chemiluminescent reaction of SO with O and with O_3, *Nature*, **204**, 992.

Ham, D. O. (1973). Contribution to the general discussion, *Faraday Disc. Chem. Soc.*, **53**, 313.

Ham, D. O. (1974). Energy limits in chemiluminescent, atom transfer reactions: Bond dissociation energy of NaF, *J. Chem. Phys.*, **60**, 1802.

Ham, D. O. and Chang, H. W. (1974). Chemiluminescence spectra of the new molecules NaF_2 and $NaCl_2$ and their implications to reaction dynamics, *Chem. Phys. Lett.*, **24**, 579.

Harding, L. B. and Goddard, W. A. (1977). Intermediates in the chemiluminescent reaction of singlet oxygen with ethylenes. *Ab initio* studies, *J. Amer. Chem. Soc.*, **99**, 4520.

Harris, H. H., Crowley, M. G., and Leventhal, J. J. (1975). Luminescence from C^+ (H_2, H) CH^+ below 20 eV, *Phys. Rev. Lett.*, **34**, 67.

Harris, H. H. and Leventhal, J. J. (1976). Ultraviolet emission in $O^+ + H_2$ reactive scattering, *J. Chem. Phys.*, **64**, 3185.

Hayashi, S., Mayer, T. M., and Bernstein, R. B. (1978). Crossed molecules beam chemiluminescence study of the metastable mercury reaction: $Hg(6^3P_0) + Br_2 \rightarrow HgBr(B) + Br$, *Chem. Phys. Lett.*, 53, 419.

Hayes, E. F. (1966). Bond angles and bonding in group IIa metal halides, *J. Phys. Chem.*, **70**, 3740.

Helm, H. (1976). Mobilities of Kr^+, $Xe^+(^2P_{1/2,3/2})$ in Kr, Xe, *J. Phys. B*, **9**, 2931.

Herbelin, J. M. and Cohen, N. (1975). HELP: A model for evaluating the feasibility of using various CL reaction systems as chemical lasers, *J. Quantum Spectr. Rad. Transf.*, **15**, 731.

Herbelin J. M. and Cohen N. (1973). The chemical production of electronically excited states in the H/NF_3 system, *Chem. Phys. Lett.*, **20**, 605.

Herbelin, J. M., Kwok, M. A., and Spencer D. J. (1977). In *Chemistry and Scale-up of the NF-system, Electronic Transition Lasers II* (Eds. L. E. Wilson, S. Suchard, and J. I. Steinfeld). MIT Press, p. 96.

Herzberg, G. (1950). *Spectra of Diatomic Molecules*, Van Norstrand, Princeton, New Jersey.

Herzberg, G. (1967). *Electronic Spectra of Polyatomic Molecules*, Van Norstrand, Princeton, New Jersey.

Hessel, M. M., Drullinger, R. E. and Broida H. P. (1975). Chemiluminescent reactions in a heat pipe oven, *J. Appl. Phys.*, **46**, 2317.

Hildenbrand, O. L. (1977). Dissociation energies of CaBr, SrBr, BaBr, BaCl from mass spectrometric studies of gaseous equilibria, *J. Chem. Phys.*, **66**, 3526.

Hsu, C. J., Krugh, W. D., and Palmer, H. B. (1974a). Pressure dependence of the BaO $A^1\Sigma - X^1\Sigma$ photon yield in the reaction of Ba(g) with N_2O and NO_2, *J. Chem. Phys.*, **60**, 5118.

Hsu, C. J., Krugh, W. D., Palmer, H. B., Obenauf, R. H., and Aten, C. F. (1974b). A new electronic band system of BaO, *J. Mol. Specy.*, **53**, 273.

Hui, K. K., Rosen, D. I., and Cool, T. A. (1975). Intermode energy transfer in vibrationally excited O_3, *Chem. Phys. Lett.*, **32**, 141.

Husain, D. and Wiesenfeld, H. R. (1975). Comments on 'gas phase reaction of Ba + N_2O. II. Mechanism of reaction', *J. Chem. Phys.*, **62**, 2010.

Jaffe, R. L., Morokuma, K., and George T. F. (1975). *Ab initio* and semiempirical study of multiple surfaces and their analytic continuation for collinear $F(^2P_{3/2}, {}^2P_{1/2}) + H_2 \rightarrow FH + H$, *J. Chem. Phys.*, **63**, 3417.

Jarmain, W. R. and McCallum, J. C. (1970). *TRAPRB: A Computer Program for Molecular Transitions*, University of Western Ontario, London, Ontario.

Hildenbrand, O. L., and Kleinschmidt, P. D. (1978). Dissociation energies of CaI, SrI and BaI from mass spectrometric studies. *J. Chem. Phys.*, **68**, 2819.

Johnson, S. E., Scott, P. B., and Watson, G. (1974). Visible Chemiluminescence from supersonic mixing metal–oxidant flames, *J. Chem. Phys.*, **61**, 2834.

Jonah, C. D. and Zare R. N. (1971). Formation of group IIa dihalides by two body radiative association, *Chem. Phys. Lett.*, **9**, 65.

Jonah, C. D., Zare, R. N., and Ottinger, Ch. (1972). Crossed-beam chemiluminescence studies of some group IIa metal oxides, *J. Chem. Phys.*, **56**, 263.

Jones, C. R. and Broida, H. P. (1973). An efficient chemiluminescent reaction, *J. Chem. Phys.*, **59**, 6677.

Jones, C. R. and Broida, H. P. (1974). Gas phase reaction of Ba with N_2O. I. Measurement of production efficiency of excited states, *J. Chem. Phys.*, **60**, 4369.

Jones, R. W. and Gole, J. L. (1977). Single collision chemiluminescent studies of the La–Cos reaction-vibrational analysis of the LaS $C^2\Pi - X^2\Sigma^+$ system and determination of D_0^0(LaS), *Chem. Phys.*, **20**, 311.

Kaldor, A., Braun, W., and Kurylo, M. J. (1974). Infrared laser enhanced reactions: O_3 + SO, *J. Chem. Phys.*, **61**, 2496.

Kaufmann, K. J., Kinsey, J. L., Palmer, H. B., and Tewarson, A. (1974a). Chemiluminescent emission spectra and possible upper-state potentials of KCl and KBr, *J. Chem. Phys.*, **61**, 1865.

Kaufmann, K. J., Kinsey, J. L., Tewarson, A., and Palmer, H. B. (1974b). Potassium iodide chemiluminescence in diffusion flames and the KI upper state potential, *J. Chem. Phys.*, **60**, 4023.

Kiefer, R., Siegel, A., and Schultz A. (1978). Chemiluminiszenzuntersuchung der Reaktion $Ba + CCl_4$, Paper A17, Verhandlungen d. Deutschen Physikal. Gesellschaft, München, May.

Kolb, C. E., Gersh, M. E., and Hershbach, D. R. (1975). A suggested mechanism for the visible chemiluminescence observed in gas phase aluminium oxidation, *Comb. Flame*, **25**, 31.

Kondratiev, V. N. (1964). *Chemical Kinetics of Gas Reactions*, Pergamon Press, New York.

Kowalski, A. and Heldt, J. (1978). Chemiluminescent studies of excited Mg* and Sr* Atoms with Cl_2 in the beam-gas arrangement, *Chem. Phys. Lett.*, **54**, 240.

Krause, H. F. (1978). *A Carbon Reaction Studied by Crossed Molecular Beams*, reported as paper 3-Ia-6 at Conference on Dynamics of Molecular Collisions, Pacific Grove, Cal.

Krause, H. F., Johnson, S. G., Datz, S., and Schmidt-Bleek F. K. (1975). Crossed molecular beam study of excited atom reactions: $Hg(6^3P_2^0)$ with Cl_2 and chlorinated methane molecules, *Chem. Phys. Lett.*, **31**, 577.

Krenos, J. (1978). On electronic emission intensity in chemiluminescent-reactions, *J. Chem. Phys.*, **68**, 343.

Krenos, J. R., Bowen, K. H., and Herscbach, D. R. (1975). Chemiluminescence in molecular beams: Statistical partitioning of electronic energy in the $Cl + K_2$ reaction, *J. Chem. Phys.*, **63**, 1696.

Krenos, J. R. and Tully, J. C. (1975). Statistical partitioning of electronic energy: Reactions of alkali dimers with halogen atoms, *J. Chem. Phys.*, **62**, 420.

Kuntz, P. J. (1976). Features of potential energy surfaces and their effect on collisions, in *Dynamics of Molecular Collisions* (Ed. W. H. Miller), Part B, Plenum Press, New York, Chap. 2, p. 53.

Kurylo, M. J., Braun, W., Kaldor, A., Freud, S. M., and Wayne, R. P. (1974). Infrared laser enhanced reactions: Chemistry of vibrationally excited O_3 with NO and $O_2(^1\Delta_g)$, *J. Photochem.*, **3**, 71.

Kurylo, M. J., Braun, W., Xuan, C. N., and Kaldor, A. (1975). Infrared laser enhanced reactions: Temperature resolution of the chemical dynamics of the $O_3^{+} + NO$ reaction system, *J. Chem. Phys.*, **62**, 2065.

Kurylo, M. J., Braun, W., Abramowitz, M., and Krauss, M. (1976). A Study of the chemiluminescence of the $Pb + O_3$ reactions, *J. Res. Natl. Bur. St.*, 80A, 167.

Kusunoki, I. and Ottinger, Ch. (1978a). *Chemiluminescence in Reactions of N^+ Ions with Hydrogen and Hydrocarbons;* I. *Rotational–Vibrational Product Excitation*, Max Planck Institut f. Strömungsforschung, Göttingen, Report 113/1978.

Kusunoki, I. and Ottinger, Ch. (1978b). *Chemiluminescence in Reactions of N^+ Ions with Hydrogen and Hydrocarbons;* II. *Dependence of Total Reaction Cross Sections on Collision Energy*, Max Planck Institut f. Strömungsforschung, Göttingen, Report 114/1978.

Kwok, M. A., Herbelin, J. M., and Cohen, N. (1976). *Collisional Quenching and Radiative Decay Studies of $NF(a^1\Delta)$ and $NF(b^1\Sigma^+)$. Electronic Transition Lasers I*, (Ed. J. I. Steinfeld), MIT Press, p. 8.

Laidler, K. J. (1955). *Chemical Kinetics of the Excited State*, Clarendon Press, Oxford.

Lechtken, P. and Turro, N. J. (1973). Thermal and photochemical generation of electronically excited organic molecules, tetramethyl-1, 2-dioxetane and naphthalene, *Pure Appl. Chem.*, **33**, 363.

Lee, P. H., Woolsey, G. A., and Slafer, W. D. (1976). Integration over cylindrical luminous volumes using a calibrated piston source, *Appl. Opt.*, **15**, 2825.

Levi, H. (1934). *Ueber die Spektren der Alkalihalogen-Daempfe*, Doctoral Dissertation, Friedr. Wilh. Univ. Berlin.

Levine, R. D. and Ben-Shaul, A. (1977). Thermodynamics of molecular disequilibrium, in *Chemical and Biochemical Applications of Lasers*, Vol. II, Academic Press, New York, p. 145.

Lindsay, D. M. and Gole, J. L. (1977). $Al + O_3$ chemiluminescence: perturbations and vibrational population anomalies in the $B^2\Sigma^+$ state of AlO, *J. Chem. Phys.*, **66**, 3886.

Linton, C. and Broida, H. P. (1976). Chemiluminescent spectra of PbO from reactions of Pb atoms, *J. Mol. Specy.*, **62**, 396.

Linton, C. and Broida, H. P. (1977). Flame spectroscopy of TiO: Chemiluminescence, *J. Mol. Specy.*, **64**, 382.

Liskow, D. H., Bender, C. F., and Schaefer, H. F. (1974). Potential energy surfaces related to the ion–molecule reaction $C^+ + H_2$, *J. Chem. Phys.*, 61, 2507.

Luria. M., Eckstrom, D. J., and Benson, S. W. (1976a. Heat pipe oven reactor (HPOR): I. A new device for flame studies; photon yields in the reaction of Na with CCl_4 and N_2O, *J. Chem. Phys.*, **64**, 3103.

Luria, M., Eckstrom, D. J., and Benson, S. W. (1976b). Heat-pipe-oven-reactor (HPOR) studies. II. Formation of excited CN in Li–NF_3–CCl_4 ternary flame system, *J. Chem. Phys.*, **65**, 1595.

Luria, M., Eckstrom, D. J., Edelstein, S. A., Perry, B. E., and Benson, S. W. (1976c). Chemiluminescence studies. V. Production of electronically excited alkali atoms in reactions of Li, Na, Cs with NF_3 and C_5 with F_2, *J. Chem. Phys.*, **64**, 2247.

Magee, J. L. (1940). The mechanism of reactions involving excited electronic states: The gaseous reactions of alkali metals and halogen, *J. Chem. Phys.*, **8**, 687.

Manos, D. M. (1976). *Crossed Beam Chemiluminescence of Group IIIb Metals*, Ph.D. Thesis, Ohio State University, Columbus, Ohio.

Manos, D. M. and Parson, J. M. (1975). Crossed molecular beam study of chemiluminescent reactions of group IIIb atoms (Sc, Y, La) with O_2. *J. Chem. Phys.*, **63**, 3575.

Manos, D. M. and Parson, J. M. (1978). Chemiluminescent reactions of group IIIb atoms with O_2: Spectral simulations and extended energy dependence, *J. Chem. Phys.*, **69**, 231.

Maya, J. and Nordine, P. C. (1976). Chemiluminescence from $Tl + F_2$ reactions, *J. Chem. Phys.*, **64**, 84.

McDermott, W. E., Pchelkin, N. R., and Benard, D. J. (1978). An electronic transition chemical laser, *Appl. Phys. Lett.*, **32**, 469.

McFadden, D. L. (1978). Chemiluminescence in gas phase alkali halide systems, in *Alkali Halide Vapors: Structure, Spectra and Reaction Dynamics* (Eds. P. Davidovits and D. L. McFadden, Academic Press.

Mendenhall, G. D. (1977). Analytical applications of chemiluminescence, *Angew. Chem.* (*Int. Ed.*), 16, 225.

Menzinger, M. (1974a). Dynamics of the electronically chemiluminescent $Ca + X_2(F_2, Cl_2, Br_2)$ reactions, *Chem. Phys.*, **5**, 350.

Menzinger, M. (1974b). Electronic chemiluminescence in $M + X_2$ reactions: Dissociation energies of the alkaline earth monohalides MX(M = Ca, Sr, Ba; X = F, Cl, Br), *Can. J. Chem.*, 52, 1688.

Menzinger, M. and Wolfgang, R. (1969). The meaning of the Arrhenius activation energy, *Angew. Chem.* (*Int. Ed.*), 8, 438.

Menzinger, M. and Wren, D. J. (1973). Hermaphroditism in chemical dynamics: The reaction $Ba + Cl_2 \rightarrow BaCl^* + Cl$, *Chem. Phys. Lett.*, **18**, 431.

Metiu, H., Ross, J., and Silbey, R. (1974). On symmetry properties of reaction coordinates, *J. Chem. Phys.*, **61**, 3200.

Metiu, H., Ross, J., and Whitesides, G. M. (1978). A basis for orbital symmetry rules, *Angew. Chem.* (*Int. Ed.*), in press.

Miller, W. B., Safron, S. A., and Hershbach, D. R. (1967). *Faraday Disc. Chem. Soc.*, 44, 108.

Mims, C. A. and Brophy, J. H. (1977). Angular distributions of chemiluminescence from $Ba + Cl_2$, *J. Chem. Phys.*, 66, 1378.

Moulton, M. C. and Hershbach, D. R. (1966). Chemiluminescence in molecular beams: Electronic excitation of alkali atoms in exchange reactions of vibrationally excited alkali halides, *J. Chem. Phys.*, **44**, 4010.

Moy, J., Bar-Ziv, E., and Gordon, R. J. (1977). Temperature dependence of the laser-enhanced reaction $NO + O_3(001) \rightarrow NO_2(^2B_{1,2}) + O_2$, *J. Chem. Phys.*, **66**, 5439.

Neoh, S. K. and Hershbach, D. R. (1975). Reactive and nonreactive modes of electronic excitation and molecular dissociation in hyperthermal collisions of alkali atoms with alkali halides, *J. Chem. Phys.*, **63**, 1030.

Nicholls, R. W. and Stewart, A. L. (1962). Allowed transitions, in *Atomic and Molecular Processes* (Ed. D. R. Bates), Academic Press, New York.

Nikitin, E. E. (1974). *Theory of Elementary Atomic and Molecular Processes in Gases*, Clarendon Press, Oxford.

Obenauf, R. H., Hsu, C. J., and Palmer, H. B. (1972). Mechanism of production of electronically excited BaO in the reaction of Ba vapor with O_2, *Chem. Phys. Lett.*, **17**, 455.

Obenauf, R. H., Hsu, C. J., and Palmer, H. B. (1972/73). Distribution of electronic states in products of elementary reactions: $Ba + N_2O$ or $NO_2 \rightarrow BaO(A^1\Sigma$ or $X^1\Sigma) + N_2$ or NO, *J. Chem. Phys.*, **57**, 5607 and *erratum: ibid.*, **58**, 2674 (1973).

Obenauf, R. H., Hsu, C. J., and Palmer, H. B. (1973). Distribution of electronic states in products of elementary reactions. II. (Ba, Ca) + ONCl → NO + BaCl($C^2\Pi$, $A^2\Pi$, $X^2\Sigma$) or CaCl ($B^2\Pi$, $A^2\Pi$, $X^2\Sigma$), *J. Chem. Phys.*, 58, 4693.

Oldenburg, R. C., Dickson, C. R., and Zare, R. N. (1975). A new electronic band system of PbO, *J. Mol. Specy.*, **58**, 283.

Oldenburg, R. C., Gole, J. L., and Zare, R. N. (1974). Chemiluminescent spectra of alkali–halogen reactions, *J. Chem. Phys.*, **60**, 4032.

O'Malley, T. F. (1971). Diabatic states of molecules—quasistationary electronic states, *Adv. Atom. Mol. Phys.*, **7**, 223.

Ottinger, Ch. (1976). *Chemoluminiszente Ionen–Molekül-Reaktionen*, Max Planck Institut f. Strömungsforschung, Göttingen, Bericht 18/1976 (Habilitationschrift).

Ottinger, Ch. (1978). *Chemiluminescence in Ion–Molecule Reactions*, *Electronic and Atomic Collisions* (Ed. G. Watel), Proceedings X. ICPEAC, Paris 1977, North Holland Publ. Co. Amsterdam, p. 639.

Ottinger, Ch. and Simonis, J. (1975). Rotational excitation of reaction products: $C^+ + O_2 \rightarrow CO^+(A^2\Pi) + O$, *Phys. Rev. Lett.*, 35, 924.

Ottinger, Ch. and Zare R. N. (1970). Crossed beam chemiluminescence, *Chem. Phys. Lett.*, **5**, 243.

Palmer, H. B. and Hsu, C. J. (1972). A reexamination of electronic chemiluminescence from TiO, *J. Mol. Specy.*, **43**, 320.

Palmer, H. B., Hsu, C. J., and Krugh, W. D. (1975). *Chemiluminescent Spectra and Light Yields for Several low-pressure Diffusion Flames of Alkaline Earth Metal Vapors*,

15th Symp. on Combustion, The Combustion Institute, Pittsburgh, PA, p. 951.

Parr, T. P., Freedman, A., Behrens, R., and Herm, R. R. (1977). Crossed molecular beam kinetics: BaO recoil velocity spectra for $Ba + N_2O$, *J. Chem. Phys.*, **67**, 2181.

Parson, J. C. (1978). Private communication.

Parson, J. M. and Conway, T. J. (1977). *Crossed Beam Chemiluminescence Reactions of Ti and V with* (O_2, SO_2, CO_2, N_2O), Paper 29 Sympos. on State-to-State Chem., ACS Meeting, New Orleans.

Pasternack, L. and Dagdigian, P. J. (1978). The reaction of metastable Ca atoms with O_2 and CO_2, *Chem. Phys.*, preprint.

Pearson, R. (1976). *Symmetry Rules for Chemical Reactions*, Wiley, New York.

Pechukas, P., Light, J. and Rankin, C. (1966). Statistical theory of chemical kinetics: Application to neutral-atom–molecule reactions, *J. Chem. Phys.*, **44**, 794.

Polanyi, J. C. and Schreiber J. L. (1974). The dynamics of bimolecular reactions, in *Physical Chemistry, An Advanced Treatise* (Ed. W. Jost), Vol. VIA, *Kinetics of Gas Reactions*, Academic Press, New York.

Polanyi, J. C. and Woodall K. B. (1972). Energy distribution among reaction products. VI. $F + H_2, D_2$, *J. Chem. Phys.*, **56**, 1563.

Polanyi, M. (1932). *Atomic Reactions*, Williams and Norgate, London.

Polanyi, M. and Schay M. (1928). Ueber hochverduennte Flammen III Beweis und Ausbau des Reaktions-und Leuchtmechanismus. Die beiden Reaktionstypen. Ueberblick ueber die ganze Untersuchung, *Z. Phys. Chem.*, **B1**, 30.

Pratt, D. W. and Broida, H. P. (1969). Microwave-optical double resonance: Frequencies and linewidths of stimulated emission transitions in $A^2\Pi_{3/2}$ state of CN, *J. Chem. Phys.*, **50**, 2181.

Preuss, D. R. and Gole, J. L. (1977a). The use of temperature dependent reaction rates to correct metal halide (oxide) dissociation energies as determined from chemiluminescence reactions. *J. Chem. Phys.*, **66**, 880.

Preuss, D. R. and Gole, J. L. (1977b). The temperature dependence of 'single collision' bimolecular beam-gas chemiluminescent reactions. I. Theory, *J. Chem. Phys.*, **66**, 2994.

Preuss, D. R. and Gole, J. L. (1977c). The temperature dependence of 'single collision' bimolecular beam-gas chemiluminescent reactions. II. Experimental studies, *J. Chem. Phys.*, **66**, 3000.

Pruett, J. G. and Torres-Filhi, A. (1977). Dark Products in Chemiluminescent Flames of $Ba + N_2O$, in: State-to-State Chemistry (Eds. P. B. Brooks and E. Hayes), ACS Symposium Series 56, ACS, Washington, DC, p. 139.

Radford, H. E. and Broida, H. P. (1963). Chemical and magnetic enhancement of perturbed lines in the violet spectrum of CN, *J. Chem. Phys.*, **38**, 644.

Redpath, A. E. and Menzinger, M. (1971). Molecular beam chemiluminescence. I. Kinetic energy dependence of the $NO + O_3 \quad NO_2^* + O_2$ reaction, *Can. J. Chem.*, **49**, 3063.

Redpath, A. E. and Menzinger, M. (1975). Molecular beam chemiluminescence. V. Reactivities of NO $(^2\Pi_{1/2})$ and $(^2\Pi_{3/2})$ fine structure components in the $NO + O_3 \rightarrow NO_2^* + O_2$ reaction, *J. Chem. Phys.*, **62**, 1987.

Redpath, A. E., Menzinger, M., and Carrington, T. (1978). Molecular beam chemiluminescence. XI. Kinetic and internal energy dependence of the $NO + O_3 \rightarrow NO_2^*$, $NO_2^{\ddagger}$ reaction, *Chem. Phys.*, **27**, 409.

Rosenwaks, S., (1976). Chemiluminescent reactions of Al atoms and halogens, *J. Chem. Phys.*, **65**, 3668.

Rosenwaks, S. and Broida H. P. (1976). Chemiexcitation transfer (in $Al + NF_3$) to high Rydberg levels of Al*, *J. Opt. Soc. Amer.*, **66**, 75.

Rosenwaks, S., Steele, R. E., and Broida, H. P. (1975). Chemiluminescence of A10, *J. Chem. Phys.*, **63**, 1963.

Rosenwaks, S., Steele, R. E., and Broida, H. P. (1976). Observation of $a^3\Pi - X^1\Sigma$ intercombination emission in AlF, *Chem. Phys. Lett.* **38**, 121.

Russek, A. (1971). Rotationally induced transitions in atomic collisions, *Phys. Rev.*, **A4**, 1918.

Sakurai, K., Adams, A., and Broida, H. P. (1976). Chemiluminescence of CaH and AlH in the reaction of the metal and formaldehyde, *Chem. Phys. Lett.*, **39**, 442.

Sayers, M. J. and Gole, J. L. (1977). Chemiluminescence from $Al + O_3$: Perturbations, populations and vibrational analysis of the AlO $A^2\Pi - X^2\Sigma^+$ transition, *J. Chem. Phys.*, **67**, 5442.

Schay, G. (1931). in *Fortschritte der Chemie, Physik und Physikal. Chemie*, Vol. 21, p. 1.

Siegel, A. and Schultz, A. (1978). Direct observation of angular distribution of the chemiluminescence of $Ba + N_2O \rightarrow BaO^* + N_2$, *Chem. Phys. Lett.*, **28**, 265.

Smith, F. T. (1969). Diabatic and adiabatic representations for atomic collision problems, *Phys. Rev.*, **179**, 111.

Smith, G. P. and Zare, R. N. (1975). Facile spin-forbidden reactions, $Ba + SO_2 \rightarrow BaO + SO$, *J. Amer. Chem. Soc.*, **97**, 1985.

Smith, K. K., Koo, J. Y., Schuster, G. B., and Kaufmann K. H. (1977). Unimolecular decay of photoexcited tetramethyldioxetane by picosecond spectroscopy, *Chem. Phys. Lett.*, **48**, 267.

Sridharan, U. C., McFadden, D. L., and Davidovits, P. (1977). Chemiluminescence from the gas phase reaction of atomic boron with alkali metal fluorides, *J. Chem. Phys.*, **65**, 5373.

Stephenson, J. C. and Freund, S. M. (1976). Infrared laser-enhanced reactions: Chemistry of $NO(v = 1)$ with O_3, *J. Chem. Phys.*, **65**, 4303.

Stolte, S. (1978). Univ. Nijmegen, Holland, private communication.

Struve, W. S. (1973). Semiempirical pseudopotential surfaces for chemical reactions: The $(AB)^+X^-$ dialkali halide systems, *Mol. Phys.*, **25**, 777.

Struve, W. S., Kitagawa, T., and Hershbach, D. R. (1971). Chemiluminescence in molecular beams: Electronic excitation in reactions of Cl atoms with Na_2 and K_2 molecules, *J. Chem. Phys.*, **54**, 2759.

Struve, W. S., Krenos, J., McFadden, D. L., and Hershbach, D. R. (1973). Contribution to the general discussion ($M_2 + X$, X_2), *Faraday Disc. Chem. Soc.*, **53**, 314.

Struve, W. S., Krenos, J. R., McFadden, D. L., and Hershbach, D. R. (1975). Molecular beam kinetics: Angular distributions and chemiluminescence in reactions of alkali dimers with halogen atoms and molecules, *J. Chem. Phys.*, **62**, 404.

Sutton, D. G. and Suchard, S. N. (1975). Electronic transition lasers: A parametric evaluation, *Appl. Opt.*, **14**, 1898.

Swearengen, P., Davis, S., and Niemczyk, T. (1977). *A measurement of the Rate Coefficient for the $Ge + N_2O$ Reaction. Electronic Transition Lasers II* (Eds. L. E. Wilson, S. Suchard, and J. I. Steinfeld, MIT Press, p. 132.

Tang, S. P., Utterback, N. G., and Frijchtenicht, J. F. (1976). Measurement of chemiluminescent reaction cross sections for $B + N_2O \rightarrow BO^* + N_2$ and $Ho + N_2O \rightarrow HoO^* + N_2$, *J. Chem. Phys.*, **64**, 3833.

Tang, S. P., Wicke, B. G., and Frijchtenicht, J. F. (1978). Studies of the chemiluminescent reaction $Ho + N_2O \rightarrow HoO^* + N_2$, *J. Chem. Phys.*, **68**, 5471.

Thrush, B. A. (1968). Gas reactions yielding electronically excited species, *Ann. Rev. Phys. Chem.*, **19**, 371.

Thrush, B. A. (1973). Determination of bond energies from chemiluminescent atom transfer reactions, *J. Chem. Phys.*, **58**, 5191.

Tully, J. C. (1974a). Collision complex model for spin forbidden reactions: Quenching of $O(^1D)$ by N_2, *J. Chem. Phys.*, **61**, 61.

Tully, J. C. (1974b). Collisions of $F(^2P_{1/2})$ with H_2, *J. Chem. Phys.*, **60**, 3042.

Tully, J. C. (1975). Reactions of $O(^1D)$ with atmospheric molecules, *J. Chem. Phys.*, **62**, 1893.

Tully, J. C. (1976). *Nonadiabatic Processes in Molecular Collisions, Dynamics of Molecular Collisions* (Ed. W. H. Miller), Part B, Plenum Press, New York.

Tully, J. C. (1977). *Collisions involving Electronic Transitions, State-to-State Chemistry* (Eds. P. R. Brooks and E. Hayes), ACS Sympos. Series 56, A.C.S., Washington, D.C., p. 206.

Turro, N. J., Lechtken, P., Shore, N. E., Schuster, G., Steinmetzer, H. Ch., and Yekta, A. (1974). Tetramethyl-1,2-dioxetane. Experiments in chemiexcitation, chemiluminescence, photochemistry, chemical dynamics and spectroscopy, *Accs. Chem. Res.*, **7**, 97.

Utterback, N. G., Tang, S. P., and Frijchtenicht, J. F. (1976). A laser-vaporization Molecular beam source, *Phys. Fluids*, **19**, 900.

Valentini, J. J., Coggiola, M. J., and Lee, Y. T. (1977). Crossed beam studies of endoergic bimolecular reactions: Production of stable trihalogen radicals, *Faraday Disc. Chem. Soc.*, **62**, 246.

West, J. B., Bradford, R. S., Eversole, J. D., and Jones C. R. (1975). Flow system for production of diatomic metal oxides and halides, *Rev. Sci. Instr.*, **46**, 164.

West, J. B. and Broida, H. P. (1975). Chemiluminescence and photoluminescence of diatomic iron oxide, *J. Chem. Phys.*, **62**, 2566.

Wicke, B. G., Revelli, M. A., and Harris D. O. (1975). On the importance of $Ba(^3D)$ as the key reactant leading to $BaO(A\text{–}X)$ chemiluminescence in the $Ba + N_2O$ reaction, *J. Chem. Phys.*, **63**, 3120.

Wicke, B. G., Tang, S. P., and Frijchtenicht, J. F. (1978). Velocity dependence of the $Pb + N_2O \rightarrow PbO(B, v = 0) + N_2$ chemiluminescent reaction, *Chem. Phys. Lett.*, **53**, 304.

Wiesenfeld, J. R. and Yuen M. J. (1976). Kinetic study of the reaction $Sn + N_2O \rightarrow SnO + N_2$, *Chem. Phys. Lett.*, **42**, 293.

Wilson, L. E., Suchard, S. N., and Steinfeld, J. I. (Eds.) (1977). *Electronic Transition Laser*, MIT Press, Cambridge, MA.

Wilson, T. (1976). Chemiluminescence in the liquid phase: thermal cleavage of dioxetanes, in *International Review of Science*, Series 2, Vol. 9, *Chemical Kinetics* (Eds. A. D. Buckingham and D. R. Hershbach), Butterworths, London, p. 265.

Woolsey, G. A., Lee, P. H., and Slafer, W. D. (1977). Measurement of the rate constant for NO–O chemiluminescence using calibrated piston source of light, *J. Chem. Phys.*, **67**, 1220.

Wren, D. J. and Menzinger, M. (1973). Quenching of the $Ba + Cl_2$ chemiluminescence: Estimate of $BaCl_2^*$ Radiative lifetime, *Chem. Phys. Lett.*, **20**, 471.

Wren, D. J. and Menzinger, M. (1974). Molecular beam chemiluminescence: Three body processes in the micro-torr region, *Chem. Phys. Lett.*, **27**, 572.

Wren, D. J. and Menzinger, M. (1975). Molecular beam chemiluminescence. VII. Enhancement of $Ba + N_2O \rightarrow BaO^* + N_2$ cross section by N_2O bending vibration: Evidence for electron transfer, *J. Chem. Phys.*, **63**, 4557.

Wren, D. J. and Menzinger, M. (1979a). Molecular beam chemiluminescence. XIII. Kinetic and internal energy dependence of the $Ba + N_2O$ and $Ba + (NO_2, N_2O_4)$ reactions, *Faraday Disc. Chem. Soc.*, to be published.

Wren, D. J. and Menzinger, M. (1979b). Kinetic energy dependence of the chemiionization reactions $M + X_2 \rightarrow MX^+ + X^-$ (M = Ca, Sr, Ba; X = F, Cl, Br), *Chem. Phys. Lett.*, to be published.

Wren, D. J. and Menzinger, M. (1979c). Molecular beam chemiluminescence. XIV. Mechanisms and dynamics of the $M + X_2 \rightarrow MX_2^*$, $MX^+ + X^-$ (M = Ca, Sr, Ba; X = F, Cl, Br) reactions, *Chem. Phys.* to be published.

Yarkony, D. R., Hunt, W. J., and Schaefer, H. F. (1973). Relation between electronic structure and the chemiluminescence arising from collisions between alkaline earth atoms and halogen molecules, *Mol. Phys.*, **26**, 941.

Yokozeki, A. and Menzinger, M. (1976). Molecular beam chemiluminescence. VIII. Pressure dependence and kinetics of Sm + (N_2O, O_3, F_2, Cl_2) and Yb + (O_3, F_2, Cl_2) reactions. Dissociation energies of the diatomic reaction products, *Chem. Phys.*, **14**, 427.

Yokozeki, A. and Menzinger, M. (1977a). Molecular beam chemiluminescence. IX. Sm + N_2O: Translational and vibrational energy dependence of cross sections, *Chem. Phys.*, **20**, 9.

Yokozeki, A. and Menzinger, M. (1977b). On the dynamical content of excitation functions: Simple linerization procedure, *Chem. Phys.*, **22**, 273.

Zahr, G. E., Preston, R. K., and Miller, W. H. (1975). Theoretical treatment of quenching in $O(^1D) + N_2$ collisions, *J. Chem. Phys.*, **62**, 1127.

Zare, R. N. (1963). *Programs for Calculating Relative Intensities in the Vibrational Structure of Electronic Band Systems*, Lawrence Radiation Laboratory, Berkeley CA., Report UCRL-10925.

Zare, R. N. (1964). Calculation of intensity distribution in the vibrational structure of electronic transitions: The $B^3\Pi_{0+u}$–$X^1\Sigma_{0+g}$ resonance series of molecular iodine, *J. Chem. Phys.*, **40**, 1934.

Zare, R. N. (1966). Molecular level-crossing spectroscopy, *J. Chem. Phys.*, **45**, 4510.

Zare, R. N. (1974). Flurescence of free radicals: Method for determination of D_0^0-limits, *Ber. Bunsen. Phys. Chem.*, **78**, 153.

Potential Energy Surfaces
Edited by K. P. Lawley

SEMIEMPIRICAL DIATOMICS-IN-MOLECULES POTENTIAL ENERGY SURFACES

JOHN C. TULLY

Bell Laboratories, Murray Hill, New Jersey, 07974, *USA*

CONTENTS

I. INTRODUCTION

A. The Potential Energy Surface

The primary focus of most experimental and theoretical inquiries in the field of chemical dynamics is the potential energy hypersurface; i.e. the electronic energy of a group of N interacting atoms as a function of their $3N-6$ relative position coordinates. Embodied in the potential energy surface characterizing a particular reactive system (Fig. 1) is information about the isolated reactant and product species, about their long-range

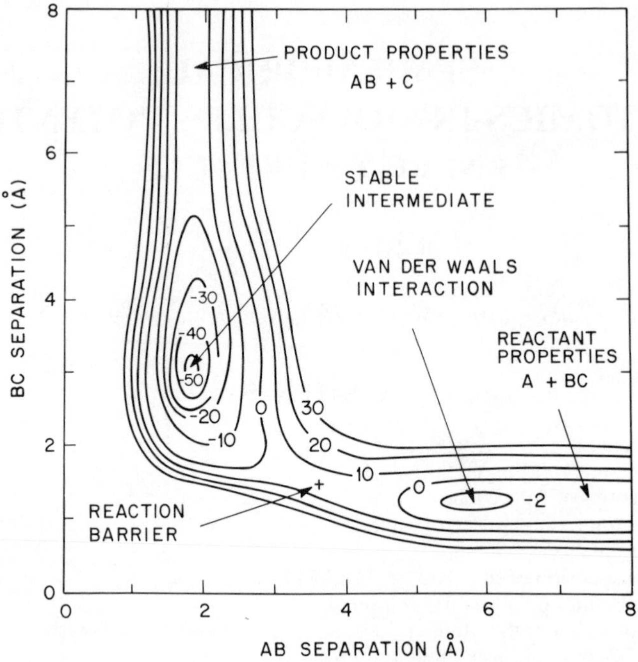

Fig. 1. Schematic illustration of the central role of the potential energy surface in chemical dynamics.

(van der Waals) interactions, about molecular deformations which lead to the breaking and forming of chemical bonds, about energy barriers which must be surmounted to achieve chemical reaction, and about the geometries and properties of transient reaction intermediates. A wide variety of experimental and theoretical approaches are converging on a single goal, qualitative and quantitative elucidation of the potential energy surface. Different techniques provide knowledge about different regions or features of potential energy surfaces. Spectroscopic and thermochemical measurements are invaluable in characterizing fragments and stable intermediates. Transport properties and elastic or inelastic differential cross-section measurements probe interactions between unrearranged reactant or product molecules, i.e. the 'entrance' and 'exit' valleys of the potential energy surface. Information about the transition-state region can be obtained from the energy or temperature dependence of reaction cross sections or, in more detail, from the specific angular and velocity patterns of scattered products produced in collisions of monoenergetic beams of reactants. Excursions off the beaten path, i.e. away from the reaction coordinate, are achievable using accelerated reactant beams or reactants prepared in specific internal states. Similarly, different theoretical approaches are appropriate for different features of the potential energy surface. For example, single determinant SCF methods can provide properties of isolated closed-shell fragments, perturbation approaches are suitable for obtaining van der Waals interactions, and extensive configuration interaction or semiempirical procedures are required to obtain accurate reaction barriers. The ultimate objective is to somehow synthesize all of this disparate information into a 'best' potential energy surface which, with continual refinement, will serve as a quantitative compendium of our knowledge of a particular chemically reactive system.

Once constructed, a potential energy surface that accurately embodies the electronic interactions characteristic of some interesting chemically reactive system can be extremely useful for many purposes. Properties such as reaction endo- or exoergicities and the electronic state correlations of reactants and products are apparent immediately from the potential energy surface. Since the forces experienced by the nuclei are given simply by the negative of the derivatives of the potential energy with respect to displacements, the potential energy surface provides the strength and directionality of the chemical forces that ultimately dictate the course of a collision event. Classical or quantum mechanical 'trajectories' tracing out the motion of the nuclei over the potential energy surface can provide both quantitative predictions of the outcome of a collision process and a qualitative understanding of how it occurred, i.e. the reaction mechanism. Dynamical calculations can provide very detailed information such as product angular and velocity distributions, the partitioning of energy among specific product internal states, etc., as well as 'crude' information such as absolute reaction rates. 'Numerical experiments' based on the potential energy surface can be performed for conditions that might be difficult to reproduce in the laboratory, e.g. reactants far from equilibrium or prepared in specific internal states.

In systems for which electronic transitions—transitions between one potential energy surface and another—can occur, we require characterization of more than one potential energy surface, one surface corresponding to each electronic state that may be involved in the collision. We require, in addition, specification of the non-adiabatic coupling (i.e. off-diagonal potential energy surface) that promotes transitions between two potential energy surfaces. With this expanded concept of the 'potential energy surface matrix' we can address a great many additional phenomena, including electronic energy transfer, charge transfer, spin-forbidden reactions, electronic-to-vibrational energy transfer, radiationless transitions, etc.

B. Theoretical Approaches

The quest to develop theoretical methods for computing potential energy surfaces began with the original definition of the (adiabatic) potential energy surface (Born and Oppenheimer, 1927) and with London's realization of the importance of this concept to the theory of chemical reactions (London, 1929). In principle, the potential energy surface can be computed by obtaining, for all possible nuclear configurations, the eigenenergy of the electronic Schrödinger equation for each fixed nuclear geometry. Even with the power of modern computers, however, *ab initio* calculations of potential energy surfaces to useful chemical accuracy are discouragingly difficult. Quantitative *ab initio* potential energy surfaces have been obtained at present for only a handful of the very simplest systems, including H_3^+, H_3, and FH_2. The computational task vastly increases in magnitude for larger systems for two reasons. First, the time required to compute a single point on a potential energy surface increases dramatically with the number of electrons in the system. An approximate rule of thumb is the time of calculation increases roughly as the fourth power of the number of electrons. Second, the number of points required to adequately characterize a potential energy surface grows with the number of nuclear degrees of freedom. Thus if 10 points are required to define a diatomic potential curve, roughly 10^3 will be required for a triatomic surface, 10^6 for a four-atom system, etc. The prognosis for obtaining quantitatively reliable potential energy surfaces by purely *ab initio* techniques for the majority of systems of chemical interest is extremely bleak for the foreseeable future. We must therefore develop procedures for systematic incorporation of various types of experimental information within our apparatus for constructing potential energy surfaces. *Ab initio* theory must still play a central role in this programme. Available experimental information will almost never be sufficient by itself. *Ab initio* calculations can provide additional input, input that is particularly valuable because it can be focused on the most crucial uncertain features of the potential energy surface.

Procedures for utilizing experimental information to construct potential energy surfaces can be categorized as either *empirical* or *semiempirical.* Empirical methods employ more or less arbitrary functional forms to represent the potential energy surface and choose adjustable parameters to fit experimental data. This approach is fraught with danger. Experimental information is usually available only for restricted regions of the potential energy surface, usually concentrated in the reactant and product valleys and perhaps in the vicinity of a stable intermediate. Empirical potential energy surfaces amount to interpolations, frequently between greatly separated points, and are extremely dependent on the arbitrary functional form employed. In fact, in recognition of this, additional parameters which can be varied to make the surface 'look right' are often employed. This approach is unsatisfactory except, perhaps, in those rare situations for which a dense body of experimental or reliable theoretical data is available.

Semiempirical theories attempt to reduce the arbitrary nature of potential surface construction by employing a specific framework which has its roots in the Schrödinger equation. The form of the potential energy surface is thereby greatly restricted and, hopefully, the resulting potential energy contours reflect in a realistic way the quantum mechanical interactions that determine chemical forces. There are, of course, serious limitations to this approach as well. The approximations to the Schrödinger equation required to derive the semiempirical framework may be inaccurate, the framework itself may be too inflexible to accurately represent the true surface, or the method may be complicated or cumbersome to apply. With future research we can hope to reduce the importance of the first two problems. The third is a price we must be prepared to pay if we are to obtain potential energy surfaces that are reliable and quantitative.

There is a bewildering galaxy of semiempirical methods for computing electronic energies. Many of these have been reviewed elsewhere (Balint-Kurti, 1975; Segal, 1977). Most of these methods are based on the molecular orbital approach (e.g. Hückel, CNDO, INDO, MINDO, $X\alpha$). Semiempirical molecular orbital methods employing a single determinant wave function are generally not suitable for obtaining potential energy surfaces for chemical dynamics studies. One reason for this is associated with the usual failure of a single configuration to correctly describe molecular fragmentation limits. In addition, single configuration methods are incapable of adequately incorporating important configuration mixing effects that might occur, for example, in the vicinity of avoided surface crossings.

These problems can be overcome, in principle, by multiconfiguration molecular orbital semiempirical formalisms or, alternatively, by semiempirical valence-bond methods. While there has been very little development to date in the first direction, there has been considerable success in the second. Semiempirical valence-bond approaches and their application to the construction of potential energy surfaces for reactive systems are the subject of this chapter. We will concentrate on the *Diatomics-in-Molecules* (DIM) method (Ellison, 1963) and closely related methods, most notably the *London–Eyring–Polanyi–Sato* (LEPS) method (e.g. Parr and Truhlar, 1971), which is equivalent to DIM applied to three interacting ^2S atoms.

The DIM method, as outlined in the next section, is simple, widely applicable and rigorously formulated. It automatically displays exact behaviour upon separation into any combination of fragments, and it describes surface crossing effects in a natural and realistic way. Two characteristics which make DIM ideally suited to calculation of potential energy surfaces for chemical dynamics studies are its flexibility to describe a wide variation of topographical features of potential energy surfaces and its realistic inclusion of the quantum mechanical interactions which determine the strength and directionality of chemical bonding. These characteristics, as mentioned earlier, are absolutely crucial and are considered in detail in Section III.

There are other semiempirical valence-bond methods that are not based on the DIM formalism. Of particular note is the 'orthogonalized Moffit' (OM) method developed for potential energy surface calculations by Balint-Kurti and coworkers (Balint-Kurti and Karplus, 1971; Balint-Kurti and Yardley, 1977). The OM method is essentially an *ab initio* valence-bond calculation, usually with a relatively restricted basis set, with empirical corrections to make the isolated atomic energies exact. Like DIM, OM incorporates proper dissociation limits and realistic treatment of configuration mixing. The method has produced some very valuable results. In fact, we will refer to an OM potential energy surface for LiFH (Balint-Kurti and Yardley, 1977) in one of our later 'case studies'. OM requires considerably more computational effort to apply, but it has one important advantage over DIM: it requires only known atomic empirical information. The diatomic information required by DIM is very often not completely available and must either be estimated or obtained by supplementary calculation.

The OM method lies somewhere between the non-empirical *ab initio* techniques and the heavily empirical DIM method. The three approaches thus complement each other. In particular, while DIM is a very powerful method—far more powerful than is generally recognized—it cannot stand alone. DIM can be reliable and predictive only when supported, at critical points, by the results of *ab initio* and/or OM studies.

The DIM formalism is outlined in the next section. The following section is a discussion of the strengths and weaknesses of the method, with respect to both inclusion of qualitative chemical bonding effects and production of quantitatively reliable potential energy surfaces, as tested against experiment or accurate *ab initio* theory. The remainder of the chapter, with the exception of a short summary, is a collection of short case studies of particular chemically reactive systems. Examples of exoergic and endoergic and of adiabatic and non-adiabatic processes are included. The examples are selected to illustrate how experimental and theoretical inquiries, with the potential energy surface serving as their link, can provide a comprehensive understanding of the dynamics of basic chemical processes.

II. THE DIATOMICS-IN-MOLECULES METHOD

A. Formulation

DIM was first proposed by Ellison (Ellison, 1963). The original technique consisted of expanding the polyatomic wavefunction in terms of canonical valence-bond functions which were assumed to be eigenfunctions of their respective atomic and diatomic fragment Hamiltonians, with eigenvalues equal to experimental energies. Polyatomic energies could then be obtained by a simple procedure requiring no electronic integral evaluations except, perhaps, overlaps. More recently the method has been reformulated in a rigorous and systematic way and extended to properly describe the directional character of chemical binding (Kuntz and Roach, 1972; Tully, 1973a, b; Steiner *et al.*, 1973; Tully and Truesdale, 1976). This more general formulation is outlined here (Tully, 1977b).

1. *Expansion of* Ψ *in Atomic Product Functions*

Consider a molecule composed of n electrons and N nuclei, $N \geq 3$. We first define a set of 'polyatomic basis functions' (pbf's) Φ_m defined by

$$\Phi_m(1, \ldots n) = \mathscr{A}_n \phi_m(1, \ldots n), \tag{1}$$

where $\mathscr{A}_n$ is the n-electron antisymmetrizer and ϕ_m is a product of N atomic functions $\zeta_m^{(K)}$ (Moffitt, 1951):

$$\phi_m(1, \ldots n) = \zeta_m^{(A)}(1, \ldots n_A)\zeta_m^{(B)}(n_A + 1, \ldots n_A + n_B) \ldots \zeta_m^{(N)}(n - n_N + 1, \ldots n). \tag{2}$$

The atomic functions $\zeta_m^{(A)}$ are taken to be antisymmetric but otherwise need not be specified explicitly.

We now expand the total wavefunction Ψ describing the motion of the electrons in the field of the N fixed nuclei,

$$\Psi(1, \ldots n) = \sum_m \Gamma_{1m} \Phi_m(1, \ldots n). \tag{3}$$

The desired energy which we will identify as one point on the potential energy surface is an element of the diagonal eigenvalue matrix $\mathbf{E}$ which satisfies the matrix equation

$$\mathbf{H\Gamma} = \mathbf{S\Gamma E} \tag{4}$$

where elements of the Hamiltonian matrix $\mathbf{H}$ are defined by

$$H_{mm'} = \langle \Phi_m | \mathscr{A}\mathscr{H} | \phi_{m'} \rangle \tag{5}$$

and elements of the overlap matrix $\mathbf{S}$ by

$$S_{mm'} = \langle \Phi_m | \Phi_{m'} \rangle. \tag{6}$$

In equation (5) we have made use of the (crucial) fact that since the electronic Hamiltonian operator $\mathscr{H}$ is symmetric with respect to interchange of any two electrons, it commutes with $\mathscr{A}_n$.

2. *Partitioning of the Hamiltonian*

The n-electron Hamiltonian operator $\mathscr{H}$ can be partitioned as follows (Ellison, 1963):

$$\mathscr{H} = \sum_K^N \sum_{L>K}^N \mathscr{H}^{(\mathrm{KL})} - (N-2) \sum_K^N \mathscr{H}^{(\mathrm{K})} \tag{7}$$

where $\mathscr{H}^{(KL)}$ and $\mathscr{H}^{(K)}$ are diatomic and atomic fragment Hamiltonian operators defined as follows: $\mathscr{H}^{(K)}$ is the Hamiltonian operator for an isolated atom K. It contains the electron kinetic energy operators, the electron–electron repulsions, and the electron–nucleus K attractions involving those electrons assigned to atom K. Thus $\mathscr{H}^{(A)}$ will involve electrons $1, 2, \ldots n_A$. Similarly, $\mathscr{H}^{(KL)}$ is the Hamiltonian operator for isolated diatomic fragment KL. Equation (7) is an *exact* expression and is the cornerstone of DIM.

Atomic and diatomic fragment Hamiltonian matrices can now be defined,

$$H_{mm'}^{(K)} = \langle \Phi_m | \mathscr{A}_n \mathscr{H}^{(K)} | \phi_{m'} \rangle \tag{8}$$

$$H_{mm'}^{(KL)} = \langle \Phi_m | \mathscr{A}_n \mathscr{H}^{(KL)} | \phi_{m'} \rangle . \tag{9}$$

The total Hamiltonian matrix $\mathbf{H}$ is then

$$\mathbf{H} = \sum_{K}^{N} \sum_{L>K}^{N} \mathbf{H}^{(KL)} - (N-2) \sum_{K}^{N} \mathbf{H}^{(K)} . \tag{10}$$

Note that the individual fragment matrices are not Hermitian since the n-electron antisymmetrizer $\mathscr{A}_n$ does not commute with the fragment Hamiltonian operators. The total Hamiltonian matrix $\mathbf{H}$ given by equation (10) is Hermitian, of course, since equation (10) is exact.

Since the atomic fragment Hamiltonian operator $\mathscr{H}^{(K)}$ involves only those electrons assigned to atom K, we can expand $\mathscr{H}^{(K)}$ in terms of a complete set of atomic functions $\zeta_m^{(K)}$:

$$\mathscr{H}^{(K)} \zeta_m^{(K)} = \sum_{l} \zeta_l^{(K)} \Delta_{lm}^{(K)} \tag{11}$$

where $\Delta_{lm}^{(K)}$ are expansion coefficients. Substitution of equation into equation (8) gives, in matrix notation

$$\mathbf{H}^{(K)} = \mathbf{S} \boldsymbol{\Delta}^{(K)} \tag{12}$$

where the elements of the expansion coefficient matrix $\boldsymbol{\Delta}^{(K)}$ can be determined solely from properties of isolated atom K. Thus if we define for isolated atom K an overlap matrix $\boldsymbol{\sigma}^{(K)}$ and Hamiltonian matrix $\mathbf{h}^{(K)}$,

$$\sigma_{mm'}^{(K)} = \langle \zeta_m^{(K)} | \zeta_{m'}^{(K)} \rangle \tag{13}$$

and

$$h_{mm'}^{(K)} = \langle \zeta_m^{(K)} | \mathscr{H}^{(K)} | \zeta_{m'}^{(K)} \rangle \tag{14}$$

and obtain the atomic energies $\boldsymbol{\varepsilon}^{(K)}$ and eigenvalues $\gamma^{(K)}$ from the equation

$$\mathbf{h}^{(K)} \boldsymbol{\gamma}^{(K)} = \boldsymbol{\sigma}^{(K)} \boldsymbol{\gamma}^{(K)} \boldsymbol{\varepsilon}^{(K)} \tag{15}$$

then using equation (11) we obtain

$$\boldsymbol{\Delta}^{(K)} = \boldsymbol{\gamma}^{(K)} \boldsymbol{\varepsilon}^{(K)} \boldsymbol{\gamma}^{(K)\dagger} \boldsymbol{\sigma}^{(K)} . \tag{16}$$

Similarly for diatomic fragments,

$$\mathbf{H}^{(\mathrm{KL})} = \mathbf{S}\boldsymbol{\Delta}^{(\mathrm{KL})} \tag{17}$$

where

$$\boldsymbol{\Delta}^{(\mathrm{KL})} = \gamma^{(\mathrm{KL})}\varepsilon^{(\mathrm{KL})}\gamma^{(\mathrm{KL})\dagger}\sigma^{(\mathrm{KL})}; \tag{18}$$

i.e. $\boldsymbol{\Delta}^{(\mathrm{KL})}$ involves information about diatomic fragment KL only. Substituting equations (12) and (17) into equations (10) and (4), we obtain finally (Tully, 1977b)

$$\boldsymbol{\Delta\Gamma} = \boldsymbol{\Gamma}\mathbf{E} \tag{19}$$

where

$$\boldsymbol{\Delta} \equiv \mathbf{S}^{-1}\mathbf{H} = \sum_{K}^{N}\sum_{L>K}^{N} \boldsymbol{\Delta}^{(\mathrm{KL})} - (N-2)\sum_{K}^{N}\boldsymbol{\Delta}^{(\mathrm{K})}. \tag{20}$$

Thus we have replaced our original eigenvalue equation, equation (4), with an entirely equivalent one, equation (19), involving the non-Hermitian matrix $\boldsymbol{\Delta}$. According to equation (20), $\boldsymbol{\Delta}$ is a sum of fragment matrices $\boldsymbol{\Delta}^{(\mathrm{K})}$ and $\boldsymbol{\Delta}^{(\mathrm{KL})}$ each of which can be constructed solely from knowledge of isolated atoms and diatomics. The DIM method consists of constructing these matrices using experimental and theoretical information about the isolated fragments in equations (16) and (18), and then obtaining the polyatomic electronic energy from equation (19). This procedure would be exact if a complete set of pbf's were used. In practice, of course, only a finite set of pbf's can be employed resulting in inexactness for two reasons. First, the expansion of the polyatomic wavefunction Ψ of equation (3) becomes approximate, just as in any calculation based on a truncated expansion. Second, expansions like equation (11) become only approximate, so equations (16) and (18) are no longer exact. The assumption that equations (16) and (18) are exact for an incomplete basis set is the fundamental approximation of DIM. It is hoped that inaccuracies introduced by this approximation can be minimized by using empirical information to construct the fragment matrices $\boldsymbol{\Delta}^{(\mathrm{K})}$ and $\boldsymbol{\Delta}^{(\mathrm{KL})}$.

B. Application of the Method

1. *Selection of Basis Functions*

In applying the general method outlined above, specific choices must be made at several stages. The first is selection of the atomic basis functions. Only the spin and symmetry of the atomic functions are needed; actual numerical specification of the basis functions is never required in DIM. This is a great advantage because it introduces substantial flexibility into the basis functions. They can be considered to expand, contract or distort in any way that preserves symmetry in order to optimally adapt to any particular molecular environment.

The main consideration in choosing atomic basis functions is to include all states required to provide a realistic description of the atom in any chemically bonded situation. For example, to adequately describe the water molecule ground state, $O(^1D)$ configurations known to be important in bonding would have to be included in addition to $O(^3P)$. To allow for sp^n hybridization in carbon compounds, the $C(^5S)$ state arising from the excited $2s^12p^3$ configuration is required. Ionic configurations should be included in situations which exhibit pronounced ionic character. Many of the reported failures of DIM can be attributed simply to inadequate selection of atomic basis functions.

Once the atomic basis functions have been selected, they must be combined to form the pbf's defined by equations (1) and (2). It is obvious that this produces more pbf's than are required. For example, for the H_3 molecule with one ^2S function employed for each atom, 8 different pbf's differing only in spin variables are possible. By selecting linear combinations of pbf's that are eigenfunctions of spin, we can reduce the 8×8 problem to a 2×2 describing doublet H_3 states and a 1×1 describing quartet H_3. Thus DIM basis functions are actually spin eigenfunctions constructed from the pbf's defined in equations (1) and (2). Techniques for accomplishing the required spin transformations are described elsewhere (Tully, 1973a; Pickup, 1973).

2. *Utilization of Empirical Information*

Empirical information is introduced in DIM through equations (16) and (18). This is the most sensitive step in applying the method, and most failures of DIM can be attributed to improper incorporation of experimental information into equations (16) and (18). In particular, the coefficient matrices $\gamma^{(K)}$ and $\gamma^{(KL)}$ are frequently taken to be the unit matrix and the $\varepsilon^{(K)}$ and $\varepsilon^{(KL)}$ taken to be experimental energies. This procedure is probably acceptable in most cases for atomic fragment information, but rarely for diatomics. In most cases diatomic states are not well represented by a simple product of atomic-like functions, $\mathscr{A}\zeta_m^{(K)}\zeta_m^{(L)}$. They are mixtures of such states, and this mixing must be reflected in off-diagonal elements of $\gamma^{(KL)}$ of equation (18). This information about the mixing of various states as a function of internuclear separation is the essential input required by the DIM method. Since this type of information is not usually available from experiment it must be obtained elsewhere. Sometimes simple valence-bond considerations are adequate, but more often theoretical calculations on the diatomic fragments are required, preferably tailored for this specific purpose.

Another complication in utilization of experimental information in DIM, in addition to non-vanishing off-diagonal elements of $\gamma^{(KL)}$, is associated with the diatomic excited-state energies $\varepsilon^{(KL)}$ in equation (18). Actual excited state wavefunctions very likely reflect important contributions from atomic product configurations that are not included in the DIM basis set. Thus the direct use of experimental energies for $\varepsilon^{(KL)}$ is incorrect. 'Diabatic' energies (e.g. Lichten, 1963) are frequently more appropriately associated with simple atomic product functions. Both the off-diagonal elements of $\gamma^{(KL)}$ and the excited-state energies might best be determined by reproducing some known features of the polyatomic system to be described; i.e. by introducing empirical information or accurate theoretical results on the polyatomic system in addition to empirical fragment information. The great advantage of this procedure over purely empirical methods lies with the rigorous quantum mechanical framework of DIM, which will hopefully allow reliable determination of a complete potential energy surface with minimal available input information.

3. *Overlap*

Once basis functions have been selected and the fragment matrices $\boldsymbol{\Delta}^{(K)}$ and $\boldsymbol{\Delta}^{(KL)}$ have been constructed from equations (16) and (18), the desired polyatomic energies and coefficients are obtained from the eigenvalue equation, equation (19), with $\boldsymbol{\Delta}$ given by equation (20). Note that $\boldsymbol{\Delta}$ is non-Hermitian, as it should be since it is an approximation to the matrix $\mathbf{S}^{-1}\mathbf{H}$. If $\boldsymbol{\Delta}$ were exactly $\mathbf{S}^{-1}\mathbf{H}$, then it would have all real eigenvalues and solving equation (19) would be equivalent to solving equation (4). However, since $\boldsymbol{\Delta}$ is approximate (via incompleteness of expansions like equation (11) in introduction of empirical information), we are no longer assured that its eigenvalues will be real. Application to H_3 (Tully and Truesdale, 1976) has shown that, for this system, direct diagonalization of $\boldsymbol{\Delta}$ is an accurate and well-behaved procedure. There may be other situations (e.g. avoided surface crossings) where this will not be the case, however; i.e. where $\boldsymbol{\Delta}$ may have complex roots. In such cases it may be preferable to neglect overlap; i.e. as an additional approximation to set the fragment overlap matrices $\boldsymbol{\sigma}^{(KL)}$ in equation (18) equal to unit matrices. This might be accomplished, for example, by employing 'symmetric orthogonalization' (Löwdin, 1970) in the diatomic calculations. Then $\boldsymbol{\Delta}$ will be automatically Hermitian. The zero-overlap procedure has been shown to give results comparable to the more rigorous non-Hermitian procedure (Tully and Truesdale, 1976).

The conventional method for treating overlap in DIM differs from both of the above methods. A Hermitian approximation to $\mathbf{H}$ is obtained from the expression

$$\mathbf{H} = \tfrac{1}{2}(\mathbf{S}\boldsymbol{\Delta} + \boldsymbol{\Delta}^{\dagger}\mathbf{S}) \tag{21}$$

where the overlap matrix **S** is computed from some assumed basis functions. As shown by Steiner *et al.*, in addition to adding arbitrariness to the method by requiring explicit specification of basis functions, this procedure can introduce serious errors (Steiner *et al.*, 1973). It should not be used.

III. DESCRIPTION OF CHEMICAL BONDING

The most important requirement of a semiempirical method for constructing potential energy surfaces is that it incorporate the chemical interactions specific to each system in a qualitatively correct way. The attractive or repulsive nature of interactions, chemical valence effects, and the directionality of chemical bonding must be described realistically. If the method accomplishes this then it should be possible, by normalization to a very limited selection of empirical data, to obtain quantitatively accurate potential energy surfaces (e.g. Kuntz, 1972). If it does not, then like arbitrary empirical procedures, there is little hope of constructing a useful potential energy surface unless an exhaustive supply of input exists.

The greatest strength of the DIM method is that it appears to correctly reflect the gross qualitative features of almost any chemical binding situation. We will illustrate this below for a variety of systems displaying very different potential energy topographies. To do this we will have to forego the usual descriptions of chemical binding and reactivity that employ molecular orbital language (e.g. Walsh, 1953; Hoffmann and Woodward, 1968), and replace them with the ultimately equivalent but perhaps less familiar valence-bond concepts (Goddard *et al.*, 1973).

A. Energetics of Bonding

The most essential kind of information that must be contained in the theory is gross energetics. Is the molecule stable or unstable? Does the transition state exhibit a large energy barrier or a deep attractive well? Is the reaction endoergic or exoergic? The last of these questions is automatically answered exactly by DIM. To illustrate how the other types of information are provided by DIM we will examine two of the simplest polyatomic systems, H_3^+ and H_3 (Ellison *et al.*, 1963).

To apply DIM in its simplest form to H_3, we construct pbf's from the three atomic ^2S basis functions, one assigned to each atom. There are two linearly independent combinations of these three states which produce a final doublet state. Labelling the $1s$ functions on each atom by A, B, and C, we can define the two pbf's,

$$\Phi_1 = (1/\sqrt{2})\mathscr{A}[\bar{A}CB - A\bar{C}B] \quad (22)$$

$$\Phi_2 = (1/\sqrt{6})\mathscr{A}[2AC\bar{B} - A\bar{C}B - \bar{A}CB] \quad (23)$$

where the bar denotes spin $-\frac{1}{2}$ and no bar $+\frac{1}{2}$. Applying the DIM formulation using these basis functions, neglecting overlap, we obtain for the energy of H_3

$$E = Q_1 + Q_2 + Q_3 - [J_1^2 + J_2^2 + J_3^2 - J_1 J_2 - J_1 J_3 - J_2 J_3]^{1/2} \quad (24)$$

where

$$Q_i = \tfrac{1}{2}[{}^1E(R_i) + {}^3E(R_i)] \quad (25)$$

$$J_i = \tfrac{1}{2}[{}^1E(R_i) - {}^3E(R_i)]. \quad (26)$$

R_1, R_2, and R_3 are the distances between atoms AB, BC, and AC, respectively. ${}^1E(R)$ and ${}^3E(R)$ are the energies of the ${}^1\Sigma_g^+$ and ${}^3\Sigma_u^+$ states of H_2. For C_{2v} geometries equations (24)–(26) reduce to

$$E = \tfrac{3}{2}\,{}^1E(R_{AB}) + \tfrac{1}{2}\,{}^3E(R_{AB}) + {}^3E(R_{AC}) \quad (27)$$

Equation (24) is the London equation (London 1929). If ${}^1E(R)$ and ${}^3E(R)$, or alternatively Q_i and J_i, are considered to be adjustable diatomic interactions, then a fairly wide range of potential energy surface topographies can be constructed. A variety of semiempirical procedures, differing in the way the diatomic interactions are chosen, have been based on equation (24) and can therefore be considered variations of the DIM approach. For example, in the LEPS method (London, 1929; Eyring and Polanyi, 1931; Sato, 1955; Kuntz *et al.*, 1966) the quantities Q_i and J_i are written in the form $Q_i'/(1 + S_i)$ and $J_i'/(1 + S_i)$ where S_i is the 'Sato parameter'. There are related procedures (e.g. Porter and Karplus, 1964) which, although not derived directly from equation (24), still incorporate the underlying valence-bond basis of the London equation; i.e. of DIM.

The simplest DIM description of H_3^+ is slightly more complicated than that for H_3. Three pbf's of overall singlet spin must be included,

$$\Phi_1 = 1/\sqrt{2}[\bar{A}B - A\bar{B}] \quad (28)$$

$$\Phi_2 = 1/\sqrt{2}[\bar{B}C - B\bar{C}] \quad (29)$$

$$\Phi_3 = 1/\sqrt{2}[\bar{A}C - A\bar{C}] \quad (30)$$

Each of these basis functions describes the pairing of the spins of the electrons on two centres, with the third centre a bare proton. The DIM expression for the energy of H_3^+ is the lowest eigenvalue of the 3×3 matrix constructed from the 3 pbf's of equations (28)–(30). It is sufficiently complicated that we will write it down here only for the special case of equilateral geometries with atomic separation R,

$$E = {}^1E(R) + 2\,{}^2E^+(R) \quad (31)$$

${}^1E(R)$, as before, is the energy of the ground ${}^1\Sigma_g^+$ state of H_2 at internuclear separation R, and ${}^2E^+(R)$ is the corresponding energy of the ${}^2\Sigma_g^+$ ground state of H_2^+.

Now that we have obtained simple DIM expressions for equilateral H_3^+, equation (31), and for C_{2v} H_3, equation (27), we are ready to begin a qualitative discussion of the way bonding and antibonding interactions are described by DIM. We can deduce immediately either from molecular orbital or valence-bond considerations that H_3^+ will be a stable molecule and H_3 is likely to be loosely bound or unstable. The molecular orbital argument is illustrated in Fig. 2. Consider 3 protons arranged in a C_{2v} geometry. By forming linear combinations of the 3 1s orbitals, one from each

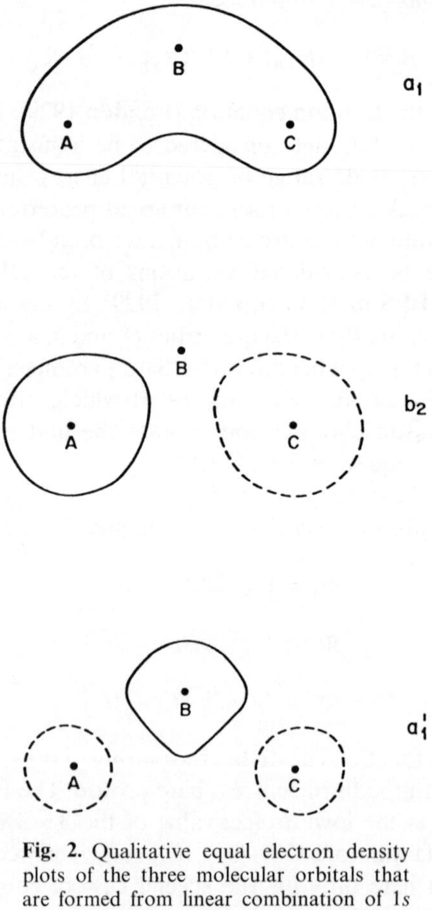

Fig. 2. Qualitative equal electron density plots of the three molecular orbitals that are formed from linear combination of 1s orbitals on each of three H atoms. Dashed lines indicate regions where the wave function is negative.

atom, we can construct 3 different molecular orbitals, labelled according to their symmetry a_1, b_2, and a'_1. The a_1 orbital is nodeless and will be strongly bonding. The b_2 orbital, with a node through the central atom, is antibonding. The a'_1 orbital with two nodes is even more strongly antibonding. In the singlet ground state of H_3^+ the two electrons are both assigned to the bonding a_1 orbital, resulting in a strongly bound, stable molecule. Since the Pauli principle prevents a third electron from occupying this orbital, the third electron in H_3 must be assigned to the antibonding b_2 orbital. H_3 is thus either weakly bound or unbound. (In fact, of course, it is unbound except for a very weak long-range van der Waals attraction.) H_3^- would be expected to be very unstable. Since H_3^+ involves only the bonding a_1 orbital, we would expect the most stable geometry of H_3^+ to be that which maximizes the overlaps of the 3 1s orbitals, i.e. equilateral. For H_3 we would expect the most stable (least repulsive) geometry to be linear so as to minimize the repulsive interaction between the two lobes of the antibonding b_2 orbital. All of these qualitative molecular orbital arguments lead to correct conclusions.

The DIM method takes account of these considerations in a way that is not as immediately apparent, but is equivalent. The molecular orbital discussions are based on two considerations. The first is the Pauli principle which prevents more than two electrons to be assigned to the same molecular orbital. The second is the concept of bonding or antibonding orbitals, determined by the qualitative shape and the number of nodes of the orbital. DIM invokes the same two effects, but they are slightly disguised. The Pauli principle is introduced via coupling of the spins of the atomic functions to construct the pbf's. For H_3^+ the basis functions of equations (28)–(30) involve coupling of only two spins, so there are no constraints arising from the Pauli principle, just as in the molecular orbital picture. However, for H_3 three doublet spins must be coupled. In the basis functions Φ_1 and Φ_2 defined in equations (22) and (23), the spin functions on atoms A and C are antiparallel (singlet) in Φ_1 and parallel (triplet) in Φ_2. The coupling between the spin functions of atoms A and B (or C and B) is neither pure singlet nor pure triplet. For example, we can rewrite equation (22) as

$$\Phi_1 = \sqrt{1/2}(1/\sqrt{2})\mathscr{A}[\bar{A}BC - A\bar{B}C] - \sqrt{3/4}(1/\sqrt{6})\mathscr{A}[2AB\bar{C} - \bar{A}BC - A\bar{B}C], \qquad (32)$$

i.e. the coupling between spins of atoms A and B (or B and C) is $\frac{1}{4}$ singlet and $\frac{3}{4}$ triplet in Φ_1. Similarly, it is $\frac{3}{4}$ singlet and $\frac{1}{4}$ triplet in Φ_2. Thus the fact that three doublet spin functions cannot simultaneously be paired, the Pauli principle, is a trivial outcome of the algebra of spin angular momentum coupling. DIM, by correctly treating the coupling of atomic spins, thus automatically incorporates all consequences of the Pauli principle.

The second important concept of molecular orbital theory, that of bonding and antibonding interactions, is introduced quantitatively in DIM via the diatomic input information. For example, in H_3^+ only the strongly bonding orbital (Fig. 2) is occupied. The bonding character of this interaction is assured in DIM because, as seen from equation (31) for equilateral H_3^+, the interaction energy is the sum of contributions from the strongly bonding ground state of H_2 and the bonding ground state of H_2^+. The addition of one electron in the antibonding b_2 orbital to form neutral H_3 introduces antibonding interactions into DIM. As shown in equation (27) for C_{2v} geometries, the interaction energy contains contributions not only from the bonding ground state but also from the repulsive lowest triplet state of H_2. This triplet state, in a molecular orbital picture, consists of one electron in a

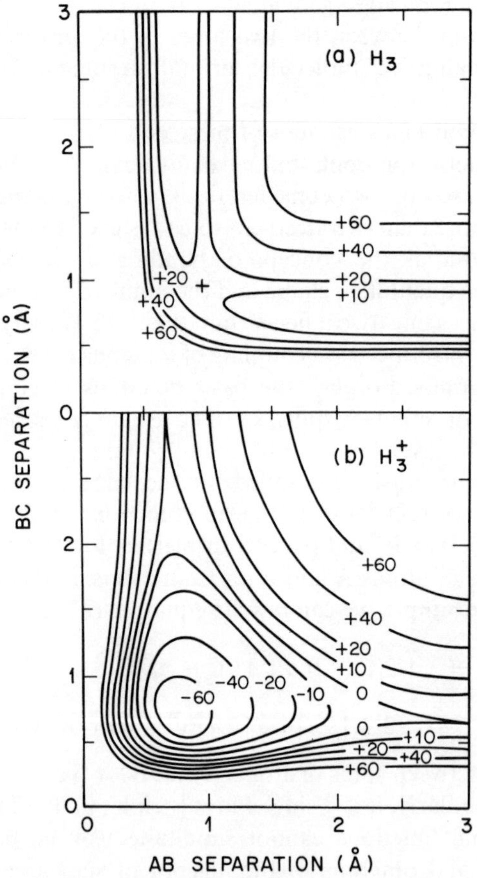

Fig. 3. Contour plots of equal potential energy for collinear H_3 (a) and H_3^+ (b).

bonding σ_g orbital and the other in the antibonding σ_u orbital. DIM also predicts geometries correctly. H_3^+ is equilateral thereby maximizing the bonding interactions, and the minimum energy of H_3 is for linear configurations for which the repulsive term ${}^3E(R_{AC})$ in equation (27) is minimized.

The DIM method thus correctly describes the qualitative bonding and antibonding interactions that determine the stability of molecules. Potential energy contours computed by DIM for H_3^+ and H_3 are shown in Fig. 3. The two potential energy surfaces are qualitatively very different. As expected, H_3^+ exhibits a deep potential well whereas H_3 is unstable relative to $H + H_2$. In fact, these potential energy surfaces are in reasonable quantitative agreement with accurate *ab initio* calculations. For example, the best *ab initio* value for the barrier to H atom exchange in the $H + H_2$ reaction (Liu, 1973) is 9.8 kcal/mol at $R_{AB} = R_{BC} = 1.76\ a_0$, compared with the DIM value of 13.0 kcal/mol at 1.78 a_0. The best *ab initio* dissociation energy of H_3^+ (Carney and Porter, 1974) is 4.49 eV at $R = 1.65\ a_0$, compared with 4.94 eV at $R = 1.73\ a_0$ for DIM. Furthermore, as shown in Table I, the DIM characterization of the avoided intersection between the ground and first excited potential energy surfaces of H_3^+ compares well with *ab initio* results (Bauschlicher *et al.*, 1973). The good agreement in these two cases was achieved with direct use of experimental energies; i.e. with no adjustment of input data. It is apparent that by slight modification of, for example, the repulsive diatomic potential curves, agreement within the accuracy of the best *ab initio* calculations is easily achievable.

TABLE I. Comparison of DIM and *Ab Initio* Descriptions of the H_3^+ Avoided Intersection

Parameter	*Ab initio*[a]	DIM[b]
Position $R_2(a_0)$	2.48	2.46
Separation ΔE at R_2 (kcal/mol)	16.6	18.1
Slope $\Delta E'$ at R_2 (kcal/mol-a_0)	35.2	34.7
Curvature $\Delta E''$ at R_2 (kcal/mol-a_0^2)	69.3	57.0

[a] Bauschlicher *et al.* (1973).
[b] Preston and Tully (1971).

B. Valence

Now that we have established that the DIM method correctly accounts for the fact that some combinations of atoms form stable molecules (e.g. H_3^+) and some are unstable (e.g. H_3), it is a simple step to show that the method correctly describes chemical valence; i.e. why CH_4 is stable but not CH_5 or NH_4 or OH_4. We illustrate this by considering the atoms F, O, N, and C which usually exhibit valences of 1, 2, 3, and 4, respectively.

The fluorine atom has a valence of 1 because it has only one unpaired electron to bond with another species, e.g. with a hydrogen atom. In the DIM formalism, the spin of the ^2P fluorine atom can couple with that of the ^2S hydrogen to form a singlet bonding interaction. But just as for the H_3 molecule, a second H atom cannot simultaneously couple with the fluorine atom as a singlet; it must be primarily a triplet repulsive interaction (equation 32). The ground state of the oxygen atom is ^3P, so a singlet H_2O molecule can be formed via two bonding OH doublet interactions; i.e. oxygen has a valence of 2. Similarly, nitrogen, with its ^4S ground state, has a valence of 3.

The ground-state configuration of carbon, like oxygen, is ^3P. Both have two unpaired $2p$ electrons. In carbon, however, a $2s$ electron can be promoted to the third vacant $2p$ level resulting in four unpaired electrons. In the isolated atom this costs 4.2 eV (the difference in energy between the excited ^5S and ground ^3P states), but this is more than made up for by the additional bonding interactions in, for example, CH_4. In the oxygen atom this $2s \rightarrow 2p$ promotion is blocked because the third $2p$ level is not vacant but contains a pair of electrons; the lowest ^5S state of oxygen is 9.1 eV above the ground state. Thus if only the ground-state ^3P configuration were employed as an atomic basis function in DIM, then carbon would be predicted to behave as a divalent species. In order to correctly describe its valence of 4, the excited ^5S state must be included among the carbon atom basis functions. Thus DIM treats the valence of carbon properly; it will exhibit a valence of 4 only if the excitation energy of the ^5S state (i.e. the $2s \rightarrow 2p$ promotion energy) is compensated by additional bonding interactions.

C. Directionality of Chemical Bonding

The potential utility of DIM for producing accurate potential energy surfaces first became apparent when it was recognized that DIM is capable of correctly describing the directional character of chemical bonds (Tully, 1973a). DIM is essentially a procedure for evaluating the energy of a polyatomic molecule from the energies of its composite diatomic and atomic fragments. Since pairwise additive interaction approximations are hopelessly inadequate in describing molecular bonding, DIM might be viewed with skepticism. Of course, DIM does not assume pairwise additive interactions. This was illustrated in the H_3 example above, which demonstrated how the approach of a third H atom toward an H_2 molecule weakens the original H_2 bond by mixing in some triplet character. The key to the description of 3,4,... body interactions in DIM lies with the fact that energies of both ground and excited fragment states are included, with the relative mixing of these states being determined globally, not pairwise. However, while non-pairwise additive interactions are present in DIM, it does not necessarily follow that they are described accurately. A sensitive test of this is the ability of the method to correctly predict geometries of polyatomic molecules; i.e. to incorporate the directional character of chemical bonding.

We illustrate this with the first-row diatomic hydrides (Tully, 1973a). In particular, consider the ground state of the water molecule. The bent geometry of water can be understood qualitatively from simple valence-bond considerations. The oxygen atom has two unpaired electrons which we can assign, for example, to the $2p_x$ and $2p_z$ orbitals. If we pair the electron of one H atom to the $2p_x$ electron, and the electron of the other H atom to $2p_z$, then to maximize overlap (maximize OH bond strength) the HOH angle would be 90°. But since the H–H interaction is repulsive ($\frac{3}{4}$ triplet, $\frac{1}{4}$ singlet), the actual bond angle will be somewhat greater than 90°.

These arguments become quantitative in DIM, but they are also somewhat less apparent. DIM employs basis functions constructed from atomic states, e.g. $O(^3P)$ and $H(^2S)$. It is not apparent from the state designation alone (^3P) that oxygen has an unpaired electron in $2p_x$ and another in $2p_z$. There are many excited states of oxygen with electrons assigned to all kinds of orbitals which couple in such a way to produce ^3P states. DIM extracts the additional required information from diatomic energies.

In this case the crucial information is contained in the two lowest $^2\Pi$ states of OH. The lower state ($^2\Pi_1$) is attractive and the upper one ($^2\Pi_2$) is repulsive. Their electronic configurations, expressed in terms of products of atomic functions, can be represented approximately by

$$\Phi(^2\Pi_1) \sim \sqrt{3/4}\,O(^3P)H(^2S) + \sqrt{1/4}\,O(^1D)H(^2S) \tag{33}$$

$$\Phi(^2\Pi_2) \sim \sqrt{1/4}\,O(^3P)H(^2S) - \sqrt{3/4}\,O(^1D)H(^2S). \tag{34}$$

Thus the lower $^2\Pi$ state has some contribution from the excited $O(^1D)$ configuration. It is easy to see how this arises. Placing the O and H atoms on the z axis, we obtain the strongest $OH(^2\Pi)$ bond by pairing the $2p_z$ electron with the $H(1s)$,

$$\Phi(^2\Pi_1) \sim \tfrac{1}{2} XZA[\alpha\alpha\beta - \alpha\beta\alpha]. \tag{35}$$

But since the $O(^3P)$ and $O(^1D)$ configurations are

$$\Phi(^3P_y) \sim XZ \begin{Bmatrix} \alpha\alpha \\ \dfrac{1}{\sqrt{2}}(\alpha\beta + \beta\alpha) \\ \beta\beta \end{Bmatrix} \tag{36}$$

and

$$\Phi(^1D_{xz}) \sim \frac{1}{\sqrt{2}} XZ[\alpha\beta - \beta\alpha] \tag{37}$$

equation (33) results immediately.

If we now approach $OH(^2\Pi_1)$ with another H atom, in order to form the ground state of water we must couple the two doublet spins in such a way to produce a singlet. To accomplish this the second H is forced to assume the following coupling with the O atom,

$$\Phi \sim \sqrt{3/4}\,O(^3P)H(^2S) - \sqrt{1/4}\,O(^1D)H(^2S). \tag{38}$$

The sign of the second term is reversed from that of equation (33). This is simply another reflection of the Pauli principle: the electrons of two H atoms cannot simultaneously be paired with the $2p_z$ electron.

If we define the z axis to be the line passing through the O and the first H atom, and then rotate the second H atom about the O atom in the xz plane, the character of the interaction of the second H with the O will change according to the rotational properties of the atomic basis functions, in this case P and D functions (Tully, 1973a). As a simple example, the interaction of the $F(^2P_z)$ state with $H(^2S)$ along the z axis would correspond to the $FH(^{1,3}\Sigma^+)$ state. If the H were rotated 90° so that it lay along the x axis, it would become an $FH(^{1,3}\Pi)$ interaction. At angles between 0 and 90°, it would be a linear combination of Σ and Π contributions. Similarly, rotating the second H atom of H_2O off the z axis requires a similar transformation which is easily accomplished using simple rotation matrices. In this case, the $O(^3P_y)$ function is invariant to rotation about the y axis, whereas $^1D_{xz}$ mixes with $^1D_{z^2}$ and $^1D_{x^2y^2}$. However, a rotation through 90° results in $^1D_{xz} \rightarrow -{}^1D_{xz}$; i.e. if the second H atom lies along the x axis, the minus sign in equation (38) will become + and the OH bond will be strong ($^2\Pi_1$), equivalent to equation (33). Thus the simple valence-bond description of the geometry of H_2O is contained in the DIM method, but only if the $O(^1D)$ excited configuration is included with $O(^3P)$ among the atomic basis functions. Applying DIM to H_2O in this way, employing no adjustable input information, results in a bond angle of 100.3° compared with the experimental value of 104.5° (Tully, 1973a). Slight adjustment of OH input information can produce an extremely accurate H_2O potential energy surface.

Similar results are achieved with the other first-row hydrides. For example, DIM correctly predicts that the ground state of CH_2 is $^3B_1$ and is bent at some undetermined but large angle (experimental value is 136°), and the first excited state is $^1A_1$ with angle of about 100° (experimental value is 102.4°). A quantitative comparison of DIM first-row dihydride potential energy surfaces with experiment is given in Table II.

TABLE II. DIM Description of Ground-State Properties of First-Row Triatomic Hydrides
(Reproduced by permission of the American Institute of Physics from Tully, 1973a.)

Molecule	Method	Configuration	Bond length (Å)	Bond angle (deg)	Atomization energy (eV)
BeH_2	DIM	$^1\Sigma_g^+$	1.37	180	6.20
	Exp.	$^1\Sigma_g^+$		180	(6.70)
BH_2	DIM	$^2A_1$	1.23	(131)[a]	7.86
	Exp.	$^2A_1$	1.18	131	8.23
CH_2	DIM	$^2B_1$	1.10	(136)[a]	8.71
	Exp.	$^3B_1$	1.08	136	(7.87–9.30)
NH_2	DIM	$^2B_1$	1.041	100.4	7.53
	Exp.	$^2B_1$	1.024	103.4	7.57
H_2O	DIM	$^1A_1$	0.979	100.3	9.37
	Exp.	$^1A_1$	0.957	104.5	9.59

[a] Adjusted to agree with experiment.

Eaker and Parr have adjusted the CH input information to obtain a reasonable fit to the energies and geometries of the CH_2 ground state and excited $^1A_1$ state. They then employed the same diatomic input unaltered in DIM treatments of CH_3 and CH_4. As shown in Table III, their results (Eaker and Parr, 1976) are encouraging. The potential energy surfaces thus obtained for ground and excited states of CH_2, CH_3, and CH_4 are probably the most accurate currently available, and should be quite satisfactory for a variety of dynamical studies.

In all of the examples cited in this section DIM provides, without any empirical adjustment of input data, semiquantitative potential energy surfaces with qualitatively correct overall shapes; e.g. reaction barriers, well depths, geometries, etc. Furthermore, it appears that with modest adjustment of input information to reproduce limited experimental or theoretical data, accurate quantitative potential energy surfaces can be obtained. However, there are situations for which the method does not appear so well suited. Most notable among these, perhaps, is H_4. The DIM (unadjusted) prediction of 69 kcal/mol above two isolated H_2 molecules for the minimum energy of square H_4 (Tully, 1977b) is in substantial disagreement with an accurate *ab initio* value of 142 kcal/mol (Rubinstein and Shavitt, 1969). While the qualitative features of the DIM and *ab initio* H_4 potential energy surfaces are similar, so great an adjustment would be required to obtain quantitative agreement that the resulting DIM surface could not be considered reliable.

This and related failures of DIM result from inaccurate representation of

TABLE III. Equilibrium Geometries and Energies of Atomization for States of CH_2, CH_3, and CH_4
(Reproduced by permission of the American Institute of Physics from Eaker and Parr, 1976)

		DIM	*Ab Initio*[a]	Expt.[a]
$CH_2(X^3B_1)$	R_e(Å)	1.083	1.081	1.078
	< HCH	127.7°	134.2°	136 ± 8°
Atomiz. energy (kcal/mol)		191.1	187	190.6
$CH_2(^1A_1)$	R_e(Å)	1.118	1.116	1.11
	< HCH	99.7	102.5	102.4
$^3B_1 \rightarrow {}^1A_1$ (kcal/mol)		12.1	9.2 ± 3	8–13
$CH_2(^1B_1)$	R_e(Å)	1.086	1.092	1.05
	< HCH	137.3°	143.8°	140 ± 15°
$^3B_1 \rightarrow {}^1B_1$ (kcal/mol)		37.0	43.8	
$CH_3(X^2A_2'')$	R_e(Å)	1.097	(1.079)	1.079
	< HCH	120°	(120°)	120°
Atomiz. energy (kcal/mol)		301.5	~230	307
$CH_3(^2A_1')$	R_e(Å)	1.127	(1.079)	1.12
	< HCH	112.1°	(120°)	120°
$^2A_2'' \rightarrow {}^2A_1'$ (kcal/mol)		139.4	124.6	132
$CH_3(^2E')$	R_e(Å)	1.148	1.172	...
	< HCH	86.4°	92°	...
$2A_2'' \rightarrow {}^2E'$ (kcal/mol)	153.1	145.6		
$CH_4(X'A_1)$	R_e(Å)	1.119	1.091	1.094
	< HCH	109.5°	109.5°	109.5°
Atomiz. energy (kcal/mol)		398.5	404.7	419

[a] References to *ab initio* and experimental studies are given in Eaker and Parr (1976).

3- and 4-centre integrals by the approximate expressions, equations (16) and (18) (Tully, 1977a). The occurrence of such failures underscores the need for corroborating theoretical or experimental evidence.

IV. CASE STUDIES OF SIMPLE CHEMICAL PROCESSES

It is apparent from the discussions of the preceding section that the great diversity and individuality displayed by the chemical elements in the ways they interact among each other will be reflected in potential energy surfaces. Hills, valleys, cliffs, and lakes of all shapes and sizes will occur. In this section we address the question of how various topographical features of potential energy surfaces determine the dynamics of molecular collision processes. We do this by citing a few examples of chemically reactive systems for which

DIM or LEPS potential energy surfaces, in combination with experimental results, have contributed to our understanding of the collision dynamics. Before beginning with the first example, however, we mention a few general considerations which will be referred to later.

The potential energy surface feature that, in cases where it exists, probably has the most direct effect on reactivity is the reaction barrier; i.e. the saddle point or point of maximum energy along the minimum energy path connecting reactants and products. The height of the reaction barrier, of course, is usually closely related (but certainly not equal) to the experimental activation energy for reaction. Furthermore the location of the barrier and the nature of the potential energy surface in the vicinity of the barrier can have a strong influence on the reaction dynamics, particularly for collision energies near threshold.

A systematic study of these effects has been carried out by Polanyi and coworkers (Kuntz *et al.*, 1966; Polanyi and Wong, 1969; Perry *et al.*, 1974; Polanyi and Schreiber, 1974; Kuntz, 1976) who performed classical trajectory simulations employing LEPS potential energy surfaces with parameters chosen to produce a wide variation of topographies. For exoergic reactions $A + BC \rightarrow AB + C$ with a single barrier, Polanyi defines three categories of potential energy surfaces according to whether energy is released mostly before, during, or after the region of maximum curvature of the reaction path. Potential energy surfaces exhibiting 'attractive energy release', i.e. for which most of the exoergicity of the reaction is converted to kinetic energy before the region of maximum curvature of the reaction path, are seen from trajectory studies to deposit a relatively large amount of energy into vibrational motion of the product AB molecule. This is illustrated in Fig. 4(a), in which the schematic trajectory is accelerated to such an extent as it rolls down from the barrier that it cannot follow the minimum energy path around the bend. Instead, its momentum carries it up the repulsive wall corresponding to a compressed AB bond, thus producing vibrational excitation. 'Repulsive energy release', Fig. 4(b), occurs when most of the heat of reaction is released after the region of maximum curvature. Trajectories can more nearly follow the minimum energy path, typically resulting in relatively low product vibrational energy and high translational or rotational energy. The intermediate case of 'mixed energy release' can result in partitioning of energy into either product internal motion or translation depending, in part, on the masses of the three atoms (e.g. Kuntz, 1976). Thus in favourable cases, the disposal of energy among product degrees of freedom for an exoergic reaction can be qualitatively understood directly from simple features of the potential energy surface.

Of great interest for endoergic reactions is the selectivity of energy requirements; i.e. is the reaction facilitated more effectively if the required energy is inserted initially into vibrational or translational motion of the reactants?

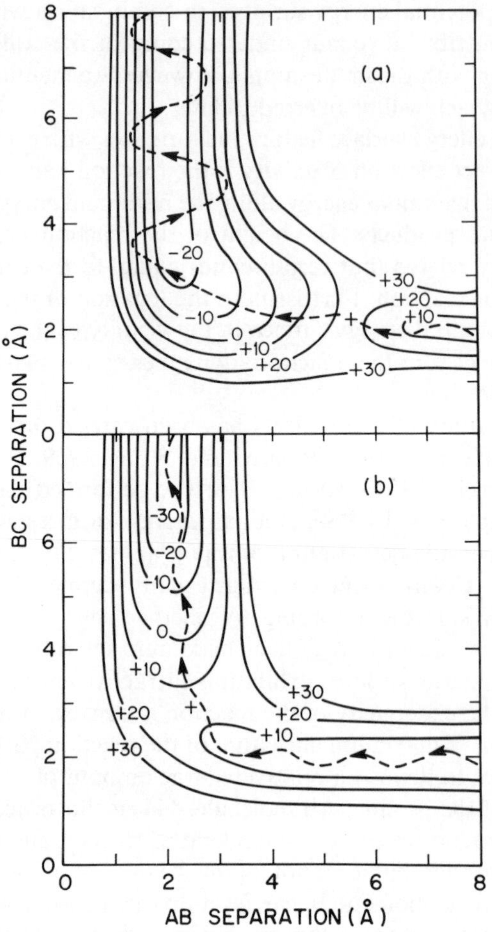

Fig. 4. Qualitative behaviour of potential energy surfaces for exothermic reactions A + BC → AB + C. (a) Attractive energy release. (b) Repulsive energy release.

The location of the barrier relative to the region of maximum curvature is again critical, as can be seen from detailed balance considerations. Thus for an endoergic reaction AB + C → A + BC with the barrier in the exit valley, reactant vibrational energy will usually be much more effective than translation in surmounting the barrier (the reverse of Fig. 4a). If the barrier is in the entrance valley (reverse of Fig. 4b), translational or rotational energy is likely to be more effective than vibration in promoting reaction.

There are features of potential energy surfaces in addition to barriers that can have major effects on the dynamics of collisions. One such feature is a potential well like that exhibited by the reaction $H^+ + H_2 \rightarrow H_2 + H^+$ (Fig. 3b). Such reactions have the possibility of proceeding via a 'collision complex' that survives for many vibrational periods. This is particularly likely if the available energy (exoergicity plus initial reactant energy) is only a fraction of the well depth and/or if there are a large number of internal degrees of freedom among which to distribute the available energy.

Another potential energy surface feature which, when present, can be crucial in determining the dynamics of a collision process is the surface crossing or region of strong non-adiabatic coupling. For reactions which require a transition from one potential energy surface to another, the electronic transition is frequently the critical step that determines whether the reaction will proceed or not. Thus like the position of the barrier for an adiabatic reaction, the location and shape of the non-adiabatic interaction region can impose specific energy requirements for non-adiabatic processes. It is frequently apparent immediately from contour plots of off-diagonal potential energy surfaces, i.e. non-adiabatic interaction strength, whether vibrational or translational energy will be more effective in promoting electronic transitions (Preston and Tully, 1971; Carrington, 1974; Chapman and Preston, 1974; Galloy and Lorquet, 1977).

A. Na + FH → NaF + H

In a pioneering experiment in 1955, Taylor and Datz (Taylor and Datz, 1955) performed the first successful crossed molecular beam study of a chemical reaction,

$$K + BrH \rightarrow KBr + H. \tag{39}$$

Since that time a number of detailed molecular beam and chemiluminescence studies of reactions of alkali atoms with hydrogen halides have been reported. Part of the interest in this family of reactions concerns the role of non-adiabatic interactions between covalent (M—XH) and ionic (M^+—X^-H) potential energy surfaces (Lacmann and Herschbach, 1970). Thus both covalent and ionic configurations must be included explicitly in a DIM study of these systems; i.e. a minimum of five configurations must be employed, the four arising from $M(^2S)X(^2P)H(^2S)$ and one from $M^+(^1S)X^-(^1S)H(^2S)$.

A DIM calculation has been carried out in this way (Tully, 1978) for the endoergic reaction

$$Na + FH \rightarrow NaF + H. \tag{40}$$

Energy contour plots of the lowest potential energy surface are shown in Fig. 5 for collinear Na—F—H. The potential surface is qualitatively similar to one for LiFH computed by the OM method (Balint-Kurti and Yardley, 1977), and also to an *ad hoc* potential energy surface for KBrH (Roach, 1970).

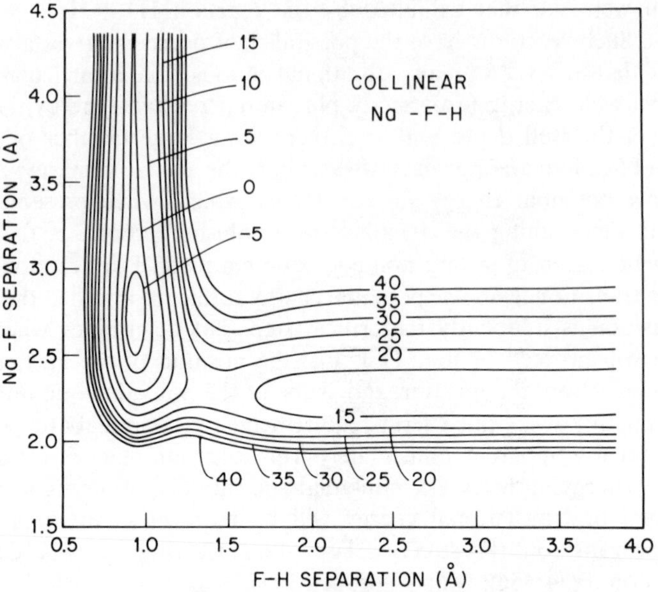

Fig. 5. Contour plot of equal potential energy for collinear Na—F—H computed by DIM. Contours are in kcal/mol relative to separated Na and HF. (After Tully, 1978.)

One of the most striking features of these surfaces is the existence of a barrier with its crest located well into the product valley; Fig. 5 is qualitatively similar to Fig. 4(a). We would therefore anticipate from the trajectory results of Polanyi and coworkers that reaction (40) will be promoted more effectively by initial HF vibrational energy than by relative translation.

This has been demonstrated experimentally for this system (Blackwell *et al.*, 1977). Vibrationally excited HF was formed by a 'pre-reaction' $F + H_2 \rightarrow HF^{\dagger}$ $(v = 1 - 3) + H$ or $F + HBr \rightarrow HF^{\dagger}$ $(v = 1 - 5) + Br$. Reaction of $HF(v = 5)$ with Na was found to proceed extremely rapidly, with a collision efficiency of order unity. Under these conditions ($v = 5$, corresponding to 51.8 kcal/mol vibrational excitation), the distribution of energy in the reactants is 3% translation, 2% rotation, and 95% vibration. Thus vibrational energy is indeed very effective in promoting reaction, consistent with the location of the barrier in Fig. 5. The threshold for reaction was observed by Blackwell et al. to be $v = 2$, corresponding to a threshold energy of $\leq$ about 21 kcal/mol. This compares favourably with the value of 18.2 kcal/mol for collinear NaFH computed by DIM.

The experiments of Blackwell *et al.* are consistent with observations of Brooks and coworkers (Odiorne *et al.*, 1971; Pruett *et al.*, 1974) on the reaction $K + HCl \rightarrow KCl + H$. These workers found that increasing the relative translational energy had little effect whereas excitation of HCl to the $v = 1$ level increased the reaction cross section by two orders of magnitude. Thus it appears that for this system, too, the reaction barrier is located in the exit valley.

Molecular beam studies of the reaction of K with HBr, reaction (39), have uncovered an unusual isotope effect. Product angular distributions are observed to peak backwards (in the recoil direction) for K + HBr at initial relative energy of 2.8 kcal/mol, predominantly forward for K + DBr at 2.8 kcal/mol, and predominantly backward for K + TBr at 1.4 kcal/mol (Martin and Kinsey, 1967; Gillen *et al.*, 1969). Roach has proposed qualitative arguments which can explain this effect if the potential energy surface displays the following two features (Roach, 1970): first, the energy of the 'corner' of the surface—the position of maximum curvature of the reaction path—must increase as the KBrH angle is decreased from 180°; second, the barrier height relative to the energy at the corner must decrease with decreasing angle. Both of these features are in fact exhibited by the DIM NaFH surface and the OM LiFH surface. There is, however, a disparity between the two theoretical treatments with respect to the behaviour of the barrier height measured with respect to the energy of reactants. For NaFH, DIM predicts a gradual increase in barrier height upon bending, similar to the sketch by Roach, whereas for LiFH, OM predicts a continual lowering of the barrier as the angle is decreased from 180° to 90°. Unanswered questions such as this could almost certainly be resolved by comparing the predictions of classical trajectory simulations employing these potential energy surfaces with results of molecular beam experiments performed as a function of initial relative translational energy.

Lacmann and Herschbach have observed the formation of K^+ ions in 5–20 eV collisions of K with HCl but observe no emission from excited states of K. In order to shed light on this result, it is useful to construct an off-diagonal potential energy surface showing the non-adiabatic interaction strength as a function of nuclear position. Non-adiabatic coupling is a vector quantity which we will define as

$$\mathbf{d}_{kl}(\mathbf{R}) = \langle \Psi_k | \nabla | \Psi_l \rangle \tag{41}$$

where Ψ_k and Ψ_l are the wave functions of equation (3) for different electronic states k and l, ∇ is the gradient with respect to nuclear coordinates, and the brackets indicate integration over electronic coordinates only. Fortunately, DIM provides a very simple approximation to $\mathbf{d}_{kl}(\mathbf{R})$ (Preston and Tully, 1971; Tully, 1973b),

$$\mathbf{d}_{kl}(\mathbf{R}) = \sum_{mm'} \Gamma^*_{mk} S_{mm'} \nabla \Gamma_{m'l} \tag{42}$$

where Γ_{ij} are the eigenvector elements of equation (19). Employing this simple procedure to the NaFH system, we obtain the off-diagonal potential

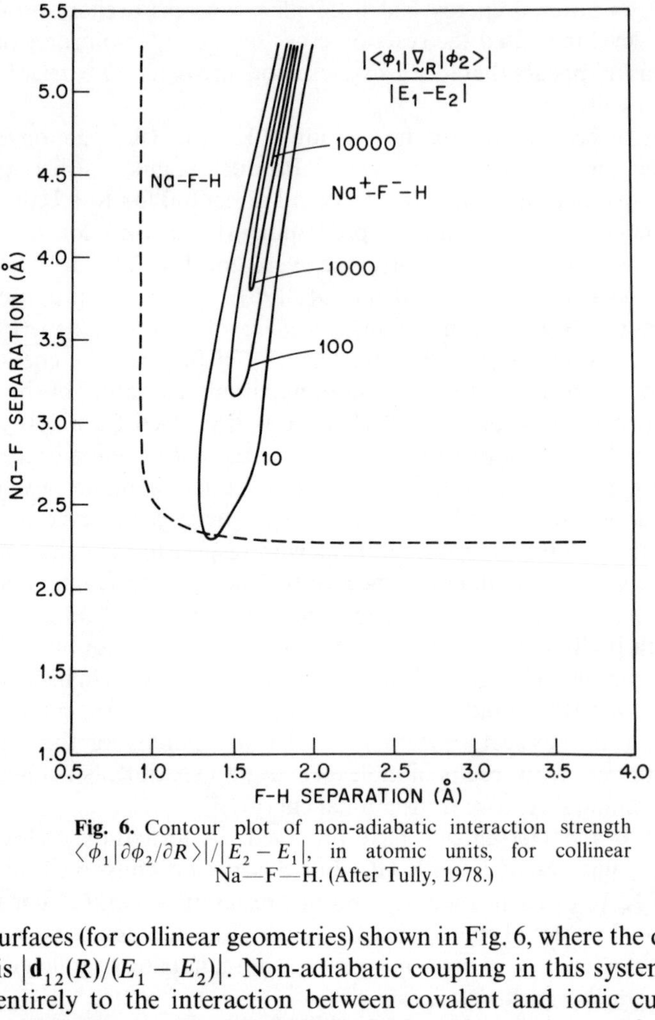

Fig. 6. Contour plot of non-adiabatic interaction strength $\langle \phi_1 | \partial\phi_2/\partial R \rangle | / | E_2 - E_1 |$, in atomic units, for collinear Na—F—H. (After Tully, 1978.)

energy surfaces (for collinear geometries) shown in Fig. 6, where the quantity plotted is $|\mathbf{d}_{12}(R)/(E_1 - E_2)|$. Non-adiabatic coupling in this system is due almost entirely to the interaction between covalent and ionic curves, as suggested by Lacmann and Herschbach. In fact, the maximum of the non-adiabatic coupling defines an avoided crossing 'seam' between the ionic and covalent states. It can be seen from Fig. 6 that electron transfer does not occur as the reactants approach; it is incorrect to think of a transition from $Na + HF$ to $Na^+ + HF^-$. The avoided crossing is traversed in the region of the barrier where the ground-state potential energy surface exhibits a smooth transition over a distance of about 1 Å from covalent to ionic character. A rough estimate of the likelihood of occurrence of an electronic transition to the first excited state in this region is provided by the quantity $|\dot{\mathbf{R}}\mathbf{d}_{12}|$, where $\dot{\mathbf{R}}$ is the velocity of the nuclei (Tully, 1977a). If this parameter is of order unity or larger, transitions are highly probable. If it is much less than unity, transitions are unlikely. For energies of 20 eV or so, $\dot{\mathbf{R}}$ is of order 10^{-2} atomic velocity units. Thus it is clear from Fig. 6 that non-adiabatic coupling in the vicinity of the barrier is too weak to promote transitions to the first excited state, consistent with Lacmann and Herschbach's absence of K* radiation in the KClH system. Non-adiabatic transitions can occur with high probability only in the upper right-hand corner of Fig. 6, where the three atoms are at relatively large separations. Since only motion perpendicular to the avoided crossing seam is effective in promoting electronic transitions, it is primarily the motion of the H atom, not Na—F dissociation, that is responsible in this case.

B. $F + H_2 \rightarrow FH + H$

The dynamics of the reaction of F atoms with H_2 and its isotopic variants has been the subject of exhaustive experimental and theoretical study. The reaction is exoergic by 32 kcal/mol and produces HF with sufficiently high vibrational excitation to be an efficient pumping mechanism for the HF laser. Thus many of the experimental and theoretical investigations of this reaction have focused on the disposal of energy among product translational and internal modes. Experimental measurements of product energy distributions have been performed by chemical laser (Parker and Pimentel, 1969; Coombe and Pimentel, 1973; Berry, 1973), infrared chemiluminescence (Polanyi and Tardy, 1969; Jonathan *et al.*, 1971; Polanyi and Woodall, 1972; Perry and Polanyi, 1976; Douglas and Polanyi, 1976), and molecular beam (Schafer *et al.*, 1970) techniques.

The potential energy surface of the ground state of FH_2 has been determined to quite high accuracy. This is particularly true for collinear configurations for which an extensive *ab initio* calculation has been performed (Bender *et al.*, 1972). A variety of semiempirical potential energy surfaces has also been proposed, of which some have been adjusted to reproduce experimental data and are thought to be at least as accurate as the best *ab initio* surface (Muckerman, 1971, 1972; Jaffe and Anderson, 1971; Wilkins, 1972; Polanyi and Schreiber, 1977).

The DIM potential energy surface for FH_2 (Tully, 1973a) is the lowest root of a 4×4 matrix constructed from the 4 doublet A' states that arise from coupling two $H(^2S)$ atoms and an $F(^2P)$. In collinear configurations it reduces to a 2×2 matrix, identical to LEPS. Since the reaction occurs almost exclusively in nearly collinear configurations, at least near threshold, LEPS provides a satisfactory representation of the potential energy surface and has been used in most dynamical studies.

A great many dynamical calculations have been reported. For collinear collisions, quantum (Schatz *et al.*, 1975), semiclassical (Whitlock and Muckerman, 1974), and classical (Whitlock and Muckerman, 1974) studies have been reported on the same LEPS surface. Three-dimensional classical trajectory studies have been carried out for several potential energy surfaces (Jaffe and Anderson, 1971; Muckerman, 1971, 1972; Wilkins, 1972; Polanyi and Schreiber, 1977). Three-dimensional quantum calculations have also been reported recently (Redmon and Wyatt, 1977).

A DIM potential energy surface for collinear FH_2 is shown in Fig. 7 (Tully, 1973a). This potential surface is very similar to the *ab initio* and semiempirical surfaces mentioned above. It can be seen from Fig. 7 that the potential energy surface is predominantly repulsive; i.e. energy release occurs mainly after the region of maximum curvature of the minimum energy path.

We would expect from the general considerations discussed earlier in this section that initial translational energy would be more effective in surmounting the barrier than would vibrational energy. This has been verified by classical trajectory studies (Polanyi and Schreiber, 1977). Fig. 8, taken from Polanyi and Schreiber, shows that the effect is quite dramatic. While this has not been demonstrated experimentally for this reaction, it has for two presumably similar 'substantially exothermic' reactions, $H + Cl_2 \rightarrow HCl + Cl$ (Ding *et al.*, 1973) and $H + F_2 \rightarrow HF + F$ (Polanyi *et al.*, 1976).

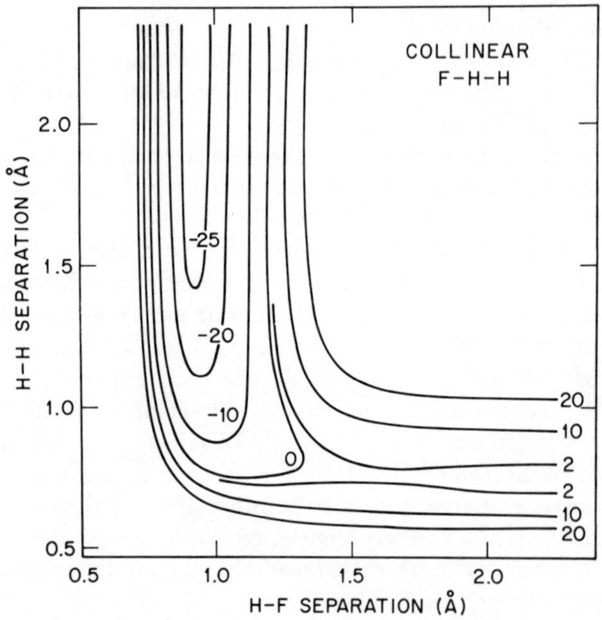

Fig. 7. Contour plot of equal potential energy for collinear FH_2 computed by DIM. Contours are in kcal/mol relative to separated $F + H_2$.

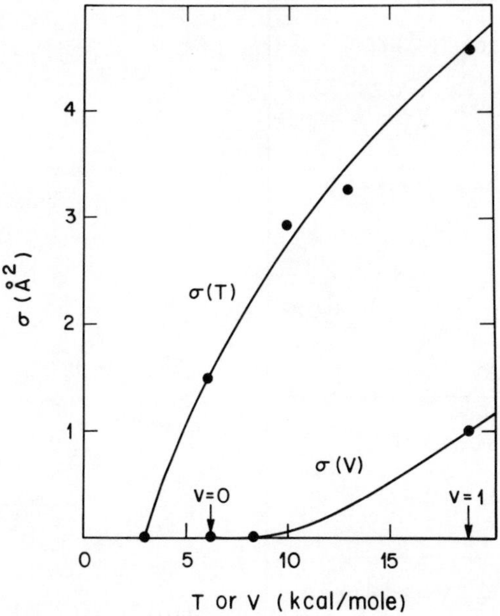

Fig. 8. Reaction cross section for $F + H_2(J = 1)$. Upper curve shows cross section σ as a function of translational energy T, with H_2 vibrational energy fixed at $V = 1$ kcal/mol. Lower curve shows σ as a function of V with T fixed at 1 kcal/mol. Note: Zero-point energy of H_2 is 6.2 kcal/mol. (Reproduced by permission of the Chemical Society from Polanyi and Schreiber, 1977.)

We might also expect from the general considerations discussed earlier that the repulsive FH_2 surface would channel most of the exoergicity of reaction into product translation and rotation, and relatively little into FH vibration. In fact, this is not the case. Trajectory calculations and experiment both show that energy is channelled efficiently into high vibrational levels of HF. A comparison between classical trajectory and experimental detailed rate constants is shown in Fig. 9. The explanation for this, as given by Polanyi and Schreiber, is that trajectories do not tend to follow the minimum energy path. Instead, they tend to 'cut the corner', corresponding to a release of repulsive energy while the new bond is still forming. Formation of the HF bond in an extended state gives rise to vibrational excitation. For a systematic and very extensive description of the detailed dynamics of the $F + H_2$ reaction, the reader is referred to the paper of Polanyi and Schreiber (Polanyi and Schreiber, 1977).

The above discussion is based on the assumption that collisions of F atoms with H_2 evolve on a single potential energy surface. This is not true,

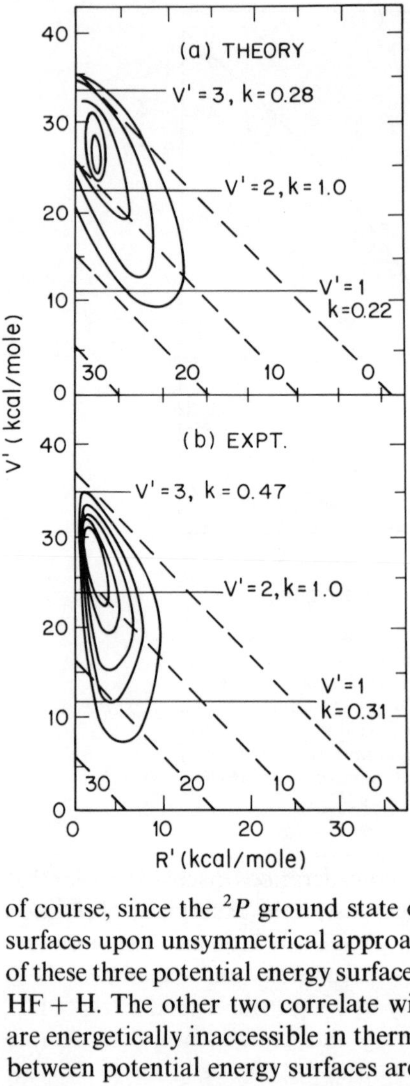

Fig. 9. Contours of equal detailed rate constant $K(V'R')$, drawn at 0.015, 0.046, 0.077, 0.11, and 0.15. Plots are normalized to same maximum value. Relative rates into different V' levels are indicated next to lines which give HF vibrational energies. Dashed lines indicate approximate values of translational energy. (a) Classical trajectory results. (b) Chemiluminescence experiments. (Reproduced by permission of the Chemical Society from Polanyi and Schreiber, 1977.)

of course, since the ^2P ground state of F splits into three different potential surfaces upon unsymmetrical approach of the H_2 molecule. Only the lowest of these three potential energy surfaces correlates with ground-state products HF + H. The other two correlate with electronically excited products that are energetically inaccessible in thermal energy collisions. Thus if transitions between potential energy surfaces are unlikely, single surface treatments for reactive events will be justified (Truhlar, 1972; Muckerman and Newton, 1972).

DIM potential energy surfaces for collinear approach of F and H_2 are shown in Fig. 10 (Tully, 1973b). Because of spin–orbit coupling, the $F(^2P)$ State splits into $F(^2P_{3/2})$ and $F(^2P_{1/2})$, the latter 1.15 kcal/mol higher in energy. Under the usually adequate assumption that two-centre terms may be neglected in the spin–orbit operator, spin–orbit interactions can be computed quite easily with the DIM formalism. The spin–orbit interaction has been included in the potential energy surfaces of Fig. 10.

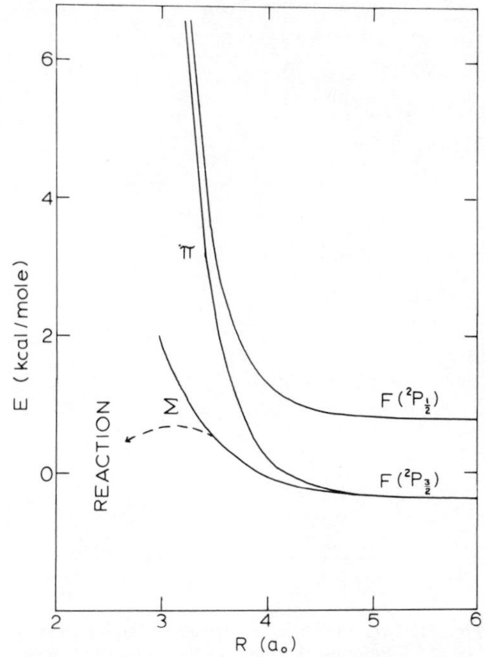

Fig. 10. Slices of the three lowest potential energy surfaces of FH_2, with spin–orbit included, computed by DIM. Curves are plotted as a function of F—H_2 distance for collinear geometries with H_2 internuclear distance fixed at 1.4 a_0. The reactive path is indicated schematically. (Reproduced by permission of the American Institute of Physics from Tully, 1973b.)

Non-adiabatic coupling between the reactive potential surface and the surface correlating with F($^2P_{1/2}$), induced by F–H_2 relative motion, is shown in Fig. 11. (The third potential energy surface is coupled to these two only by much weaker rotational coupling.) The non-adiabatic coupling, although not enormous, is not small. For room temperature collisions with relative velocity of order 2×10^5 cm/s or 10^{-3} atomic velocity units, the Massey parameter $|\mathbf{vd}_{13}/(E_1 - E_3)|$ at the peak of the non-adiabatic interaction is 0.3; i.e. not quite unity but not negligible. The non-adiabatic coupling is also fairly strongly localized even though there is no avoided surface crossing. The coupling arises from the fact that at large F–H_2 separations the wavefunctions are approximately atomic LS eigenstates, F($^2P_{3/2}$) and F($^2P_{1/2}$), whereas at short F–H_2 separations the chemical interaction dominates the spin–orbit interaction, and the wavefunctions become approximately Σ or Π in nature (for collinear FHH). This changeover between the asymptotic and chemical regions requires a recoupling of spin

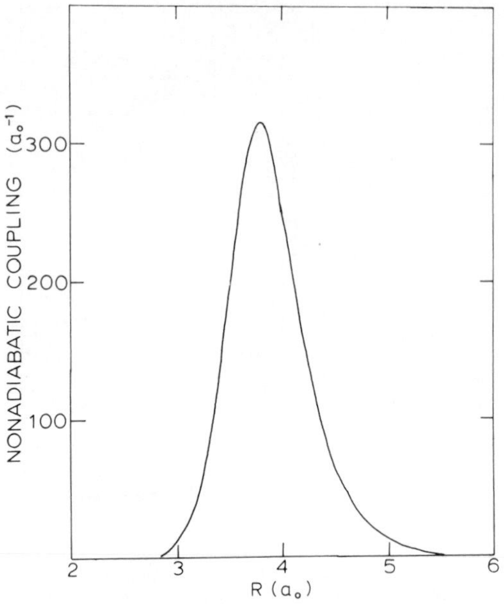

Fig. 11. Non-adiabatic coupling $|\langle\phi_1|\partial\phi_2/\partial R\rangle|/|E_3 - E_1|$ between the reactive FH_2 potential energy surface and the surface correlating with $F(^2P_{1/2})$. Curves are plotted as a function of $F—H_2$ distance for collinear geometries with H_2 internuclear distance fixed at 1.49 a_0. (Reproduced by permission of the American Institute of Physics from Tully, 1973b.)

and orbital angular momentum. The change in the nature of the wavefunction through the intermediate region is reflected in large non-adiabatic coupling, just as it is in cases of changeover from ionic to covalent nature.

As a result of this relatively strong coupling, non-adiabatic transitions do occur in $F–H_2$ collisions. Dynamical studies employing DIM potential surfaces (Tully, 1974; Kormonicki *et al.*, 1977) predict that the $F(^2P_{1/2})$ state will react with H_2 with a rate that is somewhat lower than that of $F(^2P_{3/2})$, but still significant (Fig. 12). Reactivity of the $^2P_{1/2}$ state of F with H_2 is interesting because there are two barriers to reaction, the ordinary energy barrier of about 2 kcal/mol on the reactive potential energy surface, and the barrier to accomplishing a non-adiabatic transition from the initial unreactive surface to the reactive surface. If the 2 kcal/mol energy barrier were the main hurdle, then we would expect the excited $F(^2P_{1/2})$ to utilize its extra 1.15 kcal/mol toward surmounting the barrier and thereby react faster than $F(^2P_{3/2})$. Trajectory studies using DIM potentials suggest, rather, that the transition from unreactive to reactive surfaces is sufficiently improbable that $F(^2P_{1/2})$ reacts with H_2 with considerably lower rate than

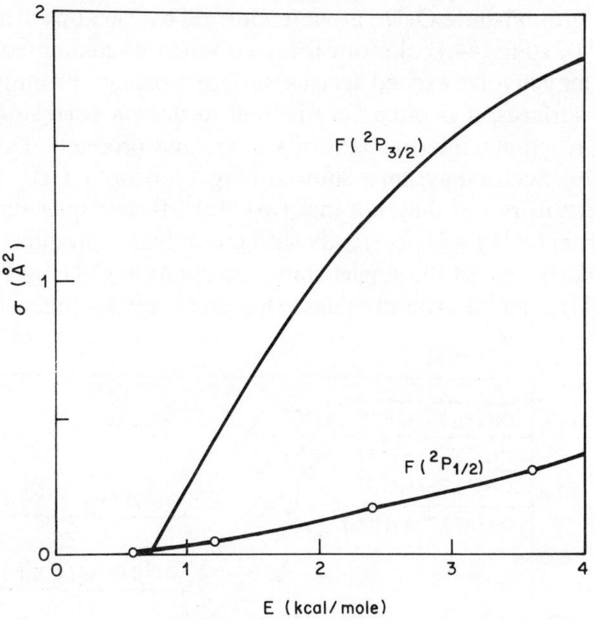

Fig. 12. Cross sections for reaction of $F(^2P_{1/2})$ and $F(^2P_{3/2})$ with H_2 (After Tully, 1974).

$F(^2P_{3/2})$ (Tully, 1974). While this prediction has not been verified in the laboratory, experimental studies of an analogous reaction $Br + HI \rightarrow HBr + I$ report a factor of 3 greater reaction rate for the $^2P_{3/2}$ state than the $^2P_{1/2}$ (Bergmann *et al.*, 1975).

If these calculations are correct, then electronic transitions play a relatively minor, although not negligible, role in $F-H_2$ reactive collisions. The extensive theoretical investigations described above based on a single potential energy surface therefore appear to be justified. This is not the case, however, for non-reactive inelastic $F-H_2$ collisions. Semiclassical (Tully, 1974; Lang *et al.*, 1975; Kormonicki *et al.*, 1977) and collinear quantum (Zimmerman and George, 1975) calculations using DIM potentials, and 3*D* quantum studies using *ab initio* potentials (Rebentrost and Lester, 1975, 1976, 1977) have established that transitions between F atom fine-structure states occur with high probability in non-reactive $F-H_2$ collisions, particularly when a near-resonance condition exists between the $^2P_{3/2}-^2P_{1/2}$ splitting and the energy of a molecular rotational transition.

C. $O + H_2 \rightarrow OH + H$

The reaction of oxygen atoms with H_2 exhibits dynamical behaviour very different from the previous two examples. We will consider here reaction of both the ground-state $O(^3P)$ atom (endoergic by 2 kcal/mol) and the first excited $O(^1D)$ state (44 kcal/mol exoergic). When discussing collision processes that may involve excited species, surface crossings, or multiple potential energy surfaces, it is often very helpful to draw a correlation diagram connecting reactants, intermediates if known, and products (Donovan and Husain, 1970). Such a diagram is shown in Fig. 13(a) for $O + H_2 \rightarrow OH + H$. It can be seen from this diagram that two of the three triplet states of H_2O that arise from $O(^3P) + H_2$ correlate with ground-state products $OH(^2\Pi) + H(^2S)$. Similarly, two of the singlet states arising from $O(^1D) + H_2$ correlate with $OH(^2\Pi) + H(^2S)$ ground states. However, only for one of these four

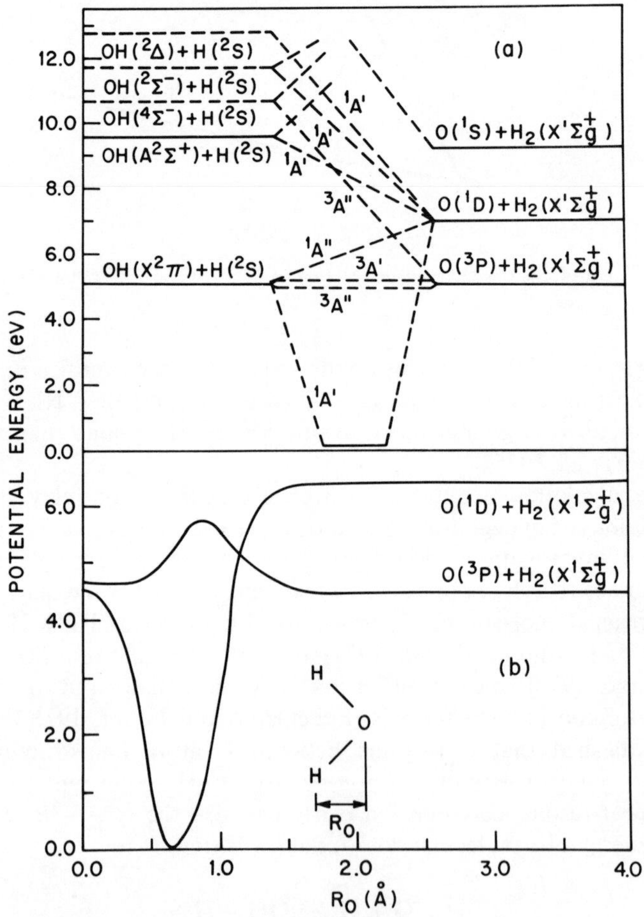

Fig. 13. (a) Correlation diagram connecting reactants $O + H_2$ and products $OH + H$. (b) Quantitative reaction path for insertion of O into H_2, computed by DIM. (After Whitlock *et al.*, 1976.)

potential energy surfaces, the lowest $^1A'$ surface corresponding to the ground state of H_2O, is anything definitive known about the interaction region. For the other states it is not known whether there are wells or barriers along the reaction path, whether favoured approach is collinear or bent, etc. In fact, except for the region of configuration space near the bottom of the well of the H_2O ground state, little is known about the lowest $^1A'$ state either. Furthermore, it is apparent from Fig. 13(a) that there are surface crossings exhibited by this system, but the locations of these are also uncertain.

It would be very valuable if a semiquantitative correlation diagram could be constructed that showed energies along some suitably defined reaction coordinate continuously connecting reactants and products. DIM affords a simple way to do this. Crude diagrams of this sort that may be helpful for qualitative discussions can be obtained using very approximate or estimated diatomic input. More refined input information can hopefully provide quantitative diagrams. Fig. 13(b) shows, for example, the minimum energy pathways for approach of the O atom to H_2, constrained to C_{2v} symmetry for the $^1A_1$ and $^3B_1$ states. These curves are from a DIM calculation employing accurate optimized input information (Whitlock *et al.*, 1976), and almost certainly represent the most accurate determination of these two potential energy surfaces, with the exception of the vicinity of the ground-state minimum. Note that Figure 13(b) shows no barrier to insertion of $O(^1D)$ into the H_2 bond, whereas a substantial barrier of about 31 kcal/mol is present for $O(^3P)$ insertion.

Figure 13(b) also shows a surface crossing between the $^1A'$ and $^3A''$ states. It is conceivable that spin–orbit interaction could induce singlet–triplet transitions between these two states. However, it is thought that the spin–orbit interaction is too small in this system (of order 100 cm^{-1}) for this to be a significant process. This is backed by experimental evidence (DeMore, 1967) and theoretical arguments (Tully, 1975) which indicate that quenching of $O(^1D)$ by H_2 results almost exclusively in reaction to form $OH + H$, with essentially no non-reactive $O(^3P) + H_2$ formation. Therefore in the following discussion we will ignore non-adiabatic effects and concentrate on independent motion on the $^3A''$ and $^1A'$ potential energy surfaces.

These two potential energy surfaces display very different landscapes resulting in very different reaction dynamics. The $^3A''$ surface describing the $O(^3P) + H_2$ reaction, according to DIM (Whitlock *et al.*, 1976), favours collinear approach. An energy barrier of approximately 13 kcal/mol is located slightly toward the exit valley, suggesting that reactant vibrational motion will be effective in promoting reaction. This is indeed found to be the case in trajectory calculations (Whitlock *et al.*, 1976). Trajectory studies of this reaction have also been carried out on an LEPS surface with the one adjustable Sato parameter chosen to reproduce the experimental activation energy of 8.9 kcal/mol (Johnson and Winter, 1977). The resulting potential energy surface is very similar to the DIM surface, with a barrier of 12.5 kcal/mol located slightly toward the product valley. The calculations of Johnson and Winter reproduce the experimental rate of reaction of H_2 ($v = 0$) from 300 K to 1000 K quite well, and also predict an enhancement of the rate by a factor of 5×10^3 for H_2 ($v = 1$) at 300 K. This is consistent with the only experimental investigation of this effect (Birely *et al.* 1975), in which an upper bound of the enhancement factor of 3.8×10^4 was determined. The dynamics of the reaction of $O(^3P)$ with H_2 thus appear qualitatively quite similar to the behaviour we have seen previously for systems that can be represented by LEPS-type potential energy surfaces.

This is not true for reaction of $O(^1D)$ with H_2 which follows the $^1A'$ potential energy surface. The reason for this, of course, is that the $^1A'$ surface exhibits a deep well corresponding to the ground state of the water molecule. Because of the presence of this well, as will be discussed below, it is of interest to determine whether the reaction proceeds via a long-lived collision intermediate. It is also of interest to examine whether the reaction proceeds by an abstraction mechanism, in which an end-on attack of the oxygen atom results in direct removal of the nearest H atom, or by insertion, in which the oxygen approaches the H_2 broadside and inserts into the H—H bond preliminary to reaction.

These questions have been investigated by trajectory studies (Whitlock *et al.*, 1979; Sorbie and Murrell, 1975, 1976). Both the DIM potential energy surface employed by Whitlock *et al.* and the empirical surface employed by Sorbie and Murrell exhibit no activation energy to perpendicular insertion. This is in contrast to *ab initio* results (Gangi and Bader, 1971) which predict a barrier of 8.6 kcal/mol, but it is consistent with the experimental finding that reaction of $O(^1D)$ with H_2 proceeds without activation energy (DeMore, 1967). The DIM surface of Whitlock *et al.* predicts a low barrier to collinear abstraction of about 1 kcal/mol, in contrast to the large barriers to abstraction obtained by Bader and Gangi, and assumed by Sorbie and Murrell. If the DIM prediction is correct, the abstraction and insertion mechanisms should be competitive.

Indeed, the trajectory studies of Whitlock *et al.* show both mechanisms to be important. Figs. 14 and 15 show typical insertion and abstraction reactions, respectively. Note that abstraction occurs by a single impulsive collision whereas insertion involves more complicated multiple collisions. The relative contributions of the two mechanisms were determined by monitoring the potential energy along each trajectory. Trajectories could be classified almost exclusively into two groups, those for which the minimum potential energy along the path was only barely below that of the products OH + H, and those for which the minimum potential energy was at least 30 kcal/mol below that of the products. The latter group, trajectories that traversed the region of the deep potential well, were identified as insertion

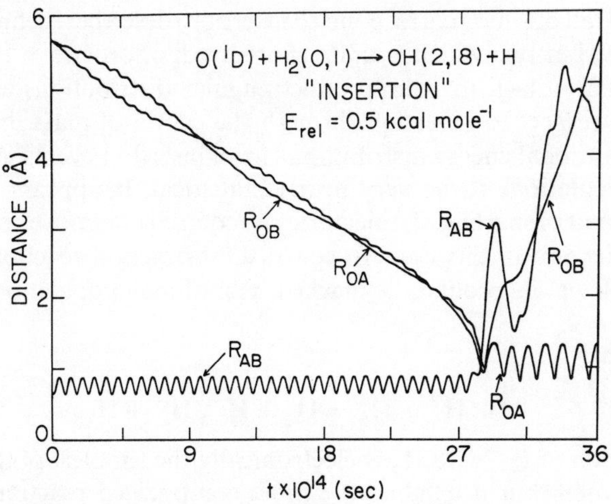

Fig. 14. Internuclear separations as a function of time along a trajectory describing an 'insertion' reaction of O with H_2. (After Whitlock *et al.*, 1979).

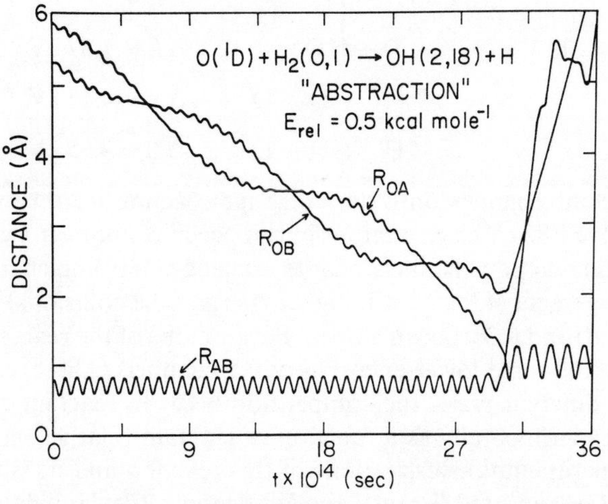

Fig. 15. Internuclear separations as a function of time along a trajectory describing an 'abstraction' reaction of O with H_2. (After Whitlock *et al.*, 1979.)

events, and the others as abstraction events. In this way, Whitlock *et al.* determined that at room temperature about 75% of reactive collisions proceeded by insertion and 25% by abstraction. This is consistent with indirect experimental evidence (DeMore, 1967) which suggests that at least 60% of reactive events occur by insertion.

The insertion trajectories of Whitlock *et al.* and the trajectories of Sorbie and Murrell (all of which correspond to insertion) describe a definite collision intermediate, but one which is quite short-lived, of order 5×10^{-14} s. It is so short-lived, in fact, that the product angular distributions are not symmetric with respect to inversion through the centre of mass. Nevertheless, product vibrational energy distributions from insertion events are predicted by both calculations to be very nearly statistical. It appears that only a modest amount of snarling of trajectories is required to remove any specificity of energy disposal in this case. In contrast, abstraction reactions observed by Whitlock *et al.* produce a marked vibrational population inversion, peaked at $v = 2$.

D. $H^+ + H_2 \rightarrow H_2 + H^+, H_2^+ + H$

The reaction of H^+ with H_2 is electronically the simplest of the examples we have chosen, but it exhibits the most complicated reaction dynamics. Consider, for example, the following isotopic arrangement:

$$H^+ + D_2 \rightarrow D^+ + HD, \qquad \Delta E = 0.04 \text{ eV} \tag{43a}$$
$$\rightarrow H + D_2^+, \qquad \Delta E = 1.85 \text{ eV} \tag{43b}$$
$$\rightarrow D + HD^+, \qquad \Delta E = 1.87 \text{ eV} \tag{43c}$$
$$\rightarrow H^+ + D + D, \qquad \Delta E = 4.55 \text{ eV} \tag{43d}$$
$$\rightarrow D^+ + H + D, \qquad \Delta E = 4.55 \text{ eV} \tag{43e}$$

At low reactant energies only the exchange reaction (43a) can occur. For energies above 1.85 eV electron transfer can occur, resulting in two additional molecular ion reaction channels, charge exchange (43b) and charge exchange with rearrangement (43c). At still higher energies the collisional dissociation channels (43d) and (43e) become open. Because one of the reactants is an ion and because the three lower-energy product channels (43a–c) each produce an ion of a different mass, the competition between reactant channels can be studied experimentally as a function of reactant relative energy. Several accurate experimental investigations of this reaction and its isotopic variations have been reported (Krenos and Wolfgang, 1970; Holliday *et al.*, 1971; Maier, 1971; Krenos *et al.*, 1971, 1974; Ochs and Teloy, 1974; Lees and Rol, 1975). As a result, very detailed and quantitative information about this reaction is known, including absolute cross sections, isotope effects, and product angular and internal energy distributions.

In addition to the complicated competition between channels, there are several other interesting facets to this reaction. As shown in Fig. 3, the potential energy surface of H_3^+ exhibits a deep well. Thus we may wonder whether the reaction proceeds through a long-lived collision intermediate and, if so, what is its lifetime and mode of breakup. Furthermore, the observation of two kinds of products (43a) and (43b, c) is associated with the presence of two interacting potential energy surfaces. Of the four reaction case studies discussed in this chapter, this is the only one for which non-adiabatic interactions between potential energy surfaces play a dominant role.

The availability of detailed experimental results for this reaction represents a challenge for theory. It is an unparalleled challenge because of the simplicity of the $H^+ + H_2$ system. H_3^+ contains only two electrons, so it is feasible to construct potential energy surfaces to whatever accuracy is needed. Accurate *ab initio* points have been obtained by several groups for restricted regions of the potential energy surfaces (Schwartz and Schaad, 1967; Conroy, 1969; Csizmadia *et al.*, 1970; Bauschlicher *et al.*, 1973; Carney and Porter, 1974). DIM potential energy surfaces and non-adiabatic coupling (Preston and Tully, 1971) agree well with the available *ab initio* results and have provided the basis for both qualitative understanding of the reaction mechanism and quantitative calculations of the dynamics.

The two lowest singlet H_3^+ potential energy surfaces are shown in Fig. 16. Also shown in this figure are three cuts across the surfaces which display the avoided crossing between the two potential surfaces when two nuclei are separated by about 2.5 a_0. If the third nucleus is very distant, the two

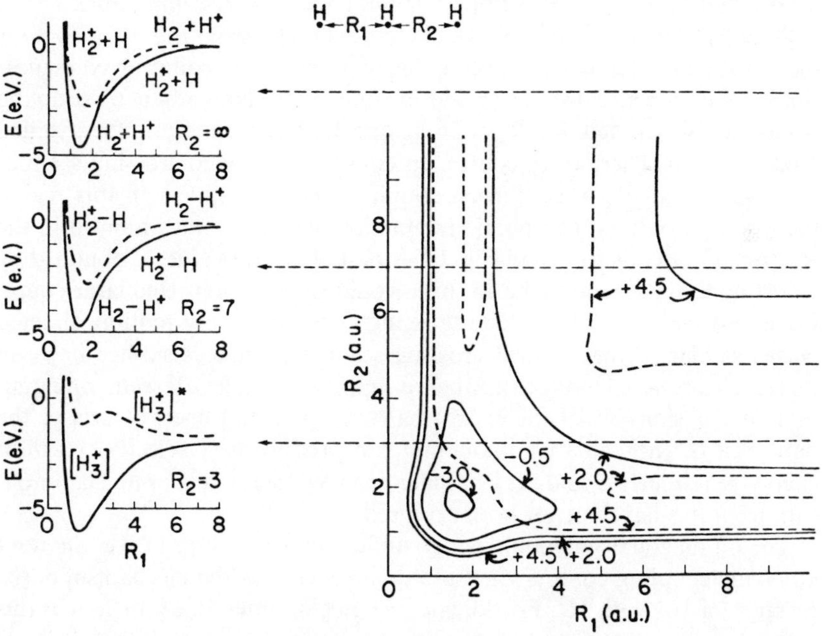

Fig. 16. DIM potential energy surfaces for collinear H_3^+. Solid line: ground state. Dashed line: first excited singlet state. Also shown are three cuts across the potential surfaces at fixed values of R_2, showing the avoided crossing at $R_1 \simeq 2.5\ a_0$.

surfaces approach very close to each other. If the third nucleus is close, the adiabatic surfaces are repelled strongly and the avoided crossing becomes unrecognizable.

Avoided crossings arise from the interaction of two different electronic configurations, in this case one corresponding to $H^+ \ldots H_2$ and the other to $H \ldots H_2^+$. If the interaction between the two electronic configurations is weak, then the resulting adiabatic potential energy surfaces will approach each other very closely and the change in electronic configuration along an adiabatic surface will occur over a very short distance; i.e. non-adiabatic coupling will be large. This is the case in H_3^+ when one of the atoms is at a large distance. Trajectories moving through the avoided crossing region then tend to hop from one adiabatic potential surface to the other, because the electrons do not have time to accomplish the configuration change that occurs over so short a distance along the adiabatic surface. Thus undergoing a transition from one adiabatic surface to another corresponds to retaining electronic configuration, whereas remaining on the adiabatic surface corresponds to changing electronic configuration. When the third atom is at short distances, the splitting between the adiabatic surfaces is large, the electronic configuration change is gradual, and trajectories usually remain on an adiabatic surface. There is an interesting intermediate region for which the probabilities of hopping or not hopping are comparable.

Regions where surface hops are likely can be seen from Fig. 17, a contour plot of the non-adiabatic coupling $\langle \phi_1 | \partial \phi_2 / \partial R_1 \rangle$ for collinear configurations, computed by DIM (Tully and Preston, 1971). Non-adiabatic coupling is largest, as expected, for $R_1 \simeq 2.5$ a_0 and R_2 large; i.e. along the avoided crossing seam. There is, of course, an equivalent avoided crossing seam at $R_2 \simeq 2.5$ a_0 and R_1 large. Strong coupling does not appear in this region because it is only non-adiabatic interaction associated with motion in the R_1 direction that is shown in Fig. 17. A plot of $\langle \phi_1 | \partial \phi_2 / \partial R_2 \rangle$, obtained by reflecting Fig. 17 through the 45° line, would show the expected large coupling along the $R_2 = 2.5$ avoided crossing. Thus it is only motion directed perpendicular to the avoided crossing seam that induces surface hops, in this case vibrational motion of either reactants or products. It seems apparent that in collisions involving more than one potential energy surface, the nature of the non-adiabatic coupling can produce very selective reactant energy requirements and very specific product energy disposal. The implications of this have not yet been explored.

The off-diagonal potential energy surface shown in Fig. 17, i.e. the non-adiabatic coupling contours, are useful in visualizing the mechanism of the reaction of H^+ with H_2. For kinetic energies of order 10 eV or less in this system, unless the non-adiabatic coupling is $\gtrsim 1$ surface hops will be of negligible importance. The contours in Fig. 17 exceed unity in two regions, the avoided crossing and at small values of R_1. The latter region occurs so

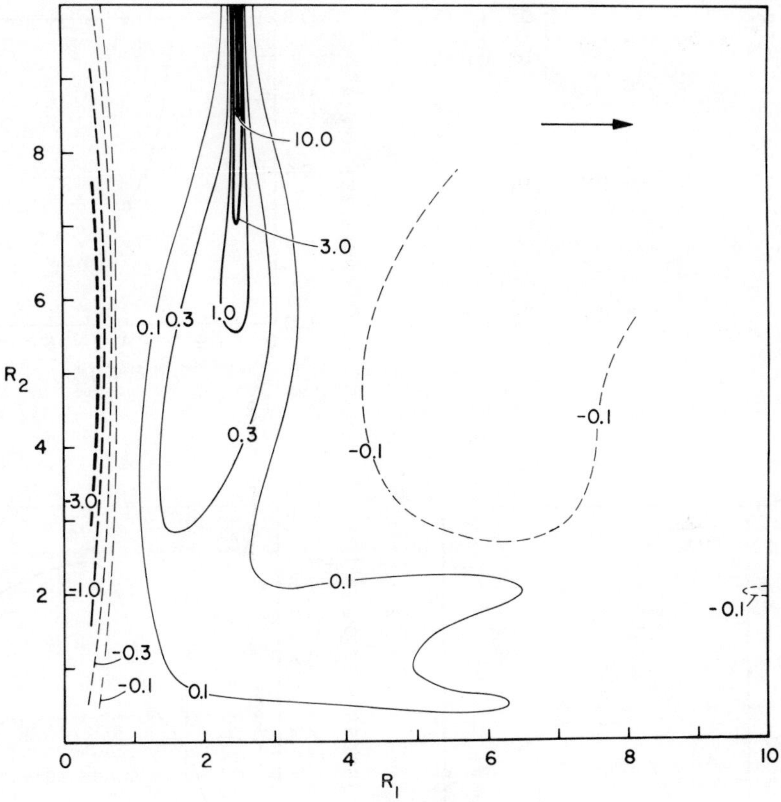

Fig. 17. Contour plot of nonadiabatic coupling $\langle \phi_1 | \partial\phi_2/\partial R_1 \rangle$ in atomic units, computed by DIM for collinear H_3^+.

high up the repulsive wall, however, that it cannot be reached. Surface hops therefore can occur only at the avoided crossing. As the H^+ and H_2 initially approach on the lower potential surface, if the vibrational state of H_2 is less than $v = 4$ its vibrational amplitude will be insufficient to reach the avoided crossing; charge exchange does not occur as the reactants approach and the system remains on the lower potential surface. Whether reaction is direct or involves trapping in the deep well, in the region where all three nuclei are close non-adiabatic coupling as shown in Fig. 17 is too small to be important. Thus surface hops, if they are important, must occur as internally excited products recede.

This qualitative picture is borne out both by dynamical calculations discussed below and by experiment. For example, products with electronic configuration $H^+ + H_2$ formed with relatively high internal energy, and therefore low translational energy, are likely to be depleted by competition

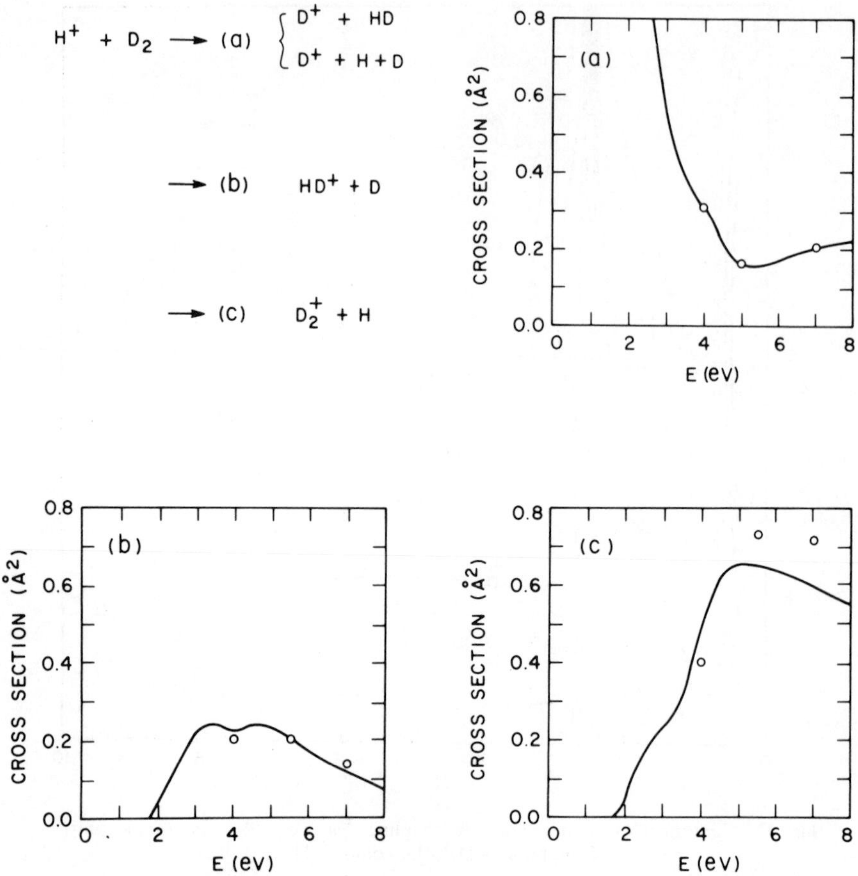

Fig. 18. Absolute cross sections for reaction of H^+ with D_2. Solid lines, experiment (Ochs and Teloy, 1974). Circles, surface-hopping trajectory calculations (Krenos *et al.*, 1974.)

with molecular ion channels $H + H_2^+$. Thus surviving $H^+ + H_2$ should exhibit relatively high translational energy. This has been confirmed both experimentally (Krenos *et al.*, 1974; Lees and Rol, 1975) and theoretically (Krenos *et al.*, 1974).

Extensive surface-hopping trajectory (SHT) calculations (Tully and Preston, 1971) have been carried out for reaction (43) and isotopic variations using DIM potential energy surfaces (Krenos *et al.*, 1974). Results are in excellent agreement with experiment for essentially every measured feature of the reaction. Fig. 18 compares absolute cross sections computed by SHT with subsequently measured results (Ochs and Teloy, 1974). Neither the experimental nor theoretical curves have been normalized in any way, and the theory contains no adjustable quantities of any kind. Details of the collision process are also predicted well by theory. Fig. 19 compares experimental and computed product translational energy distributions.

Fig. 20 shows the computed distribution of delay times for trajectories relative to the fastest one. At 3 eV initial relative energy the distribution of delay times falls off approximately exponentially, indicative of a collision intermediate of 1.4×10^{-13} s duration. At this collision energy computed

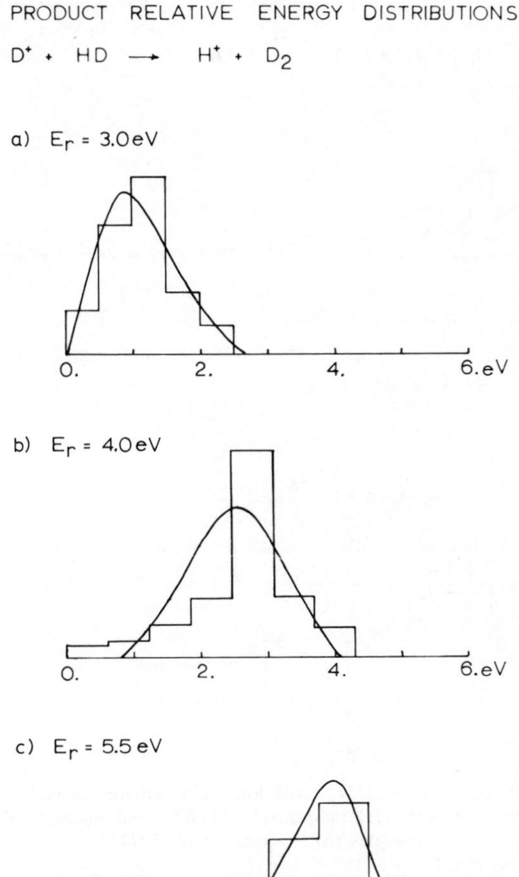

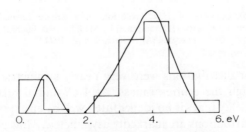

Fig. 19. Product translational energy distributions for the reaction $D^+ + HD \rightarrow H^+ + D_2$. Solid curves, molecular beam results along the collision axis, $\theta = 0°$. Histograms, SHT calculations for $\theta < 30°$. (After Krenos *et al.*, 1974.)

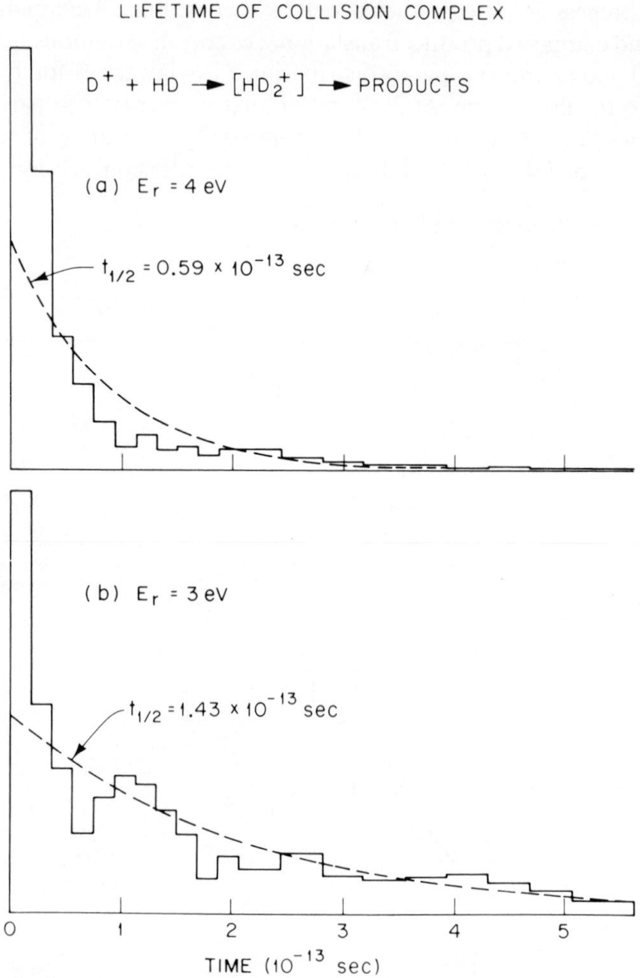

Fig. 20. Lifetime of HD_2^+ collision intermediate formed during reaction of D^+ with HD computed by SHT for two different collision energies. (After Krenos *et al.*, 1974.)

product angular distributions were very nearly symmetric with respect to inversion through the centre-of-mass. At 4 eV initial relative energy the distribution of delay times is no longer fitted well by an exponential (Fig. 20), and angular distributions are markedly asymmetric. Thus at energies below about 4 eV the reaction does indeed proceed via an intermediate collision complex, but at higher energies the mechanism becomes direct. This change in mechanism has been experimentally confirmed to occur at about 4 eV (Lees and Rol, 1975).

V. CONCLUSIONS

We have tried to illustrate *via* the examples discussed above several of the ways in which DIM potential energy surfaces can provide useful information about simple molecular collision processes. Crude DIM potential energy surfaces, constructed using incomplete diatomic input information obtained from available experiments and approximate calculations, supplemented by estimates, can be of great value in developing qualitative insight into chemical reaction dynamics. Quantitative descriptions and reliable predictions can be obtained from accurate DIM potential energy surfaces constructed from (1) high level *ab initio* valence-bond calculations of diatomic input and/or (2) accurate experimental or theoretical determinations of a few crucial points on the polyatomic potential energy surfaces.

There are some limitations to the DIM approach, of course. Perhaps the most serious limitation at present is a temporary one, unavailability of the required diatomic input data. *Ab initio* calculations for diatomic molecules to the required accuracy are now standard, so this deficiency should be corrected in time. A second limitation is uncertainty in the accuracy and reliability of the approach, and in the optimal way to apply it; e.g. how many atomic states are required to achieve a particular level of accuracy. Systematic documentation of the method is required here. A third limitation of DIM is its complexity. For each of the examples chosen above, the DIM descriptions were very simple. Since DIM is based on the coupling between atomic states and does not explicitly take account of individual electrons, the complexity does not increase if, for example, Na were replaced by Cs and F by I in the first example of the last section. However, the calculation can become quite extensive if many different atomic states must be employed, as may be the case in systems composed of many atoms. DIM, of course, still remains vastly simpler than an *ab initio* computation of the same accuracy for these cases. Nevertheless, it can become cumbersome.

A possible approach to simplifying the method for systems composed of many atoms has been proposed recently (Tully, 1977b). DIM is based on the exact partitioning of the Hamiltonian given by equation (7). There are alternative exact ways to partition the Hamiltonian upon which alternative theoretical methods can be based. For example, a *polyatomics-in-molecules* approach might be developed based on the following partitioning scheme (Tully, 1977b):

$$\mathscr{H} = \mathscr{H}^{(M_1)} + \mathscr{H}^{(M_2)} + \sum_{K'}^{M_1}\sum_{K''}^{M_2} \mathscr{H}^{(K'K'')} - M_2 \sum_{K'}^{M_1} \mathscr{H}^{(K')} - M_1 \sum_{K''}^{M_2} \mathscr{H}^{(K'')} \tag{44}$$

Thus we would consider the system to be composed of two fragments with M_1 and M_2 atoms, respectively. This method would require as input ground- and excited-state information about the polyatomic fragments M_1 and M_2 which would generally be more difficult to obtain than the diatomic information required by DIM. However, it would have the advantage that once input information was assembled, the method would be easier to apply than DIM. This and related polyatomics-in-molecules approaches might prove valuable in extending DIM to larger molecules and even to condensed-phase situations, e.g. gas–solid interactions.

As illustrated above, one of the great advantages of the DIM approach is the facility with which spin–orbit and non-adiabatic interactions can be computed. This greatly extends its range of applicability to the host of interesting phenomena that involve more than one potential energy surface.

DIM may also be valuable for investigating the dynamics of molecular processes occurring in strong laser fields. Such processes are conveniently described using 'dressed-molecule' potential energy surfaces (George *et al.*, 1977) which include interactions with the radiation field. Construction of the dressed potential energy surfaces requires knowledge of transition moments as a function of molecular geometry. It is straightforward to compute these properties by DIM (Tully, 1976) provided, of course, that the necessary diatomic input information is available.

References

Balint-Kurti, G. G. (1975). In *Molecular Beam Scattering: Physical and Chemical and Chemical Applications* (Ed. K. P. Lawley), Wiley, London, p. 137.

Balint-Kurti, G. G. and Karplus, M. (1971). *Chem. Phys. Lett.*, **11**, 203.

Balint-Kurti, G. G. and Yardley, R. N. (1977). *Faraday Disc. Chem. Soc.*, **62**, 77.

Bauschlicher, C. W. Jr., O'Neil, S. V., Preston, R. K., Schafer, H. F. III, and Bender, C. F. (1973). *J. Chem. Phys.*, **59**, 1286.

Bender, C. F., O'Neil, S. V., Pearson, P. K., and Schaefer, H. F. III (1972). *Science*, **176**, 1412.

Bergmann, K., Leone, S. R., and Moore, C. B. (1975). *J. Chem. Phys.*, **63**, 4161.

Berry, M. J. (1973). *J. Chem. Phys.*, **59**, 6229.

Birely, J. H., Kasper, J. V. V., Hai, F., and Darnton, L. A. (1975). *Chem. Phys. Lett.*, **31**, 220.

Blackwell, B. A., Polanyi, J. C., and Sloan, J. J. (1977). *Faraday Disc. Chem. Soc.*, **62**, 147.

Born, M. and Oppenheimer, J. R. (1927). *Ann. Physik*, **84**, 457.

Carney, G. D. and Porter, R. N. (1974). *J. Chem. Phys.*, **60**, 4251.

Carrington, T. (1974). *Acts. Chem. Res.* **7**, 20.

Chapman, S. and Preston, R. K. (1974). *J. Chem. Phys.*, **60**, 650.

Conroy, H. (1969). *J. Chem. Phys.* **51**, 3979.

Coombe, R. D. and Pimentel, G. C. (1973). *J. Chem. Phys.*, **59**, 251.

Csizmadia, I. G., Karl, R. E., Polanyi, J. C., Roach, A. C., and Robb, M. A. (1970). *J. Chem. Phys.*, **52**, 6205.

DeMore, W. B. (1967). *J. Chem. Phys.*, **47**, 2777.

Ding, A. M. G., Kirsch, L. J., Perry, D. S., Polanyi, J. C., and Schreiber, J. L. (1973). *Faraday Disc. Chem. Soc.*, **55**, 252.
Donovan, R. J. and Husain, D. (1970). *Chem. Revs.*, **70**, 489.
Douglas, D. J. and Polanyi, J. C. (1976). *Chem. Phys.* **16**, 1.
Eaker, C. W. and Parr, C. A. (1976). *J. Chem. Phys.* **64**, 1322.
Ellison, F. O. (1963). *J. Amer. Chem. Soc.*, **85**, 3540.
Ellison, F. O., Huff, N. T., and Patel, J. C. (1963). *J. Amer. Chem. Soc.*, **85**, 3544.
Eyring, H. and Polanyi, M. (1931). *Z. Physik. Chem. (Leipzig)*, **B12**, 279.
Galloy, C. and Lorquet, J. C. (1977). *J. Chem. Phys.* **67**, 4672
Gangi, R. A. and Bader, R. F. W. (1971). *J. Chem. Phys.* **55**, 5369.
George, T. F., Zimmerman, I. H., Yuan, J.-I., Laing, J. R., and DeVries, P. L. (1977). *Accts. Chem. Res.*, **10**, 449.
Gillen, K. T., Riley, C., and Bernstein, R. B. (1969). *J. Chem. Phys.*, **50**, 4019.
Goddard, W. A., Dunning, T. H., Hunt, W. J., and Hay, P. J. (1973). *Accts. Chem. Res.*, **6**, 368.
Hoffmann, R. and Woodward, R. B. (1968). *Accts. Chem. Res.*, **1**, 17.
Holliday, M. G., Muckerman, J. T., and Friedman, L. (1971). *J. Chem. Phys.* **54**, 1058.
Jaffe, R. L. and Anderson, J. B. (1971). *J. Chem. Phys.*, **54**, 2224.
Johnson, B. R. and Winter, N. W. (1977). *J. Chem. Phys.*, **66**, 4116.
Jonathan, N., Melliar-Smith, C. M., and Slater, D. H. (1971). *Mol. Phys.*, **20**, 93.
Kormonicki, A., Morokuma, K., and George, T. F. (1977). *J. Chem. Phys.*, **67**, 5012.
Krenos, J. R., Preston, R. K., Wolfgang, R., and Tully, J. C. (1971). *Chem. Phys. Lett.* **10**, 17.
Krenos, J. R., Preston, R. K., Wolfgang, R., and Tully, J. C. (1974). *J. Chem. Phys.*, **60**, 1634.
Krenos, J. R. and Wolfgang, R. (1970). *J. Chem. Phys.*, **52**, 5961.
Kuntz, P. J. (1972). *Chem. Phys. Lett.*, **16**, 581.
Kuntz, P. J. (1976). In, *Dynamics of Molecular Collisions* (Ed. W. H. Miller), Part B, Plenum Press, N.Y., p. 53.
Kuntz, P. J., Nemeth, E. M., Polanyi, J. C., Rosner, S. D., and Young, C. E. (1966). *J. Chem. Phys.*, **44**, 1168.
Kuntz, P. J. and Roach, A. C. (1972). *J. Chem. Soc. Faraday Trans. II*, **68**, 259.
Lacmann, K. and Herschbach, D. R. (1970). *Chem. Phys. Lett.*, **6**, 106.
Lang, J. R., George, T. F., Zimmerman, I. H., and Lin, Y. W. (1975). *J. Chem. Phys.*, **63**, 842.
Lees, A. B. and Rol, P. K. (1975). *J. Chem. Phys.*, **63**, 2461.
Lichten, W. (1963). *Phys. Rev.*, **131**, 229.
Liu, B. (1973). *J. Chem. Phys.*, **58**, 1925.
London, F. (1929). *Z. Electrochem.*, **35**, 552.
Löwdin, P. O. (1970). *Adv. Quantum Chem.*, **5**, 185.
Maier, W. B. (1971). *J. Chem. Phys.*, **54**, 2732.
Martin, L. R. and Kinsey, J. L. (1967). *J. Chem. Phys.*, **46**, 4834.
Moffitt, W. (1951). *Proc. Roy. Soc. (London)*, **A210**, 245.
Muckerman, J. T. (1971). *J. Chem. Phys.*, **54**, 1155.
Muckerman, J. T. (1972). *J. Chem. Phys.*, **56**, 2997.
Muckerman, J. T. and Newton, M. D. (1972). *J. Chem. Phys.*, **56**, 3191.
Ochs, G. and Teloy, E. (1974). *J. Chem. Phys.*, **61**, 4930.
Odiorne, T. J., Brooks, R. P., and Kasper, J. V. V. (1971). *J. Chem. Phys.*, **55**, 1980.
Parker, J. H. and Pimentel, G. C. (1969). *J. Chem. Phys.*, **51**, 91.
Parr, C. A. and Truhlar, D. G. (1971). *J. Phys. Chem.*, **75**, 1844.
Perry, D. S. and Polanyi, J. C. (1976). *Chem. Phys.*, **12**, 37, 419.

Perry, D. S., Polanyi, J. C., and Wilson, C. W. Jr. (1974). *Chem. Phys.*, **3**, 317.
Pickup, B. T. (1973). *Proc. Roy. Soc. (London)*, **A333**, 69.
Polanyi, J. C. and Schreiber, J. L. (1974). In *Physical Chemistry, An Advanced Treatise* (Ed. W. Jost), Vol. VIA, Academic Press, N.Y., Chap. 6.
Polanyi, J. C. and Schreiber, J. L. (1977). *Faraday Disc. Chem. Soc.*, **62**, 267.
Polanyi, J. C., Solan, J. J., and Wanner, J. (1976). *Chem. Phys.*, **13**, 1.
Polanyi, J. C. and Tardy, D. C. (1969). *J. Chem. Phys.*, **51**, 5717.
Polanyi, J. C. and Wong, W. H. (1969). *J. Chem. Phys.*, **51**, 1439.
Polanyi, J. C. and Woodall, K. B. (1972). *J. Chem. Phys.*, **57**, 1574.
Porter, R. N. and Karplus, M. (1964). *J. Chem. Phys.*, **40**, 1105.
Preston, R. K. and Tully, J. C. (1971). *J. Chem. Phys.*, **54**, 4297.
Pruett, J. G., Grabiner, F. R., and Brooks, P. R. (1974). *J. Chem. Phys.*, **60**, 3335.
Rebentrost, F. and Lester, W. A. Jr. (1975). *J. Chem. Phys.* **63**, 3737.
Rebentrost, F. and Lester, W. A. Jr. (1976). *J. Chem. Phys.*, **64**, 3879.
Rebentrost, F. and Lester, W. A. Jr. (1977). *J. Chem. Phys.*, **67**, 3372.
Redmon, M. J. and Wyatt, R. E. (1977). *Int. J. Quantum Chem. Symp.*, **11**, 343.
Roach, A. C. (1970). *Chem. Phys. Lett.*, **6**, 390.
Rubinstein, M. and Shavitt, I. (1969). *J. Chem. Phys.*, **51**, 2014.
Sato, S. (1955). *J. Chem. Phys.*, **23**, 2465.
Schafer, T. D., Siska, P. E., Parson, J. M., Tully, F. P., Wong, Y. C., and Lee, Y. T. (1970). *J. Chem. Phys.*, **53**, 3385.
Schatz, G. C., Bowman, J. M., and Kupperman, A. (1975). *J. Chem. Phys.*, **63**, 674.
Schwartz, M. E. and Schaad, L. J. (1967). *J. Chem. Phys.*, **47**, 5325.
Segal, G. A. (Ed.) (1977). *Semiempirical Methods of Electronic Structure Calculation*, Parts A and B, Plenum Press, N.Y.
Sorbie, K. S. and Murrell, J. N. (1975). *Mol. Phys.*, **29**, 1387.
Sorbie, K. S. and Murrell, J. N. (1976). *Mol. Phys.*, **31**, 905.
Steiner, E., Certain, P. R., and Kuntz, P. J. (1973). *J. Chem. Phys.*, **59**, 47.
Taylor, E. H. and Datz, S. (1955). *J. Chem. Phys.*, **23**, 1711.
Truhlar, D. G. (1972). *J. Chem. Phys.*, **56**, 3189.
Tully, J. C. (1973a). *J. Chem. Phys.*, **58**, 1396.
Tully, J. C. (1973b). *J. Chem. Phys.*, **59**, 5122.
Tully, J. C. (1974). *J. Chem. Phys.*, **60**, 3042.
Tully, J. C. (1975). *J. Chem. Phys.*, **62**, 1893.
Tully, J. C. (1976). *J. Chem. Phys.*, **64**, 3182.
Tully, J. C. (1977a). In *Dynamics of Molecular Collisions* (Ed. W. H. Miller), Part B, Plenum Press, N.Y., p. 217.
Tully, J. C. (1977b). In *Semiemprical Methods of Electronic Structure Calculation* (Ed. G. A. Segal), Part A, Plenum Press, N.Y., p. 173.
Tully, J. C. (1978). Unpublished results.
Tully, J. C. and Preston, R. K. (1971). *J. Chem. Phys.*, **55**, 562.
Tully, J. C. and Truesdale, C. M. (1976). *J. Chem. Phys.*, **65**, 1002.
Walsh, A. D. (1953). *J. Chem. Soc.*, 2260, 2266.
Whitlock, P. A. and Muckerman, J. T. (1974). *J. Chem. Phys.*, **61**, 4618.
Whitlock, P. A., Muckerman, J. T., and Fisher, E. R. (1976). *Theoretical Investigations of the Energetics and Dynamics of the Reactions* $O(^3P, ^1D) + H_2$ *and* $C(^1D) + H_2$, Report of Research Institute for Engineering Sciences, Wayne State University, unpublished.
Whitlock, P. A., Muckerman, J. T., and Fisher,. R. (1979). Unpublished.
Wilkins, R. L. (1972). *J. Chem. Phys.*, **57**, 912.
Zimmerman, I. H. and George, T. F. (1975). *Chem. Phys.*, **7**, 323.

Potential Energy Surfaces
Edited by K. P. Lawley

DYNAMICS OF OXYGEN ATOM REACTIONS

M. C. LIN

Chemistry Division, Naval Research Laboratory, Washington, D. C. 20375, *USA*

CONTENTS

I. INTRODUCTION

The reactions of oxygen atoms, particularly those in the ground electronic state ($2s^2\ 2p^4\ ^3P_J$, $J = 0, 1, 2$) are not only important in combustion and atmospheric chemistry, but also are important in highly exoergic propulsion and laser systems. Because of the renewed interest in hydrocarbon and alternate fuel combustion chemistry brought about by the 'energy crisis', there are considerable demands for accurate rate data and reaction mechanisms over broad ranges of temperature and pressure for detailed kinetic modelling of high-temperature combustion processes. However, at present, relatively few reactions have been studied carefully over wide ranges of experimental conditions.

Recently, there has been much theoretical and experimental interest in the reaction and relaxation of electronically excited oxygen atoms,

$O(2s^2 2p^4\,{}^1D_2)$ and $O(2s^2 2p^4\,{}^1S_0)$. This stems partly from the discovery of new chemical laser systems involving these states.

The energy-level diagram of some of the low-lying electronic states of neutral oxygen atoms is shown in Fig. 1. Among these states, only the lowest three states are chemically important. Very little chemical data are available for the higher ones. On the basis of the known kinetic data, the chemical reactivity of these three states with hydrogen or alkanes, for example, decreases in the following order: ${}^1D_2 \gg {}^1S_0 \gtrsim {}^3P_J$. This clearly indicates that the chemical reactivity of atoms and free radicals is controlled not only by energetics, but also by many other important factors, such as polarizability, ionization potential, orbital symmetry, etc.

Rate parameters for some of the reactions of oxygen atoms in these three states have recently been compiled by the U.S. National Bureau of Standards for the Climatic Impact Assessment Program (Hampson and Garvin, 1975). More recently, Schofield (1978) has up-dated some of the recommended values for the reactions of $O({}^1D)$ and $O({}^1S)$ atoms. For $O({}^3P)$, rate constants

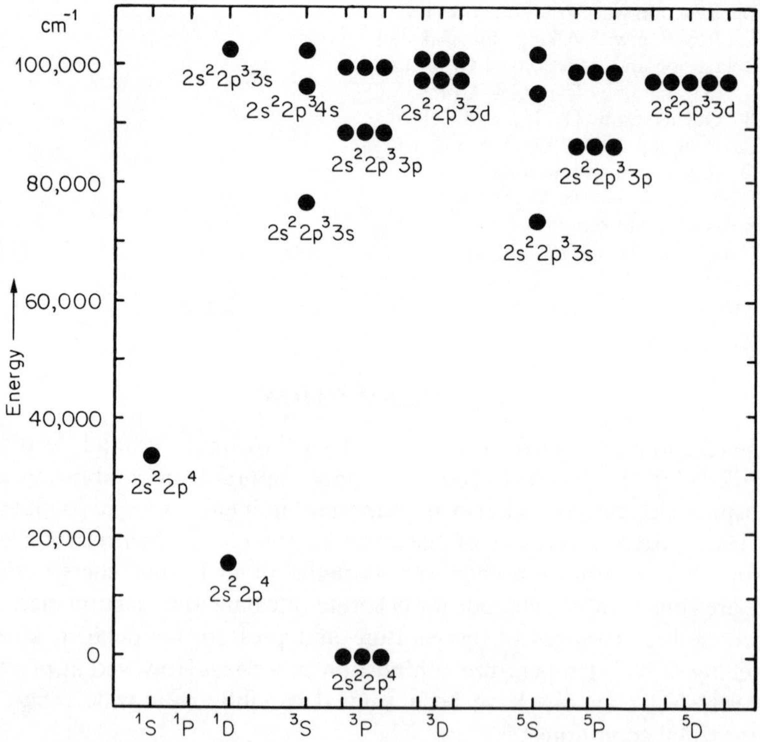

Fig. 1. An energy-level diagram of some low-lying electronic states of the oxygen atom.

for simple organic and inorganic reactions have been compiled and critically evaluated by Herron and Huie (1974) and Schofield (1973), respectively. Some of these reactions will be discussed later. In this chapter, we do not plan to critically review those overall rates and comment on their accuracies. Instead we will discuss mainly the dynamics of the reactions involving these three low-lying electronic states, with particular emphasis on the results obtained from detailed microscopic rate measurements. In cases where over-all kinetic data are useful for the interpretation of reaction mechanisms, pertinent references will also be given. The methods commonly used to generate and detect etc. O atoms in these three states are summarized in Table I.

To describe fully the mechanics of atomic rearrangement during the course of a chemical reaction from certain specified initial reactant states (E, V, R, T) to certain specified final product states (E', V', R', T'), one requires ideally a complete set of microscopic rate constants, $k(E, V, R, T/E', V', R', T')$; where E, V, R, and T represent electronic, vibrational, rotational, and translational energies, respectively. Experimentally, however, such a detailed set of rate constants is very hard to obtain even for simple three-atom systems in view of the number and possible combinations of the variables (i.e. quantum states) involved. Except for a few simple exchange reactions involving hydrogen halides for which more detailed information (such as the effects of V and T on V', R', and/or T') has been obtained by means of chemiluminescence and molecular beam experiments (Polanyi and Schreiber, 1974), only partial information at the molecular level has been measured experimentally for most reactions studied to date. This is true for most oxygen atom reactions discussed in this review.

II. REACTIONS OF $O(^3P_J)$ ATOMS

A. Reaction with Vibrationally Excited H_2 and HCl

The reaction of $O(^3P)$ atoms with hydrides generally occurs *via* the direct abstraction of H atoms producing OH radicals:

$$O(^3P) + RH \rightarrow OH(X^2\Pi) + R$$

$$R = H, Cl, OH, \text{alkyl, etc.}$$

The simplest of these metathetical processes is $O + H_2 \rightarrow OH + H$. It is one of the two key chain-carrying steps in the H_2/O_2 flame and is also an important reaction in high-temperature hydrocarbon combustion systems. The reaction is slightly endoergic ($\Delta H^0 = 1.9$ kcal/mol); accordingly, very little detailed dynamic information has been obtained experimentally. Very recently, Light (1978) studied the effect of the vibrational excitation of H_2

on the rate of OH production using the laser-induced fluorescence technique (Kinsey, 1977). The OH formed in each of the $v' = 0$ and 1 vibrational levels from the vibrationally excited H_2 reaction:

$$O(^3P) + H_2(v = 1) \rightarrow OH(v' = 0, 1) + H$$

has been measured. The production of OH in the ground state is exoergic by 9.9 kcal/mol, whereas the formation of $OH(v' = 1)$ is slightly endoergic ($\Delta H^0 = 0.3$ kcal/mol). Interestingly, however, Light found that $OH(v' = 1)$ is the major product from the above reaction. The rate constants for these two channels are $k_{v'=1} = 1 \times 10^{-14}$ and $k_{v'=0} \leq 4.7 \times 10^{-15}$ cm^3/molecule.s, with a branching ratio of ≥ 2. The reported total reaction rate for $O(^3P) + H_2(v = 1)$ is thus about 3×10^3 times higher than that for the ground-state H_2 reaction. This confirms a previously reported upper limit ($< 3 \times 10^4$) by Birely *et al.* (1975) using a different experimental technique. The effectiveness of H_2 vibrational energy in promoting the rate of the $O(^3P) + H_2$ reaction may be partly responsible for the observed deviation from linearity of the Arrhenius plot of the thermal rates at high temperatures (Schott *et al.*, 1974), although the theoretical interpretation of the non-linearity is still not certain (Gardiner, 1977; Bauer, 1978).

Theoretical studies of the $O(^3P) + H_2$ reaction have been carried out by several investigators (Gangi and Bader, 1971; Whitlock *et al.*, 1976; Johnson and Winter, 1977; Schinke and Lester, 1979). The results of these studies, except that of Schinke and Lester, have been reviewed by Tully elsewhere in this volume. These theoretical and experimental results have been discussed in more detail by Schinke and Lester, who recently carried out trajectory calculations employing a fitted *ab initio* surface obtained from an extensive CI calculation for the O–H_2 system at large numbers of nuclear geometries suitable for three-dimensional scattering (Howard and Lester, 1979). Important general conclusions from this and the previous trajectory calculations of Whitlock *et al.* (1976) and of Johnson and Winter (1977) can be summarized as follows. (1) The vibrational excitation of H_2 promotes effectively the rate of the $O(^3P) + H_2$ reaction, consistent with the experimental findings mentioned above, because the reaction barrier lies in the exit channel; (2) the computed branching ratio, $k(v = 1, v' = 1, T)/k(v = 1, v' = 0, T) > 2$, and the total thermal rates, $k(T)$, are also in good agreement with experimental data, although $k(T)$ depends rather strongly on the barrier height used which cannot be accurately calculated. A much better fit with experimental rate constants over a broad range of temperature was obtained when the barrier was first adjusted to match low-temperature data (Johnson and Winter, 1977).

The large probability (branching ratio) for the formation of OH ($v' = 1$) indicates the reaction $O(^3P) + H_2(v) \rightarrow OH(v') + H$ occurs to a significant extent with 'vibrational adiabaticity' (Ding *et al.*, 1973; Polanyi and Schreiber,

1976; Blackwell *et al.*, 1977). This is also observed experimentally in the analogous $O(^3P) + HCl(v) \rightarrow OH(v') + Cl$ system (Blackwell *et al.*, 1976, 1977; Butler *et al.*, 1978). For the forward reaction $O(^3P) + HCl(v = 1)$, a similar pronounced enhancement of the reaction rate has been reported by several groups of investigators (Arnoldi and Wolfrum, 1974; Karny *et al.*, 1975; Butler *et al.*, 1978). In a more extensive experiment recently carried out in this Laboratory (Butler *et al.*, 1978) employing two pulsed lasers (a chemical HCl laser for pumping the HCl reactant and a flash-lamp-pumped tunable dye laser for probing the OH product), it was found that the vibrational excitation of HCl to $v = 1$ and 2 enhances the rates of $O(^3P) + HCl \rightarrow OH + Cl$ by about two orders of magnitude for each vibrational quantum. This finding agrees semiquantitatively with the results of a quasi-classical trajectory calculation, based on a simple one-parameter LEPS potential, which assumed the energy barrier to be approximately situated in a symmetrical position relative to the reactants and products (Brown and Smith, 1978). The results of this calculation also support our observation that the interaction of $O(^3P)$ with $HCl(v = 1)$ results predominantly in quenching rather than reaction, contrary to the preliminary result of Arnoldi and Wolfrum (1974). Additionally, we also found experimentally that the reaction of $HCl(v = 2)$ generates almost exclusively the vibrationally excited $OH(v' = 1)$ product (Butler *et al.*, 1978).

The preferential production of $OH(v' = 1)$ from $O(^3P) + HCl(v = 2)$, which is the only process that has sufficient energy, is also consistent with the finding of Polanyi and coworkers (Blackwell *et al.*, 1977) for the reverse $Cl + OH(v')$ reaction, employing the infrared chemiluminescence method (Polanyi, 1963). In their study, the vibrational energy of the $OH^\dagger$ was found to be efficiently converted into the $HCl^\dagger$ product. The $OH^\dagger$ was produced by two different pre-reactions, $H + O_3$ and $H + NO_2$, which led to widely different average vibrational energy contents, $\langle V' \rangle = 72$ and 21 kcal/mol, respectively. The extent of vibrational adiabaticity ($\Delta V' \rightarrow \Delta V$) for the $Cl + OH^\dagger$ reaction was found to increase with increasing reagent vibrational excitation. Since the breadths of the $HCl^\dagger$ product energy distributions are comparable with those of the $OH^\dagger$ formed in both pre-reactions, the $Cl + OH$ reaction probably does not take place via a long-lived HOCl complex. The formation of such a complex is probably not likely because the stable HOCl molecule is singlet and the spin–orbit interaction between the singlet and triplet ($O + HCl$) curves for this relatively light species may be too small to be important. The formation of $H + OCl$ from $Cl + OH$, however, may occur via HOCl, and there is evidence that it probably does take place when OH is highly excited ($v' \geq 4$). According to the observed $OH^\dagger$ and $HCl^\dagger$ chemiluminescence intensities and the available Einstein radiative transition probabilities, Blackwell *et al.* (1977) concluded that about 86% of the $OH^\dagger$ removed from $v' = 1$–3 formed $HCl(v = 1$–$2)$, whereas only about 33% of

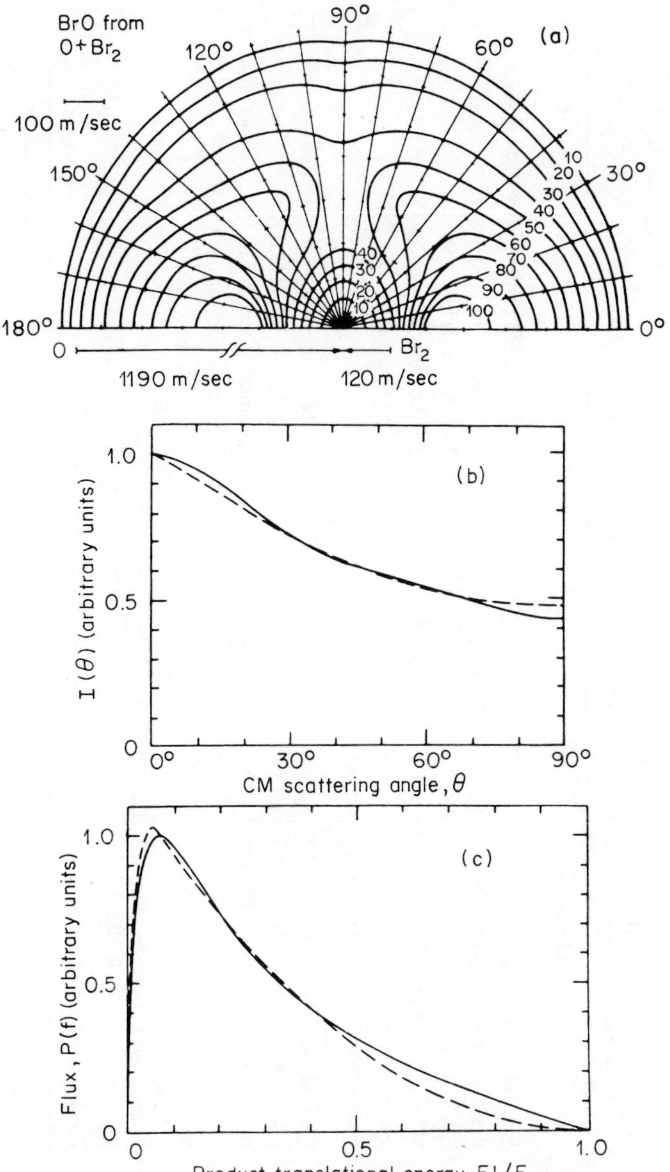

Fig. 2. Crossed molecular beam results of the $O(^3P) + Br_2$ reaction. (a) Contour map of the BrO product angle–velocity flux distribution in the centre-of-mass system. (b) Angular distribution obtained from cut through contour map at the fixed velocity of 300 m/s. (c) Product (relative) translation energy distribution as a function of E'/E_{tot}, fraction of the total available energy. Solid curves—experiment. Dash curves—statistically predicted distributions based on the RRKM-AM model. (Reprinted with permission from Parrish and Herschbach, *J. Am. Chem. Soc.*, 95, 6113. Copyright by the American Chemical Society.

TABLE II. Methods Commonly Used to Generate and Detect $O(^3P)$, $O(^1D)$, and $O(^1S)$ Atoms

Species	Generation	Detection
$O(^3P)$	1. Photodissociation: O_3($h\nu \simeq 600$ nm), NO_2 ($h\nu \geqslant 300$ nm), O_2($h\nu \leqslant 200$ nm), SO_2($h\nu < 200$ nm) 2. $Hg(^3P) + N_2O$ 3. Microwave or electrical discharge of O_2, N_2, or N_2O followed by NO titration 4. Electron-beam dissociation of O_2, N_2O	1. Resonance absorption or fluorescence ($3\,^3S_1^0$–$2\,^3P_J$, 130 nm) 2. Electron spin resonance 3. Chemiluminescence: O + CO, O + NO (including O + NO_2)
$O(^1D)$	1. Photodissociation: O_3($h\nu \simeq 200$–300 nm), NO_2 ($h\nu < 300$ nm), N_2O ($h\nu < 200$ nm), CO_2 ($h\nu < 160$ nm), O_2($h\nu \leqslant 175$ nm)	1. Resonance absorption or fluorescence ($3\,^1D_2^0$–$2\,^1D_2$, 115.2 nm) 2. Fluorescence ($2\,^1D_2 - 2\,^3P_J$, 630 nm) 3. Chemical reactions (N_2O, neopentane, etc.)
$O(^1S)$	1. Photodissociation: N_2O($h\nu \sim 120$ nm), CO_2 ($h\nu \sim 110$ nm), O_2($h\nu \lesssim 130$ nm) 2. Electron-beam dissociation of O_2	Fluorescence ($2\,^1S_0$–$2\,^1D_2$, 558 nm)

Crossed-beam experiments carried out at low collision energies (< 5 kcal/mole) for $O(^3P) + Br_2$, I_2, ICl, and Cl_2 (Parrish and Herschbach, 1973; Radlein *et al.*, 1975; Sibener *et al.*, 1978; Clough *et al.*, 1978) all indicated that the angular and velocity distributions of the products observed could be satisfactorily interpreted by means of the statistical (RRKM-AM) model derived by Herschbach and co-workers (Miller *et al.*, 1967; Safron *et al.*, 1972). Fig. 2 presents the results of the $O(^3P) + Br_2$ reaction obtained by Parrish and Herschbach (1973). The observed angular and translation energy distributions were shown to be consistent with those predicted by a statistical model assuming a 'tight' linear O—Br—Br complex that has a decay time longer than a few rotational periods, or 5×10^{-12} s. This conclusion, they believe, is concordant with the prediction of the 'electronegativity ordering rule' derived from Walsh's molecular orbital correlation scheme (Walsh, 1953). The rule favours a linear or approximately linear triplet X—Y—Z complex with the least electronegative Y atom at the centre. On the basis of this rule, Parrish and Herschbach (1973) predicted that the reactions of $O(^3P)$ with IX (X = Cl, Br) should proceed via triplet O—I—X complexes yielding the thermally neutral products OI + X (rather than the exoergic OX + I products). In the case of $O(^3P) + F_2$, in which O is less electronegative than the F atoms, the reaction is expected to take place through a symmetric OF_2 with a much higher energy barrier. With the exception of the $O + F_2$ reaction, all reactions listed in Table II have been corroborated by the results of crossed-beam studies. The predicted high activation energy for the $O(^3P) + F_2$ reaction has been confirmed experimentally by Krech *et al.* (1977) in a fast flow system using the e.s.r. technique to measure the $O(^3P)$ atoms (Table II).

The reaction of O atoms with Cl_2 has recently been studied at the higher collision energy of 9 kcal/mol (Gorry *et al.*, 1977). The angular distribution of the OCl product was found to have a backward peak that was about $\frac{2}{3}$ that of the forward one, implying that the collision complex at this higher energy is now short-lived and has a lifetime comparable to about one rotational period. In the case of $O + I_2$, a recent experiment performed at a collision energy of 7 kcal/mol (Clough *et al.*, 1978), however, indicated a more drastic change in reaction dynamics, from the long-lived complex mechanism discussed above to a backward scattering 'rebound' mechanism. This was attributed to the decrease in electronic interaction between O and I_2 at this higher collision energy (Clough *et al.*, 1978).

C. Reaction with CX Radicals

The reactions of $O(^3P)$ atoms with carbon-containing diatomic radicals:

$$O(^3P) + CX \rightarrow CO + X$$
$$X = S, Se, N, F, H$$

are all very exoergic. Carbon monoxide laser emissions from these five reactions have been demonstrated* (Rosenwaks and Smith, 1973; Lin, 1974a; Shortridge and Lin, 1974; Hsu and Lin, 1978a). With the exception of the O + CH reaction, detailed vibrational energy distributions of the CO products formed in these reactions have been measured using chemiluminescence and/or CO laser resonance absorption techniques.

The reaction of the CS radical with $O(^3P)$,

$$O(^3P) + CS(X^1\Sigma^+) \rightarrow CO(X^1\Sigma^+) + S(^3P)$$
$$\Delta H^0 = -85 \text{ kcal/mol}$$

was first proposed by Hancock and Smith (1969) to be responsible for the observed CO laser emission from the flash photolysis of CS_2 and O_2 mixtures (Pollack, 1966). Numerous studies have been made on the dynamics of the O + CS reaction (Hancock *et al.*, 1971, 1972; Tsuchiya *et al.*, 1973; Djeu, 1974; Powell and Kelley, 1974; Hudgens *et al.*, 1976; Hsu *et al.*, 1979a). Fig. 3 shows the initial vibrational energy distributions of the CO produced from the O + CS reaction using two different experimental methods (Hancock *et al.*, 1972; Hsu *et al.*, 1979a). In the experiment of Hancock *et al.*, the chemiluminescence from the $O + CS_2$ reaction carried out in a slow discharge-flow system was measured. The $CO^\dagger$ was formed in the following fast chain reactions:

$$O + CS_2 \rightarrow CS + SO$$
$$O + CS \rightarrow CO + S$$
$$S + O_2 \rightarrow SO + O$$

Loss processes such as radiative and collisional relaxations were subsequently corrected to generate the unrelaxed distribution corresponding to the nascent $CO^\dagger$ (Hancock and Smith, 1971). The distribution thus obtained has been corroborated by data determined by laser absorption measurements (Hancock *et al.*, 1971; Djeu, 1974; Hsu *et al.*, 1979a), including our results shown in Fig. 3. We have employed the flash photolysis—cw CO laser resonance absorption technique (Lin and Shortridge, 1974) to measure the vibrational energy distribution of the nascent $CO^\dagger$ formed in this and numerous other chemical laser and combustion-related reactions, some of which will be discussed in the latter part of this review. The results of our experiment, carried out in a quartz flash system ($\lambda \geq 200$ nm) using Ar-diluted mixtures of SO_2, CS_2, and N_2O** agree closely with those of Hancock

*References for earlier work on the CO laser produced from the O + CS reaction can be found in Arnold and Rojeska (1973).

**N_2O was added to relax the $CS^\dagger$ formed in the initial photodissociation of CS_2 (Hancock *et al.*, 1971). The fact that $CO^\dagger$ ($v > 15$) was still present (see Fig. 3) indicates that the relaxation was not complete.

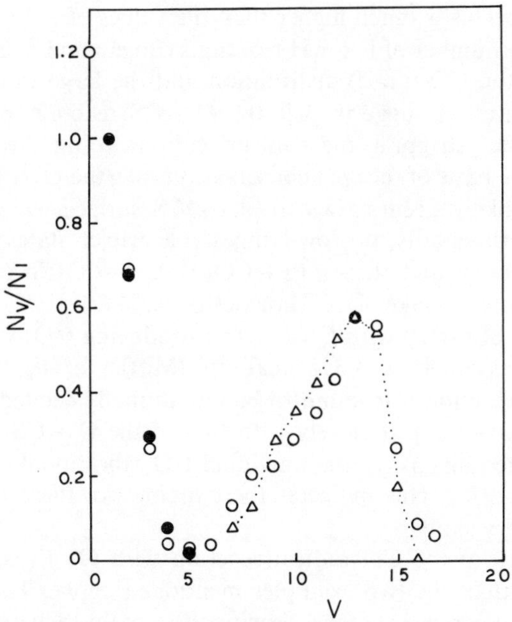

Fig. 3. Relative CO vibrational energy distributions obtained from $O(^3P) + CS$ and CS_2 reactions. Open circles—data taken from flash photolysis of SO_2, CS_2, N_2O, and Ar mixtures (0.8 : 1.0 : 5.6 : 42.6) above 200 nm; the distribution is normalized to $v = 1$ (Hsu *et al.*, 1979a). Filled circles—results of flash photolysis of NO_2, CS_2, and SF_6 (1 : 2 : 10) mixtures above 300 nm; the distribution is again normalized to $v = 1$ (Hsu *et al.*, 1979a). Triangles—data of Hancock *et al.* (1972) obtained from the O + CS reaction; the distribution is normalized to the $v = 13$ value of the distribution given by open circles. The distributions below $v = 5$ are attributable to the $O + CS_2 \rightarrow CO + S_2$ reaction discussed in Section II.D.

et al. (1972) above $v = 7$. The $CO^{\dagger}$ formed in these higher vibrational levels results mainly from the O + CS reaction. However, in our experiment we observed a significant amount of $CO^{\dagger}$ present below $v = 5$. Since the concentration of the CS radical formed in our low-energy flash (0.3 kJ) was quite small, these vibrationally colder CO molecules were most probably produced by the reaction of O atoms with undissociated CS_2 molecules. More detailed discussion on the dynamics of the $O + CS_2$ reaction will be given in the following section.

The results of Hancock *et al.* (1972) for the O + CS reaction indicate that about 80% of the total available energy ($E_{tot} = -\Delta H^0 + E_a + 5RT/2 =$ 87 kcal/mol, assuming $E_a = 0$) is channelled into the vibrational degree of freedom of CO. This is much higher than the values of ~40–70% from the HF formed in a number of F + RH reactions (Bogan and Setser, 1978). The narrowness of the CO ($v > 5$) distribution and the large extent of the CO vibrational excitation suggests that the O + CS reaction probably takes place by a direct stripping mechanism without going through an OCS complex. On the basis of the spin conservation rule, the $O(^3P) + CS(X^1\Sigma^+)$ reaction most likely occurs via a triplet OCS surface which is probably repulsive. Experimentally, no low-lying stable triplet states of OCS have ever been observed, and the $S(^3P) + CO(X^1\Sigma^+) \rightarrow OCS(X^1\Sigma^+)$ reaction has not been shown to take place (Hancock *et al.*, 1972).

The reaction of $O(^3P)$ with $CSe(X^1\Sigma^+)$ producing $CO(X^1\Sigma^+) + Se(^3P)$ is exoergic by as much as ~118 kcal/mol (Morley *et al.*, 1972). The CO formed in this reaction was found to be vibrationally excited up to $v = 20$; its distribution overlapped closely with that of the O + CS reaction when both were plotted against f_v, the fractional CO vibrational energy (E_v/E_{tot}) (Morley *et al.*, 1972). This indicates the dynamics of these two analogous reactions are very similar.

The dynamics of the CN radical reaction with $O(^3P)$ is, interestingly, more complex than the two examples mentioned above. The reaction has been shown to take place via the following three paths (Schacke *et al.*, 1973):

$$O(^3P) + CN(X^2\Sigma^+) \begin{cases} \xrightarrow{a} CO(X^1\Sigma^+) + N(^4S) + 74 \text{ kcal/mol} \\ \xrightarrow{b} CO(X^1\Sigma^+) + N(^2D) + 19 \text{ kcal/mol} \\ \xrightarrow{c} NO(X^2\Pi) + C(^3P) - 33 \text{ kcal/mol.} \end{cases}$$

This reaction system has been studied both experimentally and theoretically by Wolfrum and coworkers who investigated the effect of reagent vibrational energy on the kinetics as well as on the CO product vibrational energy distribution (Schacke *et al.*, 1973; Schmatjko and Wolfrum, 1975, 1977, 1978). Experimentally, the reaction was investigated in a discharge flow system, used in conjunction with a flash lamp for the CN radical production (employing C_2N_2). Since the $O(^3P)$ atoms, generated by the microwave discharge of O_2 or N_2 followed by NO titration, do not react with C_2N_2 rapidly, the reaction could be studied in a relatively slow flow. The disappearance of the $CN^\dagger$ radical was monitored by kinetic absorption spectroscopy using a second spectro-flash. The measured rate constants for the reactions of $CN(v \leq 7)$ with $O(^3P)$ are summarized in Fig. 4. Interestingly, the rates of the $O + CN(v)$ reactions below $v = 6$ increase very slowly with vibrational energy, whereas the rate of $O + CN(v = 7)$ increases abruptly by more than a factor of 5, compared with the lower vibrational level reactions. This sudden increase in rate was attributed to the occurrence of the endoergic

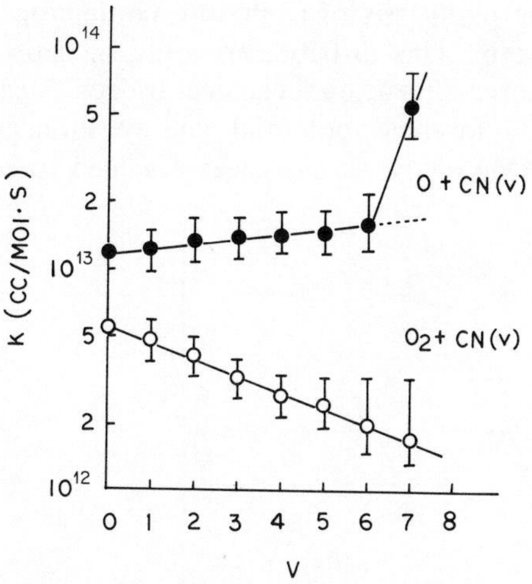

Fig. 4. Effects of the vibrational excitation of CN on $O(^3P) + CN$ and $CN + O_2$ reactions. (Reproduced by permission of Verlag Chemie from Schacke *et al.*, 1973.)

process (*c*), which is energetically accessible to the $O + CN(v = 7)$ reaction (see Fig. 5a). Similar pronounced enhancement effects have also been observed in several other reactions with high-energy barriers, including the reaction of $O(^3P)$ with H_2 and HCl discussed earlier.

Schacke *et al.* (1973) have also examined the reaction of $CN(v)$ with O_2 in the same reaction system. In this case, the reaction is known to proceed primarily by O-atom abstraction from O_2 instead of breaking the excited CN bond, *viz.*,

$$CN(v) + O_2 \rightarrow OCN + O$$
$$\Delta H^0 = -2 \text{ kcal/mol}$$

Consequently, the rate of this reaction was found to decrease with increasing vibrational excitation of the attacker, CN. This is again a very interesting observation. The closest example of this type is the reaction of O_3 with NO, in which NO is vibrationally excited (Stephenson and Freund, 1976). In this case, however, the rate of the reaction was found to be enhanced by $NO(v = 1)$ excitation, and was comparable to that observed when O_3 was vibrationally excited in the (001) mode (Gordon and Lin, 1973, 1976; Kurylo *et al.*, 1974; Bar-Ziv *et al.*, 1978; Hui and Cool, 1978).

Wolfrum and coworkers also measured the vibrational energy distribution of the CO employing the CO laser resonance absorption method alluded to earlier. The observed distribution has two distinctive peaks; one appearing at $v = 0$ and the other at $v = 9$. A typical set of distributions obtained from the photolysis of a mixture containing C_2N_2, O, and N_2 is shown in Fig. 5(b). This distribution could be satisfactorily accounted for by the use of three-dimensional classical trajectory calculations employing two different empirical potential energy surfaces (Schmatjko and Wolfrum, 1977). One of the surfaces was assumed to be repulsive; it was

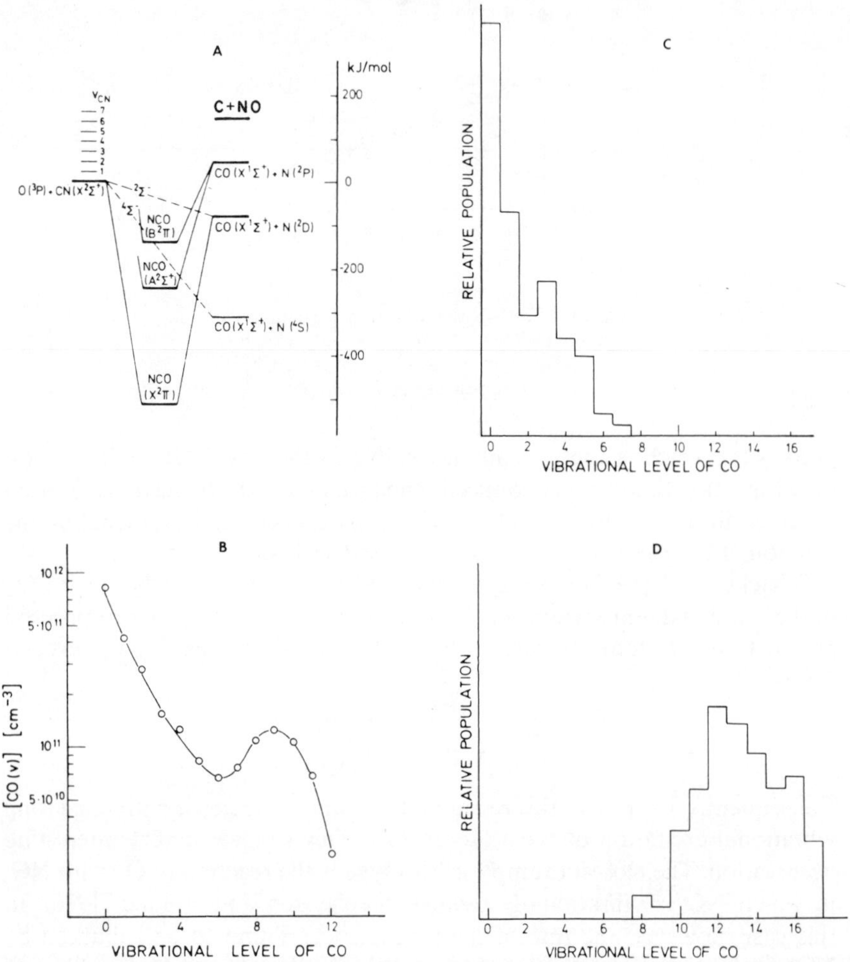

Fig. 5. (a) Schematic energy diagram of the $O(^3P) - CN(X^2\Sigma)$ system. (b) The vibrational energy distribution of the $CO^†$ formed in the $O(^3P) + CN(v)$ reaction; data obtained from the flash photolysis of C_2N_2 (6.2×10^{-9} mol/cm³) and $O(^3P)$ (8×10^{-10} mol/cm³, formed by N + NO); $P_t = 2.1$ Torr. (c) Relative CO vibrational energy distribution obtained from classical trajectory calculations for the channel: $O(^3P) + CN(X^2\Sigma^+) \rightarrow NCO(X^2\Sigma) \rightarrow CO(X^1\Sigma^+) + N(^2D)$. (d) Classical trajector results from the channel: $O(^3P) + CN(X^2\Sigma^+) \rightarrow CO(X^1\Sigma^+) + N(^4S)$. (After Schmatjko and Wolfrum, 1977). Reproduced by permission of the Combustion Institute.

used to calculate the production of $CO^\dagger$ by path (*a*). The other surface, which was constructed according to the stable NCO radical, correlates with both the reactants $O(^3P) + CN(X^2\Sigma^+)$ and the products $CO(X^1\Sigma^+) + N(^2D)$. The predicted CO vibrational distribution based on the latter surface is shown in Fig. 5c; it resembles the portion of the near statistical distribution at lower vibrational levels. This portion of the distribution was found to be less sensitive to the variation in the CN vibrational energy (V), due to the large extent of internal energy randomization in the NCO complex prior to its decomposition.

The calculated CO vibrational energy distribution for the repulsive path (*a*), as shown in Fig. 5(d), resembles that of the $O(^3P) + CS$ reaction discussed earlier. This hotter distribution was found to be rather sensitive to the change in the reagent vibrational energy, V. Analysis of their data, with the aid of trajectory calculations, reveals that about 50% of V is converted into V' (the CO product vibrational energy), indicating a fairly large extent of vibrational adiabaticity (Schmatjko and Wolfrum, 1978). Additionally, they established that roughly 20% of the overall reaction produces N atoms in the ground ^4S state and 80% in the ^2D state. This conclusion was qualitatively supported by the result of a vacuum u.v. N-atom resonance absorption measurement (Schmatjko and Wolfrum, 1977).

The reaction of the CF radical with $O(^3P)$, as mentioned previously, has been shown to generate CO-stimulated emission (Hsu and Lin, 1978a). The intensity of the emission, however, was considerably weaker than that detected in the O + CS reaction examined under similar experimental conditions, although the present system,

$$O(^3P) + CF(X^2\Pi) \rightarrow CO(X^1\Sigma^+) + F(^2P)$$

is significantly more exoergic ($\Delta H^0 = -126 \pm 2$ kcal/mol). In our experiment, the CF radical was produced by the flash photolysis of $CFBr_3$ in a quartz ($\lambda \geq 200$ nm) or in a 'Suprasil' ($\lambda \geq 165$ nm) system. The CO product, measured by CO laser resonance absorption, was observed to be vibrationally excited up to $v = 24$ with a Boltzmann temperature of 1.4×10^4 K. A surprisal analysis (see, for example, Levine and Bernstein, 1974) of the observed $CO^\dagger$ distribution gave rise to a near-straight line with a zero slope (i.e. $\lambda_v \simeq 0$), which is sharply different from the value, $\lambda_v \simeq -9$ for the $CO^\dagger$ formed in the O + CS reaction, based on the data of Hancock *et al.* (1971). The gross difference in the dynamics of these two reactions is attributable to the involvement of a relatively long-lived FCO intermediate radical in the O + CF reaction. Using the value of $D(\text{F—CO}) \simeq 59$ kcal/mol evaluated from the heat of formation, $\Delta H_f^0(\text{FCO}) \simeq -66$ kcal/mol (Gangloff *et al.*, 1975), the RRK lifetime of the FCO complex was estimated to be 10^{-12}–10^{-13} s, corresponding to a period of several vibrations. This is quite sufficient for the available reaction energy to be completely randomized among all internal degrees of freedom of this three-atom intermediate.

The CH radical is one of the most reactive species in high-temperature hydrocarbon combustion systems. Its reactions with $O(^3P)$, O_2, and NO are all very exoergic and have been shown to generate intense CO-stimulated emissions (Lin, 1973a, b, 1974a, b). The reaction of interest, $O(^3P) + CH$, is the most exoergic process among the five O + CX reactions discussed here. Many possible products can be formed in this reaction; for example,

$$O(^3P) + CH(X^2\Pi) \begin{cases} \xrightarrow{a} CO(X^1\Sigma^+) + H(^2S) + 176\ \text{kcal/mol} \\ \xrightarrow{b} OH(X^2\Pi) + C(^3P) + 22\ \text{kcal/mol} \\ \xrightarrow{c} CHO^+ + e + 2\ \text{kcal/mol} \end{cases}$$

The amount of energy released in the first step, actually, is sufficient to pump CO up to several higher electronically excited states; e.g. $a^3\Pi(\Delta H^0 = -38$ kcal/mol), $a'^3\Sigma^+$ ($\Delta H^0 = -18$ kcal/mol), and $d^3\Delta_i$ ($\Delta H^0 = -3$ kcal/mol). The relative importance of these processes is not known.

The chemiionization process (*c*) was first suggested by Calcote (1962) to be an important source of ions observed in high-temperature hydrocarbon combustion systems (for a review, see Fontijn, 1971). On the basis of the results of a theoretical study by MacGregor and Berry (1973), Dalgarno *et al.* (1973) pointed out that the chemiionization reaction $O + CH \rightarrow CHO^+ + e$ may be an important source of free electrons and positive ions in interstellar clouds. The subsequent dissociative recombination, $HCO^+ + e \rightarrow CO +$ H, may also be an efficient source of CO in both diffuse and dense clouds.

D. Reaction with Triatomic Species

In this section, we will cover the reactions of $O(^3P)$ atoms with triatomic species, for which at least some quantitative microscopic kinetic data are available. The discussion will be started with the reactions with sulphides, mainly CS_2 and OCS. Comments on the reactions with XO_2 (X = N, O, Cl) and other sulphides such as R_2S (where R = H or alkyl groups) will also be briefly given. These reactions are of interest to atmospheric chemistry. Subsequently, we will discuss the recent results obtained mainly in our own laboratory on the reactions of $O(^3P)$ with CHF, CF_2, and CH_2 and also comment on the related reaction $O + C_2O$. These reactions are related to chemical lasers and combustion chemistry.

The $O(^3P) + CS_2$ reaction is kinetically very interesting and intriguing. Although only four atoms are involved in the reaction, there are at least three known reaction paths according to Gutman and coworkers (Slagle *et al.*, 1974; Graham and Gutman, 1977):

$$O(^3P) + CS_2(X\,^1\Sigma_g^+) \begin{cases} \xrightarrow{a} CS + SO + 21\ \text{kcal/mol} \\ \xrightarrow{b} CO + S_2 + 83\ \text{kcal/mol} \\ \xrightarrow{c} OCS + S + 54\ \text{kcal/mol} \end{cases}$$

The exoergicities given here are based on the formation of ground electronic state products. Among these three reaction channels, only channel (*c*) was measured quantitatively by Gutman and coworkers. The branching ratio, $f_c = k_c/\Sigma k$, for this route was determined as varying from 0.098 ± 0.004 at 249 K to 0.081 ± 0.007 at 500 K based on experiments carried out in a fast discharge-flow system. Earlier estimates ranged from < 0.015 to 0.22 (Hancock and Smith, 1971; Suart *et al.*, 1972). A recent attempt to detect OCS in a chemiluminescence experiment by Hudgens *et al.* (1976) showed no detectable emission from the product. This however, cannot rule out its possible physical presence since the OCS molecule may be formed vibrationally cold (although it is considered as rather unusual since this channel is quite exoergic). On the basis of the intensities of the photoionized mass peaks of CS, SO, and S_2 from the $O + CS_2$ reaction performed in a crossed-beam apparatus, Slagle *et al.* (1974) crudely estimated the relative abundance of channel (*a*) to be within the range, 0.7–0.85. This rough estimate places the remaining channel (*b*) in the range of 0.05–0.2. These branching ratios are important to the kinetic modelling and the performance improvement of the CS_2/O_2 chemical CO laser system (Slagle *et al.* 1974). More experiments are needed before one can confidently use these values.

The dynamics of formation of $CS + SO$ by channel (*a*) was first investigated by Smith (1967) using kinetic absorption spectroscopy. The $O(^3P)$ atom was generated by the flash photolysis of NO_2 above 300 nm. On the basis of the measured relative vibrational populations of $CS^\dagger$ and $SO^\dagger$, Smith concluded that about 9% and 20%,* respectively, of the total available reaction energy (23 kcal/mol, adding the activation and thermal energies) were channelled into product vibrations. The dynamics of the formation of these two products were simulated by trajectory calculations assuming collinear collisions. The preliminary results of these calculations based on approximate LEPS potential energy surfaces indicate that the vibrational excitation of the $CS^\dagger$ results mainly from the repulsive release between CS and SO (Smith, 1967).

Recently, several crossed-beam studies have been made on the $O + CS_2$ reaction, following only the major product channel (*a*) (Geddes *et al.*, 1974;

*These values have been adjusted by using the newly established heat of formation for the CS radical (Okabe, 1972).

Gorry *et al.*, 1977; Clough *et al.*, 1978). There has been no report of the observation of the minor products, CO, S_2, and OCS. These studies have found quite definitely that the production of SO occurs via a direct stripping mechanism with a considerably large fraction of the reaction energy going into the products rotational degrees of freedom. This observation could be rationalized by the assumption of a bent transition state O—S—C—S. Since the CS_2 molecule is linear, the formation of a bent transition-state complex during the reaction with $O(^3P)$ is expected to result in considerable rotational excitation in both products (Gorry *et al.*, 1977).

Aside from the evidence provided by Slagle *et al.* (1974) from mass spectroscopic measurments, there has been another indirect, but strong indication from the observed $CO^\dagger$ vibrational energy distributions obtained from pulsed electrical discharge experiments (Tsuchiya *et al.*, 1973; Powell and Kelley, 1974) that channel (*b*) does occur (Kelley, 1976; Levine and Ben-Shaul, 1977). The dynamics of CO formation from this channel were first investigated by Hudgens *et al.* (1976) using the chemiluminescence method. Recently, we have also examined the production of the $CO^\dagger$ from this reaction by CO laser absorption spectroscopy. In our work, the first evidence of the occurrence of this route came from the result previously shown in Fig. 3, which was obtained from a low-energy photolysis of SO_2 and CS_2 mixtures in a quartz flash tube. The colder portion ($v < 5$) of the $CO^\dagger$ distribution shown in this figure is believed to have arised directly from $O + CS_2$. To confirm this hypothesis, we have carried out additional experiments in a Pyrex flash system using NO_2 and CS_2 mixtures. Here NO_2 was employed as the $O(^3P)$ atom source, as had been used by Smith (1967). In this system, only the $O(^3P) + CS_2$ reaction could occur. The initial CO vibrational energy distribution obtained from the Pyrex tube experiment is presented in Fig. 3. Our distributions are considerably colder and are free from the second hump peaking at $v = 7$ as reported by Hudgens *et al.* (1976). In another experiment performed at a higher flash energy, 1 kJ, we detected the presence of a small residual tail, which peaked at $v = 13$ and extended up to $v = 16$, resembling that of the O + CS reaction shown in Fig. 3. At this higher flash energy, a very small fraction of CS_2 might be dissociated near 300 nm. Since the O + CS reaction is probably at least two orders of magnitude faster than the $O + CS_2$ reaction occurring via channel (*b*), this secondary (O + CS) reaction thus becomes prominent in this high-energy flash. This secondary reaction probably also occurred in the chemiluminescence experiment which had ~ 1 ms residence time. Hudgens *et al.* (1976) interpreted their observed dual maxima as being due to two different reaction paths, one occurring by the direct attack of $O(^3P)$ on the C atom forming the 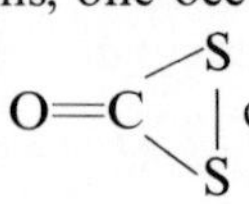 complex, and the other via an indirect attack on one of the S atoms with its subsequent migration onto the C atom. The C_{2v} OCS_2 complex may be the intermediate involved in channels (*b*) and (*c*), producing $CO + S_2$ and $OCS + S$, respectively. Our observed $CO^{\dagger}$ distribution from channel (*b*) is close to the prior statistical distribution (Kelley, 1976).

The reaction of the $O(^3P)$ atom with OCS yielding CO and SO:

$$O(^3P) + OCS(X^1\Sigma^+) \rightarrow CO(X^1\Sigma^+) + SO(X^3\Sigma^-)$$
$$\Delta H^0 = -52 \text{ kcal/mol}$$

has been investigated by CO laser absorption spectroscopy performed in the same manner as described above (Shortridge and Lin, 1975; Hsu *et al.*, 1979a). Since the reaction is quite slow at room temperature (Schofield, 1973), $CO^{\dagger}$ absorptions for all levels up to $v = 5$ were rather small. From these measurements, the vibrationally excited CO was estimated to carry about 9% of the total available energy. This is exactly the same as that carried by the $CS^{\dagger}$ in the $O + CS_2$ reaction as discussed earlier. Accordingly, the mechanism of $O(^3P) + OCS$ is probably similar to $O + CS_2$ (producing $CS + SO$), taking place mainly by direct stripping. The results of our isotope-labelled experiments using ^{18}O atoms show that CO was formed exclusively from the parent OCS molecule (Hsu *et al.*, 1979a). Recently, Clough *et al.* (1978) attempted to study the reaction in a crossed-beam machine. They could not detect any appreciable reaction at a mean collision energy of 7 kcal/mol, which is 2 kcal/mol higher than the activation energy of the reaction (Schofield, 1973). Attempts were also made to increase the internal energy of OCS by increasing its beam temperature up to 700 K; no observable change in SO signals was detected. Manning *et al.* (1976) have recently studied the effect of OCS vibrational excitation on the rate of the reaction using an energy transfer method. They also failed to observe any significant enhancement effects.

The reactions of $O(^3P)$ with NO_2, O_3, and ClO_2 have been shown to produce vibrationally excited O_2 molecules, presumably by a direct O-atom abstraction mechanism:

$$O(^3P) + NO_2 \rightarrow O_2 + NO$$
$$\Delta H^0 = -46 \text{ kcal/mol}$$

$$O(^3P) + O_3 \rightarrow 2O_2$$
$$\Delta H^0 = -94 \text{ kcal/mol}$$

$$O(^3P) + ClO_2 \rightarrow O_2 + ClO$$
$$\Delta H^0 = -60 \text{ kcal/mol}$$

Most experiments carried out on these reactions used kinetic absorption spectroscopy and only qualitative information is available with regard to the partitioning of reaction energies. The existing work on these systems has been reviewed recently (Smith, 1975; Donovan and Gillespie, 1975). For the O + ClO_2 reaction, a recent work shows it to be quite slow (Bemand *et al.*, 1973); thus the $O_2^{\dagger}$ observed in previous flash photolysis experiments must have resulted from the faster O + ClO reaction (Donovan and Gillespie, 1975).

The reactions of $O(^3P)$ with CS_2 and OCS, as described above, occur predominantly via O-atom attack on S atoms. This was also found to be true for the reactions involving alkyl sulphides (Lee *et al.*, 1976; Slagle *et al.*, 1975), viz.,

$$O(^3P) + R_1SR_2 \rightarrow [R_1R_2S{=}O] \rightarrow SO + R_1 + R_2$$

where R_1,R_2 = alkyl groups or H atoms. These reactions were found to take place with little or no activation energies. We have attempted to isolate these sulphoxide intermediates using the matrix isolation technique. They were indeed observed. In this experiment, we have also studied the reaction of H_2S with $O(^3P)$, which has been shown to occur with a higher, 3–4 kcal/mol of activation energy producing, presumably, OH + HS. In our experiment carried out at 8 K in Ar matrices, we detected an appreciable amount of the new species, HOSH (Smardzewski and Lin, 1976). This species was postulated to come from the isomerization of the H_2SO adduct within matrix cages, since the high activation energy abstraction reaction (forming HO + SH) observed at high temperatures is not expected to be important at 8 K. The occurrence of this new association path may account for the curvature in an Arrhenius plot for O + H_2S (Hollinden *et al.*, 1970).

There is strong experimental evidence that the reactions of CHF, CF_2, CH_2, and C_2O with $O(^3P)$ atoms all take place via long-lived complexes. Based on the results of HF-stimulated emission and CO laser resonance absorption measurements, we have recently concluded that the O + CHF reaction takes place via the following two major paths (Hsu *et al.*, 1979b):

$$O(^3P) + CHF(\tilde{X}\,^1A') \rightarrow [^3HFCO^{\dagger *} \rightarrow {}^1HFCO^{\dagger}]$$

$$\xrightarrow{a} HF^{\dagger} + CO^{\dagger} + \sim 190\ \text{kcal/mol}$$

$$\xrightarrow{b} H + F + CO^{\dagger} + \sim 54\ \text{kcal/mol}$$

where $^3HFCO^{\dagger *}$ and $^1HFCO^{\dagger}$ represent the vibronically excited triplet and singlet HFCO molecules, respectively. A simplified schematic energy diagram for the HFCO system is shown in Fig. 6. Although the production of $CO(a^3\Pi)$ is energetically feasible as indicated in the diagram, the result of a test designed to examine its possible presence showed it to be unimportant (Hsu *et al.*, 1979b). The $HF^{\dagger}$ formed in the reaction via channel (*a*) was found to have a near statistical vibrational energy distribution; the $CO^{\dagger}$

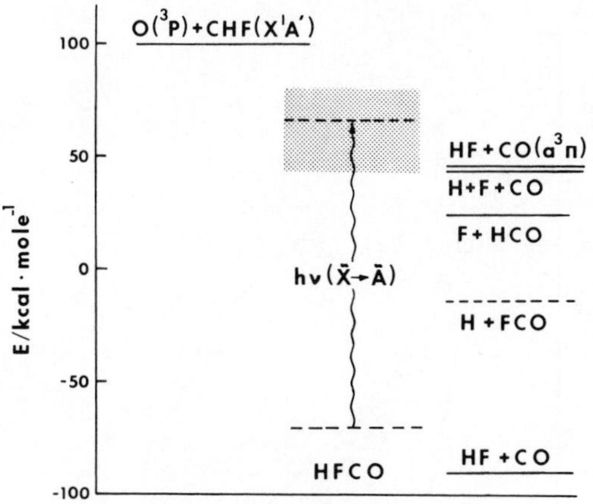

Fig. 6. A simplified schematic energy diagram for the $O(^3P) + CHF \rightarrow HFCO \rightarrow CO + HF$ (or $H + F$) reaction. (After Hsu *et al.*, 1979b.)

distribution shown in Fig. 7, however, is considerably colder. It lies between the two statistical distributions expected for channels (*a*) and (*b*) as shown. This indicates both channels are occurring concurrently. By analysis of the observed $CO^{\dagger}$ distribution using the computed statistical curves, we estimated that about 40% of the $O + CHF$ reaction occurs via the molecular elimination channel (*a*) and about 60% via the atomic production channel (*b*). In this experiment, we employed $CHFX_2$ ($X = Cl$, Br) as the photochemical sources of the CHF radical (Lin, 1973a, 1978). A similar study of the decomposition of HFCO has been carried out photolytically in the 200 nm region (see Fig. 6) by Klimek and Berry (1973). From the appearance of HF-stimulated emission and the result of a test with added D_2 (which led to DF laser emission), they came to the same conclusion that the decomposition of $HFCO^{\dagger}$ (formed by rapid intersystem crossing) occurs by both molecular and atomic elimination channels. No information on CO and the branching ratio of these two channels could be determined in this experiment.

The reaction of $O(^3P)$ and the CF_2 radical was recently investigated for the purpose of elucidating the mechanism of the $O(^3P) + C_2F_4 \rightarrow CF_2O + CF_2$ reaction (Hsu and Lin, 1977a). The question lies in the identity of the CF_2 radical produced in the reaction. Early flash photolysis experiments employing kinetic absorption spectroscopy (Michell and Simons, 1968; Tyerman, 1969) detected only the ground-state CF_2 $(^1A_1)$ radical, whereas the results of product analysis (Heicklen *et al.*, 1965; Saunders and Heicklen, 1966) suggested that the excited triplet CF_2 $(^3B_1)$ radical was also present in the reaction, as anticipated by the spin conservation rule. In our experi-

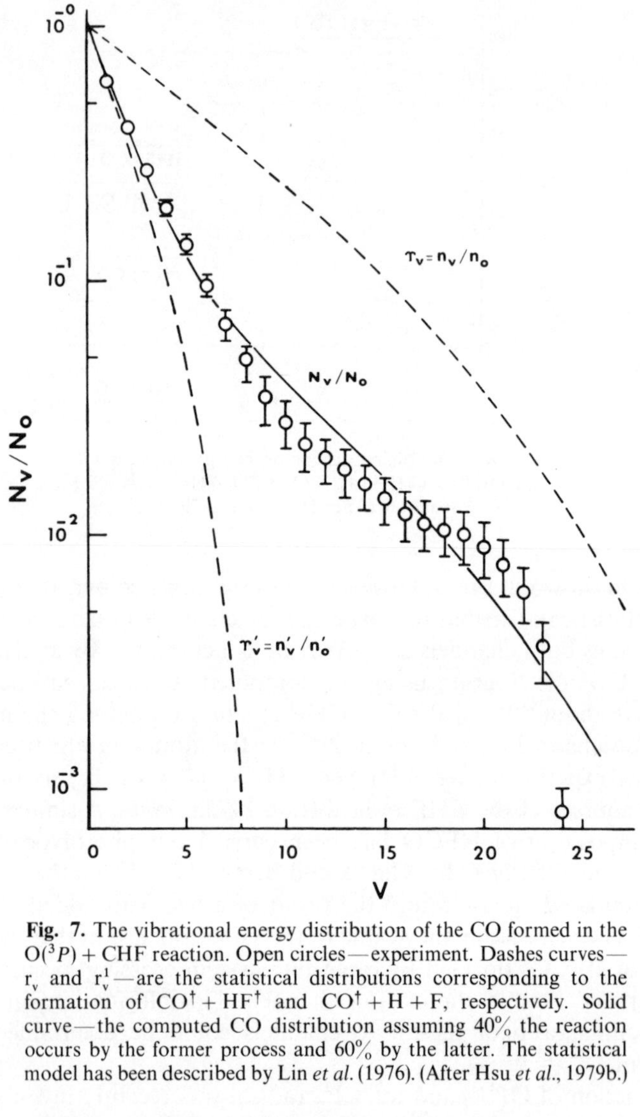

Fig. 7. The vibrational energy distribution of the CO formed in the $O(^3P)$ + CHF reaction. Open circles—experiment. Dashes curves—r_v and r_v^1—are the statistical distributions corresponding to the formation of $CO^{\dagger} + HF^{\dagger}$ and $CO^{\dagger} + H + F$, respectively. Solid curve—the computed CO distribution assuming 40% the reaction occurs by the former process and 60% by the latter. The statistical model has been described by Lin *et al.* (1976). (After Hsu *et al.*, 1979b.)

ments, we flash photolysed mixtures of NO_2 and C_2F_4 in a Pyrex tube ($\lambda \geq 300$ nm) to avoid the photodissociation of C_2F_4. The CO product formed in the following sequential reactions:

$$O(^3P) + C_2F_4 \rightarrow CF_2O + CF_2$$

$$O(^3P) + CF_2 \rightarrow CO + F_2 \text{ (or 2F)}$$

was monitored by a cw CO laser. The CO formed in this system was found to be vibrationally excited up to $v = 11$, corresponding to as much as 63 kcal/mol of energy. From the analysis of the observed $CO^\dagger$ distribution and the consideration of all possible paths and energetics associated with the $O + CF_2$ reaction, we concluded that this highly excited $CO^\dagger$ could only come from the reaction,

$$O(^3P) + CF_2(^3B_1) \rightarrow CO^\dagger + F_2$$
$$\Delta H^0 = -89 \text{ kcal/mol}.$$

The predicted CO distribution, assuming the reaction occurs via a CF_2O complex, agrees closely with the experimental ones. Our conclusion that $CF_2(^3B_1)$ is indeed produced in the $O(^3P) + C_2F_4$ reaction has been recently confirmed by Koda (1978), who detected a discrete emission from this reaction in the 470–720 nm region which was attributable to the $CF_2(^3B_1 \rightarrow {}^1A_1)$ phosphorescence.

The $O(^3P) + CH_2$ reaction is an important process in acetylene flames (Williams and Smith, 1970) since the major products of the $O + C_2H_2$ reaction have now been firmly established as $CH_2 + CO$ (see the next section). According to the results of $O + CHF$ described above and the known photochemistry of H_2CO (Calvert and Pitts, 1966; Lee, 1977), it is likely that the $O + CH_2$ reaction may take place via the following two major channels, as was proposed previously by other investigators (Arrington *et al.*, 1965; Brown and Thrush, 1967):

$$O(^3P) + CH_2 \rightarrow [^3CH_2O^{\dagger *} \rightarrow {}^1CH_2O^\dagger]$$
$$\xrightarrow{a} CO + H_2 + 179 \text{ kcal/mol}$$
$$\xrightarrow{b} CO + 2H + 75 \text{ kcal/mol}$$

Hydrogen has commonly been assumed to be a major product of this reaction, but experimentally it is not certain whether H_2 and H atoms are formed simultaneously. If H_2 is produced initially in the reaction, one would expect a much different combustion chemistry than if H atoms only, were present in the early stage of the $O + C_2H_2$ reaction.

To elucidate the mechanism of this important reaction, we have recently investigated the reaction of $O(^3P)$ with CH_2 in a quartz flash photolysis system using mixtures of SO_2 and CH_2X_2 (X = Br, I). Here, CH_2X_2 is used as the source of the CH_2 radical (Lin, 1973a, b; Hsu and Lin, 1977b). The $CO^\dagger$ formed in the reaction was found to be excited up to $v = 18$ with a vibrational temperature of $\sim 10^4$ K. Since the atomic channel (*b*) can only excite CO up to the maximal level of 13, we concluded that the molecular channel (*a*) is also occurring simultaneously. An analysis of our preliminary data indicates that both channels are approximately equally important.

The extent of the CO vibrational excitation ($T_v \sim 10^4$ K) observed in this reaction is sufficient to account for the previously reported weak CO-stimulated emissions from the $O + C_2H_2$ and the $O + CH_2$ reactions (Lin, 1973a, b). These two reactions had also been investigated earlier by Clough *et al.* (1970) and Creek *et al.* (1970) in flow systems using the chemiluminescence method. Significant CO vibrational excitation was observed in these studies.

The reaction of $O(^3P)$ atoms with C_2O ($\Delta H^0 \approx -205$ kcal/mol) has been proposed (Becker and Bayes, 1966, 1968) to be responsible for the vacuum u.v. CO ($A^1\Pi - X^1\Sigma^+$) emission detected in the $O + C_2H_2/C_3O_2$ atomic flames and in acetylene and other hydrocarbon flames (Williams and Smith, 1970). The rate constant has been measured to be $\sim 1 \times 10^{-10}$ cm^3/ molecule·s with about 1–10% of the reaction producing $CO(A^1\Pi)$ (Shackleford *et al.*, 1972). The overall reaction,

$$O(^3P + C_2O(X^3\Sigma^-) \rightarrow CO^*(A^1\Pi, d^3\Delta_i, e^3\Sigma^-, \text{etc.}) + CO(X^1\Sigma^+)$$

is believed to proceed via a long-lived C_2O_2 complex because the intensities of the $A \rightarrow X$ emission from both $C^{16}O^*$ and $C^{18}O^*$ observed in the $^{18}O(^3P) + C_2O$ reaction were found to be comparable (Bayes, 1970). However, this conclusion should be accepted with caution in view of the high pressure (> 5 Torr) used. It is now known that the cross section for energy transfer involving these highly excited states of CO is very large (Comes and Fink, 1972; Slanger and Black, 1973; Taylor and Setser, 1973; Provorov *et al.*, 1977). Fontijn and Johnson (1973) investigated the effect of inert-gas pressure on the $A \rightarrow X$ emission in the $O + C_2H_2$ reaction carried out in a discharge-flow system. They observed a strong inert-gas effect and thus concluded that the $CO(A^1\Pi)$ was produced by energy transfer reactions involving other excited states ($d^3\Delta_i$, $e^3\Sigma^-$,...) which are perturbed strongly with the $A^1\Pi$ state (Simmons *et al.*, 1969; Provorov *et al.*, 1977).

E. Reaction with Alkenes and Alkynes

The reactions of the $O(^3P)$ atom with alkenes and alkynes are significantly more complex than with alkanes, which occur only by direct H-atom abstraction. The rate constants for the reactions of $O(^3P)$ with simple organic compounds have recently been compiled (Herron and Huie, 1974; Kerr and Ratajczak, 1977) and the mechanisms reviewed (Huie and Herron, 1975).

The reactions of unsaturated hydrocarbons with $O(^3P)$ atoms can take place by both H-atom abstraction and, preferentially, by addition to unsaturated bonds forming excited adducts (Cvetanović, 1955, 1963). The subsequent unimolecular reactions of these excited adducts, which are believed to be biradicals, are complex and have been the subject of much controversy. This is true for alkenes as well as alkynes (Westenberg, 1973). The mechanism originally proposed by Cvetanović (1955, 1963) for O + alkenes is now firmly established, thanks to the improved direct mass spectrometric sampling techniques first developed by Gutman and coworkers (Kanofsky and Gutman, 1972; Kanofsky *et al.*, 1973, 1974; Gilbert *et al.*, 1976) and more recently by Blumenberg *et al.* (1977). Some of their results on the reactions of $O(^3P)$ with simple alkenes and alkynes are summarized in Table III.

Cvetanović's mechanism for the reactions of $O(^3P)$ atoms with alkenes, which is also equally applicable to alkynes (as demonstrated by the examples shown in Table III), can be schematically given as follows (Cvetanović, 1963):

TABLE III. Major Reaction Channels and Branching Ratios in the Reactions of $O(^3P)$ with Simple Alkenes and Alkynes

Reactant	Reaction channels		Branching ratios		
C_2H_4 [1,2]	$CH_3 + CHO$	(*a*)	0.95		
	$H_2 + CH_2CO$	(*b*)	0.05		
C_2H_3X [3]	$CH_3 + CXO$	(*a*)	X = F	X = Cl	X = Br
			<0.07	0.22	0.29
	$CH_2X + CHO$	(*b*)	>0.82	0.70	0.51
	$HX + CH_2CO$	(*c*)	0.10	0.08	0.20
C_3H_6 [2]	$C_2H_5 + CHO$	(*a*)	$a + b \geqslant 0.95$		
	$CH_3 + CH_3CO$	(*b*)	$a : b = 1 : 1.05$		
	$C_2H_4 + H_2CO$	(*c*)	≤ 0.03		
	$H_2 + C_3H_4O$	(*d*)	<0.02		
$1 - C_4H_8$ [2]	$C_3H_7 + CHO$	(*a*)			
	$C_2H_5 + CH_3O$	(*b*)	$a : b : c = 1 : 2 : 1.7$		
	$CH_3 + C_2H_5O$	(*c*)			
C_2H_2 [2]	$CO + CH_2$	(*a*)	0.95		
	$HC_2O + H$	(*b*)	<0.03		
	$H_2 + C_2O$	(*c*)			
CH_3C_2H [2]	$CO + CH_3CH$		0.95		
$C_2H_5C_2H$ [2]	$CO + C_2H_5CH$		0.95		
$CH_3C_2CH_3$ [2]	$CO + CH_3CCH_3$		0.95		

1. Kanosfky *et al.* (1973).
2. Blumenberg *et al.* (1977).
3. Slagle *et al.* (1975).

(i) Addition to form vibronically excited biradical adducts.

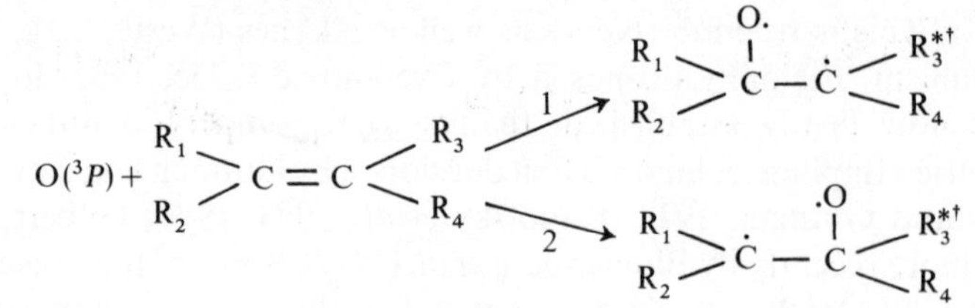

(ii) Unimolecular reactions of the excited adducts (taking one example).

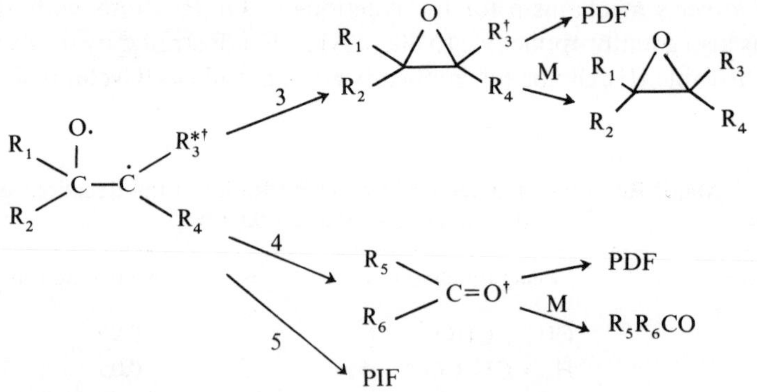

Where PDF and PIF stand for 'pressure-dependent fragmentation' and 'pressure-independent fragmentation', respectively, of the vibrationally excited ground electronic state adducts. The intersystem crossing from the vibronically excited triplet to the ground electronic singlet state at high energies is expected to be rapid. The extent of the pressure effect, of course, depends on the lifetimes of the excited species and the pressure employed in the experiment. Since the biradicals disappear rapidly via ring clossure (step 3) and isomerization (step 4 taking place via H-atom or alkyl-group migration), only very fast unimolecular decomposition processes can occur by PIF (step 5). In the case of $O + C_2H_3F$, which will be discussed later, there is strong evidence that the production of the vibrationally excited HF followed by stimulated emission occurs to some extent via step (5).

According to Cvetanović (1963), the addition reaction is 'non-stereospecific', indicating that rotation about the newly opened double bonds in the biradical intermediates does occur somewhat. For unsymmetrical alkenes, addition to the less substituted C atom is preferred, similar to the addition of alkyl or perfluoroalkyl radicals to substituted alkenes (Tedder and Walton, 1978).

We have recently utilized the above chemical activation scheme to generate hydrogen fluoride lasers (Umstead *et al.*, 1979). The reactions of $O(^3P)$ atoms with fluorinated ethenes (C_2H_3F, *cis*- and *trans*- 1,2-$C_2H_2F_2$,

1,1-$C_2H_2F_2$, and C_2HF_3) were carried out in an optical cavity consisting of a grating and a total reflecting mirror. Hydrogen fluoride laser emissions in the $v = 1 \rightarrow v = 0$ transition were detected from all five reactions. The inversion ratio of the HF formed in the $O(^3P) + C_2H_3F$ reactions was found to be $N_1/N_0 = 0.52 \pm 0.02$. This is much higher than the statistical value of 0.31. Similar higher inversion ratios ($N_1/N_0 = 0.62 \pm 0.05$) were also observed for C_2HF_3 and *cis*- and *trans*-1,2-$C_2H_2F_2$. In the O + 1, 1-$C_2H_2F_2$ case, however, $N_1/N_0 \simeq 0.42$, which is closer to the statistical value (0.31). This finding can be rationalized by means of Cvetanović's mechanism, which is supported by the results of a detailed study of these reactions by Gutman and coworkers (Slagle *et al.*, 1975; Gilbert *et al.*, 1976). Fig. 8 shows various possible routes which may produce $HF^\dagger$ from the $O(^3P) + C_2H_3F$ reaction. The branching ratios for three different sets of products shown in the diagram are the measurements of Slagle *et al.* (1975). This energy diagram is consistent with the general reaction scheme of Cvetanović, including the formation of the two aldehydic isomers, CH_2FCHO and CH_3CFO. The observation of the higher inversion ratios (greater than the statistically expected value) could be attributed to the occurrence of the direct three-centered HF elimination by step (*d*) as shown in the figure, which is a PIF process. Since this process takes place rapidly before molecular rearrangement can occur, the available reaction energy (~ 106 kcal/mol for these reactions) has not been completely randomized. These PIF processes may also be responsible for the production of formaldehyde products by C—C bond cleavage (HFCO and F_2CO) in the reactions of more heavily fluorinated ethenes as discussed in the paper by Gilbert *et al.* (1976) and several other references cited therein. The $HF^\dagger$ produced from PDF processes (via steps *a*, *b*, and *c* as labelled in Fig. 8) however, is expected to be closer to the statistical distribution because of more extensive molecular rearrangements before the four-centred elimination can take place. This is indicated by the results of the O + 1,1-$C_2H_2F_2$ reaction in which the direct three-centered HF elimination via the PIF mechanism is not possible. Accordingly, the inversion ratio observed is much lower than that for the rest of fluoroethenes and is closer to the statistical value.

There are, however, other important dynamic factors which may influence the partitioning of reaction energies in this type of four-centered elimination process, such as 'exit channel interaction' occurring in the repulsive region of the potential hypersurface where reaction products separate. A very interesting chemiluminescence study has been made by Gleaves and McDonald (1975) on the reaction of $O(^3P)$ with 3-chlorocyclohexene and 5-chloro-1-pentene from which vibrationally excited HCl molecules were produced. The measured $HCl^\dagger$ distributions are shown in Fig. 9 for comparison with the statistically predicted values. The observed distributions are considerably non-statistical and are independent of the site of HCl

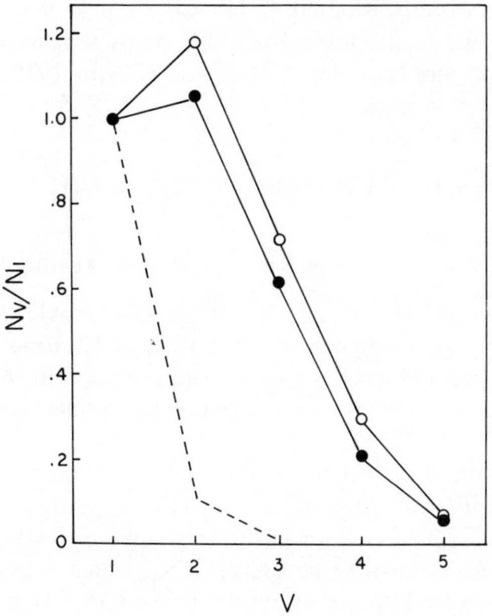

Fig. 9. Relative vibrational energy distribution of the HCl[†] formed in $O(^3P)$ + 5-chloro-1-pentene (open circles) and $O(^3P)$ 3-chlotocyclohexene (filled circles) reactions. Dashed curve—the estimated statistical HCl distribution for the latter reaction. (After Gleaves and McDonald, 1975.)

elimination with respect to the location of activation. These findings suggest the importance of the 'exit channel interaction' in these two reactions.

A similar study has been made by Moehlmann and McDonald (1973) on the reaction of $O(^3P)$ with cyclooctene. The reaction produces vibronically excited cyclooctanone, which is expected to relax rapidly to the ground electronic state before i.r. chemiluminescence can occur. The vibrationally excited cyclooctanone carries about 118 kcal/mol of internal energy, with an expected RRKM lifetime of $\sim$ 65 s (cf. 1 ms observation time) (Moehlmann and McDonald, 1973). The observed vibrational energy distribution into the three i.r. active modes of the cyclooctanone, C—H stretch, C=O stretch and CH_2 bend, was found to be in close agreement with statistically expected values.

As indicated earlier, the mechanism of alkyne reactions with $O(^3P)$ can be effectively described by Cvetanović's scheme proposed for alkene reactions. Instead of forming expoxide and aldehyde or ketone intermediates as in the alkene case, alkyne reactions are believed to take place via ketene intermediates. For the simplest reaction, $O + C_2H_2$, Haller and Pimentel

(1962) have successfully isloated $CH_2{=}C{=}O$ in a low temperature Ar matrix. Indeed one of the several possible paths that may lead to ketene has now been commonly accepted (Huie and Herron, 1975)

$$O(^3P) + HC{\equiv}CH \rightarrow [H\dot{C}{=}CH(\dot{O})]^{*\dagger} \rightarrow CH_2{=}C{=}O^\dagger$$

$$\rightarrow CH_2^\dagger + CO^\dagger + 47 \text{ kcal/mol.}$$

This mechanism is, of course, similar to Cvetanović's mechanism for $O + C_2H_4$. For higher alkynes, the migration of H atoms or alkyl groups in the initial excited adducts has also been shown to occur (Avery and Heath, 1972; Csizmadia *et al.*, 1973), as indicated by the results shown in Tables III and IV.

We have measured the extent of vibrational excitation in the CO formed in a series of $O(^3P)$ + alkyne reactions in order to verify the above mechanism. Take $O + CH_3C{\equiv}CH$, for example, the major final products formed in the reaction are known to be $CO + C_2H_4$, which is associated with an exoergicity of 118 kcal/mol. However, if $CO + CH_3CH$ is formed initially (Brown and Thrush, 1967; Kanofsky *et al.*, 1974; Herbrechtsmeir and

TABLE IV. Average Vibrational Energies of the CO Formed in $O(^3P)$ + Alkyne and Allene Reactions[a]

Reactant	Intermediate	Product	E_a	$-\Delta H^0$	$\langle V' \rangle$ expt.	$\langle V' \rangle$ calc.[k]
$HC{\equiv}CH$	CH_2CO	$CH_2 + CO$	3.0[b]	47	~5.8[g]	6.92
$CH_3C{\equiv}CH$	CH_3CHCO	$CH_3CH + CO$	2.0[c]	50	2.3 ± 0.3[h]	2.20
$CF_3C{\equiv}CH$	CF_3CHCO	$CF_3CH + CO$		~50	2.0 ± 0.3[i]	~2.2
$CH_3CH_2C{\equiv}CH$	C_2H_5CHCO	$C_2H_5CH + CO$	1.6[d]	50	0.97 ± 0.04[j]	1.18
$CH_3C{\equiv}CCH_3$	$(CH_3)_2CCO$	$(CH_3)_2C + CO$	1.8[e]	49	1.14 ± 0.07[j]	1.18
$CH_2{=}C{=}CH_2$	$\Delta{=}O$	$C_2H_4 + CO$	1.6[f]	119	6.8 ± 0.6[h]	6.79

[a] Energetics are in units of kcal/mol.
[b] Herron and Huie (1974).
[c] Herbrechtsmeier and Wagner (1974); Arrington and Cox (1975).
[d] Herbrechtsmeier and Wagner (1975a).
[e] Herbrechtsmeier and Wagner (1975b).
[f] Herbrechtsmeier and Wagner (1972).
[g] Unpublished work (Shaub *et al.*).
[h] Lin *et al.* (1976); Unstead *et al.* (1977).
[i] Hsu *et al.* (1978c).
[j] Unstead and Lin (1977).
[k] Computed according to simple statistical models of Lin *et al.* (1976).

Wagner, 1974; Arrington and Cox, 1975), as has been proposed on the basis of the $O + C_2H_2$ mechanism shown above, then the available energy for CO excitation will be considerably less. This is because the isomerization energy for CH_3CH (ethylidene) $\rightarrow C_2H_4$ (ethylene), 68 kcal/mol, released after the separation of CO and CH_3CH, is obviously not available to the CO (Lin *et al.*, 1976); *viz.*,

$$\mathrm{O(^3\mathit{P}) + CH_3C_2H \rightarrow [CH_3{-}\dot{C}{=}\overset{\overset{\cdot O}{|}}{C}H]^{*\dagger} \rightarrow CH_3CH{=}C{=}O^{\dagger}}$$
$$\mathrm{\rightarrow CH_3CH^{\dagger} + CO^{\dagger} + 50\ kcal/mol}$$
$$\mathrm{\rightarrow C_2H_4^{\dagger} + CO^{\dagger} + 118\ kcal/mol.}$$

The measured CO vibrational energy distribution, shown in Fig. 10, indeed agrees closely with that predicted statistically using $\Delta H^0 = -50$ kcal/mol, instead of the full exoergicity, 118 kcal/mol. Interestingly, the CO formed in the isomeric reaction, $O(^3P) + CH_2{=}C{=}CH_2$ (allene), which is just as exoergic ($\Delta H^0 = -119$ kcal/mol), has a distribution that is considerably hotter and agrees with the statistical one using the full amount of reaction energy (see Fig. 10). In this case, the known mechanism (Herbrechtsmeier and Wanger, 1972; Havel, 1974) indicates that C_2H_4 is formed predominantly via the following mechanism:

$$\mathrm{O(^3\mathit{P}) + CH_2{=}C{=}CH_2 \rightarrow [\dot{C}H_2{-}\overset{\overset{\cdot O}{|}}{C}{=}CH_2 \leftrightarrow \underset{\cdot}{C}H_2{-}\overset{\overset{O}{\|}}{C}{-}\underset{\cdot}{C}H_2]^{*\dagger}}$$
$$\mathrm{\rightarrow \overset{\overset{O^{\dagger}}{\|}}{\Delta} \rightarrow C_2H_4^{\dagger} + CO^{\dagger}}$$
$$\Delta H^0 = -119\ \mathrm{kcal/mol.}$$

The mass spectrometric results for the O + alkynes summarized in Table III have been verified by our CO laser absorption measurements. In Table IV, the average CO vibrational energies, $\langle V' \rangle = \Sigma N_v E_v / \Sigma N_v$ (where E_v is the vibrational energy of CO at the vth level excluding the zero-point energy) measured for the reactions are generally in good agreement with those predicted by the statistical model assuming the reactions to proceed via ketene intermediates that subsequently dissociate into CO and the corresponding biradicals.

The reaction of $O(^3P)$ with C_3O_2 (carbon suboxide) is analogous to O + allene ($O{=}C{=}C{=}C{=}O$, cf. $H_2{=}C{=}C{=}C{=}H_2$). In the $O + C_3O_2$ reaction, the $O(^3P)$ atom attacks predominantly on the centre C atom, as in the allene case (Liuti *et al.*, 1967; Williamson and Bayes, 1967; Pilz and

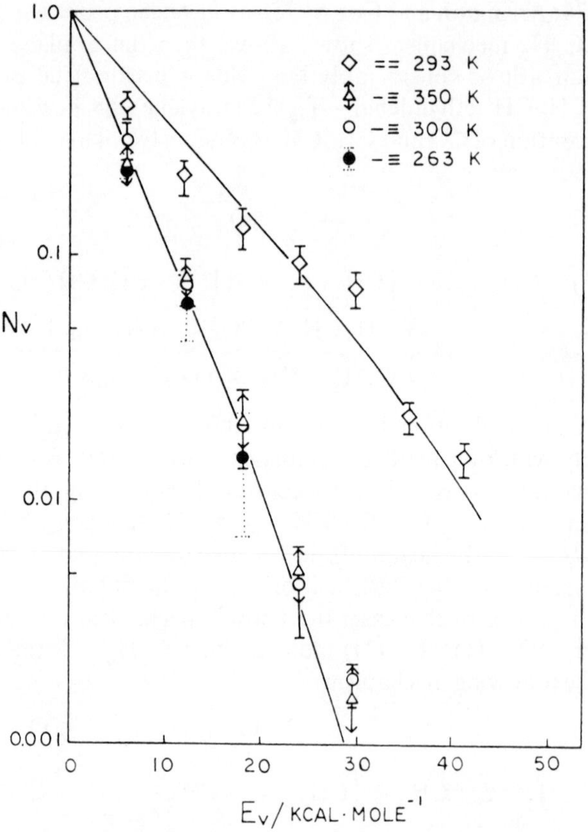

Fig. 10. Observed and calculated vibrational population distributions for the $CO^{\dagger}$ produced from $O(^3P)$ + allene and $O(^3P)$ + methylacetylene reactions at different temperatures. Solid curves—computed statistical distributions based on the mechanisms discussed in the text. (After Umstead *et al.*, 1977.)

Wagner, 1974)

$$O(^3P) + C_3O_2(X^1\Sigma_g^+) \rightarrow [O{=}\dot{C}{-}\overset{\cdot O}{\overset{|}{C}}{=}C{=}O \leftrightarrow$$

$$O{=}\underset{\cdot}{C}{-}\overset{O}{\overset{\|}{C}}{-}\underset{\cdot}{C}{=}O]^{*\dagger} \rightarrow 3CO(X^1\Sigma^+)$$

$$\Delta H^0 = -\,115\ \text{kcal/mol.}$$

The reaction, which is apparently spin forbidden as is the allene reaction, occurs rather rapidly. The spin-conserved process producing $CO_2 + C_2O$, however, has been shown recently to be negligible, i.e. $\leq 0.4\%$ (Pilz and Wagner, 1974). The dynamics of CO production from this reaction have been studied by Hsu and Lin (1978b). From the results of $O(^3P) + C_3O_2$ and $^{18}O(^3P) + C_3O_2$ experiments, it was concluded that the CO resulting from the reaction of $O(^3P)$ with the centre C atom was hotter and carried about 15 kcal/mol of vibrational energy, whereas each of the two CO's deriving from the two C=O bonds initially present in the C_3O_2 carried only about 4 kcal/mol of vibrational energy. This result implies that the reaction sequence shown above happens so fast that the total amount of available energy in the reaction has not been statistically distributed among all degrees of freedom in the C_3O_3 complex prior to its decomposition. The total amount of the vibrational energy possessed by the three CO's (19% of E_{tot}) however, is nearly the same as the amount expected statistically (Hsu and Lin, 1978b). This reaction is unique in the way that three identical molecular species are formed simultaneously in one reaction.

III. RELAXATION AND REACTIONS OF $O(^1D_2)$ ATOMS

A. Quenching by Rare-Gas Atoms

The relaxation of $O(^1D)$ atoms by rare gases has been investigated by a number of investigators employing different diagnostic methods (Preston and Cvetanović, 1966; Donovan *et al.*, 1970; Yamazaki, 1970; Castellano and Schumacher, 1972; Davidson *et al.*, 1978). The first direct, complete measurement of the deactivation reactions by the whole series of rare-gas atoms (He, Ne, Ar, Kr, and Xe) was made by Heidner and Husain (1974) employing time-resolved resonance absorption at 115.2 nm $[O(3^1D_2^0 \rightarrow 2^1D_2)]$. They found that the efficiency of deactivation by these gases decreases rapidly from Xe to He. Xe has a near gas-kinetic quenching cross section, whereas He requires $> 10^5$ collisions to effectively relax the $O(^1D)$ atom. Earlier conflicting results for lighter rare gases were attributed to experimental errors caused by diffusion (Heidner and Husain, 1974). The absolute rate constants for these quenching reactions are summarized in Table V, together with those for simple reactions involving $O(^1D)$ and $O(^1S)$ atoms.

The high quenching efficiency of Xe was ascribed to the formation of a chemical bond (XeO) and strong spin–orbit interaction (due to its large mass) (Husain and Kirsch, 1971). This interpretation was based on semi-quantitative potential energy curves derived partly from the spectroscopic data of XeO obtained by Cooper *et al.* (1961). Recently more spectroscopic data (including lifetimes) were made available (Golde and Thrush, 1974; Lorents and Huestis, 1975; Goodman *et al.*, 1977) which should aid considerably the theoretical understanding of rare-gas oxides (RgO). Theoretical calculations on various states of ArO arising from the $Ar + O(^3P, {}^1D, {}^1S)$

TABLE V. Rate Constants for Some Simple Reactions Involving O(^1D) and O(^1S) Atoms at 298 °K.[a]

Reactant	$k \times 10^{10}$ cm^3/molecule·s		
	O(^1D)		O(^1S)
He	$< 10^{-6}$		$\sim 7 \times 10^{-10}$
Ne	$5 \pm 2 \times 10^{-5}$		$3.6 \pm 1.0 \times 10^{-9}$
Ar	$3 \pm 2 \times 10^{-3}$	$(5.0 \pm 1.5 \times 10^{-3})^b$	$4.8 \pm 1.0 \times 10^{-8}$
Kr	$6.6 \pm 1 \times 10^{-2}$	$(6.4 \pm 1.3 \times 10^{-2})^b$	$2.0 \pm 0.5 \times 10^{-7}$
Xe	0.72 ± 0.14	$(0.72 \pm 0.14)^b$	$2.5 \pm 1.0 \times 10^{-5}$
O_2	0.37 ± 0.03		$2.8 \pm 0.8 \times 10^{-3}$
NO	0.4 ± 0.1		5.7 ± 0.6
N_2	0.28 ± 0.03		5×10^{-7}
CO	0.36 ± 0.05		$9.4 \pm 1.9 \times 10^{-4}$
CO_2	1.0 ± 0.2		$3.6 \pm 0.4 \times 10^{-3}$
COS	1.5 ± 0.2		
N_2O	1.2 ± 0.15		$9.4 \pm 1.9 \times 10^{-2}$
H_2	1.25 ± 0.25		$2.6 \times 10^{-6} \pm 100\%$
H_2O	2.0 ± 0.3		$5.0 \pm 100\%$
CH_4	1.5 ± 0.3	1.4^c	$2.7 \times 10^{-4} \pm 100\%$
CF_3H	0.54 ± 0.11		
CF_3Cl	1.3 ± 0.3		
CF_2Cl_2	2.6 ± 0.5	1.5^c	
$CFCl_3$	3.0 ± 0.3	2.2^c	
CCl_4	4.7 ± 0.9	3.1^c	

[a] All data summarized here are the recommended values by Schofield (1978) except those noted below.
[b] Davidson *et al.* (1978a).
[c] Davidson *et al.* (1978b).

separated-atom limits have been made by Julienne *et al.* (1976), with particular emphasis on the collision-induced O($^1S_0$)–O($^1D_2$) emission near 558 nm. Their calculations, based on the multiconfiguration, self-consistent-field technique (MCSCF) of Das and Wahl (1972), showed that all of the states were unbound. More recently, Dunning and Hay (1977) carried out more extensive *ab initio* CI calculations on various electronic states of RgO derived from the interaction of O(^3P, ^1D, ^1S) with Ne, Ar, Kr, and Xe. The resulting potential energy curves for the electronics states of these oxides are shown in Fig. 11. With the exception of the $1^1\Sigma^+$ curves of XeO and KrO, Dunning and Hay found that all other potential energy curves were repulsive. The binding energies of the $1^1\Sigma^+$ states were reported to increase in the order: NeO (unbound), ArO (essentially flat), KrO ($D_e = 5.8$ kcal/mol),

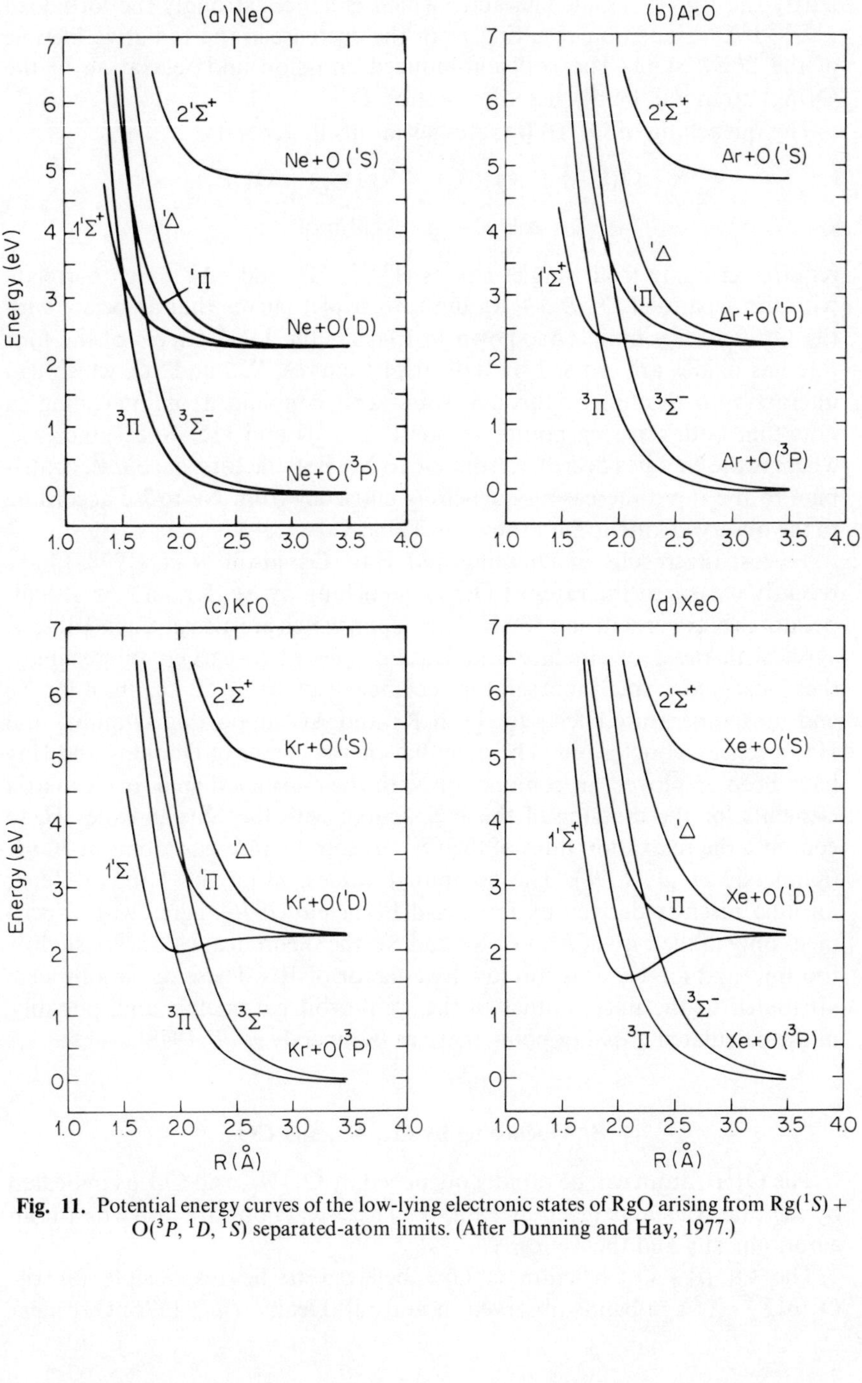

Fig. 11. Potential energy curves of the low-lying electronic states of RgO arising from $Rg(^1S)$ + $O(^3P, ^1D, ^1S)$ separated-atom limits. (After Dunning and Hay, 1977.)

and XeO ($D_e = 16.1$ kcal/mol). The $1^1\Sigma^+$ state was also found to contain partly the Rg^+O^- ionic character which enhances strongly the forbidden $2^1\Sigma^+-1^1\Sigma^-$ transition resulting from the decrease in the radiative lifetime of the $2^1\Sigma^+$ state. The collision-induced emission and relaxation of the $O(^1S_0)$ atom will be discussed in Section IV.

The quenching of $O(^1D)$ by rare-gas atoms in general:

$$O(^1D_2) + Rg(^1S_0) \rightarrow Rg(^1S_0) + O(^3P_J)$$
$$\Delta E = -45.4 \text{ kcal/mol}$$

requires crossing from singlet curves ($1^1\Sigma^+$, $^1\Pi$, and $^1\Delta$), which correlate with the separated $O(^1D_2) + Rg$ limit, to triplet curves that associate with the $O(^2P_J) + Rg$ limits. As shown in Fig. 11, the $1^1\Sigma^+$ curves of the four rare-gas oxides are crossed by both triplet curves, $^3\Sigma^-$ and $^3\Pi$, which can interact with the former through spin–orbit coupling. It is interesting to note that both crossing points (i.e. for $1^1\Sigma^+-^3\Pi$ and $1^1\Sigma^+-^3\Sigma^-$) increase, whereas the energies decrease, from Ne to Xe. Both factors, of course, contribute to the rapid increasing quenching efficiency from Ne to Xe according to the observed kinetic data shown in Table V.

To test the results of Dunning and Hay, Davidson *et al.* (1978a) have recently measured the rates of $O(^1D)$ quenching by Xe, Kr, and Ar at temperatures between 110 and 330 K. Their data, which are also given in Table V, agree with those of Heidner and Husain referred to earlier. Interestingly, they observed a small temperature coefficient of 103 ± 52 kcal/mol for Xe and no temperature effects for both Kr and Ar, supporting Dunning and Hay's general conclusions. The potential energy curves of Dunning and Hay have been employed, in conjunction with the calculated spin–orbit matrix elements for the crossing of the $1^1\Sigma^+$ curve with the $^3\Sigma^-$ and the $^3\Pi$, to compute the relaxation rates of the $O(^1D)$ atom by rare-gas atoms at 300 K (Kinnersly *et al.*, 1978). The computed values, using the Landau–Zener formula recently derived by Faist and Bernstein (1976), agree with experiment only in the case of Ar. For Kr and Xe, the theoretical values are slightly too low, and for Ne, it is too low by a factor of 10^4. These deviations were attributed to the uncertainties in the spin–orbit parameters and, possibly, in the calculated crossing-point energies (Kinnersly *et al.*, 1978).

B. Quenching by O_2, N_2, and CO

The $O(^1D)$ atom can be rapidly quenched by O_2, N_2, and CO, as indicated by the rate constants given in Table V. These reactions are important both aeronomically and theoretically.

The $O(^1D) + O_2$ reaction is now believed to be responsible for the $O_2(b^1\Sigma_g^+-X^3\Sigma_g^-)$ bands observed in auroral (Deans *et al.*, 1976; Gattinger and Jones, 1976), as well as in the twilight airglow and the day glow, near and above 100 km of the earth's atmosphere (Wallace and Hunten, 1968; Noxon, 1975). Although the overall quenching rate constant for the $O(^1D) + O_2$ reaction is now well established (see Table V), the extent of $O_2(b^1\Sigma_g^+)$ production from the reaction,

$$O(^1D) + O_2(X^3\Sigma_g^-) \rightarrow O(^3P) + O_2(b^1\Sigma_g^+)$$
$$\Delta H^0 = -7.9 \text{ kcal/mol}$$

has been a subject of much controversy in the past several years (Young and Black, 1967; Izod and Wayne, 1968; Jones *et al.*, 1970; Noxon, 1970; Snelling and Gauthier, 1971; Giachardi and Wayne, 1972; Snelling, 1974). Most recently, Lee and Slanger (1978) measured the production of $O_2(b^1\Sigma_g^+)$ from the above reaction by means of time-resolved chemiluminescence from the $v = 1$ and 0 levels of the b state following the photodissociation of O_2 at 160 nm using an H_2 laser. The quantum efficiency of converting $O(^1D)$ electronic energy into $O_2(b^1\Sigma_g^+)$ was found to be 0.77 ± 0.2, which favours the observation of Snelling (Snelling and Gauthier, 1971; Snelling, 1974). In this reaction the relative population of $O_2(b^1\Sigma^+)$ at $v = 1$ and 0 was determined to be $N_1/N_0 = 0.7$, which is slightly higher than the value, 0.5, predicted by the phase space theory (Light, 1967). The production of $O_2(a^1\Delta_g)$ from $O(^1D) + O_2(X^3\Sigma_g^-)$, though also spin-allowed, is much less important. Its rate constant has been estimated to be only as much as $\sim 5 \times 10^{-12}$ cm^3/molecule·s (Gauthier and Snelling, 1971).

The facility of the quenching of $O(^1D)$ by N_2 and CO,

$$O(^1D) + N_2(X^1\Sigma_g^+) \rightarrow O(^3P) + N_2(X^1\Sigma_g^+, v' \geq 0)$$
$$O(^1D) + CO(X^1\Sigma^+) \rightarrow O(^3P) + CO(X^1\Sigma^+, v' \geq 0)$$

although in apparent violation of spin conservation, is believed to result from the formation of the long-lived intermediates, N_2O and CO_2 (Donovan and Husain, 1970; Fisher and Bauer, 1972; Tully, 1974; Zahr *et al.*, 1975; Shortridge and Lin, 1976). For the N_2 reaction, Fisher and Bauer (1972) and Delos (1973) employed an atom + atom-like model to describe the dynamics of the singlet–triplet crossing and obtained a much smaller overall quenching cross section. The extent of N_2 vibrational excitation predicted by the model ($\sim 5\%$ of the total 45.4 kcal/mol) is also much less than the experimentally determined value, $\sim 33 \pm 10\%$ (Slanger and Black, 1974). In these calculations, the potential energy surface interaction is assumed to occur twice—once when the reactants approach and once again when they depart. This model, as pointed out independently by Tully (1974) and Zahr *et al.* (1975), is inaccurate because the potential energy surface interactions can occur more than twice in the course of a collision due to the formation of a complex which has several vibrational degrees of freedom.

Tully's model employs the Landau–Zener approximation (see, for example, Faist and Bernstein, 1976) to estimate the probability of surface crossing, and the RRKM theory (Robinson and Holbrook, 1972; Forst, 1973) to calculate the rate of the complex decomposition (e.g. N_2O, in the case of $O(^1D) + N_2$). The complex was formed by an attractive potential,—Cr^{-6}. Miller and coworkers (Zahr *et al.*, 1975), however, used a different approach. They carried out classical trajectory calculations based on crude, but qualitatively reasonable, potential energy surfaces for $O(^1D) + N_2$. Their calculations indeed showed that the cross section for complex formation was quite appreciable (~ 40 Å^2) and that 'surface hopping' (Tully and Preston, 1971) can occur fairly often within the lifetime of the N_2O complex. Interestingly, the predicted overall quenching rates ($2\text{–}7 \times 10^{-11}$ cm^3/molecule·s) and the extent of N_2 vibrational excitation (20–30%) in the $O(^1D) + N_2$ reaction by these two different models agree closely with experimental findings. Further and more concrete results have been obtained from the $O(^1D) + CO$ experiments described below.

Since CO and N_2 are isoelectronic, the theoretical models described above may also be applied to the $O(^1D) + CO$ reaction, which can be more readily and more reliably studied by means of the CO laser-probing method discussed in the previous section. The CO laser resonance absorption measurements (Lin and Shortridge, 1974; Shortridge and Lin, 1976) showed that the CO formed in the $O(^1D) + CO$ quenching reaction was vibrationally excited up to $v = 7$ (corresponding to the limit of available electronic energy) with a Boltzmann vibrational temperature of 8000 K. The surprisal analysis of the observed and the predicted statistical distributions is given in Fig. 12. These results strongly support the complex formation mechanism:

$$O(^1D) + CO(X^1\Sigma^+) \rightarrow {}^1CO_2^* \rightarrow {}^3CO_2^* \rightarrow O(^3P) + CO(X^1\Sigma^+, v' \leq 7).$$

Here ${}^1CO_2^*$ and ${}^3CO_2^*$ represent electronically excited singlet and triplet CO_2 molecules, respectively. However, one of the singlet states may be the ground electronic state. There are several low-lying singlet and triplet states known to be present in the vicinity of the $O(^3P) + CO(X^1\Sigma^+)$ limit (Winter *et al.*, 1973). It is quite possible that all these low-lying states may be initially involved in the reaction prior to a successful crossing over to the repulsive triplet states which correlate with $O(^3P_J) + CO(X^1\Sigma^+)$. However, if the lifetime of the ${}^1CO_2^*$ is as long as a period of several vibrations, then it is probably immaterial which initial states are involved. This explains in part the good agreement between the observed and calculated rate constants based on Tully's simplistic statistical model for several $O(^1D)$-atom reactions important to atmospheric chemistry (Tully, 1975). For the $O(^1D) + CO$ reaction, the calculated total quenching rate, 8.0×10^{-11} cm^3/molecule·s compares reasonably with the experimental value shown in Table V.

The complex-formation mechanism given above for $O(^1D) + CO$ is also

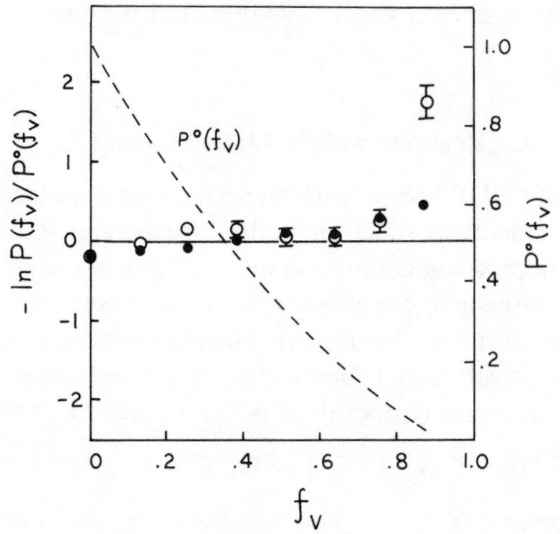

Fig. 12. Surprisal analysis of the $CO^{\dagger}$ distribution resulting from the relaxation of $O(^1D)$ by $CO(X^1\Sigma^+, v = 0)$ at 300 K. Open circles—experimental data of Shortridge and Lin (1976). Filled circles—the statistically estimated $CO^{\dagger}$ distribution based on the model described by Lin *et al.* (1976).

demonstrated by the observation that the absolute rate of $C^{16}O$ formation in the following ^{18}O-labelled experiment:

$$^{16}O(^1D) + C^{18}O \rightarrow C^{16}O + {}^{18}O(^3P)$$

is 50% of that measured in the $^{16}O(^1D) + C^{16}O$ reaction. Quite clearly, the $^{16}OC^{18}O$ complex is involved in the reaction which produces $C^{16}O$ and $C^{18}O$ with equal probabilities. Since the complete CO vibrational energy distribution ($7 \geq v \geq 0$) was obtained in this experiment, it was possible to determine the $E \rightarrow V$ energy transfer efficiency accurately. The transfer efficiency was found to vary from $17 \pm 1\%$ at 246 K to $22.5 \pm 0.8\%$ at 323 K with a slight temperature (translational energy) effect. This effect was also predicted by Zahr *et al.* (1975) for the $O(^1D) + N_2$ reaction. The observed $E \rightarrow V$ efficiency at room temperature, $21 \pm 0.5\%$, is in good accord with the theoretical value of $\sim 20\%$ by Zahr *et al.* and $30 \pm 10\%$ by Tully as mentioned earlier for $O(1D) + N_2$. Our value, however, is a factor of two lower than that of Slanger and Black (1974), $40 \pm 10\%$, based on $CO(A^1\Pi - X^1\Sigma^+)$ resonance fluorescence measurements. In these measurements, however, the $CO(v = 0)$ formed in the $O(^1D) + CO$ reaction, which was found to be as much as 33% of the total CO product population based on our extrapolation, could not be measured directly. Because of this, their $O(^1D) + N_2$ energy transfer efficiency, which was estimated to be $83 \pm 10\%$

of that for CO (Slanger and Black, 1974), would become ~ 17% if our CO value is used.

C. Reaction with N_2O, COS, and CO_2

The reaction of $O(^1D)$ atoms with these triatomic molecules also occurs very rapidly via chemical reaction and/or quenching (see Table V). The relative importance of quenching versus reaction, however, has not been quantitatively established experimentally from these reactions. For the $O(^1D) + CO_2$ reaction, a theoretical calculation made by Tully (1975) based on his statistical model mentioned earlier indicated that the major reaction channel at room temperature is the quenching of the $O(^1D)$ atom:

$$O(^1D) + CO_2(X^1\Sigma_g^+) \rightarrow O(^3P) + CO_2(X^1\Sigma_g^+)$$

At higher temperatures ($T > 1000$ K), however, the chemical reaction channel:

$$O(^1D) + CO_2(X^1\Sigma_g^+) \rightarrow CO(X^1\Sigma^+) + O_2(a^1\Delta_g)$$

becomes more important. The ratio of quenching vs. reaction at 300 K was computed to be $\sim 1.6 \times 10^{-10}/8 \times 10^{-11} = 2$. A recent attempt to detect CO from this reaction using CO laser resonance absorption, however, failed to observe any measurable amount of CO (Shortridge and Lin, 1975), implying that the rate constant for the above reactive channel is probably not more than 10^{-13} cm^3/molecule·s. It should be noted that, in this particular case, the accuracy of the predicted relative rate depends strongly on the position of the singlet–triplet crossing points on the CO_3 potential energy surfaces, which are still not well characterized theoretically and experimentally. The quenching reaction probably occurs exclusively via a long-lived CO_3 complex (that has a lifetime of more than several vibrations). The results of isotopically labelled experiments indicated that the exchange of $^{16}O(^1D)$ with ^{18}O in $C^{18,18}O_2$ can be accounted for by the following stoichiometry (Yamazaki and Cvetanović, 1964a; Baulch and Breckenridge, 1966):

$$^{16}O(^1D) + C^{18,18}O_2 \rightarrow \tfrac{1}{3}\,^{16}O(^3P) + \tfrac{2}{3}\,^{18}O(^3P) + \tfrac{1}{3}C^{16,18}O_2 + \tfrac{2}{3}C^{18,18}O_2$$

The reaction of $O(^1D)$ with N_2O probably takes place predominantly via the following two reactive channels (Donovan and Husain, 1970; Tully, 1975):

$$O(^1D) + N_2O \rightarrow 2NO, \Delta H^0 = -81 \text{ kcal/mol}$$

$$\rightarrow N_2 + O_2, \Delta H^0 = -125 \text{ kcal/mol},$$

with both channels having about equal importance. The dynamics of NO production from this reaction has been investigated by Boxall *et al.* (1972) using a flash photolysis technique. N_2O was photodissociated at wavelengths near 200 nm to generate the $O(^1D)$ atoms, and the NO formed in the subsequent rapid $O(^1D) + N_2O$ reaction was detected by the absorption in the $\gamma(X^2\Pi \rightarrow A^2\Sigma^+)$ band employing a second spectro-flash. The NO formed in the reaction was found to be excited to $v = 2$ with a characteristic vibrational temperature of ~ 5200 K. The results of an isotope-labelled experiment (using ^{15}NNO) indicate that each NO carries vibrationally $\sim 10\%$ of the total available energy (81 kcal/mol), with a major fraction of the energy going into translation. These results, aided by those obtained from an approximate SCF MO calculation suggest that the reaction probably takes place via an ONNO complex (Boxall *et al.*, 1972). The extent of NO vibrational excitation observed in the reaction, however, is much less than that expected statistically.

The $O(^1D) + COS$ reaction is dynamically very interesting. It may proceed via the following three possibly paths if spin is conserved in the reaction:

$$O(^1D) + COS \rightarrow CO + SO(a^1\Delta), \Delta H^0 = -79 \text{ kcal/mol}$$

$$\rightarrow CO + SO(b^1\Sigma^+), \Delta H^0 = -67 \text{ kcal/mol}$$

$$\rightarrow CO_2 + S(^1D), \Delta H^0 = -73 \text{ kcal/mol.}$$

This reaction has been studied by Jones and Taube (1973) in low-temperature Ar and Xe matrices. They showed that both CO + SO and CO_2 + S products were formed in the reaction. Since no CO_2S intermediate was detected in their i.r. absorption measurements, the CO_2S intermediate, if it was present at all, is probably short-lived. Their data obtained from an isotope-labelled experiment using ^{18}OCS indicate that the CO formed in the reaction was derived exclusively from the parent COS molecule with no evidence of exchange in the oxygen atoms.

The vibrational energy distribution of the CO produced in the reaction has been measured by Shortridge and Lin (1975) using the CO laser resonance absorption method. The CO was found to be vibrationally excited to $v = 9$ with an unexpectedly cold vibrational temperature of 3300 K. The average CO vibrational energy was calculated to be only 4 or 5% of the available energies, depending on the electronic state of the SO radical ($^1\Delta$ or $^1\Sigma^+$) formed in the reaction. This is rather surprising in view of the fact that the CO and CS formed in the following respective reactions:

$$O(^3P) + COS \rightarrow CO + SO, \Delta H^0 = -52 \text{ kcal/mol}$$

$$O(^3P) + CS_2 \rightarrow CS + SO, \Delta H^0 = -21 \text{ kcal/mol}$$

carry proportionally more average vibrational energy ($\sim 9\%$ in each case as indicated previously). Thus the newly formed SO bond in the $O(^1D) +$ COS reaction could probably carry proportionally more internal energy. Undoubtedly, more experimental as well as theoretical work is required in order to fully understand the dynamics of these three very fast reactions.

D. Reaction with Hydrides

The reaction of $O(^1D)$ atoms with hydrogen-containing molecules, particularly H_2, HCl, H_2O, and CH_4, is atmospherically important. These reactions, which occur mainly via reactive channels with near gas-kinetic rates (see Table V), were first shown by Norrish and coworkers (McGrath and Norrish, 1960; Basco and Norrish, 1961) to generate vibrationally excited OH radicals. For these small molecules, however, it is still not certain whether the OH radical is formed by direct H-atom abstraction or by the decomposition of vibrationally excited insertion products, $ROH^\dagger$ (where R = H, Cl, OH, and CH_3 for the reactions mentioned above). The lifetimes of these small insertion products are too short to allow them to be collisionally deactivated for final product analysis even in condensed phases. For the $O(^1D) + H_2$ reaction, for example, H_2O was observed in an experiment carried out in liquid Ar at 87 K (DeMore, 1967). However, it is difficult to distinguish whether it was the deactivated insertion product or the product of the recombination reaction of H and OH, which may be derived from both mechanisms, within the liquid Ar cages. The results of recent theoretical calculations (Sorbie and Murrell, 1976; Whitlock *et al.*, 1976) indicated that the $H_2O^\dagger$ formed by insertion has a lifetime as short as 5×10^{-14} s, a time span too short to allow successful experimental measurements. Therefore, any conclusion drawn from an analysis based on a presumed mechanism is not reliable. For larger alkanes such as neopentane, however, it is possible to at least partially deactivate the excited insertion product (neopentyl alcohol) under normal experimental conditions because of its longer lifetime (4×10^{-9} s). In this case, the relative importance of abstraction versus insertion can be reliably estimated (Paraskevopoulos and Cvetanović, 1970). In this subsection, we shall start our discussion of $O(^1D)$-atom reactions with small molecules and then go on to describe the dynamics of the decomposition of chemically activated fluoroalcohols formed by insertion reactions.

Recently, there has been much theoretical interest in the reaction of $O(^1D)$ with H_2 (Tully, 1975; Sorbie and Murrell, 1976; Whitlock *et al.*, 1976), stemming partly from its unique deviation from the much-studied hydrogen halide systems. The results of different theoretical studies have been discussed in some detail by Tully elsewhere in this volume. Hence, we will review briefly only the part of the results relevant to experiments recently carried out in this Laboratory (Smith *et al.*, 1978). As indicated in the correlation diagram shown in Fig. 13(a) of Tully's review, only two of the five singlet states arising from $O(^1D_2) + H_2(X^1\Sigma_g^+)$ correlate with $H(^2S) + OH(X^2\Pi)$, and only the lowest $^1A'$ state corresponds to the ground electronic state of H_2O. Tully (1975) employed the statistical RRKM treatment discussed earlier to compute the overall rate assuming that the reaction takes place via the two singlet surfaces that correlate with H + OH. The calculated rate constant, 2.87×10^{-10} cm^3/molecule·s, agrees closely with the observed values given in Table V. In this calculation, however, detailed product energy partitioning data were not generated. More recently, Sorbie and Murrell (1976) and Whitlock *et al.* (1976) have carried out classical trajectory calculations based on different potential energy surfaces. The former employed an empirical surface assuming the reaction to proceed solely by insertion. The direct abstraction process was ignored because of its large energy barrier (similar to Gangi and Bader's conclusion (1971)). The calculation of Whitlock *et al.* was based on the energy surfaces derived from the DIM method (see the review by Tully). Their trajectory calculations revealed that ~25% of the $O(^1D) + H_2$ reaction occurs by abstraction and ~ 75% by insertion. The abstraction reaction generates OH with negative rotational and vibrational temperatures (N_v peaking at $v = 2$), whereas the insertion process gives rise to near statistical distributions. Sorbie and Murrell's total-insertion model, however, leads to non-statistical vibrational energy distributions.

Experimentally, we have studied the partitioning of the reaction energy (44 kcal/mol) in the overall reaction,

$$O(^1D) + H_2(X^1\Sigma_g^+) \rightarrow OH(X^2\Pi) + H(^2S)$$

employing two pulsed lasers [Smith *et al.*, 1978]. The first laser (quadrupled Nd:YAG laser operating at 266 nm, FWHM ~ 100 ns) was used to dissociate O_3 to generate the $O(^1D)$ atom and the second laser (Chromatix CMX-4 flashlamp-pumped tunable dye laser operating near 300 nm, FWHM ~ 1 μs) was used to probe the OH radical formed in the subsequent $O(^1D) + H_2$ reaction. By varying the reaction pressure and the interval between the two laser pulses, it was possible to measure unrelaxed rotational energy distributions at different vibrational levels. The observed $OH(v = 0)$ rotational energy distribution is shown in Fig. 13. Although the distribution is nearly statistical, the populations at lower ($J < 10$) and higher ($J > 20$) rotational levels deviate noticeably from statistical values. The former is colder, whereas the latter, hotter. Further work on this and other simple reactions mentioned above is still underway.

The mechanism of the $O(^1D) + H_2O \rightarrow 2OH$ reaction is not well understood. It is quite likely that both abstraction and insertion mechanisms are important. In a brief study of the $^{16}O(^1D) + H_2{}^{18}O$ reaction employing the flash photolysis method, Engleman [1965] concluded that the newly formed ^{16}O—H bond carried preponderantly more vibrational energy than the old ^{18}O—H bond. The experiment, however, was carried out at a relatively

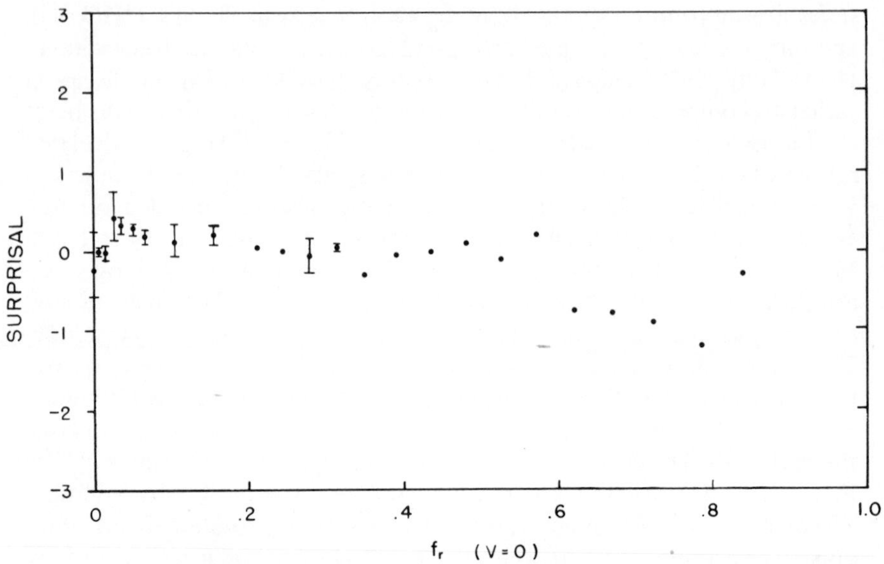

Fig. 13. Surprisal analysis of the rotational energy distribution of the $OH(v = 0)$ produced from the $O(^1D) + H_2$ reaction. (After Smith *et al.*, 1978.)

high pressure (10 torr) and at a long observation time (15 μs); more refined data are required before one can draw a definite conclusion regarding the dynamics of the reaction. The total reaction rate constant has been calculated by Tully [1975] using his RRKM model. The computed value, 3.58×10^{-10} cm^3/molecule·s, is again in close agreement with the experimental value listed in Table V, further demonstrating the usefulness of his statistical approach for predicting the rates of these complex-forming reactions.

The reaction of $O(^1D)$ with CH_4 and higher alkanes is also believed to occur by both abstraction and insertion mechanisms:

$$O(^1D) + RH \rightarrow OH + R$$
$$\rightarrow ROH^\dagger \rightarrow OH + R, \text{etc.}$$

where R represents the alkyl group. For larger alkanes, the excited alcohol intermediates, $ROH^\dagger$, also decompose readily into small alkyl radicals by C—C bond scission (Yamazaki and Cvetanović, 1964b; Paraskevopoulos and Cvetanović, 1969a, b, 1970; Lin and DeMore, 1973). Under higher pressure conditions, however, the excited insertion products can be stabilized and measured. In the $O(^1D)$ + neopentane reaction, for example, it was found that ~25% of the reaction took place via abstraction and at least 66% via insertion, based upon the amount of neopentyl alcohol measured (Paraskevopoulos and Cvetanović, 1970). In this case, molecular H_2

elimination forming pivalaldehyde accounted for only 1–2% of the total reaction. The remaining balance (~ 6–7%) perhaps disappeared via other decompositions or secondary reactions producing smaller molecular species (CO, H_2CO, etc.).

The reaction of $O(^1D)$ and CH_4 was found to yield primarily CH_3 and OH radicals as initial products (Basco and Norrish, 1961; DeMore and Raper, 1967; Lin and DeMore, 1973). Additionally, about 9% of molecular products, CH_2O and H_2, were detected in the reaction carried out in liquid Ar at 87 K (DeMore and Raper, 1967). The molecular elimination process, according to Lin and DeMore, takes place by a direct mechanism without involving the insertion product, $CH_3OH^\dagger$. On the basis of a rough RRKM calculation, however, the lifetime of the $CH_3OH^\dagger$ was found to be as short as 10^{-13} s; one therefore cannot confidently rule out $CH_3OH^\dagger$ as a possible precursor of both the $CH_3 + OH$ and the $CH_2O + H_2$ products. Further evidence for the presence of short-lived (non-RRKM) intermediates is also provided by our study of the reaction of $O(^1D)$ with partially halogenated alkanes from which hydrogen halide stimulated emissions were observed (Lin, 1971, 1972; Burks and Lin, 1978).

The our study of the $O(^1D)$ + haloalkane reactions, the insertion process was utilized to generate vibrationally excited α-halogenated alcohols. Since these alcohols are thermally unstable, they rapidly eliminate hydrogen halides yielding carbonyl compounds:

$$O(^1D) + CH_nX_{4-n} \rightarrow [CH_{n-1}X_{4-n}OH]^\dagger$$
$$\rightarrow HX^\dagger + CH_{n-1}X_{3-n}O$$

where X = Cl, F and $n = 1, 2$, or 3. The overall reactions producing HX are very exoergic ($\Delta H^0 \cong -135 \sim 155$ kcal/mol). Accordingly, HX-stimulated emissions were observed in these six and several other chlorofluoromethane reactions (Lin, 1971, 1972). In these reactions, only the insertion mechanism can satisfactorily account for the production of vibrationally excited hydrogen halides. Secondary reactions due to halogen atoms or OH radicals can be readily ruled out according to the results of our recent detailed kinetic modelling (Burks and Lin, 1978).

In order to understand the dynamics of these interesting reactions, we have recently carried out a series of experiments in a grating-tuned laser cavity (Green and Lin, 1971; Berry, 1973; Burks and Lin, 1978). The vibrational energy distribution of the HF formed in the following series of reactions has been measured:

$$O(^1D) + R_FH \rightarrow [R_FOH]^\dagger \rightarrow HF^\dagger + R'_FCFO$$
$$\Delta H^0 \simeq -155 \text{ kcal/mol}$$
$$R_F = CF_3, C_2F_5, C_3F_7, \text{and } C_7F_{15}.$$

From the observed sequence of appearance times of various HF laser lines in a vibration–rotation manifold, we were able to evaluate the inversion ratio, N_v/N_{v-1} (Burks and Lin, 1978). The vibrational population of the HF produced in the $O(^1D) + CF_3H \rightarrow HF^\dagger + CF_2O$ reaction is given in Fig. 14. It is interesting to note that the observed distribution is much hotter than the one expected statistically, probably indicating that the $CF_3OH^\dagger$ intermediate is short-lived and thus the total available reaction energy (~ 155 kcal/mol) has not been effectively randomized before the dissociation takes place. This rapid insertion–elimination process may, in fact, occur concertedly in the following manner:

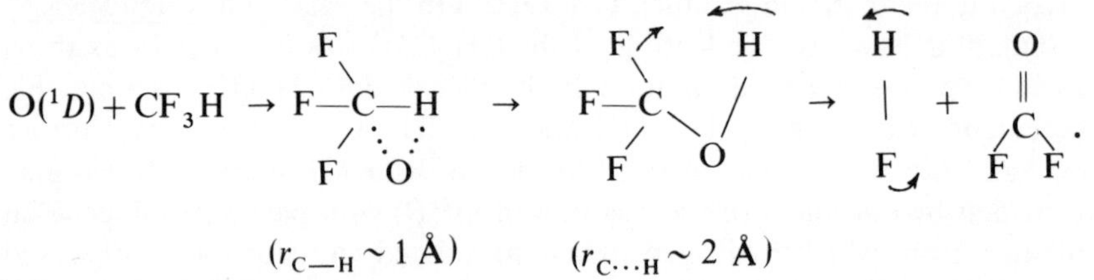

Since the C—H bond distance has been doubled from CF_3H to CF_3OH (based on $r_{C,,H}$ in CH_3OH), the C—O—H bending vibration, aided by the impact of the $O(^1D)$ attack from the side, is expected to be violently excited. This bending vibration, intuitively, coincides with the motion that

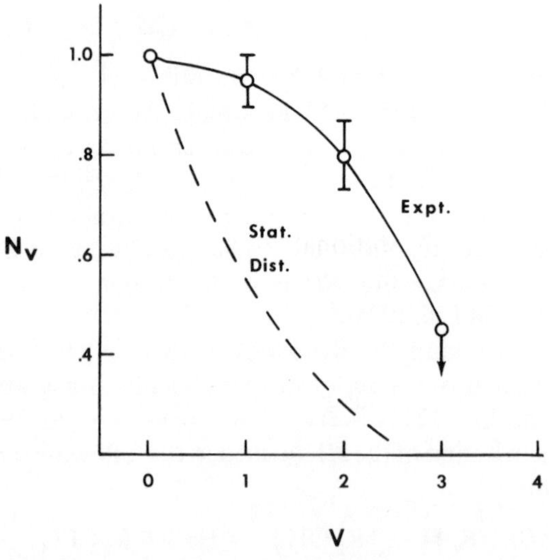

Fig. 14. The vibrational energy distribution of the HF† formed of the $O(^1D) + CF_3H$ reaction. Data taken from Burks and Lin (1978).

would result in the four-centered HF elimination reaction. Similar HF vibrational energy distributions were also observed in the reaction of $O(^1D)$ with CH_3F and CH_2F_2. In order to account for these high HF vibrational energy contents, one has to assume considerably smaller numbers of effective oscillators in our statistical (RRKM) calculations (for example, $s = 3$ instead of 12 for these six-atom systems). This evidently results from incomplete randomization of the available energies in the dissociating complexes.

Fig. 15 shows the N_1/N_0 ratios of the HF from the four homologous reactions, including CF_3H, as a function of the number of vibrational modes of the alcohol intermediates (R_FOH). The results obtained from the reaction of the larger fluoroalkanes again support the conclusion that these chemically activated $R_FOH^\dagger$ intermediates are short-lived.

This picture of the nearly concerted insertion–elimination mechanism described above for the $O(^1D) + R_FH$ reactions can also account for the observation of CH_2O and H_2 in the reaction of $O(^1D)$ with CH_4 in the gas phase as well as in liquid Ar as mentioned previously. It is not difficult to

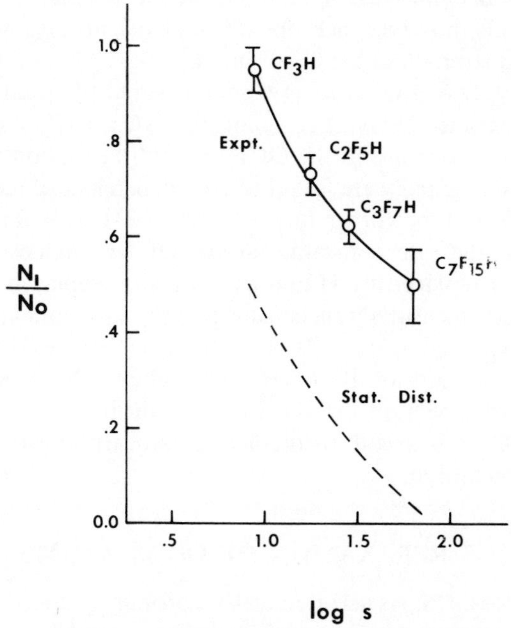

Fig. 15. The relative vibrational population of the $HF(v = 1)$ and the $HF(v = 0)$ produced from the reaction of $O(^1D)$ with a series of fluoroalkanes plotted as a function of the number of vibrational modes of the alcohol intermediates. (T. L. Burks and M. C. Lin, unpublished work).

imagine the rapid, simultaneous occurrence of the reactions

$$[CH_3OH]^\dagger \rightarrow CH_2O + H_2$$
$$\rightarrow CH_3 + OH$$

involving the newly formed, energetically non-randomized $[CH_3OH]^\dagger$ intermediate, within a time-span shorter than the RRKM lifetime ($\leq 10^{-13}$ s as quoted earlier). The same mechanism can also be applied to the reaction of $O(^1D)$ with other larger alkanes.

E. Reaction with Halocarbons

The reaction of chlorofluorocarbons with $O(^1D)$, particularly $CFCl_3$ and CF_2Cl_2, is relevant to the breakdown of man-made halocarbons and the chemistry of ClO_x in the stratosphere (Molina and Rowland, 1975). There is direct experimental evidence that the reaction of $O(^1D)$ with a C—Cl bond can occur by abstraction and possibly by other paths including physical quenching (Jayanty *et al.*, 1975; Gillespie and Donovan, 1976; Donovan *et al.*, 1976; Fletcher and Husain, 1976; Green and Wayne, 1977). Its reaction with C—F bonds, however, perhaps takes place entirely by quenching in view of the great strength of the C—F bond.

Most recently, Davidson *et al.* (1978b) measured the total reaction rates for a series of reactions involving six halocarbons (e.g. CCl_4, $CFCl_3$, $CHFCl_2$, etc.), based on the decrease in the $O(^1D) \rightarrow O(^3P)$ emission at 630 nm. All six reactions investigated were found to be independent of temperature. On the basis of these and the known rate of $O(^1D) + CH_4$ reaction (see Table V), they found that the rate constants for the $O(^1D)$ + halocarbon reactions decrease as Cl is replaced by H and as H or Cl is replaced by F. A least-squares fit of many measured rate constants to the equation: $k(C_nH_aF_bCl_c) = a\,k_H + b\,k_F + c\,k_{Cl}$, led to $k_H = 0.32 \pm 0.02$, $k_F = 0.030 \pm 0.003$, and $k_{Cl} = 0.74 \pm 0.03$, all in units of 10^{-10} cm^3/molecule·s. The given error limits are one standard deviation derived from the fitting of the adopted rates. The above equation is useful for predicting halocarbon rate constants that have not been measured.

IV. RELAXATION OF $O(^1S_0)$ ATOMS

The metastable $O(^1S_0)$ is aeronomically a prominent species. It is responsible for the distinct green emission at 558 nm, due to the $^1S_0 \rightarrow {}^1D_2$ transition, in the air glow and the aurora (Chamberlain, 1961). Recently, it has also been shown to be present in the atmosphere of Mars (Barth *et al.*, 1969, 1972).

Photochemically, it can be readily generated by the dissociation of

O_2, N_2O, and CO_2 in the vacuum u.v. region as shown in Table I (Hampson and Okabe, 1970; Ridley *et al.*, 1973; Black *et al.*, 1975a). The chemistry of the reactions of the $O(^1S)$ atom is not clearly understood, with the exception of some simple relaxation processes by rare-gas atoms and O atoms, for example. Investigation of the $O(^1S)$ chemistry is complicated mainly by the high reactivity of the $O(^1D)$ state, whose reaction usually follows immediately its production by relaxation of the ^1S state and thus confuses observations. In order to clearly separate the original ^1S state processes from subsequent contributions from the ^1D state, studies should be carried out under single-collision conditions.

Recently, the quenching of $O(^1S)$ atoms has been studied by a number of investigators who followed the $^1S_0$–$^1D_2$ emission at 558 nm. These studies were made most prominently by K. H. Welge and coworkers, and R. A. Young, T. G. Slanger, G. Black, and their coworkers. Most published data have been critically evaluated recently by Schofield (1978). Accordingly, we will not recite those numerous references here in this space. Some of the recommended kinetic data are summarized in Table V and compared with those of $O(^1D)$ reactions discussed in the preceding section. With the exception of collisional quenching by NO, NO_2, H_2O, O_3, and unsaturated hydrocarbons, the reaction and/or relaxation (which are not easily separable) of the $O(^1S)$ atom with most simple gases occur rather slowly.

It has been well documented that the forbidden $^1S_0 \rightarrow {}^1D_2$ green emission can be significantly enhanced by heavier rare-gas atoms such as Ar, Kr, and Ar (see for example, Black *et al.*, 1975b). This collision-induced emission has been successfully utilized to generate stimulated emissions from the green bands of XeO and KrO excimers (Powell *et al.*, 1974). The physics of the collision-induced $O(^1S_0$–$^1D_2)$ emission by these rare gases has been investigated by Julienne *et al.* (1976) and Dunning and Hay (1977), as mentioned in Section III. A on the quenching of $O(^1D_2)$ atoms by these gases. The computed emission rate coefficients (or 'pseudo-lifetimes') were found to agree reasonably well with experimental values. For example, the results of Dunning and Hay for the lifetimes of the $2^1\Sigma^+ \rightarrow 1^1\Sigma^+$ transitions for ArO, KrO, and XeO are 23, 4.5, and 0.31 μs, which compare closely with the experimental values, 7–10, 1.6, and 0.17–0.29 μs, respectively (Black *et al.*, 1975b; Hughes *et al.*, 1976; Welge and Atkinson, 1976).

The collisional relaxation of $O(^1S)$ by $O(^3P)$,

$$O(2^1S_0) + O(2^3P_J) \rightarrow O(2^1D_2)$$
$$\Delta E = -5.9 \text{ kcal/mol}$$

has been studied theoretically by Olson (1973). The calculations for the collision-induced spin–orbit coupling between the $^3\Pi_g$ and $^1\Sigma_g^+$ states were carried out using the *ab initio* potential energy curves of O_2 and the Landau–

Zener formula for estimating the curve-crossing probability. The calculated quenching rate at 300 K, 2.9×10^{-12} cm^3/molecule·s agrees reasonably with the experimental value of Felder and Young (1972), 7.5×10^{-12} cm^3/molecule·s, in view of the uncertainties in both the experimental data (which were rather scattered) and the theoretical curves used.

V. CONCLUDING REMARKS

In this chapter, we have described briefly the dynamics of reactions involving the three low-lying electronic states, 2^3P_J, 2^1D_2, and 2^1S_0, of the oxygen atom. Emphasis was placed primarily upon the elucidation of reaction mechanisms in terms of the data obtained from detailed microscopic rate measurements employing laser, chemiluminescence, and molecular beam techniques. Most information available to data on O-atom reactions, as is evidenced by this review, is only partial at best.

Hopefully, this brief account of the present status of O-atom reaction dynamics will stimulate more experimental and theoretical interest in these important species which relate rather intimately in one way or another to our daily life.

Acknowledgement

The author is grateful to many colleagues for permitting him to use their published and unpublished work. He is particularly indebted to Dr. Merle E. Umstead for reading the manuscript and for his valuable comments.

References

Arnold, S. J. and Rojeska, H. (973). *Appl. Opt.*, **12**, 169.
Arnoldi, D. and Wolfrum, J. (1974). *Chem. Phys. Lett.*, **24**, 234.
Arrington, C. A., Brennen, W., Glass, G. P., Michael, J. V., and Niki, H. (1965). *J. Chem. Phys.*, **43**, 525.
Arrington, C. A. and Cox, D. J. (1975). *J. Phys. Chem.*, **79**, 2584.
Avery, H. E. and Heath, S. J. (1972). *J. Chem. Soc. Faraday Trans. I*, **3**, 512.
Barth, C. A., Fastie, W. G., Hord, C. W., Pearce, J. B., Kelly, K. K., Stewart, A. I., Thomas, G. E., Anderson, G. P., and Raper, O. F. (1969). *Science*, **165**, 1004.
Barth, C. A., Hord, C. W., Steward, A. I., and Lane, A. L. (1972). *Science*, **175**, 309.
Bar-Ziv, E., Moy, J., and Gordon, R. J. (1978). *J. Chem. Phys.*, **68**, 1013.
Basco, N. and Norrish, R. G. W. (1961). *Proc. Roy. Soc. (London)*, **A260**, 293.
Bauer, S. H. (1978). *Chem. Rev.*, **78**, 147.
Baulch, D. L. and Breckenridge, W. H. (1966). *Trans. Faraday Soc.*, **62**, 2768.
Bayes, K. D. (1970). *J. Chem. Phys.*, **52**, 1093.
Becker, K. H. and Bayes, K. D. (1966). *J. Chem. Phys.*, **45**, 3967.
Becker, K. H. and Bayes, K. D. (1968). *J. Chem. Phys.*, **48**, 653.
Bemand, P. P., Clyne, M. A. A., and Watson, R. T. (1973). *J. Chem. Soc. Faraday Trans I*, **69**, 1356.

Berry, M. J. (1973). *J. Chem. Phys.*, **59**, 5229.
Birely, J. H., Kasper, J. V. V., Hai, F., and Darnton, L. A. (1975). *Chem. Phys. Lett.*, **31**, 220.
Black, G., Sharpless, R. L., Slanger, T. G., and Lorents, D. C. (1975a). *J. Chem. Phys.*, **62**, 1975.
Black, G., Sharpless, R. L., and Slanger, T. G. (1975b). *J. Chem. Phys.*, **63**, 4546.
Blackwell, B. A., Polanyi, J. C., and Sloan, J. J. (1976). *Faraday Disc. Chem. Soc.*, **62**.
Blackwell, B. A., Polanyi, J. C., and Sloan, J. J. (1977). *Chem. Phys.*, **24**, 25.
Blumenberg, B., Hoyermann, K. and Sievert, R. (1977). *16th Int. Symp. Combust.*, The Combustion Institute, p. 841.
Bogan, D. and Setser, D. W. (1978). *ACS Symp. Ser.*, **66**, 237.
Boxall, C. R., Simmon, J. P., and Tasker, P. W. (1972). *Faraday Disc. Chem. Soc.*, **53**, 182.
Brown, R. D. H. and Smith, I. W. M. (1978). *Int. J. Chem. Kinet*, **10**, 1.
Brown, J. M. and Thrush, B. A. (1967). *Trans. Faraday Soc.*, **63**, 630.
Burks, T. L. and Lin, M. C. (1978). *Chem. Phys.*, **33**, 327.
Butler, J. E., Hudgens, J. W., Lin, M. C., and Smith, G. K. (1978). *Chem. Phys. Lett.*, **58**, 216.
Calcote, H. F. (1962). *8th Int. Symp. Combust.*, Williams and Wilkins, Baltimore, p. 184.
Calvert, J. G. and Pitts, J. N., Jr. (1966). *Photochemistry*, John Wiley & Sons, New York.
Carter, C. F., Levy, M. R., Woodall, K. B., and Grice, R. (1973). *Faraday Disc. Chem. Soc.*, **55**, 381.
Castellano, E. and Schumacher, H. J. (1972). *Z. Phys. Chem.*, (N.F.) **76**, 258.
Chamberlain, J. W. (1961). *Physics of the Aurora and Airglow*, Academic Press, New York.
Clough, P. N., O'Neil, G. M., and Geddes, J. (1978). *J. Chem. Phys.*, **69**, 3128.
Clough, P. N., Schwartz, S. E., and Thrush, B. A. (1970). *Proc. Roy. Soc. (London)*, **A317**, 575.
Comes, F. J. and Fink, E. H. (1972). *Chem. Phys. Lett.*, **14**, 433.
Copper, C. D., Cobb, G. C., and Tolnas, E. L. (1961). *J. Mol. Spectr.*, **7**, 223.
Creek, D. M., Melliar-Smith, C. M., and Jonathan, N. (1970). *J. Chem. Soc. (A)*., 646.
Csizmadia, I. G., Gunning, H. E., Gosavi, R. K., and Strausz, O. P. (1973). *J. Amer. Soc.*, **95**, 133.
Cvetanović, R. J. (1955). *J. Chem. Phys.*, **23**, 1375.
Cvetanović, R. J. (1963). *Adv. Photochem.*, **1**, 115.
Dalgarno, A., Oppenheimer, M., and Berry, R. S. (1973). *Astrophys. J.*, **183**, L21.
Das, G. and Wahl, A. C. (1972). *J. Chem. Phys.*, **56**, 1769.
Davidson, J. A., Schiff, H. I., Brown, T. J., Streit, G. E., and Howard, C. J. (1978a). *J. Chem. Phys.*, **69**, 1213.
Davidson, J. A., Schiff, H. J., Brown, T. J., and Howard, C. J. (1978b). *J. Chem. Phys.*, **69**, 4277.
Deans, A. J., Shepherd, G. G., and Evans, W. F. J. (1976). *J. Geophys. Res.*, **81**, 6227.
Delos, J. B. (1973). *J. Chem. Phys.*, **59**, 2365.
DeMore, W. B. (1967). *J. Chem. Phys.*, **47**, 2777.
DeMore, W. B. and Raper, O. F. (1967). *J. Chem. Phys.*, **46**, 2500.
Ding, A. M. G., Kirsch, L. J., Perry, D. S., Polanyi, J. C., and Schreiber, J. L. (1973). *Faraday Disc. Chem. Soc.*, **55**, 252.
Dixon, D. A., Parrish, D. D., and Herschbach, D. R. (1973). *Faraday Disc. Chem. Soc.*, **55**, 385.
Djeu, N. (1974). *J. Chem. Phys.*, **60**, 4109.

Donovan, R. J. and Gillespie. H. M. (1975). *Reaction Kinetics*, Vol. 1, The Chem. Soc., Burlington, p. 14.
Donovan, R. J. and Husain, D. (1970). *Chem. Rev.*, **70**, 489.
Donovan, R. J., Husain, D., and Kirsch, L. J. (1970). *Chem. Phys. Lett.*, **6**, 488.
Donovan, R. J., Kaufmann, K., and Wolfrum, J. (1976). *Nature*, **262**, 204.
Dunning, T. H., Jr. and Hay, P. J. (1977). *J. Chem. Phys.*, **66**, 3767.
Engleman, R., Jr. (1965). *J. Amer. Chem. Soc.*, **87**, 4193.
Faist, M. B. and Bernstein, R. B. (1976). *J. Chem. Phys.*, **64**, 3924.
Felder, W. and Young, R. A. (1972). *J. Chem. Phys.*, **56**, 6028.
Fisher, E. R. and Bauer, E. (972). *J. Chem. Phys.*, **57**, 1966.
Fletcher, I. S. and Husain, D. (1976). *J. Phys. Chem.*, **80**, 1837.
Fontijn, A. (1971). *Progr. Reaction Kinet.*, **6**, 75.
Fontijn, A. and Johnson, S. E. (1973). *J. Chem. Phys.*, **59**, 6193.
Forst, W. (1973) *Theory of Unimolecular Reactions*, Academic Press, New York.
Gangi, R. A. and Bader, R. F. W. (1971). *J. Chem. Phys.*, **55**, 5369.
Gangloff, H. J., Milks, D., Maloney, K. L., Adams, T. N., and Matula, R. A. (1975). *J. Chem. Phys.*, **63**, 4915.
Gardiner, W. C., Jr. (1977). *Accts. Chem. Res.*, **10**, 326.
Gattinger, R. L. and Jones, A. V. (1976). *J. Geophys. Res.*, **81**, 4789.
Gauthier, M. and Snelling, D. R. (1971). *J. Chem. Phys.*, **54**, 4317.
Geddes, J., Clough, P. N., and Moore, P. L. (1974). *J. Chem. Phys.*, **61**, 2145.
Giachardi, D. J. and Wayne, R. P. (1972). *Proc. Roy. Soc. (London)*, **A330,** 131.
Gilbert, J. R., Slagle, I. R., Graham, R. E., and Gutman, D. (1976). *J. Phys. Chem.*, **80**, 14.
Gillespie, H. M. and Donovan, R. J. (1976). *Chem. Phys. Lett.*, **37**, 468.
Gleaves, J. T. and McDonald, J. D. (1975). *J. Chem. Phys.*, **62**, 1582.
Golde, M. F. and Thrush, B. A. (1974). *Chem. Phys. Lett.*, **29**, 486.
Goodman, J., Tully, J. C., Bondybey, V. E., and Brus, L. (1977). *J. Chem. Phys.*, **66**, 4802.
Gordon, R. J. and Lin, M. C. (1973). *Chem. Phys. Lett.*, **22**, 262.
Gordon, R. J. and Lin, M. C. (1976). *J. Chem. Phys.*, **64**, 1058.
Gorry, P. A., Nowikow, C. V., and Grice, R. (1977). *Chem. Phys. Lett.*, **49**, 116.
Graham, R. E. and Gutman, D. (1977). *J. Phys. Chem.*, **81**, 207.
Green, R. G. and Wayne, R. P. (1977). *J. Photochem.*, **6**, 371.
Green, W. H. and Lin, M. C. (1971). *J. Chem. Phys.*, **54**, 3222.
Haller, I. and Pimentel, G. C. (1962). *J. Amer. Chem. Soc.*, **84**, 2855.
Hampson, R. F., Jr. and Garvin, D. (Eds.) (1975) *Chemical Kinetics and Photochemical Data for Modelling Atmospheric Chemistry*, N.B.S. (U.S.), Tech. Note 866.
Hampson, R. F., Jr. and Okabe, H. (1970). *J. Chem. Phys.*, **52**, 1930.
Hancock, G., Morley, C., and Smith, I. W. M. (1971). *Chem. Phys. Lett.*, **12**, 193.
Hancock, G., Ridley, B. A., and Smith, I. W. M. (1972). *J. Chem. Soc. Faraday Trans. II.*, **68**, 2117.
Hancock, G. and Smith, I. W. M. (1969). *Chem. Phys. Lett.*, **3**, 573.
Hancock, G. and Smith, I. W. M. (1971). *Trans. Faraday Soc.*, **67**, 2586.
Havel, J. (1974). *J. Amer. Chem. Soc.*, **96**, 530.
Heicklen, J., Cohen, N., and Saunders, D. (1965). *J. Phys. Chem.*, **69**, 1774.
Heidner, R. F., Jr. and Husain, D. (1974). *Int. J. Chem. Kinet.*, **6**, 77.
Herbrechtsmeier, P. and Wagner, H. Gg. (1972). *Ber. Bunsen. Phys. Chem.*, **76**, 517.
Herbrechtsmeier, P. and Wagner, H. Gg. (1974). *Z. Phys. Chem.*, **93**, 143.
Herbrechtsmeier, P. and Wagner, H. Gg. (1975a). *Ber. Bunsen. Phys. Chem.*, **79**, 461, (1975b); *ibid*, **79**, 673.

Herron, J. T. and Huie, R. E. (1974). *J. Phys. Chem.*, Ref. data, **2**, 467.
Hollinden, G. A., Kurylo, M. J., and Timmons, R. B. (1970). *J. Phys. Chem.*, **74**, 988.
Howard, R. E. and Lester, W. A., Jr. (1979). To be published.
Hsu, D. S. Y., Colcord, L. J., and Lin, M. C. (1978c). *J. Phys. Chem.*, **82**, 121.
Hsu, D. S. Y. and Lin, M. C. (1977a). *Chem. Phys.*, **21**, 235.
Hsu, D. S. Y. and Lin, M. C. (1977b). *Int. J. Chem. Kinet.*, **9**, 507.
Hsu, D. S. Y. and Lin, M. C. (1978a). *Int. J. Chem. Kinet.*, **10**, 839.
Hsu, D. S. Y. and Lin, M. C. (1978b). *J. Chem. Phys.*, **68**, 4347.
Hsu, D. S. Y., Shaub, W. M., Burks, T. L., and Lin, M. C. (1979a). To be published.
Hsu, D. S. V., Shortridge, R. G., and Lin, M. C. (1979b). To be published.
Hudgens, J. W., Gleaves, J. T., and McDonald, J. D. (1976). *J. Chem. Phys.*, **64**, 2528.
Hughes, W. M., Olson, N. T., and Hunter, R. (1976). *Appl. Phys. Lett.*, **28**, 81.
Hui, K.-K. and Cool, T. A. (1978). *J. Chem. Phys.*, **68**, 1022.
Huie, R. E. and Herron, J. T. (1975). *Progr. Reaction Kinet.*, **8**, 1.
Husain, D. and Kirsch, L. J. (1971). *Trans. Faraday Soc.*, **67**, 2886.
Izod, T. P. J. and Wayne, R. P. (1968). *Proc. Roy. Soc. (London)*, **A308**, 81.
Jayanti, R. K. M., Simonaitis, R., and Heicklen J. (1975). *J. Photochem.*, **4**, 381.
Johnson, B. E. and Winter, N. W. (1977). *J. Chem. Phys.*, **66**, 4116.
Jones, I. T. N., Kaczmar, U. B., and Wayne, R. P. (1970). *Proc. Roy. Soc. (London)*, **A316**, 431.
Jones, P. R. and Taube, H. (1973). *J. Phys. Chem.*, **77**, 1007.
Julienne, P. S., Krauss, M., and Stevens, W. (1976). *Chem. Phys. Lett.*, **38**, 374.
Kanofsky, J. R. and Gutman, D. (1972). *Chem. Phys. Lett.*, **15**, 236.
Kanofsky, J. R., Lucas, D., and Gutman, D. (1973). *14th Int. Symp. Combust.* The Combustion Institute, p. 285.
Kanofsky, J. R., Lucas, D., Pruss, F., and Gutman, D. (1974). *J. Phys. Chem.*, **78**, 311.
Karny, Z., Katz, B., and Szoke, A. (1975). *Chem. Phys. Lett.*, **35**, 100.
Kelley, J. D. (1976). *Chem. Phys. Lett.*, **41**, 7.
Kerr, J. A. and Ratajczak, E. (1977). *Third Supplementary Tables of Bimolecular Gas Reactions*, Department of Chemistry, The University of Birmingham, England.
Kinnersly, S. R., Murrell, J. N., and Rodwell, W. R. (1978). *J. Chem. Soc. Faraday Trans. II*, **74**, 600.
Kinsey, J. L. (1977). *Ann. Rev. Phys. Chem.*, **28**, 349.
Klimek, D. E. and Berry, M. J. (1973). *Chem. Phys. Lett.*, **20**, 141.
Koda, S. (1978). *Chem. Phys. Lett.*, **55**, 353.
Krech, R. H., Diebold, G. J., and McFadden, D. L. (1977). *J. Amer. Chem. Soc.*, **99**, 4605.
Kurylo, M. J., Braun, W., Kaldor, A., Freund, S. M., and Wayne, R. P. (1974). *J. Photochem.*, **3**, 71.
Lee, E. K. C. (1977). *Accts. Chem. Res.*, **10**, 319.
Lee, J. H., Timmons, R. B., and Stief, L. J. (1976). *J. Chem. Phys.*, **64**, 300.
Lee, L. C. and Slanger, T. G. (1978). *J. Chem. Phys.*, **69**, 4053.
Levine, R. D. and Bernstein, R. B. (1974). *Accts. Chem. Res.*, **7**, 393.
Levine, R. D. and Ben-Shaul, A. (1977). *Chemical and Biochemical Applications of Lasers*, Academic Press, New York, p. 145.
Light, G. C. (1978). *J. Chem. Phys.*, **68**, 2831.
Light, J. C. (1967). *Faraday Disc. Chem. Soc.*, **44**, 14.
Lin, C.-L. and DeMore, W. B. (1973). *J. Phys. Chem.*, **77**, 863.
Lin, M. C. (1971). *J. Phys. Chem.*, **75**, 3642.
Lin, M. C. (1972). *J. Phys. Chem.*, **76**, 811, 1425.
Lin, M. C. (1973a). *Int. J. Chem. Kinet.*, **5**, 173.

Lin, M. C. (1973b). *Chemiluminescence and Bioluminescence*, Planum Press, N.Y., p. 61.
Lin, M. C. (1974a). *Int. J. Chem. Kinet.*, **6**, 1; (1974b). *J. Chem. Phys.*, **61**, 1835.
Lin, M. C. (1978). *J. Chem. Phys.*, **68**, 2004.
Lin, M. C. and Shortridge, R. G. (1974). *Chem. Phys. Lett.*, **29**, 42.
Lin, M. C., Shortridge, R. G., and Umstead, M. E. (1976). *Chem. Phys. Lett.*, **37**, 279.
Liuti, G., Kunz, C., and Dondes, S. (1967). *J. Amer. Chem. Soc.*, **89**, 5542.
Lorents, D. C. and Huestis, D. L. (1975). *Lecture Notes in Physics: Laser Spectroscopy*, Vol. 43, Springer, Berlin, p. 100.
MacGregor, M. and Berry, R. S. (1973). *J. Phys. B*, **6**, 181.
Manning, R. G., Braun, W., and Kurylo, M. J. (1976). *J. Chem. Phys.*, **65**, 2609.
McGrath, W. D. and Norrish, R. W. G. (1960). *Proc. Roy. Soc. (London)*, **A254**, 317.
Michell, R. C. and Simons, J. P. (1968). *J. Chem. Soc.*, **13**, 1005.
Miller, W. B., Safron, S. A., and Herschbach, D. R. (1967). *Faraday Disc. Chem. Soc.*, **44**, 108.
Moehlmann, J. G. and McDonald, J. D. (1973). *J. Chem. Phys.* **59**, 6683.
Molina, M. J. and Rowland, F. S. (1975). *Rev. Geophys. Space Phys.*, **13**, 1.
Morley, C., Ridley, B. A., and Smith, I. W. M. (1972). *J. Chem. Soc. Faraday Trans. II*, **68**, 2127.
Noxon, J. F. (1970). *J. Chem. Phys.*, **52**, 1852.
Noxon, J. F. (1975). *J. Geophys. Res.* **80**, 1370.
Okabe, H. (1972). *J. Chem. Phys.*, **56**, 4381.
Olson, R. E. (1973). *Chem. Phys. Lett.*, **19**, 137.
Paraskevopoulos, G. and Cvetanovic, R. J. (1969a). *J. Chem. Phys.*, **50**, 590; (1969b). *J. Amer. Chem. Soc.*, **91**, 7572.
Paraskevopoulos, G. and Cvetanović, R. J. (1970). *J. Chem. Phys.*, **52**, 5821.
Parrish, D. D. and Herschbach, D. R. (1973). *J. Amer. Chem. Soc.*, **95**, 6133.
Pilz, G. and Wagner, H. Gg. (1974). *Z. Phys. Chem.*, **92**, 323.
Polanyi, J. C. (1963). *J. Quantum Spectr. Rad. Transf.*, **3**, 471.
Polanyi, J. C. and Schreiber, J. L. (1974). *Physical Chemistry: An Advanced Treatise*, Vol. VIA, *Kinetics of Gas Reactions*, Academic Press, New York, p. 383.
Polanyi, J. C. and Schreiber, J. L. (1976). *Faraday Disc. Chem. Soc.*, **62**, 267.
Pollack, M. A. (1966). *Appl. Phys. Lett.*, **8**, 237.
Powell, H. T. and Kelley, J. D. (1974). *J. Chem. Phys.*, **60**, 2191.
Powell, H. T., Murray, J. R., and Rhodes, C. K. (1974). *Appl. Phys. Lett.*, **25**, 730.
Preston, K. F. and Cvetanović, R. J. (1966). *J. Chem. Phys.*, **45**, 288.
Provorov, A. C., Stoicheff, B. P., and Wallace, S. (1977). *J. Chem. Phys.*, **67**, 5393.
Radlein, D. St. A. G., Whitehead, J. C., and Grice, R. (1975). *Mol. Phys.*, **29**, 1813.
Ridley, B. A., Atkinson, R., and Welge, K. H. (1973). *J. Chem. Phys.*, **58**, 3878.
Robinson, P. J. and Holbrook, K. A. (1972). *Unimolecular Reactions*, Wiley–Interscience, New York.
Rosenwaks, R. and Smith, I. W. M. (1973). *J. Chem. Soc. Faraday Trans. II*, **69**, 1416.
Safron, S. A., Weinstein, N. D., Herschbach, D. R., and Tully, J. C. (1972). *Chem. Phys. Lett.*, **12**, 564.
Saunders, D. and Heicklen, J. (1966). *J. Phys. Chem.*, **70**, 1950.
Schacke, H., Schmatjko, K. J., and Wolfrum, J. (1973). *Ber. Bunsenges. Phys. Chem.* **77**, 248.
Schinke, R. and Lester, W. A., Jr. (1979). To be published.
Schmatjko, K. J. and Wolfrum, J. (1975). *Ber. Bunsen. Phys. Chem.*, **79**, 696.
Schmatjko, K. J. and Wolfrum, J. (1977). *16th Int. Symp. Combust.*, The Combustion Institute, p. 819.
Schmatjko, K. J. and Wolfrum, J. (1978). *Ber. Bunsen. Phys. Chem.*, **82**, 419.

Schofield, K. (1973). *J. Phys. Chem.*, Ref. data, **2**, 25.
Schofield, K. (1978). *J. Photochem.*, **9**, 55.
Schott, G. L., Gettzinger, R. W., and Seitz, W. A. (1974). *Int. J. Chem. Kinet.*, **6**, 921.
Shackleford, W. L., Mastrup, F. N., and Kreye, W. C. (1972). *J. Chem. Phys.*, **57**, 3933.
Shortridge, R. G. and Lin, M. C. (1974). *J. Phys. Chem.*, **78**, 1451,
Shortridge, R. G. and Lin, M. C. (1975). *Chem. Phys. Lett.*, **35**, 146.
Shortridge, R. G. and Lin, M. C. (1976). *J. Chem. Phys.*, **64**, 4076.
Sibener, S. J., Buss, R. J., and Lee, Y. T. (1978). *XIth Symp. Rarefied Gas Dym.*, Cannes, France.
Simmons, J. D., Bass, A. M., and Tilford, S. G. (1969). *Astrophys. J.*, **133**, 345.
Slagle, I. R., Gilbert, J. R., and Gutman, D. (1974). *J. Chem. Phys.*, **61**, 704.
Slagle, I. R., Gutman, D., and Gilbert, J. R. (1975). *15th Int. Symp. Combust.*, The Combustion Institute, p. 785.
Slanger, T. G. and Black, G. (1973). *J. Chem. Phys.*, **58**, 3121.
Slanger, T. G. and Black, G. (1974). *J. Chem. Phys.*, **60**, 468.
Smardzewski, R. R. and Lin, M. C. (1976). *J. Chem. Phys.*, **66**, 3197.
Smith, G. K., Butler, J. E., and Lin, M. C. (1978). *International Conference on Lasers*, '78, Orlando, Florida, Dec. 11–15, 1978.
Smith, I. W. M. (1967). *Faraday Disc. Chem. Soc.*, **44**, 194.
Smith, I. W. M. (1975). *Adv. Chem. Phys.*, **28**, 1.
Snelling, D. R. (1974). *Can. J. Chem.*, **52**, 257.
Snelling, D. R. and Gauthier, M. (1971). *Chem. Phys. Lett.*, **9**, 254.
Sorbie, K. S. and Murrell, J. N. (1976). *Mol. Phys.*, **31**, 905.
Stephenson, J. C. and Freund, S. M. (1976). *J. Chem. Phys.*, **65**, 1893.
Suart, R. D., Dawson, P. H., and Kimbell, G. H. (1972). *J. Appl. Phys.*, **43**, 1022.
Taylor, G. W. and Setser, D. W. (1973). *J. Chem. Phys.*, **58**, 4840.
Tedder, J. M. and Walton, J. C. (1978). *ACS Symp. Ser.*, **66**, 107.
Tsuchiya, S., Nielsen, N. and Bauer, S. H. (1973). *J. Phys. Chem.*, **77**, 2455.
Tully, J. C. (1974). *J. Chem. Phys.*, **61**, 61.
Tully, J. C. (1975). *J. Chem. Phys.*, **62**, 1893.
Tully, J. C. and Preston, R. K. (1971). *J. Chem. Phys.*, **55**, 562.
Tyreman, W. J. R. (1969). *Trans. Faraday Soc.*, **65**, 163.
Umstead, M. E. and Lin, M. C. (1977). *Chem. Phys.*, **25**, 353.
Umstead, M. E., Shortridge, A. B., and Lin, M. C. (1977). *Chem. Phys.*, **20**, 271.
Umstead, M. E., Woods, F. J., and Lin, M. C. (1979). To be published.
Wallace, L. W. and Hunten, D. M. (1968). *J. Geophys. Res.*, **73**, 4813.
Walsh, A. D. (1953). *J. Chem. Soc.*, 2266.
Welge, K. H. and Atkinson, R. (1976). *J. Chem. Phys.*, **64**, 531.
Westenberg, A. A. (1973). *Ann. Rev. Phys. Chem.*, **24**, 77.
Whitlock, P. A., Muckerman, J. T., and Fisher, E. R. (1976). *Theoretical Investigations of the Energetics and Dynamics of the Reactions* $O(^3P, {}^1D) + H_2$ *and* $C(^1D) + H_2$, Report of Research Institute for Engineering Sciences, Wayne State University, unpublished.
Williams, A. and Smith, D. B. (1970). *Chem. Rev.* **70**, 267.
Williamson, D. G. and Bayes, K. D. (1967). *J. Amer. Chem. Soc.*, **89**, 3390.
Winter, N. W., Bender, C. F., and Goddard, W. A., III (1973). *Chem. Phys. Lett.*, **20**, 489.
Yamazaki, H. (1970). *Can. J. Chem.*, **48**, 3269.
Yamazaki, H. and Cvetanović, R. J. (1964a). *J. Chem. Phys.*, **40**, 582; (1964b), *ibid*, **41**, 3703.
Young, R. A. and Black, G. (1967). *J. Chem. Phys.*, **47**, 2311.
Zahr, G. E., Preston, R. K., and Miller, W. H. (1975). *J. Chem. Phys.*, **62**, 1127.

Potential Energy Surfaces
Edited by K. P. Lawley

EXCITED-STATE POTENTIAL SURFACES AND THEIR APPLICATIONS*

C. BOTTCHER

Physics Division, Oak Ridge National Laboratory, P.O. Box X, Oak Ridge, Tennessee 37830 *USA*

CONTENTS

I. INTRODUCTION

This article will review the methods available for calculating potential energy surfaces associated with the electronically excited states of molecules. In addition, we survey briefly the applications which can be made of these surfaces in non-reactive scattering problems. Other applications, e.g. the calculation of spectroscopic constants, will not be discussed, nor will purely

*Research sponsored by the Division of Physical Research, U.S. Department of Energy under contract W-7405-eng-26 with the Union Carbide Corporation.

empirical methods of constructing surfaces.[1] As yet, little is known about reactions on excited surfaces. Most of our knowledge is still concerned with diatomics and carries over with few changes to polyatomics. However, where possible, we stress the novel features of polyatomics since most research will deal with these in the future. Being interested in collision problems we shall also try to give a reasonable account of dynamical corrections to the Born–Oppenheimer approximation.

Two major sections follow dealing respectively with the calculation of potential surfaces and applications to scattering problems. Atomic units will be used unless otherwise stated. Certain accepted abbreviations from Quantum Chemistry will be freely used:

AO	atomic orbital
MO	molecular orbital
LC	linear combination (e.g. LCAO)
SCF	self-consistent field
HF	Hartree–Fock (almost interchangeable with SCF)
MC	multiconfiguration (e.g. MCSCF)
CI	configuration interaction
B–O	Born–Oppenheimer

II. THE CALCULATION OF EXCITED-STATE POTENTIAL SURFACES

A. General Remarks

We begin by considering the methods available for calculating the adiabatic potential surfaces of electronically excited molecular states. Dynamical corrections to the adiabatic model will also be discussed. Consider a molecule M having $\mathcal{N}$ electrons and $\mathcal{K}$ nuclei with position vectors $\mathbf{r} = \{\mathbf{r}_i\}$ and $\mathbf{R} = \{\mathbf{R}_\alpha\}$, respectively. If the nuclei have masses $\{A_\alpha\}$ and charges $\{Z_\alpha\}$, the Schrödinger equation for the wavefunction Ψ of M is

$$(T + H_{\mathrm{el}})\Psi = E\Psi$$

$$T = -\frac{1}{2}\sum_{\alpha=1}^{\mathcal{K}} \frac{1}{A_\alpha}\frac{\partial^2}{\partial \mathbf{R}_\alpha^2} + \sum_{\alpha=1}^{\mathcal{K}-1}\sum_{\beta=\alpha+1}^{\mathcal{K}} \frac{Z_\alpha Z_\beta}{|\mathbf{R} - \mathbf{R}_\beta|} \tag{1}$$

$$H_{\mathrm{el}} = -\frac{1}{2}\sum_{i=1}^{\mathcal{N}} \frac{\partial^2}{\partial \mathbf{r}_i^2} - \sum_{\alpha=1}^{\mathcal{K}}\sum_{i=1}^{\mathcal{N}} \frac{Z_\alpha}{|\mathbf{R}_\alpha - \mathbf{r}_i|} + \sum_{i=1}^{\mathcal{N}-1}\sum_{j=i+1}^{\mathcal{N}} \frac{1}{|\mathbf{r}_i - \mathbf{r}_j|}.$$

We now make the Born–Oppenheimer separation in which the nuclei are initially considered as fixed.[2,3] Because of the great mass of the nuclei compared with electrons, the approximation should be an excellent one for bound states of M and for dissociating states in which the fragments have relative energies below a few KeV; the limitations of this statement are explored in II.D below. If we let

$$\Psi(\mathbf{R},\mathbf{r}) = \Phi(\mathbf{R}|\mathbf{r})\chi(\mathbf{R}) \tag{2}$$

where Φ satisfies the fixed nucleus Schrödinger equation

$$H_{\text{el}}\Phi = W(\mathbf{R})\Phi \tag{3}$$

then to order $1/A_\alpha$, χ satisfies the equation for motion on the potential surface W,

$$H_{\text{nuc}}\chi = E\chi, H_{\text{nuc}} = T + W. \tag{4}$$

It is with the solution of (3) that we are mainly concerned. The wavefunction Φ involves the spatial and spin coordinates of all $\mathcal{N}$ electrons and is antisymmetric under interchange of any two electrons. Any operator $\mathcal{O}$ which commutes with H_{el} provides another quantum number since Φ must be an eigenfunction of $\mathcal{O}$. Except for molecules containing heavy atoms (say beyond Kr) such operators would include the total spin $\mathbf{S}^2$ and its projection S_z on an axis fixed in space. For linear molecules, one also has the projection L_n of total orbital angular momentum on the molecular axis. In polyatomics high spatial symmetries (e.g. T_d, D_{3h}) are only available in limited subspaces of the nuclear coordinates, though it is often possible to discuss a collision process in terms of restricted configurations in which the molecule is linear or has some reflection symmetry.

If M is pulled apart into constituents a, b, c, ..., which may be atoms or molecules and whose isolated energies are E(a), E(b), E(c), ..., we can introduce a new quantity

$$\Delta W(\text{a,b,c},\ldots) = W - E(\text{a}) - E(\text{b}) - E(\text{c}) - \ldots \tag{5}$$

which tends to zero at infinite separation. Instead of thinking of the *potential energy* of M, one can think of ΔW, the *interaction* between a, b, c, ... In II.B we describe methods which focus on W and in II.C, methods which focus on ΔW. Little is known about the conditions under which one can add pairwise interactions to obtain ΔW, but we will make some remarks about this approximation in II.C.

In II.D we introduce the dynamical effects which arise in the real world because the nuclei change their positions with time. These effects are of two kinds. When the nuclei move rapidly or when two energy surfaces are close together, the total wavefunction is a superposition of B–O eigenstates $\chi_n\Phi_n$ connected by coupling terms $\mathscr{C}_{mn}$. Alternatively, if the electronic state of M is unstable against emission of a photon or electronic rearrangement, W acquires an imaginary part $-iG/2$. The calculation of $\mathscr{C}_{mn}$ or G is closely related to that of W. We also consider the important subject of avoided crossings.

Much of II.B–II.D applies equally to ground and excited states. Excited states are in a sense more complicated systems, e.g. they are often degenerate or they separate into degenerate states. On the other hand, they usually involve relatively diffuse valence orbitals which are amenable to approximate treatments.

B. Quantum Chemical Methods

By 'Quantum Chemical' methods we refer to those which calculate W directly by solving (3) either accurately or approximately. They may be classified under three headings: 1. *ab initio* methods, 2. model or pseudo-potential methods, and 3. other semi-empirical methods. Exact analytic solutions are only possible for one-electron diatomics (see II.D below).

1. Ab Initio *Methods*

'*Ab initio*' implies the attempted solution of (3) without using any experimental information beyond the fundamental constants. Many books and review articles[1,4,5] describe the technology and folklore or these calculations so we shall here only indicate the special features of excited states. Two procedures are available in practice, that of the self-consistent field, and configuration interaction. In the SCF method the wavefunction is approximated by a single determinant (configuration) of $\mathscr{N}$ molecular orbitals $\{\phi_i\}$ chosen so as to minimize the expectation value of the Hamiltonian. According to the Rayleigh–Ritz variational principle, this leads to the best possible wavefunction and energy of the assumed form. If each ϕ_i is a LCAO

$$\phi_i = \sum c_{is}\chi_s \tag{6}$$

the problem is reduced to a non-linear eigenvalue system for the coefficients $\{c_{is}\}$. Most difficulties come from the evaluation of two-electron molecular integrals in the AO basis. Often excited states cannot be represented by one determinant, but the SCF procedure is readily generalized to a small number of configurations, giving the MCSCF method.

One is then led naturally to consider wavefunctions which are superpositions of determinants in which occupied orbitals $k, l \leq \mathscr{N}$ are replaced by unoccupied (also called excited, virtual) orbitals $k', l' > \mathscr{N}$: this is the CI method. In an obvious notation[6]

$$\Phi = C_0 D_0 + \Sigma C_{kk'} D(kk') + \Sigma C_{klk'l'} D(klk'l') + \dots \tag{7}$$

where the coefficients $\{C\}$ are derived from a linear eigenvalue system. The singly excited determinants provide the simplest description of the valence excited states, while the doubly excited determinants describe the correlations between pairs of electrons.

The great advantage of *ab initio* calculations over semi-empirical methods is their totally predictive power. By providing a complete representation of the wavefunction, they make it possible to calculate any expectation value or transition matrix element. However, their drawbacks of complexity and lack of transparency make it desirable to look for simplifications. Most excited states consist of one or two valence electrons outside a tightly bound core, whence to a good approximation

$$W(\mathbf{R}) = W_c(\mathbf{R}) + W_v(\mathbf{R}). \tag{8}$$

Over a wide range of **R** it is often the case that $\Delta W_c \ll \Delta W_v$ so that a relatively crude representation of the core orbitals suffices to determine ΔW. If there is only one valence electron an SCF calculation is adequate, as one has only to optimize the valence orbital in an essentially single electron claculation.[7] These considerations lead naturally to the model or pseudo-potential concepts discussed in II.B.2 below.

Many-body theories such as the random phase approximation, or diagrammatic techniques have not been much used to calculate surfaces, though some interesting work has been done.[8] They are mostly applied to study correlation in atoms and molecules near their equilibrium configuration. In practice, these methods so far have been limited to systems whose ground state is a single Slater determinant, and it must be said that the same results can usually be obtained as a special case of CI.

2. *Model and Pseudo-Potential Methods*

In II.B.1 we saw that even in the *ab initio* framework, potential surface calculations are greatly simplified by separating the core and valence electrons. One would like to carry this idea to a logical conclusion and replace (3) by a Schrödinger equation for a wavefunction Φ_v involving only the valence electrons,

$$(H_v + \mathscr{U})\Phi_v = (E - E_c)\Phi_v. \tag{9}$$

The Hamiltonian of the valence electrons alone is H_v, the bare core energy is E_c, and the valence electron–core interaction is represented by $\mathscr{U}$, sometimes referred to as the 'optical potential'. Two approaches may be used to justify (9).

The first stems from Feshbach's projection technique.[9] We write the full wavefunction as

$$\Phi = \mathscr{A}(\Phi_c\Phi_v + \Sigma\Phi_c'\Phi_v') \tag{10}$$

where Φ_c, Φ'_c refer to ground and excited states of the core and $\mathscr{A}$ is the antisymmetry. By a series of manipulations, an equation of the form (9) can be derived, with $\mathscr{U}$ a non-local and energy-dependent operator. Invariably $\mathscr{U}$ is not calculated from first principles, but the theory is a guide to possible analytic forms whose parameters can be adjusted to fit empirical data.

The second approach assumes that outside some surface R_c, $\mathscr{U} = 0$ or is at least small. Then we forget (9) and solve the simpler problem

$$\begin{aligned} H_v \Phi_v &= (E - E_c)\Phi_v \ \text{(outside } R_c) \\ \frac{\partial}{\partial n} \ln \Phi_v &= B_c \ \text{(on } R_c) \end{aligned} \tag{11}$$

where the boundary function B_c is determined empirically or from an approximate theory (e.g. an SCF calculation). To turn the argument about, the same energies and wavefunctions (outside R_c) will result from (9) for any $\mathscr{U}$ which reproduces B_c. We believe that this is the simplest and most rigorous definition of the optical potential and one which shows that $\mathscr{U}$ is far from unique.

We now describe calculations based on (9). More information can be found in review articles by Weeks *et al.*[10] and Bardsley;[11] papers by Baylis[12] and Bottcher and Dalgarno[33] contain much relevent technical detail. Most applications are to systems with one or two electrons outside closed shells, though open-shell cores have been considered.[14]

The easiest to describe are the model potential methods which attempt to construct the true potential seen by an electron in the field of the core; in particular, this potential will tend to the unscreened Coulomb potential near a nucleus. For a single valence electron, such a potential would have the form

$$\mathscr{V} = V_{\text{HFL}} + V_{\text{SR}} + V_{\text{LR}} \tag{12}$$

when V_{HFL} is the local part of the HF potential, V_{SR} is an adjusted short-range potential to simulate exchange and other effects left out of V_{HFL}, and V_{LR} contain the long-range electron–core interactions.[15] We implied above that $\mathscr{U}$ is negligible outside R_c but this is an oversimplification. Intuitively, we expect that $\mathscr{U}$ contains long-range terms of which the most important is the induced polarization energy

$$V_{\text{LR}} \sim -\alpha_d / 2r^4, \tag{13}$$

α_d being the dipole polarizability of the core and r the distance from a point within R_c. Outside R_c these terms are local and unique; all derivations of $\mathscr{U}$ should predict their form correctly. Within R_c the asymptotic expansions break down and r^{-n} terms must be cut off. The parameters in V_{SR} may be fixed by spectroscopic or scattering data depending on the system. For an alkali core, one would fit the quantum defects of the Rydberg states; for an inert-gas core accurate low-energy scattering data are available. If we consider a system of $\mathcal{N}$ valence electrons outside $\mathcal{K}$ well-separated cores, the interaction is to a first approximation

$$\mathcal{U} = \sum \mathcal{V}(\mathbf{R}_\alpha - \mathbf{r}_i) \tag{14}$$

adapting the notation of (3). The methods of II.B.1 can then be applied to (9), and the procedure is almost wholly predictive.

The additive assumption (14) has to be qualified in several ways. For cores with many electrons, the core–core overlap is a significant contribution to E_c at surprisingly large separations and recent papers have devoted more attention to this point.[16] Apart from the pairwise core interactions, additional terms must be added to (14) in all but the most trivial case ($\mathcal{N} = \mathcal{K} = 1$). If we imagine two particles with charges e_1, e_2 outside a spherical core, e_1 induces a dipole moment along its position vector $\mathbf{r}_1$ which interacts with e_2, so that the full long-range interaction is

$$\mathcal{U}_{\mathrm{LR}} \sim -\frac{\alpha_{\mathrm{d}}}{2}\left[\frac{e_1^2}{r_1^4} + \frac{e_2^2}{r_2^4} + \frac{2e_1e_2\mathbf{r}_1\cdot\mathbf{r}_2}{r_1^3r_2^3}\right]. \tag{15}$$

As $r_1 \to r_2$, $\mathcal{U}_{\mathrm{LR}} \sim -\alpha_{\mathrm{d}}(e_1 + e_2)^2/2r^4$ as it should, showing that the non-spherical term is essential in a complete theory. Equation (15) was first applied for two electrons outside Ca^{2+} by Chisholm and Opik[17] and to Li_2^+ (i.e. Li^+, e outside Li^+) by Dalgarno *et al.*[18] In a system like (Li^+ + e) + He (which is easier to visualize than Li_2^+) the non-spherical term cancels out a spurious $1/R^4$ interaction which would otherwise appear between two neutral atoms.[19] Furthermore, the leading interaction between the atoms is correctly given as $-C_6/R^6$ where

$$C_6 = \alpha_{\mathrm{d}}(\mathrm{He})\langle r^2(1 + P_2)\rangle_{\mathrm{Li}} \tag{16}$$

and P_2 is a spherical harmonic referred to the LiHe axis. This expression is precisely the Unsold approximation derived by assuming that the excitation energies of the core are much greater than those of the valence electrons. Bottcher and Dalgarno[13] showed that (15) and all higher multipole analogues may be derived quantally, and when they are included in $\mathcal{U}$, atomic interactions are correctly predicted to all orders, within Unsold's approximation. Some controversy regarding the dynamical long-range forces is still outstanding.[20]

We now turn to the complementary pseudo-potential methods. The pseudo-potential may be defined as that choice of $\mathcal{U}$ which is so weak that it can be treated in first-order perturbation theory, or equivalently for which

$$\left|\int_{\mathrm{core}} \mathcal{U}\,\mathrm{d}\tau\right|$$

is a minimum. The concept probably originated in neutron-nucleus scattering theory and has proved highly successful in solid-state physics. The model potential as defined above can be very large inside the core, e.g. for Cs^+, $\mathscr{V} \to -55/r$ near the nucleus and has 5 bound s states of lower energy than the $6s$ valence state. If we soften $\mathscr{V}$, e.g. by setting

$$\begin{aligned} \mathscr{V}_{\mathrm{P}} &= \mathscr{V}(r)(r > a_l) \\ &= \mathscr{V}(a_l)(r < a_l) \end{aligned}$$

then a_l can be increased until the lowest bound state has the $6s, 6p, 5d, \ldots$ binding energy. These pseudo-orbitals are nodeless leading to difficulties if one wants to calculate matrix elements; $\mathscr{V}_{\mathrm{P}}$ is usually sensitive to l which also causes problems.

The pseudo-potential is usually introduced through HF theory. The valence orbital satisfies

$$(h + V_{\mathrm{HF}} - \varepsilon)\psi = 0 \tag{17}$$

where V_{HF} is the core interaction, including exchange. If we introduce a nodeless pseudo-orbital

$$\psi_{\mathrm{p}} = \psi + \Sigma \langle \phi_{\mathrm{c}} | \psi_{\mathrm{p}} \rangle \phi_{\mathrm{c}} \tag{18a}$$

where $\{\phi_{\mathrm{c}}\}$ are the core orbitals, then

$$[h + V_{\mathrm{HF}} + \Sigma(\varepsilon - \varepsilon_{\mathrm{c}}) | \phi_{\mathrm{c}} \rangle \langle \phi_{\mathrm{c}} |] \psi_{\mathrm{p}} = \varepsilon \psi_{\mathrm{p}}. \tag{18b}$$

The additional potential (albeit non-local) cancels the strongly attractive part of V_{HF}; in practice, $\mathscr{V}_{\mathrm{P}}$ is simulated by a plausible local potential with adjusted parameters. The true orbitals may be recovered from (18a).

Most applications of model and pseudo-potentials are concerned with closed-shell atomic cores, but the same ideas can be extended to molecular cores whose potentials are no longer spherically symmetric. If the cores have spin, then the potentials must contain a spin dependence; however, it is probably essential that the core is representable by a single determinant. To illustrate these refinements, we shall describe briefly a fairly involved application to the calculation of excited-state potential surfaces for excited atoms interacting with stable molecules, in particular the $Na(^2P) + N_2$ system.[21] The Na^+ core poses no problems nor does the long-range anisotropic part of the $e + N_2$ potential. However, it appears that the interaction of a low-energy electron with N_2, especially if the $^2\Pi_g$ resonance is correctly predicted, requires a short-range pseudo-potential which depends both on the angular momentum and energy of the electron. This form of potential raises no severe problem since the Hamiltonian is readily expanded in projection operators

$$\mathscr{P}_{lm} = |lm\rangle\langle lm|$$

which select out individual angular momentum channels. As to energy dependence, one only needs to calculate the local kinetic energy of the valence electron at the edge of the N_2 core. When increasingly sophisticated effects are incorporated, it becomes clearer that the model or pseudopotential is only a scaffolding to support the boundary condition (11). It would be interesting to carry out a programme based on (11) alone which would unify the potential methods and other scattered topics e.g. quantum defect theory. The long-range part of $\mathscr{U}$, however, is established on a firm basis and can be incorporated in any formulation.

3. *Other Semi-Empirical Methods*

The methods of II.B.2 are tractable for systems with a small number of valence electrons. We now consider semi-empirical techniques which apply to a large number of valence electrons, the extended Hückel (EH) method,[22] complete neglect of differential overlap (CNDO),[23] and diatomics in molecules (DIM).[24] The first two have been extensively applied to ground-state, near-equilibrium problems, though CNDO has been used to calculate ground-state surfaces.[25] On the other hand, DIM has been very successfully applied to excited-state surfaces. If surfaces are increasingly demanded for large polyatomics, these or related methods are probably the way to progress.

Both EH and CNDO may be regarded as approximations to SCF but with the possibility of being adjusted to fit experimental data. The SCF molecular orbitals satisfy[26]

$$h\phi_n = \varepsilon_n\phi_n, \ \phi_n = \Sigma c_{n\mu}\chi_\mu \tag{19}$$

where χ_μ is an atomic orbital, localized on a specific site. The equations are invariant on transforming to the atomic basis, except that Lagrange multipliers appear

$$h\chi_\mu = \Sigma f_{\mu\nu}\chi_\nu, \ f_{\mu\nu} = \langle\chi_\mu|h|\chi_\nu\rangle. \tag{20}$$

In the EH method we treat f as the Hamiltonian matrix which produces the correct MOs. The number of parameters contained in f is made very small by the principle of 'atomic invariance'; if μ is defined by an AO quantum number Q_μ (e.g. $2p$) and a site R_μ,

$$f_{\mu\nu} = f(Q_\mu, Q_\nu; \mathbf{R}_\mu - \mathbf{R}_\nu). \tag{21}$$

In the well-known case of benzene, one needs only $\alpha = -f(2h, 2h)$ and $\beta = f(2h, 2h')$ where $2h, 2h'$ are $2s$–$2p$ hybrid orbitals on adjacent C atoms.

Even if experimental data are lacking, one can do very well by taking $f_{\mu\mu}$ as an atomic ionization potential I_μ and

$$f_{\mu\nu} = \tfrac{1}{2}K(I_\mu + I_\nu)S_{\mu\nu} \tag{22}$$

where $S_{\mu\nu}$ is the AO overlap and $K(\simeq 1.5)$ is a universal constant (at least near equilibrium geometries). Away from equilibrium, the number of independent parameters grows and they become difficult to predict. Furthermore, the assumption that all atomic populations are exactly unity may break down; this defect is remedied in the CNDO method.

Neglect of differential overlap means that all two-electron integrals involving products $\chi_\mu(1)\chi_\nu(2)$ of different AOs are neglected. We can write the SCF Hamiltonian matrix

$$h_{\mu\nu} = H^0_{\mu\nu} + \Sigma[2(\mu n|\nu n) - (\mu n|n\nu)] \tag{23}$$

using a shorthand notation for the two-electron repulsion integral; n runs over doubly occupied MOs; H^0 is the one-electron part of h including core interactions. If we transform the two-electron integrals using (19) and neglect differential overlap, we find that

$$h_{\mu\nu} = H^0_{\mu\nu} + 2\delta_{\mu\nu}\Sigma P_{\lambda\lambda}g_{\lambda\mu} - P_{\mu\nu}g_{\mu\nu} \tag{24}$$

where $\{P_{\mu\nu}\}$ is the population-bond order matrix

$$P_{\mu\nu} = \sum_n c^*_{n\mu}c_{n\nu} \tag{25}$$

and $g_{\mu\nu}$ is a Coulomb repulsion integral

$$g_{\mu\nu} = (\mu\nu|\mu\nu). \tag{26}$$

Since $\{c_{n\mu}\}$ are the eigenvectors of h, they must be determined self-consistently. The $\{g_{\mu\nu}\}$ satisfy the invariance principle (21) and may be treated as adjustable parameters or derived from one of many prescriptions. Finally, the one-electron part may be taken as

$$H^0_{\mu\nu} = -\tfrac{1}{2}(I_\mu + I_\nu)S_{\mu\nu} \tag{27}$$

where $\{I_\mu\}$ satisfy invariance and may be adjusted if necessary.

When extended to excited states, EH and CNDO introduce many extra parameters, while the amount of experimental information diminishes. Being founded on SCF theory, they have only a limited ability to describe correlations. Nonetheless, CNDO is probably worth pursuing for large organic systems. Methods based on AOs do not fully recognize that electrons in molecules spend much of their time in the bands between pairs of atoms. The DIM procedure to be discussed next achieves its power by focusing on pairs of atoms and associated groups of electrons.

Our discussion closely follows Tully.[24] It is helpful to think of DIM as a simple LCAO–MO calculation adjusted to reproduce known diatomic potentials for specific pairs of atoms. However, a rigorous discussion begins with the observation that the Hamiltonian (3) can be broken into diatomic and atomic fragments

$$H = \sum_{\alpha>\beta} H(\alpha\beta) - (\mathscr{K} - 2)\Sigma H(\alpha) \tag{28}$$

and the matrix elements of $H(\alpha\beta)$, $H(\alpha)$ constructed in a properly chosen basis. Let

$$\Omega_{\mathrm{M}} = \mathscr{A}\mathscr{S}\prod_{\alpha} \phi_{\mathrm{M}}(\alpha) \tag{29}$$

be a product of atomic wavefunctions, antisymmetrized and spin-coupled. The exact wavefunctions are formally

$$\Phi_{\mathrm{I}} = \Sigma\Gamma_{\mathrm{MI}}\Omega_{\mathrm{M}} \tag{30}$$

whence the secular equation for the coefficients becomes

$$\mathbf{H\Gamma} = \mathbf{S\Gamma E}, \tag{31}$$

$\mathbf{E}$ being the diagonal matrix of eigenvalues. It is easily seen that since Ω_{M} is an eigenstate of $H(\alpha)$,

$$\mathbf{S}^{-1}\mathbf{H}(\alpha) = \mathbf{e}_{\alpha} \tag{32}$$

where $\mathbf{e}_{\alpha}$ is a diagonal matrix of atomic energies. To probe $H(\alpha\beta)$, we introduce the new basis

$$\Omega_{\mathrm{M}}^{\alpha\beta} = U_{\alpha\beta}\Phi_{\mathrm{M}} \tag{33}$$

whose elements are eigenstates of the conserved operators pertaining to $\alpha\beta$, *viz.* $\mathbf{S}^2, S_z$, and L_n. If all atoms $\gamma \neq \alpha, \beta$ are removed to infinity, the secular equation for the levels of $\alpha\beta$ in the basis (33) becomes

$$\mathbf{hc} = \mathbf{scw}\,(\alpha\beta \text{ understood}). \tag{34}$$

Suppose that $\mathbf{w}$ is taken from accepted data on the potentials of $\alpha\beta$ and that $\mathbf{s}, \mathbf{c}$ are known by symmetry or simple MO calculations, we can write

$$\mathbf{h} = \mathbf{U}_{\alpha\beta}^{-1}\mathbf{S}^{-1}\mathbf{H}(\alpha\beta)\mathbf{U}_{\alpha\beta} = \mathbf{scwc}^{-1}. \tag{35}$$

Since $\mathbf{h}$ involves only electrons on $\alpha\beta$, it is clear that (35) defines $\mathbf{H}(\alpha\beta)$ whether the other atoms are present or not. Thus $\mathbf{H}$ can be constructed from (28), (32), and (35); $\mathbf{S}$ is taken as the unit matrix or estimated from MO theory. Though the method is obviously not exact in a finite basis, it has proved successful for neutral and ionized molecules and for ground- and excited-state surfaces. When first proposed, DIM was criticized on the grounds that little information was available on the excited states of diatomics and that the method does not explicitly construct a wavefunction. The first objection is no longer valid, while the extension needed to calculate properties has been considered by Tully.[27]

C. Perturbation Methods

An attractive alternative to solving for the eigenenergies of the molecular Hamiltonian (3) is to start with a collection of undisturbed atoms, treating their interaction as a perturbation. Comprehensive reviews of perturbation theory with emphasis on atomic applications are given by Dalgarno and Hirschfelder *et al.*[28] More information on the topics of the present section is contained in the volume edited by Hirschfelder.[29] It is convenient to summarize some formulae here.

Let the unperturbed Schrödinger equation be

$$H_0\Psi_0 = E_0\Psi_0 \tag{36}$$

while the perturbed Hamiltonian and wavefunction are

$$H = H_0 + V, \Psi = \Psi_0 + \Psi'. \tag{37}$$

Instead of proceeding in the usual way, which is not suited to calculating exchange forces (II.C.2 below), we insert Ψ in the Rayleigh–Ritz functional. To second order in V, the energy

$$E = E_0 + \bar{V} + 2\langle\Psi_0|V - \bar{V}|\Psi'\rangle + \langle\Psi'|H_0 - E_0|\Psi'\rangle \tag{38}$$

where the first-order energy

$$E^{(1)} = \bar{V} = \langle\Psi_0|V|\Psi_0\rangle. \tag{39}$$

If E is stationary in Ψ',

$$(E - H_0)\Psi' = (V - \bar{V})\Psi_0, \tag{40}$$

whence the second-order energy

$$E^{(2)} = \langle\Psi_0|V - \bar{V}|\Psi'\rangle. \tag{41}$$

If we expand Ψ' in a set $\{\psi_t\}$ of eigenfunctions of H_0, without assuming that $\psi_0 = \Psi_0$, then

$$E^{(2)} = \sum_{t\neq 0} \frac{\langle\Psi_0|V - \bar{V}|\psi_t\rangle^2}{(E_0 - E_t)}. \tag{42}$$

For ground states $E^{(2)} \leq 0$; most calculations of $E^{(2)}$ are variational in essence and thus provide upper bounds.

The following discussion is in two parts. In II.C.1 we consider long-range interactions which arise from the permanent and fluctuating multipole moments of the interacting atoms. In II.C.2 we consider short-range forces which are strongly influenced by the Pauli exclusion principle. The problem of long-range forces is for practical purposes solved, but much less progress has been made in the short-range domain. We exclude any mention of relativistic effects, including spin–orbit coupling, which is discussed below (III.C) in the context of collision problems. Nor is small R perturbation theory considered.

1. *Long-Range Forces*

We shall make free use of spherical tensor notation[30] in this subsection.* The interaction between two non-overlapping systems (atoms or molecules) A, B may be expanded as

$$V = \Sigma V(k, l)/R^{k+l+1},$$
$$V(k, l) = \mu(2k, 2l)C^{(k+l)}(\hat{\mathbf{R}}) \times (Q_k^{\mathrm{A}} \times Q_l^{\mathrm{B}})^{(k+l)} \tag{43}$$

where $\mathbf{R}$ is the displacement between points fixed in the two systems and Q_l are multipole moments

$$Q_l = \sum r^l C^{(l)}(\theta\phi). \tag{44}$$

Another way of writing (43) is

$$V = \sum Q_k^{\mathrm{A}} \cdot F_k^{\mathrm{B}} \tag{45}$$

where

$$F_k^{\mathrm{B}} = \sum_l \mu(2k+1, 2l)(C^{(k+l)} \times Q_l^{\mathrm{B}})^{(k)}/R^{k+l+1}$$

is the 2^k-pole field at A set up by B. Without an argument $C^{(k)}$ means $C^{(k)}(\hat{\mathbf{R}})$; note that if $\mathbf{R}$ is the z-axis, $C_q^{(k)} = \delta_{q0}$. We take the unperturbed states to be products of eigenstates of A, B ignoring exchange

$$\psi_t = \Phi_{t'}^{\mathrm{A}}\Phi_{t''}^{\mathrm{B}}, \tag{46}$$

with $\Psi_0 = \psi_0$ in the notation of (42). From (43) four types of long-range interaction arise, two in first-order and two in second-order perturbation theory.

(*a*) *Permanent multipole.* If we take the ground-state expectation value of V, the operators $Q_k^{\mathrm{A}}, Q_l^{\mathrm{B}}$ in (43) are replaced by their expectation values, which define the permanent multipole moments of A, B. The most important terms are $V(0, 1)$, $V(1, 1)$, and $V(0, 2)$. In considering surfaces involving excited atoms, it should be remembered that atoms can have multipole moments. An excited hydrogen atom in the presence of another system

*The general tensor product $(A^{(k)} \times B^{(l)})_Q^K = \Sigma\langle KQ|kqlm\rangle A_q^{(k)}B_m^{(l)}$. The scalar product $A^{(k)} \times B^{(k)} = \Sigma(-1)^q A_q^{(k)}B_{-q}^{(k)}$. The Racah tensor $C_q^{(k)} = [4\pi/(2k+1)]^{1/2} Y_{kq}$. We also use the abbreviation $\mu(m, n) = (m+n!/m!n!)^{1/2}$.

will evolve into a Stark state (e.g. $2s \pm 2p_0$) with a dipole moment. Thus the asymptotic interaction of $A^+ + H(2s \pm 2p_0)$ is $\pm 3/R^2$. A P-state atom has a quadrupole moment so that the interaction of $A^+ + B(P)$ has a leading $1/R^3$ term.

(*b*) *Resonance.* Consider two like atoms, one in its ground state and the other in an optically excited state, $A(S) + A^*(P)$. Resonance with the degenerate state $A^*(P) + A(S)$ through $V(1,1)$ leads to a pair of molecular potentials with long-range tails

$$\pm f |\langle S|\mathbf{D}|P\rangle|^2/R^3,$$

f being a symmetry factor. The strong attraction in the lower state explains the stability of the important A_2^* excimers.

(*c*) *Inductive.* In the second-order sum (42) one can distinguish terms in which both A, B are excited and those in which, say, B remains in its ground state. The latter can be related to the multipoles induced in A by the proximity of B. From (45) the perturbation is essentially

$$\Sigma Q_k^{A} \cdot \langle F_k^{B} \rangle$$

where $\langle F_k^{B} \rangle$ is the field of the permanent moments of B. To illustrate the structure of the inductive terms, we consider only the dipole term. Setting $Q_1 = D, \langle F_1 \rangle = F$ for brevity, we get

$$\Delta E_{\text{ind}} = -\tfrac{1}{2}\Sigma(-1)^{m+n} F^{B}_{-m} F^{B}_{-n} \alpha^{A}_{mn} \tag{47}$$

where the tensor polarizability

$$\alpha^{A}_{mn} = 2\langle \Phi_0^{A} | D_m \mathscr{G}_0^{A} D_n | \Phi_0^{A} \rangle \tag{48}$$

and

$$\mathscr{G}_0^{A} = \sum_{t \neq 0} \frac{|\Phi_t^{A}\rangle\langle\Phi_t^{A}|}{E_t^{A} - E_0^{A}} \tag{49}$$

is the Green function of A. We can transform D to an axis fixed in A,

$$D = \Sigma \hat{D}_\mu \tau_\mu$$

where τ_μ is a tensor whose components are Wigner matrix elements $\mathscr{D}^{(1)}_{\mu m}$. Then

$$\Delta E_{\text{ind}} = -\tfrac{1}{2} F^{B} F^{B} : \Sigma \hat{\alpha}^{A}_{\mu\nu} \tau_\mu \tau_\nu \tag{50}$$

$\hat{\alpha}^{A}_{\mu\nu}$ being intrinsic to A. If A is a linear molecule, $\hat{\alpha}$ is diagonal when the z-axis is the axis of symmetry,

$$\hat{\alpha}_{00} = \alpha_{\parallel}, \quad \hat{\alpha}_{11} = \hat{\alpha}_{-1-1} = \alpha_{\perp}. \tag{51}$$

Lastly, if A is an S-state atom,

$$\alpha_{\parallel} = \alpha_{\perp} = \alpha$$

so that the leading term is

$$\Delta E_{\mathrm{ind}} \sim -\alpha^{\mathrm{A}}(D^{\mathrm{B}})^2/2R^6.$$

(*d*) *Dispersive.* Finally, we consider the second-order terms in which both A, B are excited. These are the dispersive terms of which the $1/R^6$ van der Waals term is the most important. Inserting $V(1,1)$ in (42), we get

$$\Delta E_{\mathrm{disp}} = -C/R^6 \tag{52}$$

where

$$C = 6\sum C_m^{(2)}C_n^{(2)}\langle\psi_0|(D^{\mathrm{A}}\times D^{\mathrm{B}})_m^{(2)}\mathscr{G}_0(D^{\mathrm{A}}\times D^{\mathrm{B}})_n^{(2)}|\psi_0\rangle \tag{53}$$

and $\mathscr{G}_0$ is the Green function of A + B,

$$\mathscr{G}_0 = \sum_{t\neq 0}\frac{|\psi_t\rangle\langle\psi_t|}{(E_t^{\mathrm{A}} + E_t^{\mathrm{B}} - E_0^{\mathrm{A}} - E_0^{\mathrm{B}})}. \tag{54}$$

Again transforming to axes fixed in A, B, we find

$$C = \Sigma\hat{C}_{\mu\nu\mu'\nu'}(\tau_\mu\times\tau_\nu)^{(2)}(\tau_{\mu'}\times\tau_{\nu'})^{(2)} : C^{(2)}C^{(2)},$$

$$\hat{C}_{\mu\nu\mu'\nu'} = 6\langle\psi_0|\hat{D}_\mu^{\mathrm{A}}\hat{D}_\nu^{\mathrm{B}}\mathscr{G}_0\hat{D}_{\mu'}^{\mathrm{A}}\hat{D}_{\nu'}^{\mathrm{B}}|\psi_0\rangle. \tag{55}$$

If A, B are linear molecules, we only need $\mu = \mu'$, $\nu = \nu' = \parallel$ or $\perp$ as with $\hat{\alpha}$.

Calculations of C are almost always based on the single-centre formula

$$\hat{C}_{\mu\nu\mu'\nu'} = \frac{3}{\pi}\int_0^\infty \hat{\alpha}_{\mu\mu'}^{\mathrm{A}}(i\omega)\hat{\alpha}_{\nu\nu'}^{\mathrm{B}}(i\omega)d\omega \tag{56}$$

where $\alpha(\omega)$ is a dynamic polarizability. The quantities α^{A}, α^{B} may be constructed from theoretical or experimental information on oscillator strength distributions. If the ionization potentials I_{A}, I_{B} are not too different, one has the useful London approximation,

$$\hat{C}_{\mu\nu\mu'\nu'} \simeq \frac{3}{2}\frac{I_{\mathrm{A}}I_{\mathrm{B}}}{(I_{\mathrm{A}} + I_{\mathrm{B}})}\hat{\alpha}_{\mu\mu'}^{\mathrm{A}}(0)\hat{\alpha}_{\nu\nu'}^{\mathrm{B}}(0). \tag{57}$$

If $I_{\mathrm{B}} \gg I_{\mathrm{A}}$, it is better to use the analogue of (16),

$$\hat{C}_{\mu\nu\mu'\nu'} \simeq \hat{\alpha}_{\nu\nu'}^{\mathrm{B}}(0)\langle\hat{D}_\mu\hat{D}_{\mu'}\rangle_{\mathrm{A}}. \tag{58}$$

This is the appropriate formula when A is excited and it is usually accurate to a few per cent.

The subject of long-range forces has many other ramifications, e.g. three-body interactions are significant in condensed matter. Although the importance of the leading long-range terms (polarization or van der Waals) in any given system is beyond dispute, the high-order terms must be viewed with caution. If we consider the systems mentioned in the preceding paragraph for which $I_B \gg I_A$, the second-order terms arising from $V(1, n)$ go off at least as fast as

$$\frac{\alpha^B \langle r^{2n} \rangle_A}{R^{2n+4}} \sim \frac{\alpha^B 2n! \langle r \rangle_A^{2n}}{R^{2n+4}}.$$

This series is asymptotic, the smallest term occurring about $n \sim R/\langle r \rangle_A$. In physical problems the exchange terms are almost always more important than the higher-order dispersion terms, even though the former fall off exponentially.

2. *Short-Range Forces*

If two systems A, B are brought near each other, their electrons cannot share the same cells in phase space, so they must rearrange themselves, setting up an interaction which can be repulsive or attractive; if A, B are both closed shells, the force is necessarily repulsive. In a quantal calculation these effects only appear if electrons are permuted between A, B so that the resulting interactions fall off exponentially with the separation R; this section deals with such short-range (or exchange) forces, to use the accepted terminology. Over a wide range of separations (sometimes referred to as 'intermediate') the exchange forces, though small, may be comparable with or larger than the long-range forces. It was first clearly recognized by Nikitin[31] that the short-range terms dominate in many processes involving excited species. Their importance in charge and excitation transfer has long been recognized. In constructing excited-state potential surfaces, two kinds of short-range interactions enter, one due to the overlap of valence electrons with distant cores, and the other due to core–core overlap. The first kind can be dealt with fairly concisely.

Consider a valence electron on A, in the atomic orbital $\psi_A(\mathbf{r})$, interacting with the core B through the weak pseudo-potential $\mathscr{V}_B$. The leading short-range interaction between A and B is

$$\begin{aligned} W_{SR} &\simeq \langle \psi_A | \mathscr{V}_B | \psi_A \rangle \\ &\simeq \tfrac{1}{2} a_B |\psi_A(\mathbf{R})|^2 \end{aligned} \tag{59}$$

where a_B is the e + B scattering length. This term will emerge from any model or pseudo-potential calculation. The approximations are only valid if the orbital populations on A, B are relatively unperturbed. In his original discussion, Nikitin did not derive the coefficient of ψ_A^2, but estimated it from other calculations. It is interesting that simple LCAO theory will always give $W_{SR} > 0$, although higher-order terms, included in (59) may change the sign of a_B; this is the case for all system B which exhibit a Ramsauer effect, e.g. Ar, Kr, Xe.

We now turn to the interaction between two cores whose ionization potentials are comparable. It is helpful to quote some lemmas on the asymptotic forms of exchange integrals. Let a, b be orbitals on A, B falling off like $\exp(-\lambda_a r)$, $\exp(-\lambda_b r)$ where λ_a is related to the ionization potential, $\lambda_a^2 = 2I_a$. If $\lambda_a \simeq \lambda_b$ and $f(\mathbf{r})$ is a smoothly varying function

$$\langle a|f|b\rangle \simeq \frac{\pi R^2}{(\lambda_a + \lambda_b)}(abf)_0 \tag{60}$$

where the subscript indicates that everything is evaluated at the midpoint of AB. Similarly,

$$\begin{aligned} \langle ab|1/r_{12}|ab\rangle - \text{LRT} &\simeq \langle ab|1/r_{12}|ba\rangle \\ &\simeq \frac{\pi R^2}{2(\lambda_a + \lambda_b)^3}(a^2b^2)_0 \end{aligned} \tag{61}$$

where LRT stands for long-range terms ($\sim R^{-n}$). To introduce the short-range interactions, we use the perturbation formulae (36)–(42), following Murrell and Shaw.[32] The zeroth-order wavefunction

$$\Psi_0 = \mathscr{L}(1 + \mathscr{P})\psi_0 \tag{62}$$

where $\mathscr{P}$ permutes electrons between the two centers and $\mathscr{L}$ is a normalization factor; ψ_t are the products (46). After substituting in (39) and (42), it is instructive to expand further in powers of the overlap between ψ_0 and $\mathscr{P}\psi_0$. The terms of zeroth order in overlap are

$$W_{LR}^{(1)} = V_{00}, W_{LR}^{(2)} = \sum_{t \neq 0} \frac{V_{0t}^2}{(E_0 - E_t)} \tag{63a}$$

which are the usual long-range interactions. The terms of first order in overlap are

$$\begin{aligned} W_{SR}^{(1)} &= V'_{00} - S'_{00}V_{00}, \\ W_{SR}^{(2)} &= \sum_{t \neq 0} \frac{V_{0t}}{(E_0 - E_t)}(V'_{t0} - S'_{00}V_{t0} - S'_{t0}V_{00}) \end{aligned} \tag{63b}$$

where

$$V'_{t0} = \langle\psi_t|V\mathscr{P}|\psi_0\rangle, S'_{t0} = \langle\psi_t|\mathscr{P}|\psi_0\rangle.$$

In general, $W_{SR}^{(1)}$ involves a product of single electron overlaps $\langle a_k|b_k\rangle$, $1 \le k \le N(\mathscr{P})$ where $N(\mathscr{P})$ is the number of electrons permuted. The larger $N(\mathscr{P})$ the shorter the range of W_{SR}. If we exchange one electron, as in an ion-atom system,

$$W_{\rm SR}^{(1)} \simeq \frac{2\pi R}{(\lambda_a + \lambda_b)}(ab)_0. \tag{64}$$

If two electrons are exchanged

$$W_{\rm SR}^{(1)} \simeq \frac{\pi R^2}{2(\lambda_a + \lambda_b)^3}(\rho_a \rho_b)_0 \tag{65}$$

where ρ_a is the electron density in A. One should be aware that if ionic terms are present, they will swamp other exchange terms at large enough R, e.g. if A has a negative ion state of attachment energy $J_{\rm A}$, the leading term is

$$W_{\rm SR} \simeq \frac{\langle a^- | b \rangle^2}{I_{\rm B} R^2} \simeq \frac{J_{\rm A}^{1/2}}{I_{\rm B} R^4} \exp[-2(2J_{\rm A})^{1/2} R].$$

Though slowly decaying, this expression has a small outside factor and may not be practically important. Finally, we note that $W_{\rm SR}^{(2)}$ is the exchange-polarization term which gives exponentially decaying corrections to the coefficients of the long-range (R^{-n}) interactions.[34]

While the general features of exchange forces are fairly clear, the accuracy of the leading terms (63) is uncertain. The perturbation theory is an expansion in two quantities V, S, neither of which is particularly small, so that calculating higher orders will probably not supply an answer; cancellation between direct and exchange terms is probably severe in the higher orders.[34] The nub of the difficulty is brought out by formulae such as (64) which show that one needs an accurate knowledge of the wavefunction far from both centres. A surprisingly small perturbation due to (say) B may drastically change the orbitals of A. Thus if B has a charge ζ, λ_a is modified by $\zeta/\lambda_a R$ so that (64) acquires a factor $\exp(-\zeta/2\lambda_a)$; for large $\zeta (\gtrsim 3)$ the polarization of A must also be taken into account. The importance of the wavefunction far from both centres is clarified in the alternative approach due to Holstein and Firsov.[35] Consider the integral

$$\int (\Psi_0 H \psi_0 - \psi_0 H \Psi_0) \mathrm{d}\tau$$

over the half-space $z > 0$, where A, B are located at $z = \pm \frac{1}{2} R$. Transforming to a surface integral over $z = 0$, we find that

$$W_{\rm SR} = 2 \int (\mathscr{P} \psi_0) \frac{\partial \psi_0}{\partial z} \mathrm{d}x \mathrm{d}y \tag{66}$$

which, apart from factors close to unity, is asymptotically equivalent to V'_{00}.

An interesting and widely quoted approach to the exchange forces is the electron gas model of Gordon and Kim.[36] The interaction of A, B is divided into electrostatic and electron gas contributions. The former is computed straightforwardly from the unperturbed charge densities (taken from SCF calculations) of A, B and has its own short-range terms of the form (64). The latter is made up of kinetic, exchange and correlation terms taken from the energy-density functional $\mathscr{E}(\rho)$ of an infinite uniform electron gas,

$$W_{\mathrm{EG}} = \int [\rho\mathscr{E}(\rho) - \rho_a\mathscr{E}(\rho_a) - \rho_b\mathscr{E}(\rho_b)]\,\mathrm{d}\tau. \tag{67}$$

The combined density $\rho \simeq \rho_a + \rho_b$ correct to first order in the orbital overlap. It can be shown that the leading term in (67) is approximately

$$W_{\mathrm{EG}} \simeq \frac{4.05\pi R^2}{(\lambda_a + \lambda_b)}(\rho_a\rho_b)_0^{5/6}. \tag{68}$$

This falls off less rapidly than the LCAO approximation (65), as suggested by physical arguments. Although this method is simple and fairly successful, its accuracy is difficult to assess because there is no sequence of higher approximations.*

3. *Note on the Additivity of Pairwise Interactions*

If the interaction between a cluster of atoms is treated by second-order perturbation theory (44), the interaction energy contains, in addition to contributions from each pair of atoms, three-body terms analogous to (15). Many other pointers indicate that, unless all the atoms are widely separated, one should start with electrons in MOs corresponding to bonds rather than single atoms. Then $E^{(2)}$ contains inter-bond terms as well as inter-atom terms. The pointers referred to include the success of DIM (II.B.3), the fact that many molecular properties (e.g. polarizabilities) are additive with respect to bonds rather than atoms,[3] and the rigidity of linear and planar molecules which cannot be accounted for by pairwise atomic forces.

D. Dynamical Effects

Until now we have only considered the calculation of stationary electronic states associated with given nuclear configurations. As a bridge to the discussion of collision problems, we shall now ask what complications enter the adiabatic potential surface picture when the molecular system evolves in time. Obviously, the most important effect is the breakdown of the B–O approximation when the nuclei are allowed to move (II.D.1). This leads naturally to the subjects of diabatic states and avoided crossings (II.D.2). Finally, we say a little about decaying electronic states, which have been

*Counting errors should be no more than $\mathcal{O}(1/\mathcal{N})$. If W_{EG} is small, the calculation ought to be equivalent to SCF with a local exchange potential.

inadequately treated in the chemical literature. In general, the topics of this section are not extensively discussed in textbooks or review articles, though standard texts on atomic collision theory[37] form an adequate introduction.

1. *Nuclear Motion*

Let us recall the Schrödinger equation (1) for the total wavefunction

$$(T + H_{\text{el}})\Psi = E\Psi \tag{69}$$

where T is the nuclear kinetic energy less the centre-of-mass energy. Introduce a complete set of solutions of (3)

$$H_{\text{el}}\Phi_n = W_n\Phi_n \tag{70}$$

so that we can expand

$$\Psi\Sigma\chi_n(\mathbf{R})\Phi_n(\mathbf{R}|\mathbf{r}). \tag{71}$$

When (71) is inserted in (69), it is found that different χ_n are coupled by the nuclear kinetic energy. It is convenient to introduce momentum and velocity operators,

$$\mathbf{P}_\alpha = -i\frac{\partial}{\partial \mathbf{R}_\alpha}, \dot{\mathbf{R}}_\alpha = \mathbf{P}_\alpha/A_\alpha$$

where A_α is the nuclear mass, as in (1).
After some manipulation, we find the set of coupled equations

$$(H^{(n)}_{\text{nuc}} - E)\chi_n = -\Sigma(\mathscr{C}_{nm} + \mathscr{C}'_{\text{nm}})\chi_{\text{m}} \tag{72a}$$

$$H^{(n)}_{\text{nuc}} = T + W_n \tag{72b}$$

$$\mathscr{C}_{nm} = -i\sum_\alpha \mathbf{C}^\alpha_{nm}\dot{\mathbf{R}}_\alpha, \quad \mathscr{C}'_{nm} = \langle\Phi_n|T|\Phi_m\rangle \tag{72c}$$

$$\mathbf{C}^\alpha_{nm} = \langle\Phi_n|i\mathbf{P}_\alpha|\Phi_m\rangle. \tag{72d}$$

The operators $\mathscr{C}$, $\mathscr{C}'$ describe the breakdown of the B–O approximation. If they are neglected, as in (2)–(4) above, we recover the usual B–O separation

$$\Psi = \chi_n\Phi_n, \quad (H^{(n)}_{\text{nuc}} - E)\chi_n = 0. \tag{73}$$

Thus χ_n is the vibrational wavefunction of the nuclei moving on the adiabatic potential surface W_n set up by the electrons. The terms $\mathscr{C}'_{nm}$ are usually thought to be small since they are really of second order in the nuclear velocities, and we do not consider them further ($\mathscr{C}'_{nn}$ provides a correction to W_n, usually < 0.01 eV). This leaves $\mathscr{C}_{nm}$, which (like $\mathbf{C}^\alpha_{nm}$) is real and skew-symmetric.

The expression (72c) for $\mathscr{C}_{mn}$ has the transformation properties of the nuclear kinetic energy

$$T = \Sigma \tfrac{1}{2} \mathbf{P}_\alpha \dot{\mathbf{R}}_\alpha$$

which may be written

$$T = \Sigma \tfrac{1}{2} P_\xi \dot{Q}_\xi + \tfrac{1}{2} I(Q)^{-1} \mathbf{L}^2 \tag{74}$$

where $\{Q_\xi\}$ are the $(3K-6)$ normal coordinates, $\mathbf{L}$ the total angular momentum, and $I(Q)$ the moment of inertia. It follows that

$$\mathscr{C}_{nm} = -i \Sigma C^{\xi}_{nm} \dot{Q}_\xi - i I(Q)^{-1} \mathbf{C}^0_{nm} \mathbf{L}, \tag{75a}$$

$$C^{\xi}_{nm} = \left\langle \Phi_n \left| \frac{\partial}{\partial Q_\xi} \right| \Phi_m \right\rangle, \quad \mathbf{C}^0_{nm} = \langle \Phi_n | \mathbf{L}^{\text{el}} | \Phi_m \rangle \tag{75b}$$

where $\mathbf{L}^{\text{el}}$ is the electronic orbital angular momentum. In a diatomic (75) reduces to

$$\mathscr{C}_{nm} = -i C^1_{nm} \dot{R} - i(AR^2)^{-1} C^0_{nm} L^{\text{el}}_x, \tag{76a}$$

$$C^1_{nm} = \left\langle \Phi_n \left| \frac{\partial}{\partial R} \right| \Phi_m \right\rangle, \quad C^0_{nm} = \langle \Phi_n | L^{\text{el}}_x | \Phi_m \rangle \tag{76b}$$

for nuclei moving in the 0_{yz} plane. The operator C^1 then connects states of the same $\Lambda = L^{\text{el}}$, and C^0 states differing by one unit in Λ.

In principle most atomic collision problems at low energies (i.e. at velocities $\ll$ one atomic unit, $2 \times 10^6 \text{ ms}^{-1}$) can be formulated in terms of (72a). In addition to the surfaces W_n, one needs only the couplings C^{ξ}_{nm} which are readily calculated from the wavefunctions. It is sometimes convenient to use the identity obtained by commuting $\partial/\partial Q_\xi$ with H_{el},

$$C^{\xi}_{nm} = (W_n - W_m)^{-1} \left\langle \Phi_n \left| \frac{\partial H_{\text{el}}}{\partial Q_\xi} \right| \Phi_m \right\rangle. \tag{77}$$

For collision processes involving two atoms, including elastic and exciting collisions and charge transfer, few problems remain (beyond finding enough computer time!) but as yet little can be done with larger numbers of atoms. The system (72) then becomes one of *coupled* partial differential equations which are formidable by any standards; most of chemical reaction theory is only concerned with the single partial differential equation (73).

Some remarks can be made about the atom–diatom collision A + BC, where ABC is restricted to a linear or isosceles configuration so that

$$T_{\text{ABC}} = -\frac{1}{2M}\frac{\partial^2}{\partial R^2} - \frac{1}{2\mu}\frac{\partial^2}{\partial \rho^2}. \tag{78}$$

The separation BC is denoted by ρ, and that between A and the centre-of-mass of BC by R. The expansion (71) has the form

$$\Psi = \Sigma \mathscr{F}_n(R,\rho)\Phi_n(R,\rho|\mathbf{r}) \tag{79a}$$

where $\mathscr{F}_n$ may be formally expanded in the vibrational wavefunctions of BC,

$$\mathscr{F}_n(R,\rho) = \Sigma F_{nv}(R)\zeta_v(\rho). \tag{79b}$$

Great simplification follows from an assumption of vibrational adiabaticity, in the sense that the vibrational states of BC retain their integrity during the collision. If one inserts (79b) in (72a), coupling terms appear like

$$\langle v|W_n|v'\rangle, \langle v|C^R_{nm}|v'\rangle\frac{\mathrm{d}}{\mathrm{d}R}, \left\langle v\middle|C^\rho_{nm}\frac{\mathrm{d}}{\mathrm{d}\rho}\middle|v'\right\rangle.$$

The last-written term may be dropped as really of second order. The first term can be made to vanish for $v \neq v'$ if ζ_v varies adiabatically with R; extra terms $\langle v|\partial/\partial R|v'\rangle$ then appear but are again of second order. One is left with

$$\langle v|C^R_{nm}|v'\rangle\frac{\mathrm{d}}{\mathrm{d}R}$$

and if even this coupling is dropped, the problem has been reduced to motion on a single (vibrationally adiabatic) surface

$$W_{nv}(R) = \langle v|W_n|v\rangle.$$

Such approaches, incorporating some physical insights, may be more fruitful in the short term than a frontal numerical attack on (72a).

In most physical problems the matrix elements C^ξ play a more important role than C^0, since the form can vary rapidly over a small range of nuclear configurations, while the latter tends to be slowly varying everywhere. In the next section, we inquire why and where C^ξ should change rapidly.

2. *Diabatic States and Avoided Crossings*

The calculation of molecular states usually starts with a set of atomic product states $\{\psi_t\}$ such as (46). The eigenstates of H_{el}, i.e. the usual adiabatic states, are then linear combinations

$$\Phi_n = \Sigma c_m \psi_t \tag{80}$$

where the coefficients are determined variationally. We can define diabatic states as those described by eigenfunctions ψ_t of some operator $\mathcal{O}$ other than H_{el}, e.g. any approximate Hamiltonian H_0, an unperturbed atomic Hamiltonian H_{A}, or an operator which commutes with H_0, as total orbital angular momentum commutes with H_{A}. Associated with ψ_t are the diabatic potentials

$$H_{tt} = \langle\psi_t|H_{\mathrm{el}}|\psi_t\rangle. \tag{81}$$

Common examples would be a single product of atomic eigenstates, e.g. any molecular Rydberg state of assigned l, or a single product of LCAO-MOs, e.g. $1\sigma_g^2\, 1\sigma_u^2$ for the ground state of He_2. While the adiabatic states are unique, the diabatic states of a system may be chosen in an infinite number of ways. Paradoxically, the diabatic concept is most useful in describing the behaviour of the adiabatic states.[38]

To understand the transformation (80), it is sufficient to consider only two diabatic states ψ_1, ψ_2. The eigenvalues of H_{el} are obtained from the secular equations

$$\begin{aligned} H_{11}c_1 + H_{12}c_2 &= Wc_1 \\ H_{21}c_1 + H_{22}c_2 &= Wc_2 \end{aligned} \tag{82a}$$

and they exhibit the well-known avoided crossing behaviour (Fig. 1). Diabatic states often have different approximate symmetries, but the associated adiabatic states have the same exact symmetry. The non-crossing rule, that adiabatic states of the same symmetry do not cross, has been questioned in recent years, but the only apparent exceptions are some cases in one-electron systems where hidden symmetries exist.[39] Avoided crossings are vital in

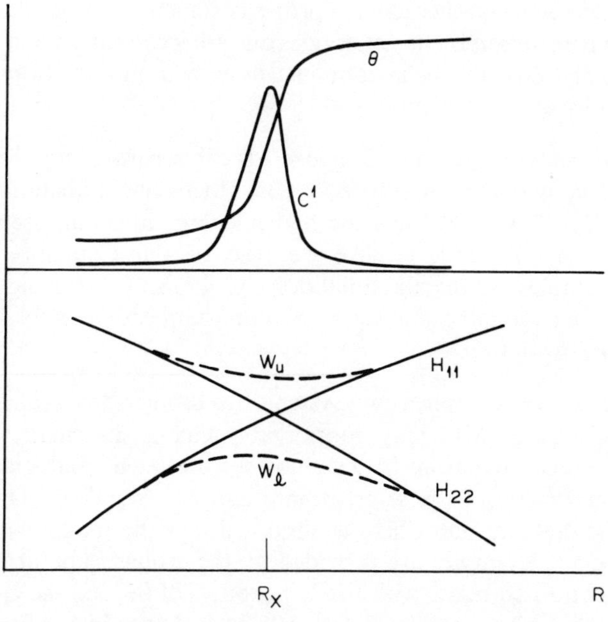

Fig. 1. Behaviour of diabatic and adiabatic potentials near a crossing point R_X (lower box). The mixing angle and radial coupling matrix element as defined in (82)–(83) are shown in the upper box.

understanding the qualitative features of surfaces, e.g. they explain almost all barriers and multiple minima. To return to (82a) the adiabatic wavefunctions may be written

$$\Phi_l = \cos\theta\psi_1 + \sin\theta\psi_2,\quad \Phi_u = \sin\theta\psi_1 - \cos\theta\psi_2,$$

$$\tan 2\theta = \frac{2H_{12}}{(H_{11} - H_{22})}. \tag{82b}$$

If $H_{\rm el}$ depends on only a single normal coordinate R, generalizations being obvious,

$$C^1_{lu} = \mathrm{d}\theta/\mathrm{d}R. \tag{83}$$

If the matrix elements are slowly varying functions of R, θ and hence C^1 will only vary rapidly if $H_{11} - H_{22}$ passes through zero, i.e. if the diabatic potentials (81) cross, say at R_x. Then θ will jump through $\pi/2$ at R_x and C^1 will exhibit a peak. On one side of R_x the adiabatic states $\Phi_l, \Phi_u \simeq \psi_1, \psi_2$, while on the other side they exchange their characters $\Phi_l, \Phi_u \simeq \psi_2, \psi_1$. In a slow encounter, assuming R is the coordinate which varies in time, a system initially in Φ_l will remain there with only a small probability of transition into Φ_u. However, in a fast encounter, the adiabatic states will be strongly mixed by C^1 as R passes through R_x. In contrast, the diabatic states preserve their character in a fast encounter, a property sometimes used as a definition. Whichever description is chosen, transitions are localized around R_x.

The avoided crossing behaviour just described almost always arises in one of two ways.

(*a*) *Ionic–covalent interaction.* Suppose AB can separate into either neutral or ionized components $A + B$, $A^+ + B^-$; these are diabatic states with $\mathcal{O} = H_A + H_B$. If $A^+ + B^-$ has the higher energy at infinite separation, it must cross $A + B$ at $R_x \simeq 1/\Delta E$ because of the Coulomb attraction. Notable examples are the alkali halides, e.g. KF/K^+F^-. It is not necessary for the negative ion to be stable, e.g. the important case of $NaN_2/Na^+N_2^-$ is discussed below (III.D).

(*b*) *Orbital promotion.* When two systems are brought from infinite to zero separation, a given MO may considerably change its character, usually because the separated atom MO cannot be correlated adiabatically with a united atom orbital of the same atomic quantum numbers. The operator defining the diabatic states here is almost always the total orbital angular momentum $\mathbf{L}^2$. Examples are provided by the ground state of He_2, where the σ_n 1*s* orbitals correlate with the 2*pσ* orbitals of Be, and the ground state of LiHe, where $\sigma 2s_{\rm Li}$ correlates with $\sigma 3p$ of B. Other instructive cases are found in the excited states of NaHe.[40]

A situation complementary to the avoided crossing, usually associated with Demkov,[41] occurs when H_{11} and H_{22} do not cross but H_{12} rises steeply for $R < R_0$. Though of less importance than curve crossings, this case is frequently found in charge-transfer problems,[42] e.g. $Na^+ + Li \rightarrow Na + Li^+$. Near R_0, θ goes from 0 to $\pi/2$ and C^1_{ln} peaks; the adiabatic states pass from an atomic character $\Phi_l, \Phi_u = \psi_1, \psi_2$ to a mixed character $\Phi_l, \Phi_u = \psi_1 + \psi_2$, $\psi_1 - \psi_2$ where the diabatic states ψ_1, ψ_2 correspond to AB^+, A^+B. In realistic calculations C^{ξ}_{mn} can exhibit more complicated behaviour, e.g. going rapidly through zero near R_x.[43]

In dealing with excited states of polyatomics, one must be alive to the possibility of a Jahn–Teller effect. This refers to the theorem that a non-linear nuclear configuration belonging to a symmetry group $\mathcal{G}$ cannot be in equilibrium if the electronic state belongs to a degenerate representation ρ of $\mathcal{G}$. Suppose we expand the electronic Hamiltonian about the nuclear configuration $\mathcal{G}$ in terms of normal coordinates Q_ξ which belong to representations of $\mathcal{G}$,

$$H_{el} \simeq H_{el}(0) + \Sigma Q_\xi F_\xi(0), \quad F_\xi = \frac{\partial H_{el}}{\partial Q_\xi}. \tag{84a}$$

If ρ is a non-degenerate representation associated with the electronic state Φ_ρ, the expectation value of (84a) becomes

$$W_\rho \simeq W_\rho(0) + Q_1 \langle \rho | F_1 | \rho \rangle \tag{84b}$$

where Q_1 belongs to the completely symmetric representation of $\mathcal{G}$. Thus it is possible by adjusting Q_1 to find a minimum of W_ρ. However, if ρ is degene-

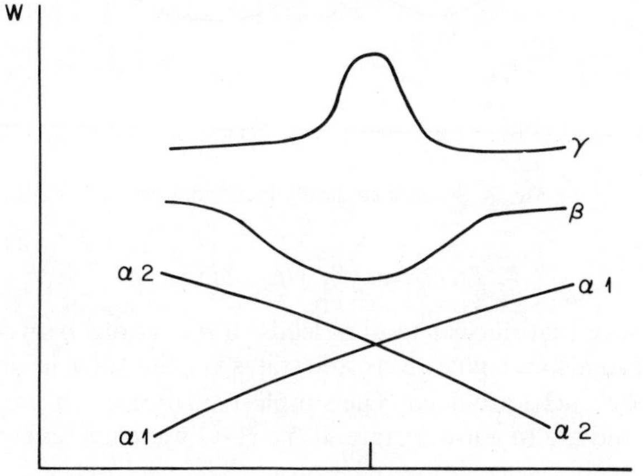

Fig. 2. Potentials associated with a doubly degenerate representation α of the molecular symmetry group. Also those associated with non-degenerate representations β, γ (cf. (84)).

rate, with substates $\rho\sigma$ which diagonalize the Q_ξ,

$$W_{\rho\sigma} \simeq W_\rho(0) + \Sigma Q_\xi \langle \rho\sigma | F_\xi | \rho\sigma \rangle. \tag{84c}$$

In this case it is not possible to remove the linear terms for all σ at once so no stable configuration exists. Fig. 2 shows a section across the potential surface for singly and doubly degenerate representations. In the latter case the two surfaces are degenerate in a $(3\mathscr{K} - 8)$ dimensional hyperspace (for a $\mathscr{K}$-atom system). In the vicinity of $Q_\xi = 0$, it is clear that the simple B–O separation breaks down, though the wavefunction may be represented by a superposition of diabatic states,

$$\Psi = \Sigma \chi_\sigma(Q) \Phi_{\rho\sigma}(Q).$$

There are a few other cases where the approximation of a single B–0 term may be completely invalid, usually associated with slightly hindered rotations or small potential barriers, e.g. in Fig. 3 the nuclear motion is shared between the wells I and II.

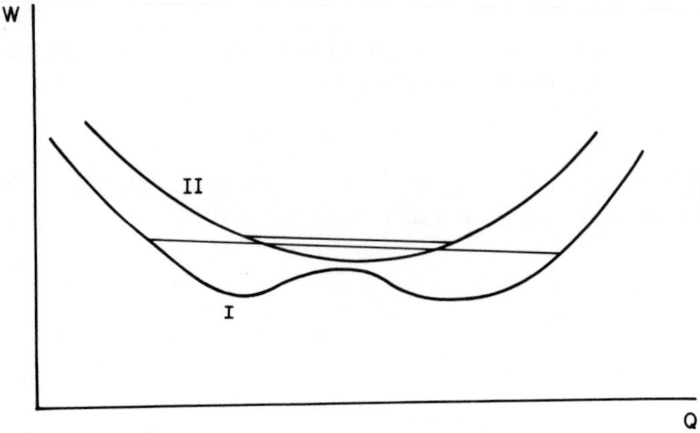

Fig. 3. Strongly coupled vibrational levels.

3. *Electronically Unstable States*

We have seen that nuclear motion leads to transitions between adiabatic states. One can also imagine electronic states which evolve in time or decay, even when the nuclei are fixed. The simplest such case is a molecular state u which can radiate to a lower state. If the B–O wavefunctions are

$$\Psi_n = \chi_n \Phi_n, \Psi_l = \chi_l \Phi_l \tag{85}$$

the rate of transitions is proportional to the square of a dipole matrix element

$$\mathbf{D}_{ul} = \langle \chi_u | \mathscr{D}_{ul} | \chi_l \rangle, \mathscr{D}_{ul} = \langle \Phi_u | \mathscr{D} | \Phi_l \rangle \tag{86}$$

where $\mathscr{D}$ is the dipole moment of all the electrons. If the nuclear motion is treated classically, the decay probability along a trajectory on the upper surface is[44]

$$= \int A_{ul}[\mathbf{R}(t)]\mathrm{d}t \tag{87}$$

where A_{ul} is the Einstein A-coefficient for a transition at one nuclear configuration. The Franck–Condon principle that the nuclei do not move during an electronic transition is implicit in this formula. The original application of (87) was to explain the formation of interstellar CH^+ and CH molecules by radiative association. In predissociation the final state is a superposition of kinematically coupled B–O states. The preceding formalism can be extended to describe processes induced by electron impact or by laser radiation.

A more complicated problem[45] is that of an excited (usually doubly excited) electronic state M_r^{**} which can autoionize into the continuum $M_c^+ + e$. The state r is thus a resonance in the scattering of electrons by M_c^+. Let the electronic wavefunction describing $e + M_c^+$ have the form

$$\Phi_c \sim f\phi_c \tag{88a}$$

when the electron is at large distances, f being the single electron scattering state and ϕ_c the core wavefunction. Then the states M_r^{**} are usually defined by the constraint

$$P\Phi_r = 0, \quad P = |\phi_c\rangle\langle\phi_c| \tag{88b}$$

i.e. P is the projection operator on to ϕ_c. At a given nuclear configuration the rate of decay from r into c is proportional to the square of a matrix element

$$V_{rc}(R_c) = \langle \Phi_r | H_{el} | \Phi_c(k_{rc}) \rangle \tag{89}$$

where the electron is ejected at the vertical energy $\frac{1}{2}k_{rc}^2 = W_r - W_c$. The nuclear motion may be treated quantally or classically, but one must in either case allow for the decaying nature of r by adding to W_r an imaginary part, making the surface complex,

$$\mathscr{W}_r = W_r - \frac{i}{2}G_r, \; G_r = 2\pi|V_{rc}|^2. \tag{90}$$

An example of a transition from r to c is Penning ionization[46]

$$He(2^3S) + H(1s) \rightarrow [He(1^1S) + H^+] + e$$

illustrated by the curve r_1 in Fig. 4. Dissociation recombination is a transition from c to r, as shown by r_2 in Fig. 4.

It is interesting to ask what happens to resonant states, e.g. r_2 and r_3 in in Fig. 4, whose energies fall below that of c so that autoionization is no

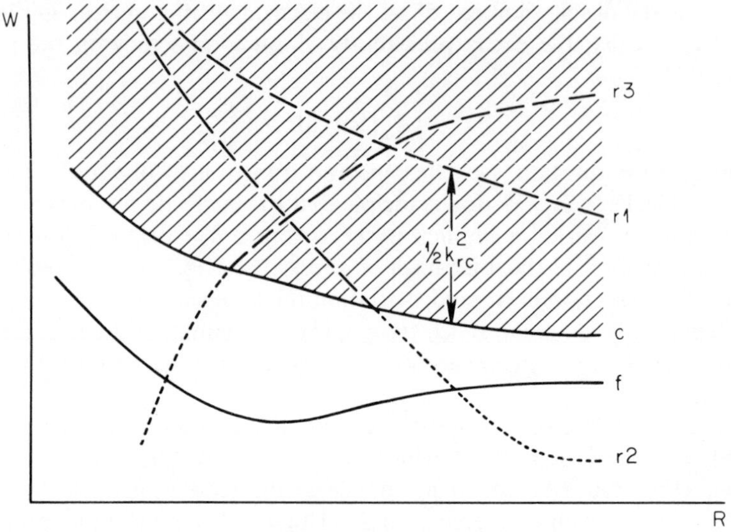

Fig. 4. Electronically decaying states. The solid lines are the potentials of stable states; c is a state of M^+ bounding the shaded $e + M^+$ continuum; r_1, r_2, r_3 are resonances of M (dashed portions), the last two becoming diabatic states below c (dotted portions); f is a stable state of M (cf. (87)–(90)).

longer possible. These may be regarded as diabatic states (with $\mathcal{O} = P$ in the notation of II.D.2) from which the system can emerge into the sunlight by crossing a final state f; they are unlikely to be electronically stable. Instead of treating the decay of r in this region as a curve-crossing problem, we could regard the process as one of internal rearrangement of the electrons (or internal autoionization). Since excess energy is absorbed by nuclear recoil rather than an ejected electron, the idea of a local width (90) is no longer valid, but each vibrational state has its own width. In certain collision problems (III.D below) these diabatic states must be explicity included.

III. APPLICATIONS TO SCATTERING PROBLEMS

From the preceding sections we see that excited-state potential energy surfaces can be calculated together with the coupling Matrix elements describing the breakdown of the B–O approximation. We now inquire how this knowledge can lead to an understanding of scattering processes involving excited species. It is not our intention to go into scattering theory in any detail but simply to survey the range of problems for which methods are available, and to draw attention to those areas where theory is still deficient. The textbooks already cited[37] may be consulted for more information on scattering formalism.

A. Formulation of Atomic and Molecular Scattering Problems

Consider a collision between systems A, B whose separation is **R** and reduced mass μ. The internal states (channels) of AB have wavefunctions $\{\phi_\alpha\}$ and wavenumbers, velocities, relative angular momenta given by $\{k_\alpha, v_\alpha, l_\alpha\}$. The amplitude $\mathscr{F}_{\alpha\beta}$ for a transition $\alpha \rightarrow \beta$ is defined by the asymptotic boundary condition

$$\Psi(\mathbf{k}_\alpha) \sim e^{i\mathbf{k}_\alpha \cdot \mathbf{R}}\Phi_\alpha + \sum_\beta \left[\frac{k_\alpha}{k_\beta}\right]^{1/2} \frac{e^{ik_\beta R}}{R} \mathscr{F}_{\alpha\beta}(\hat{\mathbf{k}}_\alpha, \hat{\mathbf{k}}_\beta). \tag{91}$$

We expand Ψ in the channel eigenstates Φ_α, which are products of the ϕ_α with spherical harmonics in $\hat{\mathbf{R}}$,

$$\Psi^\beta = \Sigma R^{-1} F_\alpha^\beta(R)\Phi_\alpha \tag{92}$$

and impose the boundary conditions

$$F_\alpha^\beta \sim \delta_{\beta\alpha} k_\beta^{-1/2} e^{-ik_\beta R} - S_{\beta\alpha} k_\alpha^{-1/2} e^{ik_\alpha R}. \tag{93}$$

Then $\Psi(\mathbf{k}_\alpha)$ may be expressed as a linear combination of Ψ^β and $\mathscr{F}_{\alpha\beta}$ as a superposition of spherical harmonics, the coefficients being multiples of $S_{\beta\alpha}$. All cross sections are then linear combinations of the partial cross sections

$$\sigma_{\beta\alpha} = \frac{\pi}{k_\beta^2} |S_{\beta\alpha} - \delta_{\beta\alpha}|^2, \tag{94}$$

the detailed form depending on the angular momentum coupling scheme in each particular case. The scattering matrix (S-matrix) elements are obtained by solving the set of equations

$$\left[\frac{d^2}{dR^2} + k_\alpha^2 - \frac{l_\alpha(l_\alpha + 1)}{R^2}\right] F_\alpha = 2\mu \sum_\beta V_{\alpha\beta} F_\beta \tag{95}$$

derived by substituting (92) in the Schrödinger equation, subject to (93).

At large collision velocities ($\gtrsim 0.05$ atomic units) the fully quantal treatment goes over to the conceptually simpler impact parameter formulation. The colliding partners are constrained to move on a straight line (or if both are charged, a Coulomb trajectory) relative path so that R is a function of time, t. The time dependent wavefunction is expanded as

$$\Psi = \Sigma c_\alpha(t)\phi_\alpha \exp(-i\textstyle\int W_\alpha dt). \tag{96}$$

Instead of (93), we put

$$c_\alpha(-\infty) = \delta_{\beta\alpha}, c_\alpha(\infty) = S_{\beta\alpha}^{\mathrm{IP}} \tag{97a}$$

and the cross sections are obtained by integrating over all impact parameters

$$\sigma_{\beta\alpha} = \textstyle\int |S_{\beta\alpha}^{\mathrm{IP}} - \delta_{\beta\alpha}|^2 d^2\mathbf{b}. \tag{97b}$$

Differential amplitudes may be obtained by performing a Hankel transformation on the elements of S^{IP}, from **b** to momentum transfer.

In the simplest problems one can imagine, elastic scattering apart, *viz.* excitation or charge transfer in a collision between two stoms, two approaches are possible. One can expand in atomic (or diabatic) states, which are usually eigenfunctions of part of the Hamiltonian H_0, so that the coupling in (95) arise from the remaining part of the Hamiltonian V. Alternatively, one can expand in adiabatic states to obtain (72a). For diatomics (72a) is equivalent to (95), with the diagonal couplings $V_{\alpha\alpha}$ given by the adiabatic interactions W_α, and the off-diagonal couplings $V_{\alpha\beta}$ by the adiabatic corrections $C^0_{\alpha\beta}$ or $C^1_{\alpha\beta}$. Many applications of both methods can be found in the recent literature; adiabatic expansions are more appropriate at low velocities[47] and diabatic expansions at high velocities,[48] though the regions of validity overlap in a substantial range $0.1 < v < 1$ atomic units.

B. Intermultiplet Transitions

Consistent application of the adiabatic formulation is probably the most satisfactory way to study low-energy atomic collisions. However, if one of the atomic partners has fine-structure levels, or if we have an atom–molecule collision in which rotational and vibrational degrees of freedom participate, a purely adiabatic approach is very cumbersome and hybrid methods are preferable. The philosophy of these methods is illustrated by the theory of intermultiplet transitions developed by Reid and Dalgarno.[49] Consider two atoms A, B in states $S_aL_aJ_a$, $S_bL_bJ_b$ which come together to form molecular states $S\Lambda$ with Hund's case (*b*) coupling. From molecular states Φ_λ we can form quasi-atomic states

$$\phi_n = \Sigma C_{n\lambda}\Phi_\lambda \tag{98}$$

where n, λ are short for all the atomic, molecular quantum numbers and $\{C_{n\lambda}\}$ are the transformation coefficients at infinite separation. We can then use (98) to evaluate matrix elements of the Hamiltonian

$$H = H_{\text{el}} + H_{\text{rel}} \tag{99}$$

assuming that the electrostatic part has matrix elements

$$\langle \Phi_\lambda | H_{\text{el}} | \Phi_{\lambda'} \rangle = W_\lambda \delta_{\lambda\lambda'} \tag{100}$$

and the relativistic part has matrix elements

$$\langle \phi_n | H_{\text{rel}} | \phi_{n'} \rangle = \Delta E_n \delta_{nn'} \tag{101}$$

where ΔE_n is the fine-structure perturbation in the free atom. The coupled equations (95) may be set up and solved in either representation. The procedure is effective because an atomic expansion is valid where the fine-structure splittings are important, while at smaller separations where the W_λ become large the fine structure may be neglected.

If the fine-structure splittings are neglected altogether, as is permissible at high enough energies, then the transformation (98) can be applied to the S-matrix elements or scattering amplitudes; this is the 'elastic' or 'exact resonance' approximation. To take a simple but important example the cross section for the fine-structure changing (FSC) transition ${}^2P_{1/2}-{}^2P_{3/2}$ in a collision with a ${}^1S$ atom (e.g. Na* + He) is given by

$$\sigma_{\mathrm{FSC}} = \tfrac{2}{9}\int |\mathscr{F}_\Sigma - \mathscr{F}_\pi|^2 \mathrm{d}\omega \tag{102}$$

where $\mathscr{F}_{\Sigma,\pi}$ are the amplitudes for elastic scattering on the ${}^2\Sigma$, ${}^2\Pi$ surfaces. The total cross section our of either ${}^2P$ state is

$$\sigma_{\mathrm{TOT}} = \tfrac{1}{3}\int (|\mathscr{F}_\Sigma|^2 + 2|\mathscr{F}_\pi|^2)\mathrm{d}\omega. \tag{103}$$

Expressions have been worked out for the pressure-induced width and shift of spectral lines (e.g. the Na D lines) in an inert-gas environment in terms of the elastic amplitudes.[19]

C. Rotational and Vibrational Excitation

In this section we review briefly processes in which rotational or vibrational energy is transferred while only one electronic potential surface is involved. This is a necessary introduction to the more general case of several surfaces which usually arises when one of the colliding species is electronically excited. Purely classical calculations are often very successful in describing energy transfer on a single surface, but they cannot describe electronic transitions, so we shall concentrate on quantal methods. These methods fall under our classification 'hybrid', though are not usually thought of in this light.

Rotational excitation has one of the largest literatures of any problem in chemical physics.[50] Most discussions set out from the formalism of Arthurs and Dalgarno in which the wavefunction is expanded in eigenfunctions $\mathrm{X}(jlJ)$ of the rotor angular momentum $\mathbf{j}^2$, the relative angular momentum l^2 and the total angular momentum $\mathrm{J}^2 = (\mathbf{j}+\mathbf{l})^2$. In an atom–rotor collision $\mathbf{j}$ has the significance just described; if two rotors collide, j is replaced by several quantum numbers. When the expansion

$$\Psi^J = \sum_{jl} R^{-1} F^J_{jl}(R)\mathrm{X}(jlJ) \tag{104}$$

is substituted in the Schrödinger equation, a coupled system of the form (95) is obtained for each J, e.g. if the potential is expanded as

$$W(R,\Theta) = W_0(R) + W_1(R)\cos\Theta + W_2(R)\mathrm{P}_2(\cos\Theta) \tag{105}$$

states $jl, j'l'$ such that $|j-j'|, |l-l'| \leq 2$ are coupled. These equations can be solved for the S-matrix elements $S^J(jl, j'l')$ and hence amplitudes and cross sections. While this procedure has been carried out in full for many cases, it is complicated and expensive especially for collision partners heavier than, say, Ar. The number of J values is at least several hundred, and for each J the number of channels is proportional to the number of rotor levels retained. Thus much effort has been devoted to finding reasonably accurate simplified schemes. Such simplifications are probably essential to making any progress with problems involving excited electronic states.

The earliest simplification was the *fixed-nucleus* (elastic, exact resonance, adiabatic) approximation, briefly mentioned in another context in III.B. If the rotor energy level separations are ignored, the coupled equations are related algebraically to those describing a collision between rotors with fixed orientations. The full S-matrix can then be obtained from the elements $S(lm, l'm')$ of the fixed nucleus S-matrix by a unitary transformation. Another way of expressing the approximation is in terms of the amplitudes $\mathscr{F}(\hat{\mathbf{n}}, \theta)$ for elastic scattering as a function of molecular orientation $\hat{\mathbf{n}}$; the rotational excitation amplitudes are given by

$$\mathscr{F}(jm \to j'm' | \theta) = \langle jm | \mathscr{F}(\hat{\mathbf{n}}, \theta) | j'm' \rangle. \tag{106}$$

The fixed-nucleus approximation is certainly valid at high energies, meaning above 1 eV, and for elastic processes $j \to j$ at all energies. At low energies its predictions for inelastic processes must be viewed cautiously.[51] When strong long-range interactions are present (e.g. when the molecule has a dipole moment) the orientation of the rotor can change appreciably during the collision so that the assumption of fixed nuclei is not valied. However, if the anisotropies of the potential W_1, W_2 are of much shorter range than W_0, we can use the method to good effect; fortunately, this condition is often satisfied in excited atom–molecule problems.

A complementary approximation is that of the *average field*, in which the dependence of the couplings on the rotor quantum numbers j, j' is retained, while an average is taken over all the other quantum numbers.[52] In effect, states with different l, l' are decoupled. Thus for each $j \simeq l$, we have as many equations as rotor levels are retained. Even at low energies, quite good results are obtained for inelastic cross sections. Many other approximation schemes can be found in the recent literature.

Vibration excitation presents no formal problem. Consider a collinear system A + BC described by the coordinates R, ρ used in (78). When the wave-function is expanded in the unperturbed vibrational wavefunctions $\zeta_\alpha(\rho)$ of BC, the coupling matrix elements are

$$V_{\alpha\beta} = \langle \zeta_\alpha | W(R, \rho) | \zeta_\beta \rangle. \tag{107}$$

A more realistic model is one in which three dimensions are included, but the anisotropy of the potential is neglected; one then has a separate set of equations for each angular momentum J (or l). Again the adiabatic approximation is obtained by taking matrix elements of the elastic scattering amplitude calculated as a function of ρ,

$$\mathscr{F}(\alpha \to \beta|\theta) = \langle \zeta_\alpha | \mathscr{F}(\rho|\theta) | \zeta_\beta \rangle. \tag{108}$$

Simultaneous rotational and vibrational excitation can be calculated with no obstacle other than complexity, though little has been done. Since vibrational energy separations usually exceed rotational ones, the most sensible approach would probably be to allow fully for the vibrational splittings while treating rotation adiabatically.

D. Collisions between Molecules and Excited Atoms

We now combine the topics of the preceding two sections. Little has been done theoretically on collisions between molecules and excited atoms apart from recent work on the $Na(^2P) + N_2$ system, some very elegant studies on collisions between $Na_2(B^1\Pi_u)$ and inert-gas atoms, and classical studies on the quenching of I_2^* and N_2^* by inert systems.

Using a tunable laser, it is possible to populate a single vibrational–rotational level of $Na_2(B^1\Pi_u)$ and thereafter to monitor the distribution of rotational states following a collision.[53] A notable feature of the rotational transfer cross sections $\sigma(\Delta j)$ is a strong asymmetry between positive and negative Δj. This effect is traceable to the dependence of the interaction on the azimuthal angle about the axis of Na_2 which occurs when the molecule is not in a Σ state. The scattering problem may be set up as in III.C except that the interaction (105) now contains a term $W_{22}(R)Y_{22}(\Theta,\Phi)$. The average Na_2 internuclear distance is large ($\sim 10\,a_0$) in the high vibrational levels of interest so that a surface for (say) Na_2He may be constructed by adding the reasonably well-known NaHe and Na*He interactions. Good accord has been found between theory and experiment.

Calculations on rotational and vibrational transfer in the $Na^* + N_2$ system have been carried out by Amaee and Bottcher[54] who began by considering the problem of rotational and intermultiplet transitions at thermal energies

$$Na(^2P_J) + N_2(j) \to Na(^2P_{J'}) + N_2(j'). \tag{109}$$

The molecule here is confined to its ground vibrational state. Potential surfaces for $Na(^2S)$, $Na(^2P)$ interacting with N_2 at its equilibrium separation have been given by Bottcher; the assumptions underlying this calculation were discussed above (II.B.2). The physically significant features of the surfaces are the small anisotropies and the large attractive exchange interactions in the $B^2\Sigma$ state. Even in the ground state the anisotropy W_2

is small at Na–N_2 separations $R > 5\,a_0$ because of the almost spherical charge distribution in the N_2 ground state. The effect is yet smaller when the Na valence orbital is excited since the N_2 is then immersed in an almost uniform electron distribution. Another consequence of near isotropy is that the projection of orbital angular momentum on the Na–N_2 axis is close to a good quantum number; thus we have only to deal with the states $X^2\Sigma$, $B^2\Sigma$, $A^2\Pi$ corresponding to the $\sigma 3s$, $\sigma 3p$, $\pi 3p$ valence orbitals. The attraction observed in the B state is due to the transfer of charge from Na* into the N_2^- resonance which is accurately represented in the pseudo-potential model. Between $R = 7$ and $20\,a_0$ the interaction is well fitted by

$$W \simeq -\frac{\zeta^2}{R} + W_{\text{exc}} \tag{110}$$

where W_{exc} is a repulsive term given by (59) with $a_{\text{B}} = 1.2\,a_0$, and $\zeta = 0.07$. Interactions in the A state are much smaller because the $\pi 3p$ orbital is directed away from the N_2. The quasi-Coulomb attraction in (110) ensures large cross sections, e.g. the experimental FSC rate for Na* + N_2 is almost twice that for Na* + Ar which is nearly isoelectronic but has no ionic states. However, it should be noted that the admixture of ionic character (Na*N_2^-) in the adiabatic states is not large ($\sim 0.5\%$).

The scattering formalism for (109) was adapted from the elastic approximation method in atom–atom collisions (III.B). Thus (103) and (104) are taken over with $\mathscr{F}_{\Sigma,\pi}$ now representing the amplitudes for a $j \to j'$ transition on the B, A surfaces. Because of the weak anisotropy, the fixed nucleus approximation should be the appropriate way to calculate $\mathscr{F}_{\Sigma,\pi}$. In support of this suggestion (*ex post facto*) rotationally inelastic processes were found to be only 10% of the total. Furthermore, direct comparison with the average-field approximation found good agreement in the least favorable case (Na(^2S) + N_2). The theoretical FSC cross section at 400 deg is close to the experimental value.

Finally, we consider the important problem of quenching with vibrational transfer.

$$\text{Na}(^2P) + N_2(v'') \to \text{Na}(^2S) + N_2(v'). \tag{111}$$

The process must proceed through crossings with the ionic state $\text{Na}^+ + N_2^-$, so we need only consider the B and $X^2\Sigma$ surfaces.[55] If the orientation dependence of the surfaces is neglected, we have only two nuclear coordinates, R the Na–N_2 separation and ρ the N–N separation. The wavefunction is expanded as

$$\Psi = \Phi_{\text{i}} \Sigma F_m(R) \zeta_m(\rho) + \Phi_{\text{f}} \Sigma G_n \zeta_n(\rho) \tag{112}$$

where $\Phi_{\text{i,f}}(R, \rho)$ are the initial and final (B, X) electronic states and $\{\zeta_m\}$ are N_2 vibrational wavefunctions. Again, we find a set of equations (95) satisfied by $\{F_m, G_n\}$ for each total angular momentum $J \simeq l$. The diagonal interactions are to sufficient accuracy $W_0(R, \rho_e)$ where ρ_e is the equilibrium N–N separation. The real problem comes in the choice of the couplings $V(im, fn)$. It follows from the small ionic component mentioned above that the adiabatic coupling between i, f is inadequate to explain the large transition probabilities and that the ionic state must be explicitly included in (112). Amaee and Bottcher[56] assumed a unit probability for the transition into the ionic state, so that the system moves on W_i for $R > R_x$ and W_d for $R < R_x$ where R_x is the crossing point (see Fig. 5). The ionic potential was taken as a Coulomb term augmented by a first-order polarization term

$$W_d = -\frac{1}{R} - \frac{D}{R^2} \tag{113}$$

where $D \simeq 5$ a.u. Without the second term, the angular momentum barrier prevents the system from reaching the crossing region (Fig. 5). Good agreement with experiment was obtained for the magnitude of the total cross section, $\simeq 36\pi a_0^2$ for initial energies between 0.1 and 0.5 eV. The predicted branching ratios from the initial state $v'' = 0$ were largest for final states $v' = 3, 4$ in accord with recent measurements.[57]

Steinfeld and his collaborators have met with some success using classical trajectory and semiclassical methods. The transfer of vibrational and rotational energy from I_2 in the $B(^3\Pi, 0_u^+)$ state to inert gases was studied

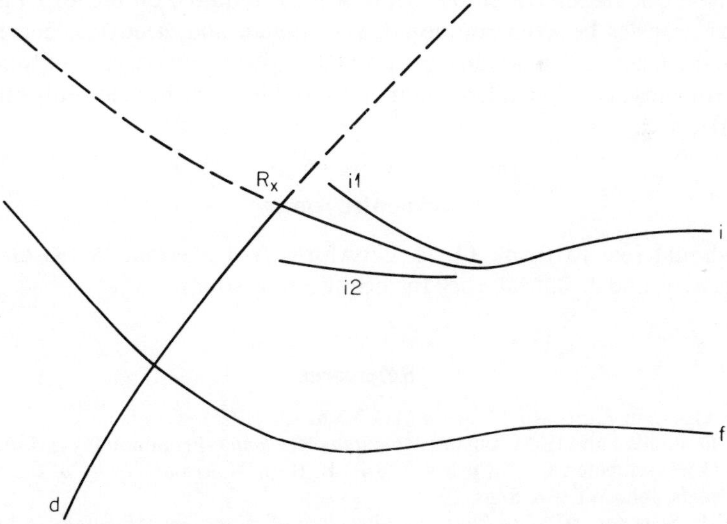

Fig. 5. Potentials associated with the Na–N_2 quenching process. The initial and final adiabatic states i, f are crossed by the ionic state d; i1, i2 illustrate the effects of adding the centrifugal barrier and polarization term (113).

using classical trajectories on a surface apprcximated by pairwise interactions.[58] Electronic transitions were incorporated through a correction based on the Landau–Zener formula. Good agreement with experiments is obtained and it is possible to eliminate dissociation of I_2 as a significant channel. The electronic deactivation of $N_2(B^3\Pi_g)$ has been explained[59] using a quasi-classical optical model, again based on Landau–Zener. The dominant channel is identified as $W^3\Delta_u$ rather than $A^3\Sigma_u^+$.

Finally we should draw attention to the practically important, but theoretically neglected, subject of the quenching of $O(^1D, {}^1S)$ by collisions with inert systems.[60] The magnitudes of these cross sections, though small (invariably $<10^{-17}\ \text{cm}^2$ and in some cases $<10^{-22}\ \text{cm}^2$) are still difficult to explain in a simple way. It seems likely that if X is an inert system, spin–orbit effects in the quasi-molecule XO* can be surprisingly large at distances well beyond the repulsive wall.

IV. CONCLUSIONS

We have seen in II that sufficient methods are available to calculate a wide range of excited-state potential surfaces to an accuracy adequate to predict cross sections. However, it is clear from III that little has been done on excited-state collisions except in atom–atom problems. The excited atom–molecule problem will probably make rapid advances in the near future; the simplified approaches to rotational excitation are vital in rendering these calculations tractable. More work is required on the formalism of energy transfer between rotational, vibrational, and electronic degrees of freedom. Chemical reactions on excited surfaces are largely untouched; the problems here are rendered more formidable by the need for fully quantal calculations.

Acknowledgement

I should like to thank O. H. Crawford, A. Dalgarno, W. R. Garrett, K. Lawley, and J. B. McGrory for helpful suggestions.

References

1. G. G. Balint-Kurti, *Adv. Chem. Phys.*, **XXX**, 137 (1975).
2. L. D. Landau and E. M. Lifshitz, *Quantum Mechanics*, Pergamon Press, 1965.
3. J. O. Hirschfelder, C. F. Curtiss, and R. B. Bird, *Molecular Theory of Gases and Liquids*, John Wiley & Sons, 1967.
4. H. F. Schaefer, III, *The Electronic Structure of Atoms and Molecules*, Addison-Wesley Publishing Co., 1972.
5. J. C. Browne, *Adv. Atom. Mol. Phys.*, **7**, 47 (1971); many articles in *Faraday Disc. Chem. Soc.* **62**, (1977) are directly relevant.

6. R. K. Nesbet, *Adv. Chem. Phys.*, **14**, 1.
7. G. Das and A. C. Wahl, *Phys. Rev.*, **A4**, 825 (1971).
8. R. F. Stewart, D. K. Watson, and A. Dalgarno, *J. Chem. Phys.*, **65**, 2105 (1976); J. C. Ho, G. A. Segal, and H. S. Taylor, *J. Chem. Phys.*, **56**, 1520 (1972).
9. H. Feshbach, *Ann. Phys. (New York)*, **5**, 357 (1958).
10. J. D. Weeks, A. Hazi, and S. A. Rice, *Adv. Chem. Phys.*, **16**, 283 (1969).
11. J. N. Bardsley, in *Case Studies in Atomic Collision Physics* (Eds. E. McDaniel and M. R. C. McDowell), Vol. IV, Elsevier, New York, 1975.
12. W. E. Baylis, *J. Chem. Phys.*, **51**, 2665 (1969).
13. C. Bottcher and A. Dalgarno, *Proc. Roy. Soc. (London)*, **A340**, 187 (1974).
14. C. Bottcher, *J. Phys. B*, **6**, 2368 (1973).
15. C. Bottcher, *J. Phys. B*, **4**, 1140 (1971).
16. W. E. Baylis, *J. Phys. B*, **10**, L583 (1977).
17. C. D. H. Chisholm and U. Opik, *Proc. Phys. Soc. (London)*, **83**, 541 (1964).
18. A. Dalgarno, C. Bottcher, and G. A. Victor, *Chem. Phys. Lett.*, **7**, 265 (1970).
19. C. Bottcher, T. C. Cravens, and A. Dalgarno, *Proc. Roy. Soc. (London)*, **A346**, 157 (1975).
20. D. Norcross and R. McCaroll, private communication. McCaroll and his collaborators have recently introduced a modified model potential method in which the valence orbitals are explicitly orthogonalized to the core HF orbitals. The method has been successfully tested on HeH and should increase the applicability and accuracy of model potential calculations.
21. C. Bottcher, *Chem. Phys. Lett.*, **35**, 367 (1975).
22. R. Hoffmann, *J. Chem. Phys.*, **39**, 1397 (1963).
23. J. A. Pople, D. P. Santry, and G. A. Segal, *J. Chem. Phys.*, **43**, S129 (1965).
24. J. C. Tully, *J. Chem. Phys.*, **58**, 1396 (1973).
25. M. J. S. Dewar, *Faraday Disc. Chem. Soc.*, **62**, 197 (1977); C. R. Claydon, G. A. Segal, and H. S. Taylor, *J. Chem. Phys.*, **54**, 3799 (1971).
26. G. G. Hall, *Proc. Roy. Soc. (London)*, **A205**, 541 (1951).
27. J. C. Tully, *J. Chem. Phys.*, **65**, 3182 (1976).
28. A. Dalgarno, in *Quantum Theory*, Vol. I, *Elements* (Ed. D. R. Bates), Academic Press, 1961; J. O. Hirschfelder, W. B. Brown, and S. T. Epstein, *Adv. Quantum Chem.* **1**, 265 (1964).
29. J. O. Hirschfelder (Ed.), *Advances in Chemical Physics*, Vol. 12, John Wiley & Sons, 1967.
30. A. R. Edmonds, *Angular Momentum in Quantum Mechanics*, Princeton University Press, 1960.
31. E. E. Nikitin, *J. Chem. Phys.*, **43**, 744 (1965).
32. J. N. Murrell and G. Shaw, *J. Chem. Phys.*, **46**, 1768 (1967).
33. M. H. Alexander and L. Salem, *J. Chem. Phys.*, **46**, 430 (1967).
34. P. Brumer and M. Karplus, *J. Chem. Phys.*, **58**, 3903 (1973).
35. S. Sinha and J. N. Bardsley, *Phys. Rev.*, **A14**, 104 (1976). This article is an updated and extended presentation of the older papers.
36. R. G. Gordon and Y. S. Kim, *J. Chem. Phys.*, **56**, 3122 (1972).
37. N. F. Mott and H. S. W. Massey, *The Theory of Atomic Collisions*, Oxford, 1965; S. Geltman, *Topics, in Atomic Collision Theory*, Academic Press, New York, 1969.
38. W. Lichten, *Phys. Rev.*, **A164**, 131 (1967); F. T. Smith, *Phys. Rev.*, **A179**, 111 (1969).
39. J. D. Power, *Proc. Roy. Soc. (London)*, **A274**, 663 (1973); a forthcoming paper by P. T. Greenland (*J. Phys. B*, **11**) corrects some errors in Power's work and goes on to calculate B-O breakdown terms.
40. C. Bottcher, *Chem. Phys. Lett.*, **18**, 457 (1973).

41. Yu. N. Demkov, *Sov. Phys. JETP*, **18**, 138 (1964).
42. C. Bottcher and M. Oppenheimer, *J. Phys. B*, **5**, 492 (1972).
43. M. Oppenheimer, *J. Chem. Phys.*, **57**, 3899 (1972).
44. D. R. Bates, *Mon. Not. R. Astron, Soc.*, **111**, 301 (1951).
45. C. Bottcher, *J. Phys. B*, **9**, 2899 (1976); A. Herzenberg and F. Mandl, *Proc. Roy. Soc. (London)*, **A274**, 253 (1963).
46. W. H. Miller, C. A. Slocum, and H. F. Schaeffer, *J. Chem. Phys.*, **56**, 1347 (1972).
47. J. Vaaben and J. S. Briggs, *J. Phys. B*, **10**, L521 (1977); C. Bottcher, *J. Phys. B*, **11**, 507 (1978).
48. M. R. Flannery, *Phys. Rev.*, **183**, 231, 241 (1969).
49. R. H. G. Reid and A. Dalgarno, *Chem. Phys. Lett.*, **6**, 85 (1970).
50. D. Secrest, *Meth. Comput. Phys.*, **10**, 243 (1971); P. McGuire, *J. Chem. Phys.*, **62**, 525 (1975).
51. S. I. Chu and A. Dalgarno, *Proc. Roy. Soc. (London)*, **A342**, 191 (1975).
52. G. Zarur and H. Rabitz, *J. Chem. Phys.*, **59**, 943 (1973).
53. H. Klar and M. Klar, *J. Phys. B*, **8**, 129 (1975).
54. B. Amaee and C. Bottcher, *J. Phys. B*, **11**, 1249 (1978).
55. C. Bottcher and C. V. Sukumar, *J. Phys. B*, **10**, 2853 (1977).
56. B. Amaee, *Ph.D. Thesis*, University of Manchester, England, 1978; B. Amaee and C. Bottcher, to be published.
57. I.V. Hertel, H. Hoffman, and K. A. Rost, *Phys. Rev. Lett.*, **36**, 861 (1976); the older literature is surveyed by H.S.W. Massey and E.H.S. Burhop, *Electronic and Ionic Impact Phenomena*, Vol. 3, Clarendon Press, Oxford, 1971.
58. M. Rubinson, G. Garetz, and J. I. Steinfeld, *J. Chem. Phys.*, **60**, 3082 (1974).
59. B. A. Garetz, J. I. Steinfeld, and L. L. Poulsen, *Chem. Phys. Lett.*, **38**, 365 (1976).
60. K. H. Welge and R. Atkinson, *J. Chem. Phys.*, **64**, 531 (1976).

Potential Energy Surfaces
Edited by K. P. Lawley

COLLISIONAL EFFECTS ON ELECTRONIC RELAXATION PROCESSES

KARL F. FREED

The James Franck Institute and The Department of Chemistry
The University of Chicago, Chicago, Illinois 60637, *USA*

CONTENTS

I. INTRODUCTION

Photochemical reaction schemes are initiated through the absorption of incident radiation by an atomic or molecular species. Subsequently, the electronically excited molecule may undergo a variety of radiationless transitions in which its electronic state changes, electronic relaxation, and/or in which it becomes vibrationally relaxed or activated, etc.[1-4] The excited molecule can then isomerize[5] or it can decompose.[6] The products of decomposition or isomerization reactions can then also undergo radiationless processes before reacting with secondary molecules in the environment, etc. The purely photophysical processes are important components of any photochemical reaction scheme. The radiationless transitions are responsible for conveying the molecule to the reactive electronic surfaces as well as for providing necessary activation energy to reactants and a stabilizing deactivation of products. These photophysical processes have become of interest in their own right, and considerable attention has been devoted to their study when photochemistry is absent.

Early studies of the photophysical radiationless processes[7] of molecular systems were carried out on molecules in condensed media, liquids, rigid matrices, and high-pressure gases. This experimental situation introduces the complication associated with the presence of the possible occurrence of a number of different photophysical relaxation processes in the same molecular system, in a fashion that mimicks the complexity of a full photochemical reaction scheme. In order to study the primary photophysical radiationless transitions, it is optimal to consider experiments in which only elementary individual processes appear. Such investigations often involve the experimental determination of radiationless transitions in isolated collision-free molecules.[8-11] For instance, collision-free experiments enable the consideration of the important phenomena of electronic relaxation and intramolecular vibrational redistribution.[12] Studies on isolated molecules have greatly contributed to our understanding of these processes, and continued progress in these areas is to be anticipated from the large number of research workers in the field. Given the understanding derived from collision-free molecule experiments, it is only natural to consider also situations in which the collisional processes play a fundamental role in the radiationless transition under consideration. In this review we concentrate on the effects of collisions on the processes of electronic relaxation in polyatomic molecules. A previous review[1] has discussed the theory of electronic relaxation processes in isolated molecules with emphasis on the description of the energy dependence of radiationless transition rates. This review is a sequel in which we describe how the presence of collisions between the excited molecule and sets of perturbing molecules alters the electronic relaxation processes of the excited molecular system. A brief description is given of those aspects of the isolated molecule theory which are necessary for the presentation of the collisional effects on radiationless processes. The interested reader is referred to the previous review article[1] and others[2-4,7] for more details on the isolated molecule case.

The theory of electronic relaxation in isolated molecules has introduced the concepts of three distinct categories of this type of radiationless transition, the small, large, and intermediate molecule limits.[1-4,7] In the small and intermediate case limits the molecule cannot undergo electronic relaxation processes in the isolated collision-free molecule, while in the large molecule limit (and some manifestations of the intermediate case) the electronic relaxation is possible in isolated molecular systems. Collisional effects on the radiationless processes have enormously different manifestations in these extreme limits. In the small and intermediate limit situations the collisions can lead to electronic relaxation.[13] On the other hand, in the large molecule case the collisions merely alter the properties of electronic relaxation process from the zero pressure limit.

In the next section we review briefly those aspects of the theory of the small, large, and intermediate case molecules which are necessary for a description of the collisional effects on radiationless transitions. Section III discusses the collisional effects on radiationless transitions in small and intermediate case molecules, while the following section considers the effects of collisions on molecules wherein electronic relaxation processes are possible in the isolated collision-free system. We note from the outset that there is a strong formal analogy between electronic and vibrational relaxation processes.[1,12] Thus much of the discussion is immediately applicable to considerations of intramolecular vibrational relaxation processes and collision-induced vibrational relaxation[14] in small and large molecules. Care, however, must be exercised in transcribing the theory between electronic and vibrational relaxation situations as the relevant molecular parameters, couplings, densities of states, initially prepared states, differ in the two cases. The highly interesting vibrational relaxation situations are also discussed as a special limit of the general theory under consideration.

II. ELECTRONIC RELAXATION IN ISOLATED MOLECULES

In the age of high-speed digital computers the electronic states of a large polyatomic molecule can be taken, in principle, to be determined by the solution of the clamped-nucleus Schrödinger equation for the electronic states of the system. In this adiabatic Born–Oppenheimer approximation the electronic Hamiltonian, $H_{el}(\mathbf{r}, \mathbf{R})$ is parametrically dependent on the positions of the nuclei, $\mathbf{R}$. A solution of this Hamiltonian,

$$H_{el}(\mathbf{r}, \mathbf{R})\phi_n(\mathbf{r}, \mathbf{R}) = E_n(\mathbf{R})\phi_n(\mathbf{r}, \mathbf{R}),$$

at each nuclear position generates the set of eigenfunctions, $\{\phi_n(\mathbf{r}, \mathbf{R})\}$, which represent the electronic wavefunctions as a function of the nuclear positions, $\mathbf{R}$. In the pre-computer days there was no real hope of obtaining reasonably accurate approximate solutions to this molecular electronic Schrödinger equation, so very drastic assumptions had to be introduced. One of the most popular involves the so-called crude Born–Oppenheimer approximation in which the properties of the molecular system are described in terms of a set of eigenfunctions of the electronic Hamiltonian for a single fixed position, $\mathbf{R}^0$, of all the nuclei. This position is often taken to be the equilibrium position of the ground electronic state of the system, but sometimes it is taken as the equilibrium position of some initially excited electronic state. It is clear that during any photochemical process, perhaps involving isomerizations and bond breaking, the molecule will traverse nuclear arrangements which differ considerably from the equilibrium position of the ground initially prepared excited state. Consequently, this crude Born–Oppenheimer approximation can be anticipated to yield a very poor representation of the nature of the potential surfaces involved in the photochemical and/or photophysical processes. In fact, calculations demonstrate that this crude Born–Oppenheimer approximation provides a *grossly inadequate* representation of even the initial reference electronic state. In particular, the predicted vibrational frequencies of that state can be in error by orders of magnitude.[15] Thus despite the fact that a considerable amount of literature is phrased in terms of this crude Born–Oppenheimer approximation, its accuracy is quite abysmal. Nevertheless, it is often possible to take theoretical expressions, involving the appropriate adiabatic Born–Oppenheimer approximation, and then afterwards introduce a crude Born–Oppenheimer basis set which is suitable for the evaluation of the particular quantities, e.g. matrix elements, of interest.[16,17] Many of the results of older works with the crude approximation can be justified by such an approach, while other results are basically consequences of symmetry restrictions, accounting for the general correctness of many of their predictions. Here we assume from the outset that the system is to be described by the appropriate adiabatic Born–Oppenheimer wavefunctions along with the adiabatic electronic potential energy surfaces, $E_n(\mathbf{R})$.

If the adiabatic Born–Oppenheimer approximation were exact, photochemistry and photophysical processes would be rather straightforward to describe. In this event, molecules would be excited by the incident radiation to some upper electronic state. Once in this electronic state, the system could radiate to a lower electronic state, or it could decompose or isomerize on the upper electronic potential energy surface. No transitions to other electronic states would be possible. The spectroscopy of the systems would also be greatly simplified as there would no longer be any phenomena such as lambda doubling,[18] etc., which lifts degeneracy of some levels of H_{el}.

There are numerous interactions which are ignored by invoking the Born–Oppenheimer approximation, and these interactions can lead to terms which couple different adiabatic electronic states. The full Hamiltonian, H, for the molecule is the sum of the electronic Hamiltonian, H_{el}, the nuclear kinetic energy operator, T_N, the spin orbit interaction, H_{SO}, and all the remaining relativistic and hyperfine correction terms, the adiabatic Born–Oppenheimer approximation assumes that the wavefunctions of the system can be written in terms of a product of an electronic wavefunction, $\phi_n(\mathbf{r}, \boldsymbol{R})$, a vibrational wavefunction, $\chi_{ni}(R)$, a rotational wavefunction, Θ_{JM} , and a spin wavefunction, χ_{spin} . However, such a product wavefunction is not an exact eigenfunction of the full Hamiltonian for the system. One consequence, noted above, is the lifting of degeneracies[18] of energy levels of these simple product wavefunctions due to couplings ignored by the Born–Oppenheimer approximation. A second consequence[7] involves the possibility for the occurrence of radiationless transition processes, in particular, electronic relaxation processes.[1–4]

A. Molecular Model

The general picture of energy levels in a polyatomic molecule can be represented schematically as in Fig. 1. A molecule has a ground electronic state, ϕ_0, with a variety of vibrational sublevels. The molecule begins in some thermally accessible vibrational level of the ground state (we omit representation of the rotational and spin sublevels at this juncture.) There are also a whole host of electronically excited states of the system. The diagram depicts one excited state of the system, ϕ_s. Isoenergetic with the low lying levels of ϕ_s is a dense manifold of vibrational levels arising from two possible sources. First, there are high-lying vibrational levels associated with the ground electronic state. Secondly, when ϕ_s is not the lowest excited electronic state of the molecule, there are also vibrational states belonging to other lower electronic states than ϕ_s. It is assumed that electronic dipole allowed transitions between the thermally accessible ground-state vibronic levels, ϕ_0, and these high-lying states $\{\phi_l\}$, are not possible because of either highly unfavourable Franck–Condon factors or because of spin selection or other symmetry considerations. Thus the primitive adiabatic Born–Oppenheimer representation of the energy levels of a polyatomic molecule in Fig. 1 would lead to the expectations of the observation of an adsorption spectrum characterized by the energy levels of ϕ_s.

The presence of the manifold of states, $\{\phi_l\}$, can and generally does modify our simple-minded expectations. This 'dense' manifold of vibronic levels leads to a series of phenomena which depend, in part, upon the densities of levels, ρ_l, within this set. For instance the small molecule limit is characterized by a very low density of levels in this set $\{\phi_l\}$. In this case the non-Born–Oppenheimer couplings can lead to the observation of perturbations in the spectrum of ϕ_s arising from the vibronic state $\{\phi_l\}$. Such perturbations are associated with the displacement of levels of ϕ_s from their anticipated

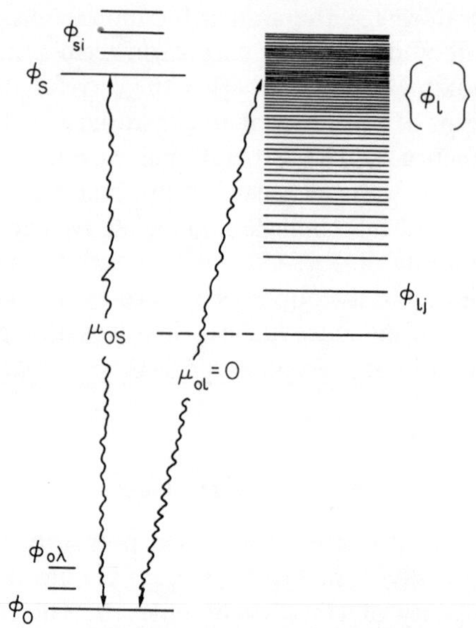

Fig. 1. Representation of the expected energy levels of a polyatomic molecule. ϕ_0 is the ground electronic state with vibronic components $\phi_{0\lambda}$. Optical excitation processes can occur from the thermally populated ϕ_0 to vibronic levels, ϕ_{si}, of electronic state ϕ_s. Isoenergetic with the $\{\phi_{si}\}$ is a dense set of vibronic levels, $\{\phi_{ij}\}$, belonging to another electronic state (or set thereof) ϕ_l. (The $\{\phi_{lj}\}$ may include high-lying levels of ϕ_0.) Optical transitions between $\{\phi_{lj}\}$ and the thermally populated $\{\phi_{0\lambda}\}$ are forbidden because of spin-selection rules and/or unfavourable Franck–Condon factors.

positions and possible in the emergence of additional lines in the absorption spectra.[19] This small molecule limit of low-level densities is an ideal case for the assignment and analysis of all spectroscopic lines.

The other extreme involves the large molecule statistical limit wherein the density of $\{\phi_l\}$ levels is extremely high. In this case the set of levels, $\{\phi_l\}$, can behave *as if* these levels were a continuum of levels on the time scales relevant to the photophysical processes in the excited electronic state. Thus these $\{\phi_l\}$ levels can act as a dissipative quasi-continuum and lead to irreversible electronic relaxation, from ϕ_s to $\{\phi_l\}$.

Between these two extremes there resides the very interesting intermediate case which covers all other possibilities. Firstly, the intermediate case may display some characteristics of both the small and large molecule limits because of strong variations in the ϕ_s–ϕ_l couplings arising from symmetry considerations, etc. A second intermediate case situation occurs where the density of $\{\phi_l\}$ levels becomes too high to hope to ever resolve and assign each of these levels individually, but where the level density is still not sufficiently high to lead to electronic relaxation in isolated molecules. This intermediate case situation is termed the 'too-many level' small molecule case and is one of the interesting situations for further consideration of collisional effects on electronic relaxation.

One of the most striking facets of radiationless transition processes is the fact that they are *all* governed by the same general molecular energy level scheme as is given in Fig. 1.[1,2] Thus this figure encompasses the description of the small molecule limit of perturbations in molecular spectroscopy to the large molecule limit of irreversible electronic relaxation as well as everything intermediate between these two limits. The phenomena are governed primarily by the details of the energy level density, ρ_l, as noted above, as well as by the decay rates, $\{\Gamma_l\}$, of the zeroth-order levels $\{\phi_l\}$ and the ϕ_s–ϕ_l couplings. Important other considerations centre about the nature of the initially prepared electronic state. The theoretical description of all of these phenomena involves a description of the interaction of radiation with a coupled many-level molecular system and of the subsequent time evolution of the prepared state of the system. A treatment of many of these facets of the theory along with discussions of much of experimental results can be found in a variety of review articles.[1–4,7] Here we merely state some of the conditions which delineate and explain the nature of the different small, intermediate, and large molecule limits.

B. The Different Limiting Cases

The important parameter characterizing the molecular limit and the nature of the observed 'radiationless processes', is given by

$$x_l = \hbar\Gamma_l/\varepsilon_l. \tag{1}$$

Here $\hbar\Gamma_l$ represents the energy uncertainty of a level with a decay rate of Γ_l, and ε_l is the average spacing between the $\{\phi_l\}$ levels. The condition for the small molecule limit is

$$x_l \ll 1, \text{ small molecule limit,} \tag{2}$$

whereas for the large molecule case it is

$$x_l \gg 1, \text{ large molecule, statistical limit.} \tag{3}$$

These limits are rather simply understood by the consideration of even a simple diatomic molecule. Suppose first that ϕ_s and ϕ_l represent two bound electronic states of the diatomic molecule. Then under isolated molecule conditions Γ_s corresponds to the radiative decay rate of the individual

ϕ_s levels. Γ_s is typically of the order of 10^{-6}–$10^{-9}\ s^{-1}$. However, in the diatomic molecule the average spacing between vibronic levels, ε, is of the order of hundreds or thousands of cm^{-1}. Thus the simple diatomic molecule example is definitely with the small molecule limit of (2).*

On the other hand we may consider a model in which ϕ_s represents a bound electronic state of the diatomic molecule, whereas ϕ_l represents a dissociative electronic state of the diatomic molecule. The dissociative state then has an energy level spacing which is zero, $\varepsilon_l \to 0$, while the radiative lifetime is non-zero (albeit possibly rather small in many cases). Thus the parameter x_l of (1) is infinite for this example, and the conditions for the 'large molecule' limit, (3), are, perforce, automatically satisfied. In this situation we can observe a radiationless transition from ϕ_s to ϕ_l which is an *irreversible non-radiative process*. Here the process of photon absorption and subsequent decomposition is simply represented as follows: first, the molecule absorbs the incident radiation and undergoes a transition from the ground electronic state, ϕ_0, to the excited electronic state, ϕ_s, which carries oscillator strength in the appropriate spectral region. The quasi-bound state ϕ_s then undergoes a non-radiative transition to the dissociative $\{\phi_l\}$ levels.

In contrast, in the small molecule limit the situation corresponds to the spectroscopist's description of perturbations in the spectra of small molecules. The molecular Hamiltonian is represented in the basis set of adiabatic Born–Oppenheimer functions, $\{\phi_s, \phi_l\}$. Coupling amongst only a few levels need be considered, namely those nearby coupled levels with couplings that satisfy the relationship

$$\frac{|v_{sl}|}{|E_s - E_l|} \gtrsim 1, \tag{4}$$

where E_s and E_l are the Born–Oppenheimer energies of ϕ_s and ϕ_l, respectively, and v_{sl} are the non-adiabatic matrix elements

$$v_{sl} = \langle \phi_s | H | \phi_l \rangle.$$

Those 'non-resonant' levels $\{\phi_l\}$, grossly violating (4), provide a small contribution and can be treated separately by perturbation theory. The diagonalization of the molecular Hamiltonian amongst the resonantly coupled states,

*The couplings between ϕ_s and ϕ_l must preserve the total angular momentum quantum number. Hence, only those rotational sublevels of ϕ_l with the appropriate rotational quantum numbers can couple to ϕ_s. Therefore, it is only necessary to consider the vibronic level spacing in our discussion of the simple diatomic molecule cases. Of course, the presence of any hyperfine structure would slightly increase the number of available ϕ_l levels for coupling to a particular spin-rovibronic state in ϕ_s. Nevertheless, this does not after the very sparse energy level distribution of states in ϕ_s with the appropriate angular momentum quantum numbers, so that the small molecule limit (2) is still safely satisfied for the two bound electronic state diatomic molecule discussed above.

satisfying (4), then leads to the molecular eigenstates, ψ_n, with energies ε_n, which are a linear superposition of the adiabatic Born–Oppenheimer states. In our two bound states simple diatomic molecule example, ignoring the minor modifications necessary to include nuclear hyperfine structure, in the usual weak coupling limit the only resonant mixing can involve a pair of levels at a time for the case that ϕ_s and ϕ_l are both singlet states. In this simple case the molecular eigenstates are represented as

$$\psi_n = a_{sn}\phi_s + a_{ln}\phi_l, \tag{5}$$

whereas when more levels of ϕ_l may be coupled, equation (5) involves the linear superposition of these few levels,

$$\psi_n = a_{sn}\phi_s + \sum_{j=1}^{N} a_{l_jn}\phi_{l_j1} \tag{6}$$

where N is a small number typically. In the absence of accidental degeneracies amongst the molecular eigenstates, $\{\psi_n\}$, the solutions (5) or (6) are the appropriate solutions of the Schrödinger equation for the isolated molecular system in a real world where spontaneous emission processes are admissible. Assuming that the ϕ_s and $\{\phi_l\}$ do not radiate to any set of common levels, it is possible to evaluate the radiative decay rates of the molecular eigenstates, $\{\psi_n\}$,

$$\Gamma_n = |a_{sn}|^2\Gamma_s + \sum_{j=1}^{N} |a_{l_jn}|^2\Gamma_l. \tag{7}$$

For many cases of interest the zeroth-order radiative decay rate of ϕ_s is much greater than that for the $\{\phi_l\}$. So under the condition that

$$\Gamma_s \gg \Gamma_l \tag{8}$$

the radiative decay rates of the molecular eigenstates are equal to

$$\Gamma_n \simeq |a_{sn}|^2\Gamma_s. \tag{9}$$

The condition that the original states, ϕ_s, be distributed amongst the molecular eigenstate implies the normalization condition

$$\sum_{n=1}^{N+1} |a_{sn}|^2 = 1, \tag{10}$$

whereupon it is clear that if more than one of the a_{sn} is non-zero, these must satisfy

$$|a_{sn}|^2 < 1. \tag{11}$$

Equations (9) and (11) yield the result that

$$\Gamma_n < \Gamma_s \quad (\text{for } \Gamma_s \gg \Gamma_l), \tag{12}$$

which implies that the radiative decay rates of the mixed, molecular eigenstates, $\{\psi_n\}$, are less than the radiative decay rate of the parent Born–Oppenheimer state, ϕ_s, containing all of the original oscillator strength. In fact, the radiative decay rates of more than one of the molecular eigenstates are often appreciable enough to enable observation of absorption or emission to more than one of the mixed molecular eigenstates. In practice this results in the observation of perturbations in the absorption spectrum where additional spectral lines are present.[20]

When the spacing between molecular eigenstates is large compared to the uncertainty widths of these levels,

$$\min|E_n - E_{n'}| \gg \tfrac{1}{2}\hbar[\Gamma_n + \Gamma_{n'}], \tag{13}$$

monochromatic excitation can only lead to the preparation of individual molecular eigenstates. Pulsed excitation, on the other hand, can lead to the excitation of a coherent superposition of a number of nearly molecular eigenstates. However, when the pulse duration, τ, satisfies the inequality,

$$\min|E_n - E_{n'}| > \hbar/\tau, \tag{14}$$

this pulsed excitation cannot excite more than one of the individual molecular eigenstates. When the pulsed duration is sufficiently short or the spacing between molecular eigenstates sufficiently small that (14) is violated, then a coherent superposition of these nearby molecular eigenstates is possible. Apart from isolated cases, the general situation (ignoring hyperfine structure) in diatomic molecules is found to satisfy (14) for nanosecond pulse durations, so even pulsed excitation is taken to prepare the molecule in single excited molecular eigenstates.

The above examples of electronic couplings in diatomic molecules sounds somewhat like motherhood and apple pie which are learned in elementary courses on molecular structure and spectroscopy. The basic principles, illustrated by these examples, however, apply to larger molecules where considerable confusion has arisen in the literature as to the applicability of either the small, large (or intermediate) case limits in specific experimental situations. As we pass to triatomic molecules, the level densities in bound electronic states increase. Thus there may be a large number of levels $\{\phi_l\}$ which can be resonantly [cf. (4)] coupled to a single level, ϕ_s. Nevertheless, when all electronic states are bound, it is quite often true that the level density is still very low, so that (2) is still safely satisfied, and the small molecule limit ensues. Then the molecular eigenstates, (6), may contain a fairly large number of terms, N of the order of 10–100 for large couplings, v_{sl}. In this case there are many additional spectral lines which may emerge in the absorption spectrum. One prime example of this complicating feature is the NO_2 molecule where the large couplings, $v_{sl} \approx 150$–$200\ \mathrm{cm}^{-1}$, and the moderate density of states leads to the appearance of a considerable number of additional levels.[20] Again, under the assumption of (8), all of the results (9)–(14) are equally valid for these small molecule limit triatomic molecules. The same considerations can even apply in an extremely large molecule. For instance, if the origins of the electronic states ϕ_s and ϕ_l are very close so that the density of states of $\{\phi_l\}$ is small at the energies E_s, the 'small molecule' limit conditions, (2), are still obeyed. This situation applies to naphthalene where the $S_1 - S_2$ splitting is ca. 3500 cm^{-1} and to 3, 4-benzepyrene where it is ca. 3800 cm^{-1}. In these cases the respective observed lifetimes of S_2, $\tau_{S_2} \approx 4 \times 10^{-8}$ s and 7×10^{-8} s are longer than those deduced from the absorption oscillator strengths, $\tau_{S_2}^{\text{rad}}$, for both molecules. The intermediate case can also arise for small $S_1 - T_1$ splittings,[21, 22] e.g. benzophenone where the splitting is ca. 3000 cm^{-1} and $\tau_{S_1} \approx 10^{-5}$ s > $\tau_{S_1}^{\text{rad}} \approx 10^{-6}$ s. Thus the behaviour of the molecule is still governed by the small molecule case, (9) through (14), when the condition (8) is obeyed. [If (8) is not met, all of the above arguments are readily modified to incorporate radiative decay processes from the $\{\phi_l\}$ levels.]

When, on the other hand, the spacing between the origins of ϕ_s and ϕ_l become large enough and/or the size of the molecule is sufficiently large so there are a large number of vibrational modes and therefore a large enough density of states, the conditions for the emergence of the large molecule statistical limit, (3), are obtained. In this situation the molecular eigenstates could, in principle, be evaluated, but they are not the most appropriate eigenfunctions for the description of the radiative decay processes of the isolated molecule. This is because the condition, (13), is violated for large sets of levels $\{\psi_n\}$, so these groups of levels no longer decay radiatively in an independent fashion.[1, 23–26] A theoretical analysis of the radiative and non-radiative decay of a group of closely coupled levels shows that the relevant quantity to be diagonalized is the effective molecular Hamiltonian,[1, 23–27]

$$H_{\text{eff}} = H - i\hbar\mathbf{\Gamma}/2, \tag{15}$$

where H is the full molecular Hamiltonian and $\mathbf{\Gamma}$ is the damping matrix. In the basis set of Born–Oppenheimer functions the latter is written as

$$\Gamma_{ij} = \Gamma_s \delta_{si} \delta_{sj} + \Gamma_l \delta_{lj} \delta_{li}, \tag{16}$$

where it is assumed that the damping matrix is diagonal amongst the $\{\phi_l\}$ levels, for simplicity. Note that (15) is a non-hermitian Hamiltonian, so its eigenvalues, E_r, are complex numbers

$$E_r = \varepsilon_r - i\hbar\Gamma_r/2. \tag{15a}$$

The real parts of the eigenvalues, E_r, give the energies of the quasi-stationary states while the negative of the imaginary parts of the eigenvalues give one half the uncertainty widths, $\hbar\Gamma_r$, of these quasi-stationary states or the lifetime from $\tau_r = \Gamma_r^{-1}$.

In the small molecule limit the effective Hamiltonian, (15), is diagonalized to a good approximation by the molecular eigenstates, $\{\psi_n\}$, of (6) with the decay rates given by (7) (in the absence of accidental degeneracies). Thus equation (15) reduces to the small molecule limit when condition, (2) is obeyed. In the large molecule limit, (3) in contrast, the nature of the eigenfunctions of the effective Hamiltonian, (15), is entirely different. The solutions correspond very much to the emergence of an eigenfunction which is very similar to ϕ_s and which is now a *resonant state* having a radiative decay rate equal to Γ_r and a non-radiative decay rate, Δ_s, given in the weak coupling limit by the familiar golden rule expression

$$\Delta_s = \frac{2\pi}{\hbar} \sum_j |v_{sl_j}| \rho_{l_j}. \tag{17}$$

The condition (3) implies that, *if* a molecule begins in state ϕ_s and then crosses over to the $\{\phi_l\}$, the final states $\{\phi_l\}$ will decay with rates Γ_l before the molecule ever has a chance to cross back to the original state ϕ_s. Thus it is the decay of the final manifold of levels $\{\phi_l\}$ which *drives* the irreversible electronic relaxation from ϕ_s to $\{\phi_l\}$.

There are, however, many situations in which the inequality (8) is obeyed for these large molecules, and the condition (3) might therefore appear to be inapplicable for these systems. However, in these cases we are primarily interested in whether or not the radiationless transition from ϕ_s to $\{\phi_l\}$ *appears* to be *irreversible* on the time scales of the experiment.[1, 24, 25] The time scales for the experiment may be limited by the radiative lifetime, Γ_s, of the state ϕ_s, or it might be limited by the detection apparatus or by the presence of collisional or other relaxation processes. Thus it is often of interest to consider the non-radiative decay properties of a system on a particular time scale, τ_e, for the experiment. The parameter,

$$x_l' = \hbar[\Gamma_l + (\tau_e)^{-1}]/\varepsilon_l, \tag{18}$$

then enables us to characterize the decay characteristics of a molecule on this experimental time scale τ_e. When x_l' obeys the condition

$$x_l' \ll 1, \tag{2'}$$

the small molecule behaviour is still manifest on the time scale τ_e. However if

$$x_l' \gg 1 \tag{3'}$$

is obeyed on the time scale τ_e, the decay characteristics of the molecule correspond to the statistical limit of states, ϕ_s, with both radiative and non-radiative decay *on the time scales*, τ_e, *of the experiment*. On these time scales for which (3′) is obeyed, the non-radiative decay rate of ϕ_s is given by equation (17) in the weak coupling limit.

C. Intramolecular Dephasing

One interesting manifestation of the intermediate case is one for which the density of $\{\phi_l\}$ states is quite large but still not large enough for the statistical limit (3) to be obeyed. In this case the molecular eigenstates in (6) have a large number of contributing terms, and it would be extremely difficult to unravel all of the individual molecular eigenstates to determine the zeroth-order Born–Oppenheimer energies and their couplings v_{sl}. This is the 'too many level' small molecule situation. It is possible to encounter cases in which the inequalities,

$$\Delta_s \gtrsim \Gamma_s \gg \Gamma_l, \tag{19}$$

are found. The quantity Δ_s of (17) is not the non-radiative decay rate of the state ϕ_s because the molecule conforms to the small molecule limit. Broad-band excitation can, in principle, produce an initially prepared non-stationary state of the molecule which closely approximates the Born–Oppenheimer state ϕ_s. The subsequent time evolution of the system is governed by the time evolution of the molecular eigenstates. For the illustrative example where the initial state of the system is ϕ_s, the state of the system at time t can be shown in this small molecule limit to be

$$\bar{\psi}(t) = \sum_n a_{ns} \exp[-i\mathscr{E}_n t/\hbar - \Gamma_n t/2]\psi_n. \tag{20}$$

The probability of observing radiative emission characteristic of the state ϕ_s at time t is proportional to the probability of finding the molecule in the zeroth-order state ϕ_s at t which is given by

$$\begin{aligned} P_s(t) &\equiv |\langle \phi_s|\bar{\psi}(t)\rangle|^2 \\ &= |\sum_n |a_{ns}|^2 \exp[-i\mathscr{E}_n t/\hbar - \Gamma_n t/2]|^2. \end{aligned} \tag{21}$$

Note that the probability of being in ϕ_s involves the absolute value squared of a complicated sum over molecular eigenstates of oscillatory damped factors.

The general behaviour of (21) could, in principle, be extremely complicated.[28] However, in the too many level, small molecule case the large number of contributing terms in the summation does provide a simplification. Again for illustrative purposes assume that the condition (8) is obeyed for this molecule and that the reason for its classification in the small molecule limit is the small value of Γ_l. However, on a short enough time scale the condition (3′) may be satisfied for time scales of the order of or less than the radiative decay rates, Γ_n, of the individual molecular eigenstates. Then, on the time scales, τ_e, the molecule *appears as if it conforms to the large molecule statistical limit*. Consequently, on these time scales the molecule appears to undergo a non-radiative decay from ϕ_s to $\{\phi_l\}$ with a non-radiative decay rate Δ_s of (17). Thus as shown by Lahmani *et al.*,[29] $P_s(t)$ of (21) for short enough times, such that (3′) is satisfied, displays an exponential decay with a decay rate given by

$$\Gamma_s^{\text{tot}}(\tau_e) = \Gamma_s + \Delta_s. \tag{22}$$

On longer time scales, where (3′) is violated, the behaviour of the system is then again given by the small molecule limit and the complicated expression (21). However, in all experiments to date the long time behaviour of these molecules, i.e. the long time behaviour of (21), is found to be adequately represented by a simple exponential decay with some average molecular eigenstate decay rate, $\bar{\Gamma}_n$. Between the short time exponential decay of (22) and the long time average molecular eigenstates exponential decay of $\bar{\Gamma}_n$, in principle, the decay given by (21) could be very complicated.[28] It could exhibit oscillations which result from quantum mechanical interference effects (quantum beats). Recent molecular beam experiments by Chaiken *et al.*[30] have observed quantum beats in biacetyl. Previous bulb experiments could not unambiguously observe these oscillatory effects, and this is presumably due to the fact that large numbers of rotational levels are involved, so that the expression (21) appropriate to these experiments presumably involves a further weighted summation of this expression over all of the individual rotational levels that are excited by the pulse of radiation, washing out any beat pattern.[31] Thus the net result is that the decay property of these intermediate case molecules, the too many level, small molecule limit, is well represented by a biexponential decay involving the short time apparent radiationless transition and the long time decay of the individual molecular eigenstates.

This short time apparent decay is a particularly interesting intramolecular dephasing process which merely accounts for the fact that the individual molecular eigenstates in the summation in (21) all have slightly different energies, $\{\mathscr{E}_n\}$. When the molecule is initially prepared in the non-stationary state ϕ_s, there is a coherent superposition of all of these molecular eigenstates with relative phases which are fixed. However, because of the differences between the energies, $\{\mathscr{E}_n\}$, these phases become different at a subsequent time, the $\exp[-i\mathscr{E}_n t/\hbar - \Gamma_n t/2]$ factors are different for each of the Ψ_n. The intramolecular dephasing of the different molecular eigenstates is what leads to the apparent exponential decay for short times governed by (3′) which *looks as if the initial state* ϕ_s is decaying into the $\{\phi_l\}$, with the subsequent time evolution then governed by that of the individual molecular eigenstates. This process is not a truly irreversible electronic relaxation since, in principle, a multiple pulse sequence could be applied to the system to produce the state ϕ_s (with diminished overall amplitude) at a subsequent time $t > 0$. The presence of a finite number of levels is what makes the dephasing (in principle) reversible. In the statistical limit the condition (3) implies the presence effectively of an infinite number of levels (i.e. continuous density of states) which can produce irreversible electronic relaxation. When an intermediate case molecule, satisfying (2) conforms to (3′) on some times scales τ_e, this molecule has not had enough time to realize that there are only a finite number of levels $\{\phi_l\}$. Because of time–energy uncertainly it 'thinks' there is a continuous density of $\{\phi_l\}$ levels on time scales τ_e, so it undergoes its exponential decay. At longer times for which (2′) becomes obeyed, the molecule 'finds out' the $\{\phi_l\}$ states are discrete, and the molecule behaves as a small molecule. Hence, the initial exponential decay of $P_s(t)$ is not an irreversible one, so the term intramolecular dephasing is useful to distinguish this process from irreversible electronic relaxation.

D. Intramolecular Vibrational Relaxtion

As mentioned in the introduction, the above discussion of the small, large, and intermediate molecule limits of electronic relaxation processes can also be utilized with very minor modifications to discuss the phenomena of intramolecular vibrational relaxation in isolated polyatomic molecules.[12] Fig. 1 is still applicable to this situation. The basis functions $\phi_0, \phi_s, \{\phi_l\}$ are now taken to be either pure harmonic vibrational states, some local mode vibrational eigenfunctions, or some alternative non-linear mode type wavefunctions. In the following the nomenclature of vibrational modes is utilized, but its interpretation as normal or local can be chosen to suit the circumstances at hand.

ϕ_0 is now the vibrationless level which carries electric dipole oscillator strength to some excited zero-order vibrational level ϕ_s. Again isoenergetic with ϕ_s is a dense manifold of other zeroth-order vibrational states, $\{\phi_l\}$ which do not carry oscillator strength from the ground vibrational level ϕ_0 (or any of the other low-lying thermally accessible vibrational levels). The zeroth-order functions $\{\phi_s, \phi_l\}$ are not exact eigenfunctions of the molecular vibrational Hamiltonian because of the presence of anharmonicities in the normal mode description, because of inter-local mode couplings and anharmonicities in the local mode description, etc. Thus there are again couplings, between ϕ_s and the dense isoenergetic manifold, $\{\phi_l\}$. These couplings can lead to a wide variety of phenomena depending on 1. the nature of initially prepared states, 2. the value of x_l of equation (1), and 3. the number of levels satisfying the near resonance criterion, (4).

In equation (1), ε_l is again taken to be the average spacing between neighbouring $\{\phi_l\}$ levels at the energy of ϕ_s. Γ_s is again the decay rate of the zeroth-order level, ϕ_s, which in isolated molecules is due to infrared emission and has typical values on the order of 10^{-3}–10^{-5} s^{-1}. In the small molecule limit the condition (2) is obeyed and the states of the system are characterized by the vibrational eigenstates of (6). In this limit the mixing amongst a few zeroth-order vibrational levels is the familiar Fermi resonance coupling. Because typical values of the couplings, v_{sl} are of the order of tens of cm^{-1}, the energy separations observed in many instances of Fermi resonance amongst lower-lying levels in polyatomic molecules safely satisfies the inequalities (13) and (14). Hence, nanosecond-type pulsed excitation can only excite single vibrational eigenstates, ψ_n of mixed parentage as in (5) or (6). With weak Fermi resonance and picosecond pulses, perhaps there are cases in which the inequality (14) is violated, so, in principle, it may be possible to coherently excite a number of vibrational eigenstates and to produce an initial nonstationary vibrational state which closely resembles the zeroth-order vibrational state, ϕ_s, which carries all of the oscillator strength from ϕ_0. The Fermi resonance is a phenomenon that is quite familiar to the vibrational spectroscopist.

On the other hand, for large enough molecules and/or high enough total vibrational energy content, the density of vibrational levels can become rather high. First, consider the case where $\rho_l = \overline{\varepsilon_l^{-1}}$ is of the order of 1–100 cm. Because of the small values of Γ_l, the small molecule limit, (2), is *still* obeyed for this situation. There are unfortunately many too many vibrational levels to every permit a complete spectroscopic analysis and sorting of each of the individual vibrational levels, so we again may term this situation the 'too many level, small molecule limit' of vibrational states (an intermediate case). At a vibrational level density of 10^3 cm^{-1} and for τ_e of the order of nanoseconds, we find that x_l' of (18) satisfies the large molecule limit of (3′) on these time scales. Hence, we again would expect to observe an exponential decay of the probability $P_s(t)$ of (21), for finding the system in the initial non-stationary state ϕ_s *if this initial non-stationary state were one that could be prepared experimentally*. Perhaps one means for producing such an initial non-stationary state would involve optical excitation from a lower electronic state to high-lying vibrational levels in the upper electronic state when the optical absorption spectra involves readily assignable peaks corresponding to the zeroth-order mode description that would be utilized in the vibrational representation of Fig. 1. Another possibility involves high overtone absorption within the ground electronic manifold.[32] On longer time scales, the criterion (2′) becomes satisfied, so the system is again in the small molecule limit with longer decay lifetimes characterized by the decay rates $\{\Gamma_n\}$. Thus again, the biexponential decay is expected to characterize the intermediate case vibrational situations *when* it is possible to initially excite a zeroth-order vibrational level, ϕ_s, of some particular mode parentage.

At much higher vibrational level densities the criterion (3) can become satisfied, whereupon the vibrational relaxation of an initially prepared ϕ_s state now involves a dissipative irreversible decay into the effective continuum of levels $\{\phi_l\}$. Questions concerning the precise nature of the prepared state, the contribution from the reversible dephasing processes at intermediate energies, and the role of intramolecular vibrational relaxation at high enough energies are central to the understanding of infrared multiphoton dissociation processes in isolated molecules.[33] The RRKM theory of unimolecular decomposition is generally expressed in terms of the assumption of very rapid intramolecular vibrational relaxation.[34] However, despite the apparent successes of this theory, a kinetically equivalent formulation invokes the opposite assumption of the *absence* of intramolecular vibrational relaxation along with the alternative assumption that it is the collisional excitation which produces its statistical distribution amongst the relevant levels of the system. Collisional processes are discussed in Section III, but we can anticipate results by nothing that at very high vibrational energy contents, corresponding to the amount of energy required to break chemical bonds, a statistical description of the initial and final states of collisions of highly vibrationaly excited molecules may be quite a reasonable one. Thus a study of intramolecular vibrational processes is of considerable interest in also furthering our understanding of unimolecular decomposition processes. To date, however, to our knowledge there has yet to be experimentally determined situations in which there is unambiguous identification of dephasing or intramolecular vibrational relaxation processes in isolated molecules.

III. COLLISIONAL PROCESSES OF SMALL AND INTERMEDIATE CASE MOLECULES

By definition irreversible electronic relaxation processes cannot occur in isolated small and too many level small (intermediate) case molecules because of the insufficient density of final levels. For long times the molecule senses the presence of a finite number of possible final levels instead of the effective continuum that is required to drive irreversible electron relaxation. When collisional processes are appended, it is clear that the continuous density of states of the colliding pair can provide the necessary driving force for irreversible relaxation. The observed magnitudes of electronic relaxation rates as well as dependencies on perturbing molecules, etc., are the aspects of the processes which are of central interest.

As an example consider the case of collision-induced intersystem crossing where the initial and final states, ϕ_s and $\{\phi_l\}$ respectively, are singlet and triplet states of the molecule. Collision-induced intersystem crossing, a spin-forbidden collision event, conjures up the image of a collisional perturber interacting via some spin–orbit or exchange interaction to enable the occurrence of the spin-forbidden transition in its collision partner. Thus it is to be expected that the rates of collision-induced intersystem crossing should increase markedly with the atomic number of the atoms in the perturbing molecule.

Unfortunately, experimental data are in total conflict with this naive picture. The collision-induced intersystem crossing rates for glyoxal[35,36] with the hydrogen molecule or helium atoms as perturbers are of the order of 5–10% of gas kinetic values (cross sections ≈ 1–2 Å^2), despite the negligibly small spin–orbit and exchange interactions available to these perturbers. As electronic relaxation processes in isolated large, statistical limit, molecules display a marked dependence on Franck–Condon factors between the singlet and triplet states,[1] it would also be expected that the collision-induced intersystem crossing rates would also be dependent on singlet–triplet Franck–Condon factors. Again the experiments on glyoxal show a lack of variation of these rates with deuteration of the glyoxal,[36] again in marked contrast with our simple anticipated results. In fact, for a wide series of collision partners the singlet to triplet rates are observed[35] to correlate well

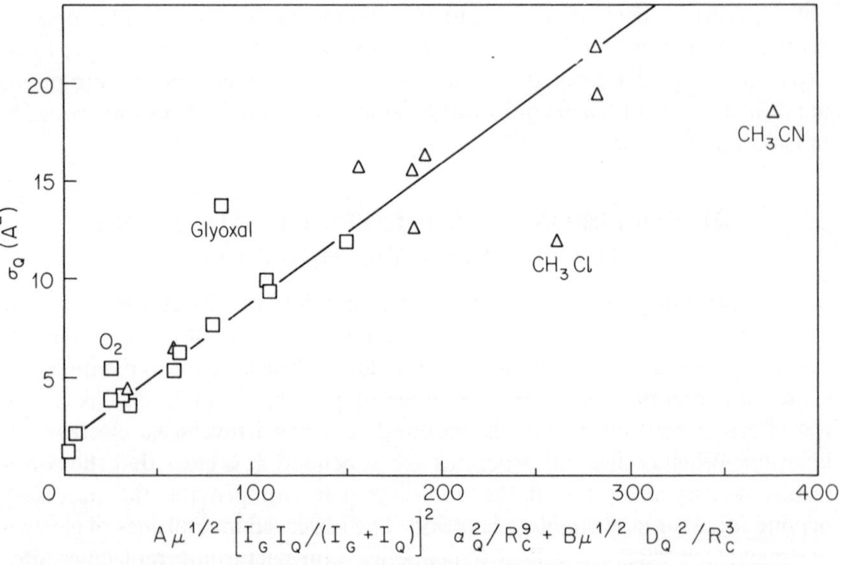

Fig. 2. Date of Beyer and Lineberger for the quenching rate of the ^1A_u state of glyoxal for a series of different perturbers in the Thayer–Yardley plot. Here μ is the reduced mass of the collision pair, I_Q and I_Q are the ionization potentials for glyoxal and the quencher, respectively, α_Q and D_Q are the dipole polarizability and electric dipole moments of the quencher, respectively, R_c is a minimum impact parameter for the collision, and A and B are constants dependent on the quenched molecule (glyoxal.) The triangles represent polar quenchers, while the squares are for non-polar ones. (Reprinted from ref. 35 by permission of the *American Institute of Physics* and the authors.)

with expected total cross sections[37] arising from the *long-range interactions* involving the perturbers' permanent electric dipole moment and dipole polarizability. This fact is illustrated in Fig. 2 where the data of Beyer and Lineberger[35] on glyoxal are presented. The abscissa involves the perturber's dipole moment, polarizability, and ionization potential. Its spin–orbit coupling constant, atomic number, are irrelevant here. The molecules CH_3Cl and CH_3CN have small dipole moments, but large quadrupoles so that Thayer–Yardley[37] type plot of Fig. 2 underestimates their long-range attractive interactions.

Parmenter has introduced an alternative correlation which also explains the perturber dependence of collision-induced intersystem crossing rates.[38] Parmenter's correlation is between these rates and the well depths of the perturbing molecules' Lennard–Jones interactions, properties again having nothing whatsoever to do with spin multiplicity changes in the collision partner. The Parmenter plot of the glyoxal data is given in Fig. 3. These facts, along with other observations, make it patently clear that some other mechanism is operative in collision-induced intersystem crossing processes of small and intermediate case molecules.

Ordinary collisional quenching processes are quite familiar. In this case the quenching rates are simply understood on the basis of the fact that the

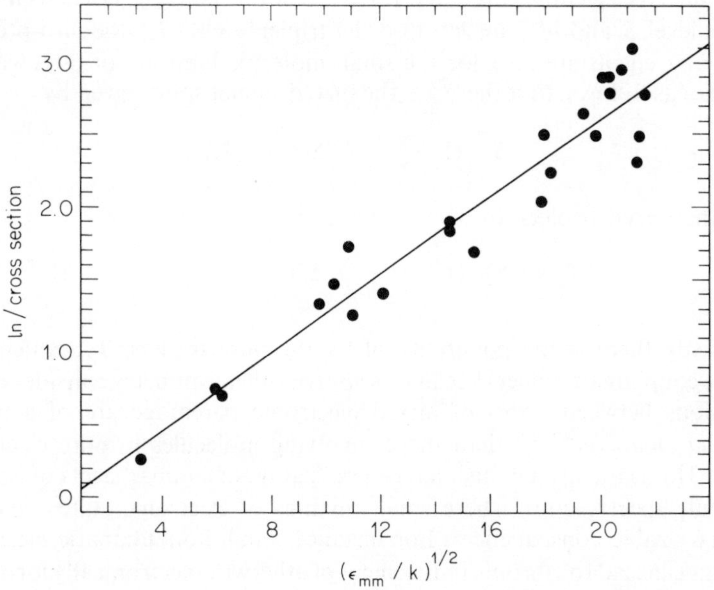

Fig. 3. Data of Fig. 2 replotted according to the Parmenter–Seaver correlation between the logarithm of the quenching cross section and the square root of the quencher well depth. (Reproduced by permission of C. Parmenter.)

molecule–perturber interaction, V, produces a coupling v_{sl}. between the initially excited molecular state, ϕ_s, and a vibronic component, ϕ_l, of another electronic state. This familiar quenching is manifested by linear Stern–Volmer kinetics over wide pressure ranges. The collision-induced intersystem crossing (and internal conversion) processes of interest to this review are explicitly those for which this traditional quenching mechanism *fails miserably*. The H_2 or He–glyoxal intermolecular interaction is much too weak to couple pure singlet and triplet glyoxal levels with cross sections of 2.0 and 1.3 Å^2, respectively. A partial listing of the molecules which to date display this non-quenching-like collision-induced electron relaxation includes methylene,[39] glyoxal,[35,36] biacetyl,[22] pyrazine,[40] CN,[41] CO,[42] formaldehyde,[43] benzophenone,[44] SO_2,[45] and pyrimidine.[46]

It is convenient to separate the discussion of collision-induced electronic relaxation into descriptions of the low and high pressure limits. A discussion of the low pressure limit can be carried out[47–50] in the limit of extremely low pressures where the probability for collisions and, therefore, for collision-induced electronic relaxation is extremely small. Under these conditions the general theoretical results of Section II provide the insight into understanding the mechanism for collision-induced electronic relaxation processes.

It has been emphasized that, when (2) is satisfied, the ϕ_s–$\{\phi_l\}$ interelectronic state coupling, v_{sl}, leads to isolated molecular eigenstates which are of mixed electronic parentage, (5)–(6). If, for instance, ϕ_s is a zeroth-order singlet level, S, and $\{\phi_l\}$ are zeroth-order triplet levels, $\{T_l\}$, the zero-pressure molecular eigenstates, (6), for the small molecule limit are of three general varieties as follows: first there are the mixed singlet states given by

$$\hat{S} = (1 - \sum_l y_l^2)^{1/2} S + \sum_l y_l T_l, \tag{23}$$

then the mixed triplets are

$$\hat{T}_l = z_l S + (1 - z_l^2 - \sum_{l' \neq l} z_{l'}^2)^{1/2} T_l + \sum_{l' \neq l} T_{l'}, \tag{24}$$

and lastly there is the possibility of having pure triplets, $T_{l'}$, which may escape coupling to singlets because of spin or other symmetry considerations. Collisions between states of mixed electronic parentage are of a *wholly different character*[47–50] than those involving molecules in pure electronic states. This is merely a collisional generalization of familiar facets of isolated molecule spectroscopy where small mixings of electronic states can lead to very sizable consequences. For instance, small non-adiabatic electronic mixings can lead to vibronic inducement of otherwise electronically forbidden radiative transitions. Likewise, the presence of these zero-pressure interelectronic state coupling lead to the occurrence of highly efficient collision-induced electron relaxation processes. In particular, the theory discussed below shows how the mixed states $\hat{S}$ 'borrow' some pure triplet state collision-induced vibrational and rotational relaxation to utilize for the collision-induced intersystem crossing in a manner which parallels intensity borrowing in isolated molecules. The intensity borrowing is generally from other electronic states with very different energies. The collisional 'borrowing', on the other hand, emerges from nearly degenerate levels.

Given that the small molecule limits. (2) and (2′), for relevant experimental time scales are both satisfied, the only real difference between a very small diatomic molecule and an intermediate-sized one like glyoxal and pyrimidine resides only in the number of coupled levels in (5) and (6), respectively. Hence, it is convenient to pursue the general analysis for the case of diatomic molecules where a pair of coupled levels, (5), is often sufficient. Then the more general case of many levels, (6), is readily generated.

Diatomic molecules can also exhibit strong collision-induced electronic relaxation. A prime example is the CN molecule[41] which undergoes collision-induced internal conversion, $A^2\Pi \rightarrow B^2\Sigma^+$ due to perturbations involving $v = 10$ of $A^2\Pi$ and $v = 0$ of $B^2\Sigma^+$ for K values of 4, 7, 8, 11, and 15. At sufficiently low pressures in active nitrogen–organic vapour flames the $A^2\Pi$ levels have a very large steady-state population relative to the isoenergetic $B^2\Sigma^+$ levels, and most of the emission is observed in the $A^2\Pi \rightarrow X^2\Sigma^+$ red-band system. The anomalously high intensity (for pressures < 1 Torr) of a few lines in the $B^2\Sigma^+ \rightarrow X^2\Sigma^+$ violet bands is due primarily to the direct formation of perturbed $B^2\Sigma^+$ levels containing some admixture of $A^2\Pi$ character. At higher presures the intensities associated with these perturbed $B^2\Sigma^+$ lines is further enhanced. Recently Lavollee and Tramer[42, 51] have observed collision-induced intersystem crossing from perturbed $A^1\Pi$ levels of CO by utilization of synchrotron excitation. These diatomic examples should provide a more quantitative test of the theoretical principles which are also applicable (with some appended summations over coupled states) to larger molecules like glyoxal.

A. Description of the Collisions

1. *The Mixed States*

We consider first the simplest case in which the coupling occurs between a pair of individual singlet and triplet rovibronic levels with electronic, vibrational, and rotational quantum numbers $SvJ\Omega M$ and $Tv\Omega JM$, respectively. Here Ω is the projection of the angular momentum along the molecule fixed axis and M is its spatial component. A Hund's case (a) basis set is employed for simplicity, and other Hund's cases are readily utilized as required.[18] (Likewise, the treatment of asymmetric tops and polyatomic symmetric tops follows directly.) The zeroth-order unperturbed energies are written as

$$E_0(SvJ\Omega) = E_0^{Sv\Omega} + B_S J(J+1) \tag{25a}$$

$$E_0(Tv'J\Omega) = E_0^{Tv'\Omega} + B_T J(J+1), \tag{25b}$$

with the $E_0^{S,T}$ the origin term energies for both states (at the given $v\Omega$). B_S and B_T are, respectively, the unperturbed singlet and triplet rotational constants, and centrifugal distortion corrections are readily appended when required.

If $v_{ST}^{vv'}$ is the coupling between these zeroth-order states, the perturbed states have energies,

$$E_{\pm}^{vv'}(J\Omega M) = \tfrac{1}{2}[E_0(SvJ\Omega) + E_0(Tv'J\Omega)] \pm \tfrac{1}{2}\{[E_0(SvJ\Omega) - E_0(Tv'J\Omega)]^2 + 4|v_{ST}^{vv'}|^2\}^{1/2}. \tag{26}$$

It is convenient to simplify the notation by utilizing ω to denote the quantum numbers $J\Omega M$, writing the latter explicitly only when necessary. The mixed level of primarily singlet character, $\hat{S}$, is expressed in this notation by

$$|\hat{S}_{vv'}\omega\rangle = \cos(\theta/2)|Sv\omega\rangle + \sin(\theta/2)|Tv'\omega\rangle \tag{27a}$$

with the perturbed energy

$$E(\hat{S}_{vv'}\omega) \equiv E_+^{vv'}(\omega) \equiv E_0(Sv\omega) + \Delta. \tag{27b}$$

Δ is defined by (27b) and (26) as the energy shift of the perturbed $\hat{S}$ level, and it may vary in sign with different values $vv'J$ and Ω. The 'mixing angle', θ, depends on v, v', J, and Ω and is given by

$$\theta = \arctan\{2v_{ST}^{vv'}/[E_0(SvJ\Omega) - E_0(Tv'J\Omega)]\}. \tag{28}$$

Likewise, the perturbed triplet state is

$$|\hat{T}_{vv'}\omega\rangle = -\sin(\theta/2)|Sv\omega\rangle + \cos(\theta/2)|Tv'\omega\rangle \tag{29a}$$

with the corresponding energy

$$E(\hat{T}_{vv'}\omega) = E_-^{vv'}(\omega) \equiv E_0(Tv'\omega) - \Delta, \tag{29b}$$

where $v_{ST}^{vv'}$ has been taken to be real for notational simplicity. When, for instance, the zeroth-order state S involves Hund's case (a) coupling (or near case (a)) and T is described by case (b) coupling (say, for a $^3\Sigma$ state), the actual zeroth-order triplet sublevels $|Tv'JNM\rangle$ with $N = J, J \pm 1$ (for a triplet) are linear combination[18] of the case (a) basis levels $|Tv'J\Omega M\rangle$. Hence, the original level $|Sv\omega\rangle$ may be coupled to any of the three spin-sublevels. Often only one of the three is close enough to give noticeable perturbations. The intermediate case (a)-case (b) character of the zeroth-order unmixed levels should be included in quantitative analysis.[18] It is ignored here to simplify the notation as its inclusion is rather straightforward.

[See (73) below, for example.] Again we should emphasize that S and T may also correspond to a pair of zeroth-order vibrational levels of a single electronic state which are coupled through the anharmonic interaction, v_{ST}—the v, v' indices here would be redundant. Thus with minor modification much of the present section can immediately be applied to describe collisional processes of Fermi resonance mixed states.

For the present it is assumed that the singlet and triplet levels only interact by pairs (i.e. one singlet with one triplet). Cases in which more levels interact are discussed afterwards. The Hamiltonian of the isolated molecule in the basis of pure spin states is written as

$$H_{\mathrm{mol}} = \sum_{\substack{Sv \\ Tv'}} \sum_{\omega = J\Omega M} |\omega\rangle \begin{pmatrix} E_0(Sv\omega) & v_{\mathrm{ST}}^{vv'} \\ v_{\mathrm{TS}}^{v'v} & E_0(Tv'\omega) \end{pmatrix} < \omega|, \tag{30a}$$

where the 2×2 matrix indices refer to zeroth-order singlet and triplet states. The sum over S, T, v, and v' runs over all singlet and triplet vibronic states. The zeroth-order states which mix are conveniently paired as in (30a), while non-mixing pure spin levels appear as $\begin{pmatrix} E_0(Sv\omega) & 0 \\ 0 & 0 \end{pmatrix}$ or $\begin{pmatrix} 0 & 0 \\ 0 & E_0(Tv'\omega) \end{pmatrix}$.

In the basis of mixed states H_{mol} becomes

$$H_{\mathrm{mol}} = \sum_{S_{vv'}} \sum_{T_{vv'}} \sum_{\omega = J\Omega M} |\omega\rangle \begin{pmatrix} E(\hat{S}_{vv'}\omega) & 0 \\ 0 & E(\hat{T}_{vv'}\omega) \end{pmatrix}. \tag{30b}$$

Here the 2×2 matrix indices designate the mixed states, while the unmixed ones remain as in (30a)

The molecule is prepared in its initial state by optical excitation (or in some cases by a subsequent collisional transition). When the incident radiation has a pulse width for which $|E_+ - E_-|$ of (3b) satisfies (14), the exciting light produces the single misled level which we take to be $|\hat{S}\omega\rangle$.

2. *The Interaction With Perturbers*

The analysis is considerably simplified by treating collisions of the molecule with a structureless perturber (called the 'atom'). The rotational and vibrational degrees of freedom of the perturber are ignored since they are easily appended if desired. The perturbers are assumed to have electronic excitation energies which are much higher in energy than the excited states of the molecule under consideration, so electronic energy transfer processes may be ignored. These interesting effects may be appended if necessary.

The distance of the atom from the centre-of-mass of the molecule is R, and the orientation angle between the vector $\mathbf{R}$ and the molecular axis is γ. Using the pure electronic state basis as in (30a), the vibronic matrix elements of molecule–perturber interactions are described by the operator

$$\mathbf{V} = \sum_{S,T,v,v'} \begin{pmatrix} V_{\mathrm{S}}^{v}(R,\gamma) & V_{\mathrm{ST}}^{vv'}(R,\gamma) \\ V_{\mathrm{TS}}^{v'v}(R,\gamma) & V_{\mathrm{T}}^{v'}(R,\gamma) \end{pmatrix} + \sum_{v \neq v'} [|Sv\rangle V_{\mathrm{S}}^{vv'}(R,\gamma)\langle Sv'| + |Tv\rangle V_{\mathrm{T}}^{vv'}(R,\gamma)\langle Tv'|]. \tag{31}$$

The first sum in (31) contains all portions, V_{S}^{v} and $V_{\mathrm{T}}^{v'}$, of the molecule–perturber interactions which are diagonal in the pure spin vibronic states Sv and Tv' and Tv' respectively. Note that $\mathbf{V}$ is still an operator with respect to the $|\omega\rangle$ for these vibronic states, $V_{\mathrm{S}}^{vv'}$ and $V_{\mathrm{T}}^{vv'}$ are the molecule–perturber interactions that can generate vibrational (and rotational) relaxation without spin change. The spin off-diagonal intermolecular couplings $V_{\mathrm{ST}}^{vv'}$ and $V_{\mathrm{TS}}^{v'v}$ are ignored below, but they could simply be reintroducd if heavy-atom effects on the collision-induced intersystem crossing became evident.[48,49] These couplings are the ones which would yield conventional-type collisional quenching processes. (For collision-induced internal conversion or collisions between anharmonically coupled vibrational levels of a single electronic state, the effects of the couplings, e.g. $V_{\mathrm{SS'}}^{vv'}$ should be more readily detected, so these terms should be retained throughout.) The important fact here is the assumption that the pure-spin energy states undergo different interactions with the perturber. These interactions are characterized by pure-spin electronic potential surfaces V_{S} and V_{T} having the elastic, V_{S}^{v} and $V_{\mathrm{T}}^{v'}$, and inelastic, $V_{\mathrm{S}}^{vv'}$ and $V_{\mathrm{T}}^{vv'}$, portions which differ for the two pure electronic surfaces.

In the absence of intramolecular interelectronic state coupling, $v_{\mathrm{ST}}^{vv'}$, the molecular–perturber interactions V_{S} and V_{T} could only lead to vibrational and rotational relaxation within the pure electronic manifolds S and T. respectively. It is this intramolecular coupling, $v_{\mathrm{ST}}^{vv'}$, which enables the occurrence of the collision-induced intersystem crossing process *when* $V_{\mathrm{ST}}^{vv'} = V_{\mathrm{TS}}^{v'v} \approx 0$. If different mixed pairs $|S_0v_0\omega_0\rangle$–$|T_0v_0'\omega_0\rangle$ and $|S_1v_1\omega_1\rangle$–$|T_1v_1'\omega_1\rangle$ are characterized by mixing arguments θ_0 and θ_1, respectively, the matrix elements of $\mathbf{V}$ in the mixed state basis are given by

$$\begin{aligned}
\langle \hat{S}_{v_1v_1'}\omega_1|\mathbf{V}|\hat{S}_{v_0v_0'}\omega_0\rangle &= \langle\omega_1|\cos(\theta_1/2)\cos(\theta_0/2)V_{\mathrm{S}}^{v_1v_0} \\ &\quad + \sin(\theta_1/2)\sin(\theta_0/2)V_{\mathrm{T}}^{v_1'v_0'}|\omega_0\rangle \\
\langle \hat{S}_{v_1v_1'}\omega_1|\mathbf{V}|\hat{T}_{v_0v_0'}\omega_0\rangle &= \langle\omega_1|\cos(\theta_1/2)\sin(\theta_0/2)V_{\mathrm{S}}^{v_1v_0} \\ &\quad - \sin(\theta_1/2)\cos(\theta_0/2)V_{\mathrm{T}}^{v_1'v_0'}|\omega_0\rangle \\
\langle \hat{T}_{v_1v_1'}\omega_1|\mathbf{V}|\hat{S}_{v_0v_0'}\omega_0\rangle &= \langle\omega_1|\sin(\theta_1/2)\cos(\theta_0/2)V_{\mathrm{S}}^{v_1v_0} \\ &\quad - \cos(\theta_1/2)\sin(\theta_0/2)V_{\mathrm{T}}^{v_1'v_0'}|\omega_0\rangle \\
\langle \hat{T}_{v_1v_1'}\omega_1|\mathbf{V}|\hat{T}_{v_0v_0'}\omega_0\rangle &= \langle\omega_1|\sin(\theta_1/2)\sin(\theta_0/2)V_{\mathrm{S}}^{v_1v_0} \\ &\quad - \cos(\theta_0/2)\cos(\theta_1/2)V_{\mathrm{T}}^{v_1'v_0'}|\omega_0\rangle.
\end{aligned} \tag{32a}$$

Amongst all the couplings in (32a), there are ones associated with purely elastic scattering which are diagonal with respect to all quantum numbers. These are written as

$$\mathbf{V}_{\text{diag}} = \sum_{\substack{S_{vv'},\\ T_{vv'}}} \sum_{\omega = J\Omega M} |\omega\rangle \begin{pmatrix} \cos^2\left(\dfrac{\theta}{2}\right) V_S^{vv} + \sin^2\left(\dfrac{\theta}{2}\right) V_T^{v'v'} & 0 \\ 0 & \cos^2\left(\dfrac{\theta}{2}\right) V_T^{v'v'} + \sin^2\left(\dfrac{\theta}{2}\right) V_S^{vv} \end{pmatrix} \langle \omega|, \tag{32b}$$

where θ is implicity dependent on the quantum numbers $vv'J\Omega$.

Available experimental results strongly suggest that the important parts of V_S and V_T involve long-range attractive interactions,[35,37,38] so these potentials may be written, for instance, in the familiar Lennard–Jones form, e.g.

$$V_S(r, R, \gamma) = \varepsilon(r, \gamma)\{[\sigma(r, \gamma)/R]^{12} - [\sigma(r, \gamma)/R]^6\}, \tag{33}$$

where r is the vibrational coordinate of the molecule and where the well depth, ε, and range, σ, are taken to have vibrational and/or orientational dependences. The long-range interactions may more accurately be represented in terms of the interaction between the multipoles on the molecule and perturbers. However, equation (33) suffices for our current purposes to demonstrate the salient physical features. Below most of the equations are written in terms of an unspecified V_S and V_T for complete generality.

3. *The Equations of Motion*

In the following, we adopt the popular semiclassical description wherein the relative motion is described classically with $\mathbf{R}(t)$ governed by a trajectory on an appropriate average potential surface $V_0(\mathbf{R})$.

We assume that the excitation process does not coherently excite pairs of mixed states as discussed in Section II.B above. Then, the molecule may be taken to initially be in some mixed molecular state, $|\hat{S}_{vv'}\omega_0\rangle$, before the collision. Its state at time t during the collision is given by

$$\bar{\psi}(t) = \sum_n a_n(t) \exp\left[-(i/\hbar) \int_{-\infty}^{t} E_n(t')dt' \right] |n\rangle, \tag{34}$$

where n runs over all possible singlet and triplet levels (mixed and unmixed). $E_n(t')$ denotes the diagonal matrix element of $\langle n|H_{\text{int}}(t')|n\rangle$ defined in (32) and (35) below. The duration of a collision is assumed to be much less than the lifetime of the excited state, so radiative, etc., decay during the collision can safely be ignored. $\Psi(t)$ obeys the time-dependent Schrödinger equation generated by the internal Hamiltonian

$$\mathbf{H}_{\text{int}} = \mathbf{H}_{\text{mol}} + \mathbf{V}_{\text{diag}} + (\mathbf{V} - \mathbf{V}_{\text{diag}}) \tag{35}$$

where H_{mol} is given by (30b), $\mathbf{V}_{\mathrm{diag}}$ by (32b), $\mathbf{V}$ by (32a) and $\mathbf{R}(t)$ is a classical trajectory for the relative molecule–perturber motion.

For relative kinetic energies sufficiently greater than any energy transfer and for weak interactions, the solution to the time-dependent Schrödinger equation is fairly accurately represented by the first-order Magnus approximation[52]

$$a_n^{M_1}(t) = \{\exp[\mathbf{M}^{(1)}(t)]\}_{n,\hat{S}_{vv'}\omega_0}, \tag{36}$$

where $\mathbf{M}_{nm}^{(1)}$ is given by

$$\mathbf{M}_{nm}^{(1)}(t) = (i\hbar)^{-1} \int_{-\infty}^{t} \mathrm{d}t' \exp\left\{(i\hbar)^{-1} \int_{-\infty}^{t'} \mathrm{d}t''[E_m(t'') - E_n(t'')]\right\} \times \langle n|H_{\mathrm{int}}(t')|m\rangle, \tag{37}$$

and, for instance

$$\begin{aligned} E\hat{S}_{vv'}\omega_0(t'') &= \langle \hat{S}_{vv'}\omega_0|H_{\mathrm{mol}} + H_{\mathrm{int}}(t'')|\hat{S}_{vv'}\omega_0\rangle \\ &= E(\hat{S}_{vv'}\omega_0) + \langle \hat{S}_{vv'}\omega_0|\mathbf{V}_{\mathrm{diag}}(t'')|\hat{S}_{vv'}\omega_0\rangle \end{aligned} \tag{38}$$

describes the time-varying energy of the initial state [cf. (32b)] during a collision characterized by classical trajectory $\mathbf{R}(t)$ for the relative motion. The coupling matrix elements between $|\hat{S}_{vv'}\omega_0\rangle$ and mixed triplets are obtained from (32b) for $\theta_1 \neq 0$ and $\theta_0 \neq 0$, while those for pure triplets are found from (32b) for $\theta_1 = 0$ and $\theta_0 \neq 0$. Notice that the coupling responsible for a mixed–unmixed $\hat{S} \rightarrow T$ transition ($\theta_1 = 0$) involves a pure triplet state molecule perturbation interaction, $\langle T_1 v_1' \omega_1 | V_{\mathrm{T}}(r, \mathbf{R}(t)) | T v_0' \omega_0 \rangle$, corresponding to vibrational and/or rotational relaxation between a pair of zeroth-order pure triplet levels. This interaction is multiplied by $\sin(\theta_0/2)$ which involves the mixing coefficient for the triplet component in the initial state $\hat{S}$. This makes it quite explicit how the mixed singlet, $\hat{S}$, 'borrows' rotational and/or vibrational relaxation processes from the pure triplet.[47–50] The collision-induced couplings between mixed singlets, $\hat{S}$, and mixed triplets, $\hat{T}$, again depend on mixing coefficients for the two states and upon differences between the pure singlet and pure triplet matrix elements of V_{S} and V_{T}, respectively. Even in the unlikely situation of equal matrix elements of V_{S} and V_{T}, the difference in mixing coefficients for different mixed states still yields a net coupling.

For groups of nearby levels the Magnus-type solution (36) is expected to be necessary.[52] This situation should also prevail in the presence of a magnetic field (see Section III.C) where the different M-sublevels are split by the field, and couplings are also introduced which are smaller or comparable to the time variations of the 'energies' like (38) during a collision. (Note that the molecule–perturber interaction also lifts the M-degeneracies.) When the spacings between these nearby levels, $|n\rangle$, become comparable or smaller than typical collisional energies, an interesting situation can ensue wherein there are a large number of curve crossings between the curves, $E_n(t)$, for these levels during the collision. A given molecule would then undergo a number of transitions during a collisional event, and the net effect of these multiple curve crossings *might* appear to resemble a quasi-statistical branching amongst the available final states. (The Magnus approach of (36) automatically treats all of these multiple crossing transitions properly.) However, the physical process is quite distinct from that generally associated with the long-lived complexes which yield statistical product distributions. In the present case the collisions can be very brief encounters, direct collisions arising from weak, long-range interactions. The magnitudes of the cross sections, their dependence on perturber and mixing coefficients, etc., can serve to distinguish the sticky collision complex from the direct multiply crossing $E_n(t)$ curves case.

For transitions to triplet levels, n, where $|E_n - E_{\hat{S}}|$ are sufficiently large, the first-order perturbation approximation,

$$a_n^{(1)}(t) \simeq \mathbf{M}^{(1)}_{n,\hat{S}_{vv'}\omega_0}(t), \tag{39}$$

suffices. Here we explicitly consider the matrix elements $\mathbf{M}^{(1)}_{n,\hat{S}}$ for n within the triplet manifold (mixed or unmixed) as our prime interest lies in first elucidating the salient physical features and, in particular, the dependence of the collision-induced intersystem crossing cross sections. $\sigma_{\mathrm{CIISC}}(\hat{S})$, upon the initial-state mixing coefficient, $\sin(\theta_0/2)$. The results discussed below using (39) are trivially generalized to (36) by likewise evaluating the requisite singlet–singlet couplings in $\mathbf{M}^{(1)}$ and then by trivially exponentiating the matrix. For cases where (39) predicts $\sigma_{\mathrm{CIISC}}(\hat{S})$ to not simply depend on $\sin(\theta_0/2)$ through an overall proportionality to $\sin^2(\theta_0/2)$, the more general case of (36) (or higher Magnus approximations)[52] of necessity have more general variations with $\sin(\theta_0/2)$.

The collisional transition probabilities are just equal to $|a_n(+\infty)|^2$. The state-to-state transition probabilities are then obtained, for a given classical trajectory with kinetic energy, E_t, by summing over initial and final molecular magnetic quantum numbers and by averaging over the initial ones,

$$P(\hat{S}_{vv'}J_0\Omega_0 \to \hat{T}'J'\Omega'|E_t) = (2J_0+1)^{-1} \sum_{M_0,M'} |a_{\hat{T}J'\Omega'M'}(+\infty)|^2, \tag{40}$$

where $\hat{T}'$ may be of mixed or unmixed electronic character and $a_n(t)$ is taken to have the initial condition $a_n(-\infty) = \delta_{n,\hat{S}J_0\Omega_0M_0}$. The cross-section is then obtained by 'summing' over all impact parameters, b, for trajectories with initial relative kinetic energy, E_t, as

$$\sigma(\hat{S}_{vv'}J_0\Omega_0 \to \hat{T}'J'\Omega'|E_t) = \int_0^\infty 2\pi b db P(\hat{S}_{vv'}J_0\Omega_0 \to \hat{T}J'\Omega'|E_t). \tag{41}$$

The thermally averaged cross section is then

$$\sigma(\hat{S}_{vv'}J_0\Omega_0 \rightarrow \hat{T}'J'\Omega'|\beta) = \int_0^\infty d(\beta E_t)\beta E_t \exp(-\beta E_t) \times \sigma(\hat{S}_{vv'}J_0\Omega_0 \rightarrow \hat{T}'J'\Omega'|E_t), \quad (42)$$

where $\beta = k_B T$ is the absolute temperature in energy units and k_B is Boltzmann's constant.

B. Types of Collisional Processes

It is useful to classify all collisional transitions $\hat{S}J_0\Omega_0 \rightarrow T'v'$ (or $\hat{T}'$)$J'\Omega'$ into two categories depending on the nature of the energy difference that appears in the exponent of (37) for the transition. Consider first the $SJ_0\Omega_0 \rightarrow \hat{T}'v'J'\Omega'$ case where this difference is written as

$$\begin{aligned} E_{\hat{S}_{vo}v_0'\omega_0}(t) - (t) - E_{\hat{T}_{vv'}\omega'}(t) &= E_0(Sv_0\omega_0) - E_0(T'v'\omega') + \Delta_0 - \Delta' \\ &\quad + \langle \omega_0 | V_S^{v_0}(\mathbf{R}(t)) | \omega_0 \rangle - \langle \omega' | V_T^{v'}(\mathbf{R}(t)) | \omega' \rangle \\ &\quad - \sin^2(\theta_0/2) \langle \omega_0 | V_S^{v_0}(\mathbf{R}(t)) - V_T^{v_0'}(\mathbf{R}(t)) | \omega_0 \rangle \\ &\quad + \sin^2(\theta_1/2) \langle \omega' | V_S^{v}(\mathbf{R}(t)) - V_T^{v'}(\mathbf{R}(t)) | \omega' \rangle \\ &\equiv \Delta E_\infty + \Delta V(S_{v_0} \rightarrow T_{v'}; t) + \delta V(S_{v_0} \rightarrow T_{v'}; t), \end{aligned} \quad (43)$$

where $\Delta V(S_{v0} \rightarrow T_{v'}; t)$ describes the time-variation of the energy difference between the unmixed levels,

$$\Delta V(S_{v0} \rightarrow T_{v'}; t) = \langle \omega_0 | V_S^{v0}(\mathbf{R}(t)) | \omega_0 \rangle - \langle \omega' | V_T^{v'}(\mathbf{R}(t)) | \omega' \rangle, \quad (44)$$

and all terms involving Δ_0, Δ', and the mixing coefficients, $\sin^2(\theta_0/2)$ and $\sin^2(\theta'/2)$ are grouped into $\delta V(S_{v0} \rightarrow T_{v'}; t)$. If the collision is such that

$$|\Delta E_\infty + \Delta V(S_{v0} \rightarrow T_{v'}; t)| \gg |\delta V(S_{v0} \rightarrow T_{v'}; t)|, \text{ all } t, \quad (45)$$

the $\delta V(S_{v0} \rightarrow T_{v'}; t)$ in (43) may be ignored. Then there is no dependence on $\sin^2(\theta/2)$ in the exponential of (37). These collisions are termed 'simply mixed' ones. On the other hand, the 'perturbed mixing' collisions are characterized by

$$|\Delta E_\infty + \Delta V(S_{v0} \rightarrow T_{v'}; t)| \lesssim |\delta V(S_{v0} \rightarrow T_{v'}; t)|, \text{ some } t, \quad (46)$$

so the exponential in (37) must contain an explicit dependence on the square of the mixing coefficients, $\sin^2(\theta/2)$.

1. *The Simply Mixed Collisions*

In the simply mixed collisions, defined by (45), the only dependence on the mixing coefficients is provided by the coupling matrix elements in (32a). *For transitions to a pure triplet* ($\theta' = 0$), for instance, the relevant matrix element is proportional to $\sin^2(\theta/2)$, so that [using (39)]

$$\sigma(\hat{S}_{vov\acute{o}}J_0\Omega_0 \to Tv'J'\Omega') = Q\sin^2(\theta_0/2), \quad \text{simply mixed collisions,} \tag{47}$$

where Q is independent of the mixing coefficient in accord with the previous theoretical arguments.[47, 48] More explicitly using the first-order perturbation approximation, this gives

$$a^{(1)}_{Tv'\omega'}(+\infty) = \sin(\theta_0/2)(i\hbar)^{-1}\int_{-\infty}^{\infty} dt \exp\left\{(i\hbar)^{-1}\int_{-\infty}^{t} dt'[\Delta E_\infty + \Delta V(S_{vo} - T_{v'};t')]\right\}\langle\omega'|V_T^{v'v\acute{o}}|\omega_0\rangle, \quad \text{simply mixed,} \tag{48}$$

which looks rather similar to a pure triplet-state transition amplitude apart from the overall $\sin(\theta_0/2)$ factor and the fact that one of the state energies in the exponent refers to a singlet. However, if it is further true that

$$|\Delta V(S_{vo} \to T_{v'};t)| \gg |\langle\omega_0|[V_S^{vo}(\mathbf{R}(t)) - V_T^{v\acute{o}}(\mathbf{R}(t))]|\omega_0\rangle, \text{ all } t, \tag{49}$$

then (48) further reduces to $\sin(\theta_0/2)$ times the amplitude for the transition $Tv'_0\omega_0 \to Tv'\omega'$ between zeroth-order pure triplets,

$$a^{(1)}_{Tv'\omega'}(+\infty) \simeq \sin(\theta_0/2)(i\hbar)^{-1}\int_{-\infty}^{\infty} dt \exp\left\{(i\hbar)^{-1}\int_{-\infty}^{t} dt'[E_0(Tv'_0\omega_0) - E_0(Tv'\omega') + \langle\omega_0|V_T^{v\acute{o}}(\mathbf{R}(t'))|\omega_0\rangle - \langle\omega'|V_T^{v'}(\mathbf{R}(t'))|\omega'\rangle]\right\} \tag{50}$$
$$\times \langle\omega'|V_T^{v'v\acute{o}}(\mathbf{R}(t))|\omega_0\rangle, \quad \text{simply mixed collisions and condition (49),}$$

exhibiting quite explicitly how the mixed initial states S 'borrows' rotational and/or vibrational relaxation processes from the pure zeroth-order triplets to provide its collision-induced intersystem crossing.[48, 49]

The case of collisions from a mixed singlet, S, to a mixed triplet, T, proceeds similarly. The coupling term from (32a) to be substituted into (37) is

$$\langle\hat{T}_{vv'}\omega'|\mathbf{V}|\hat{S}_{vov\acute{o}}\omega_0\rangle = \cos(\theta_0/2)\sin(\theta'/2)\langle\omega'|V_S^{vvo}(\mathbf{R}(t))|\omega_0\rangle - \operatorname{cs}(\theta'/2)\sin(\theta_0/2)\langle\omega'|V_T^{v'v\acute{o}}(\mathbf{R}(t))|\omega_0\rangle. \tag{51}$$

Using first-order time-dependent perturbation, (39), again and following arguments similar to the unmixed case above yields a cross section which displays the form

$$\sigma(\hat{S}_{vov\acute{o}}J_0\Omega_0 \to \hat{T}_{vv'}J'\Omega') = A\sin^2(\theta'/2)[1 - \sin^2(\theta_0/2)] + B\sin^2(\theta_0/2)[1 - \sin^2(\theta'/2)] + C\sin\theta_0\sin\theta', \quad \text{simply mixed,} \tag{52}$$

where A, B, and C are independent of the mixing coefficients. A, B, and C describe respectively, relaxation cross sections in the singlet manifold, in the triplet manifold, and cross-terms between the two. Note that part of the A-term yields a contribution independent of the initial-state mixing coefficient, $\sin(\theta_0/2)$.

2. *The Perturbed Mixing Collisions*

All collisions violating (45), the 'perturbed mixing' collisions, require the retention of terms involving mixing coefficients in the oscillating exponential of (37). Hence, this directly implies that Q in (47) and A, B, and C in (52) then develop an explicit dependence on $\sin^2(\theta_0/2)$ and $\sin^2(\theta'/2)$, while the coupling term from (32a) likewise depends on $\sin(\theta_0/2)$. The explicit forms of these variations require the detailed evaluation of the appropriate integrals for the relevant interactions, V_S and V_T, and trajectories $\mathbf{R}(t)$. For the qualitative purposes of determining the overall variation of the collision-induced intersystem crossing rate upon the magnitudes of the mixing coefficients, it is sufficient to introduce simple models.

From (43) the asymptotic separation between initial and final levels, the energy transferred to the molecule in the collision, is $|E(\hat{T}_{vv'}J'\Omega') - E(\hat{S}_{v_0v_0'}J_0\Omega_0)|$ which may be as small as a few or a few tens of wavenumbers. Because ΔV and δV in (43) are expected to be at least of this same order of magnitude, this situation corresponds to the condition (46) defining the perturbed mixing collisions wherein the overall energy difference, (43), may be very near zero for some ranges of $\mathbf{R}(t)$. Hence the exponential term in (37) can have a phase which is very slowly varying with time for certain time intervals during the collision. These stationary-phase regions give large contributions to the transition amplitude (36) or (39) and, considering the large impact parameters involved, to the overall cross section for collision-induced intersystem crossing. When $|a_n^{(1)}(+\infty)|^2$ remains small, the first order treatment of (39) suffices and these integrals with regions of slowly varying phases may be evaluated by the method of stationary phase.

The stationary-phase requirement leads to the condition that

$$E_{\hat{S}_{v_0v_0'\omega_0}}(t_0) = E_{\hat{T}_{vv'\omega'}}(t_0) \tag{53}$$

at the time of stationary phase, t_0, i.e. to the crossing of the instantaneous initial and final levels during the collision. This is assumed to be the dominant type of collision-induced intersystem crossing mechanism by Chu and Dahler,[53] but it emerges here as only a special, albeit important, contribution.

A single pass through such a region of stationary phase gives a transition probability,[54]

$$|a^{(1)}_{\hat{T}_{vv'\omega'}}(+\infty)|^2 \propto |\langle \hat{T}_{vv'}\omega'|\mathbf{V}(\mathbf{R}(t_0))|\hat{S}_{v_0v_0'}\omega_0\rangle|^2 \times \left\{\left(\frac{\partial}{\partial t}[E_{\hat{S}_{v_0v_0'\omega_0}}(t) - E_{\hat{T}_{vv'\omega'}}(t)]\right)_{t=t_0}\right\}^{-2} \tag{54}$$

In addition to the dependence of the transition matrix element on the mixing coefficients through the dependence of the matrix element of **V** as in (47) or (52), the variation of the energies in (43) with mixing coefficients also introduces dependences of the denominator in (54) upon the mixing coefficients. The collision can, however, make a second pass through the stationary-phase region (once in and once out), so the overall transition probability is

$$P_{\hat{T}_{vv'\omega'}} = 2|a^{(1)}_{\hat{T}_{vv'\omega'}}(+\infty)|^2(1 - |a^{(1)}_{\hat{T}_{vv'\omega'}}(+\infty)|^2). \quad (55)$$

For instance, for the case of transitions to pure triplets ($\theta' = 0$), it displays contribution to the overall cross section that varies as $\sin^2(\theta_0/2)f[\sin^2(\theta_0/2)] \times \{1 - \sin^2(\theta_0/2)f[\sin^2(\theta_0/2)]\}$, with f a function of $\sin^2(\theta_0/2)$. Even ignoring the dependence of f on $\sin^2(\theta_0/2)$, the dependence $\sin^2(\theta_0/2)[1 - \bar{f}\sin^2(\theta_0/2)]$ displays the type of saturation variation with the initial state mixing coefficient, $\sin^2(\theta_0/2)$, observed by Tramer and coworkers in CO.[42, 51] (If $|a_n^{(1)}(+\infty)|^2$ gets too large, the Magnus approximation (36) with (55) would still preserve the non-linear dependence on $\sin^2(\theta_0/2)$.) This saturation dependence on $\sin^2(\theta_0/2)$ is a general feature of any Landau–Zener-type[55] curve (surface) crossing contribution and is independent of the approximations invoked. Stuckelberg corrections for phase differences can readily be appended to (55), and it is, in principle, possible to note that, in reality, there are a large number of levels, the $\Omega_0 M$ and $\Omega' M'$ sublevels, whose energies may undergo numerous crossings along the trajectory $\mathbf{R}(t)$. A detailed quantitative treatment may require Monte Carlo multiple surface hopping methods.[56]

3. *Example of Perturbed Mixing Collisions*

It is interesting to note that even when a first-order treatment is valid and the saturation effect in (54) is negligible, a non-linear dependence of the transition probability emerges from (55). For the general case, the matrix element in (54) is given in (32a), and the energy difference in (43) may be re-expressed in the obvious notation as

$$\begin{aligned}\Delta E(t) = \Delta E_\infty &+ V_S^0(t)\cos^2(\theta_0/2) - V_S'(t)\sin^2(\theta'/2)\\ &+ V_T^0(t)\sin^2(\theta_0/2) - V_T'(t)\cos^2(\theta'/2).\end{aligned} \quad (56)$$

For illustration, assume that the long-range behaviour of the four interaction potentials in (56) arises from van der Waals interactions and varies as $A_{ST}[R(t)]^{-6}$. A further simplifying assumption is that one spin state need be considered in the triplet. Then the crossing point, $R_0 = R(t_0)$, satisfies the equation (ignoring differences between A and A' for simplicity)

$$0 = \Delta E_\infty + R_0^{-6}\{A_S[\cos^2(\theta_0/2) - \sin^2(\theta'/2)] - A_T[\cos^2(\theta'/2) - \sin^2(\theta_0/2)]\}. \quad (57)$$

At the crossing point we find

$$\left|\frac{\partial}{\partial t}\Delta E(t)\right|_{t=t_0} \propto \left|A_{\rm S}[\cos^2(\theta_0/2)-\sin^2(\theta'/2)] - A_{\rm T}[\cos^2(\theta'/2)-\sin^2(\theta_0/2)]\right|^{-1/6}. \quad (58)$$

We then obtain the general result that

$$|a^{(1)}_{Tvv'\omega'}(+\infty)|^2 \propto |\cos(\theta_0/2)\sin(\theta'/2)\langle\omega'|V_{\rm S}^{vv_0}(\mathbf{R}(t_0))|\omega_0\rangle - \sin(\theta_0/2)\cos(\theta'/2)\langle\omega'|V_{\rm T}^{v'v_0'}(\mathbf{R}(t_0))|\omega_0\rangle|^2 |A_{\rm S}[\cos^2(\theta_0/2) - \sin^2(\theta'/2)] - A_{\rm T}[\cos^2(\theta'/2)-\sin^2(\theta_0/2)]|^{1/3}, \quad (59)$$

which may be simplified in a few limiting cases as follows:
(i) If the initial and final states are members of the same pair, airsing from the mixing of a pair of unperturbed levels, then we have $J' = J_0$, $M' = M_0$, $\Omega' = \Omega_0$ *and* $\theta' = \theta_0$. Consequently, (59) reduces to

$$|a^{(1)}_{T_{v_0 v_0'}\omega_0}(+\infty)|^2 \propto \sin^2\theta_0|\cos\theta_0|^{1/3}|\langle\omega_0|V_{\rm S}^{vv_0}(\mathbf{R}_0) - V_{\rm T}^{v_0 v_0'}(\mathbf{R}_0)|\omega_0\rangle|^2, \quad (60)$$

which yields an underlinear dependence of $|a_T^{(1)}(+\infty)|^2$ on $\sin^2(\theta_0/2)$.
(II) If the final state is a pure triplet state, $\theta' = 0$, and we obtain

$$|a^{(1)}_{Tv'\omega'}(+\infty)|^2 \propto \sin^2(\theta_0/2)\left|\frac{A_{\rm S}+A_{\rm T}}{2} + \frac{A_{\rm S}-A_{\rm T}}{2}\cos\theta_0\right|^{1/3} \times |\langle\omega'|V_{\rm T}^{v'v_0'}(\mathbf{R}_0)|\omega_0\rangle|^2. \quad (61)$$

(iii) A transition from a pure singlet ($\theta_0 = 0$) to a mixed triplet ($\theta' \neq 0$) is similar to (61) and gives

$$|a^{(1)}_{S\to Tvv'}\omega'(+\infty)|^2 \propto \sin^2(\theta'/2)\left|\frac{A_{\rm S}+A_{\rm T}}{2} + \frac{A_{\rm T}-A_{\rm S}}{2}\cos\theta'\right|^{1/3} \times |\langle\omega'|V_{\rm S}^{v'v_0}|\omega_0\rangle|^2. \quad (62)$$

In summary, it is clear that the 'perturbed mixing' collisions provide contributions to the collision-induced intersystem crossing rates which deviate from a pure linear dependence on the initial state mixing coefficient $\sin^2(\theta_0/2)$.

Under some circumstances the collisional processes are expected to be dominated by strong short-range collisions. Here perturbation approximations like (36), (37), or (39) become untenable and more accurate methods are required. Nikitin has considered the semiclassical theory of mixed state collisions in the strong interaction regime, and we have provided an application to collision-induced intersystem crossing processes.[57] However, the greatly enhanced polarizabilities of excited electronic states imply that van der Waals interactions, which are proportional to the polarizability, are likewise increased over ground-state values.[38] This fact, coupled with the large observed values of collision-induced intersystem crossing rates and the perturber dependence thereof, lend support to the belief of the predominance of weak long-range collisions. When, however, strong, short-range interactions contribute, these must be appended in manners, for instance, similar to our exponential repulsive model.[50]

C. Magnetic Field Dependences

The introduction of an external magnetic (or electric) field leads to coupling matrix elements amongst the pure-spin singlet states, $\{|SvJ\Omega M\rangle\}$, as well as amongst the triplets, $\{|Tv'J'\Omega'M'\rangle\}$. In the zero-pressure limit the inclusion of the singlet–triplet couplings, $v_{\mathrm{ST}}^{vv'}$, leads to zero-pressure field-dependent molecular eigenstates where J and M are no longer good quantum numbers. Hence, the number of terms contributing to (6) is enhanced by the presence of the field.[48] These zero-pressure molecular eigenstates may have decay rates, Γ_n of (7), differing from the field free values.[58] This field dependence of zero-pressure decay rates and quantum yields is a rather trivial one which is briefly mentioned below. The collisional effects of the external field, on the other hand, can be quite substantial.[49,59,60] For small zero-pressure mixings, the enhanced mixing (or even demixing) caused by the field can substantially alter the collision-induced intersystem crossing rates. The field dependence of the rates can, perhaps, be utilized to provide additional information about the intramolecular couplings and molecule–perturber interactions, so it is of considerable interest to consider this phenomenon further.

As a simple example consider perturbations between a $^1\Pi$ and $^3\Sigma^+$ state. In a $|S\Lambda\Sigma\rangle$ Hund's case (a) basis one of the $^1\Pi$ Λ-doublets,[18]

$$^1\Pi^- \equiv (2)^{-1/2}[-|0,1,0\rangle + |0,-1,0\rangle], \tag{63a}$$

can interact with the $^3\Sigma_1^+$ spin component for $N = J$,

$$^3\Sigma_1^{+-} \equiv (2)^{-1/2}[-|1,0,1\rangle + |1,0,-1\rangle], \tag{63b}$$

through spin–orbit coupling. In a magnetic field of strength $\mathscr{H}$ along the space-fixed z-axis, the $^1\Pi - {}^3\Sigma_1^{+-}$ zero-pressure Hamiltonian matrix is obtained from ref. 18 as

$$\begin{pmatrix} E(^1\Pi) + B_\Pi J(J+1) & A_{10,01}^{01*} \\ \quad + \delta_\Pi^- J(J+1) - \dfrac{\mu_0 \mathscr{H} M}{J(J+1)} & F(^3\Sigma^+) + B_\Sigma[J(J+1) - 1] \\ A_{10,01}^{01} & \quad + (6)^{-1/2}[\gamma - (2)^{3/2}\bar{\gamma}] - \dfrac{2\mu_0 \mathscr{H} M}{J(J+1)} \end{pmatrix}, \tag{64}$$

where $E(^1\Pi)$ and $E(^3\Sigma^+)$ are the origins of the $^1\Pi$ and $^3\Sigma^+$ vibronic states (v and v', including spin–spin interactions in the latter case), respectively, B_Π and B_Σ are their respective rotation constants, $A_{10,01}^{01}$ is the spin–orbit matrix element (defined in Table 11 of ref. 18), μ_0 is the Bohr magneton,

$\delta_\Pi^+ - \delta_\Pi$ gives the Λ-doubling constant in the $^1\Pi$ state (analogous to $2(B_0^+ - B_0^-)$ of Table III of ref. 18 for a $^3\Pi_1$, state), and γ and $\bar{\gamma}$ are electron spin–rotation matrix elements defined in Appendix B of ref. 18. The electronic g-factor in the $^3\Sigma_1^+$ state has been taken as two for simplicity.

Using (26) the perturbed energy levels are readily found to be

$$E_\pm \frac{1}{2}\left\{E(^1\Pi) + E(^3\Sigma^+) + (6)^{-1/2}[\gamma - (2)^{3/2}\bar{\gamma}] - B_\Sigma + (B_\Pi + \delta_\Pi^- + B_\Sigma)J(J+1) - \frac{3\mu_0\mathscr{H}M}{J(J+1)}\right\}$$
$$\pm \frac{1}{2}\left\{\left[E(^1\Pi) - E(^3\Sigma^+) + B_\Sigma - (6)^{-1/2}[\gamma - (2)^{3/2}\bar{\gamma}] + (B_\Pi + \delta_\Pi^- - B_\Sigma)J(J+1) + \frac{\mu_0\mathscr{H}M}{J(J+1)}\right]^2 + 4|A^{01}_{10,01}|^2\right\}^{1/2}$$
$$\equiv \alpha_J - \frac{3\mu_0\mathscr{H}M}{2J(J+1)} \pm \frac{1}{2}\left\{\left[\beta_J + \frac{\mu_0\mathscr{H}M}{J(J+1)}\right]^2 + 4|A^{01}_{10,01}|^2\right\}^{1/2} \tag{65}$$

Hence, from (28) the mixing angle is

$$\theta(\mathscr{H}) = \arctan\left\{2|A^{01}_{10,01}|\Big/\left(\beta_J + \frac{\mu_0\mathscr{H}M}{J(J+1)}\right)\right\}, \tag{66}$$

and is dependent on the magnetic field strength $\mathscr{H}$ as well as M and J. Note that when $\mathscr{H}M$ satisfies the equality

$$\beta_J + \frac{\mu_0\mathscr{H}M}{J(J+1)} = 0, \tag{67}$$

this leads to $\theta(\mathscr{H}_0) = \pi/2$, whereupon the two perturbed levels (27a) and (29a) are now 50–50 admixtures of the zeroth-order $^1\Pi^-$ and $^3\Sigma_1^{+-}$ levels.* Thus for zeroth-order splittings, β_J, comparable to magnetic energies, $\left|\frac{\mu_0\mathscr{H}M}{J(J+1)}\right|$, the magnetic field can lead to a crossing of the zeroth-order unmixed $^1\Pi^-$ and $^3\Sigma_1^{+-}$ levels.* Note that we may have the field free situation of

$$2|A^{01}_{10,01}|/|\beta_J| \ll 1, \tag{68}$$

corresponding to very small field free mixing, and still satisfy (67). (For instance, for very weak spin–orbit coupling.)

For weak zero-field mixing, (68), and lowmagnetic fields,

$$|\beta_J| \gg \mu_0\mathscr{H}_0|M|/J(J+1), \tag{69}$$

* We assume here that the $^1\Pi^-$ and $^3\Sigma_1^{+-}$ levels do not have radiative transition probabilities to a common lower level, so off-diagonal elements of $\mathbf{\Gamma}$ in (15) need not be included. Otherwise, these off-diagonal $\Gamma(^1\Pi^-, {}^3\Sigma_1^{+-})$ terms can lead to a level anticrossing behaviour which is readily described using the effective Hamiltonian (15) instead of just the molecular one of (64).

the mixing angle (66) is approximately given by

$$\theta(\mathscr{H}) \simeq \theta(\mathscr{H}=0) - 2\frac{|A^{01}_{10,01}|}{\beta_J}\frac{\mu_0 \mathscr{H} M}{J(J+1)} + O(\mathscr{H}^2), \tag{70}$$

which contains a linear variation with $\mathscr{H}$. Thus under the conditions (68) and (69) the mixed states (27a) and (29a) have mixing coefficients $\sin^2[\theta(\mathscr{H})/2]$ that may be written in the form of

$$\sin^2[\theta(\mathscr{H})/2] \simeq \sin^2[\theta(\mathscr{H}=0)/2] - \frac{|A^{01}_{10,01}|}{\beta_J^2}\frac{\mu_0 \mathscr{H} M}{J(J+1)}\sin[\theta(\mathscr{H}=0)] + \dots \tag{71}$$

with a similar form, i.e. linear terms in $\mathscr{H}$, following even when (68) is not satisfied.

The magnetic field dependence of (66), or the special case of (71) is readily seen to affect the simply mixed collisional cross sections (47) by the introduction of contributions linear (with either sign possible) in the magnetic field for $\mathscr{H}$ small. This is in contrast to ordinary magnetic quenching processes[61] which vary as $\mathscr{H}^2$ and are pressure independent. The perturbed mixing collisions are much more complicated as the collisional energies $E_{\hat{s}\omega}(t)$ and $E_{T\omega'}(t)$ depend on M and M' by virtue of both the magnetic field and the anisotropy of the molecule–perturber interaction. There may then be many crossings amongst the perturbed magnetic sublevels during the collision so a detailed numerical treatment is necessary, and quasi-statistical effects due to multiple crossing of the $\{E_n(t)\}$ curves during a single collision might emerge (See Section III.A.2).

The above example does illustrate how the magnetic field can substantially alter already existing zero-pressure, mixing coefficients, thereby affecting collision-induced intersystem crossing rates. (A similar result may be accomplished with electric fields, but these break inversion symmetry and lead to additional couplings.) In addition, the magnetic field can lead to mixings of levels which were not coupled at zero pressure. For instance, suppose a $^1\Pi_1(J+1)$ level were close in energy to a $^3\Sigma_1^{+-}(J=N)$ level at zero pressure. The second-order Zeeman coupling mixes some $^1\Pi_1^-(J)$ into $^1\Pi_1^-(J+1)$ and some $^3\Sigma_1^{+-}(J+1)$ into $^3\Sigma_1^{+-}(J=N)$ with coefficients which are proportional to $\mathscr{H}$. The effective $^1\Pi_1^-(J+1)$–$^3\Sigma_1^{+-}(J=N)$ coupling is then linear in $\mathscr{H}$ (and $|A^{01}_{10,01}|$) again, leading to field-induced mixing of originally unperturbed levels. This then opens up new pathways, new perturbed levels, to contribute to $^1\Pi_1^- \rightarrow {}^3\Sigma_1^{+-}$ collision-induced intersystem crossing. Note that at high field strengths, second-order Zeeman energies, etc., must be appended to (64).

The other $^1\Pi$ Λ-doublet,

$$^1\Pi_1^+ = (2)^{-1/2}[|0,1,0\rangle + |0,-1,0\rangle], \tag{72a}$$

can interact with the two other $^3\Sigma^+$ spin sublevels. The latter in the (poor) case (a) representation are

$$^1\Sigma_1^{++} = (2)^{-1/2}[|1,0,1\rangle + |1,0,-1\rangle] \tag{72b}$$

and

$$^1\Sigma_0^+ = |1,0,0\rangle. \tag{72c}$$

The Hamiltonian matrix within these three levels (for a given J) is found to be

$$\begin{pmatrix} E(^1\Pi) + B_\Pi J(J+1) + \delta_\Pi^+ J(J+1) - \dfrac{\mu_0 \mathscr{H} M}{J(J+1)} & A_{10,01}^{01*} & 0 \\ A_{10,01}^{01} & \left\{\begin{array}{l} E(^3\Sigma^+) + B_\Sigma[J(J+1)-1] \\ -\dfrac{2\mu_0 \mathscr{H} M}{J(J+1)} \\ +(6)^{-1/2}[\gamma - (2)^{3/2}\bar{\gamma}] \end{array}\right\} & \begin{array}{l}[2B_\Sigma - (\bar{\gamma}/3)][J(J+1)]^{1/2} \\ +4\mu_0 \mathscr{H} M[J(J+1)]^{-1/2}\end{array} \\ 0 & \begin{array}{l}[2B_\Sigma - (\bar{\gamma}/3)][J(J+1)]^{1/2} \\ +4\mu_0 \mathscr{H} M[J(J+1)]^{-1/2}\end{array} & \begin{array}{l}E(^3\Sigma^+) + B_\Sigma[J(J+1)-1] \\ -4(3)^{-1/2}\bar{\gamma}\end{array} \end{pmatrix} \tag{73}$$

The Hund's case (b) basis for $^3\Sigma^+$ is obtained by diagonalizing the terms involving B_Σ in the 2×2 lower block of (51) to yield the $J = N \pm 1$ spin components. Often the zero field $^1\Pi^+$ level interacts with only one of the two spin sublevels, (including the contribution to spin splittings from $\bar{\gamma}$, etc.), so (73) may then be reduced to a 2×2 in this case. Nevertheless, the magnetic field then affects the levels, again leading to an alterationof mixing coefficients or the introduction of new ones in a manner analogous to the $^1\Pi^- - ^3\Sigma_1^{+-}$, case discussed above, but with richer possibilities when magnetic energies can become comparable to the spin splittings. If the $^1\Pi(J)$ level in (73) were coupled in the field free limit ($\mathscr{H} = 0$) to only one of the case (b) $^3\Sigma$ spin levels ($J = N \pm 1$), the magnetic coupling leads to a mixing with the other.

It is clear from (7) that the magnetic field can likewise alter zero-pressure radiative decay rates by changes in mixing coefficients. Similarly, as discussed by Stannard,[58] they may alter non-radiative decay rates as formulas like (7) with radiative decay rates, Γ, replaced by non-radiative ones, Δ, can be utilized for situations in which, say, pure or mixed S_1–T_1 levels can undergo electronic relaxation to S_0.

The two-level limit of (64) can be handled just as in III. A and B (with due regard to the no-longer degenerate M sublevels.) However, as $\mathscr{H}$ increases, the importance of $\Delta J \neq 0$ magnetic couplings must grow as these couplings introduce additional mixings which provide new pathways for collision-induced electronic relaxation.[48] This general situation is not simply described purely analytically even in diatomic molecules. Numerical calculations will be useful to elucidate some of the general trends. There are certain average behaviours which can be inferred from the theory, and these are discussed in Section III.D below as they utilize the simplified large molecule approximation (74) to the collision-induced intersystem crossing rate.

D. Larger Molecules

It is hopefully clear by now that even the simplest diatomic molecule limit with an atomic collision partner presents us with a number of possible types of contributing collisions to the collision-induced intersystem crossing rates. It should be emphasized that a truly quantitative treatment would have to even include any hyperfine energies as the mixing angles (28) can be greatly affected by zeroth-order hyperfine shifts which are comparable to $2|v_{\mathrm{ST}}^{vv'}|$. Thus an idealized experiment would involve the preparation of a single pure or mixed rovibronic level and the observation of the state-to-state collision-induced intersystem crossing rates. In larger molecules the only change is associated with the larger level density, producing a higher density of mixed levels and more terms possibly contributing to the mixed states (6). Otherwise, the basic phenomena and types of collisional events are the same. Here we have an increased likelihood of multiple crossings of the nearby $\{E_n(t)\}$ curves as discussed in Section III.A.2.

For a given initial vibronic level, S, (or ideally rovibronic level) it is necessary to sum over all possible contributing final levels $T_{l'}$ which may be reached in the collision event. Likewise, if more than one rovibronic level is excited initially, it is necessary to add the contributions from all of the mixed levels which are prepared by the exciting light. The collision-induced intersystem crossing cross section is of the form

$$\sigma \simeq \langle \sin^2(\theta/2) \rangle n_{\mathrm{M}} n_{\mathrm{T}} \sigma_{\mathrm{rot}}. \tag{74}$$

Here $\langle \sin^2(\theta/2) \rangle$ denotes an average mixing coefficient for the initially excited mixed robibronic levels, n_{m} is the average number of mixed singlets which are initially prepared, n_{T} is the average number of final triplet levels, T_1' (rotational, vibrational, spin, and hyperfine) which are accessible as final levels after the collision, and σ_{rot} is comparable to pure S or T electronic state rotational relaxation cross sections within a single vibronic level or to pure T-state vibrational relaxation cross sections (from one vibronic level to another) which should be comparable in magnitude. The $\langle \sin^2(\theta/2) \rangle$ dependence arises, strictly speaking, from simply mixed collisions whereas the perturbed mixing ones contribute with more general variations with mixing coefficients. If each mixed singlet contains a number of different triplets, $\sin^2(\theta/2)$ stands for $\sum_i \sin^2(\theta_i/2)$ where the sum runs over the different triplet components of a given singlet. Alternatively, we may write this as $n\langle \sin^2(\theta_i/2) \rangle$, the form given in ref. 49 where the n_{T} was inadvertently omitted. When many mixed initial levels contribute, the simple form of (74) is adequate for an initial semiquantitative estimate. The number of final levels, n_T, may be estimated as $k_B T$ times the total density of triplet vibronic sublevels. For instance, in glyoxal if we use the extreme lower limits of one mixed singlet per vibronic level, $n = 1$, and n_T calculated from *only* the vibronic density of states as, $n_T \simeq (5\ \text{cm})(200\ \text{cm}^{-1})$, then values of $\sigma_{\text{rot}} \approx 10\ \text{Å}^2$ and $\sigma \simeq 10\ \text{Å}^2$ yield the estimate from (74) of

$$\langle \sin^2(\theta/2) \rangle \sim 10^{-4}, \tag{75}$$

which is small enough that the singlet rotational spectrum is not expected to be noticeably perturbed. If a number of rotational components of an initially excited vibronic band are mixed, then the estimate of (75) becomes

$$\langle \sin^2(\theta/2) \rangle \sim 10^{-4}/n_{\text{m}}, \tag{75a}$$

so for $n_{\text{m}} \simeq 10$–100, the individual state mixing is miniscule except, perhaps, for a few favourable isolated cases. Admittedly, the estimates currently required for large molecules are rather crude. Nevertheless, they do indicate how very small mixings, e.g. (75) or (75a), can have substantial effects on collision-induced intersystem crossing transition rates. Further experimental work will require additional quantitative refinements to deal with the detailed quantitative questions. Studies on diatomic molecules are the most readily amenable to quantitative analysis, and additional research for these systems should aid in providing better estimates for larger molecules than (74).

When S and T refer to two mixed vibrational levels of the same electronic state, the zero-pressure splittings of levels of S from those of T can be substantial, e.g. tens of wavenumbers. Thus any substantial non-resonant process must have an additional term on the right in (74) to account for energetically unfavourable circumstances. Despite the differences, the experimental evidence to date[14] points to very rapid collisional transitions between Fermi resonance mixed vibrational levels in accord with the theoretical discussions.

A general understanding of the magnetic (or electric) field effects on collision-induced intersystem crossing rates in larger molecules can be deduced from the discussion in Section III.C and from the simplified approximation (74). The magnetic field is expected to have a rather minor effect on both the rotational relaxation cross section, σ_{rot}, (or the cross sections for vibrational transitions amongst the Fermi resonance strongly coupled triplets) and the number, n_T, of final triplet vibronic states which are accessible after a collision. On the other hand, the magnetic field induced breakdown of the $\Delta J = 0$ selection rules for S–T couplings, as well as the induced couplings between the spin sublevels, both imply that the number of T-sublevels coupled to a singlet rovibronic level, n_{m}, must increase with the magnetic field, $\mathscr{H}$. Some mixing coefficients $\sin^2(\theta/2)$ must decrease with $\mathscr{H}$ while other must increase (neither necessarily being monotonic or totally of either type.) The total rate, however, is dominated by the mixed S-levels largest mixing coefficients, so ultimately it is expected that the cross section be enhanced by the field. The simple example of a diatomic molecule shows that the increases in $\sin^2(\theta/2)$ with $\mathscr{H}$ can be very large; a field free value of 10^{-3}–10^{-4} can be converted to the maximum value of 1/2. The precise form requires some numerical work. Nevertheless, the final result is of the form,

$$\sigma(\mathscr{H}) \simeq \langle \sin^2(\theta(\mathscr{H})/2) \rangle n_{\mathrm{m}}(\mathscr{H}) n_{\mathrm{T}} \sigma_{\mathrm{rot}}, \tag{74a}$$

where the field dependence is made explicit. The first two, field-dependent, terms are properties of the molecule and are independent of the perturber, while the last term is perturber dependent. Equation (74a), therefore, predicts a field-dependent variation, $\sigma(\mathscr{H})/\sigma(\mathscr{H}=0)$, which is *independent of the perturbing molecule*. The number of mixed levels $n_{\mathrm{m}}(\mathscr{H})$ should increase with field strength, $\mathscr{H}$, at low magnetic fields because of the breakdown in the zero field $\Delta J = 0$ selection rules and because of field-induced couplings between the spin sublevels. At higher fields the second-order Zeeman effect is anticipated to alter $n_{\mathrm{m}}(\mathscr{H})\langle \sin^2(\theta(\mathscr{H})/2) \rangle$ through a slowly variation of this quantity with field.

E. Collision-Induced Intersystem Crossing at Higher Pressures

Traditional quenching processes are characterized by Stern–Volmer-type kinetics whereby the quenching rate is proportional to the pressure of the perturber over very wide pressure ranges. In this case the quenching rate may be understood on the basis of a thermally averaged single binary collisional quenching event. The theory of collision-induced electronic relaxation is express in terms of mixed asymptotic states and yields Stern–Volmer-like kinetics *only in the limit of low enough pressures.*[48] The remarkable feature of the theory is that it predicts[48] *very non-Stern–Volmer* quenching rates; the quenching rate is predicted to reach a high-pressure limiting value, *independent of pressure and perturbing molecule.* (If V_{ST} couplings are operative, then they produce an additional Stern–Volmer-like contribution to the quenching rates.) This type of behaviour has been observed, subsequent to the theoretical predictions, by Strickler and Rudolph[45a,b] and by Su *et al.*[45c] in the $^3B_1$ state of SO_2 and by Weisshaar *et al.*[43] in the S_1 state of formaldehyde. Knight[62] has seen some evidence for this pressure saturation effect in glyoxal, but the situation remains to be investigated further. The data on SO_2 and formaldehyde are given in Figs. 4–6.

The $SO_2(^3B_1)$ state had previously been the source of considerable conceptual difficulties. Phosphorescence quantum yields were reported which were smaller than unity. The density of lower SO_2 rovibronic levels and typical radiative decay rates places the $SO_2(^3B_1)$ level squarely in the small molecule limit, so quantum yields of less than unity violate the laws of quantum mechanics. $SO_2(^3B_1)$ was also found to have an enhanced higher

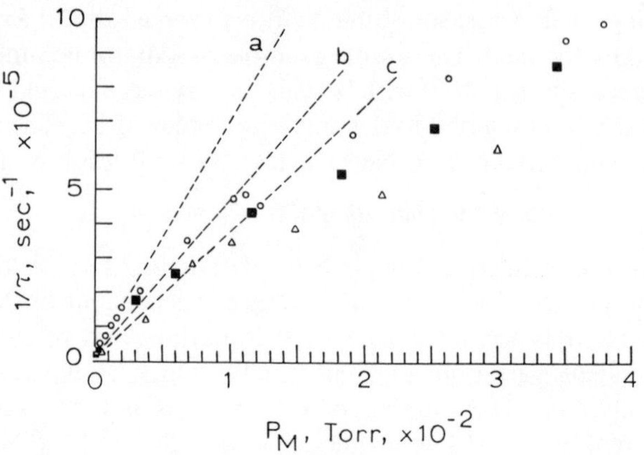

Fig. 4. Data of Su *et al.* for the inverse lifetimes of the $^3B_1$ state of SO_2 as a function of the pressure of CO_2 (circles), (squares), and N_2 (triangles). The lines a, b, and c are the linear Stern–Volmer kinetics extrapolated from the low-pressure region for CO_2, CO, and N_2, respectively. (Reproduced from ref. 45(c) by permission of *North-Holland Publ. Co.*)

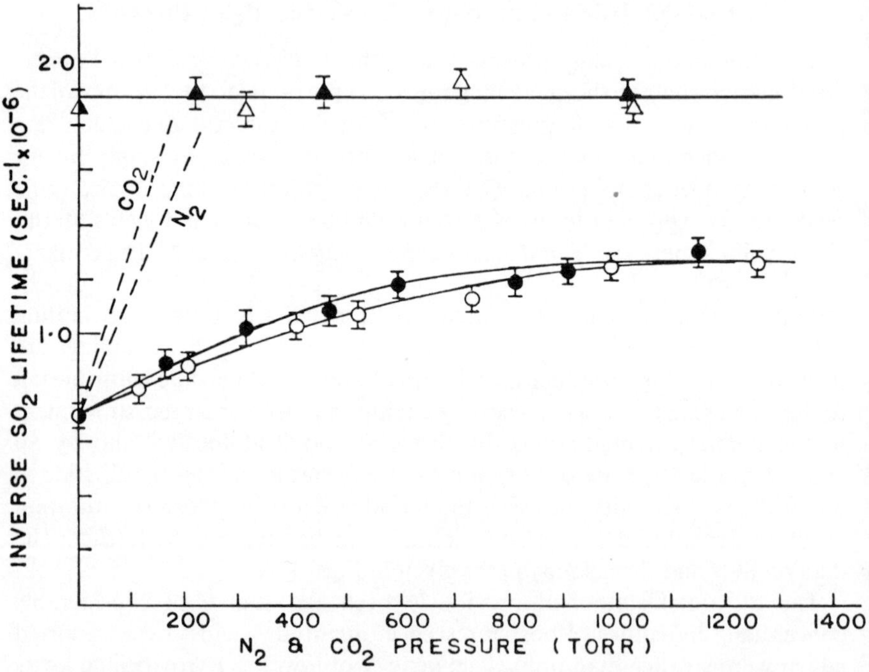

Fig. 5. Higher-pressure data of Stucker and Rudolph for the decay rate of the $SO_2(^3B_1)$ state as a function of the pressure of N_2 (open circles) and CO_2 (closed circles). Dashed lines give the linear Stern–Volmer kinetics extrapolated from the low-pressure region. (Reproduced with permission from ref. 45(b). Copyright by the *American Chemical Society*.)

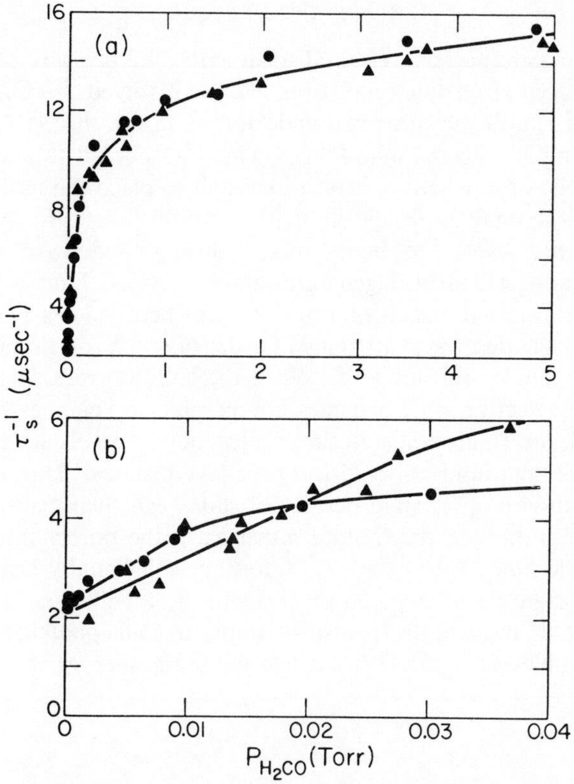

Fig. 6. Date of Weisshaar *et al.* on the pressure dependence of the decay rates of formaldehyde excited to the 4^0 (triangles) and 4^1 (circles) vibronic levels of the $S_1(\tilde{A}^1A_2)$ state. (Reproduced from ref. 43 by permission of the *American Institute of Physics* and the authors.)

bimolecular reactivity with CO and with ethylene in conflict with all reasonable expectations. The prediction[48] and observation[45] of pressure saturation effects then made it clear that the previous experiments were based on higher-pressure measurements where non-linear Stern–Volmer kinetics were appearing. Notice in Fig. 3. that this non-linear region is already apparent at pressures of 10 m Torr! Given the pressure saturation effect, the low-pressure data yield a phosphorescence quantum yield of unity and no need for enhanced reactivities of SO_2 at higher pressures.[63] Formaldehyde is a more complicated case because photochemical decomposition may proceed from the lowest vibronic levels of S_1 if the threshold is below these levels, so quantum yields of less than unity are theoretically possible from this photochemical channel. The experiments indicate the additional importance of collision-induced, internal conversion.[43]

1. *Qualitative Description*

The theoretical analysis, required to describe the pressure saturation of collision-induced electronic relaxation, is quite involved.[48] Thus it is useful to provide a simple physical rationale for its occurrence.[24] Consider an individual singlet vibronic manifold, S which is coupled to a set of triplet vibronic levels, $\{T_l\}$, where ρ_l is small enough to place the molecule in the too many level small molecule limit, the intermediate case. Zeroth-order $\{T_l^0\}$ harmonic levels are badly mixed through extensive anharmonic couplings amongst nearly degenerate vibronic levels. Thus it is expected that T_l–$T_{l'}$ vibrational relaxation rates are large between nearly isoenergetic pure-spin basis states, the rates being of order of σ_{rot}, a rotational relaxation rate within a single vibronic level. When the collision rate, k_{rot}, associated with the cross section σ_{rot}, becomes comparable to spacings between the $\{T_l\}$ sublevels, on single collision time scales the $\{T_l\}$ levels act as if they are a continuous manifold of collision-broadened states. This collisionally induced continuous $\{T_l\}$ state density of states can then drive irreversible electronic relaxation of the S state much as in the purely intramolecular large molecule limit. When the $\{T_l\}$ states are collisionally broadened, the individual S-state decay is given by (17) with ρ_{l_j} a collisionally broadened state density. However, the collisions imply a collisional induced distribution of S-sublevels, p_{si}, so (17) is converted to the average,[24]

$$\bar{\Delta}_s = \frac{2\pi}{\hbar} \sum_{i,j} p_{si} |v_{si,lj}|^2 \rho_{lj}, \tag{76}$$

when the collision-induced electronic relaxation is much slower than k_{rot}, the distribution p_{si} is a thermalized Boltzmann one, but at lower pressures the detailed kinetics must be simulated to derive p_{si}. Nevertheless, the high-pressure limit of (76) is pressure independent as the pressure-broadened state density, ρ_{l_j}, suffices to enable the replacement of the summation over j in (76) by an integration. For example, if ρ_{l_j} were modelled by simple Lorentzians,

$$\rho_{l_j} = \frac{\pi \gamma_{l_j}}{(E - E_{l_j})^2 + (\gamma_{l_j}/2)^2}, \tag{77}$$

with the binary collision pressure dependence

$$\gamma_{l_j} \propto \text{pressure} \tag{78}$$

when the spacing, ε_{l_j}, between T-levels satisfies

$$\gamma_{l_j}/\varepsilon_{l_j} \gg 1, \tag{79}$$

the large molecule limit ensues, and this statistical limit value of (76) becomes independent of the $\{\gamma_{l_j}\}$ and, hence, of the pressure. The actual situation[48] is somewhat more complicated by virtue of the fact that collisional transitions amongst the nearly degenerate $\{T_l\}$ levels can appear, but (76)–(79) provide an adequate qualitative description. The condition for the emergence of non-linear Stern–Volmer kinetics is that the pressure, P, satisfy[48]

$$\hbar k_{\text{rot}} P_0 \propto \varepsilon_{l_j}. \tag{80}$$

Between the pressures of P_0 and those where the quenching rate tends toward its high-pressure limit, the fluorescence decay of the prepared (mixed) S-state need not be a single exponential decay[64]—the multistate 'kinetics' must, in general, be analysed.

2. *Theoretical Treatment*

The consideration of the high- and intermediate-pressure regimes requires the description of multicollisional events. For instance, the appearance of (78) in (76) cannot arise from a theory involving a single collisional event. This feature, coupled with the particular mixed state characteristics of the collision-induced electronic relaxation, makes it useful to utilize a density matrix formalism. Here we outline the approach briefly, and ref. 48 can be quoted for the detailed treatment in the limit where only the simply mixed collisional contributions are explicitly treated.

The density matrix representation is conveniently introduced first in the zero-pressure limit. The effective Hamiltonian is given by (15), and U is the transformation which diagonalizes H_{eff}. The density matrix, $\rho(t)$, is defined such that the expectation value of any operator, A, at time t is given by

$$\langle A(t) \rangle = \sum_{i,j} A_{ij} \rho_{ji}(t). \tag{81}$$

For a pure state, $|\phi\rangle$, the density matrix has the form

$$\rho_{\text{pure}} = |\phi\rangle\langle\phi|, \tag{82}$$

while for a thermal ensemble it is

$$\rho_{\text{thermal}} = \exp[-\beta H]/Tr \exp(-\beta H), \tag{83}$$

where H is the relevant system's Hamiltonian and

$$Z = Tr \exp(-\beta H) \tag{84}$$

is the thermal partition function.

3. *Zero-Pressure Limit Revisited*

The time evolution of the excited states, S and T, of the system follows from the solution of the time-dependent Schrödinger equation with the full effective Hamiltonian (15). In terms of the density matrix, ρ, (in units of $\hbar = 1$) this is just[65]

$$i\frac{\partial \rho}{\partial t} = H\rho - \rho H - (i/2)(\mathbf{\Gamma}\rho + \rho\mathbf{\Gamma}) \tag{85}$$

which in matrix notation is

$$\begin{aligned} i\frac{\partial}{\partial t}\rho_{mn} &= \sum_q \{[H_{mq}\rho_{qn}(t) - \rho_{mq}(t)H_{qn}] - (i/2)[\Gamma_{mq}\rho_{qn}(t) + \rho_{mq}(t)\Gamma_{qn}]\} \\ &\equiv \sum_{kl} L_{mnkl}\rho_{kl}(t). \end{aligned} \tag{86}$$

Here, for instance, ρ_{ss} is the probability of finding the system in state $|S\rangle$ at time t, and L is the Liouville operator with matrix elements

$$L_{mnkl} = H_{mk}\delta_{nl} - H^*_{nl}\delta_{mk} - (i/2)[\Gamma_{mk}\delta_{nl} + \Gamma^*_{nl}\delta_{mk}], \tag{87}$$

where the * implies complex conjugation. In the molecular eigenstates basis, H is approximately diagonal and (87) has the value (tildes designating molecular eigenstates)

$$\begin{aligned} L_{\tilde{m}\tilde{n}\tilde{k}\tilde{l}} &\cong \delta_{\tilde{m}\tilde{k}}\delta_{\tilde{n}\tilde{l}}[E_{\tilde{m}} - E_{\tilde{n}} - (i/2)(\Gamma_{\tilde{m}} + \Gamma_{\tilde{n}})] \\ &= (UHU^{-1})_{\tilde{m}\tilde{k}}\delta_{\tilde{n}\tilde{l}} - \delta_{\tilde{m}\tilde{k}}(UHU^{-1})^*_{\tilde{n}\tilde{l}}, \end{aligned} \tag{88}$$

with the real part involving transition frequencies and the imaginary part the average decay rates. Hence, the solution to (86) is simply

$$\rho_{mn}(t) = \exp[-iL_{\tilde{m}\tilde{n}\tilde{m}\tilde{n}}t]\rho_{\tilde{m}\tilde{n}}(0), \tag{89}$$

a result which is well known from the analogous wavefunction representation.

4. *Density Matrix Theory at Non-zero Pressures*

Perturbing molecules can lead to collision-induced transitions between the states of the molecule of interest, dubbed the molecule in the following. The full theory has been presented by Fano,[66,67] but, for simplicity, we employ the impact approximation.[67,68] This approximation assumes the occurrence of only binary collisions between molecules which are uncorrelated in the precollision region. The time between collisions is taken to be much longer than the duration of a collision, so on a coarse-grained time scale, we can consider that only completed collisions have occurred. The density matrix equations (85), are now appended by the presence of a collision operator Λ.

$$i\frac{\partial}{\partial t}\rho = L\rho - i\Lambda\rho, \tag{90}$$

where[18]

$$\Lambda_{mnkl} = N \int_0^\infty v\langle \sigma_{mnkl}\rangle 4\pi v^2 \rho_v \mathrm{d}v, \tag{91}$$

N is the number density of perturbing molecules, and $4\pi v^2\rho_v \mathrm{d}v$ is the normalized Maxwell–Boltzmann velocity distribution for the relative translational motion of the collision pair. The four-indexed cross section in (91) is the thermal average over perturber internal states b of[69]

$$\sigma_{mnkl} = [\pi/(\mu v)^2] \sum_{b'} [\delta_{mk}\delta_{nl} - S_{mb';kb}S^*_{nb';lb}], \tag{92}$$

with $S_{mb';kb}$ the collision S-matrix for the transition $kb \to mb'$ in the perturber–molecule pair. (b includes the quantum numbers for the relative angular momentum of the perturber with respect to the molecule, etc.) In terms of the semiclassical formulation of Section II.A.3, the S-matrix elements of (92) are the asymptotic probability amplitudes

$$S_{n,n'} = a_n(+\infty) \quad \text{for} \quad a_n(-\infty) = \delta_{nn'}. \tag{93}$$

Other approximate formulations may be applied to the treatment of (92).

Analogous to the small molecule limit where only H of (15) need be diagonalized, when the pressure is very low, $i\Lambda$ in (3.68) is a small perturbation on L. Thus it is a good approximation to employ the zero-pressure basis set and to use the transformation that diagonalizes L alone as in the second and third lines of (88). This converts (90) to the matrix equations

$$i\frac{\partial}{\partial t}\rho_{\tilde{m}\tilde{m}}(t) = -i\Gamma_{\tilde{m}}\rho_{\tilde{m}\tilde{m}}(t) - i\sum_{\tilde{k}\tilde{l}}\tilde{\Lambda}_{\tilde{m}\tilde{m}\tilde{k}\tilde{l}}\rho_{\tilde{k}\tilde{l}}(t) \tag{94}$$

$$i\frac{\partial}{\partial t}\rho_{\tilde{m}\tilde{n}}(t) = [E_{\tilde{m}} - E_{\tilde{n}} - (i/2)(\Gamma_{\tilde{m}} + \Gamma_{\tilde{n}})]\rho_{\tilde{m}\tilde{n}}(t) - i\sum_{\tilde{k}\tilde{l}}\tilde{\Lambda}_{\tilde{m}\tilde{n}\tilde{k}\tilde{l}}\rho_{\tilde{k}\tilde{l}}(t) \tag{95}$$

where $\tilde{\Lambda}$ is defined just as in (92) and (93) except that S is replaced by $\tilde{S} = USU^{-1}$.

For simplicity, it is assumed, for now, that no coherent excitation is present, so $\rho_{\tilde{m}\tilde{n}}(0) = C_{\tilde{n}}\delta_{\tilde{m}\tilde{n}}$ is initially diagonal. Likewise, the coupling terms, that might lead to non-diagonal $\rho(t)$, are neglected here. Given these simplifications, it is only necessary to consider (94) with $\tilde{k} = \tilde{l}$ in the summation. This simplied equation is then just a master equation with exact scattering cross sections within the impact approximation. The collisional term $\sum_{\tilde{k}}\tilde{\Lambda}_{\tilde{m}\tilde{m}\tilde{k}\tilde{k}}\rho_{\tilde{k}\tilde{k}}(t)$ in (94) is then just the net rate of collisional depletion of the mixed state $|\tilde{m}\rangle$ at time t; $\tilde{\Lambda}_{\tilde{m}\tilde{m}\tilde{m}\tilde{m}}\rho_{\tilde{m}\tilde{m}}(t)$ is the rate of transitions out of the mixed state $|\tilde{m}\rangle$, and $\sum_{k\neq m}\tilde{\Lambda}_{\tilde{m}\tilde{m}\tilde{k}\tilde{k}}\rho_{\tilde{k}\tilde{k}}(t)$ is minus the collisional rate into $|\tilde{m}\rangle$

from all other mixed and unmixed states. Hence, $N^{-1}\tilde{\Lambda}_{\tilde{n}\tilde{n}\tilde{n}\tilde{n}}$ is the thermally average rate constant for the deactivation of the molecular eigenstate $|\tilde{n}\rangle$ to all other states, where N^{-1} is the perturber concentration. In general, this contains contributions from vibrational and rotational relaxation within the spin-contaminated mixed state basis for the electronic manifold to which $|\tilde{n}\rangle$ belongs along with contributions from collision-induced electronic relaxation. This initial state, $|\tilde{n}\rangle$, may be a mixed or unmixed state of primarily S or T parentage. Likewise, $-N^{-1}\tilde{\Lambda}_{\tilde{m}\tilde{m}\tilde{n}\tilde{n}}$ is the rate constant for $|\tilde{n}\rangle \to |\tilde{M}\rangle$ collisions.

The derivation[48] of the high-pressure limiting collision-induced electronic relaxation rate (76) utilizes the Green's function representation of the density matrix equations along with a projection operator formulation. While the methods follow directly as in the wavefunction representation, an introduction of these methods would take us too far afield, so the interested reader is referred to ref. 48 for the details.

IV. COLLISIONAL EFFECTS ON ELECTRONIC RELAXATION IN THE LARGE MOLECULE LIMIT

In the large molecule statistical limit electronic relaxation can occur in the isolated molecule because the $\{\phi_l\}$ density of states is sufficiently high and/or their decay rates, $\{\Gamma_l\}$, are rapid enough to drive the irreversible electronic relaxation $\phi_s \to \{\phi_l\}$ (recall Fig. 1). Collisional processes can lead to ordinary quenching of the initial state ϕ_s if the molecule–perturber interaction, V, has significant s–l coupling, $V_{sl} \neq 0$. Simple quenching is recognized by linear Stern–Volmer kinetics and is well understood from this standpoint. However, there are other collisional effects available to these large statistical limit molecules which also produce non-linear Stern–Volmer plots for the decay rate of the electronic state ϕ_s, and it is these processes to which this section is focused.

The ideal zero-pressure experiments involve the excitation of individual single vibronic levels, ϕ_{si}, and the determination of the lifetimes, τ_{si}, and quantum yields, Φ_{si}, for these levels. Then the radiative, k_{si}^{rad}, and non-radiative, k_{si}^{nr}, decay rates for these levels can be deduced from the familiar equations

$$\Phi_{si} \equiv k_{si}^{\mathrm{rad}}/(k_{si}^{\mathrm{rad}} + k_{si}^{\mathrm{nr}}) \tag{96}$$

$$\tau_{si} \equiv (k_{si}^{\mathrm{rad}} + k_{si}^{\mathrm{nr}})^{-1}. \tag{97}$$

Unfortunately, optical selection rules (and Franck–Condon factors) only enable the initial optical excitation of a small fraction of the $\{\phi_{si}\}$, corresponding to excitation of only a small fraction of the available vibrations of the molecule.[70] Collisions could then be utilized to enable transitions between these optically excitable states, $\{\phi_{si}^{a}\}$, and the forbidden ones, $\{\phi_{si}^{f}\}$, thereby enabling the determination of the lifetimes and quantum yields of the latter, as well as the interesting state-to-state collisional transition rates. But collisions are not infinitely selective, so many vibronic levels $\{\phi_{si}^{f}\}$ are created through collisions from a single ϕ_{si}^{a}.

This state-by-state analysis of collisional processes is possible only in the smallest polyatomic molecues—remember we are considering statistical limit molecules, so stable triatomic molecules, etc., are not acceptable for this limit—and only for low levels of vibrational excitation in ϕ_s. This limitation is imposed because of extensive sequence congestion and the increasing density of ϕ_s vibronic states at elevated energies. The use of pulsed supersonic beams will reduce the severity of these limitations somewhat, but there are many instances where the single vibronic level resolution is impossible. Thus we pose the more limited question of determining the nature of the pressure dependence of emission quantum yields and lifetimes as a function of the wavelength of the incident radiation.

Pressure-dependent experiments of this nature are readily performed for fluorescing states, ϕ_s.[73] However, the phosphorescing triplet states of aromatic hydrocarbons generally must be produced by excitation of a singlet with subsequent intersystem crossing[74] (or by triplet-sensitized collisions). Even if direct optical excitation to the triplets is possible, the extremely long triplet lifetimes almost preclude their study at low enough pressures to have isolated molecule conditions. Hence, the combination of very weak absorption intensities and long lifetimes implies that collisions are perforce present to complicate the radiative and non-radiative decay of these triplets. Perhaps, the most general type of experiment, that would currently be feasible for the triplets, involves the determination of the pressure and wavelength dependence of the triplet lifetime and quantum yield, a rather difficult experiment even with current technology. This type of experiment is of considerable importance in separating $S_1 \to S_0$ internal conversion from $S_1 \to T_1$ intersystem crossing rates. Theoretical predictions[1] imply that $S_1 \to S_0$ internal conversion rates in aromatic hydrocarbons increase exponentially as a function of the excess vibrational energy, ΔE_{ex}, in S_1, while the $S_1 \to T_1$ intersystem crossing rates increase only linearly. Thus the former should dominate the latter at high enough energies, ΔE_{ex}, in S_1. Experiments by Lim and coworkers[76] are in accord with these predictions, but the direct measurement of the (low) pressure and wavelength dependence of triplet yields is still necessary for a quantitative understanding and for a determination of the interesting vibrational energy dependence of the $S_1 \to S_0$, $S_1 \to T_1$, and $T_1 \to S_0$ decay rates.

Allusion has already been made to the source of non-linear Stern–Volmer kinetics, namely the variation of individual ϕ_{si} level lifetimes and quantum yields which on a coarse-grained scale translates to a variation of the zero-pressure limiting Φ_s and τ_s with ΔE_{ex}. Thus collisions can transport molecules between vibronic levels having rather different lifetimes and quantum yields, and this feature is somehow reflected in the wavelength and pressure dependence of Φ_s and τ_s. The effects of the energy dependence of radiative and non-radiative decay rates has been considered by Neporent and Mirumyants[77] who analyse a kinetic scheme which includes energy-dependent decay rates. Apart from the fact that the type of 'general' kinetic schemes, considered by them, do not yield simple closed form analytic solutions, the particular models they employ are deficient as these models do not lead to a Boltzmann distribution of vibrational populations at high pressures. Thus the implications of these models at intermediate and higher pressures are suspect.

A. Kinetic Model

In this section we discuss simple kinetic models of the pressure dependence of radiative and non-radiative decay processes of statistical limit molecules which have the following important features:[78, 79]

(a) The radiative and non-radiative decay rates are explicitly dependent on the energy, ΔE_{ex}. This dependence is chosen to be a linear variation with ΔE_{ex} in general accord with observed radiative rates and with $S_1 \rightarrow T_1$ non-radiative rates in aromatic hydrocarbons.[71]

(b) Collisional transitions transport the molecule between states of differing energy, ΔE_{ex}, and consequently varying k_{rad} and k_{nr}.

(c) We introduce a model which is amenable to *exact closed form analytical solution.* This implies that all predictions correspond to the model, albeit simple, and not to the extraneous mathematical approximations necessary to treat a highly realistic model. This condition also imposes the linearity property invoked in (a) above. It also leads us to employ a Montroll–Shuler[80] type step-ladder vibrational relaxation mechanism wherein collisions on the average result in energy gain or loss of an energy of $\hbar\omega_{eff}$. Hence the model is designed to explain general trends and not to quantitatively reproduce all experiments

(d) The model contains a rapidly increasing density of states as the vibrational energy increases.

(e) The effects of ordinary quenching processes can readily be treated in addition.

(f) A one-electronic state model[78] represents the fluorescence experiments, while a two-state version[79] can be applied to the triplets which are indirectly populated from the singlet state.

The single oscillator model is displayed in Fig. 7. The oscillator is a D-fold degenerate oscillator with energy level spacing $\hbar\omega_{eff}$. The D-fold

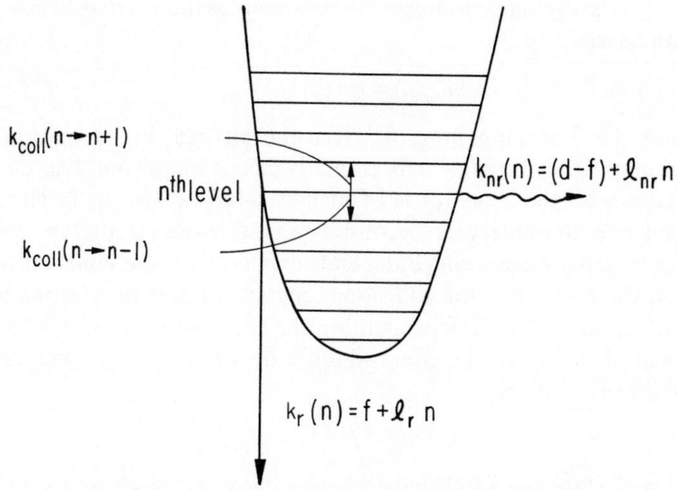

Fig. 7. The D-dimensional degenerate effective oscillator used to represent the vibronic levels of a polyatomic molecule. A step-ladder collision model is utilized and radiative and non-radiative decay rates are taken as linear functions of $\Delta E_{ex} = (n + D/2)\hbar\omega_{eff}$.

degeneracy provides the increasing level density as ΔE_{ex} increases. The model is naturally a coarse-grained one which lumps together many vibronic states into the degenerate oscillator levels. Since we expect weak quasi-elastic collisions to have the largest cross sections, these collisions can scramble these isoenergic levels and provide the coarse-graining' mechanism. The rate constants for upwards and downwards vibrational transitions are, respectively,

$$k(n \leftarrow n-1) = V_+(n+D-1) \tag{98a}$$

$$k(n \leftarrow n+1) = V_-(n+1), \tag{98b}$$

where the upward and downward coefficients, V_+ and V_-, respectively, are proportional to the pressure of the perturbing gas. V_+ and V_- must be related to each other by the principle of detailed balance which equates the rates of upward and downward transitions between levels n and $n-1$ at equilibrium. Thus if $P_{eq}(n)$ is the normalized D-fold degenerate harmonic oscillator equilibrium distribution function, detailed balance implies that

$$k(n \leftarrow n-1)P_{eq}(n-1) = k(n-1 \leftarrow n)P_{eq}(n), \tag{99}$$

or

$$V_+ = k(1 \leftarrow 0) \equiv v = V_-\zeta^{-1} \tag{100a}$$

$$\zeta = \exp(\hbar\omega_{eff}/k_B T), \tag{100b}$$

where ω_{eff} is the oscillator frequency, and n denotes the nth degenerate energy level with energy,

$$\Delta E_{\text{ex}}(n) = (n + \tfrac{1}{2}D)\hbar\omega_{\text{eff}}. \tag{101}$$

The energy level spacing, $\hbar\omega_{\text{eff}}$, is given the physical interpretation of the average energy transferred per vibrationally inelastic collision. This definition is in accord with the general use of such step-ladder models in vibrational relaxation and unimolecular decomposition theories. It therefore implies that $\hbar\omega_{\text{eff}}$ is a parameter which depends explicitly on the collision partner. Similarly, the rates, $V_{\pm}$, for collisional energy transfer of one quantum of energy, $\hbar\omega_{\text{eff}}$, are likewise dependent on the collision partner.

As mentioned in (*a*), the non-radiative decay rates $k_{\text{nr}}(n)$ are taken to increase linearly with n,

$$k_{\text{nr}}(n) = (d - f) + nl_{\text{nr}}, \tag{102}$$

where d and f are the total and radiative decay rates, respectively, of the $n = 0$ level. The parameter l_{nr} measures the incremental change in the non-radiative decay rate of the molecule per quantum of energy, $\hbar\omega_{\text{eff}}$. Since the incremental change in the non-radiative decay rates per unit energy is a property of the molecule, l_{nr} varies with collision partner because it involves the normalization energy $\hbar\omega_{\text{eff}}$. However, when this normalization is removed, the quantity $l_{\text{nr}}/\hbar\omega_{\text{eff}}$ must be a constant for a given molecule which is independent of the choice of the perturbing molecules. This criterion therefore provides a stringent test for the general applicability of the model to any particular molecular system.

It would appear that this model could also be employed to discuss cases in which k_{nr} decreases linearly with n simply by taking l_{nr} to be negative. However, this would imply that for $n > n_0 = (d - f)/|l_{\text{nr}}|$, $k_{\text{nr}}(n)$ is negative. Hence, in order to treat the case of l_{nr} negative, it is either necessary to truncate the oscillator for $n > n_0$ or to consider cases in which the probability that $n \geq n_0$ be occupied is truly negligible.

The radiative decay rates $\text{k}_{\text{rad}}(n)$ can likewise be chosen to vary linearly (increase) with quantum state,

$$k_{\text{rad}}(n) = f + nl_{\text{rad}}. \tag{103}$$

Thus, the overall decay rate of the $n = 0$ level is d, and the incremental decay rate per oscillator quantum is

$$l = l_{\text{nr}} + l_{\text{rad}}. \tag{104}$$

Once again, l_{rad} will vary with collision partner, but $l_{\text{rad}}/\hbar\omega_{\text{eff}}$ is independent of the choice of this partner.

The effects of collisional electronic quenching may readily be incorporated into the model in a trivial fashion by allowing theoverall decay rate, d, and the incremental decay rate, l, to depend linearly on pressure, i.e. they are of the form

$$d = d'P + d''$$

and

$$l = l'P + l'',$$

respectively. Here d' and l' correspond to the rate coefficients of the pressure-dependent decay rates (which give rise to the observed Stern–Volmer behaviour) while d'' and l'' account for the pressure-independent decay component.

The two-electronic state model merely arises by having two degenerate oscillators each with their values of the relevant parameters $\hbar\omega_{\text{eff}}, v, d, l, f, D,$ etc., which, in general, differ for the two electronic states in question. The additional feature is that the S-state feeds the T-state by irreversible electronic relaxation. For simplicity, choosing the two states to have a common value of $\hbar\omega_{\text{eff}}$ and taking the energy gap,

$$\Delta E_{\text{ST}} \equiv m\hbar\omega_{\text{eff}}, \tag{105}$$

the $Sn \to Tj$ state decay rate is likewise taken to vary linearly with vibrational energy as

$$k_{Sn \to Tj} = \begin{cases} \lambda + (j-m)\mu, & j \geq m \\ 0 & j < m, \end{cases} \tag{106}$$

since the intramolecular decay involves conservation of total energy.

B. The Kinetic Equations

The infinite set of kinetic equations of the model is amenable to exact analytical solution. We now consider a derivation of these kinetic equations for the one-oscillator case in order to further illustrate the physical meaning of the model. The final results are then quoted and comparisons with experiments provided.

Let $P_n(t)$ be the probability that the oscillator be in quantum state n at time t. The probability that it be in state n at time $t + \delta t$, $P_n(t + \delta t)$, must then equal the sums of the probabilities that in time δt

(1) A collision induces a transition from state $n \pm 1$ to state n, i.e.

$$(n + D - 1)V_+ P_{n-1}(t)\delta t + (n+1)V_- P_{n+1}(t)\delta t.$$

(A two-quantum jump would require two collisions and therefore would have a probability proportional to $(\delta t)^2$ which can be ignored as $\delta t \to 0$.)

(2) The oscillator remains in the state n during δt, (this is just $P_n(t)$ times one minus δt times the decay rate out of state n due to collisions and

radiative and non-radiative decay), i.e.

$$\{1 - [d + ln + (n + D)V_+ + nV_-]\delta t\}P_n(t).$$

(3) The oscillator is radiatively excited to oscillator state n (from e.g. the ground electronic state), i.e.

$$P_n^{\mathrm{irr}}(t)\delta t.$$

Thus we have

$$\begin{aligned} P_n(t + \delta t) = {} & (n + 0 - 1)V_+P_{n-1}(t)\delta t + (n + 1)V_-P_{n+1}(t)\delta t \\ & + \{1 - [d + ln + (n + D)V_+ + nV_-]\delta t\}P_n(t) \\ & + P_n^{\mathrm{irr}}(t)\delta t, \qquad n = 0, 1, 2, \ldots \end{aligned} \tag{107}$$

Rearranging (107) dividing by δt, and taking the limit $\delta t \to 0$ gives the stochastic equations[78, 81]

$$\begin{aligned} dP_n(t)/dt = {} & -[(d + Dv) + ln + v(\zeta + 1)n]Pf_n(t) \\ & + v(n + D - 1)P_{n-1}(t) + v\zeta(n + 1)P_{n+1}(t) \\ & + P_n^{\mathrm{irr}}(t), \qquad n = 0, 1, 2, \ldots, \end{aligned} \tag{108}$$

where v and ζ have been defined in (100). Equation (108) is to be solved given some initial distribution $\{P_n(0)\}$. We are interested in cases which correspond to pulsed and/or steady-state monochromatic excitations the appropriate initial condition (corresponding to selective excitation of a single oscillator level.) The pulsed case gives

$$P_n(0) = \varepsilon\delta_{nm}. \tag{109}$$

We simply assume that the steady-state illumination also populates a single oscillator level m',

$$P_n^{\mathrm{irr}}(t) = g\delta_{nm'}. \tag{110}$$

Here ε and g are basically divalued parameters which enable us to consider either only pulsed ($\varepsilon = 1, g = 0$) or only state ($\varepsilon = 0, g = 1$) excitation.

The generating function,

$$G(st) \equiv \sum_{n=0}^{\infty} s^n P_n(t), \tag{111}$$

can be introduced, where from (109) and (111) $G(st)$ initially is

$$G(s0) = \varepsilon s^m, \tag{112}$$

and the probabilities $P_n(t)$ are obtained, as usual, from $G(st)$ via

$$P_n(t) \equiv \frac{1}{n!}\left(\frac{\partial^n G(st)}{\partial s^n}\right)_{s=0}. \tag{113}$$

The equation for $G(st)$ may be obtained by multiplying (108) by s^n and summing over all $n(0 \leq n \leq \infty)$. The resultant partial differential equation may be solved by the method of characteristics. For pulsed excitation and $\varepsilon = 1$ in (112) in the solution G_m is found to be

$$G_m(st) = \frac{\exp(-\alpha t)}{[\cosh(bt)(1+zy)]^D}\left[\frac{\delta}{2}\left(\frac{\gamma}{\delta} - \frac{z+y}{1+zy}\right)\right]^m, \quad \delta^2 > 0, \tag{114}$$

where

$$\begin{aligned}
\alpha &\equiv d + vD(2-\gamma)/2 \\
\gamma &\equiv 1 + \zeta + l/v \\
\delta^2 &\equiv (\zeta - 1)^2 + 2(\zeta + 1)l/v + (l/v)^2 \\
z &\equiv z(s) \equiv (\gamma - 2s)/\delta \\
y &\equiv y(t) \equiv \tanh(bt) \\
b &\equiv \delta v/2.
\end{aligned}$$

In the case of pulsed excitation by polychromatic radiation and /or in the presence of sequence congestion, more than one level is initially populated in the excited electronic state (oscillator), and (109) is replaced by the initial conditions

$$P_m(0) = \varepsilon_m, \quad \sum_m \varepsilon_m = 1. \tag{115}$$

Since the equation for $G(st)$ is linear and homogeneous, its solution for the initial condition (115) is obtained from (113) by linear superposition as

$$G(st) = \sum_m \varepsilon_m G_m(st), \tag{116}$$

where $G_m(st)$ are given by (114).

Expressions have been given[78] for the evaluation of the total quantum yields and decay rates and their various limiting forms for high- and low-pressure situations as well as for the time-resolved emission intensity in selected wavelength regions.[82] Similar treatments are available for the two-electronic state model.[79]

C. Pressure Dependence in Statistical Limit

It is quite remarkable that the kinetic scheme of (108) produces a total emission rate which can be very well reproduced by just a double exponential decay[78] despite the infinite number of levels present with varying radiative and non-radiative decay rates. (Of course, at low and high pressures the decay becomes single exponential.) This feature is in accord with frequent experimental observations of double exponential decays in intermediate pressure regimes with only single exponentials at low and high pressures. (Other mechanisms than that implied by the kinetic scheme of Fig. 7.) These double exponential decays are often explained in terms of a two-level scheme for the electronically excited state. However, the full model predicts a rather complicated pressure and wavelength dependence of these two decay rates and their relative intensities which would be impossible to reconcile with a simple two-level model. Rather than displaying the approximate expressions for the two pressure-dependent decay rates, etc., it is instructive to consider comparisons between model predictions and experiments. The theoretical methods are illustrated in the Appendix where the previous results are generalized to treat the case of modulated steady-state excitation wherein expressions for the emission phase shift are derived.

Fluorescence experiments by Beddard *et al.*,[71] by Brown *et al.*, and by

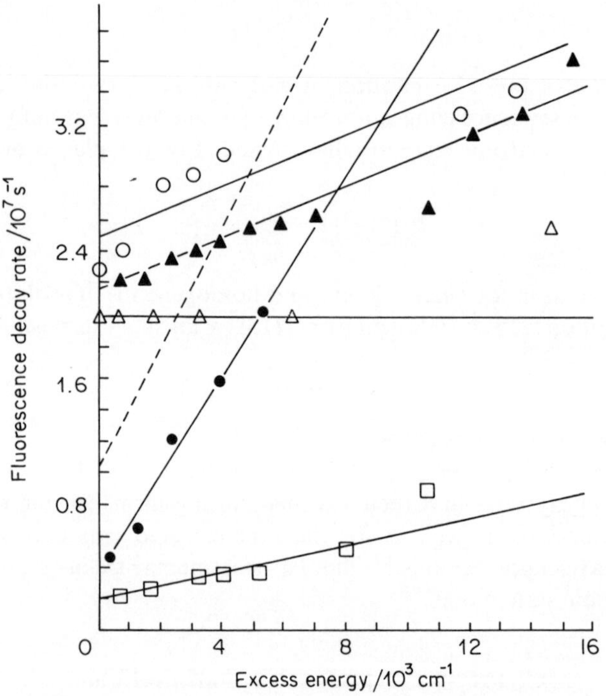

Fig. 8. Data of Beddard *et al*[71] of the fluorescence decay rates of benzene (dashed line), naphthalene (closed circles), anthracene ($\times 10^{-1}$) (open triangles), 1,2-benzanthracene (closed triangles), chrysene (open circles), and pyrene (squares) as a function of the excess vibrational energy in the excited singlet manifold. The approximate linear relation is apparent (and generally it breaks down at high enough energies.[76]) (Reproduced by permission of the Royal Society (London) from ref. 71.)

Spears[72] have been analysed in terms of the single electronic state model. Some of the data of Beddard *et al.*[71] are reproduced here because they have studied the widest set of systems. Fig. 8. presents their data for the zero pressure decay rates as a function of excess vibrational excitation, ΔE_{ex}, for a number of aromatic hydrocarbons. The observed linear variation of these decay rates with ΔE_{ex} parallels our model assumption. Since the fluorescence decay rate, $k(t)$, is time dependent in the intermediate region, Beddard *et al.*[71] compare calculated values of $k(t)$ at a particular time, $\tau^{o}{}_{b}$, after excitation with their observed values. Fig. 9. presents their date for naphthalene in argon with $\tau_{ob} = 400$ ns as a function of the argon pressure and of the wavelength of the excitation. The numbers on the curves refer to the initially excited model oscillator level obtained from

$$m + \frac{D}{2} = \Delta E_{ex}/\hbar\omega_{eff}, \tag{117}$$

where corrections for any hot band excitation because of sequence congestion cannot be applied. Note the detailed pressure and wavelength dependence which could not emerge from a simple two-vibrational level

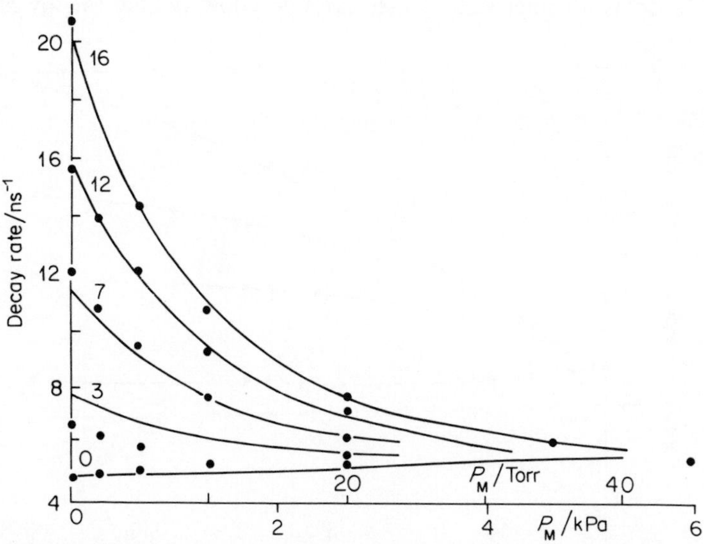

Fig. 9. Total fluorescence decay rate (at 400 ns) of naphthalene as a function of argon pressure taken from ref. 71. The points are the experimental data, while the solid lines are calculated from the one-electronic state using the parameters $D = 2$, $\hbar\omega_{eff}/k_BT = 1.5$, $l = 0.945\ ns^{-1}$, $d = 4.88\ ns^{-1}$. The numbers above the lines refer to $\Delta E_{ex}/\hbar\omega_{eff}$, the initial level of excitation of the effective oscillator. (Reproduced by permission of the Royal Society (London) from ref. 71.)

model. Fig. 10. exhibits the data of Beddard *et al.*[71] on the pressure and wavelength dependence of the fluorescence quantum yields of naphthalene in argon. Again our simple kinetic model provides a good representation of the complicated dependences.

Experiments on triplet yields[73–75] are considerably more difficult to perform, and the kinetic model perhaps contains about all (i.e. molecular parameters) we could initially hope to determine experimentally. Fig. 11. presents the calculated variation of the phosphorescence yield of naphthalene as a function of pressure for initial excitation into S_1 (or to higher singlets at 246 nm which rapidly interconvert to S_1). The quantum yield, ϕ_T, is given in units of the radiative decay rate, f', of the $j = 0$ triplet state level. Experimental data for the S_1-state model parameters are taken[79] from the model fits of Beddard *et al.*,[71] while T_1-state model parameters are approximated from the data of Soep *et al.*[75] and Beddard *et al.*[83] The $S_1 \rightarrow T_1$ model parameters are deduced from the former experiments and those of Knight *et al.*[84] The solid curve in Fig. 11 is for excitation at 308 nm and the dashed one for 246 nm. Fig. 9. of Ashpole *et al.*[73] presents experimental data for naphthalene in xenon, but plotted as ϕ_T^{-1} versus (pressure)$^{-1}$ while Fig. 8 of Ashpole *et al.*[73] provides data with other perturbers. When converted to a curve of ϕ_T versus pressure, the experimental data in Fig. 9 of Ashpole *et al.*[73] display a triplet yield which has a very low yield at the lowest pres-

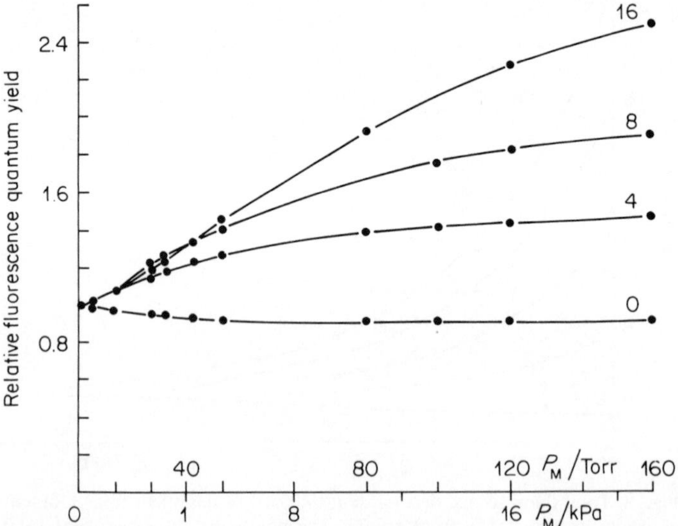

Fig. 10. Data of ref. 71 presenting the relative fluorescence quantum yield of naphthalene as a function of argon pressure. Points are experimental data, and lines give calculated results for *same* parameters as in Fig. 9. Numbers again refer to $\Delta E_{ex}/\hbar\omega_{eff}$. (Reproduced by permission of the Royal Society (London) from ref. 71.)

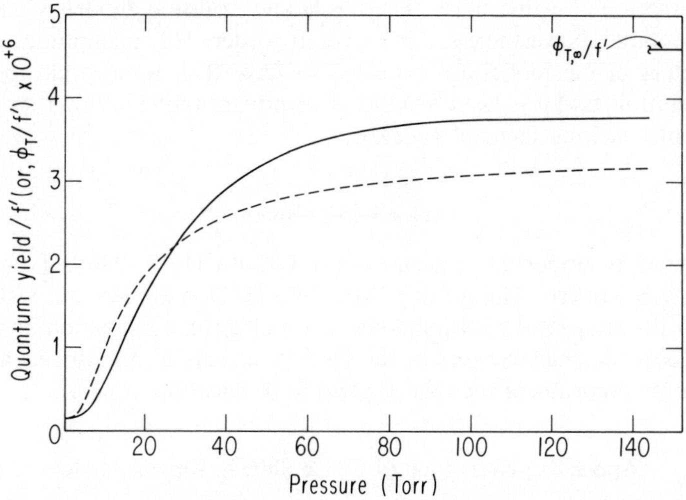

Fig. 11. Model calculations of the relative phosphorescence quantum yield of naphthalene as a function of argon pressure for 308 nm excitation (solid curve) and 246 nm excitation (dashed curve). The limiting high-pressure value is noted in the upper right by $\phi_{T,\infty}/f'$. The singlet parameters are as in Fig. 9. While the remaining ones are deduced from refs. 73–75.[79]

sure of ≈ 3 Torr, has a sharp and fairly linear rise in the yield between this low pressure and about 20 Torr, where the yield increases more slowly with pressure. The theoretical curves display these same qualitative features, except the rise in yield above 20 Torr is somewhat slower than the experimental curves, but the broad-band nature of the experiments and/or the need for a different value of v', etc., may be responsible for this difference. The triplet yield increases with increasing pressure because collisions take high-lying, more rapidly decaying triplet levels to lower, longer-lived ones. At the low-pressure limit the triplet yield is non-zero, although it is quite small because of the increase in triplet decay with increasing vibrational energy.

The study of triplet yields is a rather important problem that can provide a wealth of information and help to test theories of vibrational energy dependences of $S_1 \rightarrow S_0$ and $T_1 \rightarrow S_0$ non-radiative decay rates. This problem therefore deserves further experimental investigation. The general qualitative agreement between our simple kinetic model and the scarce available experimental data imply that the model should prove to be highly useful in the analysis of these difficult experiments. Care should be exercised in not taking the model too literally in a state-to-state sense. Theoretical predictions imply that $S_1 \rightarrow S_0$ and $T_1 \rightarrow S_0$ non-radiative decay rates should increase exponentially with ΔE_{ex}, so the assumed linear variation of the model is only an approximation. Furthermore, the step-ladder collision model is an oversimplification. Nevertheless, the overall orders of magnitude of the parameters of the model are expected to have their appropriate physical interpretation, and it is just this kind of semiquantitative information which is currently lacking for triplet yields.

Acknowledgements

This work is supported, in part, by NSF Grant CHE 77-24652. I am grateful to my coworkers, Hang Fung, Bill Gelbart, Don Heller, and Catherine Tric, on the subject of collisional effects on electronic relaxation as well as to numbers of other workers in the field for extensive discussions and for sending me preprints of their work prior to publication.

Appendix: Derivation of Phase Shift in Kinetic Model

The derivation follows just as in Section II of the Freed–Heller paper.[78] For modulated continuous excitation of the *nth* level, the source term (110) is replaced by $g \exp(i\omega t)s^m\delta_{mn}$, where ω is the modulation frequency. Using this equations (108) and (111) yield the generating function equation (for $D = 1$),

$$\frac{\partial G(st)}{\partial t} = v[s^2 - (1 + \zeta + l/v)s + \zeta]\frac{\partial G(st)}{\partial t} + (vs - d - v)G(st) + g\exp(i\omega t)s^m$$
$$\equiv P(s)\frac{\partial G(st)}{\partial s} - R(st) \tag{118}$$

The characteristic equations are

$$\frac{\mathrm{d}s}{P} = \frac{\mathrm{d}t}{-1} = \frac{\mathrm{d}G}{R}. \tag{119}$$

The s–t equation is the same as before[78] giving

$$s(t) = \frac{1}{2}\left\{\gamma + \delta\tanh\left[\frac{\delta}{2}(vt - K_1)\right]\right\}, \tag{120}$$

where

$$K_1 = vt + 2\phi/\delta, \tag{120$'$}$$
$$\phi = \operatorname{arc\,tanh}[(\gamma - 2\Delta)/\delta]. \tag{120$''$}$$

The G–t equation is

$$\frac{\mathrm{d}G}{\mathrm{d}t} - [vs(t) - d - v]G = g\exp(i\omega t)s^m, \tag{121}$$

which has the solution

$$G = \exp\left\{\int^{t}[vs(t') - d - v]\mathrm{d}t'\right\}\left\{\int_{0}^{t}\mathrm{d}t' g\exp(i\omega t')[s(t')]^{m}\right.$$
$$\left.\times \exp\left(\int^{t'}[vs(t'') - d - v]\mathrm{d}t''\right) + K_2\right\}. \tag{122}$$

The indefinite integral in the integrating factor is given by Freed and Heller[78] as

$$\int^{t}\mathrm{d}t'[vs(t') - d - v] = -\alpha^{t}_{D=1} + \ln\left\{\cosh\left[\frac{\delta}{2}(vt - K_1)\right]\right\}. \tag{123}$$

(This is actually only one branch of the integral; the final results below are valid for the general case.) Substituting (123) into (122) gives

$$G = \exp(-\alpha t)\cosh\left[\frac{\delta}{2}(vt - K_1)\right]$$
$$\cdot\left\{\int_{0}^{t}\frac{\mathrm{d}t' g\exp(i\omega t')\left[\frac{1}{2}\left(\gamma + \delta\tanh\left[\frac{\delta}{2}(vt' - K_1)\right]\right)\right]^{m}\exp(-\alpha t')}{\cosh\left[\frac{\delta}{2}(vr' - K_1)\right]} + K_2\right\}. \tag{124}$$

The initial condition for the s-equation is

$$s = \tfrac{1}{2}[\gamma - \delta\tanh(\delta K_1/2)], \tag{125}$$

while that for the G-equation is

$$G_{mm'}(t=0) = \varepsilon s^{m'} = K_2\cosh(\delta K_1/2). \tag{126}$$

Eliminating s between (125) and (126) gives the identity relation between K_1 and K_2 which is valid for all times,

$$K_2 = \frac{\varepsilon\{\frac{1}{2}[\gamma - \delta\tanh(\delta K_1/2)]\}^{m'}}{\cosh(\delta K_1/2)}. \tag{127}$$

Substitute $K_1 = vt + 2\phi/\delta$ from (120′) and K_2 from (127) into (124), then rearrange to get

$$G_{mm'}(st/\omega) = \varepsilon G_{m'}(st) + g\exp(i\omega t)\int_{0}^{t}\mathrm{d}x\exp(i\omega x)G_{m}(sx), \tag{128}$$

where G_m and $G_{m'}$ are given by (114).

As we are not interested in the pulsed case here, we can take $\varepsilon = 0$. Further, if we excited more than one state, i.e. a set of g_m, the final result (valid for a

D-dimensional oscillator model now) becomes

$$G^{\text{steady}}(st/\omega) = \exp(i\omega t)\sum_m g_m \int_0^t \mathrm{d}x \exp(i\omega x)G_m(sx), \tag{129}$$

with $G_m(sx)$ the quantity given in (114) for the pulsed case. The total steady-state yield comes from using $\omega = 0, t \to \infty, s = 1$, whereupon (129) reduces to the sum of ordinary quantum yields,

$$\Phi^{\text{steady}} = \sum_m g_m \Phi_m, \qquad \left(\sum_m g_m = 1\right),$$

$$\Phi_m = \int_0^\infty \mathrm{d}t G_m(1, t). \tag{130}$$

The case of emission from particular levels n then follows readily.

In phase shift experiments the complex steady-state yield is obtained from

$$Y(\omega s) = \sum_m g_m \int_0^\infty \mathrm{d}x \exp(i\omega x)G_m(sx). \tag{131}$$

When the emission is unresolved, the phase shift is

$$\cot\Delta\varphi = Re\,Y(\omega, 1)/Im\,Y(\omega, 1)$$

$$= \frac{\sum_m g_m \int_0^\infty \mathrm{d}x \cos\omega x G_m(1, x)}{\sum_m g_m \int_0^\infty \mathrm{d}x \sin\omega x G_m(1, x)}, \tag{132}$$

which represents only a minor change from the evaluation of quantum yields given in the pulsed case. The treatment of resolved emission again follows simply when we have $l_{\text{rad}} \neq 0$; (132) must be changed to

$$\cot\Delta\varphi = \frac{\sum_m g_m \int_0^\infty \mathrm{d}x \cos\omega x \left[fG_m(1, x) + l_{\text{rad}} \left.\frac{\partial G(s, x)}{\partial s}\right)_{s=1} \right]}{\sum_m g_m \int_0^\infty \mathrm{d}x \sin\omega x \left[fG_m(1, x) + l_{\text{rad}} \left.\frac{\partial G(s, x)}{\partial s}\right)_{s=1} \right]} \tag{133}$$

to include the increasing radiative decay rate of higher levels.

References

1. K. F. Freed, *Topics Appl. Phys.*, **15**, 23 (1976).
2. K. F. Freed, *Accts. Chem. Res.*, **11**, 74 (1978).
3. J. Jortner and S. Mukamel in *The World of Quantum Chemistry* (Eds. R. Daudel and B. Pullman), Reidel, Boston, 1974.
4. P. Avouris, W. M. Gelbart, and M. A. El-Sayed, *Chem. Rev.*, **77**, 793 (1977).
5. W. M. Gelbert, K. F. Freed and S. A. Rice, *J. Chem. Phys.*, **52**, 2460 (1970).

6. K. F. Freed and Y. B. Band, *Excited States*, **3**, 109 (1978).
7. G. W. Robinson, *Excited States*, **1**, 1 (1974).
8. G. B. Kistiakowsky and C. S. Parmenter, *J. Chem. Phys.*, **42**, 2942 (1965).
9. E. M. Anderson and G. B. Kistiakowsky, *J. Chem. Phys.*, **48**, 4787 (1968); C. S. Paramenter and A. H. White, *J. Chem. Phys.*, **50**, 1631 (1969).
10. B. K. Selinger and W. R. Ware, *J. Chem. Phys.*, **53**, 3160 (1970); C. S. Parmenter and M. W. Schuyler, *Chem. Phys. Lett.*, **6**, 339 (1970).
11. K. G. Spears and S. A. Rice, *J. Chem. Phys.*, **55**, 5561 (1971); A. S. Abramson, K. G. Spears, and S. A. Rice, *J. Chem. Phys.*, **56**, 2291 (1972); C. Guttman and S. A. Rice, *J. Chem. Phys.*, **61**, 651 (1974).
12. K. F. Freed, *Chem. Phys. Lett.*, **42**, 600 (1976).
13. L. G. Anderson, C. S. Parmenter, H. M. Poland, and J. D. Rau, *Chem. Phys. Lett.*, 232 (1971); L. G. Anderson, C. S. Parmenter, and H. M. Poland, *Chem. Phys.*, **1**, 401 (1973).
14. E. Weitz and G. Flynn, *Ann. Rev. Phys. Chem.*, **25**, 275 (1974).
15. O. Atabek, A. Hardisson, and R. Lefebvre, *Chem. Phys. Lett.*, **20**, 40 (1973).
16. W. Siebrand *Chem. Phys. Lett.*, **6**, 192 (1970); B. R. Henry and W. Siebrand, *J. Chem. Phys.*, **56**, 4058 (1972).
17. K. F. Freed and S. H. Lin, *Chem. Phys.*, **11**, 409 (1975).
18. K. F. Freed, *J. Chem. Phys.*, **45**, 4214 (1966).
19. A. E. Douglas, *J. Chem. Phys.*, **45**, 1007 (1967).
20. R. E. Smalley, B. L. Ramakrishna, D. H. Levy, and L. Wharton, *J. Chem. Phys.*, **61**, 4363 (1974).
21. P. Wannier, P. M. Rentzepis, and J. Jortner, *Chem. Phys. Lett.*, **10**, 102, 193 (1971); G. E. Busch, P. M. Rentzepis, and J. Jortner, *Chem. Phys. Lett.*, **11**, 437 (1971); *J. Chem. Phys.*, **56**, 361 (1972); D. Zevenhuijzen and R. van der Werf, *Chem. Phys.*, **26**, 279 (1977).
22. R. van der Werf and J. Kommandeur, *Chem. Phys.*, **16**, 125 (1976); R. van der Werf, E. Schutten, and J. Kommandeur, *Chem. Phys.*, **16**, 151 (1976).
23. K. F. Freed and J. Jortner, *J. Chem. Phys.*, **50**, 2916 (1969).
24. K. F. Freed, *J. Chem. Phys.*, **52**, 1345 (1970).
25. K. F. Freed, *Topics Curr. Chem.*, **31**, 105 (1972).
26. M. Bixon, Y. Dothan, and J. Jortner, *Mol. Phys.*, **17**, 109 (1969).
27. C. Tric, *Chem. Phys. Lett.*, **21**, 83 (1973); *J. Chem. Phys.*, **55**, 4303 (1971).
28. W. M. Gelbart, D. F. Heller, and M. L. Elert, *Chem. Phys.*, **7**, 116 (1975).
29. F. Lahmani, A. Tramer, and C. Tric, *J. Chem. Phys.*, **60**, 4431 (1974).
30. J. Chaiken, T. Benson, M. Gurnick and J. D. McDonald, preprint.
31. A. Villaeys and K. F. Freed, *Chem. Phys.*, **13**, 271 (1976).
32. R. L. Swofford, M. E. Long, and A. C. Albrecht, *J. Chem. Phys.*, **65**, 179 (1976); M. Berry, to be published.
33. S. Mukamel and J. Jortner, *J. Chem. Phys.*, **65**, 5204 (1976).
34. P. J. Robinson and K. A. Holbrook, *Unimolecular Reactions* Wiley, New York, 1972.
35. R. A. Beyer and W. C. Lineberger, *J. Chem. Phys.*, **62**, 4024 (1975).
36. P. F. Zittel and W. C. Lineberger, *J. Chem. Phys.*, **66**, 2972 (1977).
37. C. A. Thayer and J. T. Yardley, *J. Chem. Phys.*, **57**, 3992 (1972).
38. C. S. Parmenter and M. Seaver, to be published; M. Seaver, *Thesis*, Indiana (1978). I am grateful to Professor Parmenter and Dr. Seaver for sending me copies of their work prior to publication.
39. T. W. Eder and R. W. Carr, *J. Chem. Phys.*, **53**, 2258 (1970).
40. A. Frad, Lahmani, A. Tramer, and C. Tric, *J. Chem. Phys.*, **60**, 4419 (1974).

41. D. W. Pratt and H. P. Broida, *J. Chem. Phys.*, **50**, 2181 (1969); H. E. Radford and H. P. Broida, *J. Chem. Phys.*, **38**, 644 (1963).
42. M. Lavollée and A. Tramer, *Chem. Phys. Lett.*, **47**, 523 (1977).
43. J. C. Weisshaar, A. P. Baronavski, A. Cabello and C. B. Moore, *J. Chem. Phys.*, in press.
44. R. Naaman and R. N. Zare, *Chem. Phys. Lett.*, to be published.
45. (a) R. N. Rudolph and S. J. Strickler, *J. Amer. Chem. Soc.*, **99**, 3871 (1977); (b) S. J. Strickler and R. N. Rudolph, *J. Amer. Chem. Soc.*, **100**, 3326 1978; (c) F. Su, F. B. Wampler, J. W. Bottenheim, D. L. Thorsell, J. G. Calvert, and E. K. Damon, *Chem. Phys. Lett.*, **51**, 150 (1977).
46. K. G. Spears and M. El-Manguch, *Chem. Phys.*, **24**, 65 (1977).
47. W. M. Gelbart and K. F. Freed, *Chem. Phys. Lett.*, **18**, 470 (1973).
48. K. F. Freed, *J. Chem. Phys.*, **64**, 1604 (1976).
49. K. F. Freed, *Chem. Phys. Lett.*, **37**, 47 (1976).
50. K. F. Freed and C. Tric, *Chem. Phys.*, **33**, 249 (1978).
51. D. Grimbert, M. Lavollée, A. Nitzan, and A. Tramer, *Chem. Phys. Lett.*, **57**, 45 (1978).
52. J. C. Light, *Meth. Comput. Phys.*, **10**, 111 (1971).
53. M. Y. Chu and J. S. Dahler, *Mol. Phys.*, **27**, 1045 (1974); K. C. Kuhlander and J. S. Dahler, *J. Phys. Chem.*, **80**, 2881 (1976); *Chem. Phys. Lett.*, **41**, 125 (1976).
54. K. Lawley and J. Ross, *J. Chem. Phys.*, **43**, 2943 (1965).
55. L. D. Landau and E. M. Lifshitz, *Quantum Mechanics*, Pergamon Press New York, 1969.
56. J. C. Tully and R. K. Preston, *J. Chem. Phys.*, **55**, 562 (1971).
57. E. E. Nikitin, *Adv. Quantum Chem.*, **5**, 135 (1970).
58. P. R. Stannard, *J. Chem. Phys.*, **68**, 3932 (1978).
59. A. Matsuzaki and S. Nagakura, *Chem. Phys. Lett.*, **37**, 204 (1976); *J. Luminesc.*, **12–13**, 787 (1976); *Z. Phys. Chem.*, **101**, 283 (1976).
60. H. G. Küttner, H. D. Selzle, and E. W. Schlag, *Chem. Phys. Lett.*, **48**, 207 (1977).
61. J. H. van Vleck, *Phys. Rev.*, **40**, 544 (1932); S. Butler and D. H. Levy, *J. Chem. Phys.*, **66**, 3538 (1977).
62. A. E. W. Knight, private communication.
63. F. Su, J. W. Bottenheim, D. L. Thorsell, J. G. Calvert, and E. K. Damon, *Chem. Phys. Lett.*, **49**, 305 (1977); F. Su and J. G. Calvert, *Chem. Phys. Lett.*, **53**, 572 (1977).
64. S. Mukamel, *Chem. Phys. Lett.*, to be published; S. Mukamel and K. F. Freed, unpublished work.
65. C. P. Slichter, *Principles of Magnetic Resonance*, Harper and Row, New York, 1963; A. G. Redfield, *IBM J. Res. Dev.*, **1**, 19 (1957).
66. U. Fano, *Phys. Rev.*, **131**, 259 (1963).
67. A. Ben-Reuven, *Phys. Rev.*, **141**, 34; **145**, 7 (1966).
68. D. E. Fitz and R. A. Marcus, *J. Chem. Phys.*, **59**, 4380 (1973); **62**, 3788 (1975); W. K. Liu and R. A. Marcus, *J. Chem. Phys.*, **63**, 272, 290 (1975).
69. Y. N. Chiu, *J. Chem. Phys.*, **56**, 5741 (1972); S. H. Lin, *Proc. Roy. Soc. (London)*, **A335**, 51 (1973).
70. C. S. Parmenter, *Adv. Chem. Phys.*, **22**, 365 (1972).
71. G. S. Beddard, G. R. Fleming, O. L. J. Gijzeman, and G. Porter, *Proc. Roy. Soc. (London)*, **A340**, 519 (1974).
72. R. G. Brown, M. G. Rockley, and D. Phillips, *Chem. Phys.*, **7**, 41 (1975); K. G. Spears, *Chem. Phys. Lett.*, **54**, 139 (1978); R. P. Steer, M. D. Swords, and D. Phillips, *Chem. Phys.*, **34**, 95 (1978).

73. C. W. Ashpole, S. J. Formosinho, and G. Porter, *Proc. Roy Soc. (London)*, **A323**, 11 (1971).
74. S. J. Formosinho, G. Porter, and M. A. West, *Chem. Phys. Lett.*, **6**, 7 (1970).
75. B. Soep, C. Michel, A. Tramer, and L. Linqvist, *Chem. Phys.*, **2**, 293 (1973).
76. C. S. Huang, J. C. Hsieh, and E. C. Lim, *Chem. Phys. Lett.*, **28**, 130 (1974); **37**, 349 (1976); J. C. Hsieh, C. S. Huang, and E. C. Lim, *J. Chem. Phys.*, **60**, 4345 (1974); J. C. Hsieh and E. C. Lim, *J. Chem. Phys.*, **61**, 737 (1974).
77. B. S. Neporent and S. O. Mirumyants, *Opt. Spectr.*, **8**, 336 (1960).
78. K. F. Freed and D. F. Heller, *J. Chem. Phys.*, **61**, 3942 (1974).
79. K. H. Fung and K. F. Freed, *Chem. Phys.*, **14**, 13 (1976).
80. E. W. Montroll and K. E. Shuler, *J. Chem. Phys.*, **26**, 454 (1957).
81. V. Seshadri and V. M. Kenkre, *Phys. Lett.*, **A56**, 75 (1976); *Phys. Rev.*, **A15**, 197 (1977).
82. G. R. Fleming, O. L. J. Gÿzeman, and S. H. Lin, *J. Chem. Soc. Faraday Trans., II*, **70**, 1074 (1974); G. R. Fleming, O. L. G. Gijzeman, K. F. Freed, and S. H. Lin, *J. Chem. Soc. Faraday Trans. II*, **71**, 773 (1975); M. D. Swords and D. Phillips, *Chem. Phys. Lett.*, **43**, 228 (1976).
83. G. S. Beddard, S. J. Formosinho, and G. Porter, *Chem. Phys. Lett.*, **22**, 235 (1973).
84. A. E. W. Knight, B. K. Selinger, and I. G. Ross, *Aust. J. Chem.*, **26**, 1159 (1973).

Potential Energy Surfaces
Edited by K. P. Lawley

VIBRATIONAL AND ROTATIONAL COLLISION PROCESSES*

ANDREW E. DePRISTO† AND HERSCHEL RABITZ‡

Department of Chemistry, Princeton University, Princeton, New Jersey 08540, *USA*

CONTENTS

I. INTRODUCTION

A number of excellent review articles have recently appeared dealing with material in the same general area as this review (Secrest, 1973; Balint-Kurti, 1975; Reuss, 1975; Child, 1976; Toennies, 1976; Rabitz, 1976; Porter and Raff, 1976; Micha, 1976; Faubel and Toennies, 1977; Clark *et al*, 1977; Dickinson, 1979). The literature on vibration–rotation collision theory has mushroomed and most of these recent reviews cover the published literature extensively up until very recently. Rather than duplicate these efforts, we shall therefore take advantage of this fact and aim our article to cover the key recent ideas and developments of the past few years. In doing this we will explicitly not cite all the applications, calculations, etc., but we will cite selected references as a guide to the remaining extensive

*The authors acknowledge support from the Department of Energy, the Office of Naval Research, the National Science Foundation, and the Air Force Office of Scientific Research.
† Present address: Department of Chemistry, University of North Carolina, Chapel Hill, N.C. 27514, USA.
‡ Alfred P. Sloan Fellow, Camille and Henry Dreyfus Teacher-Scholar.

literature and to point out the problems which are feasible for study at the present time. In a similar vein, special attention will be given to the present directions of the field, and an attempt will be made to project ahead and point out new areas which deserve study. The earlier review of Takayanagi (1963) contains such a 'research shopping list' which has been a useful stimulus to many workers in the field. Unfortunately it must be added that many of the intriguing problems on that list still remain unsolved.

Before proceeding to the details of the review some background on the context of the subject is appropriate. The macroscopic properties of all real gases are a direct manifestation of collisional events at the microscopic level. The problem of proceeding forward from the Hamiltonian to the prediction of laboratory observables is just as important as inverting experimental data to determine molecular information. In both cases the dynamics of the collision must be treated. This observation brings forth the point that experiments and theory have a common goal of achieving a better physical understanding of molecular processes—an aim which is easily forgotten when examining the voluminous literature comparing experimental and theoretical results. We particularly want to stress this common general goal and Section II is devoted to some relevant observational methods. Without an observational guide it is all too easy to formulate a theory incapable of handling real problems of interest. Another pitfall is the assumption that agreement between theory and a particular experiment validates the theoretical model. If the model is physically unrealistic, then the agreement has little meaning or purpose. These comments should always be kept in mind when assessing the molecular dynamics literature.

As indicated above, theory can operate in two interrelated ways. First, starting with the Hamiltonian, appropriate dynamics can be performed and observables can be calculated. This 'forward' approach is the realm of traditional theory. In addition, theory can occupy the role of guiding the interpretation and inversion of raw experimental data back to more refined molecular information. Theory has always operated in the latter fashion to some extent, but this role has recently received additional attention. A critical matter of concern in any inversion is the reliability of the *theory and the data* along with concomitant problems of error propagation. The developments in inversion theory will be treated in Section VI, after discussing the recent advances in available methods for molecular dynamics.

In treating new methods in collision theory we shall largely confine ourselves to atom–molecule interactions, where the molecules only contain a few atoms. These 'simple' system represent the most carefully studied cases, though the desire to treat larger systems of increased complexity has always been a motivating influence in molecular dynamics. Some of the recent developments in collision theory have this goal in mind, although often only small molecules have been treated thus far.

A fundamental problem in molecular dynamics involves the determination of the relationship between the observed collisional behaviour and the structure of the underlying Hamiltonian (or intermolecular potential). There are two motivations for the study of this problem. First, if this relationship could be mapped out then reliable estimates could be given for expected behaviour when new molecular systems are encountered. It is a simple fact that even after many years of research each new molecular system seems to represent an entirely new problem in itself. Secondly, although the interactions of molecules are ultimately related to electrical forces, the actual intermolecular potentials are rarely known to the accuracy necessary for realistic dynamical predictions. This difficulty results from the many-electron nature of the problem of calculating potential surfaces. Therefore many dynamical calculations have employed some potential function which contains physically meaningful (albeit unknown) parameters such as coupling strengths. In this situation it is clear that knowledge of how the collisional observables vary with the potential parameters would be a useful aid for determining potentials from experimental data. The careful mapping-out of the input/output correlation in collision theory is expected to receive increased attention. Present developments is this area will be discussed in Section VII.

Finally in keeping with the overall intentions of this review, material and methods dating from many years past will only be treated in their modern context of usefulness. In this fashion it is hoped that the viable possibilities for current applications will become apparent. In addition, the stated limitations of the present methods should also serve to indicate new directions for future development.

II. SURVEY OF EXPERIMENTAL METHODS

Although the emphasis of this review is on theoretical developments a discussion without reference to the appropriate experimental measurements would be severely deficient. In one sense the need for theoretical studies exists due to the presence of relevant experimental observations. In another sense theory can contribute fundamentally to the detailed understanding of the physical processes at work in collisional events. It is also true that practical concerns require molecular dynamics data for appropriate design studies. The necessary information is often sparsely available from experiments alone, and theory has an important role in filling this need.

It is beyond the scope of this review to thoroughly list or detail all the possible experiments. The limited goal here is to codify the general range of possible observations and to bring out connections between them. To achieve this aim the range of experiments is listed below, starting with the most detailed and working down to more 'averaged' observations. Our classification scheme is based on the degree of state selection in an experiment.

1. *Fully State-Selected Molecular Beam Differential Cross Sections*

This technique uses crossed molecular beams of well-defined velocity and in selected internal molecular states to determine the differential cross sections

$$\sigma_{njm_j,n'j'm_{j'}}(\hat{q};E) \tag{1}$$

where j, m_j, n are respectively the initial rotational quantum number, its projection on a space-fixed axis, and the vibrational quantum number. The total centre-of-mass energy is E and $\hat{q}$ is the solid scattering angle. This data for all relevant states, energies, and angles represents the most refined measurement that could be performed; perhaps naturally, it is also the most difficult to perform and has only recently become feasible (Fluendy and Lawley, 1973; Gentry and Giese, 1977).

2. *State-Selected Total Cross Sections*

The total cross section is given by

$$\sigma_{njm_j,n'j'm_{j'}}(E) = \int \mathrm{d}\hat{q}\,\sigma_{njm_j,n'j'm_{j'}}(\hat{q};E) \tag{2}$$

This total cross section may either be measured directly by summing the angular distribution from a molecular beam experiment, or by using an appropriate angle-insensitive detection method such as laser-induced fluorescence in conjunction with a molecular beam (Wilcomb and Dagdigian, 1977).

3. *State-to-State Rate Constants*

State-to-state rates are obtained by averaging the total cross sections over a Boltzmann velocity distribution $f(v, T)$ at temperature T.

$$K_{njm_j,n'j'm_{j'}}(T) = \int_0^\infty \mathrm{d}v\, f(v, T)\, v\, \sigma_{njm_j,n'j'm_{j'}}(E) \tag{3}$$

These rates enter into the kinetic mechanisms in bulk media. In general, measurement of these rates requires successively perturbing the equilibrated system by external state-selecting forces (i.e. a resonant laser pulse) and monitoring the return to equilibrium (Steinfeld and Houston, 1978). Double resonance, laser-induced flouresecence, and/or opto-acoustic methods are all useful for these types of measurements. It is also possible to use a laser to select out a particular velocity 'group' of colliding molecules from the Doppler profile; this approach will effectively eliminate the velocity average in equation (3). The determination of the orientation dependence of $K_{njm_j,n'j'm_{j'}}$ is also feasible (Reuss, 1975; Jeyes *et al.*, 1977; Alexander and Dagdigian, 1977).

4. *Bulk Relaxation Times*

In contrast with state-to-state measurements and their inherent selective disturbance of the medium, bulk relaxation measurements extract a single overall relaxation time for the response of the system. In this case, typically, a medium is disturbed non-selectively (e.g. by shock-wave temperature jump, sudden expansion, etc.) and the overall relaxation is monitored (Lambert, 1977). These experiments may be performed by observing some property of the system, such as temperature, refractive, index, etc., that is sensitive to the energy transfer process. The relaxation time τ is in general a convolution of all the state-to-state rates $K_{njm_j,n'j'm_{j'}}(T)$. In addition, for the case of flowing gases appropriate fluid-mechanical effects must be taken into account (Rabitz and Lam, 1975). In other experiments, analogs with n.m.r. T_1 and T_2 relaxation times can be identified (Schmalz and Flygare, 1978), and these times can be related to inelastic and dephasing collision processes (Pickett, 1974, 1975). A related measurement is nuclear spin lattice-relaxation where the spin is coupled to the rotational angular momentum. Since the latter quantity is directly affected by collisions, the overall spin relaxation time is therefore a measure of the collisional processes.

5. *Vibration-Rotation Spectral Line-Shapes*

Spectral line-shapes are singled out here because they play an intermediate role between fully state-selected measurements and overall bulk relaxation times. Pressure broadening is due to collisions interrupting the normal radiative process (Rabitz, 1974). The spectral radiation essentially labels the initial collisional state, and the line-width is then a measure of collisional processes with molecules in this labelled state. Thus a series of spectral lines, either pure rotational or vib-rotational, represent *half* state-selected experiments. Line-shapes therefore potentially contain rich amounts of collisional information.

6. *Transport Properties*

Properties such as viscosity and phenomena like bulk transport of energy and angular momentum are direct consequences of molecular collisions (Curtiss, 1974). These processes can be described through collisional input into a bulk transport equation (e.g. the Boltzmann equation). Additional data can also be obtained by performing the experiments in external electric and magnetic fields (Beenakker *et al.*, 1973). In this fashion, spatial or anisotropic collisional effects can be probed.

In listing the above types of experiments, certain arbitrary classifications were made. Various combinations of the experiments can be performed to yield other types of measurements. A fundamental point is that all the observations are ultimately related to the collisional scattering matrix or amplitude. This is exactly the quantity produced by the dynamical calculations. The above classification scheme explicitly shows that the degree of averaging necessary to predict the measurable quantities increases with a decrease in state selection. Thus comparisons between theory and experiment become less direct for the more 'averaged'-type data. Nevertheless experiments of all types can provide valuable insight. As the need arises in the remainder of the review, specific references will be given to some of the above measurements.

III. INTERMOLECULAR POTENTIALS

All molecular dynamics calculations are dependent upon knowledge of an intermolecular potential. As mentioned in the introduction, obtaining an accurate surface is not a simple problem; indeed the subject is an entire field in itself (Schaefer, 1977). Only brief comments will be presented here on the theoretical methods for obtaining potential surfaces, since, for our purpose, the form of the interaction is most important.

First consider an atom A, infinitely separated from a diatomic molecule BC. The two 'particles' are assumed to have known energies, E_{A}^{∞} and E_{BC}^{∞} in this asymptotic limit. When the two objects are brought together an interaction energy, $E_{\mathrm{A,BC}}$ will occur. After the asymptotic atomic and molecular energies are subtracted the remainder is identified as the intermolecular potential V.

$$V(R, r, \gamma) = E_{\mathrm{A,BC}} - E_{\mathrm{A}}^{\infty} - E_{\mathrm{BC}}^{\infty} \tag{4}$$

The coordinates, R, r, γ, are illustrated in Fig. 1. It is assumed that during the scattering event the nuclei move sufficiently slowly to validate the use of the Börn–Oppenheimer approximation. Therefore the potential surface $V(R_i, r_i, \gamma_i)$ is mapped out at each point $R_i, r_i, \gamma_i, i = 1, 2, \ldots$ in nuclear coordinate space. This function is the potential energy occurring in the scattering Hamiltonian

$$H = T + H_0 + V \tag{5}$$

where T is the relative kinetic energy of the atom–molecule, H_0 is the Hamiltonian for the freely rotating–vibrating molecule and the atom is assumed to be structureless throughout this review. The Schrödinger

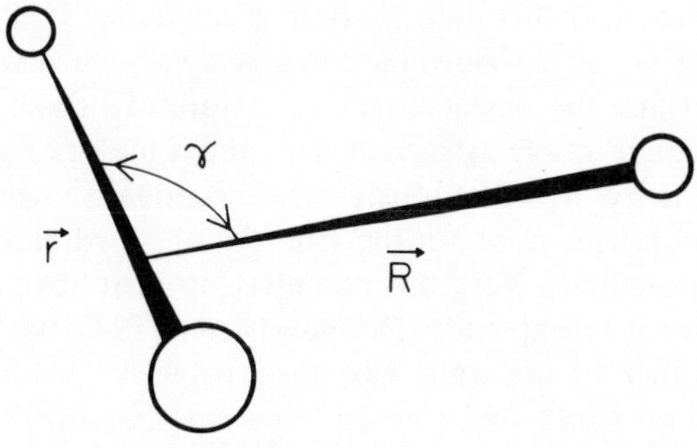

Fig. 1. Collision geometry for the atom–diatomic molecule system. The position vectors **r** and **R** pertain to the internal and relative translational degrees of freedom respectively.

equation with this scattering Hamiltonian describes the nuclear motion of the three particles.

The total interaction energy $E_{A,BC}$ can be obtained through a variety of methods developed for molecular structure calculations. At long range perturbation theory methods can be utilized, while at shorter distances Hartree–Fock and configuration–interaction methods are in principal applicable (Schaefer, 1077). The main difficulty, neglecting relativistic effects, is the large computational effort involved, and only for very simple systems are fully converged calculations just becoming available. A number of other approximate moleculat structure methods could also be applied to these problems, and most recently an electron gas theory (Kim and Gordon, 1974; Parker *et al.*, 1976) has been developed which utilizes the charge density of the free colliding particles as input. The approximate Coulomb, exchange, and correlation energies can be calculated in this fashion, and reasonable results have been obtained in a number of cases (Green *et al.*, 1975). In addition, this procedure is rather efficient to apply. Although rotational inelastic processes seem to be adequately described by this method, it is not clear at this time if this approach is accurate enough for use in vibrationally inelastic collision dynamics.

Historically, rotationally inelastic collisions have received more careful attention than the corresponding vibrational processes. This emphasis stems in part from the fact that more is known about the expected behaviour of the anisotropic part of the potential responsible for rotational coupling. In general for an atom vibrator the potential may be written as follows

$$V(R, r, \gamma) = \sum_{\mu=0} V_\mu(R, r) P_\mu(\cos\gamma) \qquad (6)$$

where P_μ is a Legendre polynomial and $\cos\gamma = \hat{r}\hat{R}$. The generally expected behaviour of V_μ is known as a function of R for r fixed (rigid-rotor approximation). Often when no other information is available $V_\mu(R, r)$ is taken as $\alpha_\mu V_0(R, r)$ where α_μ is a coordinate-independent anisotropy coefficient. Typically $V_0(R, r)$ has the characteristic behaviour of a short-range repulsive core, an intermediate-range attractive well, and a long-range inverse power decay to zero energy. Most seriously for vibrationally inelastic scattering, little information is known about the vibrational coordinate dependence of intermolecular potentials. Very few potential surfaces have been calculated at various r values. A recent study (Dougherty *et al.*, 1977) has drawn together the limited number of accurate vibration–rotation surfaces and certain common behaviour was found to exist. Nevertheless, until further accurate surfaces are available even exact dynamical calculations of vibrational inelasticity must generally be classified as model studied. At the present stage of development in molecular dynamics valuable insight can still be obtained from such studies.

IV. DYNAMICS

In this section, we outline the classical, semiclassical, and quantal formulations of molecular scattering. The choice of dynamical method depends upon the specific task at hand. In this regard, note that the calculated results are sensitive to the accuracy of the intermolecular potential—a topic treated in detail in Section VII. The consideration of dynamical methods with an emphasis on numerical and approximation techniques is separated from the kinematical developments for convenience and clarity. The two areas overlap significantly and advances in one are frequently useful in the other.

A. Classical Mechanics

An excellent exposition of classical trajectory methods has been given by Porter and Raff (1976). They discuss the rationale for using classical mechanics and also explain at considerable length the numerical procedures used in applications. A more qualitative analysis is presented here for atom–molecule collisions. The Hamiltonian is given by

$$H = \tfrac{1}{2}\mu \dot{\boldsymbol{R}}^2 + H_0(\boldsymbol{r}) + V(\boldsymbol{r}, \boldsymbol{R}) = E \tag{7}$$

where E is the total energy, the Hamiltonian for the internal motion is

$$H_0(\boldsymbol{r}) = \tfrac{1}{2} m \dot{\boldsymbol{r}}^2 + V(\boldsymbol{r}) \tag{8}$$

and the dot signifies time differentiation. The coordinates $\boldsymbol{R}$ and $\boldsymbol{r}$ describe the relative translational and internal motion respectively, μ and m are the collision and molecular reduced mass respectively and $V(\boldsymbol{r}, \boldsymbol{R})(V(\boldsymbol{r}))$ is the intermolecular (free molecule) potential. The time evolution of the collision is described by Hamilton's equations

$$\dot{\boldsymbol{p}} = -\nabla_{\mathrm{q}} H \tag{9a}$$

$$\dot{\boldsymbol{q}} = \nabla_{\mathrm{p}} H \tag{9b}$$

where $\boldsymbol{p}$ and $\boldsymbol{q}$ are the appropriate momenta and coordinates respectively.

The procedure in a classical trajectory calculation involves two steps:

1. For given initial conditions, $(\boldsymbol{p}_0, \boldsymbol{q}_0)$, propagate the equations to find the final quantities, $(\boldsymbol{p}_f, \boldsymbol{q}_f)$.
2. Vary all the initial conditions, subject to the conservation constraints, so as to cover the initially accessible phase space and then average.

A major advantage to classical techniques is that each trajectory can often be rapidly calculated. One drawback involves the need for a large number of trajectories to cover the phase space. This difficulty increases with the number of degrees of freedom. At the present time, the calculation of integral cross sections in atom–molecule and molecule–molecule systems is feasible but sometimes expensive. Basically, a classical trajectory treatment of complex systems requires a large amount of computational effort if state-selected observables are desired. The expense is greatly reduced when only less detailed information such as energy transfer moments is calculated. A more subtle problem involves the possibility of extracting quantum state-to-state quantities from the continuous classical distributions. Underlying all these features is the tacit assumption that classical mechanics gives an accurate description of the quantum mechanical world at the molecular level.

Investigations into these problems have been carried out. A reduction in the number of trajectories has recently been accomplished by use of a non-random sampling of initial conditions (Cheng, *et al.*, 1973; Suzukawa and Wolfsberg, 1978). Due to the difficulty (among others) in estimating the accuracy in non-random sampling procedures, it is not clear whether the technique is useful in general. The translation of classical distributions into quantum state transitions has been treated by two different methods. The simplest and most fundamental is the histogram binning technique (LaBudde and Bernstein, 1973; Alper *et al.*, 1978). Starting with a given internal state energy ε_i, a trajectory is calculated and the final internal energy determined. The ratio of the number of trajectories in which the final energy is between $\varepsilon_{f-1/2}$ and $\varepsilon_{f+1/2}$ to the total number of trajectories then yields the transition probability $P_{if}(b)$, where b is the impact parameter. Total cross sections are given by

$$\sigma_{if} = 2\pi \int_0^\infty P_{if}(b) b \mathrm{d}b \tag{10}$$

The other method assumes a parametrized form for σ_{if} and deconvolutes the classical energy transfer moments (Truhlar, 1976; Procaccia and Levine, 1976). We defer further discussion of this approach until Section VI on inversion procedures, where the topic arises naturally. At this time the histogram method seems to be the most accurate, at least for rotational transitions (Chapman and Green, 1977).

The argument that classical mechanics is accurate except for hydrides needs to be tested more completely. Treatment of various systems by classical and exact quantum methods using the same potential would be an ideal measure. Unfortunately, the quantal results are usually known only for hydrides or the lower quantum levels of heavier molecules. However, some comparisons for rotationally inelastic processes are available (LaBudde and Bernstein, 1973; Augustin and Miller, 1974; Brumer, 1974; Pattengill, 1975; Chapman and Green, 1977; Alper *et al.*, 1978). The classical and quantum results are in reasonable agreement provided that purely quantum symmetry effects are unimportant. By contrast, vibrationally inelastic collisions are not accurately described by classical mechanics except at very high energies (Miller, 1974). In addition, some caution is necessary in applying classical methods to the calculation of phase-sensitive quantities (Gordon, 1966; Smith *et al.*, 1976). It is clear that classical mechanics is a viable approach for treating many but not all types of collision phenomena.

We also mention that the ability to follow the time development of a collision is an important qualitative feature of classical trajectories. This aspect, which is useful in the detailed analysis of energy transfer (Barg and Toennies, 1977), chemical reactions, unimolecular decay, etc., is often obscured in fully quantum methods (Section IV.C) but retained in the semi-classical approaches.

B. Semiclassical Theory

The combination of quantal and classical mechanics can be accomplished in a variety of ways to yield semiclassical theories. One method due to Miller (1970, 1974), Marcus (1971) and Pechukas (1969) involves a complicated generalization of the JWKB approximation for elastic scattering. This approach, which has become known as classical S-matrix theory, was the subject of intensive activity in the early 1970's. The matrix elements of S are given by

$$S_{njm_j,n'j'm_{j'}} = \left[(-2\pi i)^2 \frac{\partial(n'j'm_{j'})}{\partial(q_n q_j q_{m_j})} \right]^{-1/2} \exp(i\Phi_{njm_j,n'j'm_{j'}} \hbar^{-1}) \tag{11}$$

where $\Phi_{njm_j,n'j'm_{j'}}$ is the classical action. The symbol $\partial(n'j'm_{j'})/\partial(q_n q_j q_{m_j})$ stands for the determinant of the Jacobian transformation taking the initial conjugate coordinates $(q_n q_j q_{m_j})$ into the continuous classical final values

$n_c j_c m_c$ evaluated at the stationary point $n_c j_c m_c = n'j'm_{j'}$. This stationary point is determined by varying the initial conjugate coordinates until all the final variables $n_c j_c m_c$ are integral. This requires a three-dimensional root search for *V–R, T* processes with the dimensionality increasing with the number of degrees of freedom. Clearly, kinematical approximations which effectively reduce the number of degrees of freedom would be extremely helpful in classical *S*-matrix theory.

The classical *S*-matrix approach has many advantages which include a micro-reversible *S*-matrix, a simple physical interpretation and the accurate prediction of classically forbidden processes (Miller, 1974). The major disadvantage is associated with the need for multidimensional root searches (Miller and George, 1972). For classically forbidden processes, this entails a root search in the complex plane which can be very time consuming. For classically allowed processes such as rotational transitions, the results of purely classical trajectories and classical *S*-matrix theory are very similar unless quantum symmetry effects are important (Augustin and Miller, 1974; Miller, 1974; McCurdy and Miller, 1977).

An alternate semiclassical approach involves a straightforward expansion of the total time-dependent wavefunction in the free molecule eigenfunctions $\varphi_{njm_j}(\boldsymbol{r})$ (Bates, 1962). A non-local set of coupled integrodifferential equations then results for the probability amplitudes $C_{njm_j}(t)$. Under the assumption of a classical translational path the non-local character is eliminated which yields

$$i\hbar C_{njm_j}(t) = \sum_{n'j'm_{j'}} \exp[i\omega_{nj,n'j'}t] V_{njm_j,n'j'm_{j'}}(\boldsymbol{R}(t)) C_{n'j'm_{j'}}(t) \tag{12}$$

where the transition frequencies are given in terms of the internal energy ε_{nj} as

$$\omega_{nj,n'j'} = (\varepsilon_{nj} - \varepsilon_{n'j'})\hbar^{-1} \tag{13}$$

and the potential matrix elements are

$$V_{njm_j,n'j'm_{j'}}(\boldsymbol{R}(t)) = \int \varphi^*_{njm_j}(\boldsymbol{r}) V(\boldsymbol{r}, \boldsymbol{R}(t)) \varphi_{n'j'm_{j'}}(\boldsymbol{r}) \mathrm{d}\boldsymbol{r} \tag{14}$$

The evaluation of this integral is facilitated by the expansion of the potential in equation (6). The time dependence of $\boldsymbol{R}(t)$ is governed by the classical equations of motion using quantum expectation values (i.e. Ehrenfest's theorem). It is standard but not necessary to determine the trajectory using only the spherical potential (Nielson and Gordon, 1973). Recently, the full potential has been used with good results for the He–H_2 (McCann and Flannery, 1975) and He–NH_3 (Boggs, 1978) systems. Two-molecule collisions have not been accurately treated using equation (11), to the best of our knowledge. Under the additional assumption that rotations are classical, equation (12) for *n* only has been applied to the HF–HF system (Poulsen and Billing, 1978).

Two features of equation (12) are important in applications: first, the number of coupled equations increases dramatically as higher rotor levels are included. Thus the solution of equation (12) is essentially as difficult as the solution of the corresponding fully quantal equations. Second, even an accurate solution of equation (12) violates microscopic reversibility (Augustin and Rabitz, 1978). Attempts to remove the latter difficulty have usually considered an average translational path (see, for example, Billing, 1975). In any event, this problem is not severe except in the threshold energy regime—which unfortunately is quite important for $V-T$ processes.

A different semiclassical approach involves the propagation of Gaussian wave packets (Heller, 1975) in which the spreading of the packet into other states is related to transition probabilities. Only applications to the collinear vibrational problem have been published. The extension to full three-dimensional problems would be interesting.

C. Quantum Mechanical Theory

In this section, the quantal formulation of molecular scattering is presented. Since the proliferation of indices in the formalism increases greatly with the complexity of the system (sometimes just keeping track of the indices seems as challenging as solving the coupled equations), we limit the treatment to atom–diatomic molecule collisions. More complex systems are not considered explicitly because almost all accurate calculations have been limited to atom–molecule systems. This points out that future work will probably focus on the treatment of larger systems—which allow more complicated questions and answers. However, note that within certain limitations it is now possible to handle real experimental systems; there are exceptions, but the time when collinear oscillator problems were of prime interest has passed.

The coordinate representation of the Schrödinger equation in the centre-of-mass system is

$$\left(-\frac{\hbar^2}{2\mu}\nabla_R^2 + H_0(\boldsymbol{r}) + V(\boldsymbol{r},\boldsymbol{R}) - E\right)\psi(\boldsymbol{r},\boldsymbol{R}) = 0 \tag{15}$$

where $\psi(\boldsymbol{r},\boldsymbol{R})$ is the total wavefunction. Expansion of $\psi(\boldsymbol{r},\boldsymbol{R})$ in the total angular momentum representation (Arthurs and Dalgarno, 1960; Lester, 1976) yields the set of coupled equations in matrix notation

$$\left(\mathbf{1}\frac{\mathrm{d}^2}{\mathrm{d}R^2} + \mathbf{k}^2 - \mathbf{l}^2/R^2 - 2\mu\hbar^{-2}\mathbf{V}^J(R)\right)\mathbf{U}^J(R) = 0 \tag{16}$$

where $\mathbf{U}^J(R)$ is the radial wavefunction,

$$(\mathbf{k})_{njl,n'j'l'} = (2\mu\hbar^{-2}(E - \varepsilon_{nj}))^{1/2}\delta_{nn'}\delta_{jj'}\delta_{ll'} \tag{17}$$

$$(\mathbf{l})_{njl,n'j'l'} = (l(l+))^{1/2}\delta_{nn'}\delta_{jj'}\delta_{ll'} \tag{18}$$

and

$$\begin{aligned}(\mathbf{V}^J(R))_{njl,n'j'l'} &= \sum_\mu f_\mu(jl, j'l'; J)\int \varphi^*_{nj}(r)V_\mu(r, R)\varphi_{n'j'}(r)r^2\,\mathrm{d}r \\ &\equiv \sum_\mu f_\mu(jl, j'l'; J)V^\mu_{nj,n'j'}(R) \end{aligned} \tag{19}$$

The Legendre polynomial expansion of the potential in equation (6) is used in equation (19). Also l is the orbital angular momentum, J is the total angular momentum, f_μ is a Percival–Seaton coefficient, and the internal wavefunctions were separated as $\varphi_{njm_j}(\mathbf{r}) = \varphi_{nj}(r)Y_{jm_j}(\hat{r})$. The connection to physical properties occurs through the S-matrix which is defined by the asymptotic form

$$\mathbf{U}^J(R) \overset{R\to\infty}{\longrightarrow} R\mathbf{k}^{-1/2}(h^*_l(\mathbf{k}R) - \mathbf{S}^J(E)h_l(kR))k^{-1/2} \tag{20}$$

where the Hankel function (Abramowitz and Stegun, 1965) matrix is given by

$$(h_1(\mathbf{k}R))_{njl,n'j'l'} = h_l(k_{njl,njl}R)\delta_{nn'}\delta_{jj'}\delta_{ll'} \tag{21}$$

The determination of $\mathbf{S}^J(E)$ is the basic problem (as it also is in the previously discussed semiclassical approach) since all observables are ultimately related to it. It is important to realize that sums over J and averages over scattering angle, energy, rotational state, etc. may be necessary to generate experimentally measurable quantities. For the calculation of bulk relaxation data, a Boltzmann velocity average of the cross sections and proper consideration of the kinetics is often required (Rabitz and Lam, 1975).

From a computational viewpoint, the solution of the infinite set of coupled equations (16) subject to the boundary conditions in equation (20) yields the desired S-matrix. Thus a natural goal is the development of efficient methods for solving equation (16). In practice, the S-matrix is calculated with a truncated set of coupled equations and the set is slowly increased until the S-matrix elements of interest are converged to within the desired accuracy. The tolerance will vary depending upon, among other factors, the accuracy of the intermolecular potential and the physical quantity to be calculated.

An exact solution of equation (16) involves a direct numerical integration of the coupled differential equations. In one method, a piecewise linear (Gordon, 1969) or quadratic (Rosenthal and Gordon, 1976) fit to the intermolecular potential is used. Over each interval, the local solution is a combination of either Airy functions (linear potential) or parabolic cylinder functions (quadratic potential). Matching of the derivatives and magnitudes of these local solutions allows the reconstruction of the total wavefunction and thus yields the S-matrix. If the potential is well approximated by a linear (or quadratic) function over a large interval (in R), *then only a few* intervals are necessary and Gordon's method is very efficient. This condition typically occurs for the long-range multipole potentials that are important in rotationally inelastic collisions. For vibrationally inelastic processes in atom–molecule systems, the potential is a strongly varying function of R (e.g. exponential). In this case methods which are based upon direct approximation of the wavefunction (rather than the potential) are often more appropriate. These include direct integration of equation (16) by use of the de Vogelaere algorithm (de Vogelaere, 1955) and numerical solution of the corresponding integral equation (Sams and Kouri, 1969). A technique that uses the R-matrix approach to scattering theory has also been developed recently (Light and Walker, 1976).

Rotationally inelastic processes in many atom–linear molecule systems have been treated using the above-mentioned numerical procedures. Two listings of studied systems are available in the reviews by Toennies (1976) and Rabitz (1976). Only a few calculations on more complex systems have been attempted with the most complete study occurring for H_2–H_2 collisions (Green, 1975a; Schaefer, 1979). In addition, limited results are available for the HF–HF system (DePristo and Alexander, 1977a). The only calculations on atom–symmetric top and atom–spherical top systems are those of Green (1975b) for He–NH_3 and Heil and Secrest (1979) for He–CH_4. The situation for V–R, T processes is much less satisfactory with the only accurate results occurring for the He–H_2 (Lin and Secrest, 1978) and Li^+–H_2 (Schaefer and Lester, 1975) systems. We emphasize that these results were obtained using neither dynamical nor kinematical approximations.

The previously outlined numerical procedures are inefficient when a large number of coupled equations must be solved to yield a converged S-matrix. Physically, this difficulty occurs when the total wavefunction $\psi(\mathbf{r}, \mathbf{R})$ is significantly perturbed from the molecular eigenstates because of the influence of $V(\mathbf{r}, \mathbf{R})$. Reactive scattering and strongly coupled V–R, T transitions, such as those proceeding through complex formation, are obvious examples but certainly not the only ones. Since the basis set expansion underlying equation (16) is slowly convergent, one approach is to expand the wavefunction in a locally optimal basis (Light and Walker, 1976) which leads to a smaller set of coupled equations. Another avenue of research eliminates all basis set expansions and instead focuses on the numerical solution of the partial differential equation (15) (Askar *et al.*, 1978; Rabitz *et al.*, 1978). The dimensionality of this PDE increases with the number of degrees of freedom. At the present time, it appears feasible to treat up to three coordinates efficiently. Perhaps a combination of a PDE for the strongly perturbed degrees of freedom and a locally optimal basis for the weakly perturbed channels would be most effective. Although a small set of coupled low-dimensional PDE's would have to be solved, this might be easier than solving either a high-dimensional PDE or a large number of coupled ordinary differential equations. Clearly, a variety of interesting questions and combinations could be considered. The investigation of numerical procedures should become increasingly important as the dynamics of larger and more complex systems are treated accurately.

An approximation to the exact solution of equation (16) is provided by perturbation theory. However, due to the strong-coupling nature of most $V-R, T$ processes, the results of first-order Born or distorted-wave Born approximation (DWBA) calculations are generally disappointing. The exponential approximation (Levine, 1971) attempts to include higher-order effects on the S-matrix by defining

$$\mathbf{S}^J = \exp(-i\boldsymbol{\eta}^J)\exp(-i\mathbf{A}^J)\exp(-i\boldsymbol{\eta}^J) \tag{22}$$

where

$$(\boldsymbol{\eta}^J)_{njl,n'j'l'} = \delta_{nn'}\delta_{jj'}\delta_{ll'}\eta^J_{njl} \tag{23}$$

is the phase shift due to the diagonal (Tarr and Rabitz, 1977; Bosanac and Balint-Kurti, 1975) or spherical potential (Levine, 1971). The coupling matrix is given by

$$\mathbf{A}^J = \mu\hbar^{-2}\mathbf{k}^{1/2}\left[\int_0^\infty \mathbf{U}^{J(\mathrm{DWBA})\dagger}(R)\mathbf{V}^J(R)\mathbf{U}^{J(\mathrm{DWBA})}(R)\mathrm{d}R\right]\mathbf{k}^{1/2} \tag{24}$$

where $\mathbf{U}^{J(\mathrm{DWBA})}(R)$ is the DWBA wavefunction. Equation (22) reproduces the perturbation theory solution to first order but also includes higher-order effects in an approximate manner and guarantees a unitary **S**-matrix. In the few available applications of this method the results were promising (Balint-Kurti, 1975; Bhattacharyya and Saha, 1978; Tarr and Rabitz, 1978). Further work along these lines would be quite useful; for instance, the application of these modifications of perturbation theory of two-molecule systems should greatly aid in understanding $R-R$ and $V-V$ processes.

V. KINEMATICS

For many problems, the numerical solution of either equations (7–9, 11, 12, 15) or (16) is prohibitively expensive and approximate mathematical solutions are either inaccurate and/or expensive. In addition, even when these solutions are feasible, they may contain a far greater amount of information than is desired for the ultimate observables. One approach to the resolution of these difficulties involves the use of Effective Hamiltonian (EH) procedures (Rabitz, 1976). These techniques involve either the partial or total elimination of dynamical degrees of freedom prior to calculation. Historically, rotational EH's which reduced the degeneracy of the rotational levels appeared first (Rabitz, 1972; Pack, 1974; McGuire and Kouri, 1974; DePristo and Alexander, 1976; Augustin and Rabitz, 1976b; Kouri *et al.*, 1976). We shall focus on the more recent developments and on the incorporation of EH methods with the various dynamical approaches (i.e. classical, semiclassical, and quantal).

Another approach to the mathematical and philosophical difficulties with exact calculations involves the replacement of the Schrödinger equation by a different and simpler fundamental relationship. This avenue of research has focused on the combination of dynamics with non-equilibrium statistical mechanics to handle complex molecular systems that have a high density of states. An important feature of both the statistical dynamical and EH methodology is the approximate physical modelling of the collision. This should be contasted with the mathematically approximate solutions to the exact classical, semiclassical, and quantal equations given in Section IV.

A. Effective Hamiltonian Methods

The recent review by Rabitz (1976) exhaustively presents many of the developments in rotational EH theory. In order not to duplicate that article, we assume the reader has some familiarity with basic EH theory. Two of these methods are inherently defined only in terms of matrix elements of the potential and thus have no classical analogue. First, the effective potential, EP, (Rabitz, 1972, 1976) preaverages over the degenerate m_j states to yield the equations

$$V_{nj,n'j'}(R) = \sum_{\mu} V^{\mu}_{nj,n'j'}(R) f^{\mathrm{EP}}_{\mu}(j,j') \tag{25}$$

where

$$f^{\mathrm{EP}}_{\mu}(j,j') = \exp[i\pi/2(|j-j'|+j+j')]([j][j']/[\mu]^2)^{1/4}\begin{pmatrix} j & j' & \mu \\ 0 & 0 & 0 \end{pmatrix} \tag{26}$$

the effective states are $|j\rangle$ and l is a good quantum number. Since all m_j effects have been removed, the dimensionality of the quantum-coupled Equation (16) is reduced to the *minimum*—provided both n and j are retained. Second, the decoupled l-dominant method, DLD (DePristo and Alexander, 1976), focuses on both the importance of the centrifugal barrier in long-range collisions and on the coupling strength to replace the coefficients in equation (19) by

$$f^{\mathrm{DLD}}_{\mu}(j\lambda, j'\lambda; J) = f_{\mu}(j(J-j+\lambda), j'(J-j'+\lambda); J) \tag{27}$$

where

$$|\lambda| \leq 2\min(j,j') \tag{28}$$

This decoupling in l does not completely eliminate degeneracy effects but instead implies a correlation between m_j changes (Alexander and Dagdigian, 1977). Consequently, within the DLD approximation, a few sets of coupled equations of reduced dimensionality must be solved. We mention that these methods are appropriate for different types of collisions: the EP works well for weakly anisotropic systems (Green, 1975b) while the DLD is most accurate for strong long-range anisotropies (DePristo and Alexander, 1976, 1977a). Note that the EDWBA can be used along with either of these (and other) dimensionality reducing methods. This combination of dynamical and kinematic approximations will probably be necessary for the study of large molecular systems.

The most accurate and widely applied EH procedure is the coupled states, CS, (McGuire and Kouri, 1974; Pack, 1974) approximation. The basic idea involves the replacement of the centrifugal barrier in equation (16) by the effective value $\bar{l}(\bar{l}+1)$ where $\bar{l}$ is independent of n and j. The wavefunction is transformed according to the prescription (Kouri *et al.*, 1976; DePristo and Alexander, 1977b)

$$\mathbf{F}^J(R) = \mathbf{B}^J\mathbf{U}^J(R)(\mathbf{B}^J)^\dagger \tag{29}$$

where

$$(\mathbf{B}^J)_{nj\Omega,n'j'l} = \delta_{nn'}\delta_{jj'}(-1)^{\Omega}[l]^{1/2}\begin{pmatrix} l & j & J \\ 0 & \Omega & -\Omega \end{pmatrix} \tag{30}$$

to give the equations

$$\left\{\mathbf{1}\left(\frac{d^2}{dR^2} - \bar{l}(\bar{l}+1)/R^2\right) + \mathbf{k}^2 - \mathbf{V}(R)\right\}\mathbf{F}^J(R) = 0 \tag{31}$$

Th index Ω is a projection along an appropriate spatial axis which will be considered in more detail later. The potential which is diagonal in Ω becomes

$$\begin{aligned}(\mathbf{V}(R))_{nj\Omega,n'j'\Omega'} &= (\mathbf{B}^J\mathbf{V}^J(R)(\mathbf{B}^J)^\dagger)_{nj\Omega,n'j'\Omega'} \\ &= \delta_{\Omega\Omega'}\sum_{\mu} V^{\mu}_{nj,n'j'}(R)(-1)^{\Omega}([j][j'])^{1/2}\begin{pmatrix} j & j' & \mu \\ 0 & 0 & 0 \end{pmatrix}\begin{pmatrix} j & j' & \mu \\ \Omega & -\Omega & 0 \end{pmatrix}\end{aligned} \tag{32}$$

The asymptotic solution of equation (31) then yields the CS S-matrix $S^{l,\Omega}_{nj,n'j'}$.

The role of $\bar{l}$ in equation (31) is identical to that of an impact parameter in the time-dependent semiclassical formalism. Thus a classical path approximation analagous to the CS method is possible. The independent development of the classical path CS(CPCS) was presented by Smith, Giraud, and Cooper (1976) and termed the 'peaking' approximation because of the basis for their derivation. The latter development uses the classical path–internal quantum state approach, equation (12) and assumes that the time dependence of the angular variable $\hat{R}(t)$ is very slow in the transition region and thus can be replaced by $\hat{R}_0$. After letting $\boldsymbol{R}(t) = (\hat{R}_0, R(t))$ and using equation (6), the potential matrix elements become

$$V_{njm_j,n'j'm_{j'}}(\hat{R}_0, R(t)) = \sum_{\mu} \int \varphi^*_{njm_j}(\boldsymbol{r}) V_\mu(r, R(t)) P_\mu(\bar{r}, \hat{R}_0) \varphi_{n'j'm_{j'}}(\boldsymbol{r}) \mathrm{d}\boldsymbol{r} \tag{33}$$

The *space-fixed* angle $\hat{R}_0$ in the original unprimed coordinate system now lies along the z'-axis of a new *space-fixed* coordinate system as shown in Fig. 2. We emphasize that these are space-fixed and not body-fixed coordinate systems. In the primed space-fixed coordinate system, equation (33) is diagonal in the projection quantum number Ω. The relationship between the transition amplitudes is (Heil, 1978)

$$C^{n_0 j_0 m_{j0}}_{njm_j}(t) = \sum_{\Omega} D^j_{m_j \Omega}(\hat{R}_0) C^{n_0 j_0 \Omega}_{nj\Omega}(t) D^{j_0}_{m_{j0}\Omega}(\hat{R}_0)^* \tag{34}$$

where $D(\hat{R})$ is a rotation matrix (Edmonds, 1974) and the initial state $n_0 j_0 m_{j0}$ is explicitly shown. Equation (34) shows that changes in m_j in the original coordinate system depend upon the 'peaking' angle $\hat{R}_0$. However, $\hat{R}_0$ depends upon $\bar{l}$ through the impact parameter $b = (\bar{l} + \frac{1}{2})(\mathbf{k})^{-1}_{nj\bar{l},nj\bar{l}}$ and is thus dependent upon the *dynamics of the collision*. For degeneracy averaged cross sections, this difficulty is not of concern because the rotation matrix elements sum to unity. By contrast, cross sections for m_j changes depend upon the dynamics through *both* $\hat{R}_0$ and $\bar{l}$. This fact is important in the interpretation of the time-independent quantal formulation equation (31).

A point of some controversy in the quantal CS, equation (31), involves the choice of the value of $\bar{l}$, which has been identified as the total angular

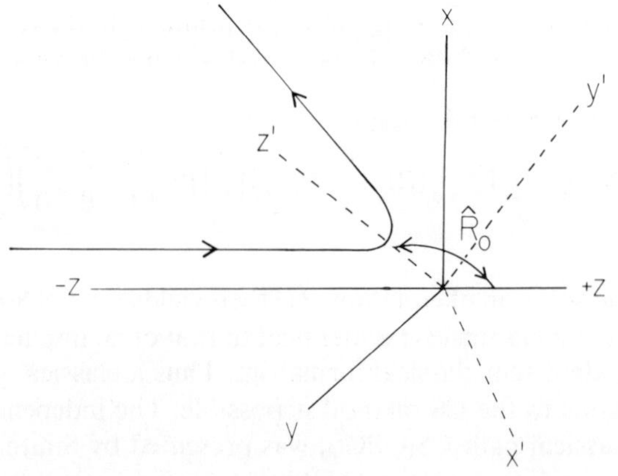

Fig. 2. Rotation of the unprimed coordinate system to the primed coordinate system as prescribed by the 'peaking' approximation.

momentum J (Pack, 1974; DePristo and Alexander, 1977b), the final orbital angular momentum l' (Shimoni and Kouri, 1977; Parker and Pack, 1977), and the initial orbital angular momentum l (Khare, 1977; Fitz, 1977; Schinke and McGuire, 1978). The reason for the disagreement is evident in the relationship analogous to equation (29).

$$S^J_{njl,n'j'l'} = \sum_{\Omega}([l][l'])^{1/2}\begin{pmatrix} l & j & J \\ 0 & \Omega & -\Omega \end{pmatrix}\begin{pmatrix} l' & j' & J \\ 0 & \Omega & -\Omega \end{pmatrix} S^{\bar{l},\Omega}_{nj,n'j'} \tag{35}$$

where the phase of S^J is obviously dependent upon the choice of $\bar{l}$. Thus the phase-sensitive quantities such as the differential and pressure-broadening cross sections are strongly dependent upon the choice of $\bar{l}$ (Green *et al.*, 1977). Note that the choice $\bar{l} = l$ or l' does not yield a unitary S-matrix in the (njl) basis. Although this might be considered strong evidence for the choice $\bar{l} = J$, it seems clear from recent work (Green *et al.*, 1977; Tarr *et al.*, 1977) that this assignment is incorrect. The choice $\bar{l} = l'$ leads to the conclusion that changes in m_j are forbidden (Shimoni and Kouri, 1977; Khare, 1977) while $\bar{l} = l$ does not. Experimental evidence for m_j conservation exists for the I_2–I_2 system (Jeyes *et al.*, 1977) while exact calculations on the HCl–He system yield results that are better approximated using $\bar{l} = l$ (Monchick and Kouri, 1978). In addition, the dynamical origin of m_j changes exhibited in equation (34) implies that strictly kinematic factors such as in equation (35) cannot completely account for reorientations. A complicating factor is that the use of equation (35) with $\bar{l} = l$ or l' yields the same formula for pressure-broadening cross sections. It may be that different assignments of $\bar{l}$ work best for various systems and different physical observables.

The purely classical analogue of the CS approximation has been developed (Augustin and Rabitz, 1976b; Tarr *et al.*, 1977) by transforming the Hamiltonian equation (7) to the body-fixed frame and neglecting the Coriolis coupling. A similar classical technique involves the replacement of the orbital angular momentum, $\boldsymbol{l}$, by its planar value (Pattengill, 1977). Both of these methods reduce the number of degrees of freedom and thus make the classical calculations somewhat less expensive. To judge from the many available applications and in agreement with theoretical analysis (Kouri *et al.*, 1976; DePristo and Alexander, 1977b), the CS approximation should be reasonable for systems dominated by short-range anisotropic forces.

The sudden approximation assumes that the time scale of the collision is small compared to some characteristic internal motion time. We include it under EH procedures because nearly all of the applications have used the infinite order sudden, IOS, approximation, which can be derived from the CS method by assuming $\varepsilon_j = 0$ (Secrest, 1975; Goldflam *et al.*, 1977b). In the case where rotations are sudden but vibrations are not then the IOS

S-matrix is

$$S^{\bar{l},\Omega}_{nj,n'j'} = \int Y^*_{j\Omega}(\theta,\varphi) S^{\bar{l}}_{nn'}(\theta) Y_{j'\Omega}(\theta,\varphi)\,\mathrm{d}(\cos\theta)\,\mathrm{d}\varphi \tag{36}$$

where $S^{\bar{l}}_{nn'}(\theta)$ is determined by solving the quantal or semiclassical coupled vibrational equations at a fixed angle θ. As in the CS method, the choice of $\bar{l}$-labelling is somewhat arbitrary but $\bar{l}=l$ or l' yields identical degeneracy averaged differential and total cross sections and pressure-broadening cross sections. An important advantage of the sudden approximation is the ability to relate different S-matrix elements (Goldflam, Green, and Kouri, 1977b) by

$$S^{\bar{l},\Omega}_{nj,n'j'}(E) = (-1)^{\Omega}([j][j'])^{1/2} \sum_L [L]^{1/2} \begin{pmatrix} j & j' & L \\ 0 & 0 & 0 \end{pmatrix} \begin{pmatrix} j & j' & L \\ \Omega & -\Omega & 0 \end{pmatrix} S^{\bar{l},\Omega}_{nL,n'0} \tag{37}$$

Equation (37) is derived from equation (36) by expanding $S^{\bar{l}}_{nn'}(\theta)$ in Legendre polynomials and identifying the special result for $j'=0$. The use of the scaling relationship in equation (37) allows for a simplification of and interrelationship between various types of cross sections. This can lead to a greater understanding of the relationship among various experimental quantities. Unfortunately, these developments are only valid when the sudden approximation is accurate—a situation that typically is limited to light reduced-mass collision systems, small rotational energy spacings, and short-range potentials. Thus the use of equation (37) to analyse and invert experimental data has not been possible (DePristo and Rabitz, 1978c; Ramaswamy *et al.*, 1979b).

A more general scaling relationship appropriate for many inelastic processes has been derived (DePristo *et al.*, 1979) by considering both the energetic level spacing and the finite collision duration. For R–T processes in neutral atom–molecule systems, the result at a given impact parameter b is

$$S^{(b)}_{jm_j,j'm_{j'}}(E_k+\varepsilon_j) = (-1)^{m_j}([j][j'])^{1/2} \sum_{Lm_L} [L]^{1/2} \begin{pmatrix} j & j' & L \\ 0 & 0 & 0 \end{pmatrix} \times \begin{pmatrix} j & j' & L \\ m_j & -m_{j'} & -m_L \end{pmatrix} A^j_L S^{(b)}_{LM_L,00}(E_k+\varepsilon_L) \tag{38}$$

where the adiabaticity factor is given by

$$A^j_L = \frac{6+(\omega_{LL-1}b/2v)^2}{6+(\omega_{jj-1}b/2v)^2} \tag{39}$$

and ω_{jj-1} is the transition frequency for the $i \to i-1$ transition. Equation (38) is applicable to downward transitions because the proper energetic considerations are incorporated by evaluating the S-matrices at the same kinetic energy above the initial state. This scaling relationship still allows for a simplification of and interrelationship between different cross sections. In essence, equation (38) can be used even when the collisions are not sudden and reduces to equation (37) when the sudden approximation holds. However, some of the phase relationships displayed by the IOS result, equation (37), are destroyed.

The EH methods detailed previously are concerned with the problem of m_j degeneracy of the rotational levels. In many experiments, information about the rotational states j is also not determined because the $R-T$ transitions are much faster than the $V-T$ transitions. A generalized EH (GEH) procedure (DePristo and Rabitz, 1978b) has recently been derived which preaverages over rotations prior to calculation of vibrational transitions. In this case, the classical path equations (12) would be used for n changes only with the potential

$$V_{nn'}(\boldsymbol{R}(t)) = \sum_{\substack{jm_j \\ j'm_{j'}}} \exp[i(\omega_{nj,n'j'} - \omega_{n0,n'0})t] f_{jm_j}^{n*}(t) V_{njm_j,n'j'm_{j'}}(\boldsymbol{R}(t)) f_{j'm_{j'}}^{n'}(t) \tag{40}$$

where $f_{jm_j}^{n}(t)$ is the amplitude for rotational state changes in the vibrational level n. The time dependence of $f_{jm_j}^{n}(t)$ is determined by solution of equation (12) with $n' = n$. Thus for $n = 0, \ldots n^*$, there are $n^* + 1$ rotational problems to be handled along with the purely vibrational equation (40). This is considerably easier than treating the full rotation–vibration coupled equations. In addition, the rotational transitions could be approximated which would yield additional savings. Applications of the GEH method are not yet available; thus it is not possible to assess the accuracy or practical utility of this approach. A similar technique to this generalized EH approach has been developed and applied based upon the optical potential theory (Gerber *et al.*, 1978; Zaristsky *et al.*, 1978). The results for the He–H_2 and He–N_2 systems were encouraging. Within the classical S-matrix formalism, the partial averaging method (Doll and Miller, 1972) also eliminates the rotational states before calculations. This effectively reduces equation (11) to only n, q_n variables. All of these techniques can be applied to systems with many rotational states but at the price of eliminating information about $R-T$ and $V-R$ processes.

B. Stochastic Methods

The treatment of complex molecular systems poses a formidable computational task because of the large number of internal states that play a role. This very fact suggests that the transitions between these densely packed states could be modelled by a stochastic process. The implementation of this idea has followed along two very different lines. The first, based upon a complete neglect of the phase information in the probability amplitudes, leads to a master equation for the probabilities (Augustin and Rabitz, 1976a, 1977, 1979). The second, based upon the Zwanzig projection operator formalism in non-equilibrium statistical mechanics (Zwanzig, 1961), leads to a pair of coupled Liouville equations—one for the variables of interest and another for the 'bath' variables (Schatz *et al.*, 1977; Eu, 1978). The latter approach is quite similar in spirit to the GEH method described previously. In fact, time-scale arguments are usually made to justify the separation into 'interesting' and 'bath' variables. Before briefly describing the two formulations, we mention a few important points. First, many of the EH techniques can be combined with these stochastic theories to reduce the number of equations to be solved. Second, one interesting use for stochastic methods involves examining the deviation from statistical behaviour in molecular collisions which has been generally approached within the information theoretic formalism (Levine and Bernstein, 1977). The two methods differ radically: information theory examines the constraints (usually just energy transfer) in a system and has typically been applied as a fitting procedure to *characterize* the deviation from a purely statistical result (see Alhassid and Levine, 1977; Eu, 1978 for different uses). By contrast, stochastic theory involves dynamical calculations with an intermolecular potential energy surface and is thus capable of *predictions* about the deviation from statistical behaviour.

The stochastic theory due to Augustin and Rabitz (1976a) leads to a non-local finite difference master equation for the transition probabilities. In practice, this equation was localized by using a classical translational path. The derivation is involved and we just state the pertinent result as

$$P_{njm_j}(t+\tau) = \sum_{n'j'm_{j'}} A_{njm_j,n'j'm_{j'}}(t+\tau, t) P_{n'j'm_{j'}}(t) \tag{41}$$

where the coefficient matrix $\mathbf{A}$ is related to the intermolecular potential. An important feature of equation (41) is that the probabilities P_{njm_j} are guaranteed to be positive. The time interval τ is chosen such that $t+\tau$ is the first time after t such that a random phase approximation in the probability amplitudes (see equation 12) holds. In practice, τ is estimated to be the minimum time long enough for a classical transition between adjacent quantum states. Extensions of this method such as using different τ values for $V-T$ and $R-T$ processes and approximating equation (41) by a Fokker–Planck equation have been considered (Ramaswamy *et al.*, 1979a). Finally, we mention that as in all classical path methods, equation (41) finds the cross sections out of each initial state separately; the use of the probability scaling relationship analogous to equation (38) is then particularly advantageous.

A different stochastic theory (Schatz *et al.*, 1977) results from the application of time-dependent projection operators (Willis and Picard, 1974) to the total density matrix $\rho(t)$. Letting $\rho_I(t)$ and $\rho_B(t)$ be the density matrices for the 'interesting' and 'bath' variables respectively, then the equations are

$$\frac{\partial \rho_I}{\partial t} = -iL_I^0\rho_I - i\langle L'\rangle_B\rho_I - \int_{-\infty}^{t} dt' K_{BI}(t, t')\rho_I(t') \tag{42}$$

$$\frac{\partial \rho_B}{\partial t} = -iL_B^0\rho_B - i\langle L'\rangle_I\rho_B - \int_{-\infty}^{t} dt' K_{IB}(t, t')\rho_B(t') \tag{43}$$

The total Liouvillian operator L is given by

$$L = L_I^0 + L_B^0 + L' \tag{44}$$

where L' is the interaction term, and

$$\langle L'\rangle_B = \mathrm{Tr}_B(L'\rho_B) \tag{45}$$

where Tr_B signifies a trace over the 'bath' variables. The memory kernels in equations (42) and (43) are complicated functions of L and the other variables (Schatz *et al.*, 1977). We emphasize that equations (42) and (43) are exact and non-local; their accurate solution is equivalent to that of equation (15). In contrast to equation (41) which works with the probabilities, equations (42) and (43) quickly varying phases and thus are just as difficult to solve as the exact quantal equation (15). Therefore, the crucial step involves their approximate solution, which is not in general guaranteed to conserve probability and can even lead to negative probabilities. Another important consideration is that the neglect of the effect of the 'interesting' variables on the 'bath' in equation (42) leads to the identification of $\langle L'\rangle_B$ as an EH.

From the above discussion it is clear that the two stochastic theories differ in basic interpretation. The formulation in equation (41) describes the *flow of probability* between internal states by a Pauli-type master equation which guarantees conservation of probability (or flux). An important feature involves the 'rates' $A_{njm_j,n'j'm_{j'}}$ which are not given by perturbation theory and thus allow equation (41) to handle strongly coupled systems. The physical assumptions behind the derivation of equation (41) are similar to those used for macroscopic systems (Zwanzig, 1961) except that ensemble averages and the convenient time-scale separation do not appear. In essence, equation (47) disregards all phase information and thus describes an *irreversible* transition from initial to final states. It is stochastic in that sense only—there being no bath or reservoir. By contrast equations (42) and (43) are stochastic only in the sense that a division into 'interesting' and 'bath' variables is involved, and they must be further approximated before being used in any practical applications. It clearly would be advantageous to combine the ideas of a 'bath' variable and those of the master equation flow of probability formulations. This might be accomplished through a projection operator formalism, a GEH technique or, possibly, an optical theory (Micha, 1976). In any event, a simple transposition of methods appropriate for macroscopic systems will almost certainly be inappropriate without special considerations for the *microscopic* collisional process.

VI. INVERSION OF EXPERIMENTAL DATA

In the previous two sections, many recent developments in the *a priori* theoretical treatment of $V-R, T$ processes were reviewed. Although these methods differ in various respects, the reliability of their predictions depends upon the availability of an accurate intermolecular potential—the calculation of which is also a formidable computational task. If a wide range of impact parameters, a large number of transitions, and a thermal average are necessary for the generation of experimentally measurable quantities, the expense of this direct approach can be enormous. Of course, once all of these transition probability amplitudes are available they can be used to calculate many quantities of experimental interest. Nevertheless, this sensitivity to the potential is a prime motivation for seeking alternate ways to investigate $V-R, T$ processes in detail.

The brief outline of experimental methods in Section II shows that an inherent loss of detail accompanies the transition from state-selected crossed molecular beam measurements to totally thermal averaged bulk relaxation experiments. Intermediate in complexity are the partially state-selected data such as measured in T_1 or T_2 experiments (Liu and Marcus, 1975). For the purpose of this review an inversion procedure is defined to be any method by which the more detailed underlying information is recovered directly from less detailed measurements. At first sight this task does not seem possible; however by combining a proper amount of theory with the experimental observations, the underlying information can be extracted.

A few illustrations suffice to show the generality of this definition of inversions. First, the finite angle and energy resolution in a molecular beam apparatus imply that the centre-of-mass (C.M.) angular scattering pattern is a convolution of the corresponding lab distribution with the velocity distribution of the beam (Pauly and Toennies, 1965); the determination of the C.M. angular distribution from the measured lab distribution (Siska, 1973; Creaser *et al.*, 1973; Shapiro *et al.*, 1977) thus entails an inversion procedure. Second, the determination of the spherically symmetric intermolecular potential from the elastic C.M. differential cross section involves an inversion (Miller, 1969; Buck and Pauly, 1971; Shapiro and Gerber, 1976; Duquette *et al.*, 1978; Siska, 1979). Third, the determination of the state-to-state rotationally inelastic rates from spectral linewidths (a measure of the total inelastic rate) necessitates an inversion technique (DePristo and Rabitz, 1978a). All of these examples involve different inversion methods, but in each case some degree of theoretical analysis is needed to extract the underlying detailed information.

For $V-R, T$ processes, a stable and unique inversion procedure for obtaining the total intermolecular potential even from perfectly accurate state-selected differential cross sections does not exist at this time and may not be possible (Devaney, 1978). As a practical alternative, a potential form with adjustable parameters is postulated and used to calculate the quantities of experimental interest. Disagreement between the calculated and measured values is reconciled by appropriate adjustment of the potential parameters. Using well-guided physical intuition to choose the initial form of the potential and with a little luck in adjusting the parameters, a reasonably accurate potential may be obtained in this manner (Dunker and Gordon, 1978; Holmgren *et al.*, 1978). This process can be prohibitively expensive if many adjustments of the potential are necessary of if a complex system with many states is being analysed. Two philosophical objections are also possible: first, and most important, the choice of the functional form can bias the results so that the fitted and true potentials differ substantially (Neilson and Gordon, 1973; Schafer and Gordon, 1973; Green and Thaddeus, 1976). Second, adjustments of the potential to incorporate the results of every experiment make a test of the theoretical predictions impossible. In Section VII, we consider a practical theoretical approach for determining the sensitivity of collision cross sections to the potential parameters.

In bulk relaxation experiments, the measured relaxation times result from a complicated convolution of dynamical and kinetic effects. Because of the tendency for multiple quantum rotational transitions, the extraction of the individual rates by the inversion of the experimental results has generally been considered an intractable problem (Carrington, 1961). Recently, it has been shown that an *a priori* scaling relationship for rotational rates is the fundamental theoretical development necessary for an inversion (DePristo and Rabitz, 1978a). The need for such a relationship to achieve an inversion has long been known; however, in the past the functional form of the relationship was *postulated* in terms of adjustable parameters that were determined by fitting to the data. This procedure is a test of the functional form rather than a measure of the physical information in the data. By contrast, the use of a physically justifiable *a priori* relationship reduces the number of unknown rates to a small number which can be determined from the experimental data. Since the utility of the results depends upon their sensitivity to the experimental uncertainty, it is also necessary to determine the stability of the inversion procedure. The application to linewidth measurements (DePristo and Rabitz, 1978a, c) yields the equation

$$W_{\lambda}(n_a j_a | n_b j_b) = \sum_{\Delta} C_{\Delta}(n_a j_a | n_b j_b) K_{0\Delta}(T) \tag{46}$$

where W_λ is the experimentally measured linewidth for the $n_a j_a \rightarrow n_b j_b$ spectral transition and the C_Δ are known coefficients. Equation (46) is a set of linear equations for the fundamental unknown rates $K_{0\Delta}$ which can be solved subject to the inequality constraint $K_{0\Delta} \geq 0$. An analysis of the propagation of experimental error from the widths W_λ to the deconvoluted rates $K_{0\Delta}$ shows that this inversion procedure is stable and yields physically meaningful results.

A simple illustration of the importance of error analysis occurs in the inversion of classical energy transfer moments to yield state-to-state cross sections (Truhlar, 1976; Procaccia and Levine, 1976; Chapman and Green, 1977). For simplicity we may assume that some set of cross sections is *exactly* given by

$$\sigma_{0n} = \pi b_{\max}^2 \exp(\alpha + \beta n) \tag{47}$$

where $b_{\max}$ is the maximum impact parameter in the classical trajectory calculation (see Section IV.A). Although there is little theoretical justification for the parametrization of the cross sections in equation (47), this is not relevant for the present purpose of illustrating the importance of statistical analysis in inversions. For the sake of illustration equation (47) is considered to be an accurate description of the state-to-state cross sections and its use in inversions is detailed. We consider $V - T$ processes and assume that the first two energy transfer moments are available from a classical calculation. The extraction of the individual cross sections involves solving the following equations for α and β

$$1 = \sum_{n=0}^{\infty} \sigma_{0n}/\pi b_{\max}^2 = e^{\alpha}(1 - e^{\beta})^{-1} \tag{48}$$

$$\varepsilon/\hbar\omega = \sum_{n=0}^{\infty} n\sigma_{0n}/\pi b_{\max}^2 = e^{(\alpha+\beta)}(1 - e^{\beta})^{-2} \tag{49}$$

where ε is the (classically calculated) first energy transfer moment and ω is the harmonic oscillator frequency. Assuming Gaussian random error in ε yields

$$(\Delta\sigma_{0n}/\sigma_{0n}) = (\Delta\varepsilon/\varepsilon)(\varepsilon + n)(1 + \varepsilon)^{-1} \tag{50}$$

where $\Delta\varepsilon(\Delta\sigma_{0n})$ is the fractional error in the average energy transfer (resulting cross sections). Equation (50) explicitly shows that the cross sections for multiple quantum transitions are less accurate than the input moments. This fact should be considered in attempting to find small transition probabilities from classical moments. In this regard, we emphasize that equation (50) only considers the uncertainty in the cross sections due to error in the input moments. The inaccuracy of the assumed functional form equation (47) will also contribute to the error in σ_{0n}. Unfortunately the latter difficulty inherently occurs whenever a parametrized form is used and there is no simple way of assessing this additional error.

Aside from the above example, all the inversion procedures aim to extract valuable data directly from laboratory observations. This should be an increasingly important realm of study and it provides a natural area for contact between experiment and theory.

VII. SENSITIVITY ANALYSIS IN MOLECULAR DYNAMICS

Scattering theory can be viewed along with most any well-defined physical problem in terms of an input/output structure (Tomovic and Vukabratovic, 1972). This situation is simply depicted in Fig. 3. The input is the Hamiltonian which contains parameters $\alpha_1, \alpha_2, \ldots$ and the output O is an implicit (and usually unknown) function of these parameters. Typically these parameters will reside in the intermolecular potential which is the chief 'unknown' in most collision problems. For classical mechanics, the initial conditions on the momenta and coordinates for the trajectories can also be considered as parameters. The goal of sensitivity analysis is to establish the sensitivity of the output scattering information with respect to the input parameters. This goal has always been implicit in any analysis of experimental data that attempts to fit the observations to an intermolecular potential. It is clear that general parameter-sensitivity questions deserve careful attention.

In the simple limit where the dynamics reduce to using first-order perturbation theory, the connection between the input and output is straightforward. Unfortunately molecular scattering processes are often nonperturbative; realistic problems usually entail strong coupling, at least

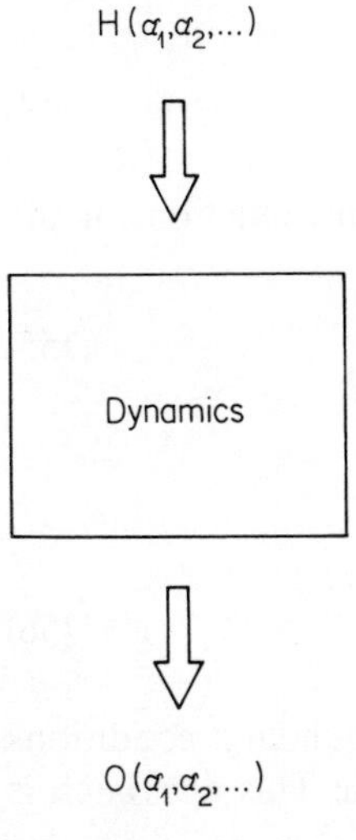

Fig. 3. Schematic illustration of the input–output nature of scattering theory. The input Hamiltonian as a function of the parameters $\alpha_1, \ldots \alpha$ is transformed into the output by the dynamics.

during a portion of the collision. In this case, the dynamics part of Fig. 3 plays the function of a scrambler of the input parameters and their detailed role. This typical situation is exactly where a systematic mathematical sensitivity treatment would be most useful. The same point is evident from work on extending the Hellman–Feynman theorem (Levine, 1966). For simplicity, we shall focus here on the sensitivity of the S-matrix, since the sensitivity of any measurables can be obtained from quantity (Hwang and Rabitz, 1979). The sensitivities are defined as

$$\mathbf{S}_{\alpha_i} = \frac{\partial}{\partial \alpha_i}\mathbf{S}, \mathbf{S}_{\alpha_i \alpha_j} = \frac{\partial^2}{\partial \alpha_i \partial \alpha_j}\mathbf{S}, \ldots \tag{51}$$

where α_i and α_j are arbitrary system parameters. The second- and higher-order mixed partial derivatives may be viewed as correlation terms since they are a measure of the joint sensitivity to more than one parameter. Often the first-order or elementary sensitivity coefficient is sufficient for many purposes, and only this case is discussed here. In a parallel fashion a classical mechanical sensitivity analysis can be developed.

There are several ways of calculating $\mathbf{S}_{\alpha_i}$ depending on how the dynamics are performed. Consider first the most rigorous treatment of working with the close-coupled scattering equations in matrix form

$$(E\mathbf{1} - \mathbf{H})\boldsymbol{\psi}(R) = 0 \tag{52}$$

The operator $\mathbf{H}$ is the usual matrix differential operator in the close-coupling equations of Section IV. Taking the partial derivative of these equations with respect to an arbitrary parameter α_i we obtain

$$(E\mathbf{1} - \mathbf{H})\boldsymbol{\psi}_{\alpha_i}(R) = \mathbf{H}_{\alpha_i}\boldsymbol{\psi}(R) \tag{53}$$

where

$$\boldsymbol{\psi}_{\alpha_i} = \frac{\partial}{\partial \alpha_i}\boldsymbol{\psi} \quad \text{and} \quad \mathbf{H}_{\alpha_i} = \frac{\partial}{\partial \alpha_i}\mathbf{H} \tag{54}$$

This equation may be solved by defining the full Green's function for the problem.

$$(E\mathbf{1} - \mathbf{H})\mathbf{G}(R, R') = \delta(R - R')\mathbf{1} \tag{55}$$

This leads to the solution

$$\boldsymbol{\psi}_{\alpha_i}(R) = \int_0^\infty dR' \mathbf{G}(R, R')\mathbf{H}_{\alpha_i}(R')\boldsymbol{\psi}(R') \underset{R\to\infty}{\longrightarrow} \boldsymbol{\psi}^{\text{out}}(R)\,{}^{\text{t}}{}_{\alpha_i} \tag{56}$$

where the latter asymptotic limit results from the boundary conditions to equation (53) above and $\boldsymbol{\psi}^{\text{out}}$ is an outgoing solution. This approach is not as formal or difficult as it might seem since $\mathbf{G}(R, R')|_{R\to\infty}$ can be expressed exclusively in terms of the *already available* solutions to the original Schrödinger equation (52). Therefore the sensitivities may be evaluated simultaneously while solving the usual scattering equations. This procedure has not been implemented at this time, but the feasibility clearly exists.

Perhaps the simplest approach to sensitivity theory is through the exponential form of the S-matrix. Let $\mathbf{S} = \exp[i\mathbf{A}]$ where $\mathbf{A}^\dagger = A$ is an appropriate Hermitian matrix. Exponential forms of this type were discussed in Section IV above. Now consider $\mathbf{S}(x) = \exp[ix\mathbf{A}]$ where $\mathbf{S}(x)$ satisfies the differential equation

$$\frac{\partial \mathbf{S}}{\partial x} = i\mathbf{A}\mathbf{S}(x),\ S(0) = 1 \tag{57}$$

and $\mathbf{S} \equiv \mathbf{S}(1)$. Take the partial derivative of this equation with respect to α_i to obtain

$$\left(\mathbf{1}\frac{\partial}{\partial x} - i\mathbf{A}\right)\mathbf{S}_{\alpha_i}(x) = i\mathbf{A}_{\alpha_i}\mathbf{S}(x) \tag{58}$$

Again applying the Green's function idea, the solution of this latter equation is simply given by

$$\mathbf{S}_{\alpha_i} = i\mathbf{S}(1)\int_0^1 \mathbf{S}(-x)\mathbf{A}_{\alpha_i}\mathbf{S}(x)\mathrm{d}x \tag{59}$$

This result is only useful if $\mathbf{A}_{\alpha_i} = \partial\mathbf{A}/\partial\alpha_i$ is known. Fortunately this quantity is known in the usual exponential S-matrix approximation. For example, taking $\mathbf{A} = \mathbf{V}$, as in the exponential Born approximation, we get a very simple result from equation (59). Since $\mathbf{S} = \exp[i\mathbf{A}]$ has already been calculated in the normal course of events the orthogonal transformation, $\mathbf{U}$, that diagonalizes $\mathbf{A}$ is known. Then the integral in equation (59) can be evaluated analytically to obtain

$$\mathbf{S}_{\alpha_i} = \mathbf{U}\mathbf{M}\mathbf{U}^\dagger \tag{60}$$

where

$$M_{pq} = [\mathbf{U}^\dagger\mathbf{A}_{\alpha_i}\mathbf{U}]_{pq}(\exp i\lambda_q - \exp i\lambda_p)/(\lambda_p - \lambda_q) \tag{61}$$

and λ_p is the pth eigenvalue of $\mathbf{A}$. Obtaining this result will require an approximate doubling of the labour involved in calculating $\mathbf{S}$ alone. Assuming an N^3 dependence in the expense with matrices of dimension N, a modest price is paid for the extra sensitivity information, but the additional insight can be extremely valuable. A recent study (Hwang and Rabitz, 1979) illustrated this point for R–T processes using a potential surface given by

$$V(R, \hat{r}\cdot\hat{R}) = V_0(R)[1 + \alpha_1 \mathrm{P}_1(\hat{r}\cdot\hat{R}) + \alpha_2 P_2(\hat{r}\cdot\hat{R})] \tag{62}$$

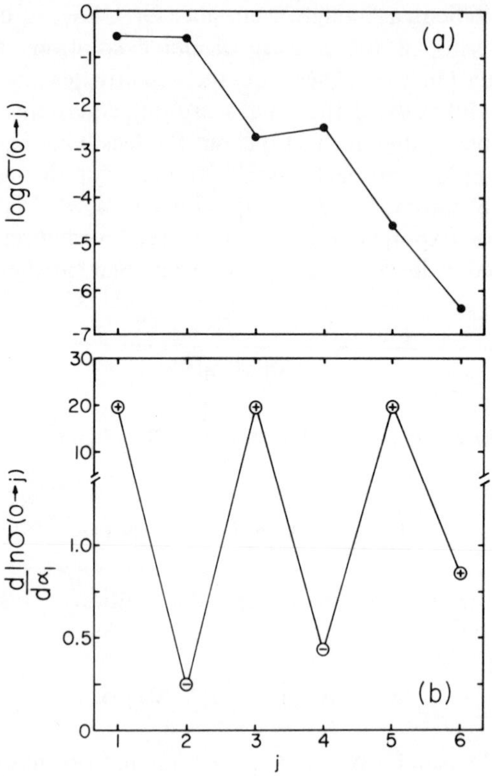

Fig. 4. Values of $\log \sigma_{0,j}$ (a) and the linear sensitivity coefficient (b) are displayed. The potential is defined in equation (62) and the + and − refer to the sign of the derivative in (b).

where $V_0(R)$ was a Lennard–Jones potential and $\alpha_1 = \alpha_2 = 0.1$. Fig. 4 illustrates the cross sections $\sigma_{0,j'}$ and sensitivity $\partial \ln \sigma_{0,j'}/\partial\alpha_1$ as a function of j'. Examination of sensitivity graphs such as this allowed the determination of how each parameter entered into the coupling scheme for a particular cross section $\sigma(j \to j')$.

The availability of collisional sensitivities can provide interesting insight into which portion of the Hamiltonian is most significant for a particular problem. In addition, the sensitivities can be used to aid inversion procedures for refining potentials such as determined from differential scattering cross section data. As another example, sensitivity theory may also be used to determine the sensitivity of averaged observations (e.g. bulk relaxation times) with respect to the underlying rate constants. Specific cases of interest here would involve situations where the relaxation time τ is a complicated non-linear functional of the kinetic rates. Finally, sensitivity analysis has been used to probe for ergodic behaviour in intramolecular energy transfer (Duff and Brumer, 1977). The general subject of sensitivity analysis should receive increasing attention in the future.

References

Abramowitz, M. and Stegun, I. A. (1965). *Handbook of Mathematical Functions*, Dover, New York.
Alexander, M. H. and Dagdigian, P. J. (1977). *J. Chem. Phys.*, **66**, 4126.
Alhassid, Y. and Levine, R. D. (1977). *J. Chem. Phys.*, **67**, 4321.
Alper, J. S., Carroll, M. A., and Gelb, A. (1978). *Chem. Phys.*, **32**, 471.
Arthurs, A. M. and Dalgarno, A. (1960). *Proc. Roy. Soc. (London)*, **A256**, 540.
Askar, A., Cakmak, A. S., and Rabitz, H. (1978). *Chem. Phys.*, **33**, 267.
Augustin, S. D. and Miller, W. H. (1974). *Chem. Phys. Lett.*, **28**, 149.
Augustin, S. D. and Rabitz, H. (1976a). *J. Chem. Phys.*, **64**, 1223.
Augustin, S. D. and Rabitz, H. (1976b). *J. Chem. Phys.*, **64**, 4821.
Augustin, S. D. and Rabitz, H. (1977). *J. Chem. Phys.*, **67**, 2082.
Augustin, S. D. and Rabitz, H. (1978). *J. Chem. Phys.*, **69**, 4195.
Augustin, S. D. and Rabitz, H. (1979). *J. Chem. Phys.*, **70**, 1286.
Balint-Kurti, G. G. (1975). *MTP Int. Rev. Sci. Phys. Chem. Ser. Two*, **1**, 285.
Barg, G. D. and Toennies, J. P. (1977). *Chem. Phys. Lett.*, **51**, 23.
Bates, D. R. (1962). *Atomic and Molecular Processes* (Ed. D. R. Bates), Academic Press, New York.
Beenakker, J. J. M., Knapp, H. F. P., and Sanctuary, B. C. (1973). *Amer. Inst. Phys.—Conf. and Symp. Ser.*, **11**, 21.
Bhattacharyya, S. S. and Saha, S. (1978). *J. Chem. Phys.*, **68**, 4292.
Billing, G. D. (1975). *Chem. Phys. Lett.*, **30**, 391.
Billing, G. D. and Poulsen, L. L. (1978). *J. Chem. Phys.*, **68**, 5128.
Boggs, J. (1978). *J. Chem. Phys.*, **69**, 2355.
Bosanac, S. and Balint-Kurti, G. G. (1975). *Mol. Phys.*, **29**, 1797.
Brumer, P. (1974). *Chem. Phys. Lett.*, **28**, 345.
Buck, U. and Pauly, H. (1971). *J. Chem. Phys.*, **51**, 1662.
Carrington, T. (1961). *J. Chem. Phys.*, **35**, 807.
Chapman, S. and Green, S. (1977). *J. Chem. Phys.*, **67**, 2317.
Cheng, V. B., Suzukawa, H. H., Jr., and Wolfsberg, M. (1973). *J. Chem. Phys.*, **59**, 3992.
Child, M. S. (1976). *Modern Theoretical Chemistry* (Ed. W. H. Miller), Vol. 2, Plenum Press, New York.
Clark, A. P., Dickinson, A. S., and Richards D. (1977). *Adv. Chem. Phys.*, **36**, 63.
Creaser, R. P., English, P., and Kinsey, J. L. (1973). *J. Chem. Phys.*, **58**, 1321.
Curtiss, C. F. (1974). *Physical Chemistry*, Vol. VIA, Academic Press, New York, p. 77.
DePristo, A. E. and Alexander, M. H. (1976). *J. Chem. Phys.*, **64**, 3009.
DePristo, A. E. and Alexander, M. H. (1977a). *J. Chem. Phys.*, **66**, 1334.
DePristo, A. E. and Alexander, M. H. (1977b). *Chem. Phys.*, **19**, 181.
DePristo, A. E. and Rabitz, H. (1978a). *J. Chem. Phys.*, **68**, 1981.
DePristo, A. E. and Rabitz, H. (1978b). *J. Chem. Phys.*, **68**, 4017.
DePristo, A. E. anh Rabitz, H. (1978c). *J. Chem. Phys.*, **69**, 902.
DePristo, A. E., Ramaswamy, R., Augustin, S. D., and Rabitz, H. (1979). *J. Chem. Phys.*, **71**, in press.
Devaney, A. J. (1978). *J. Math. Phys.*, **19**, 1526.
Dickinson, A. S. (1979). *Comp. Phys. Comm.*, in press.

Doll, J. and Miller, W. H. (1972). *J. Chem. Phys.* **57**, 5019.
Dougherty, E., Rabitz, H., Detrich, J., and Conn, R. (1977). *J. Chem. Phys.*, **67**, 4742.
Duff, J. W. and Brumer, P. (1977). *J. Chem. Phys.*, **67**, 4898.
Dunker, A. M. and Gordon, R. G. (1978). *J. Chem. Phys.*, **68**, 700.
Duquette, G., Ellis, T. H., Scoles, G., Watts, R. O., and Klein, M. L. (1978). *J. Chem. Phys.*, **68**, 2544.
Edmonds, A. R. (1974). *Angular Momentum in Quantum Mechanics*, Princeton University, Princeton, N. J.
Eu, B. C. (1978). *Chem. Phys.*, **27**, 301.
Faubel, M. and Toennies, J. P. (1977). *Adv. Atom. Mol. Phys.*, **13**, 227.
Fitz, D. E. (1977). *Chem. Phys.*, **24**, 133.
Fluendy, M. A. D. and Lawley, K. P. (1973). *Chemical Applications of Molecular Beam Scattering*, Chapman and Hall, New York.
Gentry, W. R. and Giese, C. F. (1977). *J. Chem. Phys.*, **67**, 5389.
Gerber, R. B., Zaristsky, N. C., and Minglegrin, U. (1978). *Mol. Phys.*, **35**, 1247.
Goldflam, R., Green, S., and Kouri, D. J. (1977a). *J. Chem. Phys.*, **67**, 4149.
Goldflam, R., Green, S., and Kouri, D. J. (1977b). *J. Chem. Phys.*, **67**, 5661.
Gordon, R. G. (1966). *J. Chem. Phys.*, **44**, 3083.
Gordon, R. G. (1969). *J. Chem. Phys.*, **51**, 14.
Green, S. (1975a). *J. Chem. Phys.*, **62**, 2271.
Green, S. (1975b). *J. Chem. Phys.*, **62**, 3568.
Green, S. (1976). *J. Chem. Phys.*, **64**, 3463.
Green, S., Garrison, B. J., and Lester, W. A. Jr. (1975). *J. Chem. Phys.*, **63**, 1154.
Green, S., Monchick, L., Goldflam, R., and Kouri, D. J. (1977). *J. Chem. Phys.*, **66**, 1409.
Green, S. and Thaddeus, P. (1976). *Ap. J.*, **205**, 766.
Heil, T. G. (1978). *J. Chem. Phys.*, in press.
Heil, T. G. and Secrest, D. (1978). *J. Chem. Phys.*, **69**, 219.
Heller, E. J. (1975). *J. Chem. Phys.*, **62**, 1544.
Holmgren, S. L., Waldman, M., and Klemperer, W. (1978). *J. Chem. Phys.*, **69**, 1661.
Hwang, J.-T. and Rabitz, H. (1979). *J. Chem. Phys.*, **70**, 4609.
Jeyes, S. R., McCaffery, A. J., Rowe, M. D., and Kato, H. (1977). *Chem. Phys. Lett.*, **48**, 91.
Khare, V. (1977). *J. Chem. Phys.*, **67**, 3897.
Kim, Y. S. and Gordon, R. G. (1974). *J. Chem. Phys.*, **60**, 1842.
Kouri, D. J., Heil, T. G., and Shimoni, Y. (1976). *J. Chem. Phys.*, **65**, 226.
LaBudde, R. A. and Bernstein, R. B. (1973). *J. Chem. Phys.*, **59**, 3687.
Lambert, J. D. (1977). *Vibrational and Rotational Relaxation in Gases*, Clarendon Press, Oxford.
Lester, W. A. Jr. (1976). *Modern Theoretical Chemistry* (Ed. W. H. Miller), Vol. 7, Plenum Press, New York.
Levine, R. D. (1966). *Proc. Roy. Soc. A*, **294**, 467.
Levine, R. D. (1971). *Mol. Phys.*, **22**, 497.
Levine, R. D. and Bernstein, R. B. (1977). *Modern Theoretical Chemistry* (Ed. W. H. Miller), Vol. 2, Plenum Press, New York.
Light, J. C. and Walker, R. B. (1976). *J. Chem. Phys.*, **65**, 4272.
Lin, C. S. and Secrest, D. (1979). *J. Chem. Phys.*, **70**, 199.
Liu, W. K. and Marcus, R. A. (1975). *J. Chem. Phys.*, **63**, 272, 290.
Marcus, R. A. (1971). *J. Chem. Phys.*, **54**, 3965.
McCann, K. J. and Flannery, M. R. (1975). *J. Chem. Phys.*, **63**, 4695.
McCurdy, C. W. and Miller, W. H. (1977). *J. Chem. Phys.*, **67**, 463.
McGuire, P. and Kouri, D. J. (1974). *J. Chem. Phys.*, **60**, 2488.

Micha, D. A. (1976). *Modern Theoretical Chemistry* (Ed. W. H. Miller), Vol. 7, Plenum Press, New York.
Miller, W. H. (1969). *J. Chem. Phys.*, **51**, 3631.
Miller, W. H. (1970). *J. Chem. Phys.*, **53**, 1949, 3578.
Miller, W. H. (1974). *Adv. Chem. Phys.*, **25**, 69.
Miller, W. H. and George, T. F. (1972). *J. Chem. Phys.*, **56**, 5668.
Monchick, L., and Kouri, D. J. (1978). *J. Chem. Phys.*, **69**, 3262.
Neilsen, W. B. and Gordon, R. G. (1973). *J. Chem. Phys.*, **58**, 4131, 4149.
Pack, R. T. (1974). *J. Chem. Phys.*, **60**, 633.
Parker, G. A. and Pack, R. T. (1977). *J. Chem. Phys.*, **66**, 2850.
Parker, G. A., Snow, R. L., and Pack, R. T. (1976). *J. Chem. Phys.*, **64**, 1668.
Pattengill, M. D. (1977). *J. Chem. Phys.*, **66**, 5042.
Pattengill, M. D. (1975). *Chem. Phys. Lett.*, **36**, 25.
Pauly, H. and Toennies, J. P. (1965). *Adv. Atom. Mol. Phys.*, **1**, 195.
Pechukas, P. (1969). *Phys. Rev.*, **181**, 166, 174.
Pickett, H. M. (1974). *J. Chem. Phys.*, **61**, 1923.
Pickett, H. M. (1975). *J. Chem. Phys.*, **63**, 2149. 2153.
Porter, R. N. and Raff, L. M. (1976). *Modern Theoretical Chemistry* (Ed. W. H. Miller), Vol. 2, Plenum Press, New York.
Procaccia, I. and Levine, R. D. (1976). *J. Chem. Phys.*, **64**, 808.
Rabitz, H. (1972). *J. Chem. Phys.*, **57**, 718.
Rabitz, H. (1974). *Ann. Rev. Phys. Chem.*, **25**, 155.
Rabitz, H. (1976). *Modern Theoretical Chemistry* (Ed. W. H. Miller), Vol. 2, Plenum Press, New York.
Rabitz, H., Askar, A., and Cakmak, A. S. (1978). *Chem. Phys.*, **29**, 61.
Rabitz, H. and Lam, H. (1975). *J. Chem. Phys.*, **63**, 3532.
Ramaswamy, R., Augustin, S. D., and Rabitz, H. (1979a). *J. Chem. Phys.*, **70**, 2455.
Ramaswamy, R., DePristo, A. E., and Rabitz, H. (1979b). *Chem. Phys. Lett.*, **61**, 495.
Reuss, J. (1975). *Adv. Chem. Phys.*, **30**, 389.
Rosenthal, A. and Gordon, R. G. (1976). *J. Chem. Phys.*, **64**, 1621, 1630, 1641.
Sams, W. N., and Kouri, D. J. (1969). *J. Chem. Phys.*, **51**, 4809, 4815.
Schaefer, H. F. III (1977). *Modern Theoretical Chemistry* (Ed. H. F. Schaefer), Vol. 3, Plenum Press, New York.
Schaefer, J. (1979). *J. Chem. Phys.*, in press.
Schaefer, J. and Lester, W. A., Jr. (1975). *J. Chem. Phys.*, **62**, 1913.
Schatz, G. C., McLafferty, F. J., and Ross, J. (1977). *J. Chem. Phys.*, **66**, 3609.
Schinke, R. and McGuire, P. (1978). *Chem. Phys.*, **28**, 129.
Schmalz, T. G. and Flygare, W. H. (1978). *Laser and Coherence Spectroscopy* (Ed. J. I. Steinfeld), Plenum Press, New York.
Secrest, D. (1973). *Ann. Rev. Phys. Chem.*, **24**, 379.
Secrest, D. (1975). *J. Chem. Phys.*, **62**, 710.
Shafer, R. and Gordon, R. G. (1973). *J. Chem. Phys.*, **58**, 5422.
Shapiro, M. and Gerber, R. B. (1976). *Chem. Phys.*, **13**, 235.
Shapiro, M., Gerber, R. B., Buck, U., and Schleusener, J. J. (1977). *Chem. Phys.*, **67**, 3570.
Shimoni, Y. and Kouri, D. J. (1977). *J. Chem. Phys.*, **66**, 2841.
Siska, P. E. (1973). *J. Chem. Phys.*, **59**, 6052.
Siska, P. E. (1979). Private communication.
Smith, E. W., Giraud, M., and Cooper, J. (1976). *J. Chem. Phys.*, **65**, 1256.
Steinfeld, J. I. and Houston, P. L. (1978). *Laser and Coherence Spectroscopy* (Ed. J. I. Steinfeld), Plenum Press, New York.

Suzukawa, H. H., Jr. and Wolfsberg, M. (1978). *J. Chem. Phys.*, **68**, 1423.
Takayanagi, K. (1963). *Progr. Theor. Phys. Suppl.*, **25**, 1.
Tarr, S. M. and Rabitz, H. (1978). *J. Chem. Phys.*, **68**, 642.
Tarr, S. M., Rabitz, H., Fitz, D. E., and Marcus, R. A. (1977). *J. Chem. Phys.*, **66**, 2854.
Toennies, J. P. (1976). *Ann. Rev. Phys. Chem.*, **27**, 225.
Tomovic, R. and Vukabratovic, M. (1972). *General Sensitivity Theory*, American Elsevier, New York.
Truhlar, D. G. (1976). *Int. J. Quantum Chem. Symp.*, **10**, 239.
de Vogelaere, R. (1955). *J. Res. Natl. Bur. Std.*, **54**, 119.
Wilcomb, B. E. and Dagdigian, P. J. (1977). *J. Chem. Phys.*, **67**, 3829.
Willis, C. R. and Picard, R. H. (1974). *Phys. Rev.*, **A9**, 1343.
Zaristsky, N. C., Minglegrin, U., and Gerber, R. B. (1978). *Mol. Phys.*, **35**, 1259.
Zwanzig, R. (1961). *Lectures in Theoretical Physics* (Eds. W. E. Britten, B. W. Downs, and J. Downs), Vol. 3, Interscience, New York.

Potential Energy Surfaces
Edited by K. P. Lawley

THEORETICAL STUDIES OF VIBRATIONAL ENERGY RELAXATION OF SMALL MOLECULES IN DENSE MEDIA

D. J. DIESTLER

Department of Chemistry, Purdue University, W. Lafayette, Indiana, 47907, *USA*

CONTENTS

I. INTRODUCTION

The purpose of this article is to review methods and results of theoretical investigations of vibrational relaxation (henceforth frequently abbreviated as VR) of small molecules in condensed media. Below we shall distinguish between two types of 'vibrational relaxation', namely *phase* and *energy* (or population) relaxation. By the term 'small' we shall usually mean 'diatomic', although in principle most of the approaches we shall consider are independent of the size of the relaxing molecule. The type of 'condensed medium' in which we shall be chiefly interested is that of the rare-gas solid, i.e. a monatomic lattice. Moreover, we shall generally assume that the concentration of relaxing impurity, or solute molecules is so low that interactions between them are negligible. In the applications we shall focus upon the 'prototypic' system—a single diatomic molecule imbedded as a substitutional impurity in a rare-gas lattice. This system is depicted in Fig. 1.

We can clarify the distinction between vibrational *phase* and *energy* relaxation by considering two distinct Raman-based experimental methods of studying vibrational relaxation. The first is ordinary spontaneous Raman scattering (SpRS). Consider a dilute solution of impurity molecules (solute) having a Raman-active normal mode, the relaxation of which we wish to study. Let us assume further that:

(i) the vibrations of the solute are uncorrelated with its orientations;
(ii) the orientational motion behaves classically, and
(iii) at any instant all orientations of the active molecule are equally probable.

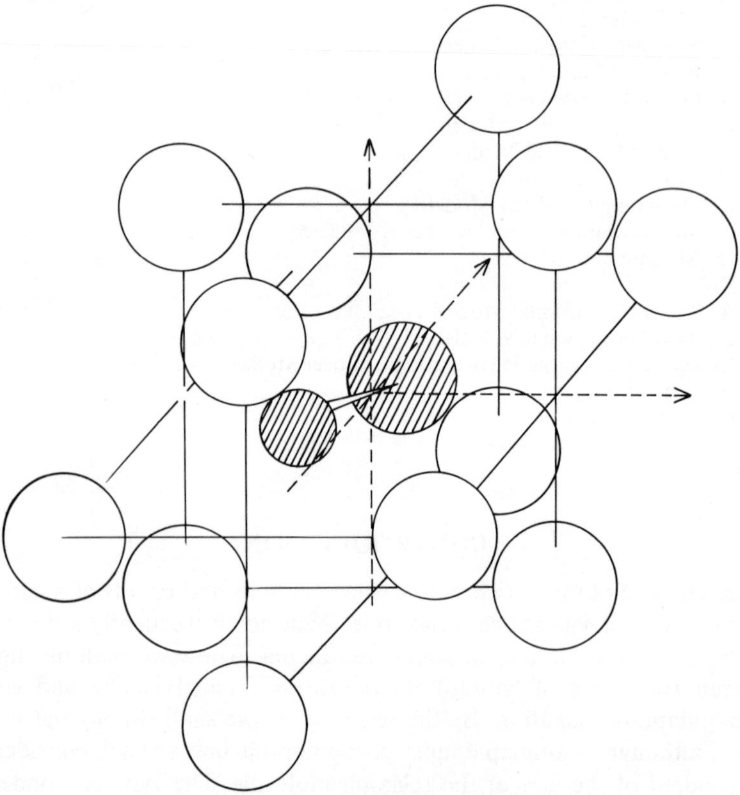

Fig. 1. A diatomic molecule occupying a substitutional site in a rare-gas crystal lattice.

Under these assumptions it can be shown (Steele, 1969; Bratos and Marechal, 1971; Bartoli and Litovitz, 1972; Nafie and Peticolas, 1972) that the isotropic component of the Stokes spectrum is expressible as

$$I_{\text{iso}}(\omega) = |(\partial\alpha_0/\partial Q)_{Q=Q_0}|^2 \int_{-\infty}^{+\infty} \mathrm{d}t \exp(i\omega t)\langle Q(0)Q(t)\rangle, \tag{1}$$

where

$$I_{\text{iso}}(\omega) = I_{\parallel}(\omega) - \tfrac{4}{3}I_{\perp}(\omega), \tag{2}$$

Q is the normal coordinate associated with the active solute mode, $I_{\parallel}$ and $I_{\perp}$ are, respectively, the intensities of scattered light having polarizations parallel and perpendicular to that of the incident light, ω is the Stokes shift, and α_0 is the *average* polarizability. Now, under the additional assumption that

(iv) the coupling of the active solute normal mode to the other degrees of freedom is weak,

the isotropic lineshape $[I_{\text{iso}}(\omega)]$ is Lorentzian. Thus a plot of the logarithm of the envelope of $\langle Q(0)Q(t)\rangle$ (obtained from the inverse Fourier transform of equation (1) *versus* t is a straight line of slope $-(2\tau)^{-1}$, where τ is called the dephasing time. Furthermore, the envelope of the function

$$|\langle Q(0)Q(t)\rangle|^2 \sim \exp(-t/\tau) \tag{3}$$

decays exponentially with the time constant τ related to the isotropic linewidth $\delta\bar{\lambda}$ (i.e. full width at half height, FWHH) by

$$\tau = (2\pi c\delta\bar{\lambda})^{-1}, \tag{4}$$

where c is the speed of light.

An alternative technique of studying vibrational relaxation is based on stimulated Raman scattering (SRS). The experiment is carried out in the time domain (as opposed to the frequency domain for SpRS) in two stages. In the first stage an intense, picosecond-duration laser pulse propagating through the solution of active solute molecules undergoes non-linear coherent Stokes scattering, thereby producing a relatively large (with respect to the usual Boltzmann population at the ambient temperature) population of molecules in the first vibrationally excited level ($v = 1$) of the active normal mode. In essence, this *is* stimulated Raman scattering. It is important to note that the vibrations of the active solute molecules are excited in phase, even though these molecules do not necessarily 'communicate' with one another. In other words, a definite relationship among the phases of the active-solute vibrations is established by the initial pulse. In the second stage of the experiment a probe pulse, delayed with respect to the initial pulse, is scattered by anti-Stokes vibrational transitions of the excited solute molecules. The *coherent* anti-Stokes intensity, observed under phase-matching conditions near the incident direction, decays at long times according to

$$I_{CAS}(t) \propto \exp(-t/\tau), \tag{5}$$

where τ is again the dephasing time [Compare equation (31]. The *incoherent* anti-Stokes intensity, I_{IAS}, observed at right angles to the incident direction, as in the SpRS technique outlined earlier, is a direct measure of the deviation $\Delta n(t)(\equiv \langle \mathbf{n}(t) \rangle - n)$ of the population $\mathbf{n}(t)$ of the $v = 1$ level from its equilibrium value $\bar{n}$. At long times

$$I_{IAS} \propto \Delta n(t) = \Delta n(0)\exp(-t/\tau'), \tag{6}$$

where τ' is the *energy relaxation* time (or vibrational lifetime).

It should be noted that the preceding analyses, which lead to Lorentzian lineshapes (in the frequency domain) or simple exponential decay curves (in the time domain) apply strictly only to systems in which the relevant vibrational transition is homogeneously broadened (see, for example, Steinfeld, 1974). For the most part, we shall restrict our attention to such systems.

In summary, we see that both the SpRS and SRS techniques yield information on the rate of vibrational *phase* relaxation, i.e. the rate of decay of the correlation function $\langle Q(0)Q(t) \rangle$. An alternative viewpoint regards $\langle Q(0)Q(t) \rangle$ as a 'memory' function. Thus its rate of decay tells one how fast the vibrations 'forget' their initial condition. In addition, the SRS technique provides directly the rate of vibrational *energy* (or population) relaxation.

Progress in the study of vibrational dephasing has been thoroughly reviewed recently in this series (Oxtoby, 1979). Experimental investigations of population relaxation of small molecules in dense media have also been extensively reviewed lately (Legay, 1978; Brus and Bondybey, 1979). Consequently, we shall choose to focus here on theoretical treatments of vibrational *energy* relaxation.

In general, two basic types of mechanisms contribute to vibrational energy relaxation (VER): radiative and non-radiative. We shall assume that the two pathways are independent (i.e. 'parallel' in the standard terminology of kinetics). Thus the VER time may be expressed as

$$\tau' = (k + \tau_r^{-1})^{-1}, \tag{7}$$

where τ_r is the radiative (natural) lifetime and k is the *non-radiative* VER rate constant. From the direct measurement of τ' (by the SRS technique for example) and an estimate of τ_r based on the natural lifetime of the *isolated* solute molecule (Legay, 1978), one can calculate k using equation (7). The determination of τ_r for molecules in solution is, in general, a non-trivial problem (Andrews and Hudson, 1978). We note that if τ_r is not known, τ'^{-1} at least provides an upper bound on k.

Stated precisely, the fundamental question we ask is: 'By what microscopic pathways is the energy contained in the intramolecular vibration of the impurity molecule dissipated non-radiatively into the medium?' To answer this question, even for the simplest type of system [i.e. the prototype (see Fig. 1)], requires analysis of a complex many-body system. For convenience, we mentally partition the entire system into two subsystems: 'relevant' and 'irrelevant'. The relevant subsystem (RS) comprises those degrees of freedom (d.f.) whose behaviour determines in large degree the outcome of the measurement of interest, i.e. $\Delta n(t)$. For example, the RS may consist simply of the intramolecular normal coordinate (Q) of the impurity. In some instances, however, in which Q interacts strongly with addition d.f. (e.g. local translations or orientations of the impurity), we may wish to augment the RS to include these additional d.f. The irrelevant subsystem (IRS) is constituted of the remaining d.f. (i.e. those not included in the RS), which we shall generally assume to be only weakly coupled to those of the RS. Thus the IRS essentially functions as a thermal bath. In the simplest case involving the prototype, in which we take Q to be the only relevant d.f., the irrelevant d.f., which collectively constitute the 'medium', consist of the following:

(i) translatory motions of the diatomic impurity in the 'cage' effectively formed by the neighbouring lattice atoms;
(ii) orientational motions of the impurity; and
(iii) oscillatory motions of the lattice atoms.

The d.f. of type (i) and (iii) may be conveniently lumped together as the normal modes of an impure lattice. We note that, in general, internal d.f. of the lattice molecules must also be taken into account. They play an utterly important part, for example, in VER in polyatomic molecular solvents.

Now VER takes place on account of a coupling between the relevant and irrelevant subsystems. In general, all of the various types of irrelevant d.f. can participate in the relaxation process. The degree of participation of a given type depends upon the strength of its coupling to the RS. In turn, the strength of that coupling is determined by the nature of the multidimensional potential energy surface (PES) of the given system. In the applications to be considered, we shall usually take the PES to be a sum of two-body interactions. However, we note that the results are often quite sensitive to the assumed forms of these interactions and that, unfortunately, reliable forms are usually lacking.

Before going on, it is worthwhile emphasizing the difference between gas-phase and condensed-phase relaxation processes in terms of the amount of information about the PES which is extractable from typical experiments. To a good approximation, relaxation processes in dilute gases can be adequately described as occurring via a sequence of bimolecular collisions. Cross sections (or probabilities) for various detailed scattering events can be defined and, in many cases, measured by molecular beam techniques (see, for example, Ross, 1966). The cross sections are related more-or-less directly to the PES of the colliding pair. In condensed phases, on the other hand, cooperative many-body effects can play a dominant role. In general, one cannot treat the relaxation process simply in terms of isolated binary encounters; the solute molecule is effectively in 'continuous collision' with surrounding solvent molecules throughout the duration of the relaxation. Hence, one cannot define, let alone measure, cross sections for specific detailed transitions between the stationary states of the system. One must be content with merely the relaxation rate constant, which, like the specific rate constant for a gas-phase reaction (see, for example, Eliason and Hirschfelder, 1959), is an average of detailed transition rates (analogous to cross sections) over initial (many-body) states of the system and a sum over final (many-body) states. Clearly, the rate constant is inherently a *macroscopic* quantity related only very indirectly to the PES.

In view of the complexity of the many-body system, one can envisage two extreme approaches to the quantitative description of VER in condensed media. At one extreme, one may simply solve the classical equations of motion (EOM) to determine the coordinates and conjugate momenta of every particle in the system as a function of time, i.e. one may determine a 'representative' trajectory of the phase point of the entire system. Then the required ensemble averages (e.g. $\langle \mathbf{n}(t) \rangle$) are expressed as time averages over this trajectory. This approach is generally known as the molecular dynamics method (see, for example, Berne and Forster, 1971). The quantum mechanical analogue of molecular dynamics, in which one computes *a priori* the detailed many-body transition rates and then sums and averages over these to obtain the non-radiative VER rate constant k, hardly seems feasible. Nor, in the opinion of this author, is such a procedure necessarily desirable. One may prefer approximate expressions for k which manifest explicit dependences upon microscopic properties (as determined by the PES) and thermodynamic variables (e.g. temperature and density). Hence, at the other extreme, instead of solving the full classical equations of motion on a computer, one can attempt to describe VER using simple models which yield, with the aid of (hopefully) reasonable approximations, closed explicit expressions for k.

In Section II we consider classical-mechanical methods of treating vibrational relaxation (VR). We discuss the application of the molecular dynamics method to VR in a diatomic molecular liquid. We also outline the stochastic classical trajectory (SCT) approach (Adelman and Doll, 1976) to VR. The SCT method takes an intermediate tack, recognizing that the impurity interacts directly only with the nearest-neighbour solvent molecules, so that its relaxation behaviour is almost entirely determined by the detailed motion of these few particles (which constitute the RS). The remaining large number of solvent molecules (the IRS) have only indirect influence. Hence, rather than solve the exact classical EOM for the entire system, one instead solves a relatively small number of Langevin-type equations describing only the RS. The effects of the IRS come in only through 'friction kernels' and random forces.

The remainder of the article is devoted to approaches from the opposite extreme (from molecular dynamics) which are based on approximate treatments of models leading to explicit closed expressions for k. In Section III we shall outline several semiclassical models. Section IV will deal with fully quantum mechanical models, with emphasis on those predicting a multiphonon mechanism of VER. All of the applications to be considered in Sections III and IV will be restricted essentially to the one-dimensional (1–d) prototypic system, with the vibration of the diatomic impurity taken as the RS and the (impure) lattice as the IRS.

In Section V we shall discuss models which attempt to account for the role of orientational motions of the impurity in VER, again restricting applications to the 2- and 3-d prototypes.

In Section VI we shall develop an extended view of the participation of local modes in VER. Like the stochastic classical-trajectory method, this extended view recognizes that the intramolecular vibration may in certain systems be strongly coupled to local motions (e.g. orientations of the impurity). If the interaction is sufficiently strong, a 'complex' or 'pseudomolecule' containing the impurity is effectively formed. An adequate description of VR should then treat the relaxation of the pseudomolecule (which becomes the RS) itself rather than simply that of the 'bare' impurity. We outline the general theory and discuss its application to the system Cl_2/Ar.

Finally, in Section VII we close with a summary and a discussion of some of the limitations of the approaches to be considered and also some possible directions for further study.

II. CLASSICAL METHODS

In this Section we consider two methods of describing VER which are based on numerical solution of the classical equations of motion (EOM).

A. Molecular Dynamics

In recent years molecular dynamics (MD) simulation has become a standard tool for the investigation of relaxation behaviour of dense systems (see, for example, Berne and Forster, 1971). MD computer 'experiments' on 3-d systems have given valuable information concerning the *external* motions of atomic and rigid molecular systems. However, any simulation which attempts to account for intramolecular vibrational motion in realistic systems encounters a twofold difficulty connected with the time scales associated with the various kinds of motion. First, since the frequencies characteristic of intramolecular vibrations are relatively higher than those associated with external motions (e.g. rotations and translations), the time scale of the internal (vibrational) motion is much shorter than that of the external motions at ordinary temperatures. Second, vibrational relaxation generally occurs on a time scale considerably longer than that of the microscopic molecular motions. Therefore, in order to mimic accurately the overall VR process, one must choose a time-step size for the numerical method sufficiently small to describe the details of the rapid vibrational motion, yet large enough to enable one to carry the computation out for a sufficiently long time to 'see' the relaxation. Even for the most favourable type of realistic system, i.e. a solution of weakly bound diatomic molecules, a full 3-d simulation of the VER appears impracticable.

In order to circumvent the time-scale difficulties discussed above, we carried out an MD study (Riehl and Diestler, 1976) of VR in a 1-d model of a diatomic molecular liquid. Although 1-d models have obvious drawbacks (e.g. orientational motions are absent), they can yield some insight into the real 3-d system of interest.

The model consists of a chain of N heteronuclear diatomic molecules AB arranged head-to-tail as follows: A—B ... A—B ... A—B. The Hamiltonian is given by

$$H = \sum_{i=1}^{N} \{p_i^2/2M + P_i^2/2\mu + \mu\omega^2(Q_i - Q_0)^2/2 + V'_{i,i+1}(x_{i+1,i}), \tag{8}$$

where

$$\begin{aligned} q_i &= (m_A r_{Ai} + m_B r_{Bi})/M \\ Q_i &= (r_{Bi} - r_{Ai}) \\ x_{i+1,i} &= r_{A,i+1} - r_{Bi} = x_{i+1,i}(Q_i, q_i, Q_{i+1}, q_{i+1}) \\ M &= m_A + m_B \\ \mu &= m_A m_B/M. \end{aligned} \tag{9}$$

In equations (8) and (9) P_i and p_i are the momenta conjugate to Q_i and q_i, respectively, and r_{Xi} is the position of atom X in molecule i. We assumed that the isolated diatomic is bound by a harmonic potential of force constant $k = \mu\omega^2$, where ω is the classical frequency. We took the interaction potential V' to be such that each atom of a given diatomic 'sees' only the nearest atom of the neighbouring diatomic. Moreover, we required that

$$V'_{N,N+1} = V'_{N,1}, \tag{10}$$

which implies that the diatomics move on a one-dimensional ring. The explicit form assumed for V' is that of the Lennard–Jones (12–6) potential

$$V'_{i,i+1} = \Delta\varepsilon[(x_{i+1,i}/\sigma)^{-12} - (x_{i+1,i}/\sigma)^{-6}]. \tag{11}$$

The more-or-less standard MD procedure is described as follows. Hamilton's EOM for the set of $4N$ coordinates and momenta $\{\dot{Q}_i = \partial H/\partial P_i, \dot{q}_i = \partial H/\partial p_i, \dot{P}_i = -\partial H/\partial Q_i, \dot{p}_i = -\partial H/\partial q_i, i = 1, 2, \ldots N\}$ are integrated numerically by a standard algorithm. For a given set of initial conditions $[Q_i(0), q_i(0), P_i(0), p_i(0), i = 1, 2, \ldots N]$, taken to correspond to a specified macroscopic condition, one computes unique sets of coordinates and momenta $\{Q_i(t_j), q_i(t_j), P_i(t_j), p_i(t_j), i = 1, 2, \ldots N\}$ at subsequent points in time t_j. These sets, stored on magnetic tape for future analysis, collectively constitute the trajectory of the phase point of the system. The required ensemble averages are calculated as time averages over the trajectory. For example, the correlation function of the intramolecular vibrational coordinate is given by

$$\langle Q(0)Q(t)\rangle = (N_s N)^{-1} \sum_{j=1}^{N_s} \sum_{i=1}^{N} Q_i(t_j)Q_i(t_j + t), \tag{12}$$

where the summation on index i is over diatomics and that on j over time origins t_j, N_s being the total number of time origins.

For the model parameters listed in Table I we (Riehl and Diestler, (1976) carried out simulations under the five different sets of macroscopic conditions given in Table II. The highest-density sample corresponds (at $T = 0$K) to a situation in which the total energy of the system is at an absolute minimum, with all atoms at their respective equilibrium positions.

TABLE I. Parameters for the one-dimensional molecular dynamics simulation Riehl and Diestler, 1976 of a heteronuclear diatomic liquid.

Parameter	Value
N	100
m_A	2.49×10^{-23} g
m_B	2.27×10^{-22} g
Q_0	2.13 Å
ω	2.51×10^{12} s^{-1}
ε	3.89×10^{-14} erg
σ	4.39 Å

TABLE II. Density (ρ), temperature (T), dephasing (τ) and energy (τ') relaxation times for various MD samples.

ρ(Å^{-1})	T(K)	τ(ps)	τ'(ps)
0.138	160	1.39	6.67
0.138	210	0.98	4.35
0.138	300	0.59	4.30
0.142	210	0.83	3.79
0.112	210	1.26	5.25

The vibrational dephasing time τ was determined by fitting the function

$$C_d(t) \equiv \frac{\langle [Q(0) - Q_0][Q(t) - Q_0] \rangle}{\langle [Q(0) - Q_0]^2 \rangle}$$

in the least-squares sense to the form $c\exp(-t/\tau)$. The VER time τ' was determined by the following procedure. The average vibrational energy of a given set (specified by a *range* of *initial* allowed vibrational energy) of diatomics is expressed as

$$\langle E^\dagger(t) \rangle = (N_s)^{-1} \sum_{j=1}^{N_s} [N_e(j)]^{-1} \sum_{k=1}^{N_e(j)} \varepsilon_k(t_j + t), \tag{13}$$

where

$$\varepsilon_k(t) = [P_k(t)]^2/2\mu + \mu\omega^2[Q_k(t) - Q_0]^2/2 \tag{14}$$

is the vibrational energy of the kth diatomic of the jth set (corresponding to the jth time origin). Now, since the vibrational energy of the diatomic is not quantized, we specified a range of allowed initial energies to correspond to a given energy level. For the samples being considered here we chose the range $3\hbar\omega \leq \varepsilon \leq 4\hbar\omega$, which brackets the level $v = 3$. Then $N_e(j)$ in equation (13) is the number of molecules at time t_j which have vibrational energy [given by equation (14)] lying in this range. The VER times listed in Table II were estimated by a least-squares fit of the function

$$E^*(t) \equiv \frac{\langle E^\dagger(t) - \bar{E} \rangle}{\langle E^\dagger(0) - \bar{E} \rangle}$$

to the form $c' \exp(-t/\tau')$. Here $\bar{E}$ is the average vibrational energy.

Of primary interest to us in this study was the dependence of τ and τ' upon thermodynamic properties such as density (ρ) and temperature (T). From Table II we reached the following conclusions:

(1) τ and τ' both decrease with increasing T at constant ρ or with increasing ρ at constant T;

(2) $\tau < \tau'$ for all combinations of ρ and T considered.

The first conclusion seems easily rationalized in terms of conventional kinetic-molecular theory. As the temperature of the system is raised at constant density, the molecules speed up, the frequency of collisions increases, and consequently the rate of relaxation increases. Likewise, as the molecules are forced closer together (increased density) at constant temperature, the time between collisions decreases and again the collision frequency increases, giving rise to an increased rate of relaxation. The second conclusion follows from more general considerations. It can be shown (see, for example, Diestler, 1976a) that for the prototypic system

$$\tau^{-1} = \tau_{\text{ph}}^{-1} + \tau'^{-1}, \tag{15}$$

where τ_{ph} is the so-called *pure* phase relaxation time. Since $\tau_{\text{ph}}^{-1} \geq 0$, it follows that $\tau^{-1} \geq \tau'^{-1}$, or $\tau \leq \tau'$.

From a detailed analysis (Riehl and Diestler, 1976) we inferred that VER in this model occurs via complicated collective motions of the system. The dispersion relation, calculated from the density–density autocorrelation function, indicates that the system behaves harmonically, even at higher temperatures. Hence, the intramolecular vibrations may be expressed in terms of the normal modes of the entire system. Energy initially associated with the internal vibrations delocalizes in a complex fashion through the concerted motions of the normal coordinates. Since the effective external force constants ($k' = 57.14\varepsilon/\sigma^2$) between atoms of *different* diatomics are comparable to the internal force constant ($k = \mu\omega^2$) the system possesses a high density of collective modes in the optical branch of the phonon spectrum having frequencies close to that of the isolated diatomic. The coupling between the internal vibration and these collective external motions is sufficiently strong to lead to rapid delocalization of the internal energy.

To assess the validity of the concerted mechanism of VER discussed above, we investigated by MD the relaxation of a *single* diatomic in a solvent of rigid diatomics. Although the statistics were quite poor, we estimated a vibrational lifetime of $\tau' > 40$ ps for sample 5 (see Table II), which is much longer than that determined for the pure diatomic liquid. The reason for the longer lifetime is that there are no collective modes of frequency comparable to that of the diatomic which can effectively participate in a concerted mechanism. Hence, an excited diatomic, of relatively high frequency, must lose its energy by slow transfer to the relatively low-frequency external centre-of-mass collective modes.

B. Stochastic Classical Trajectory Method

In performing MD simulations, one is eventually struck by the fact that so much apparently unnecessary detailed information is generated and then averaged away. The question naturally arises whether one can handle the relevant dynamics, i.e. that part which essentially determines the outcome of the experiment of interest, much more economically. This is precisely the goal of the stochastic classical trajectory (SCT) method developed originally (Adelman and Doll, 1976) for atom-surface scattering and applied recently (Shugard *et al.*, 1978) to VR in solids.

To develop the basic SCT approach, let us consider VR in the prototypic condensed system (Fig. 1). Since the diatomic impurity interacts essentially only with the nearest-neighbour lattice atoms, it is reasonable to suppose that VR is largely governed by motions of the diatomic (labelled R) and these nearest neighbours (labelled P, for 'primary'). The motion of the remaining lattice atoms (labelled Q for 'secondary') exerts only an indirect influence on VR. Now assuming that the diatomic does not interact with the secondary lattice atoms and that all interactions among lattice atoms are harmonic, one can derive the following EOM

$$\ddot{\mathbf{X}}_{\mathrm{R}}(t) = \mathbf{M}_{\mathrm{RR}}^{-1}\mathbf{F}_{\mathrm{R}}[\mathbf{X}_{\mathrm{R}}(t), \mathbf{X}_{\mathrm{P}}(t)] \tag{16a}$$

$$\ddot{\mathbf{X}}_{\mathrm{P}}(t) = -\boldsymbol{\Omega}_{\mathrm{eff}}^{2}\mathbf{X}_{\mathrm{P}}(t) - \int_{0}^{t}\boldsymbol{\Lambda}(t-t')\dot{\mathbf{X}}_{\mathrm{P}}(t')\mathrm{d}t' + \mathbf{M}_{\mathrm{PP}}^{-1}\mathbf{R}(t) + \mathbf{M}_{\mathrm{PP}}^{-1}\mathbf{F}_{\mathrm{P}}[\mathbf{X}_{\mathrm{R}}(t), \mathbf{X}_{\mathrm{P}}(t)], \tag{16b}$$

where

$$\boldsymbol{\Omega}_{\mathrm{eff}}^{2} \equiv \mathbf{M}_{\mathrm{PP}}^{-1/2}[\boldsymbol{\Omega}_{\mathrm{PP}}^{2} - \boldsymbol{\Lambda}(0)]\mathbf{M}_{\mathrm{PP}}^{1/2} \tag{17a}$$

$$\boldsymbol{\Lambda}(t) \equiv \mathbf{M}_{\mathrm{PP}}^{-1/2}\boldsymbol{\Omega}_{\mathrm{PQ}}^{2}\cos(\boldsymbol{\Omega}_{\mathrm{QQ}}t)\boldsymbol{\Omega}_{\mathrm{QQ}}^{-2}\boldsymbol{\Omega}_{\mathrm{QP}}^{2}\mathbf{M}_{\mathrm{PP}}^{1/2} \tag{17b}$$

and

$$\mathbf{R}(t) \equiv -\mathbf{M}_{\mathrm{PP}}^{1/2}\boldsymbol{\Omega}_{\mathrm{PQ}}^{2}\cos(\boldsymbol{\Omega}_{\mathrm{QQ}}t)\mathbf{M}_{\mathrm{QQ}}^{1/2}\mathbf{X}_{\mathrm{Q}}(0) - \mathbf{M}_{\mathrm{PP}}^{1/2}\boldsymbol{\Omega}_{\mathrm{PQ}}^{2}\sin(\boldsymbol{\Omega}_{\mathrm{QQ}}t)\boldsymbol{\Omega}_{\mathrm{QQ}}^{-1}\dot{\mathbf{X}}_{\mathrm{Q}}(0). \tag{17c}$$

In equations (16) and (17) $\mathbf{X}_{\mathrm{R}}$ and $\mathbf{X}_{\mathrm{P}}$ are vectors of coordinates of the impurity and primary lattice atoms, respectively. $\mathbf{F}_{\mathrm{R}}$ and $\mathbf{F}_{\mathrm{P}}$ are the (in general, anharmonic) forces between impurity and primary lattice atoms; $\mathbf{M}_{\mathrm{RR}}$, $\mathbf{M}_{\mathrm{PP}}$, and $\mathbf{M}_{\mathrm{QQ}}$ are matrices of masses of impurity, primary, and secondary $\boldsymbol{\Omega}_{\mathrm{PP}}$, $\boldsymbol{\Omega}_{\mathrm{QQ}}$, and $\boldsymbol{\Omega}_{\mathrm{PQ}}$ are matrices of frequencies characterizing the harmonic interactions within and between the primary and secondary portions of the lattice.

We emphasize that equations (16) properly describe the exact motion of the relevant subsystem, i.e. the impurity plus primary lattice atoms. The originally large set of Newton's equations has been effectively reduced to a small set of Langevin-type equations for the impurity and primary atoms. The influence of the secondary atoms is manifested through the effective frequency matrix $\boldsymbol{\Omega}_{\mathrm{eff}}^{2}$, the damping kernel $\boldsymbol{\Lambda}(t)$, and the random-force function $\mathbf{R}(t)$.

As they stand, equations (16) are apparently no easier to solve than the original full classical EOM, since the stochastic terms $\boldsymbol{\Omega}_{\mathrm{eff}}^{2}$, $\boldsymbol{\Lambda}(t)$, and $\mathbf{R}(t)$

depend upon the detailed motions of the secondary lattice atoms (see equations (17). However, by exploiting two additional features of the harmonic lattice, one can determine these stochastic terms without actually knowing the trajectories of the secondary lattice atoms. To understand how this is so, let us consider the case in which there is only a single primary lattice atom. First, one can show that for translationally invariant lattices, $\Lambda(t)$ is related to the density $g(\omega)$ of normal lattice modes by

$$g(\omega) = 2\pi^{-1}\omega^2\Lambda_c(\omega)/\{[\omega^2 - \Omega_{\text{eff}}^2 - \omega\Lambda_s(\omega)]^2 + \omega^2\Lambda_c^2(\omega)\}, \tag{18}$$

where Λ_s and Λ_c are, respectively, the sine and cosine Fourier transforms of $\pi\Lambda(t)/2$. Second, $R(t)$ can be shown (Adelman and Doll, 1976) to be a Gaussian random variable satisfying the relation

$$\langle R(0)R^{\dagger}(t)\rangle = k_B T M_{PP}\Lambda(t) \tag{19}$$

where k_B is Boltzmann's constant. Thus from a knowledge of $g(\omega)$, one can obtain Ω_{eff} and $\Lambda(t)$ through equation (18). In turn, $R(t)$ can be obtained from $\Lambda(t)$ using equation (19).

Like the MD method, the SCT method is plagued by the fact that there are several vastly different time scales inherent in the dynamics of VR. Even so, the SCT technique greatly reduces the number of equations of motion which must be explicitly solved and hence considerably extends the time scale on which one can follow VR processes.

Shugard *et al.* (1978) studied VR in the 1-d prototype. The primary portion of the lattice was taken to include only the nearest-neighbour lattice atoms, which were assumed harmonically bound to their respective equilibrium positions. The diatomic AB was permitted to move freely within its cage. The parameters of the model, given in Table III, were chosen to mimic

TABLE III. Parameters for the stochastic classical trajectory calculations on the one-dimensional model of the system Cl_2/Ar (Shugard *et al.*, 1978.)

Parameter	Value
k	1.40×10^5 erg cm^{-2}
Q_0	1.988 Å
D_e	0.4125 eV
β	2.397 Å^{-1}
Q_0'	2.228 Å
A	8.0×10^5 eV
α	5.44 Å^{-1}
R_{eq}[a]	± 3.70 Å

[a]The equilibrium separation between the centre-of-mass of the diatomic and its nearest neighbours.

the system Cl_2Ar as closely as possible. Vibrational relaxation in the electronic ground-state and $B^3\Pi$-state manifolds was simulated. In the ground state the potential binding the atoms A and B was assumed to be harmonic, i.e.

$$V = \tfrac{1}{2}k(Q - Q_0)^2,$$

while in the B state it was taken as a Morse potential

$$V' = D_e\{1 - \exp[-\beta(Q - Q_0')]\}^2$$

The interaction between the diatomic and each primary lattice atom was taken to be a repulsive exponential

$$U(r) = A\exp(-\alpha r),$$

with A and α chosen to correspond to the Ar–Ar interaction. The stochastic terms Ω_{eff}, $\Delta(t)$, and $R(t)$ were determined assuming a 3-d mode density corresponding roughly to that of Ar.

Shugard *et al.* (1978) studied the effects of variations in A, α, k, and T on both the dephasing and energy-relaxation times. In Tables IV and V we present selected results. As in the case of the MD study discussed in Section II.A, $\tau \leq \tau'$ for all cases. Furthermore, both τ and τ' decrease uniformly

TABLE IV. Dependence of τ and τ' upon α determined by the SCT method for the 1-d model of Cl_2/Ar.

$\alpha(\text{Å}^{-1})$	τ(ps)	τ'(ps)
2.84	26	54
3.78	55	404
5.44	147	950
7.56	300	>7500

TABLE V. Dependence of τ and τ' upon temperature determined by the SCT method for the 1-d model of Cl_2/Ar.

T(K)	τ(ps)	τ'(ps)
25	297	5522
50	147	950
100	34	190
150	19	60
200	16	40
500	8	4

with T at constant ρ (since the cell dimensions are the same for all simulations). This behaviour is understandable, as before in terms of elementary kinetic-molecular theory. The dependence of τ and τ' on α, however, does not seem as intuitively reasonable. This point will be discussed in another context in Section VI.B.

III. SEMICLASSICAL MODELS

In this Section we consider several treatments of VER which meld classical and quantum mechanics in an approximate description of a model. The nice feature of such approaches is that they lead to explicit expressions for the rate constant, which can be useful in interpreting VR in terms of the fundamental properties of the system.

A. General Treatments

Gouterman (1962) proposed a phenomenological theory of non-radiative transitions which is exactly analogous to the semiclassical theory of radiative transitions (see, for example, Schiff, 1955). The Hamiltonian of the system is written as

$$\mathbf{H} = \mathbf{H}^0 + \mathbf{H}' \tag{20}$$

where $\mathbf{H}^0$ is the (quantum mechanical) Hamiltonian of the free impurity and $\mathbf{H}'$ is the coupling to the medium. For $\mathbf{H}'$ Gouterman assumed the general form

$$\mathbf{H}' = \sum_{\alpha} \eta_\alpha x_\alpha F_x \cos(\omega t + \boldsymbol{\kappa}\cdot\mathbf{r}_\alpha), \tag{21}$$

where the sum on α is over impurity atoms located at positions $\mathbf{r}_\alpha$ (x_α is the x-coordinate of atom α), F_x is the x-component of the force exerted by the 'phonon' waves of the medium, and η_α is a coupling constant (taken to be proportional to the normal coordinates of the medium). In equation (21) $\boldsymbol{\kappa}$ is the phonon wave vector, whose magnitude is given approximately by

$$\omega = c_s \kappa, \tag{22}$$

where c_s is the speed of sound and ω is now the frequency of vibration of the normal modes of the medium. By an analysis which parallels that of emission and absorption of photons, one arrives at the Einstein coefficient for the spontaneous *non-radiative* transition

$$\begin{aligned} A^{s}_{n\to m} &= (4\omega^3_{nm}/3\hbar c^3_s)|\boldsymbol{\mu}^{s}_{mn}|^2, \omega_{nm} \le \omega_{\max} \\ &= 0, \omega_{nm} > \omega_{\max}, \end{aligned} \tag{23}$$

where

$$\omega_{nm} \equiv (\varepsilon_n - \varepsilon_m)/\hbar,$$
$$\boldsymbol{\mu}^{s}_{mn} \equiv e\sum_{\alpha}\eta_\alpha \langle m|\mathbf{r}_\alpha \exp(-i\boldsymbol{\kappa}\cdot\mathbf{r}_\alpha)|n\rangle,$$

n and m denote respectively the initial and final internal (zero-order) states of the impurity, and ε_n and ε_m are the corresponding energy levels. We note that equation (23) also rests on the assumption that the medium is a Debye solid having the cutoff frequency $\omega_{\max}$. Moreover, since the coupling $\mathbf{H}'$ is linear in the normal coordinates, only one-phonon emission processes are allowed. Hence, from equation (23) one concludes that only non-radiative transitions satisfying the condition

$$(\varepsilon_n - \varepsilon_m) \leq \hbar\omega_{\max} \tag{24}$$

are allowed.

Gouterman (1962) drew the following conclusions from his theory. First, one expects non-radiative transitions to be much faster in general than radiative transitions. Second, for low-energy transitions the non-radiative relaxation rate should be fairly sensitive to temperature, but less so for higher-energy transitions.

Within the context of studied of acoustical absorption in molecular crystals, Liebermann (1959) and Rasmussen (1967) developed a theory of VER which in spirit resembles the general theory of Gouterman (1962). The coupling between the intramolecular vibrational coordinate (Q) and the centre-of-mass (c.m.) displacement of the impurity, X, was taken (Rasmussen, 1967) as

$$H' = 2aQX(t)/\sqrt{3} + 4AQX(t)^3/\sqrt{3}, \tag{25}$$

where a and A are constants given explicitly in terms of the interaction potential U (and its derivatives) as

$$\begin{aligned} a &= 2U^{(2)} + 4U^{(1)}d^3, \\ A &= U^{(4)}/18 + 2U^{(3)}/9d + 2U^{(1)}/d^3, \\ U^{(m)} &\equiv (\partial^m U/\partial r^m)\big|_{r=d}, \end{aligned} \tag{26}$$

and d is the equilibrium distance between nearest neighbours. Now expressing $X(t)$ in terms of the normal modes (j) of the lattice as

$$X(t) = \sum_j Q_j \cos(\omega_j t + \phi_j),$$

one can carry out the usual analysis for a periodic time-dependent perturbation (Schiff, 1955). Neglecting the first term in equation (25) and assuming a Debye frequency distribution for the lattice, Rasmussen (1967) obtained for the VER rate constant

$$k = 9A^2N_0^4k_B^3T^3/[8hv_0mv_N^7(4\pi^2M)^3], \tag{27}$$

where N_0 is Avogadro's number, v_0 and m are, respectively, the frequency and effective mass associated with the intramolecular mode, $v_N \equiv \omega_{max}/2\pi$, and M is the molecular weight.

The theory of Liebermann (1959) and Rasmussen (1967) differs from that of Gouterman (1962) in two significant respects. First, the form of the coupling [equation (25)] indicates that both one- and three-phonon processes contribute in general to VER. Second, since the coupling constants a and A are given explicitly in terms of the interaction potential by equations (26), it is possible to calculate *a priori* the VER rate constant from an assumed known PES. Since in the Liebermann–Rasmussen model, as in all of those we are considering, the PES is taken to be simply a sum of pairwise interactions, only the two-body interaction potential U enters into the expression (27) for k.

B. A Binary-Collision Model

Sun and Rice (1965) have calculated the VER rate constant for the 1-d prototype using a semiclassical, binary-collision model. The intramolecular vibration of the diatomic impurity is treated quantum mechanically and the external lattice motions classically. The basic idea is that VER occurs via a sequence of relatively infrequent and essentially uncorrelated close encounters between the impurity and the nearest-neighbour lattice atoms. For the system considered by Sun and Rice, namely N_2/Ar, the amplitude of the internal vibration is so small compared to that of the lattice atoms that the N_2 can be regarded as a monatomic impurity, as far as the lattice motion is concerned. The rate of loss of internal vibrational energy is then given by

$$R_{n\to n-1} = \Gamma \int_{-\infty}^{-\infty} P_{n\to n-1}(v) f(v^2) \mathrm{d}v, \tag{28}$$

where Γ is the frequency of close encounters (of a kind to be specified shortly), $P_{n\to n-1}$ is the probability of transition between the diatomic's vibrational levels n and $n-1$ per collision at initial relative speed v, and f is the relative velocity distribution function.

The encounter frequency is defined as the frequency with which the distance between the c.m. of the diatomic and its nearest neighbour achieves a critical value q_0 and is equivalent to the frequency of occurrence of zeros in the function

$$\Delta = \sum_j (\alpha_{0j} - \alpha_{1j})(\varepsilon_j)^{1/2} \cos(\omega_j t + \phi_j) - q_0,$$

where α_{0j} and α_{1j} are the contributions to the c.m. and nearest-neighbour displacements of the jth normal lattice mode. By an analysis similar to that of Slater's (1959) theory of unimolecular reactions, one finds

$$\Gamma = \nu_0 \exp(-q_0^2/\alpha^2 k_B T), \tag{29}$$

where ν_0 and α^2 are given in terms of α_{0j} and α_{1j}.

The transition probability per collision $P_{n\to n-1}$, calculated using distorted-wave method (Mott and Massey, 1965) for an assumed impurity–lattice-atom interaction potential of the repulsive exponential form $e^{-\beta r}$, is given by

$$P_{n\to n-1}(v_0) = \frac{8\pi^4 \mu_m^2 \nu N}{h\mu\beta^2} \times \frac{\sinh(\pi z)\sinh(\pi z^*)}{[\cosh(\pi z) - \cosh(\pi z^*)]^2}, \tag{30}$$

where

$$z \equiv 4\pi^2 \mu_m v_0/\beta h$$
$$z^* \equiv 4\pi^2 \mu_m v_f/\beta h;$$

μ_m and μ are reduced masses

$$\mu_m = m_C(m_A + m_B)/(m_A + m_B + m_C)$$
$$\mu = m_A m_B/(m_A + m_B),$$

where the diatomic is AB and the nearest-neighbour lattice atom is C; $\nu \equiv \omega/2\pi$ is the vibrational frequency of the diatomic; v_0 and v_f are, respectively, the initial and final relative speeds of C with respect to the c.m. of AB.

The relative velocity distribution is determined by the Slater-type analysis to be

$$f(v^2) = (\pi\lambda k_B T)^{-1/2} \exp(-v^2/\lambda k_B T), \tag{31}$$

where

$$\lambda \equiv \sum_j (\alpha_{0j} - \alpha_{1j})^2 \omega_j^2$$

Using equation (28)–(31) and taking model parameters corresponding to the system N_2/Ar, Sun and Rice estimated $\tau' = 1/R_{10} \simeq 10^{-2}$ s, which seemed surprisingly long at the time. Although no measurements of τ' for VER within the electronic ground-state manifold of N_2/Ar seem to have been reported, Tinti and Robinson (1968) have found lifetimes of between 0.4 and 3.3 s for the $v' = 0$–4 vibrational levels of the electronic excited manifold ($A^3\Sigma_u^+$). Also, quite recently the vibrational lifetime of ground-state $N_2(v = 1)$ in the liquid phase at 77 K has been measured (Brueck and Osgood, 1976) to be about 100 s. Hence, the very large value of τ' calculated by Sun and Rice seems to be in relatively good agreement with experiment.

IV. QUANTUM MECHANICAL MODELS

We turn now to purely quantum-mechanical treatments of VER. Again, let us focus our attention on the prototype, taking the relevant subsystem to consist simply of the intramolecular vibration of the impurity.

A. Two-Level Models

Let us assume for the time being that the intramolecular vibrational mode has only two levels. (This assumption seems quite reasonable in case one is looking at the $1 \to 0$ transition of a diatomic molecule, or of the lowest-frequency mode of a polyatomic, having a fundamental frequency $\omega_{10} \geq k_B T/\hbar$ at the ambient temperature of the experiment.) The Hamiltonian for the system can be written

$$\mathbf{H} = \mathbf{H}^0 + \mathbf{V} = \mathbf{H}^0_{\mathrm{m}} + \mathbf{H}^0_{\mathrm{B}} + \mathbf{V}, \tag{32}$$

where $\mathbf{H}^0_{\mathrm{m}}$ and $\mathbf{H}^0_{\mathrm{B}}$ are, respectively, the Hamiltonians for the *unperturbed* intramolecular mode of the impurity (i.e. the RS) and for the unperturbed lattice or 'bath' (i.e. the IRS). $\mathbf{V}$ represents the coupling between the subsystems. The energy eigenkets of the isolated impurity mode satisfy the following relations:

$$\begin{aligned} \mathbf{H}^0_{\mathrm{m}}|i\rangle &= \varepsilon_i|i\rangle \\ \langle i|j\rangle &= \delta_{ij}, \quad i,j = 0,1 \\ \sum_{i=0}^{1} |i\rangle\langle i| &= 1. \end{aligned} \tag{33}$$

We wish to calculate the differential population $\Delta n(t)$, which for the present model is expressible as

$$\Delta n(t) = \langle \Delta \mathbf{n}(t) \rangle \equiv \langle \mathbf{n}(t) - \bar{n} \rangle, \tag{34}$$

where $\mathbf{n}$, the occupation-probability operator associated with the upper level ($i = 1$), is given by

$$\mathbf{n} = |1\rangle\langle 1| \tag{35}$$

and $\bar{n}$ is its equilibrium value. The brackets $\langle \dots \rangle$ in equation (34) signify an ensemble average. Hence, given that the system is characterized at $t = 0$ by the density operator $\boldsymbol{\rho}(0)$, we have

$$\Delta n(t) = \mathrm{Tr}\,\{\boldsymbol{\rho}(0)\Delta\mathbf{n}(t)\}. \tag{36}$$

If we assume that the impurity mode is initially (i.e. at $t = 0$) in level $i = 1$ and that the bath is in thermal equilibrium, then $\boldsymbol{\rho}$ takes the form

$$\boldsymbol{\rho}(0) = \boldsymbol{\rho}_{\mathrm{m}}(0) \otimes \boldsymbol{\rho}_{\mathrm{B}}(0), \tag{37}$$

where

$$\boldsymbol{\rho}_m(0) = |1\rangle\langle 1|, \tag{38a}$$

$$\boldsymbol{\rho}_B(0) = \exp(-\beta\mathbf{H}_B^0)/\mathrm{Tr}_B\{\exp(-\beta\mathbf{H}_B^0)\}, \tag{38b}$$

and $\mathrm{Tr}_B\{\ldots\}$ denotes a trace over the states of the bath. In equation (38b) $\beta \equiv (k_B T)^{-1}$. It follows from equations (37) and (38) that

$$\mathrm{Tr}\{\boldsymbol{\rho}(0)\} = \mathrm{Tr}_m\{\boldsymbol{\rho}_m(0)\} = \mathrm{Tr}_B\{\boldsymbol{\rho}_B(0)\} = 1. \tag{39}$$

Now using Zwanzig's (1960) projection-operator method, one can derive (Diestler, 1976a, 1976b) the following exact equation of motion for $\Delta n(t)$:

$$\Delta\dot{n}(t) = ig\Delta n(t) - \int_0^t \mathrm{d}t' K(t')\Delta n(t - t'), \tag{40}$$

where

$$\begin{aligned} g &\equiv \langle\langle \mathbf{L}\Delta\mathbf{n}\rangle\rangle \\ K(t') &\equiv \langle\langle \mathbf{L}\exp[i(1-\mathbf{P})\mathbf{L}t'](1-\mathbf{P})\mathbf{L}\Delta\mathbf{n}\rangle\rangle \\ \mathbf{P}\mathbf{X} &\equiv \Delta\mathbf{n}\langle\langle \mathbf{X}\rangle\rangle \\ \langle\langle \mathbf{X}\rangle\rangle &\equiv \langle \mathbf{X}\rangle/\langle\Delta\mathbf{n}\rangle \\ \langle \mathbf{X}\rangle &\equiv \mathrm{Tr}\{\boldsymbol{\rho}(0)\mathbf{X}\}. \end{aligned} \tag{41}$$

In equations (41) $\mathbf{X}$ denotes a general operator and $\mathbf{L}$ is the Liouville operator defined by

$$\mathbf{L}\mathbf{X} = \hbar^{-1}[\mathbf{H}, \mathbf{X}]. \tag{42}$$

Evaluating g and $K(t')$ to second order in the coupling $\mathbf{V}$ and then solving equation (40) in the van Hove (1955) weak-coupling limit, one obtains

$$\Delta n(t) = \Delta n(0)\exp(-t/\tau'), \tag{43}$$

where the VER rate is given explicitly by

$$\tau'^{-1} = \hbar^{-2}[\exp(\beta\hbar\omega_{10}) + 1] \int_{-\infty}^{+\infty} \mathrm{d}t \exp(i\omega_{10}t)\langle \mathbf{V}_{01}\mathbf{V}_{01}^{\dagger}(t)\rangle_B^0. \tag{44}$$

In equation (44)

$$\mathbf{V}_{01} \equiv \langle 0|\mathbf{V}|1\rangle, \tag{45}$$

$$\mathbf{V}_{01}^{\dagger}(t) \equiv \exp(i\mathbf{L}_B^0 t)\mathbf{V}_{01}^{\dagger} = \exp(i\mathbf{H}_B^0 t/\hbar)\mathbf{V}_{01}^{\dagger}\exp(-i\mathbf{H}_B^0 t/\hbar), \tag{46}$$

$$\langle \mathbf{X}\rangle_B^0 \equiv \mathrm{Tr}_B\{\boldsymbol{\rho}_B(0)\mathbf{X}\} \tag{47}$$

$$\omega_{10} \equiv (\varepsilon_1 - \varepsilon_0)/\hbar. \tag{48}$$

The result (44) has also been derived by Nitzan and Silbey (1974) using the cumulant-expansion technique of Kubo (1962).

B. Harmonic-Oscillator Models

A number of workers have calculated $\Delta n(t)$ for a harmonic oscillator coupled linearly (in the oscillator coordinate) to a thermal bath. In this model, the coupling takes the explicit form

$$\mathbf{V} = \mathbf{Q}\boldsymbol{\Gamma}' = (\mathbf{a} + \mathbf{a}^\dagger)\boldsymbol{\Gamma}, \tag{49}$$

where $\boldsymbol{\Gamma}$ is an arbitrary function of bath variables and $\mathbf{a}$ and $\mathbf{a}^\dagger$ are, respectively, the destruction and creation operators associated with the harmonic oscillator. In terms of $\mathbf{a}$ and $\mathbf{a}^\dagger$

$$\mathbf{H}_{\mathrm{m}}^0 = \mathbf{H}_{\mathrm{osc}}^0 = \hbar\omega(\mathbf{a}^\dagger\mathbf{a} + 1/2), \tag{50}$$

where ω is the classical frequency of the oscillator. Using the projection-operator method outlined above and taking

$$\boldsymbol{\rho}_{\mathrm{osc}}^{(0)} = |n\rangle\langle n|, \quad n = 0, 1, 2, \ldots, \tag{51}$$

Diestler and Wilson (1975) obtained

$$\tau'^{-1} = \hbar^{-2}[\exp(\beta\hbar\omega) - 1]^{-1} \int_{-\infty}^{+\infty} \mathrm{d}t \exp(i\omega t)\langle\boldsymbol{\Gamma}(0)\boldsymbol{\Gamma}(t)\rangle_{\mathrm{B}}^0. \tag{52}$$

Nitzan and Silbey (1974) arrived at an identical result via the cumulant-expansion method.

Nitzan and Jortner (1973) treated VER of a harmonic oscillator coupled to a *harmonic continuum* by an interaction of the form

$$\mathbf{V} = \mathbf{a}^\dagger\boldsymbol{\Gamma} + \mathbf{a}\boldsymbol{\Gamma}^\dagger, \tag{53}$$

where $\boldsymbol{\Gamma}$ was chosen explicitly to describe one-, two-, and many-phonon decay processes. In the case of one-phonon processes, for example, $\boldsymbol{\Gamma}$ is simply

$$\boldsymbol{\Gamma} = \sum_\nu \hbar G_\nu \mathbf{b}_\nu, \tag{54}$$

where the sum on ν is over normal modes of the continuum having associated destruction operators $\mathbf{b}_\nu$ and coupling constants G_ν. By invoking the random-phase approximation (RPA) (Rowe, 1968), which is equivalent to the assumption that the bath remains in equilibrium throughout the duration of the relaxation process, Nitzan and Jortner (1973) derived the effective linear Heisenberg equations of motion:

$$\dot{\mathbf{a}} = -i\omega\mathbf{a} - i\sum_\nu G_\nu\mathbf{B}_\nu \tag{55a}$$

$$\dot{\mathbf{B}}_\nu = -i\omega_\nu\mathbf{B}_\nu - in_\nu G_\nu^*\mathbf{a}, \tag{55b}$$

where

$$n_\nu \equiv \langle[\mathbf{B}_\nu, \mathbf{B}_\nu^\dagger]\rangle_{\mathrm{B}}^0 \tag{56}$$

and the specific forms of $\mathbf{B}_\nu$, G_ν, n_ν, and ω_ν depend upon that of $\mathbf{\Gamma}$. Note that in general ν stands for a set of phonon occupation numbers. Again, if $\mathbf{\Gamma}$ is given by equation (54), then $\mathbf{B}_\nu = \mathbf{b}_v$, $G_\nu = G_v$, $\omega_\nu = \omega_v$ and $n_\nu = 1$. Equations (55) can be solved by standard Laplace-transform techniques (see, for example, Louisell, 1973). Imposing the Wigner–Weisskopf approximation and also condition (51), one obtains finally

$$n(t) = \langle \mathbf{n}(t) \rangle = \exp(-t/\bar{\tau}')n(0) + [1 - \exp(-t/\bar{\tau}')]\bar{n} \tag{57}$$

where

$$\bar{\tau}'^{-1} = 2\pi \sum_\nu |G_\nu|^2 n_\nu \delta(\omega - \omega_\nu). \tag{58}$$

We note that equation (57) is identical in form to those obtained by Nitzan and Silbey (1974), and by Diestler and Wilson (1975). However, the precise connection between τ' given by equation (52) and $\bar{\tau}'$ by (58) is not clear, since the RPA was not employed in reaching (52). It is interesting that τ' is independent of the initial level (n) of the oscillator in all cases. Nitzan *et al.* (1974) have employed expression (58) to explore the role of multiphonon processes in VER.

Schurr (1971) considered the following 1-d model of the prototype: the diatomic impurity is represented by a particle A of mass m_A attached to a rigid wall on one side and to a chain of N identical particles of mass m_0 on the other side. The last atom of the chain is also attached to a rigid wall. All interactions are assumed harmonic. The Hamiltonian is written as

$$\mathbf{H} = \mathbf{H}^0 + \mathbf{H}', \tag{59}$$

where

$$\mathbf{H}^0 = \mathbf{H}_A + \mathbf{H}_{LC} \tag{60}$$

and $\mathbf{H}_A$ is the Hamiltonian of the unperturbed particle A and $\mathbf{H}_{LC}$ that of the unperturbed lattice. The form of the coupling is assumed to be

$$\mathbf{H}' = \sum_v F_v(\mathbf{b}_v + \mathbf{b}_v^\dagger) + (\mathbf{a} + \mathbf{a}^\dagger)\sum_v G_v(b_v + b_v^\dagger). \tag{61}$$

The first term of (61) gives rise to 'lattice relaxation' and the second to 'cascade damping'. The cascade-damping term of equation (61) is a special case of equation (49) and clearly gives rise only to one-phonon processes in the lattice. Using the Heitler–Ma (1949) perturbation theory and assuming that the particle A is initially in level n_A and that the lattice is at $T = 0$ K, Schurr calculated the cascade damping rate to be

$$\tau_{cd}^{-1} = m_0 n_A/2m_A)[\omega_{max}^2 - \omega_A^2]^{1/2}, \tag{62}$$

where ω_A is the fundamental frequency of the unperturbed particle A and ω_{max} is the cutoff frequency of the lattice. We note again that unless the fundamental frequency of the diatomic is below the cutoff, cascade damping is absent.

Unlike τ' [see equation (52) and $\bar{\tau}'$ [equation (58)], τ_{cd} [equation (62)] depends on the initial level of the diatomic impurity. The origin of this difference, discussed by Nitzan and Jortner (1973), is connected with the so-called harmonic-oscillator paradox.

C. Generalization of the Two-Level Model

The two-state result equation (44) can be generalized following the approach of Diestler (1976c). Consider the system consisting of a single solute molecule (m) and large number of solvent molecules (s). We express the total Hamiltonian as

$$\mathbf{H} = \mathbf{H}_{\mathrm{m}}^{0} + \mathbf{T}_{\mathrm{s}} + \mathbf{V}, \tag{63}$$

where $\mathbf{H}_{\mathrm{m}}^{0}$ is the Hamiltonian for the *isolated* internal vibrational mode of the solute, $\mathbf{T}_{\mathrm{s}}$ is the kinetic-energy operator for all 'external' motions, including translations and rotations of the solute, and $\mathbf{V}$ is the total potential energy of interaction between *all* molecules of the system.

Let $|i\rangle$ be an eigenket of the isolated internal Hamiltonian, i.e.

$$\mathbf{H}_{\mathrm{m}}^{0}|i\rangle = \varepsilon_i^{\mathrm{m}}|i\rangle, \tag{64}$$

where $\varepsilon_i^{\mathrm{m}}$ is the energy associated with the internal vibration. Then we may express a stationary state $|\Psi\rangle$ of the system as a linear combination of direct products of internal eigenkets $|i\rangle$ and corresponding external kets $|\chi_i\rangle$ as

$$|\Psi\rangle = \sum_i |i\rangle \otimes |\chi_i\rangle. \tag{65}$$

Requiring $|\Psi\rangle$ to be an eigenstate of the total Hamiltonian leads to the following set of coupled equations for the $|\chi_i\rangle$:

$$\mathbf{T}_{\mathrm{s}}|\chi_i\rangle + \sum_j \mathbf{V}_{ij}|\chi_j\rangle = (\varepsilon - \varepsilon_i^{\mathrm{m}})|\chi_i\rangle, \tag{66}$$

where

$$\mathbf{V}_{ij} \equiv \langle i|\mathbf{V}|j\rangle$$

and ε is the total energy of the system. As they stand, equations (66) are exact. If we neglect the off-diagonal elements of $\mathbf{V}$, however, then they decouple to give

$$\mathbf{H}_{\mathrm{s}i}^{0}|\chi_i^0\rangle = (\mathbf{T}_{\mathrm{s}} + \mathbf{V}_{ii})|\chi_i^0\rangle = (\varepsilon - \varepsilon_i^{\mathrm{m}})|\chi_i^0\rangle. \tag{67}$$

The $|\chi_i^0\rangle$ are thus eigenkets associated with the effective zero-order Hamiltonaian $\mathbf{H}_{\mathrm{s}i}^{0}$. The $|\chi_i^0\rangle$ describe the external motions in the average potential due to internal mode fixed in state $|i\rangle$. The approximation leading from equation (66) to (67) is precisely analogous to the distorted-wave Born approximation of collision theory (see, for example, Taylor, 1972). The basic idea is that the relative external motion of the molecules is affected by their internal states. Taking the effective potential $\mathbf{V}_{ii}$ into account gives a better zero-order description for the relative motion than entirely neglecting it (i.e. using the simple Born approximation).

Now rewriting the zero-order external eigenket as

$$|\chi_i^0\rangle = |\mathbf{n}_i\rangle,$$

where $\mathbf{n}_i$ denotes the collection of quantum numbers needed to specify the external state of the solvent, we can recast the Hamiltonian in terms of the complete set of states $\{|i\rangle \otimes |\mathbf{n}_i\rangle = |i\mathbf{n}\rangle\}$ as

$$\mathbf{H} = \mathbf{H}^0 + \mathbf{H}',$$

where

$$\mathbf{H}^0 = \sum_i \sum_{\mathbf{n}_i} \varepsilon^0_{i\mathbf{n}_i} |i\mathbf{n}_i\rangle\langle i\mathbf{n}_i| \equiv \sum_i \mathbf{H}^0_{si} |i\rangle\langle i| \tag{68a}$$

$$\mathbf{H}' = \sum_i \sum_{j \neq i} \sum_{\mathbf{n}_i} \sum_{\mathbf{n}_j} |i\mathbf{n}_i\rangle\langle j\mathbf{n}_j| \langle i\mathbf{n}_i|\mathbf{V}|j\mathbf{n}_j\rangle \tag{68b}$$

and the total energy of state $|i\mathbf{n}_i\rangle$ is

$$\varepsilon^0_{i\mathbf{n}_i} = \varepsilon^{\mathrm{m}}_i + \varepsilon^{\mathrm{s}}_{\mathbf{n}_i}. \tag{69}$$

Thus if the initial state of the system is given by $|\Psi(0)\rangle = |i\mathbf{n}_i\rangle$, then the coupling potentials $\langle i\mathbf{n}_i|\mathbf{V}|j\mathbf{n}_j\rangle$ give rise to transitions into other states $|j\mathbf{n}_j\rangle$ and are responsible for relaxation.

Now for reasons discussed above, let us assume that only two levels of the internal mode of the solute are involved in the relaxation process. Further, suppose the system to be described initially by the density operator

$$\begin{aligned} \boldsymbol{\rho}(0) &= |i\rangle\langle i| \otimes \boldsymbol{\rho}_{\mathrm{s}}(0) \\ \boldsymbol{\rho}_{\mathrm{s}}(0) &= \exp(-\beta\mathbf{H}^0_{si})/\mathrm{Tr}_{\mathrm{s}}\{\exp(-\beta\mathbf{H}^0_{si})\} \\ &= \exp(-\beta\mathbf{H}^0_{si})/Q_i \end{aligned} \tag{70}$$

Then calculating the transition rate in the weak-coupling limit, we get (Diestler, 1976c)

$$\begin{aligned} W_{i\to f} &= \lim_{t\to\infty} t^{-1}\mathrm{Tr}\{\boldsymbol{\rho}(t)|f\rangle\langle f|\} = \hbar^{-2}\int_{-\infty}^{+\infty} \mathrm{d}t \exp(-i\omega_{if}t) \\ &\quad \times \langle \mathbf{V}_{if}(0)\mathbf{V}^\dagger_{if}(t)\rangle^0_{\mathrm{s}}, \end{aligned} \tag{71}$$

where

$$\begin{aligned} \mathbf{V}_{if}(t) &\equiv \exp(i\mathbf{H}^0_{si}t/\hbar)\mathbf{V}_{if}(0)\exp(-i\mathbf{H}^0_{sf}t/\hbar) \\ \omega_{if} &\equiv (\varepsilon^{\mathrm{m}}_i - \varepsilon^{\mathrm{m}}_f)/\hbar \end{aligned} \tag{72}$$

and $\mathbf{H}_{si}^0$ is defined by equation (68a). It can be shown that

$$W_{f\to i} = \gamma_{if} W_{i\to f} \tag{73}$$

where

$$\gamma_{if} \equiv \exp(-\beta\hbar\omega_{if})Q_i/Q_f.$$

Also, for the two-level model we have the following relation between $W_{i\to f}$ and the lifetime of level i:

$$\tau_{if}^{-1} = (1+\gamma_{if})W_{i\to f} \tag{74}$$

Combining equations (71)–(74), we obtain finally

$$\tau_{if}^{-1} = \hbar^{-2}(\gamma_{fi}+1)\int_{-\infty}^{+\infty} \mathrm{d}t \exp(i\omega_{if}t) \times \langle \mathbf{V}_{if}^{\dagger}(0)\mathbf{V}_{if}(t)\rangle_{\mathrm{s}}^{0}. \tag{75}$$

Equation (75) reduces to equation (44) if the external states are assumed to be independent of the internal state of the solute.

D. Multiphonon Models

We (Diestler, 1974) applied the approach outlined in Section IV.C to VER in the prototypic system. We neglected rotations of the diatomic and took the lattice modes to be harmonic, i.e.

$$\mathbf{H}_{si}^0 = \sum_{v} \hbar\omega_v(\mathbf{b}_v^{\dagger}\mathbf{b}_v + \tfrac{1}{2}). \tag{76}$$

Furthermore, we assumed for the sake of simplicity that the equilibrium positions of the lattice modes simply shift as the diatomic undergoes a transition. Hence, taking the initial state i as a reference, one has for the final state

$$\mathbf{H}_{sf}^0 = \sum_{v} \{\hbar\omega_v(\mathbf{b}_v^{\dagger}\mathbf{b}_v + \tfrac{1}{2}) + \hbar\omega_v g_v^{(f)}(\mathbf{b}_v^{\dagger} + \mathbf{b}_v) + g_v^{(f)2}\hbar\omega_v\}, \tag{77}$$

where $g_v^{(f)}$ is proportional to the shift in the vth mode. Neglecting the dependence of $\mathbf{V}_{if}$ on the normal coordinates of the lattice, one obtains from equation (71)

$$W_{i\to f} \simeq \hbar^{-2}|\mathbf{V}_{if}(0)|^2 \int_{-\infty}^{+\infty} \mathrm{d}t \exp(-i\omega_{if}t)\,\mathrm{Tr}\{\boldsymbol{\rho}_{\mathrm{s}}(0)\exp(i\mathbf{H}_{sf}^0 t/h) \times \exp(-i\mathbf{H}_{si}^0 t/h)\}, \tag{78}$$

where $\mathbf{V}_{if}$ is evaluated at the equilibrium configuration of the lattice. By standard techniques of boson-operator algebra (Louisell, 1973) expression (78) may be recast as

$$W_{i\to f} = \hbar^{-2}|\mathbf{V}_{if}(0)|^2 \exp(R) \int_{-\infty}^{+\infty} \mathrm{d}t \exp[G(t)], \tag{79}$$

where

$$R \equiv -\sum_{\nu} g_{\nu}^{(f)2}(2\bar{n}_{\nu}+1) = -\sum_{\nu} g_{\nu}^{(f)2}\coth(\beta\hbar\omega_{\nu}/2)$$

$$G(t) \equiv -i\omega_{if}t + \sum_{\nu} g_{\nu}^{(f)2}[(\bar{n}_{\nu}+1)\exp(i\omega_{\nu}t) + \bar{n}_{\nu}\exp(-i\omega_{\nu}t)]$$

$$= -i\omega_{if}t + \sum_{\nu} g_{\nu}^{(f)2}\,\mathrm{csch}(\beta\hbar\omega_{\nu}/2)\cosh[i\omega_{\nu}t + \beta\hbar\omega_{\nu}/2] \tag{80}$$

and $\bar{n}_{\nu}$ is the equilibrium average occupation number of the νth normal lattice mode. Finally, evaluating the integral in equation (79) by the saddle-point approximation one arives at the expression

$$\begin{aligned} W_{i\to f} = {} & \hbar^{-2}|\mathbf{V}_{if}(\mathbf{0})|^{2}\{2\pi\sinh(\beta\hbar\omega_{\mathrm{m}}/2) \\ & \times [N_{\mathrm{m}}g_{\mathrm{m}}^{2}\omega_{\mathrm{m}}^{2}(1+\delta^{2})^{1/2}]^{-1}\}^{1/2}\exp[-N_{\mathrm{m}}g_{\mathrm{m}}^{2}(2\bar{n}_{\mathrm{m}}+1)] \\ & \times \exp\{\tfrac{1}{2}\beta\hbar\omega_{if} - \omega_{if}\ln[\delta+(1+\delta^{2})^{1/2}]/\omega_{\mathrm{m}} \\ & + N_{\mathrm{m}}g_{\mathrm{m}}^{2}(1+\delta^{2})^{1/2}[\sinh(\beta\hbar\omega_{\mathrm{m}}/2)]^{-1}\}, \end{aligned} \tag{81}$$

where

$$\delta \equiv \omega_{if}\sinh(\beta\hbar\omega_{\mathrm{m}}/2)/(N_{\mathrm{m}}g_{\mathrm{m}}^{2}\omega_{\mathrm{m}}), \tag{82}$$

ω_{if} is defined by equation (72) and ω_{m}, g_{m}, and N_{m} are the effective frequency, shift, and number, respectively, of lattice modes accepting energy in the relaxation process.

While rather complicated in appearance, equation (81) does express the VER time τ' [see equation (74)] explicitly in terms of the microscopic parameters $V_{if}(\mathbf{0})$, ω_{if}, ω_{m}, g_{m}, and N_{m}. However, as it is generally not possible to calculate these parameters *a priori*, expression (81) is really useful only in predicting general trends and in making semiempirical correlations. Indeed, it was used (Diestler, 1974) to fit semiquantitatively early experimental VER data for the system CO/Ar. The principal new feature distinguishing the present model from those discussed above [with the exception of that of Nitzan and Jortner (1973)] is that multiphonon processes arise naturally in the first order of perturbation theory. In this regard, the model is precisely analogous to earlier models for electronic relaxation in solids (Lax, 1952) and in isolated large molecules (Englman and Jortner, 1970).

Within the context of a master-equation approach to VER Lin (1974) arrived at a multiphonon-decay mechanism by another route. Considering a harmonic oscillator coupled linearly in Q to a thermal bath and taking Γ' [see equation (49)] to be

$$\Gamma' = C\exp(-\alpha\sum_{\nu}\gamma_{\nu}Q_{\nu}), \tag{83}$$

where Q_{ν} is the νth normal coordinate of the lattice, he obtained

$$W_{n\to n-1} = nC^{2}/(2\hbar\omega)\exp(S)\int_{-\infty}^{+\infty} \mathrm{d}t\,\exp[F(t)], \tag{84}$$

where

$$S \equiv \tfrac{1}{2}\sum_{v} \alpha^2 \gamma_v^2 \delta_v^{-2} \coth(\beta\hbar\omega_v/2)$$

and

$$F(t) \equiv -it\omega + \sum_{v} \frac{\alpha^2\gamma_v^2\delta_v^{-2}}{2} \operatorname{csch}(\beta\hbar\omega_v/2)\cosh[i\omega_v t + \beta\hbar\omega_v/2]. \quad (85)$$

Now comparing equations (79) and (80) with (84) and (85) and making the identifications

$$g_v^{(f)2} = \alpha^2\gamma_v^2\delta^{-2}/2$$

$$R = -S$$

we see that the expressions (79) and (84) for the rate constant are identical, except for a constant factor.

Lin *et al.* (1976) have explored the general features of $W_{n\to n-1}$ in various limits, as have Nitzan *et al.* (1975). Yakhot *et al.* (1975) have also applied this model, i.e. the exponential repulsive interaction, to the interpretation of radiative emission for excimers in solid neon.

Lin (1976) has proposed yet another multiphonon mechanism of VER based on an analogy with the Born–oppenheimer adiabatic separation of electronic and nuclear motions (see, for example, Bixon and Jortner, 1968). The basic idea is that since the internal vibrational motion of the impurity is much faster than the external motions of the solvent, the internal motion adjusts instantaneously to the external motions in exactly same fashion as the motion of electrons adjusts instantaneously to that of the nuclei. The same notion is behind the *perturbed stationary-state* method of collision theory (Mott and Massey, 1965).

In the formal treatment one replaces equation (65) with the expansion

$$\psi(Q;\mathbf{Q}_v) = \sum_i \chi_i(\mathbf{Q}_v)\phi_i(Q;\mathbf{Q}_v) \quad (86)$$

where $\mathbf{Q}_v$ denotes collectively the solvent coordinates and ϕ_i are perturbed internal stationary states of the impurity satisfying

$$[H_{\mathrm{m}}^0(Q) + V(Q,\mathbf{Q}_v)]\phi_i(Q;\mathbf{Q}_v) = \varepsilon_{\mathrm{i}}^{\mathrm{m}}(\mathbf{Q}_v)\phi_i(\mathbf{Q};\mathbf{Q}_v). \quad (87)$$

In equation (87) $\mathbf{Q}_v$ is to be regarded as a parameter. The external zero-order wavefunctions χ_i^0 are solutions of the effective Schrödinger equation

$$\{T_{\mathrm{s}}(\mathbf{Q}_v) + \varepsilon_i^{\mathrm{m}}(\mathbf{Q}_v) + \langle i|\mathbf{T}_{\mathrm{s}}|i\rangle - \varepsilon\}\chi_i^0(\mathbf{Q}_v) = 0, \quad (88)$$

which is analogous to equation (67). The coupling responsible for VER can be cast in a mixed representation as

$$\mathbf{H}' = \sum_i \sum_{j \neq i} |i\rangle\langle j| \{\langle i|\mathbf{T}_s|j\rangle - \hbar^2 \sum_v M_v^{-1} \langle i|\partial/\partial Q_v|j\rangle \partial/\partial Q_v\}, \tag{89}$$

where $|i\rangle$ denotes the perturbed stationary eigenket and M_v is the mass associated with the vth solvent coordinate. Now by procedures exactly analogous to those employed in the treatment of radiationless electronic transitions (see, for example, Fong, 1975), Lin arrived at the following expression for the transition rate

$$\begin{aligned} W_{i\to f} = {} & \hbar^{-2} \sum_v |R_v(if)|^2 \int_{-\infty}^{+\infty} \mathrm{d}t \exp(-i\omega_{if}t) \\ & \times (\omega_v/4\hbar)\{[\coth(\beta\hbar\omega_v/2)+1]\exp(i\omega_v t) + [\coth(\beta\hbar\omega_v/2)-1] \\ & \times \exp(-i\omega_v t)\} \prod_\sigma G_\sigma(t), \end{aligned} \tag{90}$$

where

$$\begin{aligned} R_v(if) &\equiv -\hbar^2 M_v^{-1} \langle i|\partial/\partial Q_v|f\rangle, \\ G_\sigma(t) &\equiv \exp\{-\Delta Q_\sigma^2(if)[\coth(\beta\hbar\omega_\sigma/2) \\ &\qquad - \operatorname{csch}(\beta\hbar\omega_\sigma/2)\cos(\omega_\sigma t - i\beta\hbar\omega_\sigma/2)]\}, \end{aligned} \tag{91}$$

and $\Delta Q_\sigma(if)$ is proportional to the shift in the equilibrium position of the normal-mode coordinate Q_σ. Lin investigated expression (90) in various limits. He found, in particular, that if the sum on v in equation (89) contains only a single non-vanishing term α and $\hbar\omega_\alpha \gg k_B T$, then equation (90) reduces to equation (79) with the identification

$$|V_{if}(\mathbf{0})|^2 = |R_\alpha(if)|^2 \omega/2\hbar.$$

In closing this Section, we note that in the so called weak-coupling limit, defined by

$$\delta \gg 1,$$

with δ given by equation (82), the multiphonon transition rate for the two-level-impurity model becomes simply

$$W_{i\to f} = c \exp[R(T)] \exp[-\omega_{if} f(T)/\omega_m] \tag{92}$$

where

$$f(T) \equiv \ln\{c'\omega_{if}/\omega_m[\bar{n}_m(T) - 1]\} - 1 \tag{93}$$

and c and c' are constants given in terms of the parameters [i.e. $V_{if}(\mathbf{0})$, $\omega_{if}, \omega_m, g_m, N_m$]. Equations (92) and (93) reveal the following characteristic features of multiphonon relaxation: (1) the energy-gap law, i.e. the rate decreases exponentially the transition-energy (gap), for a fixed effective frequency ω_m of the accepting lattice mode; (2) depending on the magnitude of c', the rate can vary strongly with temperature.

V. ROLE OF ORIENTATIONAL MOTIONS

Recent experiments (see, for example, Bondybey and Brus, 1975; Wiesenfeld and Moore, 1976) indicate that the energy-gap law predicted by the multiphonon models [see equation (92)] is strongly violated in the case of diatomic hydrides (e.g. NH, OH, HCl) in rare-gas matrices. Furthermore, observed VER rates are relatively insensitive to temperature. Bondybey and Brus (1975) suggested that rotation (or libration) of the impurity may function as a localized (more accurately, *pseudolocalized*; see Rebane (1970)) accepting mode. In support of this idea, Legay (1978) has demonstrated a nearly linear relation between log k and J_{max} (i.e. the effective final rotational quantum number of the diatomic if all of the initial vibrational energy is assumed to be converted into rotational energy). All of the multiphonon models considered in Section IV apply strictly only to the 1-d prototype, as they neglect orientational motions of the diatomic impurity. During the last couple of years several groups have developed models that account, quantitatively in some instances, for the role of orientational motions in VER. In what follows we shall describe these models in some detail.

A. Three-Dimensional Models

Berkowitz and Gerber (1977) considered the full 3-d prototype (Fig. 1), taking the impurity-lattice coupling V to be a sum of two-body interactions of the form

$$V = \sum_{i=1}^{Z} V_i(r_i, \phi_i, \rho) = \sum_i \sum_l U_l(r_i, \rho) P_l(\cos \phi_i), \tag{94}$$

where the sum on i is over the Z nearest neighbours, $\mathbf{r}_i$ is the vector distance between the centre-of-mass (c.m.) of the diatomic and lattice atom i, ϕ_i is the angle between $\mathbf{r}_i$ and the diatomic's internuclear axis, ρ is the displacement of the internuclear separation from its equilibrium value, and P_l is a Legendre polynomial. For U_l they assumed the repulsive exponential form

$$U_l(r_i, \rho) = A_l(\rho) \exp(-\alpha r_i), \tag{95}$$

where A_l is essentially linear in ρ. Now expanding r_i in terms of the displacements from equilibrium of the c.m. of the diatomic ($\mathbf{X}$) and of the lattice atoms, $\mathbf{u}_i$, one has

$$r_i \simeq r_i^0[1 + \mathbf{n}_i^0 \cdot (\mathbf{u}_i - \mathbf{X})], \tag{96}$$

where

$$\mathbf{n}_i^0 \equiv \mathbf{r}_i^0 / r_i^0$$

and $\mathbf{r}_i^0$ is the equilibrium value of $\mathbf{r}_i$, Substituting equations (95) and (96) into (94) and using the addition theorem for spherical harmonics, one obtains

$$V = \sum_i \sum_{lm} F_{ilm}(\rho, \theta, \phi) \exp[-\alpha \mathbf{n}_i^0 \cdot (\mathbf{u}_i - \mathbf{X})], \tag{97}$$

where

$$F_{ilm}(\rho, \theta, \phi) \equiv 4\pi Y_{lm}(\theta_i, \phi_i) Y_{lm}(\theta, \phi) A_l(\rho) \exp(-\alpha r_i^0)/(2l+1),$$

Y_{lm} is a spherical harmonic function, and (θ, ϕ) denote the polar coordinates specifying the orientation of the diatomic.

Taking the lattice to be harmonic Berkowitz and Gerber (1977) recast expression (97) in the usual occupation-number representation as

$$V = \sum_i \sum_{lm} F_{ilm}(\rho, \theta, \phi) \exp[-\sum_v c_v(\mathbf{b}_v + \mathbf{b}^\dagger) - c_\mathrm{L}(\mathbf{b}_\mathrm{L} + \mathbf{b}_\mathrm{L}^\dagger)], \tag{98}$$

where

$$c_v \equiv \alpha(\hbar\omega_v/2N)^{1/2} D$$
$$c_\mathrm{L} \equiv \alpha(\hbar/2\omega_v M)^{1/2}, \tag{99}$$

M is the mass of the impurity, N is the number of lattice atoms, and D is a constant independent of v. Here v refers to the bulk lattice modes and L to the (true) local mode.

One can now evaluate the VER rate using equation (71) and taking the unperturbed solvent Hamiltonian to be

$$\mathbf{H}_\mathrm{s}^0 = \mathbf{H}_\mathrm{rot}^0 + \mathbf{H}_\mathrm{L}^0, \tag{100}$$

where $\mathbf{H}_\mathrm{rot}^0$ refers to the orientations and $\mathbf{H}_\mathrm{L}^0$ to the harmonic lattice [see equation (76)]. Substituting equation (98) into (71) and explicitly evaluating the trace over the orientations, one gets

$$W_{n \to n'} = \hbar^{-2} \sum_{JM} p_{JM} \sum_{J'M'} |F_{nJM}^{n'J'M'}|^2 \int_{-\infty}^{+\infty} \mathrm{d}t \exp(-i\Omega_{nJM}^{n'J'M'} t) H(t), \tag{101}$$

where, for the time being, we take J and M to be generalized quantum numbers specifying the orientational state, with associated probability of occupation p_{JM}. The various additional symbols appearing in equation (101) are defined by

$$\Omega_{nJM}^{n'J'M'} \equiv [\varepsilon_{n'} + \varepsilon_{J'M'} - (\varepsilon_n + \varepsilon_{JM})]/\hbar, \tag{102}$$

$$\begin{aligned} H(t) \equiv \mathrm{Tr_L}[\boldsymbol{\rho}_\mathrm{L}(0) \exp\{ -\sum_{v'} c_{v'}[\mathbf{b}_{v'}(t) + \mathbf{b}_{v'}^\dagger(t)] \\ - c_\mathrm{L}[\mathbf{b}_\mathrm{L}(t) + \mathbf{b}_\mathrm{L}^\dagger(t)]\} \exp\{ -\sum_{v'} c_{v'}[\mathbf{b}_{v'} + \mathbf{b}_{v'}^\dagger] \\ - c_\mathrm{L}[\mathbf{b}_\mathrm{L} + \mathbf{b}_\mathrm{L}^\dagger]\}], \end{aligned} \tag{103}$$

$$F_{nJM}^{n'J'M'} = \sum_i \sum_{lm} \langle nJM | F_{ilm}(\rho, \theta, \phi) | n'J'M' \rangle, \tag{104}$$

ε_n and ε_{JM} are the unperturbed energy levels of the vibration and rotation, respectively, and $\boldsymbol{\rho}_\mathrm{L}(0)$ is the density operator given by equation (70), except with $\mathbf{H}_\mathrm{L}^0$ in place of $\mathbf{H}_{si}^0$. Comparing equations (79) with (101) and (103), we see that multiphonon processes are included in this model along with multiquantum rotational transitions.

Gerber and Berkowitz (1977) applied their model to the case NH, ND/Ar, making the following approximations:

(i) the true NH–Ar interaction potential is identical to that for HCl–Ar (Neilsen and Gordon, 1973);
(ii) the unperturbed orientational states of the diatomic are taken as free rigid-rotor states;
(iii) multiphonon processes take place only in the local modes, one-phonon processes in the bulk modes;
(iv) the local mode has zero width.

They obtained very good agreement with experiment, calculating $\tau' = 4.2 \times 10^{-4}$ s for NH($n = 1$)/Ar. The observed value is 1.9×10^{-4} s (Bondybey and Brus, 1975). They concluded that for the system NH, ND/Ar the dominant accepting mode is the rotation with very little excess energy taken by the lattice modes.

B. Two-Dimensional Models

Freed and coworkers (Freed and Metiu, 1977; Freed *et al.*, 1977) investigated the role of rotations in VER in the 2-d prototype (see Fig. 2). The diatomic was taken to be harmonic and the coupling potential to be of the form

$$V = \rho F(\phi, \{Q_v\}), \tag{105}$$

where the angle ϕ is defined in Fig. 2 and $\{Q_v\}$ stands for the set of normal coordinates of the lattice. Now assuming that the lattice atoms are fixed

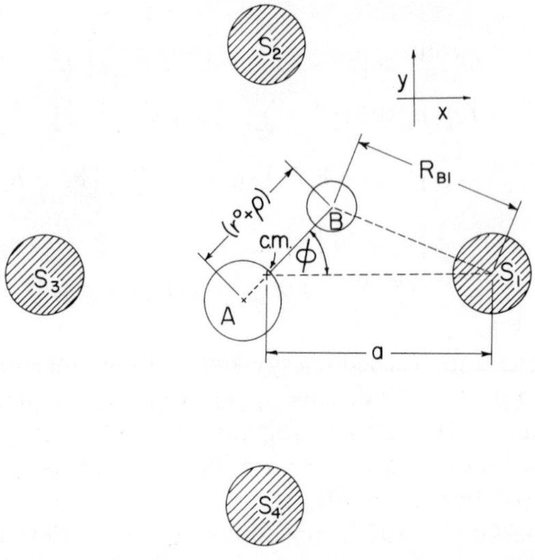

Fig. 2. Diatomic molecule AB occupying a substitutional site in a two-dimensional square lattice of rare-gas atoms S. The coordinate ρ is the displacement of the internuclear separation from its equilibrium value r_0.

at their equilibrium positions, so that only the orientational motion of the diatomic is included the 'solvent' (irrelevant subsystem) one can express the VER transition rate between vibrational levels $n = 1$ and $n = 0$ as

$$k_{\text{vib}} \equiv W_{1\to 0} = \pi\mu^{-1}\omega^{-1}\sum_{m} p_m \sum_{m'} |\langle m|F(\phi)|m'\rangle|^2 \times \delta(\hbar\omega + \varepsilon_m - \varepsilon_{m'}). \tag{106}$$

Equation (106) follows directly from equation (71), (105) and the standard relation

$$2\pi\delta(\omega) = \int_{-\infty}^{+\infty} dt \exp(\pm i\omega t).$$

In equation (106) μ and ω are the reduced mass and fundamental frequency of the diatomic; ε_m is the energy of unperturbed orientational state m, having associated occupation probability p_m.

Now let us imagine that the lattice atoms remain fixed and the diatomic is rotated about its c.m. fixed at the substitutional site. If the potential experienced by the diatomic has n-fold symmetry, we can write F as a Fourier representation

$$F(\phi) = \sum_k F_k \cos nk\phi, \tag{107}$$

where the F_k are constants. Taking the unperturbed orientational states to be free rigid-rotor states, Freed and Metiu (1977) obtained

$$k_{\text{vib}} = [\Delta/\mu\hbar^3\omega Q_r B^2(4\pi)^3] \sum_{m=-\infty}^{+\infty} \exp(-\beta m^2\hbar\bar{B})$$
$$\times \sum_k |F_k|^2\{([(nk+m)^2 - m^2\bar{B}/B - \omega/B]^2 + (\Delta/B)^2)^{-1}$$
$$+ ([(nk-m)^2 - m^2\bar{B}/B - \omega/B]^2 + (\Delta/B)^2)^{-1}\}, \tag{108}$$

where Q_r is the rotational partition function and $\bar{B}$ and B are the moments of inertia of the diatomic in vibrational levels 1 and 0, respectively. Note that the delta function in Equation (108) has been replaced by a Lorentzian of width Δ. This device effectively accounts for the coupling of the rotation to the lattice modes.

Freed and Metiu (1977) explored some of the consequences of their model numerically using parameters corresponding to Ar_2 as the substitutional impurity in a simple cubic lattice of Ar atoms. They assumed pairwise additive interactions between all Ar atoms. To determine V sufficiently accurately, they found it necessary to take into account interactions between atoms separated considerably beyond the nearest-neighbour distance (a). However, it was observed that F was well fitted by the form

$$F(\phi) = V_0 \exp[\alpha \cos n\phi],$$

from which one deduces that

$$F_0 = V_0 I_0(\alpha)$$
$$F_k = 2V_0 I_k(\alpha),\ k \neq 0 \tag{109}$$

where I_k is a modified Bessel function (Abramowitz and Stegun, 1965). Using equations (108) and (109), Freed and Metiu focused on the dependences of k_{vib} upon $(\omega/B)^{1/2}$ and upon temperature for various sets of parameters. Their principal conclusions were that

(i) $\log k_{\text{vib}}$ varies linearly with $(\omega/B)^{1/2}$ for large ω/B; and
(ii) k_{vib} is independent of temperature up to $T \simeq 20\hbar B/k_B$.

Since $(\omega/B)^{1/2} \simeq m_{\text{max}}(J_{\text{max}})$, the first conclusion is in accord with the correlation observed by Legay (1978). The second is also in agreement with the observed weak temperature dependences of k mentioned at the beginning of this Section.

We (Diestler *et al.*, 1978) recently studied the role of orientations in VER using a 2-d prototypic model similar to that of Freed and coworkers. We took $n = 4$ (see Fig. 2) and expressed the interaction of the diatomic with its nearest neighbours as

$$V = V(\rho, \phi) = \sum_{z=x,y} \sum_{i=1}^{4} U(R_{zi}), \tag{110}$$

where $U(R_{zi})$ is the interaction potential between atom $Z(=A, B)$ of the diatomic and nearest-neighbour lattice atom i. These potentials were assumed to be of the Lennard–Jones (12–6) form, i.e.

$$U(R_{zi}) = 4\varepsilon'_{zs}[(\sigma'_{zs}/R_{zi})^{12} - (\sigma'_{zs}/R_{zi})^{6}], \tag{111}$$

where the parameters ε'_{zs} and σ'_{zs} are appropriate to a given pair of atoms Z and S. Keeping the c.m. of the diatomic fixed we expanded the potential function (110) in a Taylor series as

$$V(\rho, \phi) = V_0(\phi) + \rho V_1(\phi) + \ldots \tag{112}$$

and fitted $V_i(\phi)$ to the form

$$V_i(\phi) = A_i \exp(\alpha_i \cos 4\phi) + C_i, i = 0, 1, \tag{113}$$

by a least-squares procedure. The zero-order orientational Hamiltonian was taken as that of a rigid rotor hindered by potential $V_0(\phi)$. The lowest orientational (librational) states of the hindered rotor were computed numerically by the linear variational method and the highest states by first-order perturbation theory. Effectively, the final orientational states involved in the rate expression are free.

For the VER rate constant (at $T = 0$ K) corresponding to the $n \to n-1$ transition of the diatomic, one has from equation (71)

$$k_{n\to n-1} = \pi n\mu^{-1}\omega^{-1}\sum_m |\langle 0|\mathbf{V}_1|m\rangle|^2\delta(\hbar\omega + \varepsilon_0 - \varepsilon_m), \tag{114}$$

where ε_0 is the energy of the initial librational state $|0\rangle$ and m labels the final free-rotor states. By evaluating the coupling matrix element $\langle 0|\mathbf{V}_1|m\rangle$, converting the sum on m to an integral, and performing that, we (Diestler *et al.*, 1978) approximated expression (114) as

$$k_{n\to n-1} = 4N_0^2\pi^{1/2}n\mu^{-1}\omega^{-1}A_1^2[\omega\hbar B(\hbar\omega + \varepsilon'_0)]^{-1/2} \times \left|\sum_{k=1}^{\infty}(-1)^k I_k(|\alpha_1|)\exp[-8(k-\bar{m})^2/\omega]\right|^2, \tag{115}$$

where

$$\begin{aligned} \varepsilon'_0 &\equiv \varepsilon_0 - C_0 - A_0 I_0(\alpha_0) \\ \bar{m} &\equiv [(\hbar\omega + \varepsilon'_0)/16\hbar B]^{1/2} \\ \omega &\equiv (I\kappa/\hbar^2)^{1/2}. \end{aligned} \tag{116}$$

In equation (116) κ is the effective force constant determined by fitting V_0 to the quadratic form $\kappa\phi^2/2$ about a minimum; N_0 is the normalization constant associated with the initial librational wavefunction. We emphasize that in arriving at equation (115) it was assumed that $T = 0$ K. Lattice-dynamical effects were also ignored entirely. Thus the only free parameters of the model are those associated with the interaction potential $U(R_{zi})$, i.e. ε'_{zs} and σ'_{zs}.

We (Diestler *et al.*, 1978) employed expression (115) to calculate VER rate constants for the following systems: $NH(X^3\Sigma^-)$, $ND(X^3\Sigma^-)$/Ar, Kr; HCl, DCl/Ar; $OH(A^2\Sigma^+)$, $OD(A^2\Sigma^+)$/Ne; CO/Ar. As is generally the case in model calculations of this nature, the biggest problem is that the required interaction potentials are simply not available. We did devise a very crude method of estimating ε' and σ'. However, our general approach was to vary these parameters over 'reasonable' ranges, attempting to get as good an agreement as possible with the available experimental VER data. Our major conclusions were that:

(i) for some hydride systems (e.g. NH, ND/Ar and $OH(A^2\Sigma^+)$, $OD(A^2\Sigma^+)$/Ne) orientational motions apparently play the major role in VER;
(ii) for CO/Ar orientational motions are irrelevant to VER;
(iii) for certain hydride systems (e.g. HCl, DCl/Ar; $NH(A^3\Pi)$, $ND(A^3\pi)$/Kr) it appears that orientational motions do not alone govern VER.

VI. EXTENDED VIEW OF LOCAL-MODE PARTICIPATION

Up to this point all of the models of VER which we have considered effectively take the RS to be the 'bare' intramolecular vibration of the impurity. The remaining d.f., i.e. the orientations and lattice vibrations, are lumped together to constitute a thermal bath (i.e. the IRS). The coupling between the two is then assumed weak and perturbation theory is used to derive the working-model expressions for the rate constant. Not all of the available VER data are satisfactorily accounted for by such models, as we have noted at the end of Section V.

In this Section we consider an extended view of the participation of local (or, more accurately, pseudolocal) modes in VER. The approach, which is analogous to that of the SCT method discussed in Section II.B, recognizes that in some systems the interaction of the impurity with the surrounding lattice atoms may be sufficiently strong to lead to the formation of a complex (or pseudomolecule). In fact, the formation of such complexes has been invoked (Goodman and Brus, 1976) to account for the relaxation behaviour of NH and ND in Ar–Kr alloys. Also, Ewing (1978) has explored the role of van der Waals complexes in VER in the gas phase.

Since the details of the motion of the pseudomolecule are presumably important to an accurate description of VER in such systems, one should take the pseudomolecule itself as the RS rather than simply the 'bare' vibration of the impurity. The coupling of the pseudomolecule to the remainder of the system is relatively weak and can be handled by perturbation theory, as before.

In the spirit of this idea we shall present a partial theory of local-mode participation and discuss its application to a model for the role of pseudo-molecular vibrations in VER in the 1-d prototype.

A. General Theory

We again focus on the system consisting of a single impurity (solute) molecule and a large number of solvent molecules and write the complete Hamiltonian as

$$\begin{aligned} \mathbf{H} &= \mathbf{H}_{\rm v}^{0} + \mathbf{H}_{\rm L}^{0} + \mathbf{H}_{\rm B}^{0} + \mathbf{H}_{\rm vL} + \mathbf{H}_{\rm LB}, \\ &= \mathbf{H}^{0} + \mathbf{H}_{\rm vL} + \mathbf{H}_{\rm LB}, \end{aligned} \tag{117}$$

where $\mathbf{H}_{\rm v}^{0}$, $\mathbf{H}_{\rm L}^{0}$ and $\mathbf{H}_{\rm B}^{0}$ are the zero-order Hamiltonians corresponding to the isolated intramolecular vibration (v), local mode (L) and bath (B). The bath includes all d.f. of the system except the internal impurity mode (v) and the local mode (L). The internal mode is assumed to be coupled to the local mode by the interaction $\mathbf{H}_{\rm vL}$; in turn, the local mode is coupled to the bath by $\mathbf{H}_{\rm LB}$. Direct interactions between the internal mode and the bath are neglected.

The energy eigenkets of the zero-order Hamiltonians satisfy the following relations:

$$\mathbf{H}_{\rm v}^{0}|v\rangle = \varepsilon_{v}^{\rm v}|v\rangle \; ; \; \sum_{v}|v\rangle\langle v| = 1 \tag{118a}$$

$$\mathbf{H}_{\rm L}^{0}|\lambda\rangle = \varepsilon_{\lambda}^{\rm L}|\lambda\rangle \; ; \; \sum_{\lambda}|\lambda\rangle\langle\lambda| = 1 \tag{118b}$$

$$\mathbf{H}_{\rm B}^{0}|\mathbf{n}\rangle = \varepsilon_{\mathbf{n}}^{\rm B}|\mathbf{n}\rangle \; ; \; \sum|\mathbf{n}\rangle\langle\mathbf{n}| = 1, \tag{118c}$$

where $\mathbf{n}$ stands for the *set* of quantum numbers required to specify the many-body states of the bath. Note also that, in general, λ denotes the *set* of quantum numbers required to characterize the states of the possibly several-dimensional subsystem of local modes. However, we assume that the single quantum number v is sufficient to label the states of the impurity mode. Hence, our treatment applies strictly only to a diatomic impurity.

The internal local-mode coupling is assumed to be linear in the internal vibrational coordinate (Q), i.e.

$$\mathbf{H}_{\rm vL} = \mathbf{U}_{\rm vL}\mathbf{Q}. \tag{119}$$

Now taking the bath to be a harmonic continuum and assuming that the local-mode-bath coupling is linear in the normal modes (Q_j) of the bath, one can represent $\mathbf{H}_{LB}$ as

$$\mathbf{H}_{LB} = \sum_j \mathbf{U}_j^L \mathbf{Q}_j = \sum_j \sum_\lambda \sum_{\lambda'} \sum_{n_j} \sum_{n'_j} \langle \lambda | \mathbf{U}_j^L | \lambda' \rangle \langle n_j | \mathbf{Q}_j | n'_j \rangle | \lambda \rangle \langle \lambda' | \otimes | n_j \rangle \langle n'_j |. \tag{120}$$

Anticipating deriving the rate expression for the bath initially at 0 K, we shall include only zero and one-phonon states of the bath. Further, we shall allow only for one-quantum transitions between adjacent levels of the local mode. Then, with these assumptions and the standard relation

$$\langle n_j | \mathbf{Q}_j | n'_j \rangle = (\hbar/2m_j\omega_j)^{1/2}[\sqrt{n'_j}\,\delta_{n_j,n'_j-1} + \sqrt{n'_j+1}\,\delta_{n_j,n'_j+1}],$$

where m_j and ω_j are the effective mass and frequency of the jth normal mode, equation (120) reduces to

$$\mathbf{H}_{LB} = \sum_j \sum_\lambda \{\kappa_{j\lambda} | \lambda \rangle \langle \lambda - 1 | \otimes | 0_j \rangle \langle 1_j | + \kappa_{j\lambda}^* | \lambda - 1 \rangle \langle \lambda | \otimes | 1_j \rangle \langle 0_j |\}, \tag{121}$$

where

$$\kappa_{j\lambda} \equiv (\hbar/2m_j\omega_j)^{1/2} \langle \lambda | \mathbf{U}_j^L | \lambda - 1 \rangle \tag{122}$$

We wish now to calculate the VER rate constant by solving the time-dependent Schrödinger equation approximately. We represent the state vector of the system as

$$|\psi(t)\rangle = a(t)|1, 0, \mathbf{0}\rangle + \sum_\lambda b_\lambda(t)|0, \lambda, \mathbf{0}\rangle + \sum_\lambda \sum_j c_{\lambda j}(t)|0, \lambda, 1_j\rangle, \tag{123}$$

where

$$|v, \lambda, n_j\rangle = |v\rangle \otimes |\lambda\rangle \otimes |n_j\rangle$$

are zero-order eigenstates of $\mathbf{H}^0$. Note that the form of the expansion (123) assumes implicitly that only two levels ($v = 0$ and $v = 1$) of the internal mode are involved. The assumptions that the internal mode is initially in a pure state ($v = 1$) and that the bath is at 0 K determine the initial conditions: $a(0) = 1; b_\lambda(0) = 0; c_{\lambda j}(0) = 0$. Requiring that $|\psi(t)\rangle$ satisfies the Schrödinger equation yields the following set of coupled equations:

$$\dot{a}(t) + i\omega_v a(t) = -i\hbar^{-1} Q_{10} \sum_\lambda u_{0\lambda} b_\lambda(t) \tag{124a}$$

$$\dot{b}_\lambda(t) + i\omega_\lambda b_\lambda(t) = -i\hbar^{-1} \sum_j \kappa_{j\lambda} c_{\lambda-1,j}(t) - i\hbar^{-1} Q_{01} u_{\lambda 0} a(t) \tag{124b}$$

$$\dot{c}_{\lambda-1,j}(t) + i\omega_{\lambda-1,j} c_{\lambda-1,j}(t) = -i\hbar^{-1} \kappa_{j\lambda} b_\lambda(t), \tag{124c}$$

where

$$
\begin{aligned}
\omega_{\mathrm{v}} &\equiv \varepsilon_1^{\mathrm{v}}/\hbar \\
\omega_\lambda &\equiv \varepsilon_\lambda^{\mathrm{L}}/\hbar \\
\omega_{\lambda-1,j} &\equiv (\varepsilon_{\lambda-1}^{\mathrm{L}} + \varepsilon_j^{\mathrm{B}})/\hbar \\
Q_{01} &\equiv \langle 0|\mathbf{Q}|1\rangle = Q_{10}^* \\
u_{0\lambda} &\equiv \langle 0|\mathbf{U}_{\mathrm{vL}}|\lambda\rangle = u_{\lambda 0}^*
\end{aligned}
\tag{125}
$$

and we have taken the zero-point of energy to correspond to that of the zero-order state $|0, 0, \mathbf{0}\rangle$. Now integrating equation (124c) formally and substituting the result into equation (124b), one can recast equations (124a) and (124b) as

$$
\dot{A}(t) = -i\hbar^{-1} Q_{10} \sum_\lambda u_{0\lambda} \exp[i(\omega_{\mathrm{v}} - \omega_\lambda)t] B_\lambda(t) \tag{126a}
$$

$$
\begin{aligned}
\dot{B}_\lambda(t) = {} & -i\hbar^{-1} Q_{01} u_{\lambda 0} \exp[i(\omega_\lambda - \omega_{\mathrm{v}})t] A(t) \\
& - \hbar^{-2} \sum_j |\kappa_{j\lambda}|^2 \exp[i(\omega_\lambda - \omega_{\lambda-1,j})t] \\
& \times \int_0^t \mathrm{d}t' \exp[-i(\omega_\lambda - \omega_{\lambda-1,j})t'] B_\lambda(t')
\end{aligned}
\tag{126b}
$$

where

$$
\begin{aligned}
A(t) &\equiv \exp(i\omega_{\mathrm{v}} t) a(t) \\
B_\lambda(t) &\equiv \exp(i\omega_\lambda t) b_\lambda(t).
\end{aligned}
\tag{127}
$$

Now in order to solve equation (126b), we recall that the local-mode–bath coupling has been assumed to be weak, i.e. the coupling constants $\kappa_{j\lambda}$ satisfy the relation

$$
\kappa_{j\lambda} \ll \hbar(\omega_\lambda - \omega_{\lambda-1}).
$$

Hence, equation (126b) is in a form suitable for application of the Principle of Averaging (PA) (Filatov, 1967). The inhomogeneous term in equation (126b) corresponds to a 'pumping' of the local mode by the internal mode. Now applying the PA (see, for details, Diestler and Ladouceur, 1977), one obtains

$$
B_\lambda(t) \simeq \bar{B}_\lambda(t) = -i\hbar^{-1} Q_{01} u_{\lambda 0} \exp(-\Omega_\lambda t) \int_0^t \mathrm{d}t' \exp[i(\omega_\lambda - \omega_{\mathrm{v}} - i\Omega_\lambda)t'] A(t'), \tag{128}
$$

where the bar on $\bar{B}_\lambda$ is intended to stress that it is the 'averaged' solution. In equation (128)

$$
\Omega_\lambda \equiv i\Delta\omega_\lambda + \gamma_\lambda, \tag{129}
$$

where

$$\Delta\omega_\lambda \equiv \hbar^{-2}\sum_j \frac{|\kappa_{j\lambda}|^2 P}{\omega_\lambda - \omega_{\lambda-1,j}}$$
$$\gamma_\lambda \equiv \pi\hbar^{-2}\sum_j |\kappa_{j\lambda}|^2 \delta(\omega_\lambda - \omega_{\lambda-1,j}) \tag{130}$$

and P is the principal-value operator. Here, $\Delta\omega_\lambda$ and γ_λ are, respectively, the shift and width of the local-mode level λ arising from its interaction with the bath. They are exactly analogous to the corresponding radiative quantities (Louisell, 1973).

At this point it remains to solve the coupled equations (126a) and (128). Since we have assumed that the internal mode may be strongly coupled to the local mode we cannot, in general, use perturbation theory. An exact solution would necessitate the diagonalization of the Hamiltonian of the relevant subsystem, i.e. $\mathbf{H}_v^0 + \mathbf{H}_L^0 + \mathbf{H}_{vL}$. However, in the interest of arriving at a transparent and hopefully useful expression for the VER time, we shall choose an approximate method of solution.

Now substituting expression (128) for B_λ into equation (126a), we obtain

$$\dot{A}(t) = -\hbar^{-2}|Q_{10}|^2 \sum_\lambda |u_{0\lambda}|^2 \exp[i(\omega_v - \omega_\lambda + i\Omega_\lambda)t]$$
$$\times \int_0^t dt' \exp[-i(\omega_v - \omega_\lambda + i\Omega_\lambda)]t' A(t') \tag{131}$$

If the internal local-mode coupling is sufficiently weak (i.e. $|Q_{10}||u_{0\lambda}| \leq \hbar\omega_v$), then we can again apply the PA to Equation (131), obtaining finally

$$A(t) = \exp[-i\Delta\omega_v - \Gamma_v/2]t, \tag{132}$$

where

$$i\Delta\omega_v + \Gamma_v/2 \equiv i\hbar^{-2}|Q_{10}|^2 \sum_\lambda \frac{|u_{0\lambda}|^2\{(\omega_v - \omega_\lambda - \Delta\omega_\lambda) - i\gamma_\lambda\}}{[(\omega_v - \omega_\lambda - \Delta\omega_\lambda)]^2 + \gamma_\lambda^2} \tag{133}$$

with $\Delta\omega_\lambda$ and γ_λ given by equations (130).

The vibrational energy of the internal mode is given by

$$\langle\psi(t)|\mathbf{H}_v^0|\psi(t)\rangle = |a(t)|^2 = |A(t)|^2$$
$$= \exp(-\Gamma_v t). \tag{134}$$

Hence, we identify the VER time with Γ_v^{-1} and from equation (133) find that

$$\tau'^{-1} = \Gamma_v = 2\hbar^{-2}|Q_{10}|^2 \sum_\lambda |u_{0\lambda}|^2\gamma_\lambda/[(\omega_v - \omega_\lambda - \Delta\omega_\lambda)^2 + \gamma_\lambda^2]. \tag{135}$$

It is interesting to note that in the limit in which the coupling of the local mode to the bath vanishes, the general expression (135) reduces, with the help of the relation (see, for example, Messiah, 1961)

$$\lim_{\varepsilon\to 0} \varepsilon/(x^2 + \varepsilon^2) = \pi\delta(x),$$

to

$$\Gamma_v = 2\pi\hbar^{-1}|Q_{10}|^2 \sum_{\lambda} |u_{0\lambda}|^2 \delta(\varepsilon_1^v - \varepsilon_\lambda^L). \tag{136}$$

If we identify the local mode as simply the unperturbed orientation and take the internal mode to be harmonic, then equation (136) becomes identical to equation (114) for $k_{1\to 0}$, the starting expression for the rate constant in our 2-d model for the role of rotations in VER (see Section V.B). We thus conclude that the general expression (135) takes into account dynamical participation of the bath, which was entirely ignored in our previous treatment (Diestler *et al.*, 1978).

B. Application

As a simple application of the general theory outlined in Section VI.A, we investigated (Ladouceur and Diestler, 1978) the role of pseudomolecular vibrations in VER in the 1-d prototype. In analogy to the model of Shugard *et al.* (1978), we took the RS to be a four-atom complex consisting of a *homonuclear* diatomic impurity and the two nearest-neighbour lattice atoms. The internal vibration of the diatomic was assumed to be harmonic and, in zero order, completely decoupled from the local modes. Thus in zero order the local modes are those of a triatomic pseudomolecule, the diatomic being held rigid. The motion of the c.m. of the pseudomolecule was included in the lattice, which was assumed to be harmonic in zero order. The lattice was treated as pure, the effect of the true local mode being ignored. Finally, taking the pseudolocal modes to be harmonic, we found a symmetric mode, in which the lattice atoms vibrate out of phase while the diatomic remains at rest, and an asymmetric mode of higher frequency, in which the diatomic moves out of phase with the two lattice atoms. The latter mode does not contribute to VER in the single-phonon approximation, since its frequency lies above the cutoff of the lattice.

We assumed that each atom of the diatomic interacts with only its nearest neighbours and for all atom–atom interactions we took the Lennard-Jones (12–6) form. Fitting the coupling potential [see equation (119)] according to

$$U_{vL}(Q_s) = A[\exp(-2\alpha Q_s) - \exp(-\alpha Q_s)],$$

where Q_s is the symmetric normal coordinate, we obtained

$$\begin{aligned}\Gamma_v = 4A^2/(M\hbar\omega_v) &\sum_{\lambda=0}^{\infty} (\alpha'/\sqrt{2})\lambda\gamma \\ &\times [\exp(\alpha'^2/2) - \exp(5\alpha'^2/16)/2^{\lambda-1} + \exp(\alpha'^2/8)/2^{2\lambda}] \\ &\times [(\omega_v - \lambda\omega_s)^2 + \lambda^2\gamma^2]^{-1}/\lambda!\end{aligned} \tag{137}$$

where

$$\alpha' \equiv 2\alpha(\hbar/m\omega_s)^{1/2},$$
$$\gamma \equiv (\omega_{max}^2 - \omega_s^2)^{1/2}/4,$$

M and m are the respective masses of the diatomic and lattice atoms, and ω_s is the fundamental frequency of the symmetric pseudomolecular normal mode.

As a specific application of the formula (137) we considered the Cl_2/Ar system. The atom–atom Lennard–Jones parameters were estimated by the procedure of Diestler *et al.* (1978). For the $0 \rightarrow 1$ vibrational transition in the $^3\Pi_u$ manifold, which has a fundamental frequency of 237 cm^{-1}, we estimated $\tau' = 1$ ns. This result is at least consistent with the experimental finding of Bondybey and Fletcher (1976) that VER within this manifold occurs in less than 1 μs.

In their SCT study Shugard *et al.* (1978) also chose model parameters corresponding to the Cl_2/Ar system, except that they assumed the Cl_2–Ar interaction to be the same as the Ar–Ar interaction. Also, in order to 'economize' on computer time, they took the fundamental frequency of ground-state Cl_2 to be 365 cm^{-1}, instead of the observed value of 544 cm^{-1}. At 25 K the SCT calculation yields VER time of 5.52 ns. Duplicating their parameters as closely as possible, we (Ladouceur and Diestler, 1978) estimated $\tau' = 2.9$ ns for the $1 \rightarrow 0$ transition at 0 K.

Finally, it is interesting to note that both the SCT calculations of Shugard *et al.* (1978) and our own (Ladouceur and Diestler, 1978) using equation (137) give an energy-gap law of the form

$$\ln \Gamma_v = a\omega_v + b,$$

where the slope a is negative. It thus appears that both approaches manifest the multiphonon character of the simpler quantum mechanical models considered in Section IV.

VII. SUMMARY AND DISCUSSION

The purpose of this article has been to present in some depth various approaches to the theoretical treatment of vibrational energy relaxation (VER) of molecules (solute or impurity) dissolved in dense media. We have grouped these approaches into two broad classes:

(A) those involving the (in principle, exact) solution of the equations of motion (EOM);
(B) those based on simple physical models which invoke an approximate solution of the EOM.

In most of the applications of either class A or class B methods, we have dealt with what we have defined as the prototypic physical system, namely a single diatomic molecule occupying a substitutional site in an otherwise perfect rare-gas crystal.

Two classical-mechanical methods in class A have been presented:

(1) the more-or-less standard molecular dynamics (MD) technique;
(2) the newer, less well developed stochastic classical trajectory (SCT) method.

The MD method requires the numerical solution of the classical EOM for the entire system of solute plus solvent. The great advantage of the MD method is that, in principle, there are no restrictions on the type of system to which it can be applied, except, of course, systems which are intrinsically quantum mechanical in nature. The MD method allows one to look at the motion of the system in the minutest detail. While this latter feature can often be useful in furnishing insight into the mechanisms of relaxation, it can also often be frustrating in that the MD result corresponding to the 'real' measurement must be expressed as an average over a bewildering variety of such detailed motions. Also, one quite literally pays a high price in terms of computer time for the privilege of looking at these detailed motions. But, they are admittedly very interesting and entertaining.

By recasting the full classical EOM into a smaller set of Langevin-type EOM for the relevant subsystem (e.g. impurity plus nearest-neighbour lattice atoms) the SCT method can reduce considerably the amount of computer time required to study VER. Even so, one should observe a caveat. As formulated in Section II.B, the SCT method applies only to systems in which the medium is a harmonic continuum. While this constraint is easily justifiable for VER studies of the prototype, it is not likely to be valid for liquid solutions. Additional work to enlarge the applicability of the SCT approach would be very worthwhile.

A severe problem afflicting both the MD and SCT methods concerns the vastly diverse time scales on which motions of various types take place. This problem was discussed in detail in Section II.A. Here we simply opine that it is probably feasible to employ conventional MD to simulate VR processes in 'realistic' 3-d systems, if they take place in less than about 10 ps. Shugard *et al.* (1978) claim that the SCT method is limited to systems for which the VR times are less than 10^5 ps. Hence, even though neither the MD nor SCT methods are able to simulate VER in many 'realistic' prototypic cases, e.g. N_2/Ar, they should be useful in certain non-prototypic systems, (e.g. polyatomic molecular liquids, where VER times are of the order of 10^{-12}–10^{-11} s).

We close our discussion of the class A methods with a trivial observation. Since the MD and SCT methods are inherently classical in nature, they should not apply under conditions in which quantum mechanical effects predominate. Roughly speaking, one would expect such effects to be most important at low temperature.

The class B model descriptions of VER can be subdivided into two types in accordance with the choice of the relevant subsystem (RS):

(1) those which take the RS to be simply the unperturbed intramolecular vibration of the impurity;
(2) those which augment the RS with whichever additional modes of motion (degrees of freedom) may be strongly coupled to the internal impurity mode.

The advantage of the first type of model is that it furnishes an explicit closed expression for the VER rate constant in terms of the microscopic properties of the system. Such an expression can be very useful in providing insight into the mechanism of VER. On the other hand, this expression is generally derived using an essentially uncontrolled mathematical approximation, i.e. first-order perturbation theory, so that its range of validity is uncertain. All of the models presented in Sections III–V are of the first type. We note that only the model of Berkowitz and Gerber (1977) includes the dynamical effects of both the lattice and the orientation of the impurity, i.e. the unperturbed normal modes of the lattice plus the unperturbed free rotations of the diatomic jointly constitute the medium (bath). The dynamical effects of rotational–translational coupling (RTC), known to be important in the interpretation of infrared spectra (see, for example, Friedmann and Kimel, 1965, 1967), are entirely neglected. It would be very interesting to investigate the effects of RTC on VER in a system in which the simpler models are apparently inadequate, e.g. HCl/Ar (see discussion at end of Section V.B).

In Section VI we considered class B approaches of the second type, which we pointed out are similar in spirit to the SCT method. The basic idea is to describe with high precision only the (hopefully) small number of modes of motion of the entire system which determine for the most part the course of the relaxation process. Since the impurity interacts strongly only with its nearest-neighbour solvent molecules, it seems reasonable (as a first step) to include only these few particles in the RS. Thus the RS of the type 1 methods is augmented to include the motions of the neighbouring molecules. In modelling VER more precisely we treat the relaxation of the augmented RS, which is conveniently regarded as a pseudomolecule. We emphasize that the theory of the participation of local modes in VER developed in Section VI.A is only a partial theory. The reason is that the zero-order

Hamiltonian of the RS effectively takes account of the *external* interactions of the impurity with its nearest neighbours, neglecting the coupling of the internal vibration of the impurity to these external motions. Again, the advantage realized is that one ends up with a closed expression for the VER rate constant. A complete theory would require essentially that one numerically diagonalize the augmented RS in an appropriate basis. This would clearly yield a more precise description of VER. Unfortunately, one again pays a price in terms of computer time and the lack of a simple expression by which to interpret VER.

In all of the applications of class B methods, and of the SCT method of class A, we have taken the solvent to be a rare-gas solid. The reason, of course, is that the many-body states of the solid are simply and adequately described using the harmonic approximation. Class B approaches to VER in liquid solutions seem inherently much more difficult on account of the lack of a simple, and adequate, description of the many-body states of the solvent. Several workers (Herzfeld, 1952; Litovitz, 1957; Diestler, 1975) have proposed cell models (Lennard–Jones and Devonshire, 1937). Metiu *et al.* (1977) have investigated a hydrodynamical model. Of course, if the time-scale problems discussed above can be circumvented, the MD and SCT methods may be ideally suited to the study of VER in liquids (see Section II.A). Much work remains to be done on VER in liquid solutions.

All of the class B models considered here have been applied to solutions sufficiently dilute that the impurity molecules do not interact with one another. However, the majority of experiments are performed on neat liquids, in which such interactions must play a significant role in VER. Even for apparently quite dilute solutions, however, intermolecular coupling between impurity molecules may have a profound effect on the VER behaviour. An experiment of Dubost and Charneau (1976) on CO/Ar provides an interesting illustration. Following excitation of the $v = 1$ level of $^{12}C^{16}O$ by an intense infrared laser pulse, they observed vibrational fluorescence from levels $v = 2$–8 of $^{12}C^{16}O$ and also from upper vibrational levels of the isotopes $^{13}C^{16}O$ and $^{12}C^{18}O$. Apparently, via phonon-assisted near-resonant vibrational energy transfer processes of the type

$$\mathrm{CO}(1) + \mathrm{C'O'}(n) \rightarrow \mathrm{CO}(0) + \mathrm{C'O'}(n+1) \tag{138}$$

(where the number in parentheses denote the vibrational levels and the primes distinguish between two CO molecules), the populations of molecules in vibrationally excited levels grow. Primarily on account of the radiative decay processes such as

$$\mathrm{CO}(v) \rightarrow \mathrm{CO}(v-1) + h\nu, \tag{139}$$

the populations eventually reach maxima and then decrease. Manz (1977), incorporating processes (138) and (139) into a master-equation description of the system, has succeeded in quantitatively reproducing the observations of Dubost and Charneau (1976).

Since the title of this volume is *Potential Energy Surfaces*, it is perhaps fitting to close by reiterating a point made in Section I. In all cases considered here, we have assumed the potential energy surface to be simply a sum of pairwise interactions. While that assumption may in itself be tolerable, if not reasonable, we have also, generally for lack of (knowledge of?) more reliable potentials, arbitrarily employed standard forms, e.g. Lennard–Jones (12–6) and repulsive exponential. The results of VER calculations can be quite sensitive to the estimated values of parameters associated with these standard forms. See, for example, Table IV, which shows the dependence of τ' upon α for the repulsive exponential in the model SCT calculations of Shugard *et al.* (1978). It would be very helpful to have additional reliable intermolecular potential energy surfaces for the kinds of systems, e.g. N_2/Ar, HCl/Ar, we have been considering. It is important to emphasize that the surface should be known as a function of the intramolecular vibration Q as well as of the external degrees of freedom, i.e. vector distance between the c.m. of the diatomic and the atom. Comparative studies of VER using such surfaces could yield valuable information on the conditions of validity of various approximations.

References

Abramowitz, M. and Stegun, I. A. (Eds.) (1965). *Handbook of Mathematical Functions*, U.S. Gov't Printing Office, Washington.

Adelman, S. A. and Doll, J. D. (1976). *J. Chem. Phys.*, **64**, 2375.

Andrews, J. R. and Hudson, B. S. (1978). *J. Chem. Phys.*, **68**, 4587.

Bartoli, F. J. and Litovitz, T. A. (1972). *J. Chem. Phys.*, **56**, 404.

Berkowitz, M. and Gerber, R. B. (1977). *Chem. Phys. Lett.*, **49**, 260.

Berne, B. J. and Forster, D. (1971). *Ann. Rev. Phys. Chem.*, **22**, 563.

Bixon, M. and Jortner, J. (1968). *J. Chem. Phys.*, **48**, 715.

Bondybey, V. E. and Brus, L. E. (1975). *J. Chem. Phys.*, **63**, 954.

Bondybey, V. E. and Fletcher, C. (1976). *J. Chem. Phys.*, **64**, 3615.

Bratos, S. and Marechal, E. (1971). *Phys. Rev.*, **A4**, 1078.

Brueck, S. R. J. and Osgood, R. M. (1976). *Chem. Phys. Lett.*, **39**, 568.

Brus, L. E. and Bondybey, V. E. (1979). In *Radiation-less transitions* (Ed. S. H. Lin), Academic Press, New York.

Diestler, D. J. (1974). *J. Chem. Phys.*, **60**, 2692.

Diestler, D. J. (1975). *Chem. Phys.*, **7**, 349.

Diestler, D. J. (1976a). *Chem. Phys. Lett.*, **39**, 39.

Diestler, D. J. (1976b). *Mol. Phys.*, **32**, 1091.

Diestler, D. J. (1976c). In *Topics in Appl. Phys.* (Ed. F. K. Fong), Vol. 15, Springer, Berlin, p. 169.

Diestler, D. J., Knapp, E.-W., and Ladouceur, H. D. (1978). *J. Chem. Phys.* **68**, 4056.

Diestler, D. J. and Ladouceur, H. D. (1977). *Optics Commun.*, **20**, 6.
Diestler, D. J. and Wilson, R. S. (1975). *J. Chem. Phys.*, **62**, 1572.
Dubost, H. and Charneau, R. (1976). *Chem. Phys.*, **12**, 407.
Eliason, M. A. and Hirschfelder, J. O. (1959). *J. Chem. Phys.*, **30**, 1426.
Englman, R. and Jortner, J. (1970). *Mol. Phys.*, **18**, 145.
Ewing, G. E. (1978). *Chem. Phys.*, **29**, 253.
Filatov, A. N. (1967). *Differentsial'nye Uraveniya*, **3**, 1725.
Fong, F. K. (1975). *Theory of Molecular Relaxation*, Wiley, New York.
Freed, K. F. and Metiu, H. (1977). *Chem. Phys. Lett.*, **48**, 262.
Freed, K. F., Yeager, D. L. and Metiu, H. (1977). *Chem. Phys. Lett.*, **49**, 19.
Friedmann, H. and Kimel, S. (1965). *J. Chem. Phys.*, **43**, 3925.
Friedmann, H. and Kimel, S. (1967). *J. Chem. Phys.*, **47**, 3589.
Gerber, R. B. and Berkowitz, M. (1977). *Phys. Rev. Lett.*, **39**, 1000.
Goodman, J. and Brus, L. E. (1976). *J. Chem. Phys.*, **65**, 3146.
Gouterman, M. (1962). *J. Chem. Phys.*, **36**, 2846.
Heitler, W. and Ma, S. T. (1949). *Proc. Roy. Irish Acad.*, **52A**, 109.
Herzfeld, K. F. (1952). *J. Chem. Phys.*, **20**, 288.
Kubo, R. (1962). *J. Phys. Soc. Japan*, **17**, 1100.
Ladouceur, H. D. and Diestler, D. J. (1978). Unpublished.
Lax, M. (1952). *J. Chem. Phys.*, **20**, 1752.
Legay, F. (1978). In *Chemical and Biochemical Applications of Lasers* (Ed. C. B. Moore), Vol. 2, Academic Press.
Lennard-Jones, J. E. and Devonshire, A. F. (1937). *Proc. Roy. Soc. (London)*, **163**, 53.
Libermann, L. (1959). *Phys. Rev.*, **113**, 1052.
Lin, S. H. (1974). *J. Chem. Phys.*, **61**, 3810.
Lin, S. H. (1976). *J. Chem. Phys.*, **65**, 1053.
Lin, S. H., Lin, H. P., and Knittel, D. (1976). *J. Chem. Phys.*, **64**, 441.
Litovitz, T. A. (1957). *J. Chem. Phys.*, **26**, 469.
Louisell, W. H. (1973). *Quantum Statistical Properties of Radiation*, Wiley, New York.
Manz, J. (1977). *Chem. Phys.*, **24**, 51.
Messiah, A. (1961). *Quantum Mechanics*, Vol. 1, Wiley, New York, Appendix A.
Metiu, H., Oxtoby, D. W., and Freed, K. F. (1977). *Phys. Rev.*, **A15**, 361.
Mott, N. F. and Massey, H. S. W. (1965). *The Theory of Atomic Collisions*, Oxford University Press.
Nafie, L. A. and Peticolas, W. J. (1972). *J. Chem. Phys.*, **57**, 3145.
Neilsen, W. B. and Gordon, R. G. (1973). *J. Chem. Phys.*, **58**, 414.
Nitzan, A. and Jortner, J. (1973). *Mol. Phys.*, **25**, 713.
Nitzan, A., Mukamel, S., and Jortner, J. (1974). *J. Chem. Phys.*, **60**, 3929.
Nitzan, A., Mukamel, S., and Jortner, J. (1975). *J. Chem. Phys.*, **63**, 200.
Nitzan, A. and Silbey, R. S. (1974). *J. Chem. Phys.*, **60**, 4070.
Oxtoby, D. W. (1979). *Adv. Chem. Phys.*, to be published.
Rasmussen, R. A. (1967). *J. Chem. Phys.*, **46**, 211.
Rebane, K. K. (1970). *Impurity Spectra of Solids*, Plenum Press, New York, Chap. 1.
Rihel, J. P. and Diestler, D. J. (1976). *J. Chem. Phys.*, **64**, 2593.
Ross, J. (Ed.) (1966). *Adv. Chem. Phys.*, Vol. X, Interscience Publishers, New York.
Rowe, D. J. (1968). *Rev. Mod. Phys.*, **40**, 153.
Schiff, L. I. (1955). *Quantum Mechanics*, McGraw-Hill, New York.
Schurr, J. M. (1971). *Int. J. Quantum Chem.*, **5**, 239.
Shugard, M., Tully, J. C., and Nitzan, A. (1978). *J. Chem. Phys.*, **69**, 336.
Slater, N. B. (1959). *Theory of Unimolecular Reactions*, Cornell University Press, Ithaca.

Steele, W. A. (1969). In, *Transport Phenomena in Fluids* (Ed. H. J. M. Hanley), Marcel Dekker, New York.
Steinfeld, J. I. (1974). *Molecules and Radiation*, Harper and Row, New York.
Sun, H.-Y. and Rice, S. A. (1965). *J. Chem. Phys.*, **42**, 3826.
Taylor, J. R. (1972). *Scattering Theory*, Wiley, New York, p. 270.
Tinti, D. S. and Robinson, G. W. (1968). *J. Chem. Phys.*, **49**, 3229.
van Hove, L. (1955). *Physica*, **21**, 517.
Wiesenfeld, J. M. and Moore, C. B. (1976). *Bull. Amer. Phys. Soc.*, **21**, 1289.
Yakhot, V., Berkowitz, M., and Gerber, R. B. (1975). *Chem. Phys.*, **10**, 61.
Zwanzig, R. W. (1960). *J. Chem. Phys.*, **33**, 1338.

Potential Energy Surfaces
Edited by K. P. Lawley

SPECTROSCOPY AND POTENTIAL ENERGY SURFACES OF VAN DER WAALS MOLECULES

ROBERT J. LE ROY AND J. SCOTT CARLEY

Guelph–Waterloo Centre for Graduate Work in Chemistry, University of Waterloo, Waterloo, Ontario, N2L 3G1, Canada

CONTENTS

I. INTRODUCTION

Van der Waals molecules are weakly bound complexes of atoms or molecules which characteristically have small dissociation energies and large bound lengths and are held together, not by ordinary chemical binding forces, but by 'physical' multipolar, dispersion and induction forces, or by hydrogen bonds. Significant populations of these species are found in gases at low temperature, and they are known to contribute to many physical and chemical phenomena. For example, their presence modifies the virial and transport properties of gases [1–5] and they participate in chemical reaction mechanisms,[6,7] nucleation phenomena,[8,9] spin–lattice relaxation,[10,11] and low-temperature energy transfer processes. In addition, the vibrational predissociation of selectively excited states of van der Waals molecules has been proposed as a means of achieving isotope separation.

The present discussion is concerned with the fact that the discrete spectra of van der Waals molecules are proving to be an extremely rich source of information about intermolecular forces. While spectroscopic studies of weakly bound species have long been performed using matrix isolation techniques, their usefulness has been limited by the fact that the matrix interactions have roughly the same strength as the van der Waals bonds of interest. In contrast, recently developed techniques for measuring the spectra of van der Waals molecules in the gas phase yield results completely free of these ambiguities. The rapid growth and great promise of this type of spectroscopy has aroused great interest in it and already has led to publication of a number of reviews.[12–18] Those reports focused attention mainly on the nature and quality of the spectra and on the means of obtaining them, or on the origin of the bonding. The present chapter, however, is concerned with the methods for extracting information about potential energy surfaces from such data and with the detailed nature of the potential energy functions so obtained.

While polymer clusters of various sizes have been observed mass spectrometrically[9] and discrete spectra have been reported for more than one van der Waals 'trimer' (e.g. He_2–I_2 and He_2–tetrazene),[19,20] analyses of such data on more than an elementary structural level have not yet been reported. The present work therefore considers only 'bimers', i.e. van der Waals molecules formed from two monomer species which may or may not be identical (a dimer is a special case of a bimer in which the component species are identical).

Diffuse band systems due to diatomic van der Waals molecules have long been observed in absorption and emission spectroscopy,[21,22] but intermolecular potentials were not determined quantitatively from such data until the 1970 work of Balfour and Douglas[23] on Mg_2 and of Tanaka and Yoshino[24] on Ar_2. Spectra have subsequently been observed and analysed for Ne_2, Kr_2, Xe_2, Ca_2, NeAr, and NaNe.[25–30] While these spectra are quite difficult to obtain, their analysis involves the relatively straightforward application of techniques developed for the spectroscopy of normal diatomic molecules. If both vibrational and rotational spacings are resolved, either the semiclassical RKR inversion procedure[31] or a quantum mechanical eigenvalue fitting method[32] can be used to determine the potential energy curve.[23,28,29,33,34] If only the vibrational spacings are known, one RKR equation can still be used to determine the width of the potential as a function of energy.[24,26,27] In addition, a variety of techniques may be used to model the potential when few data are available,[25] or to facilitate extrapolations beyond the observed results.[30,35,36]

The above methods are widely used and well understood, and need no elaboration here. The present discussion therefore deals with the analysis of spectra for atom–molecule or molecule–molecule complexes whose potentials depend on the relative orientation and internal bond lengths of the monomers as well as on their separation. While many phenomena reflect this dependence on the internal degrees of freedom, it is usually quite difficult to extract quantitative information about a potential energy surface from such effects. Electronic structure calculations have provided reliable information of this type for H_2–He and $(H_2)_2$,[37,38] but accurate results for systems having many electrons are extremely difficult to obtain in this manner. Analyses of relaxation times and line shapes[39–41] and of scattering cross sections for beams of state-selected molecules[42,43] have proved useful for determining the anisotropy of the potential when its isotropic or spherically averaged part is known. To date, however, the analysis of spectra of (atom–diatom) van der Waals molecules is the only technique which has simultaneously yielded information about the dependence of a potential function on all internal degrees of freedom.

Structure due to discrete transitions of species such as $(HF)_2$ or HCl–Ar began to be recognized in infrared spectra over 20 years ago.[44–46] The first (polyatomic *or* diatomic) van der Waals molecule spectra which could be assigned properly and used to test intermolecular potential models were the infrared absorption measurements for $(H_2)_2$ and H_2–Ar reported by Welsh and coworkers.[47,48] Using a model which neglected the anisotropy of the potential, Gordon and Cashion[49,50] showed that 1-dimensional (isotropic) potentials taken from the literature could reproduce the main features of these early spectra to within the experimental uncertainties. In subsequent work by Welsh and his associates,[51,52] the spectra of complexes formed with molecular hydrogen were greatly refined until effects due to the internal degrees of freedom of the diatom became sufficiently well resolved that they could be analysed.[53–57]

The recent rapid development of methods for doing spectroscopy on van der Waals molecules promises to open a wide variety of systems to scrutiny. Section II of this report describes the various experimental techniques which have been applied to this problem, outlines their limitations and summarizes the principal features of the results obtained to date. As is indicated there, the data for the van der Waals complexes involving molecular hydrogen[52] are the richest source of information about anisotropic intermolecular potentials to date, and their existence has spurred the development of the computational methods described herein. For reasons described in Section II.A, the experimental technique of long-path-length spectroscopy on samples of bulk gas which proved so successful for H_2 bimers cannot readily yield equally useful results for other species. However, Klemperer and coworkers[58] showed that complexes could be formed and their microwave and radiofrequency spectra studied in the collision-free and Doppler broadening-free environment provided by a molecular beam. Subsequent work in other laboratories showed that spectroscopy on molecular beams of van der Waals molecules could also be performed in the visible and infrared regions.[19,30,59] With the advent of these experimental techniques, high-resolution spectroscopy can in principle be performed on an extremely wide variety of van der Waals molecules. An important challenge has therefore been to devise techniques for analysing such data to determine complete potential energy surfaces.

For molecules with more than one internal degree of freedom, there is no rigorous method of inverting the observed transition frequencies and intensities to determine a potential. One must therefore adopt the approach of calculating a synthetic spectrum from an assumed potential, comparing with experiment, and then optimizing the parameters of the trial potential. The bottleneck in this procedure is the calculation of the synthetic spectrum; various methods developed for doing this are summarized and compared in Section III of this chapter.

The 'transferability' of a potential energy surface obtained from an analysis of this type, i.e. its ability to predict properties other than those from which it was determined, is a most demanding test of its quality. This capacity often depends critically on the external constraints and implicit assumptions built into the form of the potential model. A discussion of these considerations is presented in Section IV, together with a description of techniques used for performing least-squares fits to the spectroscopic data. Section V then summarizes the information about potential energy surfaces which has been obtained in this way.

Of all the existing spectroscopic studies of van der Waals molecules, only the McKellar and Welsh[52] bulk measurements of the infrared spectra of molecular hydrogen–rare gas bimers and, to a lesser extent, the Novick *et al.*[60,61] molecular beam results for isotopic Ar–HCl and Kr–ClF have yielded quantitative information about potential energy surfaces. As a result, the methods described herein were largely developed with these particular types of complexes in mind. In particular, the theory of van der Waals complexes presented in Section III is specific to the atom–diatom case. However, its generalization to the case of diatom–diatom complexes or more complex systems is in principle quite straightforward.

II. EXPERIMENTAL METHODS AND RESULTS

A. Bulk Gas (Equilibrium) Studies

The large bond lengths and small binding energies characteristic of van der Waals molecules make their gas-phase spectra much more difficult to obtain and resolve than the spectra of chemically bound species of similar composition. In the first place, the weak binding makes it relatively difficult to produce large populations of the complexes to be studied. In general, the equilibrium population of bimers may be increased simply by lowering the temperature and raising the pressure, although the fact that the partial pressure of a component cannot be greater than its equilibrium vapour pressure limits the bimer populations which may be produced in this way. However, while an increase in pressure tends to increase the equilibrium population of bimers, it also lowers the spectral resolution.

The binding energy and level spacings of a van der Waals molecule are usually much smaller than the average collision energy in the gas in thermal equilibrium. Thus the average lifetime of the molecule in any particular quantum state is simply the time between collisions τ_c. This gives rise to a 'collisional' level width of

$$\Delta E \approx \hbar/\tau_c. \tag{1}$$

For transitions in which both the initial and final states of the complex are held together by van der Waals forces, this leads to a 'collisional' line width of (in units cm^{-1})

$$\Gamma_c = 1/\pi c\tau_c = 4.2 \times 10^{-4}\sigma\rho\sqrt{T/\mu'} \tag{2}$$

where μ[amu] is the reduced mass and σ[Å^2] is the total cross section associated with collisions between the complex and a perturbing species whose number density is ρ [amagat]. Thus while increasing ρ increases the concentration of bimers, it also shortens the time between collisions and hence increases line widths. The latter consideration is crucial, since the great length of the van der Waals bond makes the moment of inertia for the end-over-end rotation of the complex relatively large so that the corresponding rotational spectrum is relatively dense. In most systems studied to date, the rotational constant B_c associated with this motion is a measure of the smallest (nuclear motion) level spacing of the complex, so that the condition

$$B_c > \Gamma_c \tag{3}$$

must be satisfied if the spectrum is to be fully resolved.

In order to delineate roughly the conditions under which equation (3) will be satisfied, a value for the cross section σ must be assumed. In view of the weak binding and small level spacings typical of these complexes, approximating it as $\sigma = \pi(2R_e)^2$, where R_e is the length of the van der Waals bond, seems reasonable (to within ca. a factor of 2). For van der Waals molecules whose infrared spectra have been studied in bulk samples in thermal equilibrium, Table I summarizes the experimental conditions and compares the resulting values of Γ_c with B_c. Note, however, that this quantity Γ_c is only a lower bound to the observed line widths, since instrumental effects can give rise to much additional broadening.

The results in Table I show that equation (3) is satisfied, so that highly

TABLE I. Characteristics of gas-phase infrared spectra of van der Waals molecules; L is the absorption path length while the other quantities are defined in the text.

Species	Ref.	T(K)	L(m)	σ(Å^2)	ρ(amagat)	B_c(cm^{-1})	Γ_c(cm^{-1})
H_2–Ar	52a	86	165	160	0.63 of H_2 0.53 of Ar	0.56	0.3
HD–Ar	52d	85	165	160	1.65 of HD 1.85 of Ar	0.82	0.8
$(HF)_2$	44d	273	1.6	100	0.105 of HF	0.22	0.02
$(HCl)_2$	62	163	6	150	0.130 of HCl	0.077	0.02
Ar–HCl	63	175	1	180	0.07 of HCl 1.1 of Ar	0.056	0.3
$(CO_2)_2$	64	192	56	210	0.92 of CO_2	0.031	0.2
O_2–Ar	65	93	122	150	2.82 of O_2 2.16 of Ar	0.020	0.7
N_2–Ar	66	87	122	190	2.68 of N_2 2.21 of Ar	0.067	0.9
$(O_2)_2$	16, 67	87	152	150	2.31 of O_2	0.045	0.3
$(N_2)_2$	68	77	122	170	3.38 of N_2	0.088	0.5

resolved spectra can readily be obtained, for two types of systems. In the first, complexes formed by molecular hydrogen, the small mass of hydrogen makes B_c sufficiently large that individual transitions could be resolved despite the large Γ_c values. The second type, here represented by $(HF)_2$ and $(HCl)_2$, consists of complexes containing a monomer which has a sufficiently large transition probability that the spectra can be observed in gases at sufficiently low density (small ρ) that τ_c will be large and Γ_e small. In both cases a long optical path is used so that ρ can be made as low as possible. While HCl–Ar has the large dipole moment required for intense transitions in the far infrared, the relatively short path length used by Boom *et al.*[63] prevented their measurements from satisfying equation (3). However, it seems clear that high-resolution spectra of $(CO_2)_2$ or complexes formed by N_2 and O_2 will not be obtainable from conventional long-path-length spectroscopy of bulk samples in thermal equilibrium. On the other hand, the extremely high sensitivity of the new detection technique of laser photoacoustic spectroscopy[69] may someday allow even these spectra to be observed at sufficiently low density that they become fully resolved.

A detailed analysis to determine a potential energy surface requires both a fully resolved spectrum and proper assignment of the observed transitions. For the complexes formed by molecular hydrogen, the relatively weak anisotropy makes the spectra qualitatively similar to those for weakly bound diatomic molecules. Taken together with the large rotational spacings of the diatom, this allowed unique assignments to be made for all of the fully resolved transitions. In contrast, for complexes formed from HCl or HF the strong potential anisotropy gives rise to much more complicated energy level patterns and the small (relative to H_2) diatom rotational constants mean that transitions correlating with different rotational states of the diatom are interspersed. In addition, the intense allowed transitions of the halide monomer obscure parts of the bimer spectra. As a result, even the fully resolved lines of the HF and HCl dimer spectra have not been spectroscopically assigned.

For $(CO_2)_2$ and the complexes formed by N_2 and O_2,[64–68] the loss of information due to lack of resolution has meant that the data have only been analysed in terms of simple structural models. While the reported structures seem quite reasonable and probably represent the best conclusions obtainable from the evidence, they must be regarded as somewhat tentative. For example, while Mannik *et al.*[64] showed that their $(CO_2)_2$ spectrum was consistent with a locked T structure, there now seems to be some doubt[70] about whether this is really the equilibrium configuration. Similar structural models have also been used to interpret the main features in the spectra of bimers formed by HCl with HCl, DCl, HBr, DBr, and HI.[62] However, until these spectra are fully resolved and properly understood, such interpretations must be deemed inconclusive.

In conclusion, the molecular hydrogen complexes studied by Welsh and coworkers[52] are the only polyatomic van der Waals molecules observed under equilibrium conditions whose spectra have been fully resolved and properly assigned so that detailed potential energy surfaces could be obtained from them.[54–57] The only other complexes for which adequate resolution can be obtained by applying traditional techniques to bulk (equilibrium) samples are those for which the associated monomer transition is sufficiently strongly allowed that the experiments could be done at very low pressure. However, the complexity of the resulting spectra makes them very difficult to assign; this is the case for the infrared measurements on $(HF)_2$ and $(HCl)_2$.[44,62] A successful analysis of such data will require the performance of extensive (and expensive, see Section III) model calculations for a variety of realistic trial potentials until the observed lines can be properly identified. This type of study has proven sufficiently forbidding that it has not yet been undertaken. In future, however, measurements of bimer spectra using molecular beam techniques (described below) may facilitate making such assignments and hence make these bulk data an important source of information about intermolecular potentials.

B. Spectroscopy in Molecular Beams

1. *General*

It has long been known that molecular clusters may be formed by the rapid cooling which occurs when a gas at high pressure expands adiabatically into a vacuum through a supersonic nozzle.[9] The nozzle conditions may be adjusted to favour bimer formation while discriminating against the production of higher polymers; for example, in expansions of Ar, dimer mole fractions in excess of 0.5 have been reported.[9b] The dominant characteristic of molecular beams produced using supersonic nozzles is the very narrow spread of the beam molecule velocities. This distribution corresponds to an effective translational temperature which can be made as low as a fraction of a degree, while the temperatures loosely associated with the distribution over internal vibration–rotation states of the beam molecules are also quite low. The virtual absence of collisions within such a beam and the relative translational temperature make it an ideal medium in which to do spectroscopy, since both Doppler broadening and the collisional line width (Γ_c of equation (3)) are negligible. This means that the spectral resolution obtainable is limited only by the characteristics of the radiation source, the transit time of the molecules crossing the light beam, and the natural lifetimes of the initial and final states of the transition being studied. At the same time, the cooling in the expansion ensures that most species present in the beam are in their lowest-energy quantum states. This greatly simplifies the spectra of complex molecules, and hence greatly facilitates the task of making proper spectroscopic assignments.

To date, three different methods for studying the spectra of van der Waals molecules in molecular beams have been developed. While all three use supersonic nozzle expansions to prepare the bimers, very different techniques are used to ascertain when light is absorbed. These methods and the results obtained from them are described below.

2. *Rabi Spectroscopy*

Molecular beam electric resonance spectroscopy, or Rabi spectroscopy, has long been used to study the microwave and radiofrequency transitions of stable species.[71] Since 1972 Klemperer and his associates have also been using it to study van der Waals molecules.[58] This method can only be applied to molecules which have an electric dipole moment sufficiently large to allow them to be selectively focused on the entrance slit of the mass spectrometer detector. The focusing field is divided into two segments, and in the resonance region which separates them the beam molecules are subjected to tuneable microwave or radiofrequency radiation. Since the dipole moment of a molecule in general changes with its internal state, molecules which absorb radiation are not refocused onto the detector slit, causing a net decrease in the mass spectrometer signal.

The Rabi technique has been used to observe radiofrequency and microwave transitions in various isotopic forms of $(HF)_2$, Ar–HCl, Ar–HF, Ar–ClF, Kr–ClF, Ar–OCS, HF–ClF, HF–HCl, and $(H_2O)_2$.[13,58,60,61,72–77] In addition to the usual information about level spacings, the Stark shift coefficients obtained by studying the effect of an applied electric field on the radiofrequency spectrum determine the dipole moment of the complex in various quantum states. Since the projection of the monomer dipole moment onto the axis of the complex is often the dominant component of this observed dipole, this in turn yields information about the relative orientation of the polar monomer within the complex. At the same time, the nuclear quadrupole coupling constants determined in this analysis yield an independent measure of this relative orientation.[58,60]

To date, virtually all of the bimer states observed in such experiments correlate with the ground vibration–rotation level of the monomer and correspond to the van der Waals bond having only its zero-point stretching energy. This limitation has been implicitly attributed to the very low temperature associated with the distribution of population over the internal states;[60] however, this population distribution could be modified somewhat by appropriate changes in source conditions. A more serious source of difficulty may be the problem of focusing excited bimers onto the detector, since internal rotation of the polar monomer in the complex will tend to make the overall dipole moment quite small. In any case, the small variety of states observed will inhibit the determination of extensive information about potential energy surfaces for these systems. However, the direct measure of the relative internal orientation of the monomer, a type of information not yet obtainable from other spectroscopic experiments, at least partially compensates for this lack of data.

For most of the molecules studied using the above technique, the data analysis has only been carried to the point of providing structural information and values of the permanent dipole moment of the bimer in various quantum states. The only exception is the analysis by Holmgren *et al.*[78] of the data for the various isotopic forms of Ar–HCl. However, while their discerning use of the data for the different isotopes allowed them to map out a portion of the potential surface, it was not possible to determine uniquely the nature of the potential anisotropy. Thus, while providing very hingh-quality structural information about these complexes, the microwave and radiofrequency data do not yet appear able to determine uniquely defined potential energy surfaces. Nonetheless, the information obtained should facilitate assignment of the spectra obtained in bulk samples where a wide range of states are observed, and hence may allow overall potential surfaces to be obtained from the combined results.

While the electric resonance techniques discussed above can only be applied to strongly polar molecules, Reuss and Verberne[79] very recently showed that the complementary, yet much more difficult technique of molecular beam magnetic resonance spectroscopy[71] could yield the radiofrequency spectrum of the hydrogen dimer $(H_2)_2$. Under the conditions of the experiment, all of the (very few) bound states of this species are populated. The binding energy of this complex is sufficiently small that only one vibrational level associated with stretching of the van der Waals bond exists, and its zero-point energy is *ca.* 90% of the potential well depth. In spite of the model dependence which this very large amplitude motion introduces into the analysis, these measurements should greatly facilitate delineation of the potential energy surface for this system.

3. *Laser-Induced Fluorescence*

A second way of doing spectroscopy on van der Waals molecules in molecular beams is the technique of laser-induced fluorescence, as applied by Smalley, Levy, and Wharton.[17–20] In this method, the beam intersects a continuous-wave dye laser whose tuning range includes a strongly allowed electronic transition of the monomer. If molecules in the beam absorb light they will signal this fact by spontaneously fluorescing (as long as the fluorescence lifetime is shorter than the lifetimes associated with decay via radiationless processes). The spectrum may then be obtained as a plot of the total detected fluorescence vs. the exciting wavelength.

This technique has so far been applied to HeI_2, He_2–I_2, He–NO_2, Ar–I_2, He–tetrazene, and He_2–tetrazene, as well as to the diatomic species Na–Ar and Na–Ne.[19, 20, 29, 30, 80–83] In the analyses of the spectra of the polyatomic complexes, a rigid asymmetric top model was applied to He–I_2 and a rigid symmetric top model to He–tetrazene and He_2–tetrazene. While the latter yielded well-defined structures,[20] the He–I_2 analysis could only define the perpendicular distance from the He to the axis of the I_2 bond.[82] Thus these spectra have not yet provided much quantitative information about potential energy surfaces.

For He–I_2 and Ar–I_2, the observed transitions correspond to the I_2 molecule in the complex going from the lowest vibrational level of its ground electronic state into a wide range of vibrational levels v' of the excited $B(^3\Pi^+_{0u})$ state. Analysis of the fluorescence for a number of these He–I_2 bands showed that virtually all of it consisted of light emitted by free B-state I_2 molecules in a vibrational level $(v' - 1)$ one lower than that associated with the bimer transition.[80] This implied that the fluorescence must have been produced by the vibrational predissociation mechanism:

$$I_2(B, v')\text{–He} \rightarrow I_2(B, v' - 1) + \text{He} \rightarrow I_2(X, v'') + \text{He} + h\nu \tag{4}$$

This in turn means that: (i) the associated B-state vibrational spacings of ca. 100 cm^{-1} are greater than the van der Waals bond strength, and (ii) the vibrational predissociation lifetime is shorter than either the fluorescence lifetime of the complex or its lifetime for the electronic predissociation process:

$$I_2^*(\mathbf{B}, v')\text{–He} \rightarrow 2\text{I}(^2p_{3/2}) + \text{He} \tag{5}$$

(In contrast, this electronic predissociation mechanism appears to be the dominant one for[83] Ar–I_2.) Subsequent measurement of the fluorescence excitation spectrum line widths for He–I_2 yielded values for these predissociation lifetimes for a range of v' levels.[84]

Since the average I_2 bond length increases with the B-state vibrational quantum number v', the differences between the bimer spectra associated with the various I_2 bands reflect the dependence of the potential on the I_2 bond length. This is precisely the part of the potential which governs vibrational predissociation of the complexes and vibrational inelasticity in He + I_2 collisions. Some indication of this dependence on the diatom bond length could be obtained by repeating the asymmetric top analysis of ref. 82 for a wide range of B-state v' levels. However, more sophisticated analyses for these systems will have to await further theoretical work, since none of the present techniques (see Section III) can accurately predict level spacings for systems where the monomer rotational spacings are very small.

4. *Bolometer Spectroscopy*

For a number of years cryogenic bolometers (a kind of microcalorimeter) have been used to measure the flux of scattered particles in molecular beam scattering experiments.[85] The bolometer is a doped silicon chip operating at ca. 4 K which responds to the total energy (kinetic plus internal plus condensation energy) associated with particles sticking to its surface. Gough, Miller, and Scoles[59,86] have recently showed that its sensitivity to the internal energy of the incident molecules makes it an ideal device to use for detecting light absorption by the molecules in a beam. If the excited state of the absorbing molecule has a long lifetime, the energy of the absorbed photon is simply transferred to the detector, giving rise to a 'positive' signal; this was the case in their studies of the infrared spectra of CO and NO.[86] If the excited state loses energy by fluorescence before reaching the detector, there will still be a substantial positive signal since the fluorescence will almost certainly populate a variety of low-lying excited states in addition to the level excited by the initial absorption step. On the other hand, if the excited-state molecule predissociates before reaching the detector, there will be a net 'negative' signal, since most of the fragments will be scattered out of the highly collimated beam, leading to a net decrease in particle flux at the detector.

It is this 'predissociation mode' of detection which has been used in these authors' recent study[59] of the infrared spectra of the van der Waals dimer $(N_2O)_2$. Unfortunately, very little structural or intermolecular force information appears to be obtainable for this case, since the spectrum appears to be a continuum. Upper and lower bounds on the vibrational predissociation lifetime were established by the time of flight between the point of irradiation and the detector, *ca.* 10^{-4} s., and a value of 10^{-12} s. obtained by attributing the total observed band width to lifetime broadening.

While the most recent technique described in this section, bolometer spectroscopy may prove to be the most useful for determining information on intermolecular forces. In particular, in the infrared region where the upper state of the complex correlates with its monomer being in an excited vibrational state, fluorescence lifetimes are long so that fluorescence becomes too diffuse for sensitive detection. The capability of bolometer spectroscopy in this region is particularly important, since the two levels joined by the transition are associated with the same electronic state of the complex, a fact which should facilitate the analyses. At the same time, this technique can also be used for molecules that fluoresce at visible wavelengths, although it may be somewhat less sensitive for those cases than fluorescence excitation techniques.

III. THEORY FOR ATOM–DIATOM COMPLEXES

A. Coordinates, Classification, and Hamiltonian

1. *Coordinates and Potential Expansion*

The theory of atom–diatom complexes may be formulated using either space-fixed or body-fixed coordinates. While most calculations to date have used the former, the latter provide a somewhat more natural description of species with strongly hindered internal rotation, and studies of the related atom–diatom scattering problem suggest that they may have some computational advantages.[87,88] The two coordinate systems are illustrated in Fig. 1; primed quantities refer to the body-fixed axes while unprimed quantities refer to the space-fixed case. The vector of length r which joins the heavy to the light component of the diatom is denoted $\boldsymbol{r}$; the vector of length R running from the diatom centre-of-mass to the atom is denoted $\boldsymbol{R}$, and

$$\theta = \cos^{-1}(\hat{\boldsymbol{R}} \cdot \hat{\boldsymbol{r}}) \tag{6}$$

where $\hat{\boldsymbol{R}} = \boldsymbol{R}/R$, and $\hat{\boldsymbol{r}} = \boldsymbol{r}/r$.

The body-fixed axes $\{x', y', z'\}$ are defined such that z' is parallel to

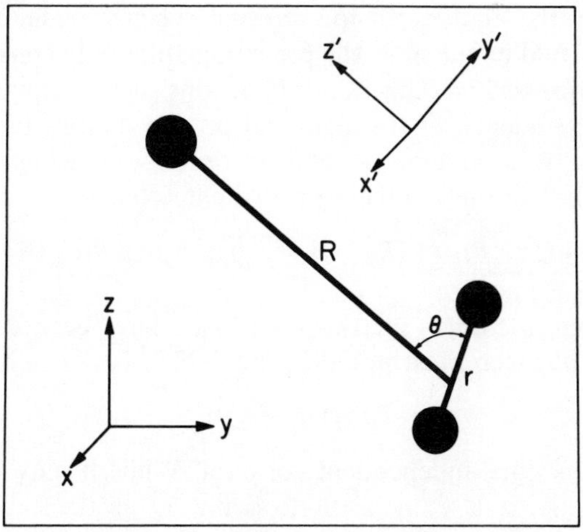

Fig. 1. Coordinate systems for an atom–diatom van der Waals molecule; unprimed (primed) axes identify the space-fixed (body-fixed) reference frame.

$\boldsymbol{R}$ and y' lies in the $\{\boldsymbol{R}, \boldsymbol{r}\}$ plane. In the space-fixed frame, the dynamics of the system are described by R, r, and the polar angles specifying the orientations of $\hat{\boldsymbol{R}}$ and $\hat{\boldsymbol{r}}$; in the body-fixed frame the relevant coordinates are R, r, θ and the three Euler angles defining the orientation of $\{x', y', z'\}$ relative to $\{x, y, z\}$.

No matter which coordinate system is used to describe its dynamics, the potential energy of the atom–diatom complex is most readily represented using the relative coordinates (R, r, θ). The relative weakness of the van der Waals bond suggests that the total potential energy of the system should be subdivided as

$$V_{\text{tot}}(R, r, \theta) = V_{\text{d}}(r) + V(R, r, \theta) \tag{7}$$

where $V_{\text{d}}(r)$ is the intramolecular potential energy of an individual diatom and $V(R, r, \theta)$ is the intermolecular potential of interest. It is often convenient to model the anisotropy of the latter using an expansion in terms of Legendre polynomials:

$$V(R, r, \theta) = \sum_{\lambda} V_{\lambda}(R, r) P_{\lambda}(\cos\theta) \tag{8}$$

At small R many terms must be retained in this expansion if good convergence is to be achieved and it cannot easily represent the coulomb singularities which occur when the nucleus of the atom approaches one of those of the diatom. Alternate ways of expressing the potential in these regions have been suggested.[89,90] However, at the distances (R values) associated with the bound states of van der Waals molecules, equation (8) should be a rapidly converging series; for the H_2–inert gas complexes, this convergence was verified by the analysis of the infrared spectra.[54] Hence, while other formulations would be possible, the present chapter only considers potential anisotropies expressed via Legendre expansions such as equation (8).

Since the dependence of the potential on the diatom bond length r is relatively weak, it should be possible to describe it using a power series expansion in the diatom stretching coordinate $\xi(r)$:

$$V(R, r, \theta) = V(R, \xi, \theta) = \sum_{\lambda} \sum_{k} \xi^k P_{\lambda}(\cos\theta) V_{\lambda k}(R) \tag{9}$$

While a number of choices for this coordinate have been considered,[57,91] the one which has seen most practical use[54,55,57] is

$$\xi(r) \equiv (r - r_0)/r_0 \tag{10}$$

where r_0 is an isotope-independent constant. While it may in principle be chosen arbitrarily, it is convenient to define r_0 as the expectation value of r for the ground state of the most abundant isotope of the diatom, so that the expectation value of $\xi(r)$ will be zero for complexes correlating with this diatom state. While the ensuing discussion could be formulated equally readily in terms of other definitions of $\xi(r)$, equation (10) is the definition assumed in the remainder of this report.

Within the electronic Born–Oppenheimer approximation, isotopic substitution does not change an intermolecular potential. However, a change in isotope for one of the atoms of the diatom in Fig. 1 causes a shift of the diatom centre-of-mass. Since bound-state or scattering calculations are most conveniently carried out using coordinates centred at the diatom centre-of-mass, an expansion of the potential in the new coordinates must be obtained. Two general methods have been devised[91,92] for transforming the Legendre expansion of an atom–diatom potential into a new Legendre expansion about a shifted (by isotope substitution) centre-of-mass.

2. *The Hamiltonian*

After separating out the motion of its centre-of-mass, the Hamiltonian for the bimer written in space-fixed coordinates is:

$$\begin{aligned}\mathbf{H}(\boldsymbol{R},\boldsymbol{r}) &= -(\hbar^2/2\mu)\nabla_R^2 + V(R,r,\theta) + \mathbf{H}_\mathrm{d}(\boldsymbol{r}) \\ &= -(\hbar^2/2\mu)R^{-1}(\partial^2/\partial R^2)R + \mathbf{l}^2/2\mu R^2 + V(R,r,\theta) + \mathbf{H}_\mathrm{d}(\boldsymbol{r}) \qquad (11)\end{aligned}$$

where $\mu = m_\mathrm{a}m_\mathrm{d}/(m_\mathrm{a}+m_\mathrm{d})$ is the atom plus diatom reduced mass, with m_a and m_d being the masses of the atom and diatom, respectively, and the eigenfunctions of the $\mathbf{l}^2$ operator are spherical harmonics:

$$\mathbf{l}^2 Y_{lm_l}(\hat{\boldsymbol{R}}) = l(l+1)\hbar^2 Y_{lm_l}(\hat{\boldsymbol{R}}) \qquad (12)$$

The last term in equation (11) is the vibration–rotation Hamiltonian for the isolated diatom,

$$\begin{aligned}\mathbf{H}_\mathrm{d}(\boldsymbol{r}) &= -(\hbar^2/2\mu_\mathrm{d})\nabla_r^2 + V_\mathrm{d}(r) \\ &= -(\hbar^2/2\mu_\mathrm{d})r^{-1}(\partial^2/\partial r^2)r + \mathbf{j}^2/2\mu_\mathrm{d}r^2 + V_\mathrm{d}(r) \qquad (13)\end{aligned}$$

where μ_d is its reduced mass and the eigenfunctions of the $\mathbf{i}^2$ operator are also spherical harmonics:

$$\mathbf{j}^2 Y_{jm_j}(\hat{\boldsymbol{r}}) = j(j+1)\hbar^2 Y_{jm_j}(\hat{\boldsymbol{r}}) \qquad (14)$$

The eigenfunctions of $\mathbf{H}_\mathrm{d}(\boldsymbol{r})$ may then be written as $r^{-1}\phi_{vj}(r)Y_{jm_j}(\hat{\boldsymbol{r}})$ where $\phi_{vj}(r)$ satisfies the usual radial Schrödinger equation

$$\{-(\hbar^2/2\mu_\mathrm{d})[\mathrm{d}^2/\mathrm{d}r^2 - j(j+1)/r^2] + V_\mathrm{d}(r) - E_\mathrm{d}(v,j)\}\phi_{vj}(r) = 0 \qquad (15)$$

and $E_\mathrm{d}(v,j)$ are the vibration–rotation eigenvalues of the isolated diatom. The diatom potential $V_\mathrm{d}(r)$ may in general be accurately determined by inversion of spectral data;[31,32,34] on substituting it into equation (15), the latter may be solved using standard numerical methods[93] to yield the exact eigenvalues and eigenfunctions of $\mathbf{H}_\mathrm{d}(\boldsymbol{r})$.

In the body-fixed coordinate system, the Hamiltonian is identical to (11) except that $\boldsymbol{R}, \hat{\boldsymbol{R}}, \boldsymbol{r}$, and $\hat{\boldsymbol{r}}$ are expressed relative to the primed axes and the angular momentum operator for the rotation of $\boldsymbol{R}$ (i.e. $\mathbf{l}$) is written as the difference between the total angular momentum operator $\mathbf{J}$ and that associated with the rotation of $\hat{\boldsymbol{r}}$ (i.e. $\mathbf{j}$), yielding

$$\mathbf{H}(\boldsymbol{R}, \boldsymbol{r}) = -(\hbar^2/2\mu)R^{-1}(\partial^2/\partial R^2)R + (\mathbf{J}-\mathbf{j})^2/2\mu R^2 + V(R, r, \theta) + \mathbf{H}_{\mathrm{d}}(\boldsymbol{r}) \quad (16)$$

In this case, the projection of $\boldsymbol{j}$ on the body-fixed axis z' is denoted by the 'helicity' quantum number Ω, so that the eigenfunctions of $\mathbf{j}^2$ become $Y_{j\Omega}(\hat{\boldsymbol{r}})$. By conservation of angular momentum, the projection of the total angular momentum $\mathbf{J}$ on this axis is also equal to $\Omega\hbar$, and necessarily $|\Omega| \leq \min(J, j)$.

The quantum numbers associated with the various types of motion of the system for both the space-fixed and body-fixed coordinate systems are summarized in Table II. It is helpful to note that these two-coordinate systems presuppose different zeroth-order pictures of the bimer. In the space-fixed case, the description of the system in terms of the rotations of $\hat{\boldsymbol{R}}$ and $\hat{\boldsymbol{r}}$ suggests a zero-anisotropy model in which the diatom rotates and vibrates perfectly freely inside a complex which itself is simultaneously and independently rotating and vibrating. This picture becomes exact in the limit when the radial strength functions $V_{\lambda k}(R)$ of equation (9) are identically zero for all $(\lambda, k) \neq (0, 0)$. In contrast, use of the body-fixed coordinates suggests a picture in which the complex is a perfect symmetric top rotating in space with a portion of it simultaneously undergoing completely unhindered internal rotation. For this second model to hold exactly, the segment of the bimer undertaking internal rotation must itself be a perfect spherical top such as methane (as in Ar–CH_4). Thus this limiting behaviour

TABLE II. Quantum numbers associated with various types of motion of an atom–diatom van der Waals complex

Coordinate system	Quantum number	Type of motion
Either:	J, M	Total angular momentum and its projection on the space-fixed z axis
	v	Stretching of r
	n	Stretching of R
Space fixed:	j, m_j	Free rotation of $\hat{r}$ in space
	l, m_l	Free rotation of $\hat{\boldsymbol{R}}$ in space
Body fixed:	Ω	Projection of J and j on the body-fixed z' axis
	j, Ω	Free rotation of $\hat{r}$ relative to the body-fixed axes

cannot be realized for an atom–diatom model merely by setting $V_{\lambda k}(R)$ functions equal to zero. On the other hand, the relatively small mass of one or more of the component atoms makes a number of the complexes which have been studied (e.g. HCl–Ar, $(HF)_2$, He–I_2) only very slightly asymmetric tops, so this picture may be nearly achieved. Note too that in either coordinate system the internal rotation quantum number j may be associated with the vibrational quantum number 'v_2' for the bending motion of the complex.

3. *Classification*

As pointed out by Bratož and Martin,[94] it is convenient to classify complexes according to the relative strength of the parts of the potential which couple together the various types of motion. This classification determines both the rigidity of the complex and the difficulty of performing accurate calculations on it. For the coupling between radial (in R) and angular motion, Ewing[15] has suggested the modified classification scheme shown in Table III. There $\Delta\bar{V}_a$ is the average effective anisotropy, i.e. the sum of all $\lambda \neq 0$ $V_{\lambda k}(R)$ terms in equation (9) averaged over the ranges of R and ξ associated with the complex; $\Delta E_d(j)$ is the rotational level spacing of the isolated diatom, and $\Delta E_c(l)$ the spacing between the levels associated with the uncoupled end-over-end rotation of the complex. For a homonuclear diatom, parity conservation causes these level spacings to correspond to $\Delta j = \pm 2$ and $\Delta l = \pm 2$, while for a heteronuclear diatom they correspond to $\Delta j = \pm 1$ and $\Delta l = \pm 1$, respectively.

For weak-coupling complexes, the diatom rotational index j is almost a good quantum number, although there may be a thorough mixing of states associated with different l values. The space-fixed coordinate system provides the most natural description of this type of bimer. For strong-coupling complexes, however, j is also thoroughly mixed and the vector $\boldsymbol{j}$ will tend to precess about $\hat{\boldsymbol{R}}$, so that use of the space-fixed coordinate system becomes less appropriate. For complexes which are only slightly asymmetric tops, the integer Ω associated with the projection of $\boldsymbol{J}$ (and of $\boldsymbol{j}$)

TABLE III. Classification of internal rotation cases for atom–diatom complexes (after Ewing[15]).

Type	Condition	Example
Free (internal) rotation	$\Delta\bar{V}_a = 0$	
Weak coupling	$\Delta E_c(l) \lesssim \Delta\bar{V}_a \ll \Delta E_d(j)$	H_2–Ar, $(H_2)_2$
Strong coupling	$\Delta E_c(l) < \Delta E_d(j) \lesssim \Delta\bar{V}_a$	Ar–HCl, $(HF)_2$
Semirigid	$\{\Delta E_c(l), \Delta E_d(j)\} \ll \Delta\bar{V}_a$	Kr–ClF, He-I_2

on the body fixed axis z' is expected to be an approximate constant of motion. For the lowest states of HCl–Ar, this approximate conservation of Ω is confirmed by the small changes in the eigenvalues calculated by Holmgren *et al.*[78] when they added $\Omega \neq 0$ terms to their wave function expansion. On the other hand, the most accurate calculations performed for this system have all used space-fixed coordinates.[55,95,96]

The only calculations so far reported for a semirigid complex are some preliminary results for Kr–ClF, mentioned by Klemperer.[97] They were performed within the BOARS or adiabatic approximation using body-fixed coordinates. In view of the well-defined structures of complexes of this type, it might seem appropriate to use the type of normal mode analysis customarily applied to chemically bound polyatomic molecules. However, the length and large stretching amplitude of the van der Waals bond should still give a significant advantage to the specialized techniques described herein. Indeed, in very recent work, one of these techniques (solution of the close-coupled equations in body-fixed coordinates) has also been used to compute eigenvalues for the lowest levels of H_2O.[98] Thus further development of the methods described herein may also contribute significantly to our understanding of 'normal' polyatomic molecules.

The classification scheme of Table III is concerned with how strongly the angle-dependent part of the potential couples the various rotational states of the diatom. It is similarly appropriate to ask how strongly the r-dependent (i.e. $k > 0$) terms of the potential equation (9) couple their various vibrational states. The vibrational predissociation observed for He–I_2 shows that this coupling is not negligible and that it may have very strong selection rules.[80,84] However, diatomic molecule vibrational level spacings are generally sufficiently large (from *ca.* 100 cm^{-1} for this He–I_2 case to 4000 cm^{-1} for H_2–Ar) that this coupling will have very little effect on the observed bimer energy levels.

B. Coupled Channel Formulation

1. *In Space-Fixed Coordinates*

a. Zero-coupling limit and general considerations. The set of coupled equations whose solution yields the eigenvalues and eigenfunctions of a van der Waals bimer is identical to the set which describes the elastic and inelastic scattering of its constituents. The latter problem has attracted considerable attention in the last decade, and a number of reliable methods for solving the relevant equations have been developed. Detailed descriptions and critical comparisons of these procedures may be found elsewhere.[99–104] The present section attempts only to show how the equations describing the bound states of the bimers are obtained and used, in order to clarify both the nature of the approximations made in calculations using this approach and the relationships between it and other methods.

In the zero-coupling limit where the potential depends only on R, so that $V(R, r, \theta) = V(R)$, the Hamiltonian of equation (11) becomes separable and its eigenfunctions may be written as

$$[R^{-1}\chi_{nl}(R)Y_{lm_l}(\hat{\boldsymbol{R}})] \times [r^{-1}\phi_{vj}(r)Y_{jm_j}(\hat{\boldsymbol{r}})]. \tag{17}$$

Here, $\chi_{nl}(R)$ is an eigenfunction of the diatomic-like radial Schrödinger equation

$$\{-(\hbar^2/2\mu)\mathrm{d}^2/\mathrm{d}R^2 + l(l+1)\hbar^2/2\mu R^2 + V(R) - E_{\mathrm{c}}(n, l)\}\chi_{nl}(R) = 0, \tag{18}$$

while the other functions are as defined above.

For truly bound states, defined by $E_{\mathrm{c}} \leq V(\infty)$, the index n is simply the vibrational quantum number associated with the stretching of R, while for the continuum of unbound states it may be associated with the asymptotic kinetic energy $[E_{\mathrm{c}} - V(\infty)]$. Arthurs and Dalgarno[105] pointed out that since the total angular momentum $\boldsymbol{J} = \boldsymbol{j} + \boldsymbol{l}$ must always be conserved, the angular part of the product in (17) should be replaced by the total angular momentum wave functions $\mathscr{Y}_{jl}^{JM}$ which are simultaneous eigenfunctions of $\mathbf{J}^2, \mathbf{J}_z, \mathbf{j}^2$, and $\mathbf{l}^2$:

$$\mathscr{Y}_{jl}^{JM}(\hat{\boldsymbol{R}}, \hat{\boldsymbol{r}}) = \sum_{m=-j}^{j} C(j, l, J; m, M-m, M)Y_{jm_l}(\hat{\boldsymbol{r}})Y_{lm_l}(\hat{\boldsymbol{R}}), \tag{19}$$

where $C(j, l, J; m_j, m_l, M)$ are the familiar Clebsch–Gordan coupling coefficients.[106] More appropriate eigenfunctions of the system in this zero-coupling limit are therefore the orthonormal functions

$$\Psi^0_{\alpha,J,M} = [r^{-1}\phi_{vj}(r)][R^{-1}\chi_{nl}(R)]\mathscr{Y}_{jl}^{JM}(\hat{\boldsymbol{R}}, \hat{\boldsymbol{r}}) \tag{20}$$

and the corresponding eigenvalues are

$$E^0(\alpha) = E^0(v, j, n, l) = E_{\mathrm{d}}(v, j) + E_{\mathrm{c}}(n, l) \tag{21}$$

This result corresponds to the limiting case of independent stretching and rotation of $\boldsymbol{R}$ and $\boldsymbol{r}$.

For a model problem in which the diatom in the complex is both rigid and homonuclear so that $V_{\lambda k}(R) \equiv 0$ both for $k > 0$ and for odd values of n, Fig. 2 shows the relevant potential curves and $E^0(\alpha)$ values for a number of the states for which $J = 2$. The curves shown are the centrifugally distorted effective potentials (see equation (18)), $[V(R) + l(l+1)\hbar^2/2\mu R^2]$, and their asymptotes are separated by the appropriate diatom level spacings $6B_{\mathrm{d}}$ and $14B_{\mathrm{d}}$, where B_{d} is the rotational constant of the rigid diatom. These curves comprise all of the $j < 6$ channels which correspond to $J = 2$. The effect of the θ-dependent (i.e. $\lambda > 0$) terms in the potential is to mix together these various zeroth-order states, giving rise to shifts in the associated energy

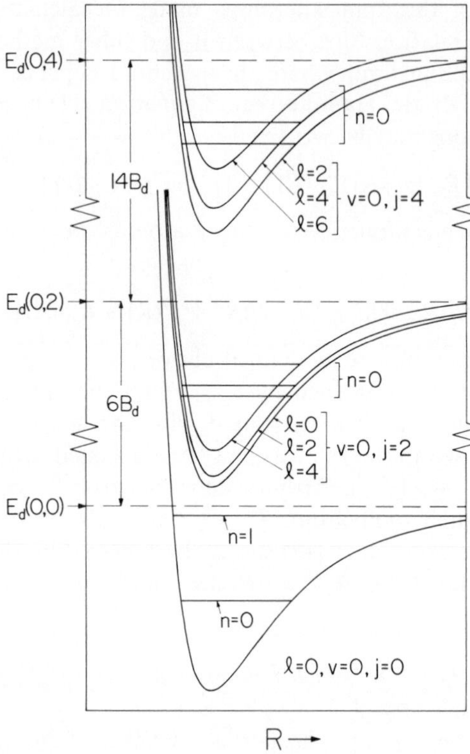

Fig. 2. Potential curves and zero-coupling limit level energies for the angular channels contributing to even parity $J = 2$ states of an atom–homonuclear diatom system.

levels. An example of the effect of this 'turning on' of the coupling is provided by the model for H_2–Ar illustrated in Fig. 3, where the $v = 1, j = 2, n = 0$ levels corresponding to different values of l are seen to be split by the anisotropy into sublevels associated with different values of the total angular momentum quantum number J. The curve shown there is the vibrationally averaged (over r) effective isotropic potential for $v = 1, j = 2, l = 0$. Note that since the states being mixed by the coupling include unbound n-levels, for large values of the diatom rotational B_d the levels associated with $j = 2$ and 4 in Fig. 2 will be metastable with respect to predissociation to continuum states with $j = 0$.

The model problem of Fig. 2 provides a convenient illustration of the distinction between the weak- and strong-coupling cases outlined in Table III. For weak coupling, the spacings between the energy levels $E_c(n, l)$ are typically much smaller than those between the diatom level energies $E_d(v, j)$.

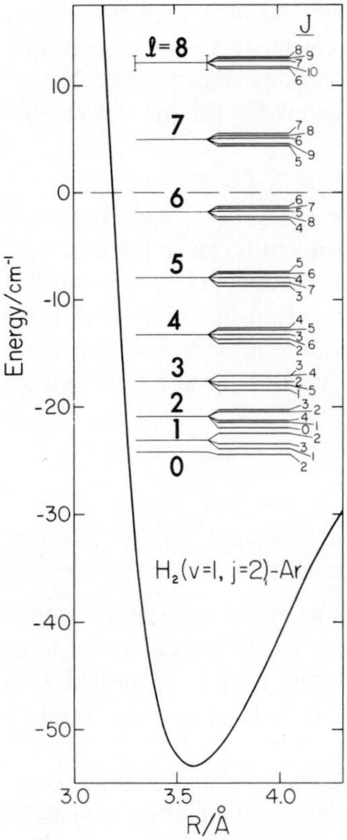

Fig. 3. Effect of potential anisotropy in splitting the l-levels of a model $H_2(v=1, j=2)$–Ar potential. Reproduced by permission of the American Institute of Physics.[54]

In practice, this means that the asymptotes of the curves for different (v, j) are displaced by an amount which is large relative to the depth of the isotropic potential $V(R)$. For strong-coupling complexes, B_d is much smaller than this well depth so that the effective isotropic potentials for different j's are nested. In this case, many of the levels associated with $j = 2$ or 4 would lie below the $j = 0$ potential asymptote, and would be truly bound rather than metastable.

The rigid-rotor model implies that: (i) the potential has no terms with $k > 0$, and (ii) that the diatom radial wave functions $\phi_{vj}(r)$, and hence the properties of the rotor, vary with v but not with j. In this limit there is absolutely no coupling between levels associated with different values of v. However, if either of these conditions is relaxed such coupling can occur. Since diatom vibrational spacings are typically much larger than van der Waals bond strengths, this coupling makes all states corresponding to non-zero values of v metastable with respect to continuum states associated with smaller v values. Fortunately, Franck–Condon overlap considerations[107] and the relatively large size of diatom vibrational spacings ensure that this coupling is usually sufficiently weak that it will have a negligible effect on bimer energy levels for which $v > 0$. As a result, this type of coupling has been neglected in all existing calculations of bimer eigenvalues. Nevertheless, it is precisely this coupling that is responsible for the vibrational predissociation of van der Waals molecules.

b. Coupled-channel equations and solutions. In the general case in which the diatom is non rigid and r- (or ξ-) and θ-dependent terms in the potential are non-zero, J (and M) remain good quantum numbers of the system but v, j, and l do not. Since the total angular momentum functions $\mathscr{Y}_{jl}^{JM}$ form a complete angular basis and the eigenfunctions of equation (15) a basis for motion along r, the eigenfunctions of the overall Schrödinger equation

$$[\mathbf{H}(\boldsymbol{R}, \boldsymbol{r}) - E_\alpha^J]\Psi_\alpha^{JM} = 0 \tag{22}$$

may be expanded in the form

$$\Psi_\alpha^{JM}(\boldsymbol{R}, \boldsymbol{r}) = r^{-1}R^{-1}\sum_{v''}\sum_{j''}\sum_{l''}\phi_{v''j''}(r)\mathscr{Y}_{j''l''}^{JM}(\hat{\boldsymbol{R}}, \hat{\boldsymbol{r}})G_{v''j''l''}^{J\alpha}(R) \tag{23}$$

where the $G(R)$ are radial channel wave functions and α is an index identifying a particular solution of (22). Since (23) must reduce to (20) in the zero-coupling limit,[56] the index α labelling a given state may be identified with the $\{v, j, n, l\}$ indices of the level with which it correlates at zero coupling (see, e.g. Fig. 3).

The expansion (23) is exact if the sum includes all j'' and l'' and the complete set of all bound and continuum eigenfunctions of equation (15). Substituting (11) and (23) into (22), multiplying from the left by $(r^{-1}\phi_{v'j'}(r)\mathscr{Y}_{j'l'}^{JM} \times (\hat{\boldsymbol{R}}, \hat{\boldsymbol{r}}))^*$ and integrating over r and the angular coordinates then yields the set of coupled equations of interest:

$$\begin{aligned}\{(\hbar^2/2\mu)[-\mathrm{d}^2/\mathrm{d}R^2 + l'(l'+1)/R^2] + E_\mathrm{d}(v', j') - E_\alpha^J\}G_{v'j'l'}^{J\alpha}(R)\\ + \sum_{v''}\sum_{j''}\sum_{l''}V(v', j', l'; v'', j'', l''; J|R)G_{v''j''l''}^{J\alpha}(R) = 0\end{aligned} \tag{24}$$

The elements of the potential matrix appearing in (24) are defined as

$$\begin{aligned}V(v', j', l'; v'', j'', l''; J|R) \equiv \int\int\int \phi_{v'j'}(r)^*\mathscr{Y}_{j'l'}^{JM}(\hat{\boldsymbol{R}}, \hat{\boldsymbol{r}})^* V(R, r, \theta)\phi_{v''j''}(r)\\ \mathscr{Y}_{j''l''}^{JM}(\hat{\boldsymbol{R}}, \hat{\boldsymbol{r}})\mathrm{d}r\mathrm{d}\hat{\boldsymbol{R}}\mathrm{d}\hat{\boldsymbol{r}}\end{aligned} \tag{25}$$

If the potential is expanded as in equation (9) this reduces to

$$\begin{aligned}V(v', j', l'; v'', j'', l''; J|R) = \sum_\lambda\sum_k\left(\int_0^\infty \phi_{v'j'}(r)^*\xi^k(r)\phi_{v''j''}(r)\mathrm{d}r\right)\\ \times f_\lambda(j', l'; j'', l''; J)V_{\lambda k}(R)\end{aligned} \tag{26}$$

where $f_\lambda(j', l'; j'', l''; J)$ are the well-known Percival–Seaton coefficients,[108] and the necessary integrals are readily computed as long as the potential $V_d(r)$ of the isolated diatom is known.

The close-coupling method for calculating the energies and other properties of van der Waals bimers requires the direct numerical integration of equation (24), subject to application of the appropriate boundary conditions. For this to be practical, the sums over v'', j'', and l'' must be truncated as early as possible, since the computational effort grows as the cube of the total number of such channels. In all studies to date, only diagonal $v'' = v' = v$ coupling has been considered. In most cases the diatom was also treated as a rigid rotor, so that the sum over k in equations (9) and (26) was truncated at $k = 0$ and the integral over r in (26) replaced by the Kronecker delta $\delta_{v'',v'}$. Considerable further simplification arises from the symmetries of the angular wave functions which cause the $f_\lambda(j', l'; j'', l''; J)$ coefficients of equation (26) to be zero unless: (i) $(j' + j'' + \lambda)$ and $(l' + l'' + \lambda)$ are both even, (ii) $|j' - j''| \leq \lambda \leq j' + j''$ and $|l' - l''| \leq \lambda \leq l' + l''$, and (iii) $(-1)^{j'+l'} = (-1)^{j''+l''}$. In addition, if the diatom bound in the complex is homonuclear, only even values of λ appear in the potential (9), and hence in (26). However, since the sums over j'' and l'' still include an infinite number of terms, they are truncated by selection of an upper bound j_{max} to the allowed range of j'' values, while retaining all allowed l'' values. In general, this j_{max} is chosen so as to include all 'open' channels (i.e. those for which $E_d(v'', j'') < E_\alpha^J$) together with 'closed' (v'', j'')-channels correlating with dissociation limits less than or equal to that for bound state in question, but excluding most or all of the other closed channels for which $E_d(v'', j'') > E_d(v, j)$.

It is this truncation of the j'' sum at j_{max} and of the v'' sum at the diagonal $v' = v'' = v$ term which makes coupled-channel calculations approximate rather than exact. Perturbation theory arguments show that these truncations introduce little error when $E_d(v'', j'') \gg E_d(v, j)$, and it seems reasonable that they should also rule out contributions from channels for which $E_d(v'', j'') \ll E_d(v, j)$.[54] Moreover, it is always possible to repeat a calculation for a series of increasing j_{max} values until convergence is achieved; see, e.g. Table III in the study of Ar–HCl by Dunker and Gordon.[95]

The only truly bound states of van der Waals molecules occur at energies where no open channels exist; this usually corresponds to the requirement that $E_\alpha^J < E_d(0, 0)$, although for complexes formed from ortho-H_2 or para-D_2, this condition becomes $E_\alpha^J < E_d(0, 1)$. The boundary conditions in this case are that all $G_{v'j'l'}^{J\alpha}(R) \to 0$ at both $R \to 0$ and $R \to \infty$. Dunker and Gordon have developed computational techniques which are particularly appropriate for this case.[110] In contrast, the radial wave function $G_{v'j'l'}^{J\alpha}(R)$ associated with an open channel obeys the scattering boundary condition:

$$G_{v'j'l'}^{J\alpha}(R) \underset{R \to \infty}{\sim} \delta_{j'j}\delta_{l'l}\delta_{v'v}\mathrm{e}^{-i[k_{vj}r - l\pi/2]} - (k_{vj}/k_{v'j'})^{1/2} S^J(v', j', l'; v, j, l)\mathrm{e}^{i(k_{v'j'}r - l'\pi/2)} \tag{27}$$

where $k_{v'j'}^2 = (2\mu/\hbar^2)[E_\alpha^J - E_d(v', j')]$ and S^J is an element of the scattering matrix. All bimer states for which the wave function contains an open-channel contribution are in fact metastable with respect to predissociation into the products associated with this open channel (e.g. all of the $j = 2$ and $j = 4$ levels for the model problem shown in Fig. 2). Thus these bimer levels are simply another manifestation of the compound state resonances of atom–molecule scattering, which attracted considerable theoretical attention a few years ago.[111–119]

For the case of only one open channel, the S^J matrix element associated with it may be expressed as $S^J = e^{2i\eta_J}$, where the (real valued) scattering phase shift $\eta_J(E)$ is expected[113] to have Breit–Wigner form

$$\eta_J(E) = \eta_J^0 + \arctan\left[\tfrac{1}{2}\Gamma_\alpha^J/(E_\alpha^J - E)\right] \tag{28}$$

near the energy E_α^J of a metastable level (compound state resonance) of width Γ_α^J. Values of E_α^J and Γ_α^J may therefore be determined by performing calculations at a range of energies and fitting the results to equation (28). An alternate procedure for determining E_α^J and Γ_α^J which makes use of one or more additional non-physical open channel(s), has been devised by Shapiro.[119]

For more than one (real physical) open channel, resonance energies and widths may still be determined from the energy dependence of the S^J matrix. However, all of the existing close-coupling calculations correspond to either no open channels[56,95,110] or only one open channel,[113–119] and a detailed procedure for doing this has not yet been reported. One difficulty which could arise is that S^J matrix elements corresponding to different initial state boundary conditions may predict slightly different energies and widths for the same metastable level. This would be analogous to the situation for single-channel orbiting resonances (quasibound levels) where it has been suggested that different experiments may yield slightly different energies for a given resonance.[120] However, such differences are expected to be small relative to the overall level width.

2. *In Body-Fixed Coordinates*

To date, no close-coupling calculations for the bound states of van der Waals molecules have been performed using the body-fixed reference frame. However, since this approach may prove useful in future, it seems appropriate to summarize it here.

In order to define the zeroth-order problem, it is first necessary to re-express the operator $(\mathbf{J} - \mathbf{j})^2$ appearing in (16) as

$$\begin{aligned}(\mathbf{J} - \mathbf{j})^2 &= \mathbf{J}^2 + \mathbf{j}^2 - 2\mathbf{J}\cdot\mathbf{j} \\ &= \mathbf{J}^2 + \mathbf{j}^2 - 2\mathbf{J}_z'\mathbf{j}_z' - \mathbf{J}_+\mathbf{j}_- - \mathbf{J}_-\mathbf{j}_+\end{aligned} \tag{29}$$

where $\mathbf{J}_{\pm} = \mathbf{J}_x \pm i\mathbf{J}_y$ and $\mathbf{j}_{\pm} = j_x \pm i\mathbf{j}_y$ are the familiar angular momentum raising and lowering operators. Since $\boldsymbol{l} = (\boldsymbol{J} - \boldsymbol{j})$ is orthogonal to z', the vectors $\boldsymbol{J}_{z'}$ and $\boldsymbol{j}_{z'}$ are identical, so that the product of the corresponding operators may be simply replaced by $\mathbf{J}_{z'}^2$. Making use of (29), the Hamiltonian (16) becomes

$$\mathbf{H}(\boldsymbol{R}, \boldsymbol{r}) = [-(\hbar^2/2\mu)R^{-1}(\partial^2/\partial R^2)R + (\mathbf{J}^2 + \mathbf{j}^2 - 2\mathbf{J}_{z'}^2)/2\mu R^2] + \mathbf{H}_\mathrm{d}(\boldsymbol{r}) + V(R, r, \theta) + (\mathbf{J}_+\mathbf{j}_- + \mathbf{J}_-\mathbf{j}_+)/2\mu R^2 \tag{30}$$

Here, the coupling between the different modes of motion is effected both by the r and θ dependence of the potential and by the last term in equation (30). In the zero-coupling limit when both of these contributions are neglected the eigenfunctions of the system are simply

$$[r^{-1}\phi_{vj}(r)][R^{-1}\chi_{nJj\Omega}(R)]Y_{j\Omega}(\hat{\boldsymbol{R}})D^J_{\Omega M}(\hat{\boldsymbol{R}}, \hat{\boldsymbol{r}}) \tag{31}$$

where $D^J_{\Omega M}$ is the usual symmetric top wave function[106], $\chi_{nJj\Omega}(R)$ is a solution of

$$\{-(\hbar^2/2\mu)\mathrm{d}^2/\mathrm{d}R^2 + [J(J+1) + j(j+1) - \Omega^2]\hbar^2/2\mu R^2 + V(R) - E_\mathrm{c}(n, J, j, \Omega)\}\chi_{nJj\Omega}(R) = 0 \tag{32}$$

and the corresponding bimer eigenvalues are

$$E^0(v, j, \Omega, n, J) = E_\mathrm{d}(v, j) + E_\mathrm{c}(n, J, j, \Omega) \tag{33}$$

This result differs qualitatively from that for the space-fixed approach in that the solutions of this uncoupled problem depend on the total angular momentum quantum number J and its projection Ω on the body-fixed axis.

When the coupling terms are introduced, v, j, and Ω are no longer good quantum numbers, but the eigenfunctions of the Hamiltonian (30) may be expressed as:

$$\Psi^{JM}_\beta(\boldsymbol{R}, \boldsymbol{r}) \equiv r^{-1}R^{-1}\sum_{v''}\sum_{j''}\sum_{\Omega''=-J}^{J} \phi_{v''j''}(r)Y_{j''\Omega''}(\hat{\boldsymbol{r}})D^J_{\Omega''M}(\hat{\boldsymbol{R}}, \hat{\boldsymbol{r}})G^{J\beta}_{v''j''\Omega''}(R) \tag{34}$$

where the index β is associated with the zero-coupling limit quantum numbers $\{v, j, J, \Omega, n\}$. Substituting (34) into (30) and taking the inner product with $[r^{-1}\phi_{v'j'}(r)Y_{j'\Omega'}(\hat{\boldsymbol{r}}')D^J_{\Omega' M}(\hat{\boldsymbol{R}}, \hat{\boldsymbol{r}})]$ then yields the body-fixed frame-coupled equations:

$$\{(\hbar^2/2\mu)[-\mathrm{d}^2/\mathrm{d}R^2 + (J'^2 + J' + j'^2 + j' - \Omega^2)/R^2] + E_\mathrm{d}(v', j') - E^J_\beta\} \times G^{J\beta}_{v'j'\Omega'}(R) + \sum_{v''}\sum_{j''} V(v', j', \Omega'; v'', j'', \Omega' | R)G^{J\beta}_{v''j''\Omega'}(R) + h^{Jj'}_{\Omega',\Omega'-1}G^{J\beta}_{v'j'\Omega'-1}(R) + h^{Jj'}_{\Omega',\Omega'+1}G^{J\beta}_{v',j',\Omega'+1}(R) = 0 \tag{35}$$

where

$$h^{Jj'}_{\Omega',\Omega'\pm 1} = [(J \pm \Omega' + 1)(J \mp \Omega')(j' \mp \Omega')(j' \pm \Omega' + 1)]^{1/2}/R^2 \qquad (36)$$

and, assuming the potential expansion of equation (9),

$$\begin{aligned} &V(v', j', \Omega'; v'', j'', \Omega' | R) \\ &\quad \equiv \int\int \phi_{v'j'}(r)^* Y_{j'\Omega'}(\hat{r})^* V(R, r, \theta) \phi_{v''j''}(r) Y_{j''\Omega'}(\hat{r}) \mathrm{d}r \mathrm{d}\hat{r} \\ &\quad = \sum_{\lambda} \sum_{k} \left(\int_0^{\infty} \phi_{v'j'}(r)^* \xi^k \phi_{v''j''}(r) \mathrm{d}r \right) \mathrm{d}_{\lambda}(j', j'', \Omega') V_{\lambda k}(R) \end{aligned} \qquad (37)$$

where

$$d_{\lambda}(j', j'', \Omega') = [(2j'' + 1)/(2j' + 1)]^{1/2} C(j'', \lambda, j'; \Omega', 0, \Omega') C(j'', \lambda, j'; 0, 0, 0) \quad (38)$$

Again, the boundary conditions for all closed channels are that $G(R) = 0$ at both $R \to 0$ and $R \to \infty$, while for open channels the latter is replaced by an appropriate scattering boundary condition which defines the S^J matrix element.[87, 88, 104]

As in the space-fixed representation, the sums over v'' and j'' in equations (34) and (35) must be truncated before the coupled equations can be solved, and the restrictions that $v' = v'' = v$ and $j', j'' \leq j_{max}$ are equally appropriate here. For any given choice of j_{max}, exactly the same number of coupled equations must be solved in either approach. The body-fixed equations are somewhat simpler because: (i) the d_{λ} coefficient of equation (38) is easier to evaluate than the f_{λ} coefficients of equation (26), and (ii) while the j and l couplings in the space-fixed equation (24) involve simultaneous sums over v'', j'', and l'', the sum over Ω'' in the body-fixed case never involves more than two off-diagonal terms, while coupling terms which are off-diagonal in one of Ω'' or (v'', j'') are necessarily diagonal in the other. A more important consideration is the fact that in some cases Ω may be nearly a good quantum number; when this is true, results of reasonable accuracy could be obtained from substantially fewer coupled equations than would be required for an equivalent calculation in the space-fixed representation. While Ω was not even approximately conserved in the $H + H_2$ scattering calculations of Choi *et al.*,[88] this may merely reflect the fact that this system is very unlike a symmetric top. In contrast, Ar–HCl is only a very slightly asymmetric top[60] and (approximate) bound state[78, 121] and scattering[122] calculations suggest that Ω is very nearly conserved for it, at least for low J. Note, that the approximation that Ω is a good quantum number arises from neglect of the last term in the Hamiltonian of equation (30); its deletion is the origin of the 'centrifugal decoupling' approximation which has proven very useful in scattering calculations.[123]

C. The Secular Equation Method

1. *Outline of the Method*

In the close-coupling method described above, it is necessary to integrate numerically the set of coupled equations (24) or (35) at a series of trial energies until all boundary conditions and continuity requirements are satisfied. While capable of yielding results of any desired accuracy, the expense of this method tends to discourage its use for the repetitive calculations required by a fit to determine potential parameters from experimental data. The present section describes an alternate approach to the solution of these coupled equations, one which can yield the same high degree of accuracy at much less computational expense, while providing a simpler physical picture of the effect of the coupling. To date, this approach has only been applied to the coupled equations in the space-fixed representation, so this is the coordinate system used below. However, its application to the body-fixed equation (35) should be quite straightforward. A very physical derivation of this method was presented in ref. 54; a more formal viewpoint is taken here in order to emphasize the generality and flexibility of the approach.

The core of the secular equation method is the expansion of each of the radial channel wave functions $G^{J\alpha}_{v''j''l''}(R)$ of equations (23) and (24), in terms of some complete set of basis functions $\psi_{\alpha''}(R)$, where $\alpha'' = \{v'', j'', n'', l''\}$ and n'' labels the members of the basis set:

$$G^{J\alpha}_{v''j''l''}(R) = \sum_{n''} A^J_{\alpha\alpha''} \psi_{\alpha''}(R) \tag{39}$$

The same set of basis functions may be used for each channel and for each bimer eigenstate, or different basis sets may be chosen for each. While not essential, it is usually convenient to require the functions associated with a given radial channel to be orthonormal, so that

$$\int_0^\infty \psi_{v''j''l''n''}(R)^* \psi_{v''j''l''n'}(R) \mathrm{d}R = \delta_{n',n''}, \tag{40}$$

and unless otherwise stated this is assumed throughout the following.

Substituting (39) into the coupled equations (24), multiplying by $[\psi_{v'j'l'n'}(R)]^*$ and integrating over R turns the problem of finding the eigenvalues of the set of coupled differential equations into a straightforward matrix diagonalization problem. This transformation could clearly be accomplished using any set of functions which are orthonormal on the interval $0 \leq R \leq \infty$ and satisfy the appropriate boundary conditions. However, for this approach to be attractive two conditions must be met: (i) the necessary radial matrix elements should be easy to evaluate, and (ii) the number of basis functions required in the expansion (39) should be as few as possible. In addition, it is highly desirable that these basis functions should be related to some simple physical picture of the system.

In order to satisfy condition (i), it has been found most convenient to define the basis functions as the discrete eigenfunctions of a radial Schrödinger equation for a carefully chosen basis generating isotropic potential $V_b(R)$:

$$\{(\hbar^2/2\mu)[-\mathrm{d}^2/\mathrm{d}R^2 + l''(l''+1)/R^2] + V_b(R) - E_b(n'', l'')\}\psi_{\alpha'}(R) = 0 \quad (41)$$

This is equivalent to rewriting the overall Schrödinger equation for the system as:

$$\{[-(\hbar^2/2\mu)R^{-1}(\partial^2/\partial R^2)R + \boldsymbol{l}^2/2\mu R^2 + V_b(R) + \mathbf{H}_d(\boldsymbol{r})] + \Delta V(R, r, \theta) - E^J_\alpha\}\Psi^{JM}_\alpha = 0 \quad (42)$$

where

$$\Delta V(R, r, \theta) \equiv V(R, r, \theta) - V_b(R) \quad (43)$$

and the wave function is expanded as

$$\Psi^{JM}_\alpha(\boldsymbol{R}, \boldsymbol{r}) = r^{-1}R^{-1}\sum_{v''}\sum_{j''}\sum_{l''}\sum_{n''} A^J_{\alpha\alpha''}\phi_{v''j''}(r)\mathscr{Y}^{JM}_{j''l''}(\hat{\boldsymbol{R}}, \hat{\boldsymbol{r}})\psi_{\alpha''}(R) \quad (44)$$

The zeroth-order eigenfunctions corresponding to the individual terms in this sum (44) are clearly exact eigenfunctions of the portion of the Hamiltonian (42) in square brackets and have eigenvalues

$$E^0_{\alpha''} = E_d(v'', j'') + E_b(n'', l'') \quad (45)$$

Substituting (44) into (42), multiplying in turn by the complex conjugate of each of the terms in the expansion (44) and integrating over $\boldsymbol{R}$ and $\boldsymbol{r}$ then yields the set of linear equations:

$$\sum_{\alpha''} A^J_{\alpha\alpha''}(H_{\alpha'\alpha''} - E^J_\alpha\delta_{\alpha''\alpha'}) = 0 \quad (46)$$

where, using the definitions of equations (25) and (26), the matrix elements of the full Hamiltonian are

$$\begin{aligned} H_{\alpha'\alpha''} &= E^0_{\alpha''}\delta_{\alpha'\alpha''} + \int_0^\infty \psi_{\alpha'}(R)^*\{V(v', j', l'; v'', j'', l''; J|R) \\ &\quad - V_b(R)\delta_{v',v''}\delta_{j',j''}\delta_{l',l''}\}\psi_{\alpha''}(R)\mathrm{d}R \\ &= E^0_{\alpha''}\delta_{\alpha'\alpha''} + \sum_\lambda\sum_k f_\lambda(j', l'; j'', l''; J)\left(\int_0^\infty \phi_{v'j'}(r)^*\xi^k\phi_{v''j''}(r)\mathrm{d}r\right) \\ &\quad \times\left(\int_0^\infty \psi_{\alpha'}(R)^*\{V_{\lambda k}(R) - \delta_{\lambda,0}\delta_{k,0}V_b(R)\}\psi_{\alpha''}(R)\mathrm{d}R\right) \end{aligned} \quad (47)$$

Solving these equations is clearly equivalent to finding the eigenvalues and eigenvectors of the (symmetric) matrix representative of the full Hamiltonian. Inclusion of enough terms in the sums over v'', j'', l'', and n'' will allow any desired accuracy to be achieved.

As indicated in the discussion of coupled-channel calculations, the first two factors appearing in the sums in equation (47) are quite easy to generate. Moreover, since the radial wave functions $\psi_{\alpha''}(R)$ are readily obtained[93] solutions of a simple one-dimensional Schrödinger equation, the integral over R also presents no difficulty. Thus requirement (i) for the radial basis set is easily satisfied by the eigenfunctions of equation (41). At the same time we have the physical picture of a set of zeroth-order eigenstates with eigenvalues given by equation (45), which are mixed by the coupling function $\Delta V(R, r, \theta)$. Since the secular equation method is simply an alternate procedure for solving the coupled equations (24), for any given angular and v'' basis, inclusion of enough terms in the sum over n'' in equations (44) and (46) should yield agreement with close-coupling results. Thus the only remaining problem is the determination of an effective isotropic potential $V_b(R)$ for which this expansion converges rapidly.

2. *Choice of the Basis Generating Potential $V_b(R)$*

a. General considerations and vibrational averaging. Probably the most obvious way of choosing $V_b(R)$ would be to set it equal to the purely isotropic component of the potential expansion of equation (9), $V_{00}(R)$. However, it would be equally valid to have it equal *any* other simple spherical potential. It is therefore convenient to introduce the symbol $U_c(R)$ to represent the central potential used in defining the basis generating potential $V_b(R)$. Possible choices of $U_c(R)$ would include the Morse or harmonic oscillator functions, for which the eigenvalues and eigenfunctions are known analytically. However, the fact that their eigenstates are not intimately related to the properties of the bimer means both that very large basis sets would be required to yield highly accurate results and that the basis states do not provide a realistic physical picture of the system.

The definition of $U_c(R)$ used in most calculations to date[54, 55, 57] is based on the zero-coupling picture described in Section III.B.1, in which the diatom is allowed to rotate and vibrate perfectly freely inside the complex. Since the vibrational frequency of the free diatom is much greater than that associated with stretching of the van der Waals bond, it seems reasonable to define $U_c(R)$ as the expectation value of the spherically averaged part of the potential, $V_0(R, r)$ of equation (8), for the diatom state correlating with the bimer level under consideration:

$$\begin{aligned} U_c(R) = \bar{V}_0(v, J \mid R) &\int_0^\infty \phi_{vj}(r)^* V_0(R, r) \phi_{vj}(r) \mathrm{d}r \\ &= \sum_k V_{0k}(R) \left(\int_0^\infty \phi_{vj}(r)^* \xi^k \phi_{vj}(r) \mathrm{d}r \right) \end{aligned} \tag{48}$$

Defining $V_b(R)$ in terms of this $U_c(R)$ means that an independent set of basis functions must be generated for each l'' channel associated with a given bimer state, and that different sets of l''-basis functions must be generated for bimer states corresponding to different diatom (v, j) levels. In addition, radial basis functions associated with different l'' values will in general all be non-orthogonal. On the other hand, this definition provides basis functions which reflect fully both the centrifugal stretching of the complex with increasing l'' and the effects on the potential due to vibrational and centrifugal stretching of the diatom. This should (hopefully) facilitate rapid eigenvalue convergence with increasing values of the basis size parameter, n_{max}. Note that for a rigid-rotor potential model, $\bar{V}_0(v, j|R) = V_{00}(R)$, the purely isotropic component of the potential expansion equation (9).

Both of the above-mentioned stretching effects are quite important for the molecular hydrogen–inert gas complexes. For the $v = j = n = 0$ levels of H_2–Ar, centrifugal stretching causes the expectation value of R to increase from 4.0 to 4.5 Å as l increases from 0 to 8, while the well depth of the effective isotropic potential increases by 3% as v goes from 0 to 1. For $n = 0$ levels of hydrogen–inert gas complexes, neglect of coupling to basis functions corresponding to $n'' > 0$ was justified in ref. 54 on the basis of perturbation theory arguments and numerical tests of the effect of including $n'' = 1$ terms. Dunker and Gordon[110] confirmed the validity of this truncation by showing

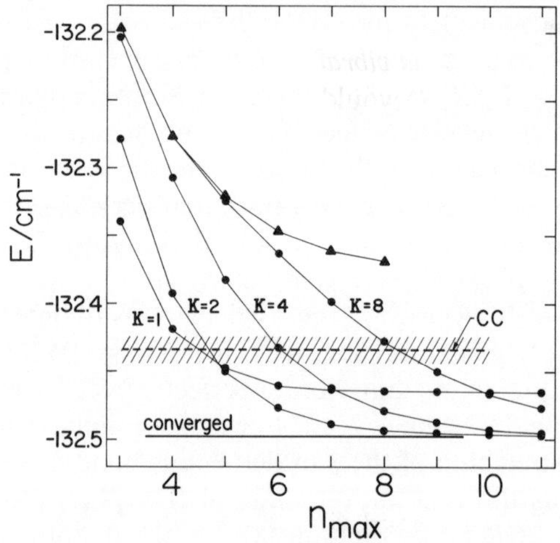

Fig. 4. Dependence on radial basis size of secular equation method eigenvalues for ground state Ar–HCl; ▲ simple central-potential basis; ● infinite-wall bases for various κ values; CC denotes the close-coupling energy of ref. 95. (Reproduced by permission of the Chemical Society (London) from LeRoy *et al.*[55a].)

(see their Table II) the excellent agreement between $n_{max} = 0$ secular equation eigenvalues and their close-coupling results for a semirealistic model H_2–Ar potential. While this extremely rapid convergence (after a single term!) of the radial basis function expansion cannot be expected to hold generally, it does demonstrate the advantage of a judicious choice of $V_b(R)$.

The equation (48) definition of $V_b(R)$ was successfully used in all of the secular equation method analyses of the spectral data for the weak-coupling hydrogen–inert gas complexes.[54,55,57] However, it was found to be inadequate when applied a model for the strong-coupling HCl–Ar complex,[55] even when up to nine bound levels of the effective isotropic potential were included in the basis. This conclusion is illustrated in Fig. 4 where eigenvalues obtained in this manner for the ground state of Ar–HCl are plotted as triangular points. Since the secular equation method is variational, their apparent convergence to an energy 0.07 cm^{-1} above Dunker and Gordon's[95] close-coupling result (horizontal dashed line) demonstrates the inadequacy of this basis. This failure implies that the set of *all* bound states of the effective isotropic potential do not form a sufficiently complete basis to represent accurately the true wave function over the range of R for which the latter has non-negligible values.

b. Infinite-wall basis sets. To remove the above difficulty, a modified central potential $U_c^w(R)$ was introduced[55,96] which is defined as

$$U_c^w(R) = U_c(R) + \Delta U(R_w | R) \tag{49}$$

where

$$\begin{aligned} \Delta U(R_w | R) &= 0 \quad \text{for} \quad R < R_w \\ &= \infty \quad \text{for} \quad R \geq R_w \end{aligned} \tag{50}$$

and R_w is typically chosen to lie beyond the outer turning point of the level in question. Such modified potentials have a complete set of discrete eigenfunctions on the interval $0 \leq R \leq R_w$, and for R_w values which are not too large, a set of these functions displays a much wider variety of behaviour *on this interval* than would the same number of eigenfunctions of $U_c(R)$. It was found[55] that R_w could conveniently be defined in terms of the asymptotic properties of the normalized radial wave function ψ_α of the unmodified potential $U_c(R)$; in particular, $R_w = R_w(\kappa)$ was defined as the maximum distance at which $\psi_\alpha(R)$ attains a magnitude of $10^{-\kappa}$, in reduced units of $(R_e)^{-1/2}$. Ground-state HCl–Ar eigenvalues obtained using these 'infinite-wall' basis functions for various values of κ and n_{max} appear as the round points in Fig. 4. Their convergence to a value 0.06 cm^{-1} below the close-coupling result indicates that the latter is in error by somewhat more than the quoted ± 0.01 cm^{-1} uncertainty (shaded region).

The results in Fig. 4 illustrate the effects of various choices of the wall position parameter κ. If R_w is not placed sufficiently far into the classically forbidden region for the state in question, as occurs for $\kappa = 1$, the real wave function cannot be properly represented in a physically important region of space and convergence to a false limit may result. However, if R_w is too large, the required basis set becomes unacceptably large. Moreover, as $R_w \to \infty$ (or $\kappa \to \infty$) the infinite-wall basis becomes identical to that generated from $U_c(R)$ itself and has the same limitations; the onset of this behaviour is shown in Fig. 4 by the growing agreement between the round and triangular points with increasing κ. In any case, it seems clear that choice of an appropriate value of R_w (say, that defined by $\kappa = 2$ or 3) allows this method to yield good results with basis sets of manageable size.

c. Single-l basis sets. A more subtle, but quite important improvement in the definitions of $V_b(R)$ given above arises from the use in equation (41) of a single effective l'' value, say l_b, for generating all of the basis functions associated with a given bimer state.[96] This is equivalent to defining

$$V_b(R) = U_c(R) + \Delta U(R_w|R) + [l_b(l_b+1) - l''(l''+1)]\hbar^2/2\mu R^2. \quad (51)$$

While this definition makes the basis-generating potential change with l'', consideration of equation (41) shows that it makes both the radial basis functions $\psi_{\alpha''}(R)$ and the associated energies $E_b(n'', l'')$ independent of l''. The only additional effort required by use of these single-l basis sets is computation of matrix elements of R^{-2} between the various radial basis functions. Note that this modification of the basis-generating potential is not related to the introduction of a single average l in the coupled states approximation[104, 123] of molecular scattering theory, since in the present situation no approximation is made to the full Hamiltonian.

In practice it is usually most convenient to define this effective l_b as the l-value identifying the bimer state in question in the zero-coupling limit. For weak-coupling (i.e. molecular hydrogen) complexes, this allows proper account to be taken of the centrifugal distortion of the complex with increasing l. At the same time, the larger reduced mass μ of strong-coupling complexes such as Ar–HCl makes these centrifugal distortion effects relatively small, so that failure of this zero-coupling l-value to be the same as the dominant-l'' contribution to the overall wave function would have no serious consequences.

While it does not change the Hamiltonian or the dimension of the matrix which must be diagonalized, use of the single-l basis substantially reduces the computation time required for generating the Hamiltonian matrix. It means that only one set of radial basis functions is required for calculating any one bimer eigenvalue and it makes all radial basis functions associated with different n''-values orthogonal. In the calculations for the strong-anisotropy HCl–Ar complexes, this change decreased computation times by roughly an order of magnitude while giving results of the same accuracy as those obtained from many-l basis function calculations for the same value of n_{max}.[96] Moreover, for the weak-coupling hydrogen–inert gas complexes, use of the single-l basis in some cases measurably improved the accuracy of the $n_{max} = 0$ eigenvalues.

D. The Adiabatic or BOARS Approximation

In the early theoretical studies of compound state resonances (i.e. bimer states with an open-channel component in their wave function), a number of decoupling approximations were developed as means of obtaining resonance energies and widths without solving a set of coupled equations. The best of these procedures is the adiabatic or best-local approximation developed by Levine, Johnson, Muckerman, and Bernstein.[112,113,115,117] A rederivation of this method using body-fixed rather than space-fixed coordinates has recently been presented by Holmgren *et al.*[121] who gave it the more descriptive title 'Born–Oppenheimer Angular Radial Separation' (BOARS) method. In spite of the different coordinate systems used and the fact that Holmgren *et al.*[121] were interested in states with no open-channel component, the two methods are fundamentally the same. Since the physical content of the more recent derivation seems somewhat more transparent, its viewpoint is taken here, albeit using space-fixed rather than body-fixed coordinates.

As is suggested by the BOARS label, the assumptions underlying this approach are precisely analogous to those motivating the Born–Oppenheimer separation of nuclear and electronic motion in the total Schrödinger equation. The key assumption here is that the vibrational–rotational motion of the diatom in the complex is much more rapid than the stretching of the van der Waals bond. This allows the Schrödinger equation for the former to be solved separately to yield eigenvalues and eigenfunctions which are only parametrically dependent on R. In this method, the diatom is treated as a rigid rotor so that the overall potential $V(R, r, \theta)$ is implicitly assumed to be vibrationally averaged over the associated diatom state, yielding

$$V(R,\theta) = \sum_{\lambda} \bar{V}_{\lambda}(v,j\,|\,R)\,P_{\lambda}(\cos\theta) \tag{52}$$

where $\bar{V}_{\lambda}(v,j\,|\,R)$ is a generalization of equation (48) to the case of $\lambda \neq 0$.

As a first step in this procedure, the Hamiltonian of equation (11) is vibrationally averaged (over r) and then written as the sum of a radial and an angular part,

$$\mathbf{H}(\boldsymbol{R},\hat{r}) = -(\hbar^2/2\mu)R^{-1}(\partial^2/\partial R^2)R + \mathbf{H}_0(R\,|\,\hat{\boldsymbol{R}},\hat{r}) \tag{53}$$

where the rotational Hamiltonian

$$\mathbf{H}_0(R|\hat{\boldsymbol{R}},\hat{\boldsymbol{r}}) = \boldsymbol{l}^2/2\mu R^2 + \boldsymbol{j}^2/2I_{\rm d} + V(R,\theta) \tag{54}$$

is parametrically dependent on R, and $I_{\rm d}$ is the diatom moment of inertia. The main approximation in this method is the assumption that the eigenfunctions of the Hamiltonian of equation (53) can be represented by the simple product

$$R^{-1}\chi_\alpha^J(R)\Phi_{jl}^{JM}(R|\hat{\boldsymbol{R}},\hat{\boldsymbol{r}}) \tag{55}$$

where $\Phi_{jl}^{JM}(R|\hat{\boldsymbol{R}},\hat{\boldsymbol{r}})$ is an eigenfunction of the angular Hamiltonian $\mathbf{H}_0(R|\hat{\boldsymbol{R}},\hat{\boldsymbol{r}})$:

$$\mathbf{H}_0(R|\hat{\boldsymbol{R}},\hat{\boldsymbol{r}})\Phi_{jl}^{JM}(R|\hat{\boldsymbol{R}},\hat{\boldsymbol{r}}) = U_{jl}^J(R)\Phi_{jl}^{JM}(R|\hat{\boldsymbol{R}},\hat{\boldsymbol{r}}) \tag{56}$$

At any chosen value of R, equation (56) can be solved by expanding its eigenfunctions as

$$\Phi_{jl}^{JM}(R|\hat{\boldsymbol{R}},\hat{\boldsymbol{r}}) = \sum_{j''}\sum_{l''} C_{j''l''}^{jl}(R)\mathscr{Y}_{j''l''}^{JM}(\hat{\boldsymbol{R}},\hat{\boldsymbol{r}}), \tag{57}$$

substituting (57) into (54), taking the inner product with $\mathscr{Y}_{j'l'}^{M}$, and diagonalizing the resulting matrix of elements

$$(H_0)_{j''l''j'l'} = [l''(l''+1)\hbar^2/2\mu R^2 + j''(j''+1)\hbar^2/2I_{\rm d}]\delta_{l'',l'}\delta_{j'',j'} + \sum_\lambda \bar{V}_\lambda(R)f_\lambda(j',l';j'',l'';J) \tag{58}$$

The diagonal energies $U_{jl}^J(R)$ thus obtained are points on the effective isotropic potentials yielded by this approach. Note that at any given R, the Φ_{jl}^{JM} functions defined by the eigenvectors of the matrix of equation (58) span the space generated by the $\mathscr{Y}_{jl}^{JM}$ functions from which they are formed.

Applying the Hamiltonian operator (53) to expression (55) and taking the inner product with Φ_{jl}^{JM} then yields the simple one-dimensional Schrödinger equation

$$\{-(\hbar^2/2\mu){\rm d}^2/{\rm d}R^2 + U_{jl}^J(R) + T_{jl}^J(R) - E_\alpha^J\}X_\alpha^J(R) = 0 \tag{59}$$

where the term

$$\begin{aligned} T_{jl}^J(R) &\equiv -(\hbar^2/2\mu)\int\int \Phi_{jl}^{JM}\frac{{\rm d}^2}{{\rm d}R^2}(\Phi_{jl}^{JM}){\rm d}\hat{\boldsymbol{r}}{\rm d}\hat{\boldsymbol{R}} \\ &= (\hbar^2/2\mu)\sum_{j'',l''}\left|\frac{{\rm d}}{{\rm d}R}C_{j''l''}^{j,l}(R)\right|^2 \end{aligned} \tag{60}$$

is precisely analogous[121] to the diagonal correction for nuclear motion in the electronic Born–Oppenheimer approximation.[124] Once the $U_{jl}^J(R)$ and $T_{jl}^J(R)$ curves are determined by solution of equation (56) at a number of R values, equation (59) may be solved using standard methods[93] to yield the adiabatic or BOARS eigenvalues, E_α^J. Holmgren *et al.*[121] showed that only a fraction of the values of $U_{jl}^J(R)$ and $T_{jl}^J(R)$ required for the numerical integration of equation (59) need be obtained directly by diagonalization of the H_0 matrix (58), and that generating the remainder by numerical interpolation does not significantly affect the accuracy of the method.

The above approach could in principle be made exact by expanding the total wave function as

$$\Psi_\alpha = R^{-1} \sum_{j'',l''} F_{j''l''}(R)\Phi_{j''l''}^{JM}(R|\hat{\boldsymbol{R}}, \hat{\boldsymbol{r}}) \tag{61}$$

and solving the coupled equations obtained upon substituting this expression into equation (53). This procedure reduces to the adiabatic approximation outlined above upon neglect of the off-diagonal terms

$$\int\int \Phi_{jl}^{JM} \frac{d^2}{dR^2}(\Phi_{j''l''}^{JM}) d\hat{\boldsymbol{R}} d\hat{\boldsymbol{r}} = \sum_{j',l'} C_{j'l'}^{jl}(R)^* \frac{d^2}{dR^2}(C_{j'l'}^{j''l''}(R)) \tag{62}$$

The corresponding diagonal term, for which $(j, l) = (j'', l'')$, is clearly the $T_{jl}^J(R)$ function of equation (60). Thus the R-dependence of the angular mixing coefficients $C_{j''l''}^{jl}(R)$ is responsible both for the inaccuracy of the adiabatic approximation due to non-zero values of the integrals of equation (62) and for the magnitude of the positive definite $T_{jl}^J(R)$ function. The differences between eigenvalues obtained on including and excluding $T_{jl}^J(R)$ from equation (59) therefore give a realistic measure of the uncertainty in the BOARS approximation level energies. Nonetheless, Holmgren *et al.*[121] showed that even for cases in which such differences are as large as several cm^{-1}, the relative energies of a set of rotational levels obtained consistently in one way or the other could be quite accurate. They also pointed out[121] that for the lowest level on the lowest adiabatic potential $U_{jl}^J(R)$ yielded by a given angular basis, the solutions of equation (59) obtained with and without the $T_{jl}^J(R)$ term, respectively, are upper and lower bounds to the true eigenvalue (within that angular basis). Their comparisons with close-coupling calculations showed that the upper-bound approach yielded slightly better eigenvalues, although the wave functions obtained in the lower-bound calculations had somewhat more realistic angular behaviour. Note however that these bounding arguments do not apply to excited levels of the lowest effective potential or to any of the levels of the upper adiabats associated with the given angular basis.

E. Treatment of Quasibound and Metastable Levels

Van der Waals bimers may undergo predissociation via two different mechanisms. The first of these affects quasibound or orbiting resonance levels which lie above the dissociation limit of the associated effective isotropic potential $\bar{V}_0(v, j|R)$ but behind a potential energy barrier which inhibits fragmentation. This type of predissociation also occurs for spherically symmetric potentials and has been extensively studied in that context (see, for example, refs. 120 and 125). The second mechanism affects metastable bimer levels which lie below the asymptote of $\bar{V}_0(v, j\,|\,R)$ but can predissociate via coupling to an open-channel component of their total wave function. In the language of scattering theory these levels are known as compound state resonances[111], and when the open-channel wave function corresponds to a lower diatom vibrational level the phenomenon is called vibrational predissociation. In the infrared spectra of hydrogen–inert gas complexes, line broadening due to quasibound level predissociation was observed a number of years ago,[52] but while the upper states of those transitions are susceptible to compound state predissociation, broadening due to it was not resolved until McKellar's[126] very recent work on HD–Ar. In this case, the observed predissociation is due to rotational energy of the diatom (HD) being converted to relative translational energy of the fragments. In contrast, the line broadening observed in the infrared spectrum[59] of $(N_2O)_2$ and in the visible spectrum[84] of He–I_2 is virtually all due to vibrational predissociation.

Calculation of detailed line shapes for transitions involving levels which predissociate by these mechanisms would require solution of the coupled-channel problem for a range of energies spread across the line profile. A procedure for doing this has been suggested by Shapiro.[119] However, his approach is computationally very expensive and the information about intermolecular potentials obtainable from the existing measurements was not sufficient to justify its use in the data analyses for the hydrogen–inert gas systems. As a result, effects due to compound state predissociation were neglected, while quasibound levels were treated using the approximate techniques described below.

For quasibound levels supported by purely isotropic radial potentials, a number of methods for calculating the resonance energy and width have been compared critically.[120,125] These studies concluded that for the narrower (and hence spectroscopically observable) resonances lying below the maximum of the potential energy barrier, the level energies are most conveniently determined using the Airy function boundary condition method and the corresponding widths obtained using semiclassical methods. In the secular equation method calculations for the hydrogen–inert gas complexes,[54,55,57] this approach was used for determining the energy E_α^0 and radial basis function $\psi_\alpha(R)$ representing the predissociating state, while the level width was assumed to be that associated with this radial basis function.

The error due to this approximation was assumed to introduce into the resulting level energy an uncertainty whose magnitude varies linearly with the calculated width. This in turn defined a statistical weight for transitions involving quasibound levels which hopefully prevented them from having an exaggerated effect on the data analysis.

A number of questions are raised by such a procedure. In the first place, it predicts the same width for all J sublevels associated with a given l, even though some are displaced to significantly higher energies than others (see, for example, Fig. 3). More seriously, it is not clear that merely defining the dominant radial basis function by the Airy function boundary condition will necessarily allow the energies of these diffuse levels to be computed in the same manner as those lying below the potential asymptote. However, it does provide a convenient and systematic framework for attempting to extract at least part of the information contained in the transitions involving quasibound levels.

The upper states of all the observed infrared transitions of the hydrogen–inert gas complexes may undergo compound state predissociation. Even if vibrational predissociation is ruled out, many of these levels can still predissociate by converting the energy of (internal) rotation of the diatom into translational motion along R. In their close-coupling calculations for these species, Dunker and Gordon[56] excluded this possibility by simply excluding open channels from the angular basis. In contrast, the secular equation method calculations[54,55,57] included those angular channels, but attempted to represent the coupling to them by the coupling to the corresponding $n = 0$ bound state radial basis functions. While the latter approach appears slightly better, no quantitative justification was offered for either approach. Fortunately, recent calculations[127] have indicated that the compound state resonance widths for the H_2 complexes are *ca.* 10^{-2} cm^{-1}. Since errors due to neglect of coupling to the continuum should be smaller than this width, such errors are unlikely to have prejudiced seriously the reported data analyses for these species. However, compound state predissociation widths for bimers formed by HD are much larger than the experimental uncertainties,[126,127b] so that detailed line-shape calculations may be required to analyse their spectra.

Levy and Wharton and coworkers[80,84] have performed careful studies of the vibrational predissociation of electronically excited He–I_2 complexes and determined absolute lifetimes for the predissociation of complexes corresponding to a wide range of I_2^* vibrational levels. Moreover, methods for calculating such lifetimes have been divised by Ashton and Child,[128] Beswick and Jortner,[129] and Ewing[107]. However, since no information on intermolecular potentials has yet been obtained from such data, these methods are not discussed in the present report.

F. Discussion and Conclusions

It is generally accepted that the close-coupling method, i.e. the direct numerical integration of the coupled equations of equation (24) or (35),[99–104] is in principle an exact procedure for calculating the eigenvalues and other properties of van der Waals molecules. The only real limitation on its accuracy arises from the computational expense of including many coupled channels or using an extremely small integration mesh. In practice, this means that this method may be readily applied to both weak- and strong-coupling complexes (see Table II), but becomes prohibitively expensive for semirigid species treated using space-fixed coordinates. This technique was used by Dunker and Gordon, both in their detailed analysis of the spectral data for the weak-coupling hydrogen–inert gas complexes,[56] and in their predictions of the spectroscopic properties of a number of previously proposed potentials for the strong-coupling Ar–HCl and Ar–DCl bimers.[95] However, for strong-coupling complexes its cost[121] would tend to discourage use of this method in the repetitive calculations associated with the determination of a potential energy surface from experimental data.

As indicated earlier, the secular equation method appears capable of achieving the same accuracy as coupled-channel calculations for all problems to which the latter has been applied, at substantially less computational expense. This conclusion is based on comparisons of predictions for bound states of both weak-[96,110] and strong-coupling[55,96] complexes, as well as predictions for compound state resonances coupled strongly to an open channel.[127] In particular, comparisons with close-coupling results[95,121] for Ar–HCl showed that the infinite-wall, single-l version of the secular equation method yielded results of better accuracy at only *ca.* 1% of the computational effort.[55,96] It therefore appears that the secular equation method should be thought of as a particularly efficients algorithm for solving the coupled equations. Thus while results obtained by direct numerical integration of the coupled equations are an invaluable touchstone, it appears that such calculations are more efficiently performed using the secular equation method.

The adiabatic or BOARS method is different in kind from the two discussed above in that it is in principle inexact. In particular, the approximations inherent in this procedure mean that for each state it yields two estimates of the energy which differ by roughly the expectation value of the $T(R)$ function of equation (60), weighted by the square of the radial wave function $X_{\alpha}^{J}(R)$ of equation (59). For the lowest levels of the potential energy surfaces considered by Holmgren *et al.*,[121] for which case these estimates are strict upper and lower bounds to the true eigenvalues, these differences ranged from 1.5 to 4.0 cm^{-1}. On the other hand, the relative spacings of neighbouring levels on a given surface, all calculated in the same way with respect to the inclusion of $T(R)$, were often accurate to better than 0.004 cm^{-1}, while the angular properties of the ground-state wave function seemed fairly insensitive to the inclusion or exclusion of $T(R)$. However, each of the states considered by Holmgren *et al.*[78,121] was the lowest level associated with a particular value of the total angular momentum J, and it is not known whether or not these conclusions are also valid for excited levels.

It is the R-dependence of the angular behaviour of the potential, as reflected in the R-dependence of the expansion coefficients $C(R)$ of equation (57), which is the origin of the uncertainties in the BOARS method eigenvalues. One illustration of this functionality is the fact that the differences between the upper and lower bounds obtained by Holmgren *et al.*[121] were largest for their potential IV, the case for which the angular behaviour of the potential energy surface changed most drastically with R (see Fig. 1 of ref. 95), and were smallest for their potential II. This R-dependence will in general be large for cases in which the radial strength functions $\bar{V}_\lambda(R)$ change sign on the R-interval between the classical turning points associated with the radial wave function $X^J_\alpha(R)$, since this sign change implies a change in the preferred relative orientation of the diatom within the complex. An alternate view of this source of the eigenvalue uncertainty is in terms of the way the radial potential energy profile varies with θ. Since the BOARS method uses a single 'best local' function[113] $X^J_\alpha(R)$ to represent the radial part of the wave function at all angles, significant uncertainties would tend to be introduced if, say, the position of the repulsive potential wall shifts with θ. The above arguments probably explain the reason for the relatively high accuracy of the BOARS (i.e. adiabatic) level energies for the model problems considered by Levine, Johnson, Muckerman, and Bernstein,[113,115,117] since in all of the cases they considered, the coupling function $V_2(R)$ did not change sign and was relatively slowly varying over the critical range of R.

The most quantitative tests of the BOARS procedure are those performed by Holmgren *et al.*[121] for the lowest levels of a set of model Ar–HCl potential energy surfaces. Their results showed that for this strong-coupling case the use of body-fixed coordinates has a significant advantage, since neglect of off-diagonal coupling in the helicity quantum number Ω had only a small effect on the eigenvalues. Consequently, results of reasonable quality can be obtained with much smaller angular bases than would be required for equivalent calculations using the space-fixed representation. In this reduced basis, their BOARS calculations for a given set of Ar–HCl levels required *ca.* 1.14% of the time required by a comparable close-coupling calculation using space-fixed coordinates. However, the inclusion of the $\Omega' \neq \Omega$ coupling, as required for obtaining high precision, increased the BOARS computation time by roughly a factor of 10 to *ca.* 1% of that required for the close-coupling calculations. Strictly speaking, it is the latter value which should be used in comparisons with the other two methods, since when the $\Omega' \neq \Omega$ coupling is included, the space-fixed and body-fixed formulations have the same number of angular channels for a given choice of

j_{max}. Since the coupled-channel and secular equation methods may also be formulated using the body-fixed reference frame, they too would achieve substantial savings in computation time if $\Omega' \neq \Omega$ coupling could be ignored.

In conclusion, therefore, it appears that for calculations of comparable scope on strong-coupling complexes, the secular equation and BOARS procedures involve roughly the same computation time and are roughly two orders of magnitude faster than the close-coupling methods. Since substantial uncertainties are usually associated with the BOARS eigenvalues, the secular equation calculations using infinite-wall, single-l basis functions currently appear to be the best way of treating these species.

IV. ANALYSIS OF SPECTRAL DATA

A. Modelling the Potential Energy Surface

1. *General Considerations*

It has long been recognized that potential energy functions determined from analysis of experimental data are often partially dependent on the analytic form of the function chosen to model the interaction. To minimize such effects, the functional form(s) for the radial strength functions $V_{\lambda k}(R)$ of equation (9) should be chosen to be as realistic and as flexible as possible. At the same time, practical considerations place some restrictions on the range of forms. In the first place, the quality and extent of the data may be insufficient to allow more than a few parameters to be determined. A second consideration is the fact that computational costs are less if all of the $V_{\lambda k}(R)$ functions for a given species may be expressed as (different) linear combinations of a limited number of functions of R. For example, when using the $LJ(m, 6)$ potential form

$$\begin{aligned} V_{\lambda k}(R) &= C_m^{\lambda k}/R^m - C_6^{\lambda k}/R^6 \\ &= \varepsilon^{\lambda k}[6(R_e^{\lambda k}/R)^m - m(R_e^{\lambda k}/R)^6]/(m-6), \end{aligned} \tag{63}$$

where $\varepsilon^{\lambda k}$ and $R_e^{\lambda k}$ are the well depth and equilibrium distance, computational effort is minimized if the same value of m is used for all λ and k.

The inverse-power form characteristic of contributions to long-range intermolecular forces[124] makes it perfectly natural to express the attractive and multipolar contributions to the radial strength functions in terms of a common set of inverse-power terms. However, use of a single functional form to represent the dominant short-range contribution to the various $V_{\lambda k}(R)$ terms may require justification; fortunately, this appears to have been provided by recent *ab initio* results. In particular, in Hariharan and Kutzelnigg's results (ref. 130, as quoted by Tang and Toennies[131]) for H_2–He and H_2–Ne, the repulsive contribution to the potential was represented by a simple exponential, $Ae^{-\beta R}$, and the exponent parameter β was found to be independent of the diatom bond length r. In addition, the values of β were approximately the same for the parallel ($\theta = 0$ or π) and perpendicular ($\theta = \pi/2$) relative orientations of the diatom. This justifies use of a common exponent parameter β in the short-range contributions to all of the radial strength functions $V_{\lambda k}(R)$ for a single species, at least for the hydrogen–inert gas systems. Note, however, that introduction of this type of constraint is largely a matter of computational convenience.

2. *Long Range* (*in R*) *Behaviour*

Most experiments are not highly sensitive to the intermolecular potential in the asymptotic region where the leading inverse-power term dominates the long-range part of the interaction energy. This is certainly true of the spectroscopic measurements considered here. Semiclassical arguments indicate that these data are only sensitive to the potential on the interval between the inner and outer turning points for the observed levels. Even for the hydrogen–inert gas complexes for which the observed levels were followed from the ground state to rotational predissociation, this only includes the interval $0.9 \lesssim R/R_e \lesssim 1.5$, where R_e is the equilibrium distance of the spherically averaged potential. Thus if a potential obtained from spectroscopic data is to have the correct long-range behaviour, it must usually be imposed on the model as an external constraint.

Fortunately, a variety of *ab initio* and semiempirical methods have been devised for calculating the coefficients of the inverse-power multipolar contributions to the long-range potential and its anisotropy.[132,133] These methods are most easily applied to interactions between closed-shell species. Thus 'theoretical' values of the long-range potential constants for van der Waals bimers may often be available; whenever they are such results should be incorporated into the potential energy function model.

The importance of taking advantage of theoretically known long-range potential coefficients has been graphically demonstrated for the H_2–Ar system. In the first full analysis of its spectroscopic data,[54] the $V_{\lambda k}(R)$ functions were all assumed to have the LJ$(m, 6)$ form of equation (63) and the fits yielded values of $\varepsilon^{\lambda k}$ and $R_e^{\lambda k}$ for the $(\lambda, k) = (0, 0)$, $(0, 1)$ and $(2, 0)$ radial strength functions. It was pointed out at the time that these effective isotropic and anisotropic potentials should not be expected to have the asymptotically correct C_6 coefficients, as indeed they did not, because the data are more sensitive to the potential near R_e^{00} than to the long-range region. A more sophisticated potential which had the same number of free parameters while being constrained to have the correct C_6^{00}, C_6^{01}, and C_6^{20} coefficients was later obtained from similar fits to the same data, which achieved exactly the same (excellent) quality of fit.[55] While the spectroscopic data could not distinguish between them, these two potentials had distinct differences;

in particular, their ε^{00} values differed by 2.6% and their ε^{20} values by 7%. However, the spectroscopic potential which was constrained to have the correct isotropic C_6 coefficient yielded excellent predictions of the low-energy total-scattering cross-section measurements of Toennies *et al.*,[134] while analogous predictions obtained from the $LJ(\mathrm{m}, 6)$ potential were quite poor. Similar findings were also obtained for the H_2–Kr and H_2–Xe systems.[134] Thus forcing the potential form to have the correct limiting behaviour in a region to which the spectroscopic data are not sensitive greatly improved the attractive branch of the potential obtained from fits to these data. This dramatically demonstrates the importance of making use of theoretical or semiempirical long-range potential constants whenever they are available.

3. *Diatom Stretching Dependence and the Collapsed-Diatom Limit*

The anharmonicity of its pair potential $V_{\mathrm{d}}(r)$ and the nature of centrifugal forces means that the average bond length of a diatomic molecule increases with vibrational or rotational excitation. As a result, data for complexes formed from a given diatom in different vibration–rotation states contain information about the diatom bond-length dependence of the intermolecular potential, although such effects may be too small to be readily resolved if only centrifugal distortion is involved. Moreover, within the (electronic) Born–Oppenheimer approximation, the vibrational levels of a second isotope of the diatom may be qualitatively considered to be levels of the first, associated with intermediate (non-integer) values of the vibrational quantum number. Thus the existence of data for more than one isotopic form of a bimer can substantially enhance the information about the R- and r-dependence of the potential.

To a first approximation, the range of r probed by data for complexes corresponding to a given set of diatom (v,j) levels corresponds to the range of expectation values of r for the isolated diatom in these levels. This implies that the data used in the H_2–Ar analyses referred to above[55,55] are most sensitive to the potential on the interval $0.77 \lesssim r \lesssim 0.82$ Å (i.e. $0 \lesssim \xi \lesssim 0.071$), while the inclusion of D_2–Ar data in the analysis extended the lower limit of this region to $r \approx 0.76$ Å (or $\xi \approx -0.010$). On this relatively narrow interval the potential energy surface was found to be effectively linear in r (or ξ) and only $k = 0$ and 1 terms in the potential expansion of equation (9) could be determined from the spectroscopic data.[54,55] However, higher-order terms certainly exist and their determination would substantially extend the utility of such surfaces. A procedure for estimating them has been devised; following ref. 57a, its description here is phrased in terms of the properties of the hydrogen–inert gas complexes, but the arguments clearly apply equally well to bimers formed by any diatomic molecule.

From the viewpoint of the Ar atom in an H_2–Ar bimer, the electronic charge distribution of the H_2 diatom is simply a distorted version of the spherical charge distribution of the isoelectronic He atom. Thus, it seems reasonable to expect that as $r \to 0$ (i.e. $\xi \to -1$), the interaction potential between Ar and H_2 should smoothly approach that between Ar and He. For the long-range (in R) isotropic and anisotropic potential constants,

$$C_6(r,\theta) = C_6^0(r) + C_6^2(r)P_2(\cos\theta), \tag{64}$$

this behaviour is clearly demonstrated in Fig. 5. The points seen there for $\xi > -1$ correspond to the semiempirical upper and lower bounds on $C_6(r,\theta=0)$ and $C_6(r,\theta=\pi/2)$ calculated by Thakkar,[135] while those at $\xi = -1$ represent his C_6^0 value for He–Ar and the constraint that the anisotropy must disappear at $r=0: C_6^2(r=0)=0$. The smooth extrapolations of the H_2–Ar results to the He–Ar limit attests to the validity of this procedure. The solid curves in the figure were obtained by fitting cubic polynomials through the points shown.

In view of the above, fits to the H_2–Ar spectroscopic data to determine parameters of the $(\lambda, k) = (0,0), (0,1)$, and $(2,0)$ components of the potential were repeated[57] with parameters of $V_{02}(R)$ being simultaneously defined

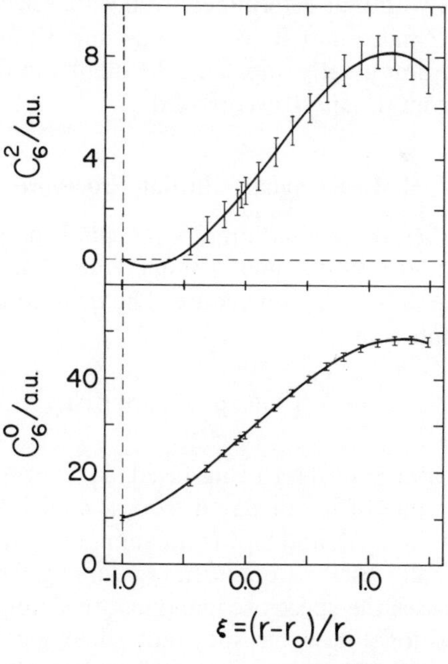

Fig. 5. Dependence of Thakkar's[135] semiempirical C_6^λ coefficients for H_2–Ar on the diatom stretching coordinate.

by the requirement that the $\xi = -1$ isotropic potential have the same depth and equilibrium distance as the known[136] He–Ar potential:

$$V_0(R, r=0) = V_{00}(R) - V_{01}(R) + V_{02}(R) - \dots = V(R; \text{He–Ar}) \quad (65)$$

This constraint effectively defined parameters of the $V_{02}(R)$ function which could not be obtained from the spectroscopic data alone; similarly, the related requirement that

$$V_2(R, r=0) = V_{20}(R) - V_{21}(R) + V_{22}(R) - \dots = 0 \quad (66)$$

effectively defined two of the parameters characterizing $V_{22}(R)$. As might be expected, since the spectral data are not directly sensitive to these $k \geq 2$ contributions to the potential,[54] inclusion of this 'collapsed-diatom limit' constraint had no net effect on the quality of the fit. However, it allowed a much more sophisticated potential energy surface to be obtained from the same data.

The improved quality of the surface yielded by this constraint is attested to in two ways. In the first place, the relative strength of the new $V_{21}(R)$ radial strength function for H_2–Ar is in much better agreement with those for H_2–Kr and H_2–Xe.[57] However, a more important development is the fact that the parameter ε^{20} of the $V_{20}(R)$ anisotropy strength function is now[57] in good agreement with that determined from state-selected total-scattering cross sections by Zandee and Reuss.[42] Thus imposition of the collapsed-diatom limit constraint greatly improved the quality of the potential energy surface obtained from the spectroscopic data.

B. Least-Squares Fitting Procedure

As indicated earlier, determination of a potential energy surface from the spectroscopic data for a polyatomic van der Waals bimer requires use of a non-linear least-squares fitting procedure. The quantity to be minimized in such fits is the χ^2 function.

$$\chi^2 = \sum_{i=1}^{N_d} [Y_i(\exp - Y_i(\boldsymbol{p})]^2/(u_i)^2 \quad (67)$$

where N_d is the number of data being fitted, Y_i (exp) the ith experimental datum, $Y_i(\boldsymbol{p})$ the value of it calculated from a model defined by the trial parameters $\boldsymbol{p} = \{p_1, p_2, \dots\}$, and $(u_i)^2$ is the sum of the squared uncertainties in the calculated and observed quantities (assuming the two to be uncorrelated). In most cases the spectroscopic data are simply the frequencies of the observed transitions; however, they may sometimes also include quantities such as line widths, relative transition intensities,[56] or structural information such as[78] the expectation value of $\cos\theta$ or $\cos^2\theta$. Indeed, if the relevant $(u_i)^2$ values are all accurately known, data obtained from a variety of experiments (e.g. spectroscopy, scattering cross sections, virial coefficients,... etc.) could be analysed simultaneously.

When a minimum of χ^2 as a function of the parameters of the model is located, a quantitative measure of the goodness-of-fit is required. For a perfect model, the expected value of χ^2 is $N_d - N_p$, where N_p is the number of parameters being varied. Thus the dimensionless root-mean-square residual (DMR) about the predictions of the model,

$$\text{DMR} = [\min(\chi^2)/(N_d - N_p)]^{1/2} \tag{68}$$

is a convenient measure of the quality of fit. For a perfect model, the expected value of this quantity is unity. However, it is impossible to tell whether a large value of DMR is due to systematic error in the data or to an inadequate model.

The real test of an optimized potential energy surface is its ability to predict accurately a wide variety of experimental phenomena. To facilitate such comparisons, a quantity called the 'dimensionless standard deviation' (DSD) has been introduced:[55]

$$\text{DSD} = [\min(\chi^2)/(N_d - 1)]^{1/2} \tag{69}$$

A DSD value of less than unity means that the average deviation of the calculated property from its observed value is smaller than the estimated combined uncertainty u_i. The distinction between the DMR and DSD indicators is the fact that the former is only used when actually fitting to determine the potential parameters, while the latter is used when comparing various predictions of a given set of data. In either case, two models which yield different DSD (or DMR) values which are both less than or equal to one are statistically indistinguishable. Note, however, that the above remarks regarding the expected optimum values of χ^2, DMR, and DSD lose much of their significance if there is real uncertainty regarding the absolute magnitudes of the statistical weights $1/(u_i)^2$.

The most efficient χ^2 minimization methods require evaluation of the partial derivatives $\partial Y_i(\boldsymbol{p})/\partial p_j$ of each calculated datum with respect to each of the parameters being varied. The most straightforward way of evaluating these derivatives is simply by divided differences. However, this would require that in each cycle of iteration all of the $Y_i(\boldsymbol{p})$ values be calculated either $2N_p + 1$ or $N_p + 1$ times, depending on whether the derivatives are obtained using symmetric or asymmetric differences about the trial $\boldsymbol{p}$ values. Such calculations are quite time consuming, so this approach should be avoided whenever possible.

If the experimental observable is a transition frequency or energy level difference, these partial derivatives may be evaluated much more efficiently[54,56] using the Hellman–Feynman theorem:

$$\partial E_\alpha/\partial p_j = \langle \psi_\alpha^{JM} | \partial V(R,r,\theta)/\partial p_j | \psi_\alpha^{JM} \rangle \tag{70}$$

The wave function Ψ_α^{JM} is usually be obtained at the same time as the eigenvalue E_α^J, and evaluation of the integrals associated with equation (70) is normally a relatively inexpensive procedure. Moreover, in the secular equation method most contributions to equation (70) are linear combinations of the same terms appearing in the Hamiltonian matrix elements of equation (47), so evaluation of these partial derivatives requires little computational effort beyond that required for determining the level energies E_α^J.[54] This is especially true if all of the radial strength functions are linear combinations of the same limited number of functions of R. Of course, for equation (70) to be used, it must be possible to readily determine the derivatives of the potential $\partial V(R,r,\theta)/\partial p_j$. While in principle quite straightforward, their evaluation may be quite time consuming in cases where the potential for the species under consideration is obtained by re-expanding the potential for a different isotope about a new diatom centre-of-mass.[91,92] As a result, while equation (70) was used in much of the data analysis for H_2– and D_2–inert gas complexes,[54–57] divided differences were used in the Holmgren *et al.*[78] fits to the data for the various isotopes of Ar–HCl.

C. Correlation Effects and the Choice of Fitting Parameters

It was shown above that imposing appropriate constraints on the potential form could yield significant improvements in the potential energy surface obtained from a fit to a given set of data. The purpose of the present section is to point out that the quality of the information obtained from such fits may also depend on which parameters are used as the independent variables in the χ^2 minimization procedure. This point is most easily made using a specific example.

In the first full analysis of the infrared spectrum of the hydrogen–inert gas complexes,[54] the radial strength functions $V_{\lambda k}(R)$ were all chosen to have the LJ$(m,6)$ form of equation (63). In addition, the traditional approach[39–41,56,111–117] of characterizing non-isotropic contributions to the potential by scaling factors $a_m^{\lambda k}$ and $a_6^{\lambda k}$ multiplying the long- and short-range parts of the isotropic potential was adopted, yielding

$$V_{\lambda k}(R) = a_m^{\lambda k}(C_m^{00}/R^m) - a_6^{\lambda k}(C_6^{00}/R^6). \qquad (71)$$

However, the latter has since been shown to be an inappropriate way of characterizing these radial strength functions at distances near or shorter than the equilibrium value of R. This conclusion was based: (i) on the large statistical uncertainties in the fitted values of these factors, (ii) on the very high degree of correlation between the $a_m^{\lambda k}$ and $a_6^{\lambda k}$ parameters for a given (λ,k), and (iii) on the irregular trends among the parameter values obtained for H_2–Ar, H_2–Kr, and H_2–Xe complexes. A further illustration of the shortcomings of this multiplicative scaling factor parameterization was provided by the Dunker–Gordon[56] analysis of the data for hydrogen–inert gas complexes. Although these authors used the more sophisticated Morse–spline–van der Waals form for the radial strength functions, the strength of the anisotropy in the bowl of their potential well was defined by scaling factors A_r and A_a multiplying the repulsive and attractive parts of the isotropic potential, respectively. The high statistical correlation between these parameters made it very difficult to distinguish between their effects on the calculated spectrum. Coupled with the fact that the close-coupling method used for their calculations is relatively expensive, this led Dunker and Gordon to forego attempting to locate the A_r and A_a values corresponding to the overall minimum of χ^2. Thus these correlation effects effectively prevented them from determining fully optimized potential energy surfaces for these systems.

An alternate way of characterizing these radial strength functions is in terms of their own overall length and energy scaling parameters, $R_e^{\lambda k}$ and $\varepsilon^{\lambda k}$ (see equation (63)).[55] This choice is particularly appropriate if $V_{\lambda k}(R)$ has a zero on the interval where Ψ_α is non-negligible, since the position of this zero will be of crucial importance in defining the magnitudes of various matrix elements of $V_{\lambda k}(R)$. For the same LJ$(m,6)$ model for the hydrogen–inert gas complexes mentioned above, this second choice yielded:[55] (i) much smaller relative statistical uncertainties in the fitted parameters, (ii) much less correlation between the parameters for a given (λ, k), and (iii) parameters which varied in a plausible and systematic way among H_2–Ar, H_2–Kr, and H_2–Xe. The two fits yielded exactly the same potential energy surface and differed only in the parameters used to define the $V_{01}(R)$ and $V_{20}(R)$ functions. However, the uncertainties of 11%, 8%, 29%, and 20% in a_m^{01}, a_6^{01}, a_m^{20}, and a_6^{20}, respectively, were replaced by[55] uncertainties of 0.6%, 4.2%, 1.6%, and 9.8% in R_e^{01}, ε^{01}, R_e^{20}, and ε^{20}. Thus it seems clear that a parameterization which treats the $V_{\lambda k}(R)$ as independent functions each with its own energy and length scaling parameters is much better than one based on the traditional multiplicative factors such as the $a_m^{\lambda k}$ and $a_6^{\lambda k}$ coefficients of equation (71).

V. POTENTIAL ENERGY SURFACES DETERMINED FROM BIMER SPECTRA

A. Molecular Hydrogen–Inert Gas Complexes

1. *Spectroscopic Potential Energy Surfaces*

Analyses of the McKellar–Welsh data for the hydrogen–inert gas bimers[52] have been reported by three groups. Petelin[53] presented a simple semi-quantitative analysis which mainly showed that an anisotropic potential model was required for explaining the observed transition frequencies. Dunker and Gordon[56] undertook a more quantitative study and determined families of anisotropic potentials which reproduced the experimental absorption profiles reasonably well. Their potentials consisted of a well-defined isotropic part plus an angle-dependent component defined by a linear equation interrelating acceptable values of multiplicative long- and short-range anisotropy strength coefficients. However, they did not attempt to fully optimize the agreement between the calculated and observed transition frequencies, and hence were unable to recommend a single best potential for each case. Moreover, while they noted that the potential for a complex formed from vibrationally excited hydrogen was different from that for a complex formed from ground-state H_2, they made no attempt to systematize these observations in order to determine the ξ-dependence of the interaction energy. On the other hand, Dunker and Gordon[56] were the only ones to undertake a quantitative analysis of the observed absorption intensities. From comparisons of their calculated absorption profiles with the experimental spectrum, they determined significant new information about the angle-dependent dipole moment functions of these species.

The most quantitative analyses of these data in terms of their ability to define unique potential energy surfaces are those of Le Roy and co-workers.[54,55,57] Their results are described here in some detail, in order to demonstrate the quality of the information which may be obtained from the spectra of van der Waals molecules. In their work,[54,55,57] the frequencies of all non-overlapping and uniquely assigned transitions for both the H_2 and D_2 isotopes of a complex were fitted to a synthetic spectrum generated from a model potential expanded in terms of the double sum (over λ and k) of equation (9). The stretching coordinate ξ of equation (10) was defined using $r_0 = 0.7666348$ Å, the expectation value of r for the ground vibration–rotation state of H_2. This choice of r_0 means that if the expansion in the stretching coordinate stops at $k_{max} = 1$, the $(\lambda, k) = (0, 0)$ radial strength function corresponds to the actual isotropic potential for $H_2(v = 0, j = 0)$–Ar. The expectation values of ξ required to define the effective potentials for the observed states of the hydrogen bimers are summarized in Table IV. All calculations used the $n_{max} = 0$ version of the secular equation method, a procedure whose accuracy was demonstrated (see Section III above) to be substantially greater than the experimental uncertainties. The fitting procedure was 'objective' in that all fitted parameters were varied simultaneously and that the uncertainties in the reported optimized values included the effects of correlation.

The radial strength functions $V_{\lambda k}(R)$ for the potential surfaces discussed below may all be expressed in terms of the flexible 'generalized Buckingham–Corner' or GBC form:

$$V(R) = AR^{-m}\exp(-\beta R) - [C_6/R^6 + C_8/R^8 + \ldots]D(R) \qquad (72)$$

TABLE IV. Expectation values of ξ, ξ^2, and ξ^3 for levels of the ground electronic state of H_2 and D_2, where $\xi \equiv (r - r_0)/r_0$ and $r_0 = 0.7666348$ Å.

		H_2			D_2		
v	j	$\langle\zeta\rangle_{vj}$	$\langle\zeta^2\rangle_{vj}$	$\langle\zeta^3\rangle_{vj}$	$\langle\zeta\rangle_{vj}$	$\langle\zeta^2\rangle_{vj}$	$\langle\zeta^3\rangle_{vj}$
0	0	0.0	0.013441	0.000295	− 0.009790	0.009492	− 0.000132
0	1	0.001505	0.013462	0.000356	− 0.009041	0.009485	− 0.000108
1	0	0.066702	0.045381	0.007041	0.036853	0.029839	0.002484
1	1	0.068270	0.045652	0.007263	0.037624	0.029917	0.002555
1	2	0.071397	0.046207	0.007712	0.039162	0.030075	0.002696
1	3	0.076070	0.047073	0.008395	0.041465	0.030321	0.002910

Here, $D(R)$ is a damping function which approximately corrects for the effects of electron overlap on the multipole expansion for attractive van der Waals forces; it was chosen to have the functional form.

$$D(R) = \exp[-h_1(h_2R_0/R - 1)^{h_3}] \quad \text{for } R \leq h_2R_0$$
$$= 1 \quad \text{for } R > h_2R_0 \tag{73}$$

If the distance R_0 corresponds to the minimum of $V(R)$, choosing (h_1, h_2, h_3) equal to (4, 1, 3) yields the standard Buckingham–Corner (BC) damping function,[137] (1.0, 1.25, 1.9) yields the Hepburn *et al.*[138] HFD1 form, and (1.0, 1.28, 2.0) corresponds to the Ahlrichs *et al.*[139] HFD2 damping function. Use of equation (72) allows the correct long-range potential constants to be included in the potential form without introducing the spurious additional inflection points which often appear in segmented splined potentials such as the 'MSV' form used by Dunker and Gordon[56] (e.g. see their Figs. 3, 7, and 9). At the same time, an appropriate choice of certain parameters can reduce equation (72) to a much simpler form such as the LJ(m, 6) or exponential-6 function; in particular, the former is obtained if $m = 12$ and $\beta = 0 = h_1 = h_2 = h_3 = C_8$, and the latter if $m = 0 = h_1 = h_2 = h_3 = C_8$.

As written, the independent parameters defining the function of equation (72) are: $A, m, \beta, C_6, C_8, \ldots$ etc., plus the parameters defining the damping function $D(R)$. The variable m and the damping function parameters were used to define a given type of model potential, while the others, $A, \beta, C_6, C_8, \ldots$ etc., were treated as quantities to be determined from the analysis.[57] As discussed in Section IV.C, the fits were much better behaved when two of the remaining parameters, say A and C_8, were replaced by the energy ε and distance R_e defining the minimum of $V(R)$. This was readily done, since

$$A = \{\varepsilon(8 - R_eD_1) - 2D_0C_6/(R_e)^6\}(R_e)^m e^{\beta R_e}/(\beta R_e + m - 8 + R_eD_1) \tag{74}$$

$$C_8 = \{(6 - R_eD_1 - \beta R_e - m)C_6/(R_e)^6 + \varepsilon(\beta R_e + m)/D_0\} \times (R_e)^8/(\beta R_e + m - 8 + R_eD_1) \tag{75}$$

where $D_0 \equiv D(R = R_e)$ and $D_1 \equiv [\partial D(R)/\partial R]_{R=R_e}$. As a result, the independent parameters became ε, R_e, β, and C_6. Of course, equations (74) and (75) correspond to the GBC($m = 0, \beta;6, 8$) functions in which only R^{-6} and R^{-8} terms contribute to the attractive part of $V(R)$; however, analogous expressions are readily obtained for cases in which additional terms are included in this expansion. More generally, it is convenient to express A and the coefficient of the highest inverse-power term (here R^{-8}) in terms of the appropriate ε and R_e values, while the coefficients of lower-power terms are either treated as parameters or held fixed at values obtained in other ways. In view of the arguments of Section IV.A, the various exponent parameters $\beta^{\lambda k}$ were all set equal to β^{00}, while the long-range potential constants $C_6^{\lambda k}$ were defined by the semiempirical results of Thakkar[135] or by the analogous results of Langhoff *et al.*[133b]

Potential surfaces with three different levels of sophistication have now been obtained from fits to the hydrogen–inert gas data.[54, 55, 57] In all cases only $\lambda = 0$ and 2 angle-dependent terms were included in the potentials, since ref. 54 concluded that $\lambda = 4$ radial strength functions have no discernable effect on the existing data for these species. For the H_2, D_2–Ar system, the parameters defining the three resulting surfaces are listed in Table V. The subscript on the label identifying each potential is k_{max}, the highest power of ξ appearing in the expansion of equation (9). The quantities in parentheses are the 95% confidence limit uncertainties in the optimized variables; parameters quoted without uncertainties were either taken from other sources and held fixed, or were defined in terms of the other parameters.

The $LJ_1(12, 6)$ surface of Table V, taken from ref. 55, differs slightly from that originally reported in ref. 54 because a different statistical weight was used for transitions involving quasibound levels. It has neither the correct long-range C_6 constants nor the correct limiting behaviour as $r \to 0$. As a result, the 'region of validity' of this surface (see Table V) is approximately bounded by the classical turning points of the observed bimer levels. On the other hand, it reproduces the spectroscopic data virtually as well as either of the more sophisticated surfaces.

The Buckingham–Corner-type $BC_1(6, 8)$ potential[57] is also linear in ξ and hence does not have the correct $r \to 0$ behaviour, but was constrained to have realistic long-range C_6 coefficients. As a result, its region of validity includes all R values beyond the inner turning point at $R \approx 3.2$Å. As implied by the name, its radial strength functions are defined by equation (72) with $m = 0$ and damping function parameters $(h_1, h_2, h_3) = (4, 1, 3)$, while the characteristic distance associated with $D(R)$ is $R_0 = R_e^{00}$, the equilibrium distance of the purely isotropic $(\lambda, k) = (0, 0)$ component of the potential. This 'intermediate' surface differs from that of ref. 55 only in that a slightly better C_6^{21} value was assumed.

The $BC_3(6, 8)$ surface[57] has the correct C_6 coefficients and attains the

TABLE V. Optimized potential energy parameter for H_2, D_2–Ar determined from fits to spectroscopic data;[54,55,57] quantities in parentheses are the 95% confidence limit uncertainities in the fitted parameters.

Parameter	$LJ_1(12, 6)$	$BC_1(6, 8)$	$BC_3(6, 8)^a$
$\varepsilon^{00}(cm^{-1})$	52.13(0.33)	50.84(0.49)	50.87(0.50)
R_e^{00}(Å)	3.5563(0.0046)	3.5737(0.0053)	3.5727(0.0051)
β^{00}(Å^{-1})	—	3.691(0.103)	3.610(0.099)
C_6^{00}(cm^{-1}Å^6)	—	136400	134500
$\varepsilon^{01}(cm^{-1})$	30.4(1.3)	30.8(1.9)	34.2(1.1)
R_e^{01}(Å)	3.833(0.026)	3.852(0.023)	3.769(0.026)
C_6^{01}(cm^{-1} Å^6)	—	177400	115800
$\varepsilon^{02}(cm^{-1})$	—	—	1.904
R_e^{02}(Å)	—	—	4.3575
C_6^{02}(cm^{-1} Å^6)	—	—	3051
C_6^{03}(cm^{-1} Å^6)	—	—	– 25270
$\varepsilon^{20}(cm^{-1})$	6.87(0.71)	7.34(0.90)	5.72(0.46)
R_e^{20}(Å)	3.821(0.059)	3.813(0.045)	3.743(0.049)
C_6^{20}(cm^{-1} Å^6)	—	14590	13500
$\varepsilon^{21}(cm^{-1})$	4.01	4.45	21.6(9.7)
R_e^{21}(Å)	4.118	4.109	3.9487
C_6^{21}(cm^{-1} Å^6)	—	50680	29600
$\varepsilon^{22}(cm^{-1})$	—	—	13.914
R_e^{22}(Å)	—	—	4.0267
C_6^{22}(cm^{-1} Å^6)	—	—	5705
C_6^{23}(cm^{-1} Å^6)	—	—	– 10395
DMR	0.90	0.87	0.83
Region of	$3.2 \lesssim R \lesssim 5.2$	$3.2 \lesssim R$	$3.2 \lesssim R$
validity (in Å)	$0.6 \lesssim r \lesssim 1.0$	$0.6 \lesssim r \lesssim 1.0$	$0 \leq r < 1.9$

[a]Recommended potential surface of ref. 57.

correct He–Ar well depth and equilibrium distance[136] in the $r \to 0$ limit. Its region of validity therefore extends to $r = 0$; while the corresponding upper bound on r also increases, it is bounded by the limit on the range of validity of the cubic expansion for the C_6 coefficients (see Fig. 5). In this case, the $\varepsilon^{\lambda k}$ and $R_e^{\lambda k}$ and $R_e^{\lambda k}$ parameters for $k = 2$ were defined by the collapsed-diatom limit constraint of equations (65) and (66), while Thakkar's[135] results were used to define $C_6^{\lambda k}$ coefficients up to $k = 3$. In addition, this surface has one more fitted parameter than the first two. In the earlier work,[54,55] the data were found to be relatively insensitive to the $V_{21}(R)$ radial strength function (defining the linear stretching dependence of the anisotropy), so its parameters were defined by the ratios

$$\varepsilon^{21} = \varepsilon^{20}(\varepsilon^{01}/\varepsilon^{00}) \tag{76}$$

$$R_e^{21} = R_e^{20}(R_e^{01}/R_e^{00}) \tag{77}$$

However, when the collapsed-diatom limit was imposed on the model, the first of these constraints had to be released if the excellent quality of fit was to be retained. This change yielded an ε^{21} value in better accord with the corresponding parameters[55,57] for H_2, D_2–Kr and H_2, D_2–Xe, and the associated ε^{20} parameter attained a value in better agreement with that determined from state-selected total-scattering cross sections by Zandee and Reuss.[42]

The differences among the fitted parameters for the three surfaces of Table V reflect the effects both of changes in the form of the radial strength functions and of imposing realistic limiting behaviour on the potential. The only effect of this increasing sophistication on the quality of fit was to improve it slightly. Indeed, the fact that the converged DMR values for all three cases are slightly less than unity suggests that McKellar and Welsh's[52] estimated experimental uncertainties (of ca. $\pm 0.03\ \text{cm}^{-1}$) may be

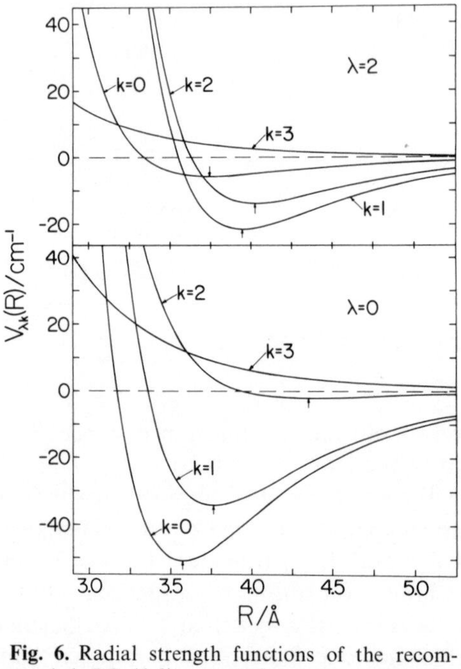

Fig. 6. Radial strength functions of the recommended $BC_3(6, 8)$ potential energy surface for H_2–Ar; vertical arrows denote the minima of the various curves.

slightly pessimistic. As a more quantitative test of the model dependence of the resulting parameters, Carley[57] performed fits to the H_2, D_2–Ar data at the intermediate ($k_{max} = 1$, correct C_6) level of sophistication using a variety of different GBC-type potentials. In all cases, the resulting parameters agreed to within the quoted uncertainties with those for the $BC_1(6, 8)$ function considered here. This appears to indicate that this type of analysis can yield a potential energy surface which is fairly model independent.

The radial strength functions which define the recommended $BC_3(6, 8)$ potential energy surface for H_2, D_2–Ar are plotted in Fig. 6, while Figs. 7 and 8 illustrate how the overall isotropic and angle-dependent radial strength functions

$$V_\lambda(R, \xi) = \sum_{k=0}^{3} \xi^k V_{\lambda k}(R) \tag{78}$$

depend on the diatom stretching coordinate ξ. In Fig. 7, the well depth and equilibrium distance of the spherical ($\lambda = 0$) part of the potential are plotted

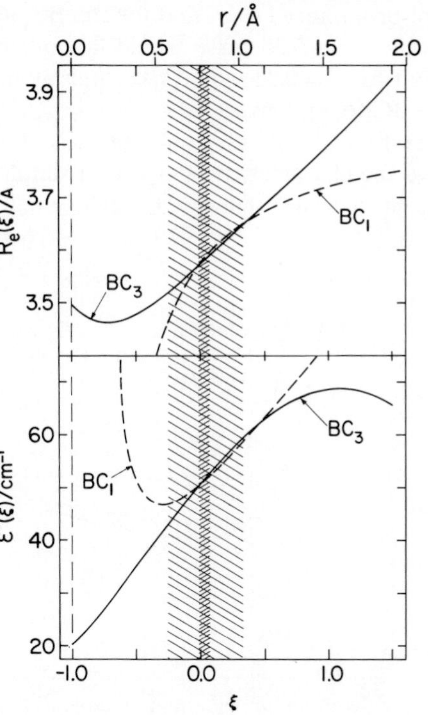

Fig. 7. Dependence on the diatom stretching coordinate of the well depth and equilibrium distance for the isotropic part of two H_2–Ar potential energy surfaces.

vs. ξ for both the BC_3 (6, 8) (solid curves) and BC_1 (6, 8) (dashed curves) surfaces of Table V; the points at $\xi = -1$ denote the He-Ar parameters of Smith *et al.*[136] The narrow shaded region seen here denotes the range of expectation values of ξ for the diatom levels associated with the bimer states defining the spectroscopic transitions used in the analysis: $-0.010 \leq \xi \leq 0.071$. However, the full range of ξ to which these data are sensitive is represented by the broad shaded region which is defined by the distance between the inner and outer turning points for the highest of these diatom states, the $v = 1, j = 2$ level of H_2. While the solid and dashed curves in Fig. 7 are quite similar over the range of ξ to which the spectroscopic data are most sensitive, the marked differences between them outside of this interval demonstrates the improvement in the stretching dependence of the potential achieved by imposition of the collapsed diatom limit.

In Fig. 8, $V_2(R, \xi)$ is plotted for ξ fixed at values corresponding to the inner and outer turning points of H_2 ($v = 1, j = 2$), -0.26 and $+0.33$, respectively, and at the expectation values of ξ for H_2 in its $(v, j) = (0, 1)$ and (1, 2) states, 0.001 and 0.071, respectively. The stretching dependence seen here is much more pronounced than that for the BC_1 (6, 8) surface. For the latter, the analogous $\xi = -0.26$ curve would lie quite near the $\xi = 0$ curve in Fig. 8, while that for $\xi = 0.33$ would like roughly half way between the $\xi = 0.071$ and 0.33 BC_3(6, 8) curves.

While it is interesting to note that most of the radial strength functions seen in Fig. 6 have the characteristic shape of a simple spherical potential curve, it is important to realize that this behaviour may only be a property

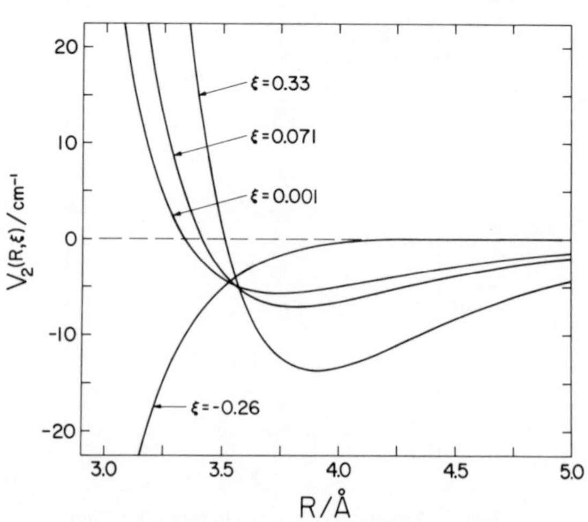

Fig. 8. Overall anisotropy strength function of the recommended BC_3 (6, 8) potential energy surface for H_2–Ar, at different values of the diatom stretching coordinate.

of this particular type of complex. In particular, if the power series expansion in the stretching coordinate ξ had been centred at the point $\xi = -0.26$ instead of at $\xi = 0$, the $V_{20}(R)$ radial strength function would have corresponded to the $\xi = -0.26$ curve in Fig. 8, which has two zeros for $R > 4.0$ Å. The resulting function clearly could not be parameterized readily in terms of the ε, R_e, β, and C_6 variables of equations (72)–(75) and Table V. However, devising a more appropriate set of fitting parameters for functions of this kind should present little difficulty.

Fits to the H_2, D_2–Kr and H_2, D_2–Xe spectroscopic data[52] were performed in the same way as those for H_2, D_2–Ar, using potential models with the same three levels of sophistication.[54, 55, 57] However, while effects associated with replacing the $LJ_1(m, 6)$ form by a $BC_1(6, 8)$ model were precisely analogous to those found for H_2, D_2–Ar,[55] some convergence difficulty was encountered when the collapsed-diatom limit constraint was applied to these systems.[57] The $k = 0$ and 1 components of the resulting $BC_3(6, 8)$ functions are very similar to those of the $BC_1(6, 8)$ surfaces reported earlier[55] and the $k = 2$ strength functions are qualitatively similar to those of the $BC_3(6, 8)$ potential for H_2, D_2–Ar. However, the dimensionless mean residual (DMR) values for the fits to the $BC_3(6, 8)$ model are 31% and 64% higher than those obtained using the simpler potentials forms. This difficulty is probably due to insufficient flexibility in the ξ-dependence of the $BC_3(6, 8)$ model used and should be able to be overcome through use of a more sophisticated potential form. However, at the present time the $BC_1(6, 8)$ potentials of ref. 55b, whose region of validity is restricted to the interval $-0.26 \lesssim \xi \lesssim 0.33$, are the current 'best' potential energy surfaces for H_2, D_2–Kr and H_2, D_2–Xe.

For H_2, D_2–Ne, both Dunker and Gordon[56] and Le Roy and coworkers[54, 57] independently concluded that no information about the potential anisotropy and very little about its stretching dependence could be extracted from the existing spectra.[52b] However, these data are still very sensitive to the isotropic or spherically averaged part of the potential and could be very helpful in determining it. Carley[57] therefore devised a $BC_3(6, 8)$ surface for this system by using fits to the spectra to determine the isotropic strength function parameters ε^{00}, R_e^{00}, and β^{00} while the stretching parameter ε^{01} was varied manually and the other parameters were defined by external constraints. In particular: (i) the stretching-dependent isotropic and anisotropic C_6 coefficients were taken from Thakkar's[135] semiempirical results; (ii) on the basis of a rough extrapolation from the results for H_2–Ar, Kr, and Xe, the distance R_e^{01} was defined as 1.1 R_e^{00}; (iii) the anisotropy parameters ε^{20} and R_e^{20} were based on the results of Tang and Toennies,[131] and (iv) the parameters ε^{21} and R_e^{21} were defined by equations (76) and (77). The parameters defining this surface are listed in Table VI. In view of the insensitivity of the data to the diatom bond length, the collapsed-diatom limit was

TABLE VI. Parameters of the $BC_3(6, 8)$ potential surface for H_2, D_2–Ne from ref. 57.[a]

(λ, k)	$\varepsilon^{\lambda k}(\mathrm{cm}^{-1})$	$R_e^{\lambda k}(\mathrm{\AA})$	$\beta^{\lambda k}(\mathrm{\AA}^{-1})$	$C_6^{\lambda k}(\mathrm{cm}^{-1}\,\mathrm{\AA}^6)$
(0, 0)	27.7(1.7)	3.255(0.026)	6.02(0.45)	39610
(0, 1)	10(?)	3.58	6.02	31160
(2, 0)	2.26	3.54	6.02	3681
(2, 1)	0.82	3.89	6.02	7722

[a]In addition, C_6^{02}, C_6^{03}, C_6^{22}, C_6^{23} are, respectively 47.7, −6487, 1322, and −2719 $\mathrm{cm}^{-1}\,\mathrm{A}^6$, and DMR = 1.00 for the fit.

not applied to these data, so $k > 1$ contributions to this potential consist only of the $C_6^{\lambda k}/R^6$ terms.

2. *Tests of the Spectroscopic Potentials*

The spectroscopic potentials for hydrogen-Ar, Kr, and Xe have been tested with regard to their ability to predict a variety of other experimental phenomena.[57] Particularly satisfying in this regard are the comparisons with the differential scattering cross sections of Rulis *et al.*[140] and the low-energy integral scattering cross sections of Toennies *et al.*[134] In both cases, the data can be properly interpreted in terms of a spherically averaged one-dimensional potential between the inert-gas atom and a ground-state H_2 molecule.[134, 140] Since the differences between the vibrationally averaged (over r) potentials for ground-state ortho- and para-hydrogen are much smaller than the uncertainties in the spectroscopic potentials, such differences could be ignored when making these comparisons. The assumption is further justified by the fact that Toennies *et al.*[134] found no noticeable differences between cross sections obtained using pure para-H_2 and those yielded by beams of normal-H_2.

For H_2–Ar, Kr, and Xe, the isotropic potentials obtained by Rulis *et al.*[140] agree with the vibrationally averaged spherical parts of the $BC_1(6, 8)$ spectroscopic potentials to within the mutual uncertainties. More important though, the DSD values (see equation (69)) for the predictions generated from the spectroscopic potentials are virtually identical to those for potentials derived from these scattering data. Thus the differential cross-section measurements cannot distinguish between potentials obtained from them and the spectroscopic potentials. Similar conclusions are yielded by the comparisons with the low-energy scattering cross sections of Toennies *et al.*[134] In this case, the predictions of the $BC_1(6, 8)$ spectroscopic potential for H_2–Ar were in sufficiently close agreement with their data that these authors did not attempt to improve on it, while their data for H_2–Kr and H_2–Xe cannot discriminate between potentials derived from them and the best spectroscopic potentials.[134]

The above comparisons confirm the reliability of the spherical part of the spectroscopic potentials. More important though, they point out that independent analyses of three completely different types of experimental phenomena have yielded, within the quoted uncertainties, exactly the same spherical potentials for H_2–Ar, Kr, and Xe. This attests to the reliability of all three techniques and indicates that the results are among the best known of *all* (spherical) pair potentials.

While the scattering cross sections of Rulis *et al.*[140] and Toennies *et al.*[134] effectively see only the spherical average of the interaction potential, collision phenomena such as relaxation cross sections or cross sections for beams of state-selected molecules are also sensitive to its angle dependence. However, model studies of the n.m.r. longitudinal relaxation time T_1/ρ, the two-level rotational relaxation time τ_{rot}, and the anisotropy of state-selected total cross sections $\Delta\sigma/\sigma$ for H_2–He showed that[57,141] for a fixed anisotropy strength function $V_2(R)$, virtually the only feature of the isotropic potential these properties are sensitive to is the position of its repulsive wall. Thus it is impossible to determine reliably both the isotropic and angle-dependent parts of a potential surface using only these properties. Moreover, the utility of these properties for determining or testing an estimate of $V_2(R)$ is critically dependent on the reliability of the isotropic potential used in the calculations, or more particularly, on the position of its repulsive wall.

In Table VII, the depths and minimum positions of the anisotropy strength functions determined by Foster and Rugheimer[39b] from T_1/ρ data and by Zandee and Reuss[42] from $\Delta\sigma/\sigma$ measurements are compared with the

TABLE VII. Anisotropy strength function parameters for $H_2(v=0, j=1)$–Ar, Kr, and Xe determined from n.m.r. relaxation T_1/ρ,[39b] from total cross-section anisotropies $\Delta\sigma/\sigma$,[42] and from the spectra of van der Waals molecules.[57]

Parameters		T_1/ρ	$\Delta\sigma/\sigma$	Spectroscopy
H_2–Ar:	$\varepsilon^2(\text{cm}^{-1})$	8.9(2)	5.03(1.02)	5.87(0.46)
	R^2(Å)	3.53(0.1)	3.72(0.08)	3.76(0.05)
	R_e^0(Å)	3.34^a	3.574^b	3.576(0.005)
H_2–Kr:	$\varepsilon^2(\text{cm}^{-1})$	9.2(2)	6.39(1.06)	8.34(0.54)
	R_e^2(Å)	3.82(0.1)	3.85(0.08)	3.84(0.10)
	R_e^0(Å)	3.65^a	3.719^b	3.721(0.006)
H_2–Xe:	$\varepsilon^2(\text{cm}^{-1})$	10.4(2)	8.05(1.04)	9.75(0.36)
	R_e^2(Å)	4.05(0.1)	3.95(0.08)	3.95(0.12)
	R_e^0(Å)	3.89^a	3.934^b	3.938(0.006)

^{a}Isotropic potential taken from Helbing *et al.*[142]
^{b}Isotropic potential is the $(\lambda, k) = (0,0)$ component of the $BC_1(6,8)$ potential of ref. 55.

corresponding parameters of the $BC_3(6,8)$ spectroscopic potentials.[57] Since the T_1/ρ and $\Delta\sigma/\sigma$ experiments effectively see only ground-state ortho-hydrogen, the spectroscopic parameters shown correspond to the vibrationally averaged function $\bar{V}_\lambda(0,1|R)$ of equation (48). The fact that the T_1/ρ parameters differ significantly from the spectroscopic values only for the one case (H_2–Ar) for which the equilibrium distance R_e^0 of the assumed isotropic potential is significantly in error supports the arguments of the preceding paragraph regarding the sensitivity of these relaxation times and cross sections to the isotropic potential. With this one exception, the R_e^2 values obtained in all three studies are in excellent agreement, while the only ε^2 values which do not agree to within the mutual uncertainties are the $\Delta\sigma/\sigma$ and spectroscopic results for H_2–Kr and Xe. Moreover, even these latter discrepancies are not very large and may well lie within the range of uncertainties arising from use of the approximate DWBA method in performing the $\Delta\sigma/\sigma$ calculations and from neglect of the effect of orbiting resonances in the $\Delta\sigma/\sigma$ analysis.[57] Thus while the agreement is less satisfying than that for properties depending only on the isotropic part of the potential, the results in Table VII do suggest that the anisotropies of the potential energy surfaces determined from the van der Waals molecule spectra are essentially correct.

To date, analysis of the spectroscopic data for the hydrogen–inert gas complexes is the only way in which the diatom bond-length dependence of a potential energy surface has been determined from experimental data. As a result, there are no results with which the spectroscopic $k > 0$ radial strength functions may be quantitatively compared. However, good agreement was obtained when the stretching-dependent spherical parts of the original[54] $LJ_1(m, 6)$ surfaces for H_2–Ar, Kr, and Xe were used to predict the matrix shift and centre-of-mass frequency for transitions of H_2 and D_2 dissolved in solid Ar, Kr, and Xe.[143] Moreover, the large disagreement with experiment of the analogous predicted[143] matrix frequency shift for HD in solid Ar is probably the simply to the fact that the change in the effect spherical potential due to displacement of the diatom centre-of-mass[91,92] was not taken into account in the calculations. In one additional test of the stretching dependence of the isotropic potentials,[144] good agreement was found when the stretching dependence of the $LJ_1(12,6)$ potential for H_2–Ar was used to predict the pressure shifts of Raman transition frequencies for H_2.[145]

The only test of the stretching dependence of the potential anisotropy is provided indirectly by the comparisons with the anisotropy parameters for H_2–Ar obtained by Zandee and Reuss[42] from their total cross-section anisotropy measurements. Of the systems considered in Table VII, the parameters obtained from the $\Delta\sigma/\sigma$ analysis are believed to be most accurate for H_2–Ar, since the shallower isotropic potential and corresponding weaker potential anisotropy should minimize possible effects due to orbiting resonances or to the inadequacies of the DWBA method. Thus the marked difference between their $\Delta\sigma/\sigma$ value, $\varepsilon^{20} = 5.0(\pm 1.0)\,\text{cm}^{-1}$, and the values $6.9(\pm 0.7)$ and $7.3(\pm 0.9)\,\text{cm}^{-1}$ associated with the earlier[54,55] LJ_1 (12, 6) and BC_1 (6, 8) spectroscopic potentials (see Table V) was somewhat disturbing. The source of this discrepancy appears to be the fact that the spectroscopic data are most sensitive to the effective anisotropy of bimers formed from vibrationally excited hydrogen, while the $\Delta\sigma/\sigma$ measurements see only $H_2(v=0, j=1)$. The angle dependence of the $LJ_1(12, 6)$ and $BC_1(6, 8)$ surfaces is only weakly dependent on the diatom bond length, so their ε^{20} and R_e^{20} values mainly reflect the effective anisotropy for the $v = 1$ diatom. This conclusion is partly based on the fact that the $BC_1(6, 8)\, V_2(r, \xi)$ curves corresponding to $\xi = 0$ and 0.071 are both virtually identical to the $BC_3(6, 8)$ curve for $\xi = 0.071$ seen in Fig. 8. Imposition of the collapsed-diatom limit constraint on the H_2, D_2–Ar analysis yielded a potential surface with a much larger anisotropy dependence. The agreement between the vibrationally averaged effective anisotropy strength $\bar{V}_2(0, 1\,|\,R)$ for the resulting $BC_3(6, 8)$ surface and the potential anisotropy determined by Zandee and Reuss[42] (see Table VII) therefore demonstrates that this anisotropy stretching dependence is reasonably correct.

It has been suggested[92] that the ability to predict transition frequencies for bimer isotopes other than those whose spectra were used to determine a potential energy surface would be a most severe test of the quality of such a surface. In this regard, refs. 55 and 92 implied that the $LJ_1(12, 6)$ and $BC_1(6, 8)$ surfaces for H_2, D_2–Ar had serious deficiencies in that neither appeared capable of yielding predictions in good agreement with McKellar's[52d] observed HD–Ar spectrum. However, it now appears that the calculations on which the latter conclusions were based are unreliable. In particular, it has been shown[96] that while $n_{\text{max}} = 0$ or 1 secular equation calculations are perfectly reliable for H_2–Ar or D_2–Ar, but they are not adequate for HD–Ar, and that[127] coupling to the continuum cannot be neglected when treating the upper states of the observed transitions for this species. Thus the reported[55,92] calculations for HD–Ar do not represent valid tests of the $LJ_1(12, 6)$ and $BC_1(6, 8)$ potential surfaces[54,55] for this system.

B. Isotopic Ar–HCl

Apart from the hydrogen–inert gas complexes, Ar–HCl is the only system for which a detailed potential energy surface has been obtained from the spectra of van der Waals molecules. The experimental data for this system consist of the microwave and radiofrequency transitions which Novick *et al.*[60] observed using molecular beam electric resonance techniques. The number of fully resolved and assigned transitions is quite small, totalling only 14 lines for the four isotopes studied, and they only involve levels lying very close to the ground state of the bimer. However, the information about the nuclear quadrupole coupling constants obtainable from these data yields effectively model-independent estimates of the expectation value of $\cos^2 \theta$ for the observed states. Similarly, the values of the bimer dipole moments obtained from the measured[60] Stark shift coefficients provide estimates, albeit somewhat more model-dependent ones, of the expectation value of $\cos \theta$. Thus in spite of the limited variety of observed states, the nature of these measurements means that they contain substantial information about the potential energy surface.

The first study to make quantitative use of the spectroscopic data for the Ar–HCl system was that of Dunker and Gordon,[95] who used close-coupling calculations to predict the spectroscopic properties of four previously proposed[40] potential surfaces. These predictions were in only rough qualitative agreement with experiment; in particular, the expectation values of $\cos \theta$ and $\cos^2 \theta$ were distinctly too large and isotope shifts of the angular and radial properties were incorrect both in magnitude and direction. This showed that the Nielsen–Gordon[40] surfaces are not reliable and emphasized the importance of the spectroscopic data for testing available potentials.

A much more detailed analysis of the Ar–HCl data has recently been reported by Holmgren *et al.*[78] who used non-linear least-squares fits to the data to optimize the parameters of two models for the potential energy surface. Although these authors performed their calculations using the approximate BOARS procedure, they showed quite convincingly that the weaknesses of this method should not introduce significant errors into the quantities they calculate. Another novel feature of their analysis is the fact that the fits simultaneously treated four different types of properties: the rotational constant B_0 and centrifugal distortion constant D_J for each of four isotopes, plus the expectation values of $\cos \theta$ and $\cos^2 \theta$ for two of these isotopes. In the spirit of the discussion of Section IV.C, these authors were also careful in selecting the most appropriate parameters to be freely varied in the fits.

Since all of the observed states of the isotopes of Ar–HCl correlate with the diatom in its ground vibration–rotation level, the potential models used[78] were independent of the diatom bond length. The fact that HCl is not homonuclear also means that terms corresponding to odd values of λ had to be included in the Legendre expansion for the potential which therefore became:

$$V(R;\theta) = V_0(R) + V_1(R)P_1(\cos \theta) + V_2(R)P_2(\cos \theta) + \dots \tag{79}$$

The analysis showed that at least two angle-dependent terms had to be included in this expansion, and since the data were adequately explained by them no higher-order angular terms were introduced.

All of the radial strength functions were assumed to have the exponential (α, m) form

$$V_\lambda(R) = \varepsilon\{me^{\alpha(1-R/Re)} - \alpha(R_e/R)^m\}/(\alpha - m) \tag{80}$$

where $m = 6$ for $\lambda = 0$ and 2 and $m = 7$ for $\lambda = 1$. The value of one isotropic parameter (either ε^0 or α^0) was taken from other sources, and two other parameters were defined by the constraints $R_e^2 = R_e^1$ and $\alpha^2 = \alpha^1$. The fitted values of the remaining parameters for the two optimized surfaces are compared in Table VIII; while the strength of the $\lambda \neq 0$ functions is here characterized by ε^1 and ε^2 rather than by the variables quoted by Holmgren *et al.*,[78] these constants and their uncertainties are precisely equivalent to the reported quantities. As indicated by the DMR values, these two surfaces cannot be distinguished on the basis of the quality of fit.

Since all of the observed levels lie within 1 cm^{-1} of the ground state, the data can contain very little information about the absolute binding energies of these levels, and hence about ε^0. However, they should provide a very reliable estimate of the position of the potential minimum and of the overall shape of the potential surface in the neighbourhood of this minimum. As a result, in spite of the large differences between the ε^λ values for these two surfaces, the appropriately scaled (to take account of uncertainty regarding the absolute binding energy) potential energy contour diagrams for them shown in Figs. 2 and 3 of ref. 78 are remarkably similar. Thus although the differences between the parameters in Table VIII might appear to suggest the contrary, the shape of the Ar–HCl potential surface in the region probed by the spectroscopic data is in fact fairly well known. This demonstrates

TABLE VIII. Parameters of the two fitted potential surfaces for Ar–HCl reported by Holmgren *et al.*;[78] the uncertainties here correspond to one standard deviation.

Parameter	Surface Ib	Surface IIb
$\varepsilon^0(cm^{-1})$	166(5)	133[a]
$R_e^0(cm^{-1})$	3.85(0.05)	3.81(0.05)
α^0	13.5[a]	20(2)
$\varepsilon^1(cm^{-1})$	29.4(19.8)	41.2(20.3)
$R_e^1(cm^{-1})$	4.26(0.02)	4.259(0.007)
α^1	10.1(0.4)	10.36(0.06)
$\varepsilon^2(cm^{-1})$	34.6(16.4)	29.8(17.1)
DMR	0.81	0.90
Region of validity (in Å)	$3.4 \lesssim R \lesssim 4.5$	$3.4 \lesssim R \lesssim 4.5$

[a] Parameter held fixed at given value.

the important results: (i) that it is the overall shape of a potential energy surface which is important, rather than the characteristics of the individual radial strength functions, and (ii) that cases may arise in which the former is well known even though the latter are not.

When the isotropic parts of these two potentials were used[78] to predict the differential scattering cross sections of Farrar and Lee,[146] the results for surface IIb were in much better agreement with the cross sections calculated from the isotropic potential determined from the scattering data. However, this might have been expected since its ε^0 parameter was held fixed at the Farrar and Lee value, and its R_e value is relatively closer to that for the scattering potential. It should be pointed out too, that this agreement does not resolve the discrepancy between the isotropic well depths suggested by the molecular beam experiments,[146] 133 cm^{-1}, and by analysis of the temperature dependence of the infrared absorbance,[46,63,147] 385–524 cm^{-1}, since the only reason the bimer spectra are at all sensitive to the ε^0 parameter is because of the simplicity of the analytic form chosen for the $V_\lambda(R)$ functions.

In the first independent test of the anisotropy of Ar–HCl potentials determined from bimer spectra, Kircz *et al.*[148] used unpublished early versions of the potentials of Holmgren *et al.*[78] to predict the collisional broadening of HCl rotational lines, for comparison with experiment. The very poor agreement which these comparisons showed should not be surprising, since these line widths depend on the potential outside the range of R to which the bimer spectra are sensitive. At the same time, the magnitude of the differences between the anisotropy strength functions of Holmgren *et al.*[78] and those which Kircz *et al.*[148] determined from the line-width data suggests that the latter exaggerate the angle dependence of the potential in the region near its minimum. Thus it is clear that further experimental input is required before a unique overall anisotropic potential can be determined for Ar–HCl. However, it seems likely that near the potential minimum at $\theta = 0$ and $R \approx 4$ Å, the shape and angle dependence of the surfaces of Holmgren *et al.*[78] are essentially correct.

VI. CONCLUDING REMARKS

From the results in Section V.A, it seems clear that the discrete spectra of atom–diatom van der Waals molecules are capable of yielding potential energy surfaces of unparalleled sophistication and accuracy. To date, the availability of sufficiently extensive and accurate data has restricted realization of this goal to the weak-coupling molecular hydrogen–inert gas bimers,[52] which fortuitously are precisely those systems which are most easily treated by the theory.[54–57] However, the development of the new experimental methods described in Section II should ensure the appearance of good data for many other systems in the near future. At the same time, the improved form of the secular equation method[55,96] described in Section III, and in appropriate cases also the approximate BOARS method,[78,121] should make detailed analyses of data for weak- or strong-coupling complexes both tractable and reliable. Indeed, the results for the Ar–HCl system show that significant information about the shape of a potential energy surface may sometimes be obtained even when the data are not sufficiently extensive to allow the potential to be uniquely defined.

One obvious future development is the generalization of the methods described in Section III to the case of diatom–diatom complexes. Although additional complexities certainly arise, the problems encountered should be ones of implementation rather than of formulation. Once this is done, an analysis of the fully assigned existing infrared[52] and radiofrequency[79] transitions for $(H_2)_2$ and $(D_2)_2$ should prove quite fruitful, as would a study of the dimer $(HF)_2$ for which fully resolved infrared data[44d] and fully resolved and assigned microwave and radiofrequency measurements[58] are available.

The main limitation of the present theory of van der Waals bimers is the fact that none of the methods described in Section III has yet been proved capable of yielding reliable results for semirigid complexes such as He–I_2 or Kr–ClF. Klemperer[97] has described the results of a preliminary study of the latter system performed using the BOARS method, but the reliability of those calculations has not been demonstrated. However, the fact that Shapiro and Balint-Kurti[98] were able to perform close-coupling calculations on the ground-state H_2O molecule treated as O–H_2, albeit at considerable computational expense, suggests that further improvements in the present methods may make treatment of semirigid complexes quite feasible. In particular, associating the basis-generating potential $V_b(R)$ of the secular equation method with the profile of the potential surface at the preferred relative orientation of the complex may allow the size of the required radial basis set to be kept quite small. This would at least partially compensate for the very large number of angular channels which has to be considered for such species.

Acknowledgements

The authors are very grateful to Dr. J. E. Grabenstetter and to Professor F. R. McCourt for both their helpful discussions during the development of this report and their incisive comments on the manuscript. We also gratefully acknowledge the financial support of the National Research Council of Canada.

References

1. T. L. Hill, *Statistical Mechanics*, McGraw-Hill, New York, 1956, Chap. 5.
2. D. E. Stogryn and J. O. Hirschfelder, *J. Chem. Phys.*, **31**, 1531 (1959); *ibid*, **31**, 1545 (1959); *ibid*, **33**, 942 (1960).
3. G. Glockler, C. P. Roe, and D. L. Fuller, *J. Chem. Phys.*, **1**, 703 (1933).
4. E. A. Guggenheim, *Mol. Phys.*, **10**, 401 (1966).
5. S. K. Kim and J. Ross, *J. Chem. Phys.*, **42**, 263 (1965).
6. S. W. Benson, *The Foundation of Chemical Kinetics*, McGraw-Hill, New York, 1960.
7. D. L. King, D. A. Dixon, and D. R. Herschbach, *J. Amer. Chem. Soc.*, **96**, 3328 (1974).
8. (a) J. P. Hirth and G. M. Pound, *Condensation and Evaporation*, Pergamon Press, Oxford, 1963; (b) W. J. Dunning, *Disc. Faraday Soc.*, **30**, 9 (1960); (c) J. Frenkel, *Kinetic Theory of Liquids*, Oxford University Press, Oxford, 1946.
9. (a) R. E. Leckenby, E. J. Robbins, and P. A. Trevalion, *Proc. Roy. Soc. (London)*, **A280**, 409 (1964); (b) D. Golomb, R. E. Good, and R. F. Brown, *J. Chem. Phys.*, **52**, 1545 (1970); (c) T. A. Milne, A. E. Vandegrift, and F. T. Greene, *J. Chem. Phys.*, **52**, 1552 (1970); (d) O. F. Hagena and W. Obert, *J. Chem. Phys.*, **56**, 1793 (1972); (e) N. Lee and J. B. Fenn, *Rev. Sci. Instr.*, **49**, 1269 (1978).
10. (a) C. C. Bouchiat, M. Bouchiat, and L. C. L. Pottier, *Phys. Rev.*, **181**, 144 (1969); (b) M. A. Bouchiat, J. Brossel, and L. C. Pottier, *J. Chem. Phys.*, **56**, 3703 (1972).
11. F. A. Franz and C. Volk, *Phys. Rev. Lett.*, **35**, 1704 (1975).
12. G. E. Ewing, *Angew. Chem. Internat. Edit.*, **11**, 486 (1972).
13. W. Klemperer, *Ber. Bunsen. Phys. Chem.*, **78**, 128 (1974).
14. G. E. Ewing, *Accts. Chem. Res.*, **8**, 185 (1975).
15. G. E. Ewing, *Can. J. Phys.*, **54**, 487 (1976).
16. B. L. Blaney and G. E. Ewing, *Ann. Rev. Phys. Chem.*, **27**, 553 (1976).
17. R. E. Smalley, L. Wharton, and D. H. Levy, *Accts. Chem. Res.*, **10**, 139 (1977).
18. D. H. Levy, L. Wharton, and R. E. Smalley, in *Chemical and Biochemical Applications of Lasers*, Vol. II, Academic Press, New York, 1977, Chap. 1.
19. R. E. Smalley, D. H. Levy, and L. Wharton, *J. Chem. Phys.*, **64**, 3266 (1976).
20. R. E. Smalley, L. Wharton, D. H. Levy, and D. W. Chandler, *J. Chem. Phys.*, **68**, 2487 (1978).
21. W. Finkelnburg and Th. Peters in *Handbuch der Physik* (Eds. Flügge), Vol. 28, Springer-Verlag, Berlin, 1957, p. 181.
22. H. E. Gunning, S. Penzes, H. S. Sandhu, and O. P. Stransz, *J. Amer. Chem. Soc.*, **91**, 7684 (1969), and references cited therein.
23. W. J. Balfour and A. E. Douglas, *Can. J. Phys.*, **48**, 901 (1970).
24. Y. Tanaka and K. Yoshino, *J. Chem. Phys.*, **53**, 2012 (1970).
25. (a) Y. Tanaka and K. Yoshino, *J. Chem. Phys.*, **57**, 2964 (1972); (b) Y. Tanaka, K. Yoshino, and D. E. Freeman, *J. Chem. Phys.*, **59**, 564 (1973); (c) G. C. Maitland, *Mol. Phys.*, **26**, 513 (1973); (d) R. J. Le Roy, M. L. Klein, and I. J. McGee, *Mol. Phys.*, **28**, 587 (1974).
26. Y. Tanaka, K. Yoshino, and D. E. Freeman, *J. Chem. Phys.*, **59**, 5160 (1973).
27. D. E. Freeman, K. Yoshino, and Y. Tanaka, *J. Chem. Phys.*, **61**, 4880 (1974).
28. (a) W. J. Balfour and R. F. Whitlock, *Chem. Commun.*, **1971**, 1231; (b) W. J. Balfour and R. F. Whitlock, *Can. J. Phys.*, **53**, 472 (1975).
29. R. E. Smalley, D. A. Auerbach, P. S. H. Fitch, D. H. Levy, and L. Wharton, *J. Chem. Phys.*, **66**, 3778 (1977).

30. R. Ahmad-Bitar, W. P. Lapatovich, D. E. Pritchard, and I. Renhorn, *Phys. Rev. Lett.*, **39**, 1657 (1977).
31. (a) For a general review of the PKR method, see: E. A. Mason and L. Monchick, *Adv. Chem. Phys.*, **12** (*Intermolecular Forces*, Wiley–Interscience, New York, 1967), p. 329. See also: (b) J. Tellinghuisen, *J. Mol. Spectr.*, **44**, 194 (1972); (c) A. S. Dickinson, *J. Mol. Spectr.*, **44**, 183 (1972).
32. W. M. Kosman and J. Hinze, *J. Mol. Spectr.*, **56**, 93 (1975).
33. E. A. Colbourn and A. E. Douglas, *J. Chem. Phys.*, **65**, 1741 (1976).
34. C. R. Vidal and H. Scheingraber, *J. Mol. Spectr.*, **65**, 46 (1977).
35. (a) W. C. Stwalley, *Chem. Phys. Lett.*, **7**, 600 (1970); (b) K. C. Li and W. C. Stwalley, *J. Chem. Phys.*, **59**, 4423 (1973).
36. (a) R. J. Le Roy, *J. Chem. Phys.*, **57**, 573 (1972); (b) R. J. Le Roy, in *Molecular Spectroscopy 1* (a Specialist Periodical Report of the Chemical Society (London), Eds. R. F. Barrow, D. A. Long, and D. J. Miller) 1973, Chap. 3, pp. 113–76.
37. (a) B. Tsapline and W. Kutzelnigg, *Chem. Phys. Lett.*, **23**, 173 (1973); (b) P. J. M. Geurts, P. E. S. Wormer, and A. van der Avoird, *Chem. Phys. Lett.*, **35**, 444 (1975); (c) F. Mulder, A. van der Avoird, and P. E. S. Wormer, *Mol. Phys.*, **37**, 159 (1979).
38. (a) C. F. Bender and H. F. Schaefer, III, *J. Chem. Phys.*, **57**, 217 (1972); (b) G. A. Gallup, *J. Chem. Phys.*, **66**, 2252 (1977); (c) G. A. Gallup, *Mol. Phys.*, **33**, 943 (1977).
39. (a) J. W. Riehl, J. L. Kinsey, J. S. Waugh, and J. H. Rugheimer, *J. Chem. Phys.*, **49**, 5276 (1968); (b) K. R. Foster and J. H. Rugheimer, *J. Chem. Phys.*, **56**, 2632 (1972); (c) J. W. Riehl, C. J. Fisher, J. D. Baloga, and J. L. Kinsey, *J. Chem. Phys.*, **58**, 4571 (1973).
40. (a) W. B. Nielsen and R. G. Gordon, *J. Chem. Phys.*, **58**, 4131 (1973); (b) *ibid*, **58**, 4149 (1973).
41. R. Shafer and R. G. Gordon, *J. Chem. Phys.*, **58**, 5422 (1973).
42. L. Zandee and J. Reuss, *Chem. Phys.*, **26**, 345 (1977).
43. H. J. Loesch, *Adv. Chem. Phys.*, **42** (1979, this volume), Chap. 9.
44. (a) G. A. Kuipers, *J. Mol. Spectr.*, **2**, 75 (1958); (b) D. F. Smith, *J. Mol. Spectr.*, **3**, 473 (1959); (c) W. F. Hergert, N. M. Gilar, R. J. Lovell, and A. H. Nielsen, *J. Opt. Soc. Amer.*, **50**, 1264 (1960); (d) J. L. Himes and T. A. Wiggins, *J. Mol. Spectr.*, **40**, 418 (1971).
45. (a) H. Vu and B. Vodar, *Compt. Rend.*, **248**, 2082 (1959); (b) G. C. Turrell, H. Vu, and B. Vodar, *J. Chem. Phys.*, **33**, 315 (1960); (c) B. Vodar, *Proc. Roy. Soc. (London)*, **A255**, 44 (1960).
46. (a) D. H. Rank, B. S. Rao, and T. A. Wiggins, *J. Chem. Phys.*, **37**, 2511 (1962); (b) D. H. Rank, P. Sitaram, W. A. Glickman, and T. A. Wiggins, *J. Chem. Phys.*, **39**, 2673 (1963); (c) D. H. Rank, W. A. Glickman, and T. A. Wiggins, *J. Chem. Phys.*, **43**, 1304 (1965).
47. A. Watanabe and H. L. Welsh, *Phys. Rev. Lett.*, **13**, 810 (1964).
48. A. Kudian, H. L. Welsh, and A. Watanabe, *J. Chem. Phys.*, **43**, 3397 (1965).
49. R. G. Gordon and J. K. Cashion, *J. Chem. Phys.*, **44**, 1190 (1966).
50. J. K. Cashion, *J. Chem. Phys.*, **45**, 1656 (1966).
51. (a) A. Kudian, H. L. Welsh, and A. Watanabe, *J. Chem. Phys.*, **47**, 1553 (1967); (b) A. K. Kudian and H. L. Welsh, *Can. J. Phys.*, **49**, 230 (1971).
52. (a) A. R. W. McKellar and H. L. Welsh, *J. Chem. Phys.*, **55**, 595 (1971); (b) A. R. W. McKellar and H. L. Welsh, *Can. J. Phys.*, **50**, 1458 (1972); (c) A. R. W. McKellar and H. L. Welsh, *Can. J. Phys.*, **52**, 1082 (1974); (d) A. R. W. McKellar, *J. Chem. Phys.*, **61**, 4636 (1974).

53. A. N. Petelin, *Opt. Spectr.*, **38**, 21 (1975).
54. R. J. Le Roy and J. Van Kranendonk, *J. Chem. Phys.*, **61**, 4570 (1974).
55. (a) R. J. Le Roy, J. S. Carley, and J. E. Grabenstetter, *Faraday Disc. Chem. Soc.*, **62**, 169 (1977); (b) J. S. Carley, *Faraday Disc. Chem. Soc.*, **62**, 303 (1977).
56. A. M. Dunker and R. G. Gordon, *J. Chem. Phys.*, **68**, 700 (1978).
57. (a) J. S. Carley, *Ph.D. Thesis*, University of Waterloo (1978); (b) J. S. Carley and R. J. Le Roy, unpublished work.
58. T. Dyke, B. J. Howard, and W. Klemperer, *J. Chem. Phys.*, **56**, 2442 (1972).
59. T. E. Gough, R. E. Miller, and G. Scoles, *J. Chem. Phys.*, **69**, 1588 (1978).
60. (a) S. E. Novick, P. Davies, S. J. Harris, and W. Klemperer, *J. Chem. Phys.*, **59**, 2273 (1973); (b) S. E. Novick, K. C. Janda, S. L. Holmgren, M. Waldman, and W. Klemperer, *J. Chem. Phys.*, **65**, 1114 (1976).
61. S. E. Novick, S. J. Harris, K. C. Janda, and W. Klemperer, *Can. J. Phys.*, **53**, 2007 (1975).
62. M. Larvor, J.-P. Houdeau, and C. Haeusler, *Can. J. Phys.*, **56**, 334 (1978).
63. E. W. Boom, D. Frenkel, and J. van der Elsken, *J. Chem. Phys.*, **66**, 1826 (1977).
64. L. Mannik, J. C. Stryland, and H. L. Welsh, *Can. J. Phys.*, **49**, 3056 (1971).
65. G. Henderson and G. E. Ewing, *J. Chem. Phys.*, **59**, 2280 (1973).
66. G. Henderson and G. E. Ewing, *Mol. Phys.*, **27**, 903 (1974).
67. C. A. Long and G. E. Ewing, *J. Chem. Phys.*, **58**, 4824 (1973).
68. C. A. Long, G. Henderson, and G. E. Ewing, *Chem. Phys.*, **2**, 485 (1973).
69. (a) W. H. Smith, G. Stella, and J. Gelfand, Paper RN13 at the *31st Symposium on Molecular Spectroscopy*, Columbus, Ohio (1976); (b) S. Bragg and W. H. Smith, Paper WG3 at the *33rd Symposium on Molecular Spectroscopy*, Columbus, Ohio (1978).
70. N. Brigot, S. Odiot, S. H. Walmsley, and J. L. Whitten, *Chem. Phys. Lett.*, **49**, 157 (1977).
71. (a) N. F. Ramsey, *Molecular Beams*, Clarendon Press, Oxford, 1956; (b) P. Kusch and V. W. Hughes, *Handbuch der Physik* (Eds. Flügge), Vol. 37, Springer-Verlag, Berlin, 1959, p. 1.
72. S. J. Harris, S. E. Novick, and W. Klemperer, *J. Chem. Phys.*, **60**, 3208 (1974).
73. S. J. Harris, S. E. Novick, W. Klemperer, and W. E. Falconer, *J. Chem. Phys.*, **61**, 193 (1974).
74. S. J. Harris, K. C. Janda, S. E. Novick, and W. Klemperer, *J. Chem. Phys.*, **63**, 881 (1975).
75. S. E. Novick, K. C. Janda, and W. Klemperer, *J. Chem. Phys.*, **65**, 5115 (1976).
76. K. C. Janda, J. M. Steed, S. E. Novick, and W. Klemperer, *J. Chem. Phys.*, **67**, 5162 (1977).
77. (a) T. R. Dyke and J. S. Muenter, *J. Chem. Phys.*, **57**, 5011 (1972); (b) *ibid.*, **60**, 2929 (1974); (c) T. R. Dyke, K. M. Mack, and J. S. Muenter, *J. Chem. Phys.*, **66**, 498 (1977).
78. S. L. Holmgren, M. Waldman, and W. Klemperer, *J. Chem. Phys.*, **69**, 1661 (1978).
79. J. Verberne and J. Reuss, *Chem. Phys.*, 1979, to be published.
80. M. S. Kim, R. E. Smalley, L. Wharton, and D. H. Levy, *J. Chem. Phys.*, **65**, 1216 (1976).
81. R. E. Smalley, L. Wharton, and D. H. Levy, *J. Chem. Phys.*, **66**, 2750 (1977).
82. R. E. Smalley, L. Wharton, and D. H. Levy, *J. Chem. Phys.*, **68**, 671 (1978).
83. G. Kubiak, P. S. H. Fitch, L. Wharton, and D. H. Levy, *J. Chem. Phys.*, **68**, 4477 (1978).
84. K. E. Johnson, L. Wharton, and D. H. Levy, *J. Chem. Phys.*, **69**, 2719 (1978).
85. (a) M. Cavallini, G. Gallinaro, and G. Scoles, *Z. Naturforsch.*, **A22**, 413 (1967);

(b) M. Cavallini, L. Meneghetti, G. Scoles, and M. Yealland, *Rev. Sci. Instr.*, **42**, 1759 (1971).
86. (a) T. E. Gough, R. E. Miller, and G. Scoles, *Appl. Phys. Lett.*, **30**, 338 (1977); (b) T. E. Gough, R. E. Miller, and G. Scoles, *J. Mol. Spectr.*, **72**, 124 (1978).
87. R. T. Pack, *J. Chem. Phys.*, **60**, 633 (1974).
88. (a) B. H. Choi, R. T. Poe, and K. T. Tang, *J. Chem. Phys.*, **69**, 411 (1978); (b) *ibid*, **69**, 422 (1978).
89. N. C. Blais and D. G. Truhlar, *J. Chem. Phys.*, **65**, 5335 (1976).
90. R. T. Pack, *Chem. Phys. Lett.*, **55**, 197 (1978).
91. W.-K. Liu, J. E. Grabenstetter, R. J. Le Roy, and F. R. McCourt, *J. Chem. Phys.*, **68**, 5028 (1978).
92. H. Kreek and R. J. Le Roy, *J. Chem. Phys.*, **63**, 338 (1975).
93. (a) J. W. Cooley, *Math. Comput.*, **15**, 363 (1961); (b) J. K. Cashion, *J. Chem. Phys.*, **39**, 1872 (1963); (c) R. J. Le Roy, University of Waterloo Chemical Physics Research Report CP-110 (1978).
94. S. Bratož and M. L. Martin, *J. Chem. Phys.*, **42**, 1051 (1965).
95. A. M. Dunker and R. G. Gordon, *J. Chem. Phys.*, **64**, 354 (1976).
96. J. E. Grabenstetter and R. J. Le Roy, unpublished work.
97. W. Klemperer, *Faraday Disc. Chem. Soc.*, **62**, 179 (1977).
98. M. Shapiro and G. G. Balint-Kurti, *J. Chem. Phys.*, **71**, 1461 (1979).
99. (a) B. R. Johnson and D. Secrest, *J. Math. Phys.*, **7**, 2187 (1966); (b) D. Secrest, *Meth. Comput. Phys.*, **10**, 243 (1971).
100. (a) W. A. Lester, Jr. and R. B. Bernstein, *J. Chem. Phys.*, **48**, 4896 (1968); (b) W. A. Lester, Jr., *Meth. Comput. Phys.*, **10**, 211 (1971).
101. (a) R. G. Gordon, *J. Chem. Phys.*, **51**, 14 (1969); (b) R. G. Gordon, *Meth. Comput. Phys.*, **10**, 81 (1971).
102. W. N. Sams and D. J. Kouri, *J. Chem. Phys.*, **51**, 4815 (1969).
103. A. C. Allison, *J. Comput. Phys.*, **6**, 378 (1970).
104. W. A. Lester, Jr., in Part A of *Dynamics of Molecular Collisions* (Ed. W. H. Miller), Plenum Press, New York, 1976, p. 1.
105. A. M. Arthurs and A. Dalgarno, *Proc. Roy. Soc. (London)*, **A256**, 540 (1960).
106. See e.g. M. E. Rose, *Elementary Theory of Angular Momentum*, Wiley, New York, 1957.
107. (a) G. Ewing, *Chem. Phys.*, **29**, 253 (1978); (b) G. Ewing, *J. Chem. Phys.*, **70**, in press.
108. I. C. Percival and M. J. Seaton, *Proc. Camb. Phil. Soc.*, **53**, 654 (1957).
109. R. B. Bernstein, A. Dalgarno, H. Massey, and I. C. Percival, *Proc. Roy. Soc. (London)*, **A274**, 427 (1963); corrigendum, see footnote 25 of ref. 56.
110. A. M. Dunker and R. G. Gordon, *J. Chem. Phys.*, **64**, 4984 (1976).
111. (a) D. A. Micha, *Chem. Phys. Lett.*, **1**, 139 (1967); (b) D. A. Micha, *Phys. Rev.*, **162**, 88 (1967).
112. R. D. Levine, *J. Chem. Phys.*, **49**, 51 (1968).
113. (a) R. D. Levine, B. R. Johnson, J. T. Muckerman, and R. B. Bernstein, *Chem. Phys. Lett.*, **1**, 517 (1968); (b) ibid, *J. Chem. Phys.*, **49**, 56 (1968).
114. D. J. Kouri, *J. Chem. Phys.*, **49**, 4481 (1968).
115. J. T. Muckerman, *J. Chem. Phys.*, **50**, 627 (1969).
116. M. von Seggern and J. P. Toennies, *Z. Physick*, **218**, 341 (1969).
117. J. T. Muckerman and R. B. Bernstein, *J. Chem. Phys.*, **52**, 606 (1970).
118. W. N. Sams and D. J. Kouri, *J. Chem. Phys.*, **52**, 2556 (1970).
119. M. Shapiro, *J. Chem. Phys.*, **56**, 2582 (1972).
120. R. J. Le Roy and W.-K. Liu, *J. Chem. Phys.*, **69**, 3622 (1978).

121. S. L. Holmgren, M. Waldman, and W. Klemperer, *J. Chem. Phys.*, **67**, 4414 (1977).
122. S. M. Tarr, H. Rabitz, D. E. Fitz, and R. A. Marcus, *J. Chem. Phys.*, **66**, 2854 (1977).
123. H. Rabitz in Part A of *Dynamics of Molecular Collisions* (Ed. W. H. Miller), Plenum Press, New York, 1976, p. 33.
124. J. O. Hirschfelder and W. J. Meath, *Adv. Chem. Phys.*, **12**, *Intermolecular Forces*, Wiley–Interscience, New York, 1967, p. 3.
125. R. J. Le Roy and R. B. Bernstein, *J. Chem. Phys.*, **54**, 5114 (1971).
126. A. R. W. McKellar, private communication (1979).
127. (a) J. E. Grabenstetter and R. J. Le Roy, *Chem. Phys.*, in press. (b) G. C. Corey, J. E. Grabenstetter, and R. J. Le Roy, unpublished work (1979).
128. (a) C. J. Ashton, *Part II B.A. Thesis*, Oxford University (1976); (b) M. S. Child, *Faraday Disc. Chem. Soc.*, **62**, 307 (1977).
129. (a) J. A. Beswick and J. Jortner, *J. Chem. Phys.*, **68**, 2277 (1978); (b) *ibid*, **69**, 512 (1978).
130. P. C. Hariharan and W. Kutzelnigg, *Progr. Report Lehrstuhl für Theoretische Chemie*, Ruhr-Universität Bochum, Federal Republic of Germany (1977).
131. K. T. Tang and J. P. Toennies, *J. Chem. Phys.*, **68**, 5501 (1978).
132. (a) A. Dalgarno and W. D. Davison, *Adv. Atom. Mol. Phys.*, **2**, 1 (1966); (b) A. Dalgarno, *Adv. Chem. Phys.*, **12**, *Intermolecular Forces*, Wiley–Interscience, New York, 1967, p. 143.
133. (a) G. A. Victor and A. Dalgarno, *J. Chem. Phys.*, **53**, 1316 (1970); (b) P. W. Langhoff, R. G. Gordon, and M. Karplus, *J. Chem. Phys.*, **55**, 2126 (1971); (c) G. Starkschall and R. G. Gordon, *J. Chem. Phys.*, **54**, 663 (1971); (d) J. S. Cohen and R. T. Pack, *J. Chem. Phys.*, **61**, 2372 (1974); (e) K. T. Tang, J. M. Norbeck, and P. R. Certain, *J. Chem. Phys.*, **64**, 3063 (1976).
134. (a) J. P. Toennies, W. Welz, and G.Wolf, *J. Chem. Phys.*, **71**, 614 (1979); (b) G.Wolf, *Dissertation*, Max-Planck-Institut für Strömungsforschung, Report 20 (1976).
135. A. Thakkar, *The Intramolecular Bond Length Dependence of the Anisotropic Dispersion Coefficients for Hydrogen Molecule–Noble Gas or Alkali Atom Intractions*, contributed paper at Sixth Canadian Symposium on Theoretical Chemistry, Fredericton, N. B., 1977.
136. K. M. Smith, A. M. Rulis, G. Scoles, R. A. Aziz, and V. Nain, *J. Chem. Phys.*, **67**, 152 (1977).
137. R. A. Buckingham and J. Corner, *Proc. Roy. Soc. (London)*, **A189**, 118 (1947).
138. J. Hepburn, G. Scoles, and R. Penco, *Chem. Phys. Lett.*, **36**, 451 (1975).
139. R. Ahlrichs, R. Penco, and G. Scoles, *Chem. Phys.*, **19**, 119 (1977).
140. A. M. Rulis, K. M. Smith, and G. Scoles, *Can. J. Phys.*, **56**, 753 (1978).
141. (a) W.-K. Liu and F. R. McCourt, *Chem. Phys.*, **19**, 137 (1977); (b) **27**, 281 (1978); (c) **37**, 75 (1979).
142. R. Helbing, W. Gaide, and H. Pauly, *Z. Physik.*, **208**, 215 (1968).
143. J. Vitko, Jr. and C. F. Coll, III, *J. Chem. Phys.*, **69**, 2590 (1978).
144. R. J. Le Roy, unpublished work, 1972.
145. (a) A. D. May, V. Degen, J. C. Stryland, and H. L. Welsh, *Can. J. Phys.*, **39**, 1769 (1961); (b) V. G. Cooper, A. D. May, and B. K. Cupta, *Can. J. Phys.*, **48**, 725 (1970).
146. J. M. Farrar and Y. T. Lee, *Chem. Phys. Lett.*, **26**, 428 (1974).
147. A. W. Miziolek and G. C. Pimentel, *J. Chem. Phys.*, **65**, 4462 (1976).
148. J. G. Kircz, G. J. Q. van der Peyl, J. van der Elsken, and D. Frenkel, *J. Chem. Phys.*, **69**, 4606 (1978).

Potential Energy Surfaces
Edited by K. P. Lawley

SCATTERING OF NON-SPHERICAL MOLECULES

HANSJÜRGEN LOESCH

Fakultät für Physik, der Universität Bielefeld, 4800 Bielefeld, West Germany

CONTENTS

I. INTRODUCTION

Molecular beam experiments on elastic atom–atom scattering have proven to be a universally applicable tool for the determination of spherically symmetric (isotropic) interaction potentials and there is currently considerable interest in establishing this successful technique as a standard method for the deduction of the more complicated non-central potentials of atom–molecule or even molecule–molecule systems as well. The first step beyond the familiar scattering by isotropic potentials (for recent reviews see Buck, 1975; Pauly, 1975) towards increasing complexity of the collision process is the scattering of an atom by a linear, rigid-rotor molecule. The replacement of one of the atoms by such a molecule leads to a departure of the interaction potential from sphericity (anisotropic potential) causing three fundamentally new scattering phenomena. First, the rotational state of the molecule may change during the course of the encounter with respect to the magnitude of the angular momentum as well as to its orientation in space. Second, the scattering becomes dependent on the initial relative orientation between the relative velocity of the colliding particles and the intermolecular axis. Third, the well-known structures of the isotropic scattering cross sections, like the glory undulations, the rainbow peaks, and the rapid (diffraction) oscillations, appear more or less quenched or vanish completely, even in the absence of inelastic events. All three phenomena reflect sensitively the anisotropy of the interaction potential. They represent the basic experimental pieces of information needed for the deduction of anisotropic potential properties and comprise the central subject covered by this review article.

In addition to the rotational degree of freedom molecules also possess vibrational ones. However, within the energy range usually covered by neutral–neutral experiments (thermal to ≈ 1 eV) the collisionally induced change of vibrational states—if energetically possible at all—plays only a minor role (Weitz and Flynn, 1974; Amme, 1975; Suzukawa *et al.*, 1978). The most dominant processes by far are the elastic and rotationally inelastic events; this permits the great majority of molecular scattering experiments to be interpreted in terms of rigid-rotor molecules. Within this framework the most general collision process between an atom or a molecule, considered to be structureless, A and a linear rigid rotor BC (electronically excited states are not considered) is given by

$$\mathrm{A} + \mathrm{BC}(j, m) + E \rightarrow \mathrm{A} + \mathrm{BC}(j', m') + E' \qquad (1)$$

where the j, m denote the rotational quantum numbers and E the relative kinetic collision energy; the primed quantities refer to the final state. This process is fully described by the differential scattering cross section:

$$I(j', m'/j, m; \vartheta, \phi, E) \tag{2}$$

where ϑ, ϕ are the deflection angles in the centre-of-mass coordinate system. These highly detailed quantities represent the most complete information on the molecular scattering process, and their knowledge is in principle the basis for the determination of the intermolecular forces, the ultimate goal of a scattering experiment.

The main emphasis of this review is on the experimental determination of the potential anisotropy from the scattering of neutrals in their electronic ground state. (For recent progress in ion–molecule scattering see the review of Faubel and Toennies, 1977.) All literature is omitted in which molecular scattering data are interpreted in terms of isotropic interactions. For a review on this subject see Buck (1975).

The first beam experiments with the aim of gaining some insight into the angular dependence of intermolecular forces date back to the years 1964 and 1965; Bennewitz *et al.* (1964) exploited the m-dependence of the integral cross section while Toennies (1965) used the small-angle scattering of molecules in completely specified initial (j, m) and final (j', m') states. In 1968 (Olson and Bernstein, 1968; Cross, 1968a) and 1970 (Cross) the quenching of structures in the total integral and differential cross sections, respectively, was recognized as a useful means of deducing potential anisotropies. This method has recently received considerable theoretical (Buck and Khare, 1977; Klassen *et al.*, 1978; Pack, 1978) and experimental (Keil *et al.*, 1978; Valbusa, 1978; Buck *et al.*, 1978c; Reed and Wharton, 1977) attention. Integral cross sections unresolved with respect to j and j' for large energy transfers have been demonstrated to be useful for the deduction of the anisotropy of the repulsive interaction (Loesch, 1976). Schepper *et al.* (1978, 1979) have recently discovered structures in the unresolved differential energy transfer cross section which directly provide the repulsive anisotropy. The first state-resolved differential cross sections measured over a wide range of deflection angles were reported in 1977 (Buck *et al.*, 1977; Gentry and Giese; 1977). Their rich content of detailed information on the potential anisotropy has been exploited by Huisken (1978) and Buck *et al.* (1979).

The scope of this review will be mainly on the results of the last five years or so up to the summer of 1978. However, for the sake of completeness, some of the older work is also briefly discussed. The article begins with some short remarks on theoretical methods for the calculation of cross sections. They represent the basis for trial and error calculations which are commonly used to deduce anisotropic potentials from scattering data (Section II). In Section III important experimental techniques are surveyed. In Section IV the correlations between particular cross-section features and potential properties known thus far are compiled and those domains of the potential which are most sensitively probed by these data are indicated. In Section V the experimental results are surveyed, and experimentally determined anisotropy parameters or other conclusions about the angular dependence of the potential are compiled in Table III.

II. THEORETICAL REMARKS

A. Potential hypersurfaces

The quantum treatment of the molecular collision problem leads to a many-body Schrödinger equation describing the simultaneous motion of the electrons and nuclei. To solve this complicated equation one fundamentally assumes the validity of the Born–Oppenheimer approximation, i.e. the separability of the electronic and nuclear motions. The electronic equation can then be solved with 'clamped' nuclei; the resulting energy eigenvalues, which depend only on the relative nuclear positions, act then as multidimensional potential hypersurfaces (PHS) for the nuclear motion. The *ab initio* solution of the complete scattering problem then proceeds via two steps: (a) the solution of the electronic Schrödinger equation to determine the PHS and (b) the calculation of the nuclear dynamics on these PHS.

Systems of particles with closed electronic shells usually have a ground-state PHS energetically far separated from the excited-state PHS. In this case the Born–Oppenheimer approximation is an excellent approach. If, however, open-shell particles are involved, for example, one collision partner has a non-vanishing electronic orbital angular momentum, the situation changes, and energetically closely lying adjacent hypersurfaces may appear. Some of them may strongly interact with each other leading to local breakdowns of the Born–Oppenheimer approximation. The effect of interacting PHS on energy-transferring collisions has been studied by Nikitin (1975). Another aspect of open-shell systems is their chemical reactivity. It is well known (Wilkins, 1975; Leone *et al.*, 1975; Quigley and Wolga, 1975; Macdonald *et al.*, 1975; for a review see Smith, (1976) that potentially reactive PHS strongly enhance vibrationally inelastic processes, but little is known about their influence on rotationally inelastic encounters.

We restrict our brief discussion to ground-state PHS of systems consisting of a closed-shell atom in the ground state (^1S) interacting with a linear molecule in the $^1\Sigma$ state. These are the cases on which most theoretical scattering studies are based. The potential energy is then a function of the three relative position coordinates r, R, γ, which are shown in Fig. 1. The PHS consists of an isotropic (γ-independent) and an anisotropic (γ-dependent) part. This becomes evident if the function is expanded in Legendre poly-

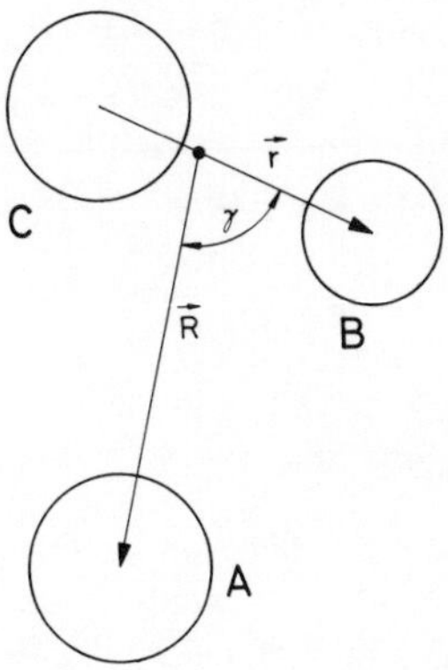

Fig. 1. Relative position coordinates of the atoms in an A + BC collision.

nominals:

$$V(r, R, \gamma) = \sum_{n=0}^{\infty} V_n(r, R) P_n(\cos \gamma). \tag{3}$$

The term with $n = 0$ represents the isotropic and the residual sum the anisotropic component. The spherical average of the latter vanishes.

In the majority of theoretical scattering studies at thermal collision energies the molecule is treated as a rigid rotor. This means that the vibrational degree of freedom is considered to have no effect on the scattering. This assumption is supported by considerable experimental and theoretical evidence even at collision energies substantially beyond the excitation threshold (Weitz and Flynn, 1974; Amme, 1975; Mariella *et al.*, 1974; Bennewitz and Buess, 1978; Suzukawa *et al.*, 1978). In the following the rigid-rotor model is adopted, and the r-dependence of the interaction energy is dropped.

In Fig. 2 a typical PHS for an atom–rigid-rotor system is represented in a contour map as a function of R (length of the radius vector) and γ (polar angle). It consists of three distinct regions:

(a) the potential well region in the vicinity of the dash-dotted line which connects the local potential energy minima $\varepsilon(\gamma)$ at the minimum separation $R_m(\gamma)$;
(b) the long-range (van der Waals) attractive region ($R > 1.5\, R_m(\gamma)$);
(c) the short-range repulsive domain ($R < R_m(\gamma)$).

The interaction leading to the attractive branch (b) is dominated by the correlation energy between the atomic and molecular electrons. It is usually calculated via perturbation theory, and the following expressions are

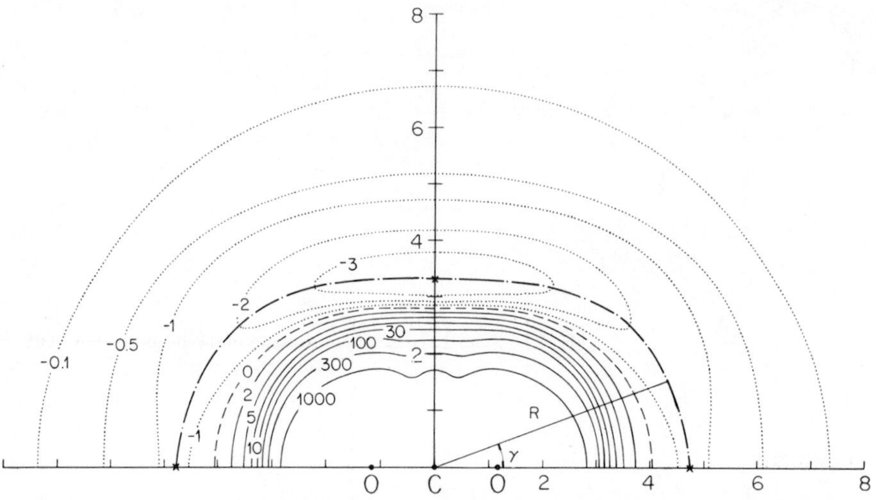

Fig. 2. Contour map of a rigid-rotor PHS for He + CO_2. The distances and energies are given in Å and meV, respectively. The lines connect points of equal potential energy with $V > 0$ (solid lines), $V = 0$ (dashed line), and $V < 0$ (dotted lines). The dash-dotted line connects the minima along fixed γ, $\varepsilon(\gamma)$, located at $R = R_m(\gamma)$. The absolute minimum of $\varepsilon(\gamma)$ is -3.9 meV occurring at the T-shaped configuration ($\gamma = \pi/2$) with $R_m = 3.31$ Å. A saddle point appears at $\gamma = 0$ and $\gamma = \pi$ (for symmetry reasons) with $\varepsilon = -1.1$ meV abd $R_m = 4.77$ Å. (Reproduced by permission of North Holland Publ. Co. from Keil *et al.*, 1978.)

obtained:

$$\underset{(R\to\infty)}{V(R,\gamma)} = -C^{(6)}R^{-6} - C^{(7)}R^{-7} - C^{(8)}R^{-8} - \dots \tag{4}$$

with

$$\begin{aligned} C^{(6)} &= C_0^{(6)} + C_2^{(6)}P_2 \\ C^{(7)} &= C_1^{(7)}P_1 + C_3^{(7)}P_3 \\ C^{(8)} &= C_0^{(8)} + C_2^{(8)}P_2 + C_4^{(8)}P_4 \end{aligned} \tag{5}$$

where each of the coefficients $C_i^{(n)}$ represents a sum of dispersion and induction contributions.

For reviews on this subject see Buckingham (1967) and Margenau and Kestner (1969); for recent progress in calculating the various constants see Langhoff *et al.* (1971), Cohen and Pack (1974), Tang (1976), Parker and Pack (1976), Nielson *et al.* (1976), and Meyer (1976).

The short-range repulsive region (c) originates mainly from electrostatic interaction and electron exchange, both of which are well described by the standard sell-consistent field (SCF) approximation.

The theoretically most complicated region is located around the well (a). Here, all interactions mentioned above play an equivalent role. However, perturbation theories fail in the region in which the electron clouds overlap substantially (Margenau and Kestner, 1969), and the SCF approximation is not applicable since it does not allow for correlation effects. For an *ab initio* determination much more involved quantum mechanical techniques, like the configuration interaction (CI) approach, must be used. For a detailed discussion of *ab initio* methods the reader is referred to the article by Tully in this book and to reviews by Hinze (1974) and Balint-Kurti (1975a).

The large computational effort required for CI calculations has limited the applicability of the method to systems with a total number of electrons up to approximately 20 (for a list of recent studies see Faubel and Toennies, 1977). In case of heavier systems one has to rely on theoretical models. Two of them have recently been used for the interpretation of molecular scattering experiments.

One of these is the electron gas (EG) model developed by Gaydaenko and Nikulin (1970), Nikulin (1971), Gordon and Kim (1972), and modified by Rae (1973), Cohen and Pack (1974), and Kim and Gordon (1974b). The surprisingly good results given by this approximation for closed-shell atom–atom (Gordon and Kim, 1972; Schneider, 1973; Kim and Gordon, 1974a, b; Cohen and Pack, 1974) interactions have encouraged its application to atom–molecule systems. At present a variety of such PHS exist: $Ar + N_2$ (Kim, 1978), Ar + HCl (Green, 1974), He + HCN (Green and Thaddeus, 1974), He + CO (Green and Thaddeus, 1976), He + HCl (Green and Monchick, 1975), $He + H_2CO$ (Green *et al.*, 1975), He, $Ar + CO_2$ (Parker *et al.*, 1976), He, Ar + HF (Detrich and Conn, 1976), Ar + NO (open shell) (Nielson *et al.*, 1977), Ar + CO (Parker and Pack, 1978a). The other model is a very promising new method to calculate PHS which utilizes *ab initio* SCF results to describe the repulsive region and an expansion similar to equation (4) to approximate the attractive part of the surface (Tang and Toennies, 1977, 1978). Simple corrections to equation (4) provide a realistic behaviour of the model within the intermediate region. This technique has been applied so far to a series of atom–atom systems as well as to He, $Ar + H_2$ (Tang and Toennies, 1977, 1978).

The reliability of any PHS based on approximate methods can only be tested by its comparison with either an 'exact' *ab initio* surface, which is presently not possible for heavier systems, or by comparing the potential-based predictions of observable quantities with corresponding experimental data. As demonstrated in Section IV the various cross sections obtained from beam scattering experiments with anisotropic molecules will be well qualified for this purpose. (Systems for which such a comparison has been performed are given in Table III marked by a T in the column 'data analysis'.)

An alternative to the above comparison technique is to use the scattering data directly for the deduction of an empirical PHS. Since in the case of anisotropic potential scattering inversion methods are not known to date, the determination of such a PHS must be accomplished by trial and error calculations which work as follows (cf. Fig. 3). First an adequate empirical PHS incorporating free parameters is constructed (several frequently used analytical model potentials are compiled in Table II); then a set of parameters is assumed and a suitable scattering theory is applied to calculate differential (or integral) cross sections. The latter may be compared to experimental cross sections which emerge from direct detector signals via a data reduction procedure. However, the most frequently used technique is, to subject the calculated cross sections to further computations which simulate as perfectly as possible all apparative imperfections and compare the results directly to the detector signals. In either case, if no satisfactory agreement has been achieved, a new set of parameters is chosen and the loop starts again until an acceptable fit is obtained. The feasibility of such a technique within a given economic frame depends vitally on the computa-

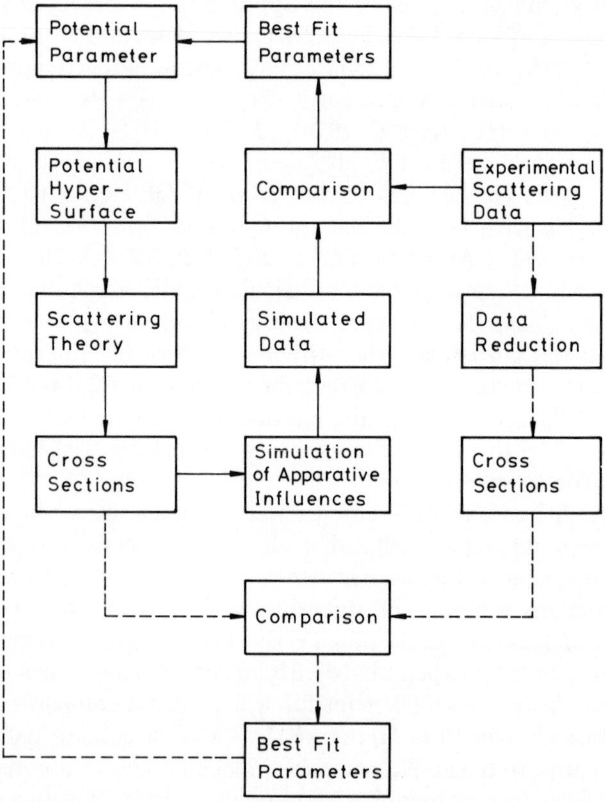

Fig. 3. Schematic flow diagram of a trial and error procedure.

tion time of one cycle. In Section II.B several adequately 'fast' approximate scattering theories are discussed.

These trial and error techniques are standard for the deduction of empirical spherically symmetric potentials from elastic scattering data (see the reviews on elastic scattering in Section IV. A) but have also been recently applied to anisotropic PHS (relevant systems are given in Table III marked by F in the column 'data analysis'). The results obtained are reliable only for the PHS area sensitively probed by the particular experiment; extrapolations beyond this domain may deteriorate rapidly. More universally valid PHS are achieved when various experimental data sensitive to different regions are incorporated in the trial and error procedure. For example, Neilsen and Gordon (1973) report a PHS for Ar + HCl constructed on the basis of infrared linewidths, vibration–rotation line shifts and linewidths, and proton magnetic relaxation times. The inelastic scattering data of Ding and Polanyi (1975) are well in accord with predictions based on this PHS (Polanyi and Sathyamurthy, 1978), and with a small correction of R_m agreement with the rainbow scattering data of Farrar and Lee (1974) could also be established (Dunker and Gordon, 1976). An analysis of the molecular beam electric resonance spectroscopy experiments of Novick *et al.* (1973) on the ArHCl molecule, however, suggested further corrections. Recently, Holmgren *et al.* (1978) reported a PHS which rationalized both, the bound state and 'elastic' scattering properties of ArHCl. Additional constraint on this PHS can be expected to arise from the results of inelastic scattering experiments.

B. Scattering Theory

The conclusive step in treating the scattering problem comprises the solution of the quantum equation of motion for the nuclei and the determination of cross sections. Two exact formal solutions have been presented. The one dates back to the work of Arthurs and Dalgarno (1960), who described the problem in a space-fixed coordinate system, the other to Curtiss and coworkers (Curtiss and Adler, 1952; Curtiss, 1953; see also Pack, 1974; McGuire, 1975), who used a body-fixed frame. In either case the theory leads for each partial wave $J(\mathbf{J} = \mathbf{l} + \mathbf{j}$ the total angular momentum given by the orbital and molecular angular momenta $\mathbf{l}$ and $\mathbf{j}$ respectively) to an infinite set of coupled differential equations which, of course, must be truncated to finite size to permit a numerical solution. Convergence of the resulting cross sections with respect to an increasing number of involved rotational states j is usually achieved with the incorporation of only the energetically accessible j-states ($j \le j_{max}$). The reduced finite set is denoted as close-coupling equations (CC). The converged cross sections are considered as exact quantum results and are frequently used as standards against which more approximate methods are tested (see below).

The main drawback to the CC method arises from the fact that usually the number N of coupled wave functions (N is roughly given by $(j_{max}+1)^2$) for each partial wave J is large, thus requiring computation times of the order of days with modern computers. As the computation time is roughly proportional to N^n, with $2 \leq n \leq 3$ depending on the number of equations and the algorithm employed, only systems involving low collision energies and light molecules are presently qualified for application of this method. For a detailed discussion of computation times and of recent CC results see Faubel and Toennies (1977).

Because of the prohibitively large computation times required for solving the CC equations a variety of approximate methods have been developed. Some of them start directly from the CC set and reduction of computation time is accomplished by partial or even complete decoupling of the equations. Others are hybrids of quantum and classical mechanics and range from classical path approximations, i.e. incorporate classically treated relative motions, to the full classical treatment of the collision process. Approximate methods are discussed in detail in recent articles by Secrest (1973), Gordon (1973), Micha (1974), Balint-Kurti (1975), Miller (1974, 1975), Lester (1975), Rabitz (1976), and Toennies (1976). We therefore may restrict ourselves to touching only briefly some frequently used methods. These may be listed as follows:

The coupled-state method (CS) developed by McGuire (1973, 1975) and McGuire and Kouri (1974). Partial decoupling of the CC set in the body-fixed frame is achieved by neglecting the intermultiplet coupling (neglection of the Coriolis force) and replacing the orbital angular momentum l by a constant value $\bar{l}$, generally the total angular momentum J, but other choices are possible (Kouri and McGuire, 1974). The CS method is applicable to a large class of systems characterized by dominant short-range and weak long-range anisotropies (McGuire, 1976; Kouri *et al.*, 1976), as for example: $Ar + N_2$ (McGuire, 1977), $He + Na_2$ (Berbenni and McGuire, 1977), $Li^+ + H_2$ (Kouri and McGuire, 1974), $He + H_2$ (McGuire 1975).

The effective potential approximation (EPA) introduced by Rabitz (1972). In this method an effective Hamiltonian with an effective PHS is defined, the latter in such a way that the potential matrix elements which couple the wave functions of the CC set are solely dependent on j and no longer on l. This definition incorporates a summation over all projection quantum numbers m and consequently no m-dependent cross sections may be calculated. The resulting set of coupled equations contains only $(j_{max}+1)$ equations. The accuracy of the EPA is dependent on the anisotropy of the PHS, yielding moderately good agreement with CC calculations for systems with small anisotropy, such as He, $H_2 + H_2$ (Zarur and Rabitz, 1973, 1974; Green, 1975a; Alexander and McGuire, 1976), and $H + H_2$ (Chu and

Dalgarno 1975a), and poor agreement for strongly anisotropic collision partners like He + HCN (Green, 1975), H + CO and He + N_2 (Chu and Dalgarno, 1975, 1975b) at collision energies approaching threshold. (For further details see Rabitz, 1975, 1976.)

The l-dominant approximation (LDA) (DePristo and Alexander, 1975a) provides a reduction of the number of CC equations by a factor of two since all wave functions with $l > J$ are neglected. This method is valid for collisions in which large impact parameters and a strongly anisotropic long-range attraction prevail (DePristo and Alexander, 1975a). A further reduction of the dimensionality is obtained if the CC equations are block diagonalized into sets with equal alignment between J and j characterized by the expression $\lambda = l + j - J$, as the preferred coupling arises from $\lambda = \lambda'$ potential matrix elements. This technique is called decoupled LDA (DLDA) (DePristo and Alexander, 1975b). The number of sets of coupled equations and their dimensionality is here equivalent to the CS approximation. Comparisons between CC and LDA (Ar + N_2 and $Li^+ + H_2$: DePristo and Alexander, 1975a and 1975c) and DLDA results (H^+ + CN and HF + HF: DePristo and Alexander, 1976, 1977) exhibit good agreement within the range of validity (large J).

The p-helicity decoupling approximation (PHDA) (Tamir and Schapiro, 1975) utilizes the p-helicity formalism (Jakob and Wick, 1959; Klar, 1969, 1971, 1973) to generate a new, exact set of coupled equations. These are simplified by neglecting all terms that couple different helicity states (states of different m_j-projection with respect to the relative momentum vector $\boldsymbol{p}$). Excellent agreement between CC and PHDA results has been demonstrated for Ar + N_2 at thermal energies (Tamir and Schapiro, 1976; Schapiro and Tamir, 1976). Computation times comparable to classical trajectory calculations are reported (Schapiro and Tamir, 1976).

The distorted wave approximation (DWA) makes use of the assumption that, in the case of a weakly anisotropic potential, most coupling terms in the CC set may be neglected. It is essentially a first-order perturbation method with respect to the non-diagonal potential matrix elements while the diagonal terms are fully accounted for (Takayanagi, 1965). For strongly anisotropic molecules, i.e. all except H_2, the DWA provides unreliable, generally too large cross sections (Jacobs and Reuss, 1977). For hydrogenic systems the agreement between CC and DWA results is good (Reuss and Stolte, 1969). The method has been extensively applied to the calculation of orientation-dependent integral cross sections (Reuss, 1975).

The infinite-order sudden approximation (IOSA) has recently received considerable attention (Parker and Pack, 1978; Pack, 1978; Buck *et al.*, 1978c; Buck and Khare, 1977). It was first suggested by Takayanagi (1963) and was then developed by Pack and coworkers (Tsien and Pack, 1970, 1971; Pack, 1972; Tsien *et al.*, 1973). Secrest (1975) and Hunter (1975) generalized the IOSA to arbitrary PHS and showed its equivalence to an early approximation of Curtiss (1968). The IOSA makes use of the CS assumption of constant orbital angular momentum $l = \bar{l}$ as well as of the additional approximation of degenerate molecular rotational states. These simplifications lead eventually to a complete decoupling of the CC set and reduce the inelastic problem to the elastic scattering of an atom by an 'isotropic' potential $V_\gamma(R) = V(R, \gamma)$, where γ acts as a parameter. Inelastic cross sections are evaluated essentially by calculating matrix elements of the 'elastic' scattering amplitude—which must be known as a function of γ—for the transition between the desired initial and final rotational states (Parker and Pack, 1978). The constant angle corresponds classically to the neglect of the angular motion of the (reduced) particle as it travels through the PHS (Dickinson and Richards, 1978). The assumption of degenerate j-states corresponds classically to a rotor with space-fixed orientation and hence implies the sudden collision limit which requires collision durations to be short compared to the rotational period and amounts of energy transfered, ΔE, small compared to the collision energy E. A further restriction of the validity of the IOSA arises from the fact that the centrifugal potential term may be chosen to be correct either for the initial or the final state but not both. This sensitively affects the weak coupling regime where l is large, while the potential dominant coupling regime is hardly influenced. Therefore, one may expect good results for molecules with strongly anisotropic short-range repulsion and weekly anisotropic long-range attraction. Comparisons with CC calculations confirm this general behaviour (Tsien *et al.*, 1973; Pack, 1975, 1977; Pfeffer and Secrest, 1977; Bowman and Leasure, 1977). Besides the surprisingly good accuracy, the most striking advantage of the IOSA is the short computation time required for inelastic cross sections, which is of the order of that required for elastic scattering calculations. These properties make the IOSA well qualified to be used extensively in trial and error calculations for the interpretation of molecular scattering data. For a review on this subject see Parker and Pack (1978).

The classical path approximations comprise a hierarchy of widely used methods which are based on the assumptions that (a) the relative motion of the colliding particles can be described by classical mechanics, and (b) the required trajectories may be determined by either ignoring the interaction potential completely (straight trajectories) or by including only the isotropic component $V_0(R)$ of the PHS (curved trajectories) while (c) their internal motion may be treated quantum mechanically. Simplification (b) decouples the relative motion from the internal degrees of freedom and is therefore valid only as long as the trajectory is not significantly affected by the inelastic processes, i.e. $\Delta E/E \ll 1$. The relation between time and space along a trajectory provides a time-dependent potential energy and leads eventually to a time-dependent Schrödinger equation for the internal motion of the molecule. A solution of this equation for the determination of cross sections is usually attempted via perturbation series for the time evolution operator (and the S-matrix) up to second order such as the time-dependent Born series (see for example: Cross and Gordon, 1966; Rabitz and Gordon, 1970a, b; Saha *et al.*, 1973; Verter and Rabitz, 1973, 1974; Cross, 1977) as well as the most frequently used exponential series of Magnus (1954) (see also Pechukas and Light, 1966) using straight and curved trajectories (see for example Cross, 1967, 1968, 1968a, 1969, 1970; Levine, 1968; Saha *et al.*, 1974; Saha and Guka, 1975; Bhattacharyya *et al.*, 1977). An important role is played by the sudden limit of the above approximations which has been introduced to rotationally inelastic scattering by Kramer and Bernstein (1964) (see also Bernstein and Kramer, 1966; Fenstermaker and Bernstein, 1967; Fenstermaker *et al.*, 1969). In this limit the use of the first-order term of the Magnus series is usually referred to as the time-dependent sudden approximation (SA) for curved and as the high-energy approximation (HEA) for straight trajectories. Both the SA and HEA have been extensively used for the analysis of differential, total differential, and total integral cross sections (see Sections IV, B and C). In the sudden limit one essentially calculates scattering amplitudes for the collision of a (reduced) projectile with a *space-fixed* (clamped) molecule, quite in contrast to the IOSA where γ is considered to be fixed. The inelastic cross sections are eventually deduced from matrix elements of these amplitudes for the desired rotational transitions. The existence of a tight relation between the IOSA in first order with the first-order Magnus approximation has been revealed by Pack (1977). Comparisons of integral cross sections obtained from the latter method with CC results (Tsien *et al.*, 1973; Pack, 1975) indicate good agreement for small energy transfers. A recent comparison between total differential cross sections obtained for various SA showed that the use of straight trajectories leads to appreciable errors (Buck and Khare, 1977). For more details and a more extensive list of references see the review article of Balint-Kurti (1975).

The classical S-matrix theory, which has been developed by Miller (1970, 1971), makes a quite different use of classical trajectories. For recent reviews see Miller (1974, 1975). In contrast to the classical path theories discussed above, all degrees of freedom of the system are treated classically. The only quantum mechanical element contained in this theory is the linear superposition principle. First, all trajectories leading from a specific initial to a specific final state are computed. (In classically forbidden regions they become complex valued.) Then the action integrals are evaluated along each trajectory to obtain the proper phases of the S-matrix elements, which are eventually superimposed to give the scattering amplitude for the process under consideration. With some simplification this semiclassical method has been applied to full three-dimensional scattering problems (Miller, 1971; Doll and Miller, 1972; Miller and Raczkowski, 1973; Raczkowski and

Miller, 1974; Augustin and Miller, 1974). Comparisons with CC calculations show good agreement (Miller, 1975). It has been shown by Dickinson and Richards (1976) that if classical perturbation theory is valid for calculation of the action change of the rotor, the classical S-matrix theory becomes identical to the strong coupling correspondence principle approximation (Dickinson and Richards, 1974, 1976, 1977).

Classical mechanics as a method for calculating cross sections for molecular collision processes via classical trajectories (CT) was first developed for the study of rearrangement processes (Bunker, 1971) but has become a standard method in the field of non-reactive scattering too. Classical mechanics has three compelling advantages: (1) the computation times involved are among the shortest of all approximate methods, (2) the PHS is taken into account exactly, and (3) the dynamics is not exact but, due to the correspondence principle, becomes more precise as the system behaves more and more classically. This means roughly: one may expect classical mechanics to be a good approximation if (1) the De Broglie wavelength of the colliding system is small compared to the relevant dimensions of the PHS, (2) the quantized amount of energy transfered ΔE is much less than the total energy E, and (3) interference effects are washed out due to the lack of adequate experimental resolving power. Several more precisely formulated conditions are given by Barg *et al.* (1976). These conditions are usually met in scattering experiments at thermal collision energies, with the possible exception of systems with H or H_2 as scattering partner. A comparison of differential cross sections calculated with the CT and the converged CC approximations for the rather unfavourable case of $Li^+ + H_2$ at $E = 0.6$ and 1.2 eV (Barg *et al.*, 1976) show excellent agreement except for the rainbow region, which is dominated by quantum interference effects.

Further comparisons between CT calculations and quantum approximations for integral cross sections ($Li^+ + H_2$: LaBudde and Bernstein, 1973; Barg *et al.*, 1976; He + CO, H_2: Augustin and Miller, 1974a; $Li^+ + CO$: Thomas *et al.*, 1978; Ar + N_2 : Thomas, 1977; Pack, 1975; H + CO, He + HCN: Brumer, 1974; Ar + HCl: Polanyi *et al.*, 1977; He + CO, HCl, Ar + HCl, H_2 + CS, COS: Chapman and Green, 1977) exhibit convincing agreement. Exceptions are systems in which the near-inversion symmetry of the PHS quantum mechanically suppresses odd Δj and favours even Δj transitions (He + CO: Augustin and Miller, 1974a; H_2 + CS, COS: Chapman and Green, 1977). In the case of homonuclear molecules, in which odd Δj transitions are forbidden, this problem is avoided by an adequate boxing procedure (LaBudde and Bernstein, 1973).

III. EXPERIMENTAL METHODS

In this section we give a survey of basic experimental concepts and some important apparatus components which have been employed to investigate the scattering from molecules. Furthermore, the relations between the measured signals on the one hand and the fundamental cross sections on the other hand are provided. Both the concepts and the relations can most conveniently be discussed with the aid of an ideal molecular beam experiment which is defined below.

A. An Ideal Scattering Experiment

A schematic design of such an experiment is sketched in Fig. 4. A unidirectional beam of molecules BC with well-determined velocity ($\boldsymbol{v}_{BC}$) and internal state (j, m) intersects a unidirectional and velocity selected ($\boldsymbol{v}_A$) beam of atoms A. The molecules scattered into the angles Θ and Φ of the laboratory frame (LAB) are then analysed with respect to their final states (j', m') and detected. The quantization axis to which the m-states refer is determined by a week external orientation field. The number of molecules BC scattered per second into the solid angle $d\Omega$ subtended by the detector is given by:

$$d\dot{N}(j', m'; \Theta, \Phi, E) = n_A n_{BC}(j, m) v \, d\tau \times I(j', m'/j, m; \vartheta, \varphi, E) \frac{\partial(\vartheta, \varphi)}{\partial(\Theta, \Phi)} d\Omega \tag{6}$$

where E is the relative collision energy $E = 1/2 \mu v^2$, with the reduced mass $\mu = m_A m_{BC}/(m_A + m_{BC})$($m_A$ and m_{BC} are the masses of A and BC, respectively)

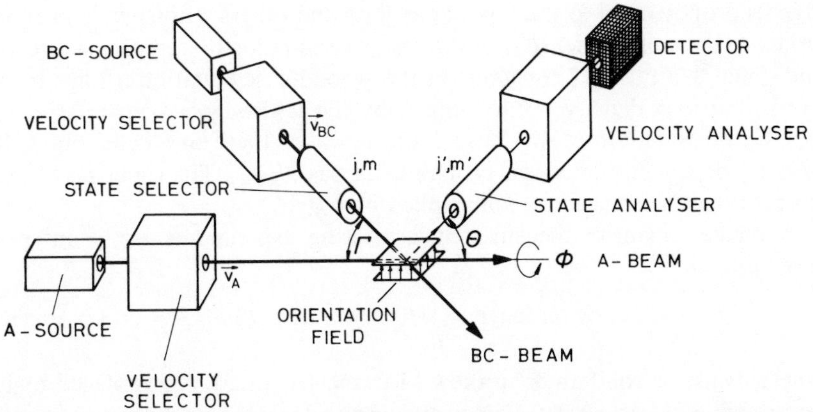

Fig. 4. Schematic perspective view of an ideal apparatus qualified to surface differential scattering cross sections of the general process $A + BC(j, m) + E \rightarrow A + BC(j', m') + E'$. The velocity *and* state-selected molecule beam intersects the velocity-selected atom beam at an angle Γ. The molecules scattered by the angles Θ and Φ are subsequently either velocity analysed (without state analyser) in a velocity-change (VC) experiment or state analysed in a state-selective experiment prior to their detection (Section III.D).

and the relative velocity $\boldsymbol{v} = \boldsymbol{v}_A - \boldsymbol{v}_{BC}$, n_A and $n_{BC}(j, m)$ denote the number densities of A and BC in the initial state, $d\tau$ the intersection volume, and $\partial(\vartheta, \varphi/\partial(\Theta, \Phi)$ the Jacobian for the transformation of the deflection angles ϑ and φ in the centre-of-mass frame (CM) to the corresponding angles Θ and Φ in the LAB frame. The Jacobian is given by:

$$\frac{\partial(\vartheta, \varphi)}{\partial(\Theta, \Phi)} = \frac{v_1'^2}{u_1'^2 \cos \beta} \tag{7}$$

where v_1' and u_1' are the LAB and CM velocities of the detected particle (BC) after the collision, and $\beta = \sphericalangle(\boldsymbol{v}_1', \boldsymbol{u}_1')$. The Newton diagram, Fig. 5, illustrates the kinematic situation of the scattering experiment for detection in the plane of the two beams ($\Phi = 0^0$) and visualizes the relations between the various LAB and CM quantities. For general transformation equations see Helbing (1968), Warnock and Bernstein (1968), and Pauly and Toennies (1968). The recoil velocity u_1' of the detected particle may be calculated from energy conservation:

$$u_1' = m_A/(m_A + m_{BC})v(1 - \Delta E/E)^{1/2}, \tag{8}$$

where ΔE denotes the energy transferred during the collision. For a rigid rotor ΔE is given by:

$$\Delta E = E - E' = B[j'(j' + 1) - j(j + 1)]; \tag{9}$$

E' denotes the recoil energy and B the rotational constant of the rotor.

The relation between the flux entering the detector and the measured signals depends on the devices used. There are detectors which generate signals proportional to the flux of particles and others with signals proportional to the number density. In the first case equation (6) represents directly the signal except for a constant; in the second case equation (6) has to be converted to a density by dividing it by the LAB velocity $v_1' = v_1'(j'/j, \boldsymbol{v}_A, \boldsymbol{v}_{BC}, \Theta, \Phi)$. In addition to these limiting cases there are detectors with velocity-dependent detection probabilities, say $P_D(\boldsymbol{v}_1')$. The signals are then proportional to equation (6) multiplied by $P_D(\boldsymbol{v}_1')$.

A further quantity measured in scattering experiments is the integral cross section:

$$Q(j'm'/jm; v) = \int_{4\Pi} I(j'm'/jm; \vartheta, \varphi, E)\, d\omega \tag{10}$$

where $d\omega$ is the solid angle in the CM frame. (As usual, E is replaced by v.) In principle, Q can be determined independently, i.e. no differential experiments necessary, by using a state-selective detector which measures the global (4Π) flux of molecules emitted by the scattering volume. The resulting signal is, according to equation (6), proportional to the desired quantity $Q(j'm'/jm; v)$. For experimental realizations see Section III. D.3.

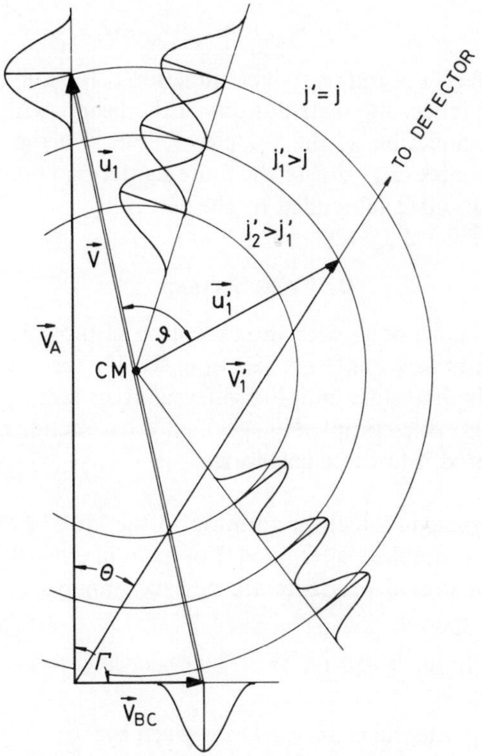

Fig. 5. Newton diagram depicting the kinematic situation of an ideal scattering experiment. The detector is positioned in the plane of the two beam velocity vectors $\boldsymbol{v}_A$ and $\boldsymbol{v}_{BC}(\Phi = 0)$. Each circle around the centre-of-mass (CM) is the locus of all possible arrow tips of the recoil velocity vector $\boldsymbol{u}'_1$ of the detected particle 1 (A or BC) the modulus of which is determined by the individual angular momentum transfer $j \rightarrow j'$ (equations (8) and (9)). Each set of CM coordinates consisting of u'_1 and the deflection angle ϑ correlates uniquely with the corresponding set v'_1 and Θ in the laboratory frame. In a realistic experiment the beams possess velocity distributions (indicated by the bell-shaped curves around the most probable velocities) and angular profiles which lead to velocity distributions of the scattered particles even for a fixed angular momentum transfer. These are illustrated again by bell-shaped curves for two typical CM angles and three distinct angular momentum transfers.

In a real experiment the beams are, of course, neither monoenergetic nor unidirectional. Then n_A and n_{BC} (j, m) in equation (6) must be replaced by:

$$n_A \rightarrow n_A f_A(\mathbf{v}_A)\,d\mathbf{v}_A$$
$$n_{BC}(j, m) \rightarrow n_{BC}(j, m) f_{BC}(\mathbf{v}_{BC})\,d\mathbf{v}_{BC} \tag{11}$$

and integrated over $\mathbf{v}_A$ and $\mathbf{v}_{BC}$. The functions $f(\mathbf{v})$ are normalized by $\int f(\mathbf{v})d\mathbf{v} = 1$. They reflect the distributions of the beams with respect to the modulus and the direction of the velocity vectors. Further integration of equation (6) is also necessary due to the finite scattering volume $d\tau \rightarrow V$ and the solid angle $d\Omega \rightarrow \Delta\Omega$ subtended by the detector.

B. Cross Sections

Most experiments to date, with the exception of those in which electrostatic state selectors *and* analysers (Section III. D) are employed, do not determine an individual differential or integral cross section but particular sums and averages over completely specified cross sections. These can be conveniently ordered into three categories:

Oriented cross sections in which in addition to the initial j either the initial or final m quantum number is retained. For example, the BC-beam is state selected but the scattered products are not state analysed. Then the data yield:

$$I(j, m; \vartheta, \varphi, E) = \sum_{j', m'} I(j', m'/j, m; \vartheta, \varphi, E). \tag{12}$$

The corresponding integral cross section is given by:

$$Q(j, m; v) = \int_{4\Pi} I(j, m; \vartheta, \varphi, E) \sin\vartheta\, d\vartheta\, d\varphi. \tag{13}$$

A well-studied quantity is the cross-section anisotropy A:

$$A = \frac{Q(j, m_1; v) - Q(j, m_2; v)}{Q(j, m_2; v)} \tag{14}$$

where m_1 and m_2 represent two different projection quantum numbers.

State-to-state cross sections which are summed and averaged over the final and initial magnetic substates. Only the j, j'-dependence is conserved. These are given by:

$$I(j'/j; \vartheta, E) = \frac{1}{2j+1} \sum_{m'm} I(j', m'/j, m; \vartheta, \varphi, E). \tag{15}$$

Here the deflection angle φ is eliminated due to the averaging over the initial m-states. The corresponding integral cross section is

$$Q(j', j, v) = 2\Pi \int_0^{\Pi} I(j'/j; \vartheta, E) \sin \vartheta \, d\vartheta. \quad (16)$$

In VC experiments (cf. Section III) in which quantum states remain unresolved it may be convenient to introduce the energy transfer cross section j, which is differential with respect to the solid angle *and* the final translation energy E′:

$$J(E', \vartheta, E)\Delta E = \sum_{\{j, j'\} \to E', E' + \Delta E} I(j'/j, \vartheta, E) \not{f}(j). \quad (17)$$

The sum includes all j, j'-combinations providing final translational energies into the interval $E', E' + \Delta E$ transmitted by the velocity analyser; the function $\not{f}(j)$ describes the initial state population. J is a (quasi) continuous function of E' (Beck, 1970). Since final velocities are measured, it is more adequate to express J as a function of u_1' rather than E'. The two functions are then related by:

$$J(u_1', \vartheta, E) = J(E', \vartheta, E) \cdot \frac{dE'}{du_1'}. \quad (18)$$

Total cross sections in which neither initial nor final states are explicitly selected. The measured quantities are therefore cross sections summed over all populated initial and all possible final states. The differential cross section is then given by:

$$I(\vartheta, E) = \sum_{j, j'} I(j'/j; \vartheta, E) \not{f}(j); \quad (19)$$

the corresponding integral cross section by:

$$Q(v) = \sum_{j, j'} Q(j', j; v) \not{f}(j). \quad (20)$$

C. Experimental Components

1. *State Analysers and Selectors*

The most commonly used devices for selecting and analysing low j, m ($j \lesssim 3$)-states of polar molecules are electric quadrupole or hexapole fields. The multipolarity chosen depends on whether the molecule exhibits a quadratic or linear Stark effect (Pauly and Toennies, 1968; Fluendy and Lawley, 1973). Due to the focusing properties of these fields a drastic increase of state-selected molecules over their concentration in the unselected beam is obtained; Bennewitz *et al.* (1964) report a factor of 200. Another technique involving fields may be applied to non-polar molecules with non-vanishing nuclear spin. Moerkerken *et al.* (1970) employed for the m-state selection of ortho-H_2 with $j = 1$ a standard magnetic Rabi field configuration with inhomogenous A and B fields as well as a homogenous C field on which an RF field is superimposed. The molecules leaving the A-field in a specific m_I-state (I = nuclear spin) enter the C-field where the RF is tuned to cause $\Delta m_I = \pm 1$ transitions for the desired molecular m-state. A modulation of the RF power eventually produces, after deflection of the particles in the B-field and subsequent collimation by a slit, a modulation of all molecules in an individual $j = 1, m$-state.

Particularly promising and increasingly utilized state selectors are i.r. lasers. This technique for preparing a beam in an individual quantum state was first employed by Odiorne *et al.* (1971) and Pruett *et al.* (1974). These authors used a pulsed HCl laser to selectively populate $v = 1, j$-states of an HCl beam. At present CW lasers are more commonly being used for state selection. Gough *et al.* (1977) have studied the $v = 0, j = 1 \rightarrow 1, 2$-transition in a CO beam using a diode laser. Brooks (1978), Lee (1978), and Karny *et al.* (1978) employ chemical HCl(HF) lasers to prepare HCl(HF) beams in specific $v = 1, j$-states. The latter authors exploit additionally the fact that linearly polarized radiation aligns molecules to investigate polarization effects. Toennies (1978) and coworkers achieve vibrational excitation of a CO beam by shining the light from a CO laser into the pinhole of a nozzle beam source. The developement of tunable i.r. lasers like the F-centre laser will expand the applicability of this technique drastically (Scoles, 1978). However, pulsed i.r. lasers may yet make a comeback due to the development of intense pulsed molecular beam sources (Gentry and Giese, 1978) (see below).

Optical pumping has recently been successfully applied to state select a Na_2 beam (Bergmann *et al.*, 1978). The laser line was tuned to the $v = 0$, $j = 28 \rightarrow v' = 6, j' = 27$ transition between the ${}^1\Sigma_g^+$ and the $B^1\Pi_u$ electronic states of Na_2. The excited state radiatively decays with a lifetime of 7×10^{-9} s (Demtröder *et al.*, 1976) into all accessible states. A laser power of 10 mW is sufficient to depopulate the lower level via this optical pumping process by 95% (laser-induced depopulation—LID). Thus by modulating the laser power, a Na_2 beam is provided in which a specific v, j-state is marked by the modulation frequency.

Infrared luminescence (IRL) from vibrationally excited molecules has been exploited to selectively detect inelastic transitions (Ding and Polanyi, 1975). This technique requires, however, a substantial concentration of initially excited molecules to produce a detectable luminescence. The measured i.r. intensity atributed to individual $v'j' \rightarrow v, j$ transitions reflects the number of molecules in the v', j'-state. If the residence time of an excited molecule within the detection volume is large compared to the lifetime of the state, the detected signal is proportional to the particle flux, for the opposite case, the signal becomes proportional to the number density of excited molecules. All intermediate cases are, of course, also possible. The method allows the measurement of state-to-state cross sections if the initial state is selectively excited, for example, by use of an i.r. laser (Lang *et al.*, 1977).

The laser-induced fluorescence technique (LIF), common now for the state analysis of products from reactive scattering, represents a sensitive state-analysing detector (Zare and Dagdigian, 1974; Dagdigian *et al.*, 1974; for a review see Kinsey, 1977). It has recently been applied to rotationally inelastic scattering (Wilcomb and Dagdigian, 1977; Bergmann *et al.*, 1978). The molecules in a specific v', j'-level of the electronic ground state are excited into a v'', j''-level of an electronically excited state by properly tuned laser radiation. The intensity of the fluorescence light emitted from the excited state can be used to deduce the population of the initial v', j'-state. The LIF features a velocity characteristic of the detection probability which is dependent on the laser power. For example, at low power only a small fraction of molecules within the probed volume is excited and hence the detection probability becomes proportional to v^{-1} (density detector); at sufficiently high power all molecules in the proper state are excited, resulting in a velocity-independent detection probability (flux detector) (Bergmann *et al.*, 1978).

Fluorescence light emitted from molecules ionized and excited by a high-energy electron beam ($\gtrsim 1000$ eV) can also serve to state-selectively detect inelastic processes (Scott *et al.*, 1973, 1974) (electron beam induced fluorescence technique). The only application of the method to date has been to N_2. The electron beam strongly excites the $B^2\Sigma_u^+$ state of N_2^+, which quickly decays to $\chi^2\Sigma_g^+$, emitting the first negative system of nitrogen. The line intensities of this system are related to the density of molecules in v', j'-states before the collisional ionization (Muntz, 1962).

2. *Detectors*

In addition to the state-selective detection methods discussed above, the standard detection devices known from elastic or reactive scattering are also used in molecular scattering experiments. Alkali atoms or alkali-containing molecules are detected by their ionization on a hot metal surface like W, Ir, Pb–W, etc. (Fluendy and Lawley, 1973) with an efficiency up to 100%. Low-temperature bolometers (Cavallini *et al.*, 1971) have frequently been used in the context of total differential cross-section experiments (see for example Bickes *et al.*, 1975; and Bassi *et al.*, 1976). Most recently they have been utilized to determine the diffraction oscillation damping in H-molecule systems (Valbusa, 1978). The detector is particularly sensitive for the latter atom since a weak flow of 0_2 onto the bolometer increases the effieiency by a factor of 20 (Bassi *et al.*, 1974). Also widely used are mass spectrometers with electron impact ion sources (see for example: Lee *et al.*, 1969; Loesch and Beck, 1971; Kuppermann *et al.*, 1973; Buck *et al.*, 1975).

The advantage of this technique is its universality, its drawback the low efficiency of $< 10^{-3}$.

3. *Velocity Selectors and Analysers*

Velocity selection is generally achieved by the standard rotating slotted disc velocity selector based on the Fizeau design (Hostettler and Bernstein, 1960; Pauly and Toennies, 1968). This device may also serve as velocity analyser. In the case of paramagnetic atoms (hydrogen) velocity selection (and polarization) can be accomplished by utilizing focusing magnetic hexapole fields (Bassi *et al.*, 1976).

Frequently, time-of-flight (TOF) techniques are used in inelastic scattering experiments for velocity analysis (for a review see Raith, 1976). There are two different ways of doing this. The first employs a disc with a few equally spaced narrow slots to chop the scattered beam into short packages. Their time-of-flight distribution is then recorded, for example by a multichannel analyser (see e.g. McDonald *et al.*, 1972; Loesch, 1976). The disadvantage of this method is that the velocity resolving power and the transmission are correlated. The latter is usually below 10% and decreases rapidly with increasing resolving power. This drawback is avoided in the second method, in which the beam is chopped by a pseudorandom sequence of slit and tooth widths. The resulting detector signal is cross-correlated with the modulating sequence to produce the TOF spectrum (Sköld, 1968; Gläser and Gompf, 1969; Meyer, 1974; Hirschy and Aldridge, 1970; Huisken, 1978). The effective transmission for such sequences is close to 50%, independent of the resolving power. This method is particularly suited for experiments with signal counting rates small compared to the background rates. Then the gain in measuring time compared to the single-pulse technique may become as large as $N/4$, where N is the number of elements of the pseudo-statistical sequence (~ 100). In the opposite case, the single-pulse technique is superior (Meyer, 1974). A novel technique of velocity analysis employs Doppler profiles of a suitable absorption line of the detected particle measured by LIF (see above and Bergmann *et al.*, 1978a). Kinsey (1977a) has demonstrated that the full three-dimensional velocity distribution of particles emerging from a small intersection volume can be recovered from Doppler profiles measured as a function of the direction of the incident light beam utilizing a three-dimensional Fourier transformation procedure. The basic experimental set up may be illustrated with the aid of Fig. 4. The detector, velocity analyser, and state analyser are replaced by the laser beam, which is directed into the intersection volume, and the fluorescence light intensity is recorded via a photomultiplier as a function of the angles θ and Φ and the wavelength of the light. A gain in the rate of signal acquisition of approximately 10^4 compared to conventional beam techniques is expected. An application of the Fourier transform Doppler spectroscopy to a scattering experiment has not yet come to my attention.

4. *Beam Sources*

Since the advent of supersonic nozzle beams in the last two decades, they are now widely used in molecular scattering experiments. Their properties have recently been reviewed by Anderson (1974), and the advantages of their utilization in inelastic scattering experiments is thoroughly treated by Faubel and Toennies (1977a). We mention therefore only briefly their most important features relevant to the scattering from molecules.

The beams are generated by expanding the beam gas from an initial pressure P_0 (a few 100 Torr up to several atmospheres) through a thin-walled orifice with a diameter d of the order of 0.1 mm to a final pressure of $P < 10^{-2}$ Torr. During the (isentropic) expansion the collisions between the gas molecules lead to a gradual transfer of the kinetic energy of the thermal motion into directed flow motion and hence to a cooling of the temperature of the translational degrees of freedom. The density within the beam decreases inversely proportional to the square of the distance from the nozzle. Therefore a certain distance exists (a few mm) beyond which this temperature remains constant, as no further substantial energy transfer occurs (freezing radius). A few mm downstream from this point the beam is usually skimmed and collimated by a slim conical skimmer. Its velocity distribution (number density) may from that point on be approximated by (Anderson *et al.*, 1966):

$$n(v)\,\mathrm{d}v = \mathrm{const}\cdot v^2 \exp\{-[(v-u)^2/v_0^2]\}\,\mathrm{d}v \tag{21}$$

where u is the final flow velocity and v_0 the most probable velocity associated with the terminal beam temperature T:

$$v_0 = (2kT/m)^{1/2}$$

(m is the mass of the beam particles.) The full width at half maximum of the velocity distribution for $v_0 \ll u$ is given by

$$\Delta v_{1/2}/u \approx 1.66/S.$$

Here the speed ratio $S = u/v_0$ characterizes the beam temperature T. The flow velocity u can be calculated from the conservation of the enthalpy of the gas during the expansion. This gives, for $v_0 \ll u$,

$$u = (2C_p T_0/m)^{1/2} \tag{22}$$

where C_p is the specific heat of the gas at constant pressure and T_0 the nozzle temperature. The quantitative treatment of the expansion according to a thermal conduction model (Habets, 1977) leads, for an atomic gas, to the speed ratio:

$$S = 867.8\left(P_0\frac{d}{2}T_0^{-4/3}\right)^{6/11}$$

(P_0 in Torr, d in cm and T_0 in K).

The beam intensity in the forward direction $\dot{N}(s^{-1}sr^{-1})$ is given by

$$\dot{N} = \frac{k}{\phi}\dot{N}_{\text{tot}}$$

where $\dot{N}_{\text{tot}}$ is the total gas flow energing from the nozzle. The peaking factor k is equal to 2 for an atomic gas and 1.35 for a diatomic gas (Habets, 1977). The upper limit of $\dot{N}_{\text{tot}}$ is governed by the capacity of the gas handling system which maintains the final pressure P. Since $\dot{N}_{\text{tot}} \propto P_0 d^2$, S may be increased for constant $\dot{N}_{\text{tot}}$ by decreasing the nozzle diameter d. This procedure is limited, however, by the onset of extensive clustering beyond a critical beam temperature T. Brusdeylins *et al.* (1977) have recently reported the following speed ratios: 225 (He), 50 (Ne), 25 (Ar), 28 (N_2), 10 (CO_2), 10 (CF_4), 8 (C_2H_6), 5 (HCl). The gas handling system employed had a pumping speed of 10 Torr 1/s. It should be noted that the 'natural' widths of the velocity distributions obtained above may already be sufficiently narrow to permit scattering experiments without an additional velocity selector.

During the adiabatic cooling of the translational degrees of freedom of the molecules, relaxation collisions occur which tend to equilibrate the energy among all available degrees of freedom. Since translational to rotational energy transfer cross sections are of the order of gas kinetic ones (except for H_2), substantial cooling of the rotation is frequently observed. Temperatures ranging from 200 K for H_2 (Gallagher and Fenn, 1974; Verberne *et al.*, 1978) to 30 K for CO_2 (Silvera and Tommasini, 1976; Loesch, 1976), 14 K for COS (Meerts *et al.*, 1978), and 9–15 K for N_2 (Scott *et al.*, 1974) have been reported. There is evidence that lower rotational levels relax more completely than higher levels. This leads to a far-reaching tail of high j-levels in the final distribution which can be empirically taken into account by a second rotational temperature (Ding and Polanyi, 1975; Kukolich *et al.*, 1974, Verberne *et al.*, 1978). Vibrational relaxation plays only a minor role since the cross sections for translation to vibration are usually much smaller than for translation to rotation (Mariella *et al.*, 1974; Bennewitz and Buess, 1978).

Another important aspect of nozzle beams is the possibility of generating enhanced translational energies using the seeded beam method (Abuaf *et al.*, 1967). This consists of expanding an inert carrier gas (H_2, He, etc.) together with a small partial pressure of the desired beam substance (seed). For the case of high dilution (seed/carrier $\lesssim 1\%$) the seed emerges from the nozzle with about the same velocity as the carrier, thus altering the kinetic energy of the seed by a factor of Mass(seed)/Mass (carrier). Depending on the mass of the seed, and the degree of dilution, translational energies in the eV range can be achieved. An approximate flow velocity may be calculated by replacing C_p and m in equation (22) by adequate mean values of the gas mixture. A decrease in the velocity of the seed is also possible by using a carrier gas heavier than the seed. Relaxation of the translational and rotational degrees of freedom also lead to narrow velocity distributions and low rotational temperatures for seeded beams (Malthan, 1976; Meerts *et al.*, 1978). A particular advantage of the method is that these effects can be obtained at pressures of the seed sufficiently low to avoid clustering, which might take place extensively in a pure beam. Due to the low terminal rotational temperatures the beam molecules are distributed only among the few principally occupied states, resulting in a drastically increased signal in state-selected experiments. Gain factors of 2 to 3 orders of magnitude have been reported for a CsF(Xe) mixture (Malthan, 1976).

The most recent development in this field is a beam source which generates very intense supersonic molecular beam pulses of about 10 μs duration (Gentry and Giese, 1978). The principle of operation of this source is given in the legend to Fig. 6. The repetition rate is currently of the order of 1 Hz which, together with the short pulse duration, results into a very low duty factor of about 10^{-5}. At first glance this looks like a severe drawback of the method. However, the low duty factor permits a drastic increase of the particle density within one pulse as compared to the continuous version *without* overloading the pumps. A gain factor of 100, achieved mainly by enlarging the nozzle diameter (0.6 mm) (Gentry and Giese, 1977a), has been reported for light gases like H_2 and He. Furthermore, a substantial reduction of the signal noise becomes possible (a) by blocking the scalers after all correlated signal pulses have been counted and (b) by reducing the partial pressure of the detected species within the detector enclosure by the subsequent closing of the detector gate value. A noise reduction of several orders

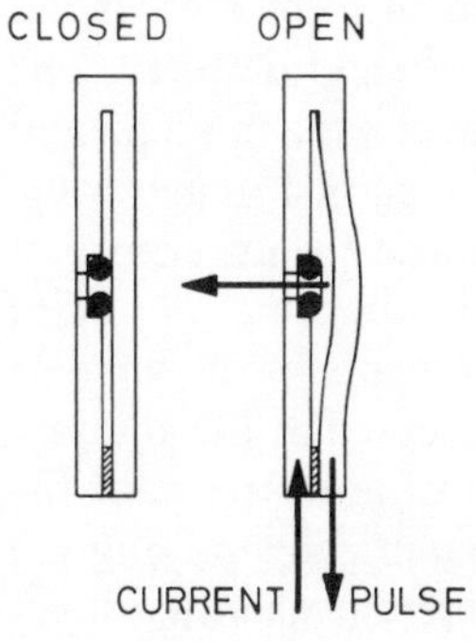

Fig. 6. Schematic cross section through the pulsed nozzle beam source of Gentry and Giese (1978). Left, valve in closed position with the flexible bar resting on the 0-ring. Right, the repulsive forces of the antiparallel flowing current pulse flexes the bar and allows the gas to emerge through the 0-ring (0.6 mm diameter) for a short period of time. The bar is rigidly clamped at both ends. (Reproduced by permission of the American Institute of Physics from Gentry and Giese, 1978.)

of magnitude has been reported (Gentry and Giese, 1977) which, together with the intensity gain (10^4 if two pulsed beams are used), may easily compensate for the loss in signal-to-noise ratio due to the low duty factor. Specific advantages of the pulsed beams are: (a) time-of-flight measurements can be performed without low-transmission choppers; and (b) in experiments requiring pulsed laser beams the factor 100 in particle density leads directly to a corresponding rise of the signal.

D. Experimental Determination of Cross Sections

1. *Completely Specified Cross Sections*

Experiments with the aim of measuring differential scattering cross sections for completely specified transitions $j, m \rightarrow j', m'$ belong to the early studies on energy transfer collisions (Toennies, 1965, 1966; see also the review of Pauly and Toennies, 1965). The basic experimental set ups are similar to the one given in Fig. 4; however, the LAB angles are kept fixed at $\Theta = \Phi = 0$ (forward scattering) and no velocity analysis is utilized. The electrostatic quadrupole state filters are set to transmit different j, m-states, namely, the selected initial (i) and the analysed final (f) states. Directly measured are the flux of molecules in the initial ($f = i$) and in the final state (f) with and without target beam, $\dot{N}_i$, $\dot{N}_f$, $\dot{N}_{i0}$, and $\dot{N}_{f0}$, respectively. These four quantities define an apparative cross section I_{AP}:

$$\frac{\dot{N}_f}{\dot{N}_i} - \frac{\dot{N}_{f0}}{\dot{N}_{i0}} = nlI_{AP}(f|i) \tag{23}$$

which is related to the differential cross section equation (2) by:

$$I_{AP}(f|i) = \frac{|\boldsymbol{v}|}{|\boldsymbol{v}_{BC}|} \int_{4\pi} I(f|i;\vartheta, \varphi, E) \cdot \bar{\eta}(f|i;\vartheta, \varphi) \mathrm{d}\omega \tag{24}$$

In equation (23) n and l denote the target particle density and the length of the target, respectively. The weight function $\bar{\eta}(f|i;\vartheta, \varphi)$ describes the probability that a particle in its final state f, scattered into the CM angles ϑ and φ, will be transmitted through the state analyser and detected. The bar indicates for velocity average. The exact knowledge of $\bar{\eta}$ is a fundamental requirement for an interpretation of I_{AP} and has to be gained numerically. The width of $\bar{\eta}$ is somewhat dependent on the system and the experimental conditions, and is of the order of a few degrees (Borkenhagen *et al.*, 1976; Malthan, 1976). Henrichs *et al.* (1977) report measurements of the angular dependence of I within the range of deflection angles accepted by the analyser field. They recovered this piece of information by closing off parts of the aperture of the analyser field at both ends, thus systematically varying the transmission function $\bar{\eta}$. The resulting I_{AP} reflects different narrow ranges of deflection angles and eventually provides the desired angular distribution of $I(f\,|\,i; \vartheta, \phi, E)$. The systems which have been studied up until now are: TlF + rare gases and a series of simple nonpolar and polar molecules (Toennies *et al.*, 1965, 1966), CsF + rare gases and several molecules (Borkenhagen *et al.*, 1976; Malthan, 1976; Henrichs *et al.*, 1977), and Kcl + Kr (Meyer and Toennies, 1978).

2. *Oriented Cross Sections*

Two different types of scattering experiments have been reported which are designed to study the dependence of integral and differential cross sections on the initial (or final) molecular orientation. They may be grouped according to the measured quantities thus:

(a) velocity dependence of the cross-section anisotropy A (equation (14));
(b) total differential cross sections for initially state-selected molecules as a function of the deflection angle (equation (12)).

Scattering experiments of type (a) were historically the first which yielded directly the anisotropy of the attractive interaction of a molecular system (Bennewitz *et al.*, 1964). They have been adequately reviewed (Reuss, 1975) and there is no need to go into details. We give therefore only a brief description of the experimental method, which is basically an attenuation experiment with either a state selector or analyser in connection with a velocity-selecting device (Fizeau-type velocity selector or highly expanded nozzle beam) (cf. Fig. 4 with $\Theta = \Phi = 0$). Inhomogeneous electric fields (Bennewitz *et al.*, 1964; Stolte *et al.*, 1972) and the Rabitype magnetic field configuration (Moerkerken *et al.*, 1970) have been employed for state selection. After preparation of an individual j, m-state the molecules enter the orientation field. Here the beam is attenuated by the target gas (or beam) as a result of all possible encounters which remove particles from it. In the ideal case the integral cross section, summed over all accessible final states, is given by the ratio of the beam intensities $\dot{N}$ and $\dot{N}_0$, with and without target gas:

$$\dot{N}/\dot{N}_0 = \exp[-nlQ(j, m, \beta, v)]$$

where n and l denote the density and length of the target and β the angle between the field vector within the orientation field and the relative velocity vector $\boldsymbol{v}$; m is referred to the orientation field vector. If Q is measured for two different field directions β_1 and β_2, a general cross-section anisotropy may be constructed:

$$A_0 = \frac{Q(j, \mathrm{m}; \beta_1, v) - Q(j, m; \beta_2, v)}{Q(j, m; \beta_2, v)}. \tag{25}$$

The $Q(j, m;\beta, v)$ can be replaced by β-independent quantities $Q(j, m';v)$ if the relative velocity $\boldsymbol{v}$ is introduced as a new quantization axis. Then

$$Q(j, m;\beta, v) = \sum_{m'} (d^j_{mm'}(\beta))^2 Q(j, m';v)$$

where $d^j_{mm'}(\beta)$ is the usual rotation matrix. For $j = 1$ and $m = 0$ the asymmetry A_0 becomes largest at $\beta_1 = 90^0$ and $\beta_2 = 0^0$ and one obtains the simple form:

$$A = \frac{Q(1, 1;v) - Q(1, 0;v)}{Q(1, 0;v)} \tag{26}$$

where the m'-states refer now to $\boldsymbol{v}$. For other states A becomes more complex (Bennewitz *et al.*, 1969a). The (small) differences between the two cross sections which are actually observable arise from the fact that at the moment of nearest approach for $m' = 0$ and $m' = 1$ the perpendicular and parallel orientation, respectively, of the molecule with respect to $\boldsymbol{v}$ prevail (Zandee and Reuss, 1977a) thus permitting the collision partner to probe two slightly different parts of the PHS.

In a real experiment the relative velocity is not homogeneous and unidirectional but distributed according to the velocity distributions of beam and target. Consequently, the directly measured cross section anisotropy A_e represents an average of the general form A_0 over β_1, β_2, and v. Further experimental imperfections requiring corrections, are for example, the finite acceptance angle of the detector and imperfect state selection. The velocity dependence of $Q(j, m;v)$ or A is usually not recovered from the data. For an analysis of the experimental results theoretically determined cross-section anisotropies A are subjected to proper averaging procedures and compared to A_e using trial and error techniques. Experimental results are available for TlF + rare gases (Bennewitz *et al.*, 1964, 1969*a*), CsF + Ar and He (Bennewitz *et al.* 1969a, b; Bennewitz and Haerten, 1969), NO + rare gases and simple molecules (Stolte *et al.*, 1973; Schwarts *et al.*, 1973; Kessener and Reuss, 1975), and H_2 + rare gases and simple molecules (Moerkerken *et al.*, 1973, 1975; Zandee *et al.*, 1976, 1977; Zandee and Reuss, 1977a, b).

Experiments of type (b) are basically the differential analogues of the above studies. The corresponding cross sections $I(j, m;\vartheta, \varphi, E)$ (see equation (12)) are obtained employing a basic experimental set up similar to the one given in Fig. 4 except for the state-analysing devices, which are eliminated. Tsou *et al.* (1977) utilized in their investigation of LiF–Ar, which is the only one of this kind to date, a highly expanded (seeded) LiF nozzle beam to prepare the initial velocity and an electrostatic quadrupole filter to select the initial state $j = 1, m = 0$. The flux of particles at the angles Θ and Φ is, according to equation (6), related to $I(j, m;\vartheta, \varphi, E)$ if properly summed over all available final states and averaged over all apparative uncertainties. For the recovery of the angular dependence of I from the recorded signals see the remarks on total cross sections in Section III. D.4. Analogous to A one may also define a differential cross section anisotropy,

$$A_d = \frac{I(j, m, \beta_1; \vartheta, \varphi, E) - I(j, m, \beta_2; \vartheta, \varphi, E)}{I(j, m, \beta_2; \vartheta, \varphi, E)}, \tag{27}$$

which can be transformed to a coordinate system with $\boldsymbol{v}$ as quantization axis in the same way as indicated above. For $j = 1$ and $\beta_1 = 0°, \beta_2 = 90°$ one obtains the expression

$$A_d = \frac{I(1, 1; \vartheta, \varphi, E) - I(1, 0; \vartheta, \varphi, E)}{I(1, 0; \vartheta, \varphi, E)} \tag{28}$$

to which the same comments on corrections of experimental imperfections apply as mentioned above.

3. *State-to-State Cross Sections*

Two basic experimental techniques are used to study differential and integral state-to-state cross sections. The one is the most universally applicable and widely used velocity-change (VC) method, the other the state-selective (SS) technique. They differ in the way the inelastic event is detected. While the VC method exploits the change of relative velocity correlated with an energy transfer, the SS technique employs a state-selective detection device (cf. Section III.C.1). In the following the two methods and their relation to the basic cross sections are discussed.

a. The velocity-change method. This method can most conveniently be explained using the Newton diagram (Fig. 5). For the depicted ideal experimental condition every j', j-combination refers to a well-defined CM recoil velocity $u'_1(j'/j, v)$ (equations (8) and (9)). A velocity scan carried out with a velocity analyser and a detector located at the angles Θ and Φ yields therefore intensity only at the discrete velocities $v'_1(j'/j, \boldsymbol{v}_A, \boldsymbol{v}_{BC}, \Theta, \Phi)$ corresponding to the intersections of the LAB velocity vector and the circles around the centre-of-mass with radii $u'_1(j'/j, v)$. The unique relation between v'_1, Θ, and Φ of the LAB and j', j, and ϑ of the CM frame permits eventually the direct determination of the cross section $I(j'/j; \vartheta, E)$ from the measured intensities via equation (6). In the latter equation $I(j', m' | j, m; \vartheta, E)$ is to be replaced by $I(j'/j; \vartheta, E)$. The basic experimental set up is obtained if the state analyser in Fig. 4 is removed and all other components remain unchanged.

In a realistic experiment the VC method suffers considerably from the beam velocity distribution and the resolving power of the velocity analyser. The LAB velocity v'_1 for a single j', j-combination no longer is well defined but is distributed around some most probable value. Therefore, the energy resolving power of the experiment with respect to two just distinguishable energy transfers ΔE_i is drastically reduced. The quantitative description of this problem is based on the equation

$$\begin{aligned} dS(v'_{10}, \Theta, \Phi) = n_A n_{BC}\, d\tau\, d\Omega \sum_{j',j} f(j) \int_{\boldsymbol{v}_A} \int_{\boldsymbol{v}_{BC}} f_A(\boldsymbol{v}_A) f_{BC}(\boldsymbol{v}_{BC}) \\ \times vT[v'_1(j'/j, \boldsymbol{v}_A, \boldsymbol{v}_{BC}, \Theta, \Phi), v'_{10}] I(j'/j, \vartheta, E) \\ \times \frac{\partial(\vartheta, \varphi)}{\partial(\Theta, \Phi)} \cdot P_D(v'_1)\, d\boldsymbol{v}_A\, d\boldsymbol{v}_{BC} \end{aligned} \tag{29}$$

which gives the relation between the signal dS measured by the detector positioned at the angles Θ and Φ and the differential cross sections. In equation (29) the velocity analyser is characterized by the transmission probability $T(v'_1, v'_{10}$ is the most likely transmitted velocity; for all other quantities see Section III.A. The integrations over $d\tau$ and $d\Omega$ have been suppressed. Since only the CM frame is of physical significance it is useful to study the signal as a function of u'_1 along those combinations of the LAB variables v_1, Φ, and Θ which correspond to a fixed CM deflection angle $\bar{\vartheta}$ of a representative Newton diagram (for example, the one drawn for the most probable beam velocities (Schepper *et al.*, 1979)). The result of a computation carried out for the in-plane scattering ($\Phi = 0$) of K by N_2 is shown in Fig. 7 (Schepper, 1978). Here, the relative full width at half maximum (FWHM) of the signal velocity distribution $\Delta_{1/2}u'_1/\bar{u}_1$ is displayed as a function of $\bar{\vartheta}$ for various relative energy transfers $\Delta E_i/E$ ($\bar{u}_1$ and E are the representative elastic recoil velocity and collision energy, respectively). The most conspicuous feature of Fig. 7 is that for each $\Delta E_i/E$ there exists a region of minimal $\Delta_{1/2}u'_1/\bar{u}_1$ and hence of best energy resolving power. This effect is due to the fact that a change in u'_1 caused by a change in v_A is approximately compensated for by the corresponding shift of the centre-of-mass. The kinematic situation is schematically illustrated by Fig. 5, where the various velocity distributions are visualized by bell-shaped curves. At large angles the maxima corresponding to the angular momentum transfers $j \to j' = j, j_1$, and j_2 are well resolved, while at small angles all peaks overlap substantially due to the concerted action of all apparative imperfections, and no separation is possible. A region of optimal resolving power occurs rather generally in in-plane scattering experiments and is particularly marked if the collision energy predominantly originates in the detected beam. Another possibility for achieving an improved resolving power is to rotate the detector in a plane perpendicular to the plane of the beams ($\Phi = 90°$). Then, over the ϑ range kinematically covered, the velocity distribution of the undetected beam is exactly eliminated, and only the angular profile remains to cause a deterioration of the resolution (Greene *et al.*,

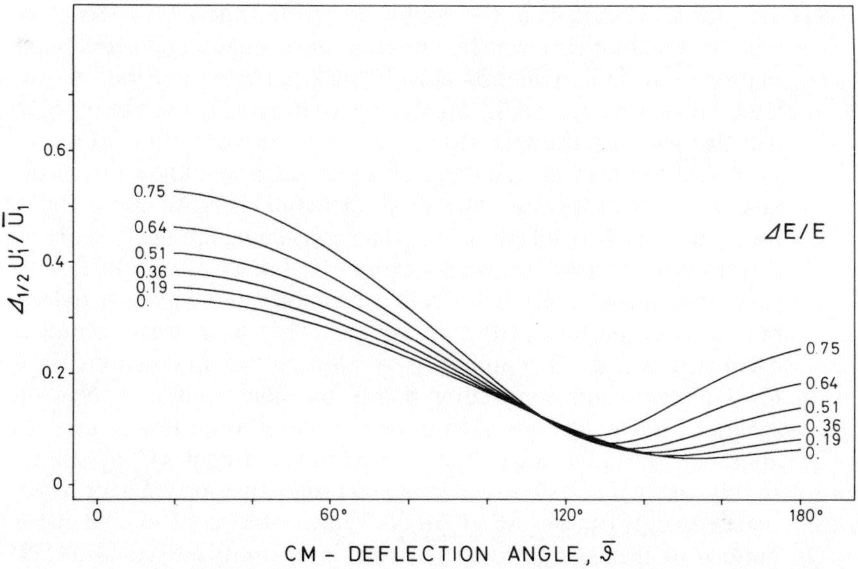

Fig. 7. The FWHM of the reduced recoil velocity $u_1'/\bar{u}_1$ calculated at various energy transfers $\Delta E/E$ along constant CM angle $\bar{\vartheta}$ of a representative Newton diagram as a function of $\bar{\vartheta}$ (Schepper, 1978). The curves are calculated for K + CO colliding at $E = 1.24$ eV. Beam data: $v_K = 3742$ m/s, FWHM 16.9%; $V_{CO} = 762$ m/s, FWHM = 19.3%, FWHM of the angular profile 5.7°, FWHM of the velocity analyser transmission function 10%.

1969a; Beck and Förster, 1970; Armstrong *et al.*, 1975). A detailed treatment of the energy resolving power in VC experiments is given in the review of Faubel and Toennies (1977).

In systems with $E \gg B$ contributions from various energetically closely lying j', j-combinations overlap, leading to a complete disappearance of the quantum structure in the velocity distributions. However, for molecules with $E \sim B$, such as hydrogen and isotopic variants at thermal collision energies, only one or a small number of states may be involved, thus permitting resolved experiments with modern beam machines (cf. Section V.A).

A futher problem which appears in all differential scattering experiments and which is by no means specific to the VC technique concerns the transformation of the experimental data from the LAB frame into the physically significant CM frame and the recovery of the angular dependence of the basic cross sections. In an ideal scattering experiment this is easily achieved as mentioned above. In all other cases it means a determination of the integrand in equation (29) using trial and error procedures. The following two methods may be used. While (a) refers only to unresolved experiments, (b) can also be applied to partially or even completely state-resolved experiments.

(a) This method starts from the double differential energy transfer cross section, equation (18), which is inserted into equation (29). The summation in the latter equation must be replaced by an integration over the LAB velocity v'_1, and the Jacobian to be inserted is given by $(v'_1/u'_1)^2$. The free parameters contained in J are then determined by fitting the simulated to the measured data. This method is well known from the analysis of reactive scattering data (Entemann, 1967). An approximate procedure which avoids the rather time-consuming fitting of a globally defined cross section has been reported by Loesch (1976). In this, to each observation angle Θ is fitted a function $j_\Theta(u'_1, E)$ which reflects only a small portion of the u'_1, ϑ-plane. The entire cross section is eventually recovered graphically by plotting the J_Θ-functions on a u'_1, ϑ-diagram and connecting points of equal values. A Newton diagram drawn for the most probable beam velocities is used to allocate an (average) angle ϑ to each pair of coordinates Θ, u'_1.

(b) In this case state-to-state-type cross sections (equation (15)) for hypothetical energy transfer ΔE_i, $I(\Delta E_i, \vartheta, E)$ are employed. The LAB distributions are then simulated by inserting this quantity into equation (29) and summing over the various transfers ΔE_i instead of j' and j. The free parameters (one set for each ΔE_i) are, in principle, again determined by fitting the simulated and measured distributions. The data analysis may be simplified by introducing θ-dependent cross sections $I_\Theta(\Delta E_i, E)$ which are fitted to each LAB velocity distribution individually. The deduction of the global $I(\Delta E_i, \vartheta, E)$ is accomplished in a way similar to that described in (a) (Beck and Förster, 1970; Armstrong *et al.*, 1975). The relation between I and the state-to-state cross sections is given by:

$$I(\Delta E_i, \vartheta, E) = \sum_{\{j,j'\} \to \Delta E_i, \Delta E_{i+1}} I(j'/j, \vartheta, E) f(j)$$

where the sum includes all j', j-combinations leading to energy transfers within the interval $\Delta E_i, \Delta E_{i+1}$. It should be noted that in experiments in which the velocity distributions are measured for a fixed angle $\bar{\vartheta}$ of a representative Newton diagram (Schepper *et al.*, 1979) the best-fit functions $J_{\bar{\vartheta}}(u'_1, E)$ (a) and $I_{\bar{\vartheta}}(\Delta E_i, E)$ (b) already represent the global cross sections.

The VC technique was first used by Blythe *et al.* (1964) to investigate inelastic collisions between $K + D_2$. Then Beck and Förster (1970) applied it to $K + CO_2$. In the meantime approximately 30 systems, including non-alkali collision partners as well (Farrar *et al.*, 1973; Loesch, 1976; Blais *et al.*, 1977), have been studied. Recent improvements in molecular beam techniques have led to state-resolved data for hydrogenic systems (Buck *et al.*, 1977, 1978; Gentry and Giese, 1977, 1977a; Farrar *et al.*, 1977) and to the discovery of structures in the unresolved energy-transfer cross sections (Schepper *et al.*, 1978, 1979).

b. State-selective method. There are three types of state-selective detection techniques which have been used to measure rotationally inelastic state-to-state cross sections. These are the laser-induced fluorescence (LIF), the infrared luminescence (IRL), and the electron beam induced fluorescence (EBF) techniques (cf. Section III.C). Basically the same apparative design as schematically shown in Fig. 4 is used except for the replacement of the velocity and state analyser, as well as the detector, by the optical device. As molecules are state-selectively detected, the energy resolving power of the experiments plays only a minor role; the magnitude of the transferred quanta is of no importance.

The LIF and EBF detectors have proven to be sufficiently sensitive to be utilized in wide-angle scattering experiments (Bergmann *et al.*, 1978; Scott *et al.*, 1973). The relation between the measured signal and the differential cross section is given by equation (6) with state-to-state cross sections inserted, properly averaged over all experimental uncertainties. Furthermore equation (6) has to be corrected for the velocity dependence of the detection probability. The recovery of the angular dependence of CM cross sections from the data may be accomplished simply by direct transformation using a representative Newton diagram or in a way similar to the one described by method (b) (see above, for one single ΔE_i). The IRL and also the LIF detectors have been applied to the direct determination of integral cross sections (Ding and Polanyi, 1975; Wilcomb and Dagdigian, 1977). In both methods, the LIF and IRL, the collection of fluorescence or luminescence light over a solid angle as close to 4π as possible (extensive use of mirrors) must be attempted. According to equation (6) (with state-to-state cross sections inserted) the measured light intensity then represents the integral cross section equation (16), provided the (LAB) velocity dependence of the detection probability is properly accounted for.

Data from state-selective experiments are now available for HCl and HF + rare gases and some molecules (IRL: Ding and Polanyi, 1975), LiH + Ar (LIF: Wilcomb and Dagdigian, 1977), Na_2 + Ne and He (LIF: Bergmann *et al.*, 1978), and N_2 + Ar (EBF: Scott *et al.*, 1973).

4. *Total Cross Sections*

This type of cross section requires neither a state-selecting nor analysing device. The basic experimental configurations are identical to those used in elastic atom–atom scattering, and we refer to recent review articles covering this field (see Section IV.A). We would like to remark that it is not in general possible to recover the angular dependence of the total integral cross sections properly from the data. This is due to the fact that the final relative velocities, and consequently the LAB–CM transformation equations, depend on the specific transitions. In cases in which inelastic encounters are negligible the elastic transformation procedure is a good approximation. However, as the inelastic cross sections increase this approach deteriorates. For this situation one then calculates model-based state-to-state cross sections, transforms them individually into the LAB frame, and compares the simulated and experimental data within the framework of a trial and error procedure (Section II.A).

IV. CROSS SECTIONS AND ANISOTROPY

The variety of scattering cross sections (Section V) which are experimentally available contain the angular dependence of the PHS in a more or less complex form. The only general method to date of exploiting their content of potential information is by trial and error procedures (Section II). A sensible application of this technique, however, requires the knowledge of at least an approximate correlation between cross section features and PHS properties *and*, what is also of importance, the location of specific areas of the PHS which are most sensitively probed by the given experiment. The latter would in principle permit the stepwise determination of the entire PHS by the properly selected utilization of those experimental data which represent different PHS regions. It is the aim of this Section to point out, as far as is known, the interrelations between cross sections and anisotropy and to localize the sensitively probed areas. We try to achieve these goals with the aid of computational results and simple models.

A. Scattering from Isotropic Potentials

The total scattering from isotropic and anisotropic potentials exhibits, to a certain extent, very similar features, such as glories in the integral cross sections as well as rainbows and diffraction oscillations in the differential cross sections. The reason for this is that $V_0(r)$, the isotropic term in the potential expansion of equation (3), usually represents the main contribution to the PHS and therefore dominates both the elastic and inelastic scattering. The most pleasant properties of the isotropic potential scattering are that the physical reasons for all the above features can be clearly visualized and easily understood and that the correlations of these structures with the interaction potential have been extensively studied in detail. To take advantage of these properties we summarize briefly the results of the isotropic scattering as far as they are related to the subject of this article. For a more detailed literature we refer to the following important articles and reviews: Pauly and Toennies (1965, 1968), Bernstein (1966), Bernstein and Muckermann (1967), Schlier (1969), Beck (1970), Toennies (1974), Pauly (1975), and Buck (1975).

The scattering from an isotropic potential is governed by the deflection function $\vartheta = \vartheta(b)$ which relates the impact parameter b of a trajectory and its final deflection angle ϑ. The general shape of $\vartheta(b)$ can be derived by qualitatively determining ϑ for a few trajectories with typical b. As illustrated in Fig. 8 trajectories starting at large b ($b \gg R_m$) cross only the weak attractive asymptotic potential region and are therefore only little deflected. For smaller b the probed forces increase, resulting in scattering into larger and larger angles. This behaviour changes as b becomes sufficiently small to permit influence on the trajectories by the onsetting repulsion. This leads

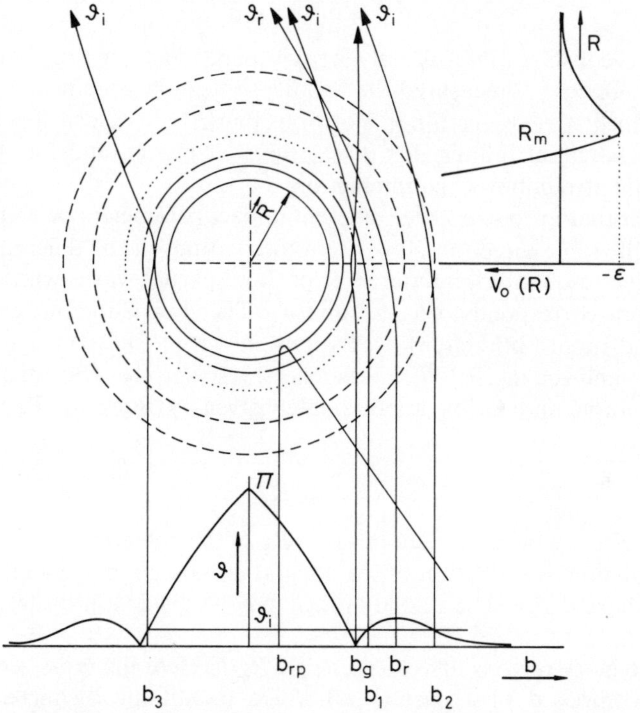

Fig. 8. Upper part: Traces of a few typical trajectories travelling through an isotropic potential. The latter is illustrated by contours with $V_0 > 0$ (solid lines), $V_0 = 0$ (dash-dotted line) $V_0 = -\varepsilon$ (dotted line), the minimal potential energy at $R = R_m$, and $V_0 < 0$ (dashed lines). A schematic cross section through the potential mountain is shown in the right-hand part. The three trajectories starting with impact parameters $b = b_1$, b_2, and b_3 lead to the same deflection angle ϑ_i, which gives rise to various interference phenomena. Lower part: Schematic deflection function $\vartheta = \vartheta(b)$. The relative maximum at $b = b_r$ is responsible for the classical rainbow scattering at $\vartheta = \vartheta_r$.

to a maximal deflection angle ϑ_r the rainbow angle at b_r, and eventually to a vanishing net deflection at b_g. For $b < b_g$ the repulsion dominates the shape of the trajectories, producing increasing scattering angles with decreasing b until at $b = 0$ backward scattering is obtained.

The classical differential cross section is given by

$$I(\vartheta, E) = \sum_{i=1}^{3} \frac{b_i}{\sin \vartheta} \left| \frac{d\vartheta}{db_i} \right|^{-1}$$

where i characterizes the three single-valued branches of $\vartheta(b)$ between $0 \leq b \leq b_g$, $b_g \leq b \leq b_r$, and $b \geq b_r$. The maximum of the deflection function at $\vartheta = \vartheta_r$ generates the well-known rainbow singularity (as $\vartheta_r^{-1}\, d\vartheta/db_i = 0$). To first order ϑ_r is given by

$$\vartheta_r = \text{const} \times \varepsilon/E \tag{30}$$

where the constant (≈ 2) is only weakly dependent on the shape of the potential. Once ϑ_r is measured the well depth ε can be obtained.

From the above discussion it is obvious that $I(\vartheta, E)$ reflects for $\vartheta < \vartheta_r$ the repulsive potential branch, for $\vartheta \lesssim \vartheta_r$ the well region, and for $\vartheta \ll \vartheta_r$ the asymptotic attraction of the interaction.

The fact that for $\vartheta < \vartheta_r$ three different impact parameters provide scattering into the same direction gives rise to oscillations in the differential cross section due to quantum interferences of those partial waves whose angular momentum correspond to b_1, b_2, and b_3. The slow (supernumerary rainbows) and rapid (diffraction) oscillations shown in Fig. 10 originate from the b_1–b_2 and the (b_1, b_2)–b_3 interferences, respectively. The angular width $\Delta\vartheta$ of the rapid undulation, approximately given by (Buck and Pauly, 1968);

$$\Delta\vartheta \approx \frac{\pi\hbar}{\mu v R_m}, \tag{31}$$

provides directly the minimum position R_m of the potential.

The potential information of the integral cross section is contained in its velocity dependence. The velocity range may be divided into three regions:

(a) The glory regions at velocities $v \lesssim \varepsilon R_m/\hbar$. Here the cross section $Q(v)$ is composed of a part $Q_{ng}(v)$ which monotonically increases as v decreases (resonances are not considered) and an oscillating part $Q_g(v)$. The main contributions to Q_{ng} are due to partial waves which probe the wide-range attraction. The glory undulations arise from interferences between partial waves corresponding to b_g which suffer no net deflection due to the compensatory influence of the repulsion and attraction and those of large b. The position and amplitude of the oscillations thus contain information on the well region. A semiclassical treatment of the glory effect provides in first order the position v_N

of the glory extrema

$$(N - 3/8)\pi = a\,\frac{2\varepsilon R_{\mathrm{m}}}{\hbar v_N} \tag{32}$$

The constant $a(\approx 0.4)$ depends slightly on the model potential used. Conventionally $N = 1, 2, 3, \ldots$ for the maxima and $N = 1.5, 2.5, 3.5 \ldots$ for the minima (Fig. 9).

(b) The high-velocity (repulsive) region at $v \gg \varepsilon R_{\mathrm{m}}/\hbar$ where $Q(v)$ decreases monotonically with increasing v. Here, $Q(v)$ is predominantly determined by the repulsive forces.

(c) The transition region at intermediate velocities. $Q(v)$ is here only weakly sensitive to either the attraction or repulsion.

B. Total Cross Sections

In atom–rigid-rotor scattering perturbations of the 'isotropic' integral and differential cross sections appear due to the potential anisotropy and the rotationally inelastic processes. These effects have been approximately studied in a series of early papers by Olson and Bernstein (1968, 1969), Cross (1968a), Miller (1969) (integral cross sections), and Cross (1970) (differential cross sections). The authors found for the integral cross sections a shifting of the glory extrema to lower velocities and a strong quenching of their amplitudes. Olson and Bernstein (1969) have derived simple formulae for the glory quenching ratio for a PHS of type IV, for example:

$$1 - Q_{\mathrm{g}}/Q_{\mathrm{g}}^{0} = 0.0394\left(\frac{2\varepsilon R_{\mathrm{m}}}{\hbar v}\right)^{2}(3.61a_2^2 - 3.78a_2b_2 + b_2^2 + 1.91b_1^2) \tag{33}$$

where Q_{g}^{0} and Q_{g} are the glory amplitudes calculated for the isotropic part of the potential and the complete PHS, respectively. Equation (33) indicates a v^{-2} dependence of the degree of quenching and demonstrates its direct relation to the anisotropy parameters. The monotonic part of the cross section proved not to be affected by the anisotropy. The perturbations of the differential cross section consist of a slight shift of the classical rainbow angle and of more or less severe shifting and damping of the main and supernumerary rainbow amplitudes and of the rapid diffraction oscillations. The classical cross section is not altered.

The theoretical studies cited above start with approximations like the DWA or classical path methods, but the results are eventually obtained in the rather poor high-velocity limit. A reliable determination of anisotropy parameters from quenching phenomena requires more precise scattering theories which, additionally, must be computationally sufficiently fast to permit trial and error calculations within sensible computation times. Recently, it has been shown that various sudden approximations (Section

II.B) meet these requirements (Buck and Khare, 1977; Klaassen *et al.*, 1978; Pack, 1978; Buck *et al.*, 1978).

Klaassen *et al.* (1978) utilized the time-dependent SA to study glory quenching (Potential type II). The reliability of this approximation is supported by satisfactory agreement with DWA results and by the convergence of the glory structure to the correct limit as the potential becomes isotropic. The latter is not the case in the older studies. The glory amplitudes exhibit considerable sensitivity to variations of the anisotropy parameters (Fig. 9) and confirm the general findings reported in the earlier papers. As indicated by equation (33) the degree of quenching depends on a combination of the complete set of anisotropy parameters which would predict for $b_2/a_2 \approx 2$ and $b_1 = 0$ no net quenching. This is supported by the results of Klaassen *et al.* (1978), who observed a damping of less than 10% for $b_2/a_2 \approx 2.15$ in (Ar + NO)-like systems (Fig. 9). Experimentally, this small amount of damping would be indistinguishable from the unquenched case and might lead to the incorrect conclusion that the PHS is isotropic. A similar effect

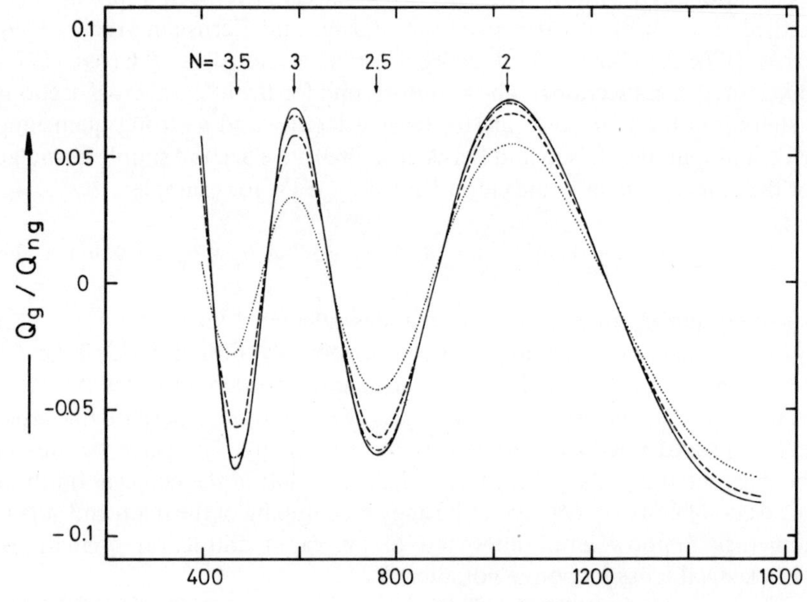

Fig. 9. Damping of the relative glory amplitude for an (Ar + NO)-like system calculated with the model potential II (Klaassen *et al.*, 1978). The parameters used are: $a_2 = b_2 = 0$ (isotropic case: solid line); $a_2 = 0.239$ and $b_2 = 0.239$ (dotted line), $b_2 = 0.3583$ (dashed line), and $b_2 = 0.478$ (dashed-dotted line). The curve for the latter set ($b_2 = 2a_2$) is quenched by less than 10% and nearly indistinguishable from the isotropic curve. All glory amplitudes exhibit an increasing degree of quenching with increasing index number N. (Reproduced by permission of North-Holland Publ. Co. from Klaassen *et al.*, 1978.)

has also been reported for the rainbow amplitude. Buck *et al.* (1978) found that its quenching depends on the parameter combination $\alpha_{eff} = a - 0.4b$ (PHS VIII). These results suggest that the anisotropies of repulsion and attraction compensate within the sensitively probed PHS area. And indeed, the strength of the angular dependence of a type II potential, characterized by the first derivative of $V(R, \gamma)$ with respect to γ,

$$\partial V/\partial\gamma = 3\varepsilon(R_m/R)^6 \cos\gamma \sin\gamma[(R_m/R)^6 b - 2a],$$

is particularly small or zero for $R \sim R_m$ with the combination $b \sim 2a$ and at $R \sim 1.04R_m$ for $b \sim 2.5a$. This strongly supports the conclusions, which might also be drawn from the isotropic scattering, that glory and rainbow quenching probe the well region at $R \sim R_m$ and at R values somewhat shifted to the inflection point, respectively.

An illuminating and stimulating study of the mechanism of rainbow and rapid undulation quenching has been presented by Pack (1978). His computations are based on the IOSA using the PHS VI, which explicitly contains the angular dependence of the well depth and separation. His results clearly indicate that the parameter a which controls the ε-anisotropy is responsible for the rainbow quenching and b for the diffraction oscillation damping (cf. Fig. 10). These findings correspond to what one would naively expect from the first-order results, equations (30) and (31), of the isotropic scattering. As the position of the rainbow $\vartheta_r \sim 2\varepsilon(\gamma)/E$ becomes γ-dependent, the averaging of the rainbow amplitude over the molecular orientation would eventually lead to its quenching and broadening. The degree of the effect would depend on the magnitude of a. The same argumentation applies to the damping of the rapid oscillations.

It appears feasible to extract a and b—or their equivalents if other PHS are used—from experimental data. In heavy systems ($Ar + CO_2$) the rapid oscillations may not be observed due to inadequate experimental resolving power. Then, by assuming a sensible b value the parameter a can be extracted. On the other hand, in light systems ($He + CO_2$), where the rainbow structure is disturbed by the widely spaced diffraction oscillations, b can be determined, provided a reasonable a value is chosen.

In addition to the above-mentioned perturbations of the isotropic cross sections a further feature in the differential cross section may arise for large anisotropies—a second rainbow at wide angles (Harris and Wilson, 1971; Buck *et al.*, 1975b). Extensive calculations with the potential of type V utilizing a classical path approximation in the sudden limit (straight-line trajectories) similar to the one proposed by Cross (1967) demonstrate strong dependence of this feature on the anisotropies. The results indicate that large attractive and small repulsive anisotropies are prerequisites for the appearance of a second rainbow (Buck *et al.*, 1975b).

We will not conclude this subsection without indicating some caution

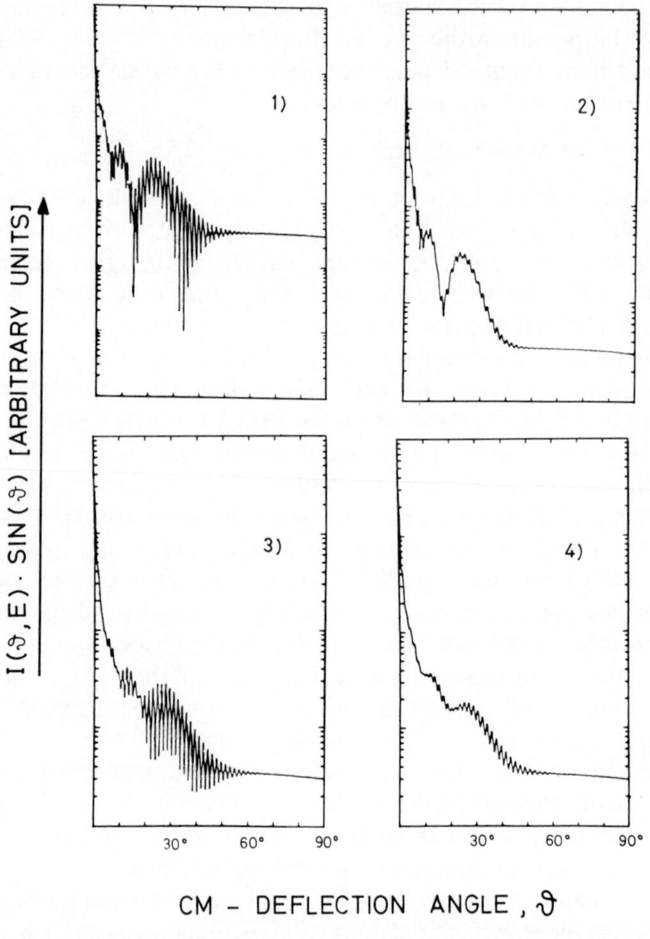

Fig. 10. Damping of the rainbow and diffraction oscillation amplitudes for an ($Ar + CO_2$)-like system calculated with the model potential VI. Part (1) refers to the unquenched isotropic case with rapid oscillations and primary and secondary rainbows; (2) the parameters $a = 0, b = 0.2$ lead to nearly complete quenching of the rapid oscillations and no damping of the rainbow amplitudes; (3) the set $a = -0.6$, $b = 0$ results in strong quenching of the rainbow maximum and no quenching of the fast undulations; and set (4), with $a = -0.6$ and $b = 0.2$ provides strong quenching of all structures. The remaining parameters are: $E = 69$ meV, $\varepsilon = 17.2$ meV, and $R_m = 3.88$ Å. (Reproduced by permission of North-Holland Publ. Co. from Pack, 1978.)

to be observed in using the method of obtaining anisotropy parameters discussed above. In addition to the physical reasons for the quenching and shifting of the various structures there are unfortunately also apparative ones. In fact, all experimental imperfections tend to reduce the amplitudes and therefore simulate quenching. These influences must be carefully eliminated in order to get reliable results. This can be done by subjecting theoretical results to further computations which simulate apparative imperfections before comparing them with the experimental data. Comparative measurements with atom–atom systems of approximately equal masses may help to test the success of such calculations (Keil *et al.*, 1978). Furthermore, in realistic systems, besides the anisotropy-induced quenching, vibrationally inelastic scattering (Kramer and LeBreton, 1967; Helbing and Rothe, 1968; Helbing, 1969; LeBreton and Kramer, 1969) and chemical reactions (Greene *et al.*, 1966; Ross and Greene 1970) may also cause a damping of cross-section structures. One has, therefore, to check carefully whether one of these mechanisms may occur in the system under investigation.

C. State-to-State (Energy Transfer) Cross Sections

Theoretical studies concerning the fundamental collision process of translational to rotational energy transfer have mostly been aimed at the development of approximate scattering theories (Section II.B), comparison between them, and their application to a few systems with known PHS. Some of these theories are presently at a sufficiently high level with respect to reliability and computation speed to permit investigations in which potential properties (particularly the anisotropy) are varied to study their influence on this process. (see for example $Ar + CO_2$ (Loesch, 1976; Preston and Pack, 1977), $K + CO_2$ (Förster, 1975). $Li^+–H_2$ (Barg *et al.*, 1976), $Ar + HCl$ (Buck and McGuire, 1976; Polanyi and Sathyamurthy, 1978), $Ar + N_2$ (Alexander, 1977, 1978), $Ne + HD$ (Huisken, 1978), and $Kr + H_2$ (Jacobs and Reuss, 1977). Schepper *et al.* (1979) have developed a classical rigid-shell model which provides considerable insight into the rotational excitation process at enhanced energies.

In the following some computational results for heavy systems with $B \ll E$ are used to establish relations between translational to rotational energy transfer cross sections and PHS properties. In a first step towards this goal the potential domains are determined which are most sensitively probed by the inelastic encounters. The procedure is based on a series of CT calculations with potential VI carried out for a variety of parameters. The results compiled in Figs. 11, 12 are obtained for the parameter sets given in Table I, which correspond to a highly anisotropic $(Ar + CO_2)$-like systems. We think that the conclusions drawn from these calculations should be of rather general validity because the PHS is realistic, parts of the findings are supported by other studies, and it is in accord with predictions obtained from the Massey criterion. The latter are discussed below.

The computations provide the following important facts:

(a) As shown in Fig. 11 there exist strong correlations between the impact parameter b and the deflection angle ϑ (see also: Thomas, 1977; Polanyi and Sathyamurthy, 1978). They are very similar to the deflection functions known from the isotropic case (Fig. 8) except for two points. First, to each b corresponds a small band of ϑ values instead of a

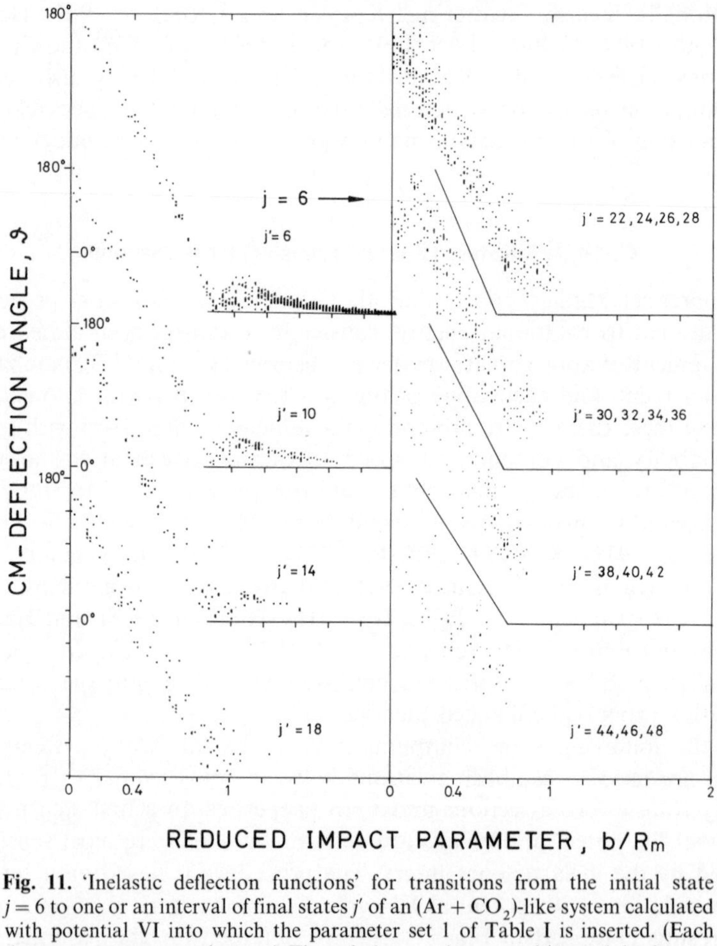

Fig. 11. 'Inelastic deflection functions' for transitions from the initial state $j = 6$ to one or an interval of final states j' of an $(Ar + CO_2)$-like system calculated with potential VI into which the parameter set 1 of Table I is inserted. (Each dot represents one trajectory.) The relative maximum at $b/R_m \sim 1$ leads to a rudimentary rainbow peak or at least to a steep shoulder in the differential cross section. This feature gradually disappears as j' increases.

single angle, and, secondly, for any j' a maximal impact parameter exists beyond which no scattering occurs. The similarities suggest using these deflection functions to allocate b-intervals and sensitively probed PHS areas in the same way as for the isotropic potential scattering (cf. Section IV. A).

(b) For small j' the deflection function exhibits rainbow and glory scattering (extremal and zero deflection angle, respectively, for intermediate b) which can be clearly seen in the left-hand portion of Fig. 11, although the b–ϑ relations are somewhat blurred. With increasing j' the maximal b decreases, thus gradually eliminating first the rainbow and then the

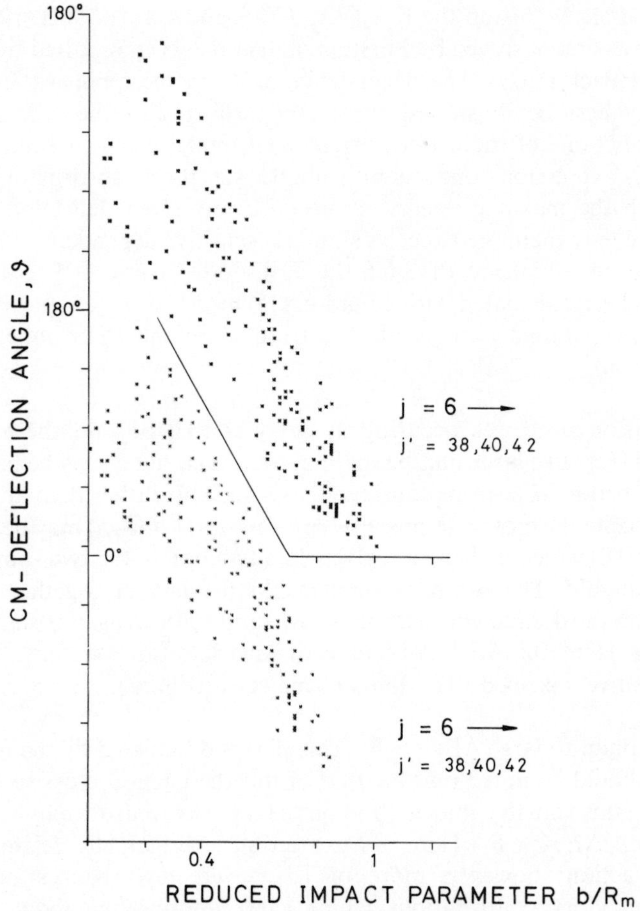

Fig. 12. The same as in Fig. 11 except that the upper portion has been calculated using the parameter set II of Table I. The larger well depth results in an increased forward scattering due to glory contributions.

glory scattering. For small to intermediate j' values the scattering is distributed between $0^\circ \leq \vartheta \leq 180^\circ$, and the intensity at small angles becomes increasingly due to glory scattering.

(c) The most conspicuous feature at intermediate to large j' is the disappearance of scattering into angles below a critical value. The position of this lower boundary of deflection angles, ϑ_c, depends on the anisotropy of the repulsion and the depth of the potential well. At a fixed j', ϑ_c decreases as the well is deepened, thus providing even at large j' more forward intensity due to (glory) collisions with impact parameters around b_g. This effect, which is illustrated by Fig. 12, was of particular importance for the interpretation of the $Ar + CO_2$ (Loesch, 1976) and the $K + CO_2$, COS, and CS_2 (Förster, 1975) data. It was first observed by Förster and has also been reported by Preston and Pack (1977). The dependence of ϑ_c on the repulsive anisotropy may best be illustrated by a comparison with the corresponding predictions of the classical rigid-shell model given in equation (35). This expression connects in a simple way the anisotropy of the shell with the maximal energy transfer for any given deflection angle ϑ. One may therefore expect a similarly sensitive dependence to be valid also for a realistic PHS (cf. the dash-dotted curve in Fig. 13, which has been calculated with equation (35) for an average repulsive anisotropy parameter of $\beta = 0.74$, attributed to the PHS under investigation).

In collision processes occuring at energies well beyond the excitation threshold ($E \gg B$) a large number of final rotational states may be populated, and j' (or better ΔE) can be considered as a variable rather than an index in characterizing the cross sections. For constant j, $I(j'/j\,;\vartheta, E)$ may be replaced by $I(\Delta E\,; \vartheta E)$ which is then, at a given E, a function of the two-dimensional ΔE, ϑ-manifold. The computational results (b) and (c), together with (a), now permit the desired correlation of $\Delta E, \vartheta$ and PHS areas. It is summarized in Fig. 13. Here, the $\Delta E/E$, ϑ-plane is divided into three sections, to which most sensitively probed PHS domains may be attributed.

(a) Medium to large ΔE; $\vartheta > \vartheta_c$. This part is dominated by the repulsion. It should be noted that because of this the integral cross section also contains valuable information on the repulsive anisotrophy.

(b) *Small* $\Delta E\,; \vartheta \ll \vartheta_r$. Here, the scattering is principally related to the attraction; however, more or less severe disturbances may arise from glory contributions which carry information about the well region.

(c) Small to intermediate $\Delta E; 0 \leq \vartheta \leq \vartheta_r$. This fraction of the cross section is governed by the well region. It reaches into area (a) because

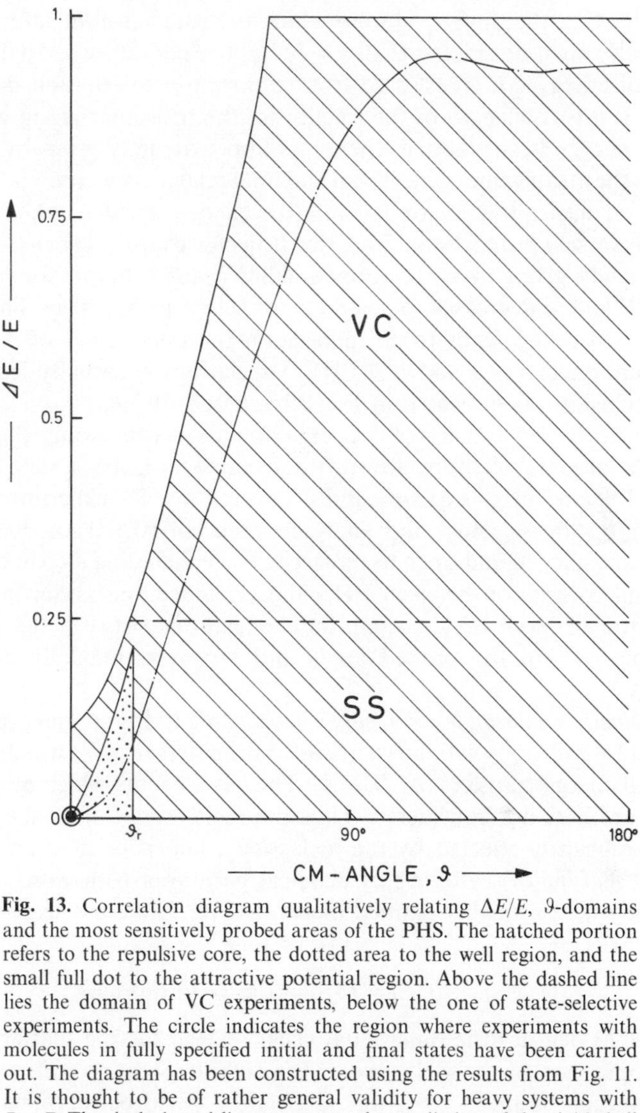

Fig. 13. Correlation diagram qualitatively relating $\Delta E/E$, ϑ-domains and the most sensitively probed areas of the PHS. The hatched portion refers to the repulsive core, the dotted area to the well region, and the small full dot to the attractive potential region. Above the dashed line lies the domain of VC experiments, below the one of state-selective experiments. The circle indicates the region where experiments with molecules in fully specified initial and final states have been carried out. The diagram has been constructed using the results from Fig. 11. It is thought to be of rather general validity for heavy systems with $B \ll E$. The dash-dotted line represents the prediction of the rigid-shell model (equation (35)).

glory collisions with relatively small impact parameters around b_g are capable of rather large transfers. The residual part with $\vartheta_r \leq \vartheta \leq 180°$ contains information on the repulsion.

These findings are in accord with conclusions drawn from the Massey criterion (Massey and Burhop, 1952). Applied to rotational inelastic scattering it states that the time, τ_i, during which the rotor and the atom interact substantially must be smaller than $\tau = h/\Delta E$, the period associated with the amount of energy ΔE transferred from translation to rotation during the collision. If d is the length of the fraction of the trajectory along which the potential energy is substantial, then τ_i is approximately given by $\tau_i \sim d/\bar{v}$, where $\bar{v}$ is the mean value of the initial and final relative velocity. The criterion provides an upper bound for the relative energy transfer, $\Delta E/E \leq \bar{v}h/Ed$, which is inversely proportional to d. For trajectories with large b ($b \gg b_r$) d is approximately given by b. Towards smaller b, for example for $b = b_r, b_g$, or b_{rp} of Fig. 8, d decreases to $d \sim r_m, \sigma$, or $0.1r_m$, respectively. This means that a decrease in b leads to an increase in the upper limit of $\Delta E/E$. The above mentioned $b \to \vartheta$ and $b \to$ PHS correlations eventually lead to the same conclusions as indicated in Fig. 13 (Loesch, 1976).

Region (a) is the domain of VC experiments because only these intermediate to large transfers are currently detectable for heavy systems (Section III.D.3). Region (b) is experimentally covered by SS experiments using electrostatic state selectors and analysers (Section III.D.1), and region (c) is widely unexplored and open to research. The concluding step in establishing the full correlation between PHS and cross sections is devoted to the presentation of more or less quantitative relations between the potential anisotropies within the probed areas and the experimentally detectable quantities.

The inelastic small-angle scattering is, according to Fig. 11, predominantly generated by collisions with large impact parameters ($b > b_r$) which produce only small energy transfers ($\Delta E/E \ll 1$). The glory events which might cause transfers are here not considered. Thus the relative motion of the particles is not significantly affected by the inelasticity, and it is quite adequate to describe this type of scattering by classical path approximations, provided a path may be reasonably defined (Section II.B). The cross section is then given by

$$I(j', m'/j, m; \vartheta, E) = P(j', m'/j, m; b, E) I(\vartheta(b), E),$$

where $I(\vartheta, E)$ denotes the differential cross section for the scattering from the isotropic part of the PHS and P the rotational transition probability of the molecule induced by a time-dependent perturbation. The latter is caused by the projectile as it travels along the classical path with impact parameter b determined by the isotropic part of the PHS. The probability P is usually calculated in the sudden limit (Section II.B) utilizing the HEA the first order (Toennies, 1966; Malthan, 1976). The m-dependence is retained since the only measurements of this type of cross section have been carried out with electrostatic fields, the focusing properties of which depend on m (Toennies, 1966; Malthan, 1976; Henrichs *et al.*, 1977; Meyer and Toennies, 1978).

If the PHS consists only of one single anisotropic term of the order n in equation (3), the theory provides the following selection rules for inelastic transitions:

$$\Delta j = j' - j = \begin{cases} \text{even for even } n \\ \text{odd for odd } n \end{cases}$$

and, in first order, additionally

$$|j' - j| \leq n \leq j' + j.$$

Fof Δm-transitions no general selection rule applies, except for the perpendicular and parallel orientation of the quantization axes with respect to the relative velocity (Bernstein and Kramer, 1966). On the basis of the first-order straight-line trajectory approximation the transition probability P can be expressed by analytic formulae (Fenstermaker *et al.*, 1969) using the simple potential: $V(R, \gamma) = V_0(R)[1 + a_n P_n(\cos\gamma)]$. For the allowed transitions, $P(j', m'/j, m; b, E)$ proves to be proportional to the anisotropy parameter $|a_n|^2$. Usually only the attractive PHS branch is taken into account for the interpretation of small-angle scattering data (neglection of glory collisions). Then, if only terms with $n = 1$ and $n = 2$ are involved, the experimentally determined cross sections for $\Delta j = 1$ and $\Delta j = 2$ transitions separately provide the anisotropy parameters $|a_1|^2$ and $|a_2|^2$, respectively, due to the above selection rule.

Inelastic rainbow scattering appears to be a common feature and should occur whenever the spherically averaged PHS would generate it. In addition to $Ar + CO_2$ (cf. Fig. 11) it has been found computationally for many systems, for example: $Ar + HCl$ (Buck and McGuire, 1976), $Ar + N_2$ (McGuire, 1977; Alexander, 1977), $Ar + TlF$ (Van de Ree, 1971), $Li^+ + H_2$ (McGuire, 1974; Barg *et al.*, 1976), and $K + CO_2$ (Förster, 1975; Scholtheis, 1976). There has been no experimental activity in this direction, probably due to the lack of appropriate high-resolution VC or SS machines. However, the combined LIF and LID technique of Bergmann *et al.* (1978) appears to be very promising for this field.

The available computational results suggest the following interrelations between rainbow scattering and PHS properties:

(a) The position of the rainbow is approximately given by $\vartheta_r \sim 2 \times \varepsilon/E$, the same relation as for the isotropic case (equation (30)). The ε to be inserted is, however, the maximal well depth found on the PHS rather than a spherically averaged value (Barg *et al.*, 1976; Scholtheis, 1976). The rainbow angle ϑ_r is only weakly dependent on Δj. If a shift is observed, it tends towards smaller angles (Barg *et al.*, 1976; Buck and McGuire, 1976', quite in contrast to the first-order prediction of

Truhlar (1973) that a shift in the opposite direction, according to $\Delta\vartheta_r/\vartheta_r \sim \frac{1}{2}\Delta E/E$, should occur.

(b) The amplitude of the rainbow is considerably influenced by the shape of the PHS. Alexander (1977) found that a strongly anisotropic attraction leads to pronounced rainbow peaks while a strongly anisotropic repulsion tends to quench them drastically. Some shift and quenching may arise from the initial rotation of the molecule (Barg *et al.*, 1977; Buck and McGuire, 1976).

The wide-angle scattering can be approximated by a classical model if the influence of the well region on the trajectories is eliminated by the assumption $E, E' \gg \varepsilon$. Then the scattering is governed solely by the steep repulsive core of the PHS. This physical situation suggests—quite in analogy to the rigid-sphere scattering in the isotropic case—replacing the hard core by an angular-dependent rigid shell. The dynamics on this PHS collapse to reflections of the (reduced) projectile from the rigid surface with conservation of angular momentum, energy, and the tangential component of the momentum as the basic dynamic assumptions. Such rigid-shell models have been used for the calculation of rotational excitation probabilities (Kolb *et al.*, 1972; Tsait *et al.*, 1973; Kolb and Elgin, 1977), for the interpretation of molecular alignment in nozzle beams (Sinha *et al.*, 1974), and recently, for the rationalization of differential rotational energy-transfer cross sections (Schepper *et al.*, 1979. According to the model the relative energy transfer $\Delta E/E$ caused by a (reduced) projectile which hits the initially non-rotating rigid shell at the position $\boldsymbol{R}$($\boldsymbol{R}$ = vector from the centre-of-mass of the molecule to the impact point) given by (Tsait *et al.*, 1973; Schepper *et al.*, 1979)

$$\Delta E/E = 4(\hat{\boldsymbol{n}}\hat{\boldsymbol{p}})^2 \frac{\mu/\Theta(\boldsymbol{R} \times \hat{\boldsymbol{n}})^2}{(1 + \mu/\theta(\boldsymbol{R} \times \hat{\boldsymbol{n}})^2)^2}, \tag{34}$$

where the unit vectors $\hat{\boldsymbol{n}}$ and $\hat{\boldsymbol{p}}$ point in the direction perpendicular to the surface at the impact point and along the initial momentum, respectively; μ is the reduced mass of the colliding system, and Θ is the moment of inertia of the molecule. It is interesting to note that the masses of the collision partners enter equation (34) only through the ratio μ/Θ. For the limiting case, $\mu/\Theta\ (\boldsymbol{R} \times \hat{\boldsymbol{n}})^2 \ll 1$, $\Delta E/E$ becomes proportional to μ/Θ. Furthermore, for given μ/θ the energy transfer is governed by the factor $(\boldsymbol{R} \times \hat{\boldsymbol{n}})^2$. This product of two body-fixed vectors is only dependent on the geometric shape of the rigid surface. As a function of $\boldsymbol{R}$ it runs through extrema which are of particular importance because they generate observable (integrable) singularities in the energy transfer cross section $J(E', \vartheta, E)$ for fixed ϑ at distinct final recoil energies $E' = E'_s$ (Schepper *et al.*, 1979, see also Thomas, 1977).

One of the simplest surfaces which may be used to describe a linear molecule with inversion symmetry is a rotationally symmetric (around the internuclear axis) ellipsoid, the centre of which coincides with the centre-of-mass. Then, two singularities of $I(E', \vartheta, E)$ appear at constant ϑ, one for the elastic events at $E'_{s1}/E = 1$ and the other for the inelastic ones at

$$E'_{s2}/E = \left[\frac{(1 - \beta^2 \sin^2 \vartheta)^{1/2} + \beta \cos \vartheta}{1 + \beta}\right]^2 \tag{35}$$

where β characterizes the anisotropy of the ellipsoid:

$$\beta = \mu/\Theta(c - a)^2.$$

Here, c and a are the lengths of the long and short semiaxis of the ellipsoid, respectively. At a given angle ϑ only recoil energies E' in the interval $E \geq E' \geq E'_{s2}$ occur provided $\beta < 1$. For $\beta > 1$ the singularities also exist, but energy transfers of $\Delta E/E = 1$ become possible (Beck, 1978).

A linear molecule without inversion symmetry may be described within this rough approximation by shifting the centre of the ellipsoid by an amount Z_0 with respect to the centre-of-mass. Then three singularities appear, the first of them again in the elastic channel as before. The two others occur at values which reflect the asymmetry in the $(\boldsymbol{R} \times \hat{\boldsymbol{n}})^2$ product due to the different length of any two lever arms $\boldsymbol{R}$ pointing in opposite directions. The longer arm leads to the singularity at the largest energy transfer, the other to the singularity at an intermediate transfer. A computational result is shown in Fig. 14 where $J(\Delta E/E, \vartheta, E)$ is plotted for $\vartheta = 150°$ for $(K + N_2)$- and $(K + CO)$-like systems. In a real scattering experiment, because of the non-perfect energy and angle resolving power, the (integrable) singularities degenerate to finite peaks with the positions of their maxima close to E'_{si}. Such peaks have indeed been observed for $K + N_2$ and $K + CO$ colliding at enhanced energies (Fig. 22, cf. Section V.E) (Schepper *et al.*, 1979).

A further feature common to elastic and inelastic scattering is the diffraction oscillation pattern of the differential cross sections (Section IV.A). In the elastic case the spacing of two adjacent maxima is given by equation (31). It has been shown in a case study for $He + Na_2$ that this formula also applies for inelastic scattering if modified for the change in the wave number corresponding to the energy transfer (McGuire, 1977; Berbenni and McGuire, 1977). The dependence of the position and spacing of the maxima on the potential anisotropy is widely unknown at present.

It should be noted that—in addition to the direct collision processes discussed above—the formation of long-lived collision complexes may also occur in inelastic scattering. The production of such complexes appears to depend predominantly on the ratio $E/\varepsilon = K$. For $K < 5$ Van de Ree (1971) found indications for complex formation and reported an increasing

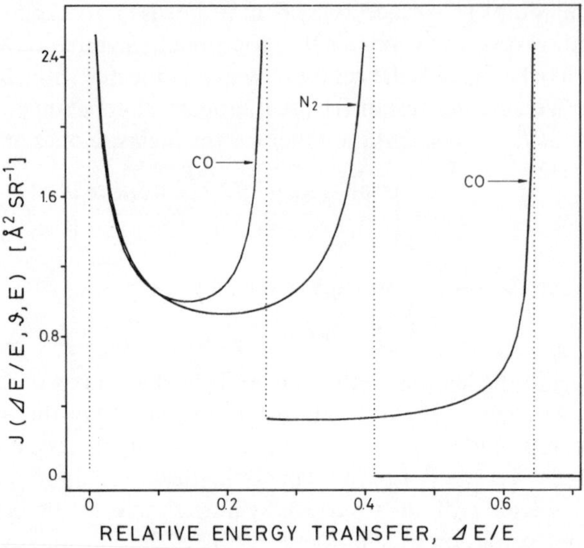

Fig. 14. Double differential energy-transfer cross sections calculated on the basis of the rigid-shell model for (K + CO)- and $(K + N_2)$-like systems at $\vartheta = 150°$. The parameters used are given in Table III (Schepper, 1978). The (integrable) singularities, reflect directly the anisotropy of the rigid shell (equation (35)).

fraction of indirect events towards smaller E. This has also been observed by Schlier (1978) for $H^+ + H_2$. The calculations on $Ar + CO_2$, however, showed no evidence for the existence of complexes at $K \sim 4$, not even for the largest energy transfers. This is probably due to the fact that encounters with $\Delta E/E \leq 1$ take place only for small impact parameters with relatively small orbital momentum $\boldsymbol{l}$. After the transfer of an essential fraction of $\boldsymbol{l}$ into the rotation $\boldsymbol{j}'$ of the molecule, the residual rotational barrier is too low to prevent the separation of the system, and no subsequent collisions occur. If the structure of the PHS permits substantial energy transfers for relatively large impact parameters, complex formation may become an important mechanism.

D. Oriented Cross Sections

1. *Cross-Section Anisotropy*

As pointed out in Section IV.B, molecular systems may exhibit more or less damped glory structures in the integral cross section. If the molecule is oriented these features persist, but as the PHS area sensitively probed by the atomic projectile depends slightly on the specific orientation, the resulting glory structures for two different orientations are essentially identical except for a small shift in the velocity scale, which, of course, reflects the anisotropy. Consequently, the relative cross-section difference A exhibits zero close to the mean extremal velocities of the two glory structures and extrema at their mean zeros (Zandee and Reuss, 1977b).

A qualitative velocity dependence of $A(v)$ is shown in Fig. 15. Here, similar to the atomic case (cf. Section IV.B), glory, transition, and repulsive regions can be clearly identified. From this argumentation it is obvious that the probed PHS areas correspond roughly to those discussed in the elastic case. Extensive investigations of this problem have been carried out by Zandee and Reuss (1977b) for the H_2–rare-gas systems.

Computations of A have been most frequently performed using the HEA (Bennewitz *et al.*, 1964, 1969a; Bennewitz and Haerten, 1969; Kuipers and Reuss, 1973) and DWA (Franssen and Reuss, 1973, 1974; Zandee and Reuss, 1977b), and, recently, also the time-dependent SA (Kuipers and Reuss, 1974; Klaassen *et al.*, 1978). Extensive DWA calculations (small anisotropies) have provided the following approximate expressions for the glory amplitude a and the peak height b of the transition region of A (PHS type II) (Zandee and Reuss, 1977b):

$$a = 0.25(b_2 - 1.5a_2)$$

$$b = 0.1(2a_2 - b_2).$$

The appearance of this kind of parameter combination is characteristic of

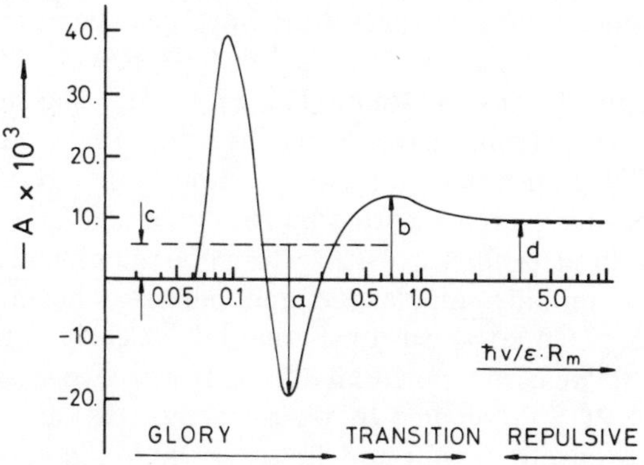

Fig. 15. Schematic velocity dependence of the cross-section anisotropy A. The various amplitudes are given in the text. The velocity scale is, analogous to the isotropic scattering, divided into three portions; glory, transitions, and repulsive (high-velocity) regions. (Reproduced by permission of North-Holland Publ. Co. from Zandee and Reuss, 1977b.)

the fact that the well region is sensitively probed (cf. Section IV.B). For $b_2/a_2 = 1.5$ the A-glory should be quenched; this has recently been nicely demonstrated by Klaassen *et al.* (1978) for (Ar + NO)-like potentials. The authors also found that for systems with a stable linear configuration a maximum of A appears at even glory indices, whereas for those with a stable T configuration a minimum occurs, and *vice versa* for odd indices.

Within the Born approximation one can calculate the limits for pure repulsive and pure attractive PHS, corresponding to the high-velocity limit and the non-glory contributions of A, respectively. The results are:

$$c = \frac{3}{50} a_2 \qquad \text{(non-glory)}$$

and

$$d = \frac{9}{220} b_2 \qquad \text{(high-velocity limit).}$$

These formulae impressively demonstrate the intimate correlation between the quantity $A(v)$ and the anisotropy of the PHS.

2. *m-Transitions*

Due to the development of new experimental techniques, particularly of the LIF (Alexander *et al.*, 1977), considerable theoretical interest in m-dependent cross sections has arisen. Questions regarding the anisotropy dependence of quantities such as polarization and depolarization of scattered molecules as well as individual Δm-transition (reorientation) cross sections have been considered. Recent theoretical work has been principally based on CC calculations: Monchick (1977) studied He + HCl, Alexander (1977, 1978) $Ar + N_2$, and Jacobs and Reuss (1977) $Kr + H_2$. The L-dominant and potential dominant limits have been investigated by Alexander and Dagdigian (1977). Expressions for cross sections with arbitrarily oriented quantization axis have been reported by Alexander *et al.* (1977). The sensitivity of the angular dependence of the vector polarization and of the depolarization ratio for initially unpolarized and polarized beams, respectively, with respect to potential anisotropies has been clearly demonstrated by Alexander (1978). He used a potential of type II with two sets of parameters which provided PHS dominated by the attractive (LR) and repulsive (SR) anisotropy, respectively. Jacobs and Reuss (1978) found substantial differences in the differential reorientation cross sections calculated for LR and SR potentials and an additional one with equal attractive and repulsive anisotropies.

An interesting behaviour of the integral cross section for $\Delta j = 0$ has been reported for $(Ar + N_2)$-like systems (Alexander, 1977) which is probably also characteristic for other atom–molecule systems in the energy range considered. For the LR potential the largest Δm-cross section occurs for $m = m' = 0$ (V is here the quantization axis). This has also been observed using the approximate L-dominant decoupling scheme (Alexander and Dagdigian, 1977). For the SR potential the final cross sections appear with a much wider m'-distribution which peaks at transitions with $m, m' = 0, 0$. It is, however, not statistical as predicted by the simple potential dominant decoupling scheme (Alexander and Dagdigian, 1977).

V. EXPERIMENTAL RESULTS

The concluding section is devoted to a survey of scattering experiments and their results with respect to anisotropic PHS, anisotropy parameters, or any other information on the deviation of the interaction from sphericity. Some of the experiments are cited although they are not analysed in terms of potential surfaces, but in these cases such analysis can be expected in the near future. The PHS used for the interpretation of scattering data are compiled in Table II except for the analytical representations of *ab initio* PHS. The best-fit parameters or other evidences about the potentials are presented in Table III together with some experimental details. In the following only a few remarks about the experimental configurations and conditions are given and we refer to Section III with respect to this subject.

A. Hydrogenic Systems

The most complete set of scattering data is available for H_2 and isotopic variants. Experimentally, this is due to (a) the nuclear moment of ortho-H_2, which permits the alignment of the internuclear axis in magnetic fields and therefore A measurements, and (b) to the large rotational constant which is a prerequisite for the resolution of individual quantum states in VC experiments. From a theoretical point of view hydrogenic systems are a favourite because *ab initio* calculations of PHS and cross sections can be performed with the smallest computational effort compared to all other collision partners. This makes H_2 a prototype for fertile comparisons between experiment and theory.

The velocity dependence of the cross-section anisotropy A has been extensively investigated by the Nymegen group for H_2 + rare gases, H_2, and some further molecules (Moerkerken *et al.*, 1973, 1975; Zandee *et al.* 1976, 1977; Zandee and Reuss, 1977a, b). In these experiments ortho-H_2, completely relaxed to the $j = 1$ level by supersonic expansion, was oriented using Rabi-type magnetic fields. The experimental cross-section anisotropy A_e was simulated numerically using the DWA on the basis of the PHS II and IX and compared to the data in a trial and error procedure. Two parameters were varied, a_2 and b_2 or A_2/A_0 and $a_2^{(8)}$ for PHS II or PHS IX, respectively (the molecular scattering partners are considered to be structureless). The $V_2(R)$ terms obtained proved to be similar in shape to the $V_0(R)$ terms, with well depths of about 10% of that of V_0 and a minimum position shifted slightly to larger distances as compared to R_m. These results compare fairly well with those obtained for H_2–rare gases from i.r. absorption measurements (LeRoy and van Kranendonk, 1974) except for a discrepancy in the well depths of V_2 of 20% to 50%.

The VC method has recently been successfully applied to measure state-to-state differential cross sections for the following systems: $HD(j=0)+Ne \rightarrow HD(j=1)+Ne$ (Buck *et al.*, 1977, 1978a; Huisken, 1978), $HD(0)+He \rightarrow HD(1)+He$ (Gentry and Giese, 1977a; Farrar *et al.*, 1977), $HD(0)+HD(0) \rightarrow HD(0;1;2)+HD(2;1;0)$ (Gentry and Giese, 1977), and $HD(0)+D_2(j) \rightarrow HD(1)+D_2(j)$ (Buck *et al.*, 1978, 1978a; Huisken, 1978). The final velocities of the scattered HD molecules were analysed with a pseudostatistical chopper (Buck *et al.* and Huisken), a single-pulse chopper (Farrar *et al.*) (together with continuous nozzle beams), and with the pulsed-beam technique (Giese and Gentry). The preparation of the $j=0$ state of HD was achieved by supersonic expansion. Buck *et al.* (1978a) and Huisken (1978) successfully measured for the first time state-to-state differential cross sections over a wide range of deflection angles. Their results for $HD(j=0)+Ne$ and $D_2(j) \rightarrow HD(j'=0,1)+Ne$ and $D_2(j)$, which exhibit well-resolved diffraction oscillations, have been used to determine potential anisotropies and/or to test available PHS. In this context it is important to note that despite the identical electronic structure of H_2 and the isotopic variant HD the two molecules behave drastically differently with respect to inelastic scattering. This is due to the displacement of the centres-of-charge and -mass which create odd terms in the expansion equation (3) of the PHS for HD. The first three terms of the Legendre polynomial for the displaced case are given by (Kreek and LeRoy, 1975):

$$\begin{aligned} V_0^d &= V_0 + 0(\delta^2/R^2) \\ V_1^d &= -\delta(\tfrac{6}{5}V_2/R + V_0' + \tfrac{2}{5}V_2') + 0(\delta^2/R^2) \\ V_2^d &= V_2 + 0(\delta^2/R^2) \end{aligned} \tag{36}$$

The functions $V_i(R)$ are the coefficients of the expansion of the H_2-atom PHS. The prime indicates differentiation with respect to R, and δ is the displacement of the two centres. The observed $0 \rightarrow 1$ transitions are due to the V_1^d term, which contains, in addition to the desired V_2 coefficient, also the large isotropic potential portion V_0. A test or the determination of a V_2 term requires therefore the exact knowledge of the isotropic component V_0.

Huisken (1978) and Buck *et al.* (1979) used for the analysis of their HD + Ne data an isotropic potential deduced from the total differential cross section of the system $D_2 + Ne$ (Buck *et al.*, 1978b) and a V_2 term of type IX, consisting of the theoretical model potential of Tang and Toennies (1978) to describe the well and attractive regions and the exponential repulsive branch with a free steepness parameter β. The quantity β could be determined via a trial and error procedure employing the CS approximation. The resulting best-fit V_2 term compares favourably with the one derived from the A measurements (PHS of type III with $m = 15$) over the entire R range (Fig. 17).

Furthermore, a series of test calculations have been performed to check the accuracy of a PHS consisting of the V_0 term of Buck *et al.* (1978b) and the *ab initio* V_2 term of Birks *et al.* (1975) and of the potential of Tang and Toennies (1978). The anisotropy parameter $a_2^{(8)}$ (cf. PHS type IX) contained

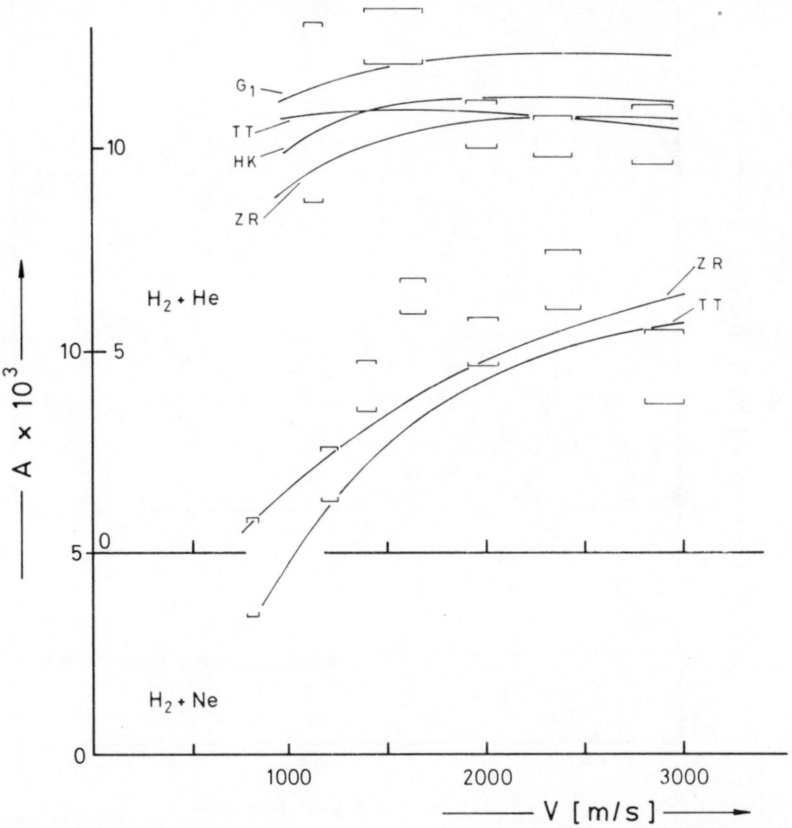

Fig. 16. Comparison of experimental cross-section anisotropies for $H_2 + He$ and Ne (from Zandee and Reuss, 1977a by permission of North-Holland Publ. Co.) with those calculated for the best-fit potential of Zandee and Reuss (ZR), the model potential of Tang and Toennies (1978, by permission of the American Institute of Physics (TT), and for the *ab initio* PHS of Hariharan and Kutzelnigg (1977) (HK) and of Geurts *et al.*, (1975) (G_1).

in the latter PHS, considered as free, could be determined to be $a_2^{(8)} = 0.26$ in a very sensitive way by a comparison of computed and measured cross-section anisotropies (Tang and Toennies, 1978). The resulting nearly perfect fit is exhibited in Fig. 16. The results of the test calculations with respect to the differential cross sections are shown in Fig. 18. The elastic part is well reproduced with either PHS indicating a correct isotropic term. The monotonous as well as the oscillatory portion of the inelastic cross section

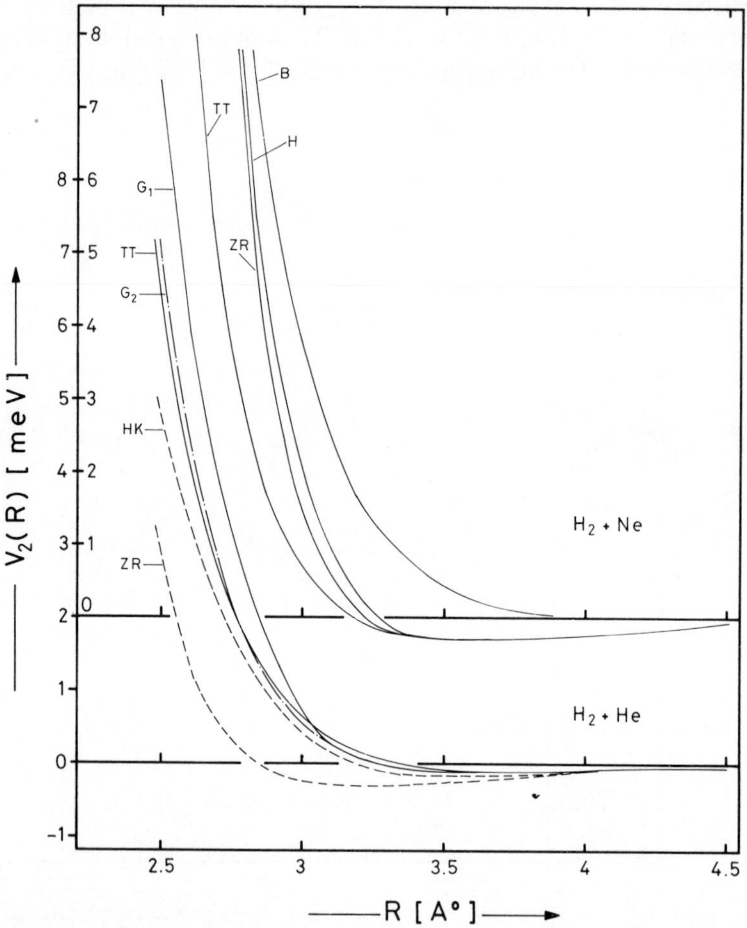

Fig. 17. The V_2 terms of $H_2 + He$ and $H_2 + Ne$ reported by: Hariharan and Kutzelnigg (1977) (HK), Geurts *et al.* (1975, 1977) (G_1 and G_2), Tang and Toennies (1978, by permission of the American Institute of Physics) (TT), and Zandee and Reuss (1977a, by permission of North-Holland Publ. Co.) (ZR) for $H_2 + He$, and by Tang and Toennies (1978) (TT), Zandee and Reuss (1977a) (ZR), Huisken (1978) (H), and Birks *et al.* (1975) (B) for $H_2 + Ne$.

are also nicely simulated except for the interchange of the maxima and minima. This 'phase error' proved to be an artifact of the CS approximation, as was demonstrated with the aid of a CC calculation using the best-fit potential of Huisken (Fig. 18). The small discrepancies at large angles occurring for the Tang and Toennies PHS are not essential as they vanish if the higher-order terms of the expansion equation (36) are taken into account (Buck, 1978). One may therefore conclude that this potential is reliable in the repulsive region around $R \sim 2.65$ Å ($V_0(R) \sim E$) and in the well area between 3.5 Å $< R <$ 4.6 Å (Zandee and Reuss, 1977b) which are most sensitively probed by the inelastic data and the A measurements, respectively.

The results of Giese and Gentry on HD + He have not yet been analysed in detail with respect to a PHS. They only report that the order of magnitude

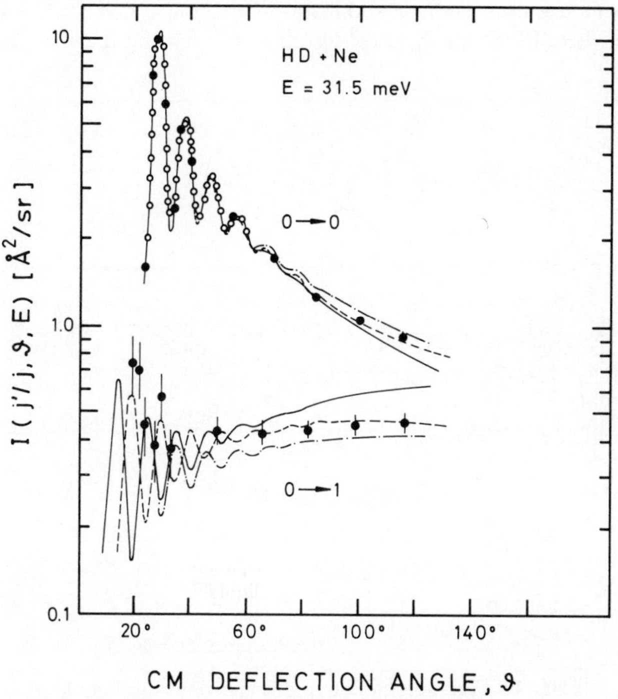

Fig. 18. Differential state-to-state cross sections for the transitions $j = 0 \rightarrow j' = 0$ and 1 in HD + Ne collisions (Huisken, 1978; Buck *et al.*, 1978a, 1979). The various lines are calculated with the potentials of Schleusener (V_0) and Birks *et al.* (1975) (V_2) (dash-dotted line), and of Tang and Toennies (1978, by permission of the American Institute of Physics) (solid-line) using the CS approximation. The dashed line is a CC calculation employing the best-fit PHS of Huisken (1978).

of the transition probabilities and their increase with increasing scattering angle and collision energy are in qualitative agreement with theoretical predictions. The data of Farrar *et al.* on HD + He were compared to CC calculations with the PHS of Shafer and Gordon (1973) and fair agreement was found. In contrast to these data the A measurements for H_2 + He have been extensively exploited to test PHS. Fig. 17 shows the best fit V_2 term of Zandee and Reuss (1977b) together with those of Tang and Toennies (1978), Shafer and Gordon (1973) (empirical), Hariharan and Kutzelnigg (1977) (*ab initio*), and Geurts *et al.* (1975, 1977) (*ab initio*). The strong departure of the anisotropic term of Zandee and Reuss from all others may be due to the unrealistically small R_m value of 3.09 Å used for their analysis. The most sensitively probed region is reported to be the well region (around $R_m \sim 3.4$ Å). Here the V_2 terms differ drastically; for example, ε and R_m values of V_2 vary from 0.17 meV and 3.57 Å (Hariharan and Kutzelnigg) to 0.079 meV and 3.80 Å (Tang and Toennies). Despite this fact, the A values calculated with the various PHS are in good agreement with the data of Zandee and

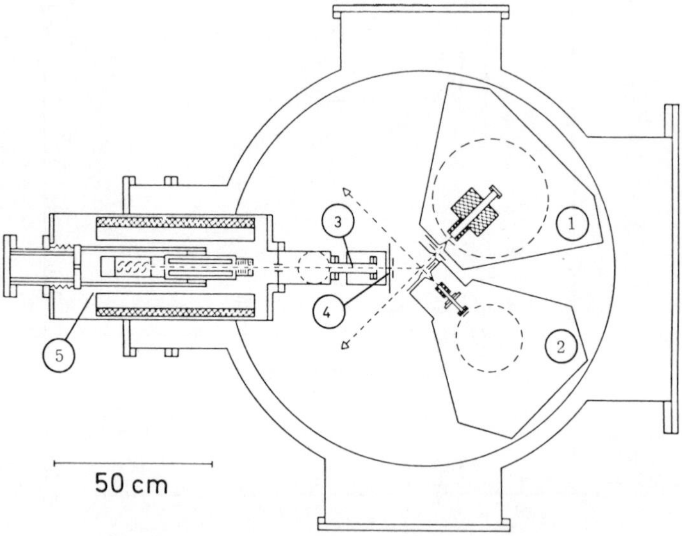

Fig. 19. Cross section through the apparatus used for the VC experiments of Buck *et al.* (1977, 1979) and Huisken (1978). The numbers refer to the following parts: (1) and (2) differential pumping chambers and nozzle beam sources (3) gate valve (4) pseudorandom chopper wheel (5) quadrupole mass spectrometer detector with electron bombardment ion source mounted in a separate UHV housing. Cross-hatched areas represent liquid-cooled surfaces (Huisken, 1978).

Reuss (1977b) (Fig. 16). More discriminating evidence may be expected from comparisons with inelastic scattering data.

The HD–D_2 measurements (Fig. 19) (Buck *et al.*, 1978, 1978a; Huisken 1978) have recently been compared to cross sections calculated with the CC approximation using the *ab initio* PHS of Meyer (1978). The results agree well with the experimental data except for the oscillatory region of the inelastic cross section. The discrepancy is reported to be due to insufficient knowledge about the PHS and does not reflect apparative uncertainties (Buck, 1978). An analysis of the HD–HD measurements has not yet been reported.

It should be mentioned that there are several total integral (Helbing *et al.*, 1968, Butz *et al.*, 1971a, b) and differential (Winicur *et al.*, 1970; Kuppermann *et al.*, 1973; Bickes *et al.*, 1975; Rulis *et al.*, 1977; Rulis and Scoles, 1977; Bauer *et al.*, 1978) cross-section measurements in which glories and diffraction oscillations have been well resolved. However, due to the relatively small anisotropy of the homonuclear hydrogen, no damping was found which proved to be sufficiently large to justify a determination of anisotropy parameters. Without exception spherically symmetric potentials were used for an interpretation of the scattering features.

B. Alkali Halide Systems

A great number of experimental investigations have been devoted to alkali halide systems involving atoms and molecules as collision partners. The reason for this preference is experimental: (a) alkali halides can be readily detected via surface ionization, and (b) their large electric dipole moment permits state selection and orientation by means of electrostatic quadrupole fields. Information on the potential anisotropy is available from A measurements, single quantum transition data, energy-transfer cross sections, and quenching phenomena, all of which will be discussed in the following in this order.

Results on cross-section anisotropies A were first reported by Bennewitz *et al.* (1964) for TlF + rare gases. The anisotropy was calculated using the HEA (cf. Section II.B) on the basis of PHS I. A comparison of the calculated and measured quantities provided for Ar and Kr a consistent a_2 value of 0.235 ± 0.041 while no unique parameter could be found for He and Ne, probably due to the neglect of the repulsion (Bennewitz *et al.*, 1969a). Under improved experimental conditions Bennewitz *et al.* (1969a, b; Bennewitz and Haerten, 1969) reexamined TlF + Ar and Kr and determined a_2 values for GsF + Ar and He. In the case of GsF + He the introduction of a repulsive term (PHS II) was necessary in order to obtain a satisfying fit of the data. The calculations of cross sections were carried out using the HEA. Partially integrated differential cross sections for specific quantum transitions (equation (24)) have been measured for rare gases and molecules plus TlF (Tonnies *et al.*, 1965, 1966), CsF (Borkenhagen *et al.*, 1976; Malthan, 1976; Henrichs *et al.*, 1977), and KCl (Meyer and Toennies, 1978). The use of electrostatic fields as state selectors and analysers limited the investigations to transitions with small Δj ($\Delta j = 1, 2$) starting from low j levels ($j = 1, 2$). The results of the more recent studies are sufficiently reliable for testing theoretical long-range anisotropic potentials. An adequate theoretical description of the dynamics is provided by the HEA frequently used in first order. Using semiempirical expressions for the dispersion and induction potentials (equation (4)) up to the R^{-7} terms Malthan (1976) found agreement with data within 30% for the $\Delta j = 1$ transitions in CsF + Ne, Ar. The computed cross sections for $\Delta j = 2$ incorporating the a_2 value of Bennewitz and Haerten (1969) are too large by a factor of 4 to 6. With the inclusion of higher order terms up to R^{-9} the agreement is expected to become better (Meyer and Toennies, 1978). Dickinson and Richards (1978) report agreement with the $\Delta j = 1$ data of Malthan (1976) within 40% using a classical path approximation with fixed γ (similar to the IOSA) and suggest that an improvement of the theoretical results may be achieved with the inclusion of the repulsive potential branch. Henrichs *et al.* (1977) deduced the angular dependence of the cross sections of CsF + Ar and Kr by varying the transmission function of the state analyser (Section III.D.1). The authors found the results to be consistent with the assumption that the $j = 2 \rightarrow j' = 1$ transition is predominantly due to the R^{-7} potential term.

Numerous VC experiments have been performed for alkali halides interacting with rare gases or molecular targets. They can be categorized according to the ratio of translational energy E to internal energy E_{int} of the alkali halides.

(a) Seeded rare-gas beams were crossed with thermal, effusive beams of CsI (Loesch and Herschbach, 1972), CsF (King *et al.*, 1973), and CsF, CsI, CsCl, CsBr, and RbI (King, 1974) leading to $E/E_{\text{int}} \gg 1$.
(b) In triple-beam experiments vibrationally excited KBr was generated via the reaction $K + Br_2 \rightarrow KBr + Br$ and shot onto thermal target beams of CH_3OH (Donohue *et al.*, 1972), Ar, and CO_2 (Chou *et al.*, 1973), polar molecules (Donohue *et al.*, 1973), non-polar molecules (Crim *et al.*, 1973), and polyatomic molecules (Crim *et al.*, 1974). The energy ratio is here $E/E_{\text{int}} \ll 1$.
(c) Two thermal effusive beams were crossed resulting in $E/E_{\text{int}} \sim 1$. The systems studied were CsI + Ar, CsCl + Ar, and CsI + Xe (Armstrong *et al.*, 1975; Greene *et al.*, 1977).

The experiments of type (a) exhibit a surprising feature. Here, with a substantial integral cross section of 5 Å^2, the available energy is nearly completely transferred into the internal degrees of freedom (ballistic effect). The inverse phenomena has been found for similar systems at the initial conditions (b). A CT study of Ar + vibrationally excited KBr on the basis of three different empirical PHS has demonstrated (Matzen and Fisk, 1977) that both depth and shape of the well region sensitively influence the energy-transfer behaviour. Double impact collisions proved to be an important mechanism for the deactivation. (Such multiple collisions may well be responsible for the ballistic effect as indicated by Loesch and Herschbach (1972). No further attempt was made to deduce additional potential information.

The experimental results of case (c) are consistent with the assumption that most encounters lead to the formation of complexes which live sufficiently long to permit the randomization of the available energy among the degrees of freedom of the complex. Recently, evidence has been found that also the initial conditions (a) and (b) for atomic partners may lead to complexes which decay statistically (McGinnis and Greene, 1978). Complex formation has also been reported for polar target molecules in type (b) experiments while non-polar molecules appear preferably to interact directly.

The velocity dependence of total cross sections has been measured for a variety of alkali halide–atom/molecule systems (Cross *et al.*, 1966; Hessel and Kusch, 1968; Bennewitz *et al.*, 1969b; Richman and Wharton, 1970; David *et al.*, 1973; Dehmer and Wharton, 1974). The absence of glory structures in all these systems (presumably due to inelastic and reorientation processes) for trajectories with impact parameters around b_g (Dehmer and Wharton, 1974) did not permit the determination of anisotropy parameters. On the other hand, high-resolution total differential cross sections measured for LiF + Ar and Kr (Reed and Wharton, 1977) exhibit usable quenching of the diffraction undulations and of the rainbow amplitude (LiF + Kr). The data were analysed in terms of opacity functions, which give for each partial wave the probability for the loss of elastically scattered particles due to Δj-and/or Δm-transitions. Using first-order transition probabilities of the HEA the V_2 term has been extracted from the opacity function.

Detailed information about the potential anisotropy can be expected from the total differential cross-section experiments with oriented LiF molecules which have recently been performed by Tsou *et al.* (1977). The total differential cross-section anisotropy A_d, (see equation (28), however, with $I(1,0;\vartheta,\varphi,E)$ replaced by the degeneracy averaged $I(1;\vartheta,E)$) was found to be $< 8\%$ for Ar + LiF and hence much larger than known A values. Furthermore, as the small-angle portion of $A_d(\vartheta,\varphi,E)$ reflects the potential at large distance, it is less subjected to severe rotational quenching than A, which already contains strongly inelastic trajectories due to the relatively small impact parameters around b_g (cf. Section IV.C). This might play an important role, particularly in the investigation of highly anisotropic molecules with quenched glories. A detailed interpretation of the A_d data in terms of potential anisotropies has not yet been offered.

C. Alkali Hydrides

These molecules are characterized by (a) a large electric dipole moment, (b) a large rotation constant, and (c) spectra in the visible and/or near ultraviolet corresponding to $\Sigma \rightarrow \Sigma$ transitions. They may therefore be state selected in electric quadrupole fields and state-selectively detected via LIF. These techniques have been utilized by Wilcomb and Dagdigian (1977) to determine integral state-to-state cross sections for LiH $(j = 1) + \mathrm{Ar} \rightarrow$ LiH $(j' = 2\text{–}5) + \mathrm{Ar}$ at an elevated collision energy of 0.8 eV (LiH is produced by passing H_2 through melted Li within the oven; this also produces a seeded beam effect (Dagdigian, 1976)). The data appear to be sensitive to the repulsive potential anisotropy (cf. Section IV.C); however, a detailed analysis has not yet been reported.

D. Hydrogen Halides

The exceptionally large probability for radiative vibrational $(\Delta v = 1)$ transitions and their large vibrational spacings make these molecules particularly qualified for state-selective detection by means of the IRL technique. This method has been utilized by Ding and Polanyi (1975) in their molecular beam investigation of the rotational–translational energy transfer between HCl and HF + rare gases and some molecules. The required population of the radiant $v = 1$ state was achieved by heating the supersonic nozzle beam source up to temperatures around 1900 K. The experiments were carried out at energies ranging from thermal up to elevated values of 1.4 eV (seeded beams). No complete rotational relaxation was achieved, resulting in a residual j-distribution $N(j)$ with $j \leq 15$. The luminescence intensity can be related to the change, $\Delta N(j)$, of the populations of a state with target beam on and off due to rotationally inelastic collisions. $\Delta N(j)$ is correlated to the integral cross section by

$$\Delta N(j') = \sum_{j \neq j'} N(j) Q(j'/j; E) - N(j') \sum_{j} Q(j/j'; E) \tag{37}$$

The first term describes the repopulation of the j' state from all the initial states, the second term its depopulation into all available states. Due to the relatively large number of initial states there are always many more unknown cross sections in equation (37) than known population changes. Ding and Polanyi (1975) attempted a solution of equation (37) by adjusting the free parameter $C(E)$ of a simple exponential ansatz (Polanyi and Woodall,

1972; Polanyi and Sathyamurthy, 1978) for the cross-section matrix:

$$Q(j'/j;E) = C_0(2j'+1)\left(\frac{E'}{E}\right)^{1/2} \exp[-C(E)|\Delta E(j',j)|]$$

$\Delta E(j',j)$ is the difference between the final and initial rotational energies (equation (9)) and C_0 a constant. The resulting best-fit parameter $C(E)$ served as the basic experimental result for testing several available PHS (cf. Section II. A) via CT calculations (Ding and Polanyi, 1975; Polanyi and Sathyamurthy, 1978). As can be expected from general considerations (cf. Section IV. C), the inelastic cross sections proved to be sensitive only to the repulsive anisotropy. Since all PHS studied behave similarly in this region, it was not possible to choose the best among them. The extensive computations carried out for Ar + HCl provided equally good fits for the entire set of surfaces.

Olson and Bernstein (1968, 1969) applied their theory of glory quenching to the experimental results of Helbing and Rothe (1968) for Li + HCl, DCl, HBr, and HI. They deduced, for a potential of type IV, upper limits of the repulsive anisotropy parameters b_1 and b_2. The attractive anisotropy a_2 was calculated from polarizabilities. Rainbow damping was observed for Ar + HCl (Farrar and Lee, 1974), but the data were only analysed in terms of spherical potentials.

E. Other Diatomic Molecules

Scattering experiments have also been performed for a variety of diatomic molecules other than those already mentioned. Collision partners were predominantly rare gases, but alkali atoms and simple molecules have also been reported.

A series of cross-section anisotropies have been measured for NO. Orientation was achieved by selecting the $j = m = \Omega = 3/2$ state in an electric six-pole field. This selection technique is based on the nearly linear Stark effect of the $^2\Pi_{3/2}$ ground state. Glories have been observed for NO + Ar, Kr, and Xe (Schwartz *et al.*, 1973), thus providing via a DWA calculation for PHS II the attractive and repulsive anisotropy parameters. For N_2 and SF_6 (Kessener and Reuss, 1975), and CO_2, CS_2, N_2O, CCl_4, and CF_3H (Stolte *et al.*, 1972, 1973) as targets, glories could not be resolved; consequently only the attractive anisotropy (PHS I) could be extracted. (The structure of the molecular targets was always neglected.)

The first state-to-state differential cross sections for a molecule other than hydrogen were determined by Bergmann *et al.* (1978) (Fig. 21). In a promising experiment the authors combined the LID technique to 'state select' (depopulate) the $v = 0, j = 28$ state of Na_2 with the LIF method to selectively detect *this* state. Collision partners were He and Ne. With the LID laser 'on' the LIF signal S_{on} is proportional to the sum of all LAB cross sections leading from the *populated* initial states to the detected final state (background signals are neglected for simplicity):

$$S_{on} \simeq \sum_{v,j\neq 0,28} I^{LAB}(v'=0, 28/v, j; \Theta, E) f(v,j) P(V'_{Na_2})$$

$f(v, j)$ characterizes the population of the initial states, and $P(V'_{Na_2})$ corrects for the velocity dependence of the sensitivity of the LIF detector. With the LID laser 'off' the sum of all (including $v=0, j=28$) initial states to $v'=0$, $j'=28$ is measured. The difference between the two signals provides the differential state-to-state cross section directly, in this case $I^{LAB}(0.28/0.28; \Theta, E)$. The preliminary results shown in Fig. 20 exhibit intense inelastic scattering over the entire angular range indicating a strongly anisotropy PHS. This is in qualitative agreement with recent CS calculations (Berbenni and McGuire, 1977). Further cross-section measurements with $\Delta j \neq 0$ are in progress (Bergmann *et al.*, 1979).

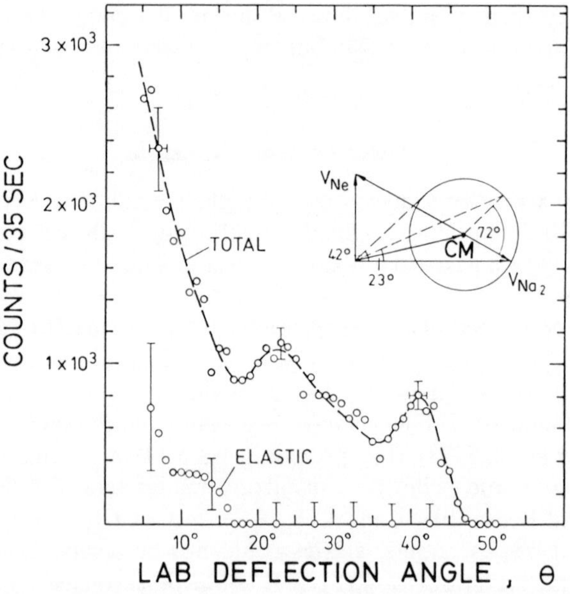

Fig. 20. Angular distributions of Na_2 scattered by Ne. The $v=0$, $j=28$ state of Na_2 is selectively detected via LIF. The curve marked 'total' incorporates all processes which end in the detected state while the curve marked 'elastic' represents the pure elastic scattering; $Na_2(v=0, j=28) + Ne \rightarrow Na_2(v=0, j=28) + Ne$ (Bergmann *et al.*, 1978, reproduced by permission of the American Physical Society).

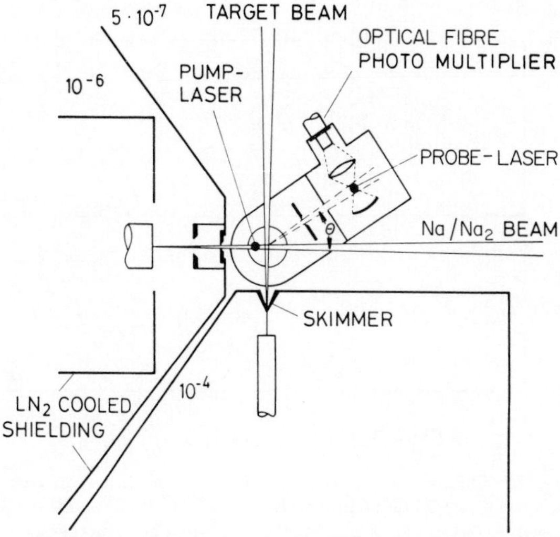

Fig. 21. Schematic cross section through the apparatus used in the state selective scattering experiment of Bergmann *et al.* (1978). The pump laser 'state selects' the Na_2 beam via LID while the scattered molecules are state selectively detected by the probe laser utilizing LIF. The fluorescent light is collimated onto an optical fibre and guided to the photomultiplier. The operating pressures indicated within the various pumping chambers are given in Torr. Reproduced by permission of the American Physical Society.

Differential cross sections ($5^\circ \leq \Theta \leq 20^\circ$) for $N_2 + Ar$ into individual rotational states measured using the EBF method have been reported by Scott *et al.* (1973). The time-dependent SA was employed to extract anisotropy parameters on the basis of PHS III. However, no consistent fit could be achieved, probably due to the potential ansatz used, which was much too simple to describe the repulsive interaction realistically.

Recently, Schepper *et al.* (1978, 1979) measured differential energy transfer cross sections for N_2 and $CO + K$ at elevated collision energies between 0.34 and 1.24 eV. The inelastic events were detected utilizing a somewhat modified VC technique (Fig. 23). The authors measured velocity distributions at constant mean CM deflection angles $\bar{\vartheta}$ rather than at constant observation angles Θ. Fixed $\bar{\vartheta}$ was accomplished by computer-controlled simultaneous variation of Θ and V_1' such that, according to the Newton diagram for the most probable beam velocities, $\bar{\vartheta}$ remained constant. This method has the striking advantage that structures appearing in $J(E', \vartheta, E)$ as a function of E' can be directly measured within one experimental run through *and* within the regions of best velocity resolving power (Section III.D.2). To gain the

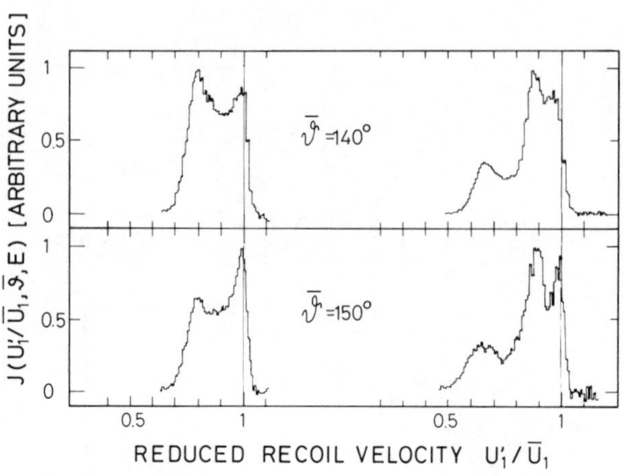

Fig. 22. Differential energy-transfer cross sections measured for two mean CM deflection angles for K + N_2 (left) and K + CO (right) colliding at $E = 1.24$ eV. The distinct structures, two peaks for N_2 and three peaks for CO, are interpreted as washed out remnants of the classical cross-section singularities provided by the rigid-shell model (cf. Fig. 14) (Schepper *et al.*, 1979).

same information with the standard method of measuring, velocity distributions at constant Θ data for numerous closely lying Θ would be required. In principle both techniques are equivalent; however, that of Schepper *et al.* is better adjusted to the physical problem of resolving structures in the E'-dependence of J. A sample of two measurements is shown in Fig. 22. The remarkable structures—two peaks for N_2, three for CO—have been directly traced back to the geometrical anisotropy of the repulsive potential branch employing the classical rigid-shell model described in Section IV.C. The interpretation of the structures as arising from vibrational excitation could be convincingly ruled out by the fact that the peaks are still present below the excitation threshold. This type of VC experiment appears to be one of the most promising techniques in determining repulsive anisotropies. Further diatomic ($^{12}C\ ^{18}O$, O_2, NO) and triatomic (CO_2) molecules are being investigated (Beck *et al.*, 1979).

Quenching of total differential cross sections has been observed for several atom–diatomic systems such as Ar + N_2 (Bickes and Bernstein, 1969; Anlauf *et al.*, 1971), Ar and Kr + N_2 and O_2 (Tully and Lee, 1972), Ar and Kr + N_2 (Cavallini *et al.*, 1971*b*), and Hg + I_2 (Wilcomb *et al.*, 1976), but the data have only been interpreted in terms of spherical potentials. Anisotropic PHS for He + N_2, O_2, CO and NO from diffraction oscillation quenching have been announced by Keil *et al.* (1979), but their results are not yet available to us.

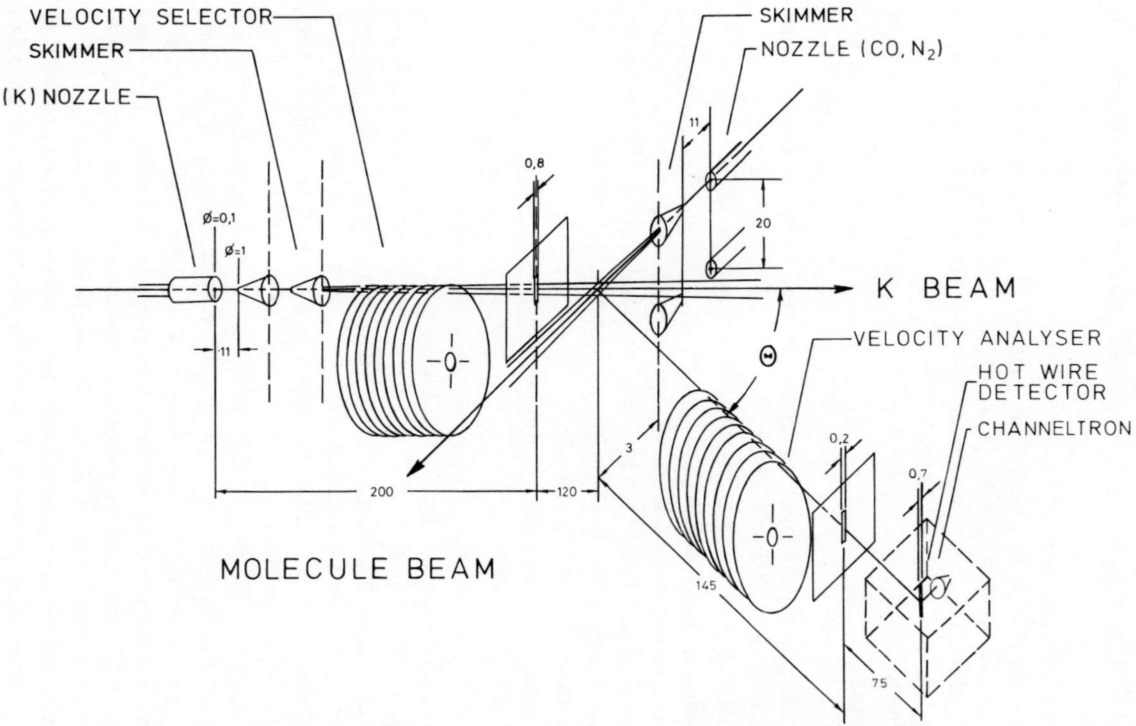

Fig. 23. Schematic perspective diagram of the apparatus used for the VC experiments of Schepper *et al.* (1979). The LAB angle Θ and the velocity transmitted by the analyser are scanned simultaneously in such a way that the CM deflection angle ϑ given by a representative Newton diagram remains constant. The two molecule beam nozzles operate alternately (0.1 Hz) to generate a chopped scattering signal without changing the background pressure. Dimensions are given in mm (Schepper, 1978).

F. Triatomic molecules

Information about the potential anisotropy of triatomic molecules originates predominantly in energy transfer cross-section measurements determined via the VC technique. Some arise from glory and diffraction pattern quenching.

Experimental results from VC measurements are available for: $K + CO_2$ (Beck and Förster, 1975), $K + COS$, CS_2, and SO_2 (Förster, 1975), $Ar + CO_2$ (Loesch, 1976), Ar, $Kr + N_2O$, CO_2 (Farrar *et al.*, 1973), and rare gases $+ CS_2$ (Blais *et al.*, 1977), and from quenching data for HD, D_2, $He + CO_2$ (Butz *et al.*, 1971a), and $He + CO_2$ (Keil *et al.*, 1978).

The energy transfer behaviour of all systems which have been investigated is very similar.

(a) The rotationally inelastic processes occurs with gas kinetic cross sections of the order of 40 Å^2.
(b) Transfer of considerable amounts of energy, up to 70–90% of the collision energy, is observed.
(c) The cross sections peak in the forward direction with substantial intensity for small and even for medium energy transfer. This property is particularly prominent for the K systems and shows a tendency to increase from CO_2 to CS_2.
(d) No evidence is found for vibrational excitation, which is energetically allowed in most cases.

All the above experimental findings could be qualitatively, partially even quantitatively, rationalized by CT calculations in which simple empirical or electron gas potentials were utilized. The dependence of the integral cross section on the final rotational quantum number was used to extract, by trial and error procedure, the anisotropy of the repulsive core of a two-centre L–J(12–6) potential (PHS VII) (Loesch, 1976). The sensitivity of this quantity with respect to the (repulsive) anisotropy is illustrated in Fig. 24. With the resulting best-fit potential the differential cross sections could at least be simulated qualitatively. A more quantitative fit of the angular distributions was achieved by Preston and Pack (1977) (cf. Fig. 25). They employed a PHS consisting of a repulsive part based on the electron gas model and an attractive portion in which experimental polarizabilities, Padé approximant methods, and combining rules were incorporated (Parker *et al.*, 1976). Depending on the way in which the two potential branches were joined they could generate a surface with a deep (34.4 meV) and shallow (5.5 meV) well at the T-shaped and linear configurations, respectively, and another surface with a rather constant well depth (~ 17 meV). The calculations clearly demonstrated that intense forward scattering (c) could only

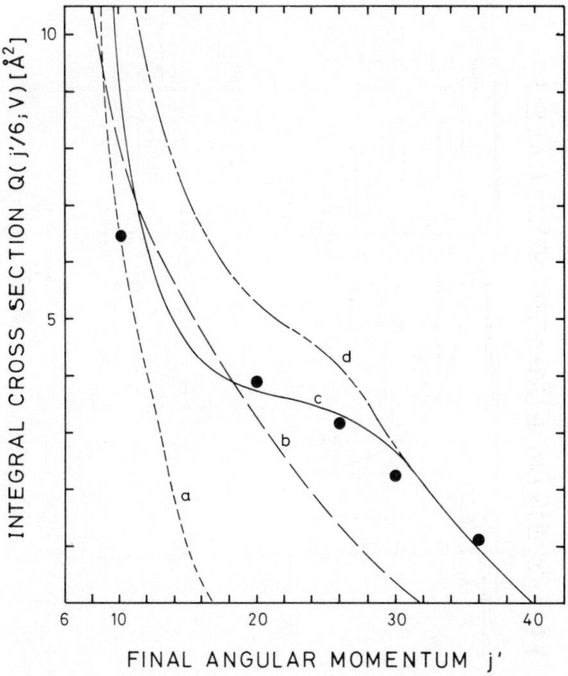

Fig. 24. Integral cross section of $Ar + CO_2$ colliding at $E = 0.069$ eV as a function of the mean final angular momentum (Loesch, 1976). The lines are calculated via classical trajectories with a potential of type VII for the parameters: l(Å), ε(meV), and R_m(Å) = 0.29052, 7.63, and 4.1 (a); 0.58105, 8.431, and 4(b); 1.1621, 9.63, and 3.8(c); and 1.74315, 10.4, and 3.7(d). (Reproduced by permission of North-Holland Publ. Co. from Loesch, 1976.)

be achieved with the deep well potential. This correlation has also been found by Förster (1975) for $K + CO_2$ utilizing a PHS similar to type VI but with a γ-dependent $R(\gamma) = R + \beta^{-1}P_2$ and constant R_m. The property (c), therefore, represents marked evidence for the existence of a deep potential well at the T-shaped configuration with a tendency to increase in depth from $Ar + CO_2$ to $K + CO_2$ to $K + CS_2$. In Section IV. C this behaviour is traced back to inelastic glory scattering. It is interesting to note that the potential well deduced from pure scattering arguments has been found spectroscopically for one system, namely, the van der Waals molecule ArCOS (Harries *et al.*, 1975).

Point (d) is supported by the results of a recent CT calculation based

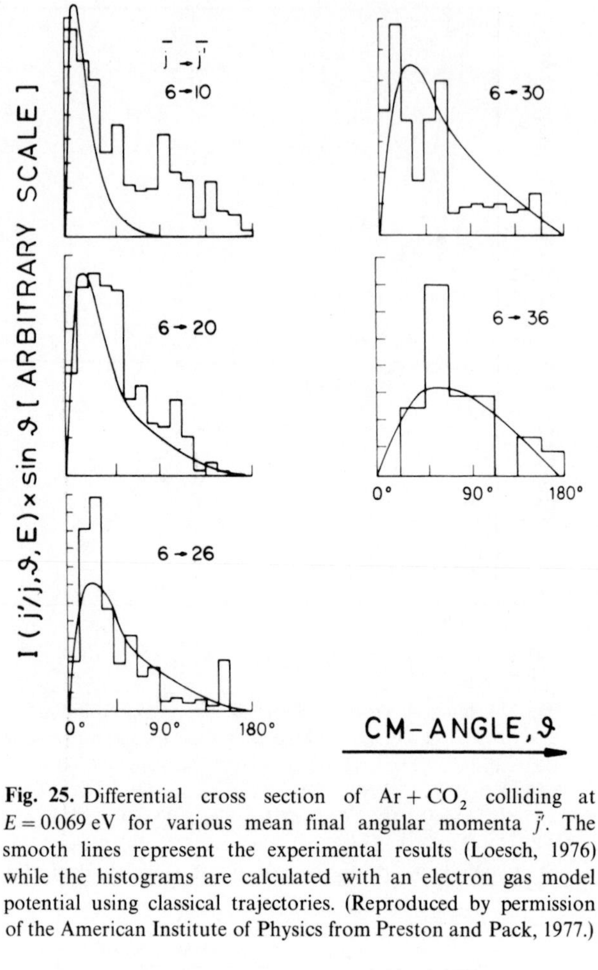

Fig. 25. Differential cross section of $Ar + CO_2$ colliding at $E = 0.069$ eV for various mean final angular momenta $\bar{j}'$. The smooth lines represent the experimental results (Loesch, 1976) while the histograms are calculated with an electron gas model potential using classical trajectories. (Reproduced by permission of the American Institute of Physics from Preston and Pack, 1977.)

on an additive empirical PHS (Suzukawa *et al.*, 1978). The computations were performed for rare gases + CO_2 colliding at energies up to 10 eV, negligible vibrational excitation was reported below 1 eV.

Damping of the glory amplitudes has been observed for the systems He, H_2, and HD + CO_2. The effect was used to estimate the anisotropy of a potential type II ($a_2 = b_2$) by means of a very crude scattering calculation (Butz *et al.*, 1971a). Damping of the diffraction oscillations has recently been reported by Keil *et al.* (1978) for He + CO_2. The total differential cross sections were analysed in terms of potential parameters by employing a trial and error procedure using the IOSA and the modified Morse–spline–van der Waals PHS given in Table II (type X). The spline coefficients $S_i(i =$

1–4) are fixed by smoothness conditions at the spline points $\rho_1 = 1 + (\ln 2)/\beta$ and $\rho_2 = 1.6$. The C_n are given by the long-range dispersion coefficients of Pack (1976). The free parameters ε_0, a, R_m (90) b, and β could be determined uniquely; the fit obtained is shown in Fig. 26. The fitting procedure was tested by applying it to the isotropic system He + Ar; the vanishing of the anisotropy parameters in this case confirmed their physical significance for He + CO_2. Further empirical PHS have been announced for He + N_2O (and C_2N_2) (Parker *et al.*, 1979).

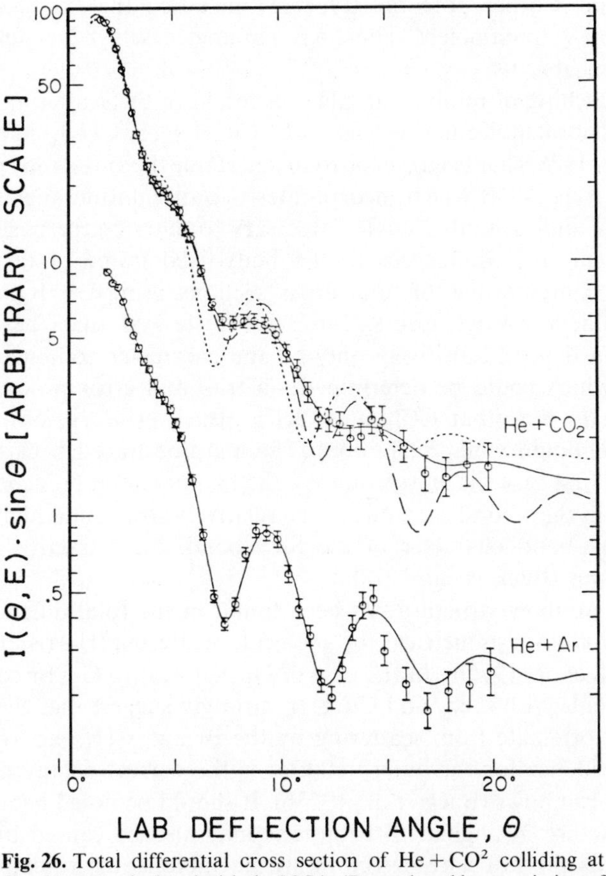

Fig. 26. Total differential cross section of He + CO^2 colliding at $E = 64$ meV calculated with the IOSA. (Reproduced by permission of North-Holland Publ. Co. from Keil *et al.*, 1978.) The solid line represents the best fit obtained with a PHS of type X using the parameters given in Table III. The dotted and dashed lines refer to scattering from two different isotropic model potentials. The lower portion shows the differential cross section of He + Ar calculated for the potential of Keil *et al.* (1979).

G. Polyatomic Molecules

Anisotropy parameters for atom–polyatomic molecule systems have been determined at yet solely by total differential cross-section measurements. A VC experiment has been performed for $K + NH_3$ (Kusunoki, 1977), but no potential anisotropy was deduced.

Strong quenching of diffraction undulations of the systems $H + C_2H_6$ and C_3H_4 has been reported very recently by Valbusa (1978) (see also Este *et al.*, 1978, and for apparative details, Bassi *et al.*, 1976). The data were analysed by trial and error calculation using the IOSA with a modified Buckingham–Corner potential (type IV including the R^{-10} term and a γ-dependent β parameter). The final parameter sets have not yet been communicated to us.

The quenching of rainbow amplitudes found in the scattering of sodium from tetrahedral molecules such as $SiCl_4$, $C(CH_3)_4$, $Si(CH_3)_4$, and $Sn(CH_3)_4$ (Buck *et al.*, 1978*c*) has been exploited to determine the anisotropy parameters of a PHS (type VIII) which incorporates two orientation angles i.e. γ and χ the polar and azimuthal angles necessary to describe the position of the impinging atom with respect to the body-fixed frame of the molecule. Extensive computations of total cross sections using the IOSA and the time-dependent SA (cf. Cross, 1967, 1968) showed that the quenching phenomena depend sensitively only on the parameter combination $\alpha_{\text{eff}} = a - 0.4b$, which could be determined via trial and error procedures. It is interesting to note that CCl_4 and $GeCl_4$ (Buck *et al.*, 1975a) feature no rainbow damping while $SiCl_4$ does. This may be traced back to the fact that in the first case the anisotropies of the two potential branches compensate while in the second case, due to the relatively strong and highly polarizable double-bond character of the SiCl bond, the attractive anisotropy predominates (Buck *et al.*, 1978).

Double rainbow structure has been found in the total differential cross sections of atom–symmetrical top molecule scattering (Harris and Wilson, 1971; Buck *et al.*, 1975b). In the case of $Na + (CH_3)_3\text{—C—Br}$ comparative measurements with CBr_4 and $C(CH_3)_4$ strongly suggest that the two rainbows may originate from scattering by the Br and CH_3 end, respectively. Further evidence for this interpretation is the correct energy dependence of the two rainbows (Buck *et al.*, 1975b). It should be noted here that additional structure in angular distributions may also be caused by chemical reactions, as the hot wire detectors are not able to discriminate between product molecules and the non-reactively deflected alkali atoms (Airey *et al.*, 1967; Greene *et al.*, 1969a; Sloane *et al.*, 1972). The above evidence, however, excludes such an interpretation for the cases discussed. This permits a reliable analysis of the data in terms of anisotropic PHS. Buck *et al.* (1975b) performed trial and error calculations using the time-dependent SA (Cross,

1967, 1968) with a potential of type V which eventually led to the determination of the free potential parameters.

Acknowledgements

I would like to thank Dr. Denise Caldwell for carefully reading the manuscript, Miss Brigitte Falkenburg for performing the calculations presented in Figs. 11 and 12, and Prof. D. Beck and Dr. W. Schepper for communicating their results prior to publication.

V. TABLES

TABLE I. Parameters of PHS VI used for the CT calculations shown in Fig. 11. The resulting PHS gives the same values for $\varepsilon(0^0)/\varepsilon(90^0)$, $R(0^0)/R(90^0)$ at $V(R, \gamma) = 0$, and the spherical average of $\varepsilon(\gamma)$ as the empirical PHS of Loesch (1976).

Set	ε(meV)	R_m(Å)	a	b	E(meV)	j
I	13.5	4.21	−0.3817	0.2456	110	6
II	27					

TABLE II. Some potential model functions used for the analysis of molecular scattering experiments. The $P_l(\cos\gamma)$ and $Y_l^m(\gamma,\chi)$ denote the Lengendre polynomials and spherical harmonics, respectively.

Type	*Van der Waals potential* $V(R,\gamma)$	Typical system	References
I	$-\frac{C^{(6)}}{R^6}(1+a_2P_2)$	Ar + TlF	Bennewitz *et al.* (1964)
	Modified Lennard–Jones potentials $V(R,\gamma)$		
II	$\varepsilon\left[\left(\frac{R_m}{R}\right)^{12}(1+b_2P_2)-2\left(\frac{R_m}{R}\right)^6(1-a_2P_2)\right]$	He + CsF	Bennewitz *et al.* (1969b)
III	$\frac{\varepsilon}{m-6}\left[6\left(\frac{R_m}{R}\right)^m(1+b_2P_2)-m\left(\frac{R_m}{R}\right)^6(1+a_2P_2)\right]$	Ne + H_2	LeRoy and van Kranendonk (1974)
IV	$\varepsilon\left[\left(\frac{R_m}{R}\right)^{12}(1+b_1P_1+b_2P_2)-2\left(\frac{R_m}{R}\right)^6(1+a_2P_2)\right]$	Li + HCl	Olson and Bernstein (1969)
V	$\varepsilon\left\{\left(\frac{R_m}{R}\right)^{12}-2\left(\frac{R_m}{R}\right)^6+\sum_{i=1}^{3}\left[\left(\frac{R_m}{R}\right)^{12}b_j-2\left(\frac{R_m}{R}\right)^6a_i\right]P_i\right\}$	Na + $(CH_3)_3$ CBr	Buck *et al.* (1975)
VI	$\varepsilon(\gamma)\left[\left(\frac{R_m(\gamma)}{R(\gamma)}\right)^{12}-2\left(\frac{R_m(\gamma)}{R(\gamma)}\right)^6\right]$	He + CO_2	Parker and Pack (1978)
	$\varepsilon(\gamma)=\varepsilon(1+aP_2);\ R_m(\gamma)=R_m(1+bP_2),\ R(\gamma)=R$		
	or		
	$\varepsilon(\gamma)=\varepsilon(1+a\cos^2\gamma);\ R_m(\gamma)=R_m;\ R(\gamma)=R-\beta^{-1}\cos^2\gamma$	K + CO_2	Foster (1975)
VII	$\varepsilon\sum_{i=1}^{2}\left[\left(\frac{R_m}{R_i}\right)^{12}-2\left(\frac{R_m}{R_i}\right)^6\right]$	Ar + CO_2	Loesch (1976)
	$R_i=(l^2+R^2+(-1)^i 2Rl\cos\gamma)^{1/2}$		
	$V(R,\gamma,\chi)$		

	$\alpha_{eff} = a - 0.4b$	Na + $SiCl_4$	Buck *et al.* (1978)
	Modified Buckingham–Corner potential $V(R,\gamma) = V_0(R) + V_2(R)P_2$		
IX	$V_0(R) = A_0 \exp(-\beta R) - \left(\frac{C^{(6)}}{R^6} + \frac{C^{(8)}}{R^8} + \dots\right) D(R)$		
	$V_2(R) = A_0 A_2/A \exp - (\beta R) \quad \left(a_2{}^{(6)}\frac{C^{(6)}}{R^6} + a_2{}^{(8)}\frac{C^{(8)}}{R^8} + \dots\right) D(R)$ q	Ar + H_2	Zandee and Reuss (1977)
	$D(R) = \begin{cases} \exp[-a(bR_m/R - 1)^c] & \text{for } R < bR_m \\ 1 & R \geq bR_m \end{cases}$		
	Morse–spline–Van der Waals potential $V(R,\gamma) = \varepsilon(\gamma) f[\rho(R,\gamma)]$		
X	$\varepsilon(\gamma) = \varepsilon(1 + aP_2), \rho(R,\gamma) = \frac{R}{R_m(\gamma)}; R_m(\gamma) = R_m(90°)\left(\frac{1 + b\sin^2\gamma}{1+b}\right)^{1/2}$		
	$\begin{cases} \exp[2\beta(1-\rho)] - 2\exp[\beta(1-\rho)] & \text{for } \rho \leq \rho_1 \\ (\rho_2 - \rho_1)[S_1(\rho_2 - \rho_1) + S_3] + \\ + (\rho - \rho_1)[S_2(\rho - \rho_1)^2 + S_4] & \text{for } \rho_1 < \rho < \rho_2 \\ -\frac{1}{\varepsilon(\gamma)}\sum_{i=1}^{3}\frac{C^{(2i+4)}}{(\rho R_m(\gamma))^{2i+4}} & \text{for } \rho \geq \rho_2 \end{cases}$	He + CO_2	Keil *et al.* (1978)
	Rigid-shell potential $V(R,\gamma)$		
XI	$\begin{cases} 0 & \text{for } R > R(\gamma) \\ \infty & R = R(\gamma) \end{cases}$ with: $\left[\frac{R(\gamma)}{a}\sin\gamma\right]^2 + \left[\frac{R(\gamma)\cos\gamma Z^0}{c}\right]^2 = 1$	K + CO, N_2	Schepper *et al.* (1979)

TABLE III. Survey of scattering experiments for which fitting or test calculations with anisotropic potentials have been or will soon be performed. Systems are given in the order of their appearance in the text (Section V).

System	Collision energy (meV)	Exp. data	State selection: method	State selection: prep. states	State analysis: method	State analysis: anal. states	Data analysis
$H_2 + He$	9.7–54	*A*	RF	$j = 1, m = 0$			F
							T
							F
HD + He	19.3	$I(j'/j; \vartheta, E)$	NB	$j = 0$	VC	$j' = 0, 1$	T
	30.9	$I(j'/j; \vartheta, E)$	NB	$j = 0$	VC	$j' = 0, 1$	
$H_2 + Ne$	6–74	*A*	RF	$j = 1, m = 0$			F
							F, T
HD + Ne	31.5 35.7	$I(j'/j; \vartheta, E)$	NB	$j = 1$	VC		F
							T T
$H_2 + Ar$	6.3–15.3	*A*	RF	$j = 1, m = 0$			F
$H_2 + Kr$	6.5–84	*A*	RF	$j = 1, m = 0$			F
$H_2 + Xe$	6.5–85	*A*	RF	$j = 1, m = 0$			F

Type	Potential: anisotropy parameters	other parameters	References
II	$a_2 = 0{\cdot}19$[1], $b_2 = 0{\cdot}25$	$\varepsilon = 1.68$, $R_m = 3.09$[2]	Zandee and
AI[13]			Reuss (1977a)
TM			Tang and Toennies (1978)
E[7]			Farrar *et al.* (1977)
			Gentry and Giese (1977)
II	$a_2 = 0{\cdot}17$, $b_2 = 0.29$	$\varepsilon = 3.20$, $R_m = 3.09$[3]	Zandee and Reuss (1977b)
TM			Tang and Toennies (1978)
IX	$V_0(R)$: empirical[8]	$C^6 = 5.06$	
		$C^8 = 21.58$ ref. 9	Huisken (1978)
	$\beta = 4.8$	$C^{10} = 113.84$	
	$a_2^6 = 0.094$	$a = 1, b = 2.29, c = 2$	
	$a_2^{10} = 0.26$ ref. 9		
	$A_2 = 7125$	$R_m = 2.065$	
AI[14]			
TM 9			
II	$a_2 = 0.12$, $b_2 = 0.14$	$\varepsilon = 6.30$, $R_m = 3.34$[3]	
Ix	$A_2/A_0 = 0.17$		
	$a_2^{(6)} = 0.107$[5]	$A_0 = 3.53 \times 10^3$[4]	
	$a_2^{(8)} = 0.17$	$\beta = 3.692$[4]	
		$C^{(6)} = 16.9$[3]	
		$C^{(8)} = 126.5$[4]	Zandee
		$a = 4.1, b = 1, c = 1.9$[6]	and
II	$a_2 = 0.09$, $b_2 = 0{\cdot}08$	$\varepsilon = 6.80$, $R_m = 3.65$[3]	Reuss (1977a)
IX	$A_2/A_0 = 0.17$		
	$a_2^{(6)} = 0.108$[5]	$A_0 = 3.21 \times 10^3$[4]	
	$a_2^{(8)} = 0.18$	$\beta = 3.462$[4]	
		$C^{(6)} = 24.1$[3]	
		$C^{(8)} = 234.3$[4]	
		$a = 4.1, b = 1, c = 1.9$[6]	
II	$a_2 = 0.12$, $b_2 = 0.11$	$\varepsilon = 7.43$, $R_m = 3.9$[3]	
IX	$A_2/A_0 = 0.14$		
	$a_2^{(6)} = 0.108$[5]	$A_0 = 1.23 \times 10^4$[4]	
	$a_2^{(8)} = 0.20$	$\beta = 3.668$[4]	
		$C^{(6)} = 41.0$[3]	
		$C^{(8)} = 124$[4]	
		$a = 4.1, b = 1, c = 1.9$[6]	

(contd.)

TABLE III *(contd.)*

System	Collision energy (meV)	Exp. data	State selection: method	State selection: prep. states	State analysis: method	State analysis: anal. states	Data analysis
HD + HD	35,0	$I(j'_1,j'_2/j_1,j_2;\vartheta,E)$	NB	$j_1=j_2=0$	VC	$j'_1,j'_2=0,1$	
$HD + D_Y$	45,4	$I(j'/j;\vartheta,E)$	NB	$j=0$	VC	$j'=0,1$	T
$H_2 + H_2$	8.8–35	A	RF	$j=1, m=0$			F
$H_2 + N_2$ $H_2 + CO$ $H_2 + CH_4$ $H_2 + CCl_4$ $H_2 + CF_4$ $H_2 + SF_6$ $H_2 + C(CH_3)_4$	10–93	A	RF	$j=1$ $m=1$			F
TlF + Ar TlF + Kr TlF + Xe	8.4–70.6	A	EQ	$j=1,$ $j=m$			F
CsF + Ar	12.4–64.4	A	EQ	$j=1,2,$ $m=0$			F
	89	$I(j'm'/j,m;$ $\vartheta,\phi,E)$	EQ	$j=2,3$ $m=0$	EQ	$j'=1,2,3,$ $m'=0$	T
	91	$I(j',m'/j,m;\vartheta,\phi,E)$	EQ	$j=2,$ $m=0$	EQ	$j'=1,$ $m'=0$	
CsF + Ne	78	$I(j',m'/j,m;$ $\vartheta,\phi,E)$	EQ	$j=2,3,$ $m=0$	EQ	$j'=1,2,3,$ $m'=0$	T
CsF + Kr	$I(j',m'/j,m;$ $\vartheta,\phi,E)$	EQ	$j=2,$ $m=0$	EQ	$j'=3,$ $m'=0$		T
	84	$I(j',m'/j,m;$ $\vartheta,\phi,E)$	EQ	$j=2,$ $m=0$	EQ	$j'=1,$ $m'=0$	
CsF + He	5–20	A	EQ	$j=1,$ $m=0$			F
KBr + Ar	33	$J(E',\vartheta,E)$	CR		VC		T
LiF + Ar	98–484	$I(\vartheta,E)$	NB	$T_r = 575$ K $T_r = 1185$ K			T
	1147 186	$I(j,m;\vartheta,\phi,E)$	EQ	$j=1,$ $m=0$			
LiF + Kr	223–385	$I(\vartheta,E)$	NB	$T_r = 575$ K $T_r = 1185$ K			T
LiH + Ar	800	$Q(j'/j;v)$	EQ	$j=1$	LIF	$j'=2–5$	

Type	Potential: anisotropy parameters	Potential: other parameters	References
			Gentry and Giese (1977)
I[15]			Buck *et al.* (1978)
I	$a_2 = 0.18, b_2 = 0.25$	$\varepsilon = 3.37, R_m = 3.34$[16]	Zandee and Reuss (1977b)
	$a_2 = 0.196, b_2 = 0.278$	$\varepsilon R_m = 20.2$	
	$a_2 = 0.182, b_2 = 0.244$	$\varepsilon R_m = 21.4$	
	$2a_2 - b_2 = 0.142$	$\varepsilon R_m = 25.3$	
I	$a_2 = 0.145, b_2 = 0.145$	$\varepsilon R_m = 28.1$	Zandee *et al.* (1977)
	$a_2 = 0.080, b_2 = 0.09$	$\varepsilon R_m = 29.7$	
	$a_2 = 0.074, b_2 = 0.089$	$\varepsilon R_m = 36.8$	
	$a_2 = 0.07, b_2 = 0.07$	$\varepsilon R_m = 47.4$	
	$a_2 = 0{\cdot}23 \pm 0.01$		Bennewitz *et al.* (1969a)
	$a_2 = 0.28 \pm 0.02$	$C^{(6)} = 350$	Bennewitz and Haerten (1969)
Eq. (4)	$C_0^{(6)} = 137, C_2^{(6)} = 41$		
	$C_1^{(7)} = 998$		Malthan (1976)
			Henrichs *et al.* (1977)
Eq. (4)	$C_0^{(6)} = 32.5, C_2^{(6)} = 9.8$		
	$C_1^{(7)} = 243$		Malthan (1976)
Eq. (4)	$C_0^{(6)} = 370, C_2^{(6)} = 111$		Meyer and
	$C_1^{(7)} = -842, C_3^{(7)} = -572$		Toennies
	$C_2^{(8)} = 4747, C_4^{(8)} = 997$		(1978)
	$C_1^{(9)} = -4513, C_3^{(9)} = -4544$		
	$C_5^{(9)} = -1218$		
			Henrichs *et al.* (1977)
II	$a_2 = 0.28, b_2 = 0.9 \pm 0.2$	$\varepsilon = 91.1, R_m = 5.90$	Bennewitz and Haerten (1969)
E			Matzen and Fisk (1977)
Eq. (4)	$C_0^{(6)} = 63.7, C_2^{(6)} = 52.7$	$\varepsilon = 25.6, R_m = 3.8$	Reed and
	$C_1^{(7)} = 94.9, C_3^{(7)} = 91.1$		Wharton (1977)
Eq. (4)			Tsou *et al.* (1977)
Eq. (4)	$C_0^{(6)} = 92.2, C_2^{(6)} = 77.4$	$\varepsilon = 33.1, R_m = 3.7$	Reed and
	$C_1^{(7)} = 152, C_3^{(7)} = 138$		Wharton (1977)
			Wilcomb and Dagdigian (1977)

(*contd.*)

TABLE III *(contd.)*

System	Collision energy (meV)	Exp. data	State selection: method	State selection: prep. states	State analysis: method	State analysis: anal. states	Data analysis
HCl + He	35–217			T_r = 200 K		j' = 0–14	
				and			
HCl + Ar	260–1304	$Q(j',j;v)$	NB	T_r = 500–1000 K	IRL	j' = 0–16	T
HCl + Kr	217–1435					j' = 0–16	
HF + Ar	348–1087					j' = 0–12	T
HCl + Li	31–490						
DCl + Li	31–490						
HBr + Li	33–536	$Q(V)$					F
HI + Li	34–552						
NO + Ar							F
NO + Kr							
NO + Xe	22–48	A	EH	$j = 3/2$			
				$m = 3/2$			
				$\Omega = 3/2$			
$NO + CO_2$							
$NO + CS_2$							
$NO + N_2O$							
$NO + CCl_4$							F
$NO + SF_6$							
NO + Li	29–471	$Q(v)$					F
Na_2 + He	98	$I(j'/j;\vartheta,E)$	LID	$v = 0$	LIF	$v = 0$	
Na_2 + Ne	190	$I(j'/j;\vartheta,E)$		$j = 28$		$j' = 28$	
CO + K	340						
		$J(E',\vartheta,E)$	NB	T_r = 30 K	VC		F
N_2 + K	1240						
COS + K	89–142						
CS_2 + K	94–151	$J(E',\vartheta,E)$			VC		
SO_2 + K	102						
CO_2 + HD	2.6–150	$Q(v)$					F
$CO_2 + D_2$	3.4–195						
	3.4–195	$Q(v)$					F
CO_2 + He	64	$I(\vartheta,E)$					F
							F

Potential:			
Type	anisotropy parameters	other parameters	References
EG[10] and			Ding and Polanyi (1975)
EG[11]			Polanyi and Sathyamurthy (1978)
EG[12]			Polanyi and Sathyamurthy (1979)
	$a_2 = 0.12, b_1 \leq 0.71, b_2 \leq 0.61$	$\varepsilon = 21.4, R_m = 4.02$	Helbing and Rothe
IV	$a_2 = 0.13, b_1 \leq 0.67, b_2 \leq 0.63$	$\varepsilon = 21.4, R_m = 4.02$	(1968)
	$a_2 = 0.06, b_1 \leq 0.75, b_2 \leq 0.46$	$\varepsilon = 28.4, R_m = 4.00$	Olson and Bernstein
	$a_2 = 0.03, b_1 \leq 0.74, b_2 \leq 0.43$	$\varepsilon = 26.1, R_m = 4.3$	(1968, 1969)
II	$a_2 = 0.22 \pm 0.02, b_2 = {}^{0\ 24}_{0\ 44} \pm 0.01$	$\varepsilon R_m = {}^{50\ 8}_{42\ 0} \pm 0.6, R_m = {}^{3\ 93}_{3\ 85}$	Schwartz *et al.* (1973)
	$a_2 = 0.24 \pm 0.01, b_2 = {}^{0\ 35}_{0\ 42} \pm 0.01$	$\varepsilon R_m = {}^{58\ 4}_{51\ 5} \pm 0.3, R_m = {}^{4\ 05}_{3\ 96}$	
	$a_2 = 0.22 \pm 0.01, b_2 = {}^{0\ 36}_{0\ 34} \pm 0.01$	$\varepsilon R_m = {}^{71\ 2}_{60\ 9} \pm 1, R_m = {}^{4\ 3}_{4\ 2}$	
I	$a_2 = 0.29 \pm 0.02$		Stolte *et al.* (1973)
	$a_2 = 0.20 \pm 0.03$		
	$a_2 = 0.25 \pm 0.05$		
	$a_2 = 0.14 \pm 0.01$		Kessener and Reuss (1975)
	$a_2 = 0.18 \pm 0.01$		
IV	$a_2 = 0.6, b_1 \leq 0.54, b_2 \leq 0.45$		Olson and Bernstein (1968, 1969)
			Bergmann *et al.* (1978)
XI	$c - a = 0.29 \pm 0.03$ Å		
	$z_0 = 0.12 \pm 0.02$ Å		Schepper *et al.* (1979)
	$c - a = 0.27 \pm 0.02$ Å		
			Förster (1975)
II	$a_2 = b_2 = 0.27$	$\varepsilon = 8.9, R_m = 3.3$	Butz *et al.* (1971b)
	$a_2 = b_2 = 0.27$	$\varepsilon = 8.9, R_m = 3.3$	
II	$a_2 = b_2 = 0.27$	$\varepsilon = 3.7, R_m = 3.5$	Butz *et al.* (1971b)
X	$\varepsilon = 2.98, a = -0.64$ $R_m(90°) = 3.31, b = -0.518$ $\beta = 4.59$	$C^{(6)} = 68.4$ $C^{(8)} = 398$	Keil *et al.* (1978)
VII	$l = 1.1621$ Å	$\varepsilon = 96.3, R_m = 3.8$	Loesch (1976)

(*contd.*)

TABLE III *(contd.)*

System	Collision energy (meV)	Exp. data	State selection: method	State selection: prep. states	State analysis: method	State analysis: anal. states	Data analysis
CO_2 + Ar	69–190	$J(E', \vartheta, E)$ $Q(j'/j, V)$	NB	$T_r = 30$ K	VC		T
CO_2 + K	105–126	$J(E', \vartheta, E)$			VC		T
Na + $SiCl_4$, $C(CH_3)_4$, $Si(CH_3)_4$	160–220	$I(\vartheta, E)$					F
CBr_4 + Na	125–250	$I(\vartheta, E)$					F
Na + $(CH_3)_3CI$, $(CH_3)_3CBr$, $(CH_3)_3CCl$, $(CH_3)I$, CBr_4	170–231	$I(\vartheta, E)$					F

References: (1) Foster and Rugheimer (1972), (2) Amdur and Malinauskas (1965), (3) Helbing *et al.* (1968), (4) LeRoy and van Kranendonk (1974), (5) Langhoff *et al.* (1971), (6) LeRoy *et al.* (1977), Hepburn *et al.* (1975), (7) Shafer and Gordon 1973), (8) Buck *et al.* (1979), (9) Tang and Toennies (1978), (10) Buck and McGuire (1976), Neilsen and Gordon (1973), Dunker and Gordon (1976), (11) Green (1974), Green and Monchick (1975), (12) Detrich and Conn (1976), (13) Geurts *et al.* (1975), Hariharan and Kutzelnigg (1977), (14) Birks *et al.* (1975), (15) Meyer (1978), (16) Dondi *et al.* (1972), (17) Parker *et al.* (1976).

	Potential:		
Type	anisotropy parameters	other parameters	References
EG[17]			Preston and Pack (1977)
VI	$\beta^{-1} = 1.16; a = -0.75,$	$\varepsilon = 56, R_m = 4{,}6$	Förster (1975)
VIII	$\alpha_{eff} = 0.90 \pm 0.07$	$\varepsilon = 20.4 \pm 0.8, R_m = 4.6 \pm 0.2$	
	$\alpha_{eff} = 0.20 \pm 0.10$	$\varepsilon = 18.2 \pm 0.6, R_m = 5.5 \pm 0.2$	Buck *et al.* (1978)
	$\alpha_{eff} = 0.44 \pm 0.06$	$\varepsilon = 18.5 \pm 0.8, R_m = 6.3 \pm 0.2$	
	$\alpha_{eff} = 0.70 \pm 0.07$	$\varepsilon = 23.1 \pm 0.8, R_m = 8.3 \pm 0.3$	
V	$b_1 = b_2 = b_3 = 0$ $a_1 = 0.2, a_2 = 0.65, a_3 = 0$	$\varepsilon = 39, R_m = 7.2$	Buck *et al.* (1975)
V	$a_1 = 0.6, a_2 = 0.4,$ $a_3 = 0.15, b_i = 0$	$\varepsilon = 28, R_m = 6.4$	Buck *et al.* (1975)
	$a_1 = 0.5, a_2 = 0.6, a_3 = 0.5,$ $b_3 = 0.9, b_1 = b_2 = 0$	$\varepsilon = 24.7, R_m = 6.7$	
	$a_1 = 0.45, a_2 = 0.6, a_3 = 0.4,$ $b_3 = 0.7, b_1 = b_2 = 0$	$\varepsilon = 20.3, R_m = 6.6$	
	$a_1 = 0.5, a_2 = 0.1,$ $a_3 = -0.2, b_j = 0$	$\varepsilon = 26.2, R_m = 5.1$	
	$a_1 = 0.2, a_2 = 0.5,$ $a_3 = 0, b_i = 0$	$\varepsilon = 36.2, R_m = 6.3$	

Key to state selection and analysis: RF = Rabi field technique, EQ = electrostatic quadrupole, EH = electrostatic exapole, NB = nozzle beam, LID = laser-induced depopulation, VC = velocity-change method, LIF = laser-induced uorescence, IRL = infrared luminescence, CR = chemical reaction.

Key to data analysis: F = fitting procedure, T = test calculation.

Key to potentials: EG = electron gas, E = empirical, AI = *ab initio*, roman numerals refer to potentials of Table II.

Dimensions: ε(meV), R_m(Å), A_0 (eV), β(Å^{-1}), $C^{(i)}$(eV Å^i).

References

Abuaf, N., Anderson, J. B., Andres, R. P., Fenn, J. B., and Marsden, D. G. H. (1967). *Science*, **155**, 997.

Airey, J. R., Greene, E. F., Reck, G. P., and Ross, J. (1967). *J. Chem. Phys.*, **46**, 3295.

Alexander, M. H. (1977). *J. Chem. Phys.*, **67**, 2703.

Alexander, M. H. (1978). *Chem. Phys.*, **27**, 229.

Alexander, M. H. and Dagdigian, P. J. (1977). *J. Chem. Phys.*, **66**, 4126.

Alexander, M. H., Dagdigian, P. J., and DePristo, A. E. (1977). *J. Chem. Phys.*, **66**, 59.

Alexander, M. H. and McGuire, P. (1976). *Chem. Phys.*, **12**, 31.

Amdur, J. and Malinauskas, A. P. (1965). *J. Chem. Phys.*, **42**, 3355.

Amme, R. C. (1975). *Adv. Chem. Phys.*, **28**, 171.

Anderson, J. B. (1974). In *Molecular Beams and Low Density Gas Dynamics* (Ed. P. Wegner), Dekker, New York.

Anderson, J. B., Andres, R. P., and Fenn, J. B. (1966). *Adv. Chem. Phys.*, **10**, 275.

Anlauf, K. G., Bickes, R. W., and Bernstein, R. B. (1971). *J. Chem. Phys.*, **54**, 3647.

Armstrong, W. D., Conley, R. J., Creaser, R. P., Greene, E. F., and Hall, R. B. (1975). *J. Chem. Phys.*, **63**, 3349.

Arthurs, A. M. and Dalgarno, A. (1960). *Proc. Roy. Soc. (London)*, **A256**, 540.

Augustin, S. D. and Miller, W. H. (1974). *J. Chem. Phys.*, **61**, 3155.

Augustin, S. D. and Miller, W. H. (1974a). *Chem. Phys. Lett.*, **28**, 149.

Balint-Kurti, G. G. (1975). *International Review of Science, Physical Chemistry* (Eds. A. D. Buckingham and C. A. Coulsen) Ser. 2, Butterworths, London, p. 285.

Balint-Kurti, G. G. (1975a). *Adv. Chem. Phys.*, **30**, 137.

Barg, G. D., Kendall, G. M., and Toennies, J. P. (1976). *Chem. Phys.*, **16**, 243.

Bassi, D., Dondi, M. G., Tommasini, F., Torello, F., and Valbusa, U. (1976). *Phys. Rev.*, **A13**, 584.

Bassi, D., Tommasini, F., and Scoles, G. (1974). *J. Chem. Phys.*, **62**, 600.

Bauer, W., Shobatake, K., Toennies, J. P., and Walaschewski, K. (1978). *J. Chem. Phys.*, **68**, 3413.

Beck, D. (1970). 'Enrico Fermi', in *Proceedings of the International School of Physics Course XLIV* (Ed. Ch. Schlier), Academic Press, New York, p. 1.

Beck, D. (1978). Private communication.

Beck, D. and Förster, H. (1970). *Z. Phys.*, **240**, 136.

Beck, D., Ross, U., and Schepper, W. (1979). To be published.

Bennewitz, H. G. and Buess, G. (1978). *Chem. Phys.*, **28**, 175.

Bennewitz, H. G., Gengenbach, R., Haerten, R., and Müller, G. (1969a). *Z. Physick*, **226**, 279.

Bennewitz, H. G. and Haerten, R. (1969). *Z. Physick*, **227**, 399.

Bennewitz, H. G., Haerten, R., and Müller, G. (1969b). *Z. Physik*, **226**, 139.

Bennewitz, H. G., Kramer, K. H., Paul, W., and Toennies, J. P. (1964). *Z. Phys.*, **177**, 84.

Berbenni, E. and McGuire, P. (1977). *Chem. Phys. Lett.*, **45**, 84.

Bergmann, K., Engelhardt, R., Hefter, U., Hering, P., and Witt, J. (1978). *Phys. Rev. Lett.*, **40**, 1446.

Bergmann, K., Engelhardt, R., Hefter, U., and Witt, J. (1979). To be published.

Bergmann, K., Hefter, U., and Hering, P. (1978a). *Chem. Phys.*, **32**, 329.

Bernstein, R. B. (1966). *Adv. Chem. Phys.*, **10**, 75.

Bernstein, R. B. and Kramer, K. H. (1966). *J. Chem. Phys.*, **44**, 4473.

Bernstein, R. B. and Muckerman, J. T. (1967). *Adv. Chem. Phys.*, **12**, 389.

Bhattacharyya, S. S., Saha, S., and Barua, A. K. (1978). *J. Phys. B: Atom. Mol. Phys.*, **10**, 1557.

Bickes, R. W. and Bernstein, R. B. (1969). *Chem. Phys. Lett.*, **4**, 111.
Bickes, Jr. R. W., Scoles, G., and Smith, K. M. (1975). *Can. J. Phys.*, **53**, 435.
Birks, J. W., Johnston, H. S., and Schaefer, H. F. (1975). *J. Chem. Phys.*, **63**, 1741.
Blais, N. C., Cross, J. B., and Kwei, G. H. (1977). *J. Chem. Phys.*, **66**, 2488.
Blythe, A. R., Grosser, A. E., and Bernstein, R. B. (1964). *J. Chem. Phys.*, **41**, 1917.
Borkenhagen, U., Maltan, H., and Toennies, J. P. (1976). *Chem. Phys. Lett.*, **41**, 222.
Bowman, J. M. and Leasure, S. C. (1977). *J. Chem. Phys.*, **66**, 288.
Brooks, P. R. (1978). Private communication.
Brumer, P. (1974). *Chem. Phys. Lett.*, **28**, 345.
Brusdeylins, G., Meyer, H.-D., Toennies, J. P., and Winkelmann, K. (1977). *AIAA Progress in Astronautics and Aeronautics*, in press.
Buck, U. (1975). *Adv. Chem. Phys.*, **30**, 313.
Buck, U. (1978). Private communication.
Buck, U., Gestermann, F., and Pauly, H. (1975a). In *50 Jahre Max-Planck-Institut für Strömungsforchung*, p. 392.
Buck, U., Gestermann, F., and Pauly, H. (1975b). *Chem. Phys. Lett.*, **33**, 186.
Buck, U., Huisken, F. Pauly, H., Pust, D., and Schleusener, J. (1975). In *50 Jahre Max-Planck-Institut für Strömungsforschung*, p. 380.
Buck, U., Huisken, F., and Schleusener, J. (1978). *J. Chem. Phys.*, **68**, 5654; and *J. Chem. Phys.*, to be published.
Buck, U., Huisken, F., and Schleusener, J. (1978a). *Book of Abstracts of MOLEC II*, Brandbjerg Hojskole, Danmark, p. 67.
Buck, U., Huisken, F., and Schleusener, J. (1978b). *J. Chem. Phys.*, to be published.
Buck, U., Huisken, F., and Schleusener, J. (1979). To be published.
Buck, U., Huisken, F., Schleusener, J., and Pauly, H. (1977). *Phys. Rev. Lett.*, **38**, 680.
Buck, U. and Khare, U. (1977). *Chem. Phys.*, **26**, 215.
Buck, U., Khare, V., and Kick, M. (1978c). *Mol. Phys.*, **35**, 65.
Buck, U. and McGuire, P. (1976). *Chem. Phys.*, **16**, 101.
Buck, U. and Pauly, H. (1968). *Z. Phys.*, **208**, 390.
Buckingham, A. D. (1967). *Adv. Chem. Phys.*, **12**, 107.
Bunker, D. L. (1971). *Meth. Comput. Phys.*, **10**, 287.
Butz, H. P., Feltgen, R., Pauly, H., and Vehmeyer, H. (1971a). *Z. Phys.*, **247**, 70.
Butz, H. P., Feltgen, R., Pauly, H., Vehmeyer, H., and Yealland, R. M. (1971b). *Z. Phys.*, **247**, 60.
Cavallini, M., Dondi, M. G., Scoles, G., and Valbusa, U. (1971a). *Chem. Phys. Lett.*, **10**, 22.
Cavallini, M., Meneghetti, L., Scoles, G., and Yealland, M. (1971b). *Rev. Sci. Instr.*, **42**, 1759.
Chapman, S. and Green, S. (1977). *J. Chem. Phys.*, **67**, 2317.
Chou, M. S., Crim, F. F., and Fisk, G. A. (1973). *Chem. Phys. Lett.*, **20**, 464.
Chu, S.-I. and Dalgarno, A. (1975). *J. Chem. Phys.*, **63**, 2115.
Chu, S.-I. and Dalgarno, A. (1975a). *Astrophys. J.*, **199**, 637.
Chu, S.-I. and Dalgarno, A. (1975b). *Proc. Roy. Soc. (London)*, **A342**, 191.
Cohen, J. S. and Pack, R. T. (1974). *J. Chem. Phys.*, **61**, 2372.
Crim, F. F., Bente, H. B., and Fisk, G. A. (1974). *J. Phys. Chem.*, **78**, 2438.
Crim, F. F., Chou, M. S., and Fisk, G. A. (1973). *Chem. Phys.*, **2**, 283.
Cross, Jr., R. J. (1967). *J. Chem. Phys.*, **47**, 3724.
Cross, Jr., R. J. (1968). *J. Chem. Phys.*, **48**, 4838.
Cross, Jr., R. J. (1968a). *J. Chem. Phys.*, **49**, 1976.
Cross, Jr., R. J. (1969). *J. Chem. Phys.*, **51**, 5163.
Cross, Jr., R. J. (1969). *J. Chem. Phys.*, **52**, 5703.

Cross, Jr., R. J. (1977). *Chem. Phys.*, **25**, 165.
Cross, Jr., R. J., Gislason, E. A., and Herschbach, D. R. (1966). *J. Chem. Phys.*, **45**, 3582.
Cross, Jr., R. J. and Gordon, R. G. (1966). *J. Chem. Phys.*, **45**, 3571.
Curtiss, C. F. (1953). *J. Chem. Phys.*, **21**, 2045.
Curtiss, C. F. (1968). *J. Chem. Phys.*, **49**, 1952.
Curtiss, C. F. and Adler, F. T. (1952). *J. Chem. Phys.*, **20**, 249.
Dagdigian, P. J. (1976). *J. Chem. Phys.*, **64**, 2609.
Dagdigian, P. J., Cruse, H. W., Schultz, A., and Zare, R. N. (1974). *J. Chem. Phys.*, **61**, 4450.
David, R., Spoden, W., and Toennies, J. P. (1973). *J. Phys. B*, **6**, 897.
Dehmer, P. M. and Wharton, L. (1974). *J. Chem. Phys.*, **61**, 4204.
Demtröder, W., Stetzenbach, W., Stock, M., and Witt, J. (1976). *J. Mol. Spectr.*, **61**, 382.
DePristo, A. E. and Alexander, M. H. (1975a). *J. Chem. Phys.*, **63**, 3552.
DePristo, A. E. and Alexander, M. H. (1975b). *J. Chem. Phys.*, **64**, 3009.
DePristo, A. E. and Alexander, M. H. (1975c). *J. Chem. Phys.*, **63**, 5327.
DePristo, A. E. and Alexander, M. H. (1976). *J. Phys. B: Atom. Mol. Phys.*, **9**, 2712.
DePristo, A. E. and Alexander, M. H. (1977). *J. Chem. Phys.*, **66**, 1334.
Detrich, J. and Conn, R. W. (1976). *J. Chem. Phys.*, **64**, 3091.
Dickinson, A. S. and Richards, D. (1974). *J. Phys. B: Atom. Mol. Phys.*, **7**, 1916.
Dickinson, A. S. and Richards, D. (1976). *J. Phys. B: Atom. Mol. Phys.*, **9**, 515.
Dickinson, A. S. and Richards, D. (1977). *J. Phys. B: Atom. Mol. Phys.*, **10**, 323.
Dickinson, A. S. and Richards, D. (1978). *J. Phys. B: Atom. Mol. Phys.*, **11**, 1085.
Ding, A. M. G. and Polanyi, J. C. (1975). *Chem. Phys.*, **10**, 39.
Doll, J. D. and Miller, W. H. (1972). *J. Chem. Phys.*, **57**, 5019.
Dondi, M. G., Valbusa, U., and Scoles, G. (1972). *Chem. Phys. Lett.*, **17**, 137.
Donohue, T., Chou, M. S., and Fisk, G. A. (1972). *J. Chem. Phys.*, **57**, 2210.
Donohue, T., Chou, M. S., and Fisk, G. A. (1973). *Chem. Phys.*, **2**, 271.
Dunker, A. M. and Gordon, R. G. (1976). *J. Chem. Phys.*, **64**, 354.
Entemann, E. A. (1967). *Ph.D. thesis*, Harvard University.
Este, G. O., Knight, G., Carraciolo, G., Valbusa, U., Marchetti, S., and Scoles, G. (1978). *Book of Abstracts of Molec II*, Brandbjerg Hojskole, Danmark, p. 69.
Farrar, J. M., Burgmans, A. L. J., Parson, J. M., Walker, R. B., and Lee, Y. T. (1977). *VI. Int. Symp. Mol. Beams*, Noordwijkerhout, The Netherlands.
Farrar, J. M. and Lee, Y. T. (1974). *Chem. Phys. Lett.*, **26**, 428.
Farrar, J. M., Parson, J. M., and Lee, Y. T. (1973). *IV. Int. Symp. Mol. Beams*, Cannes, France, p. 215.
Faubel, M. and Toennies, J. P. (1977). *Adv. Atom. Mol. Phys.*, **13**, 229.
Faubel, M. and Toennies, J. P. (1977a). *Max-Planck-Institut für Strömungsforschung*, Bericht Nr. 103/1977.
Fenstermaker, R. W. and Bernstein, R. B. (1967). *J. Chem. Phys.*, **47**, 4417.
Fenstermaker, R. W., Curtiss, C. F., and Bernstein, R. B. (1969). *J. Chem. Phys.*, **51**, 2439.
Fluendy, M. A. D. and Lawley, K. P. (1973). *Chemical Applications of Molecular Beam Scattering*, Chapman and Hall, London.
Förster, H. (1975). *Thesis*, University of Freiburg, Germany.
Foster, K. R. and Rugheimer, J. H. (1972). *Chem. Phys.*, **56**, 2632.
Franssen, W. and Reuss, J. (1973). *Physica*, **63**, 313.
Franssen, W. and Reuss, J. (1974). *Chem. Phys.*, **77**, 203.
Gallagher, R. J. and Fenn, J. B. (1974). *J. Chem. Phys.*, **60**, 3492.

Gaydaenko, V. I. and Nikulin, V. K. (1970). *Chem. Phys. Lett.*, **7**, 360.
Gentry, W. R. and Giese, C. F. (1977). *Phys. Rev. Lett.*, **39**, 1259.
Gentry, W. R. and Giese, C. F. (1977a). *J. Chem. Phys.*, **67**, 5389.
Gentry, W. R. and Giese, C. F. (1978). *Rev. Sci. Instr.*, **49**, 595.
Geurts, P. J. M., van der Avoird, A., Mulder, F., and Wormer, P. E. S. (1977). Private communication.
Geurts, P. J. M., Wormer, P. E. S., anh van der Avoird, A. (1975). *Chem. Phys. Lett.*, **35**, 444.
Gläser, W. and Gompf, F. (1969). *Nucleonik*, **12**, 153.
Gordon, R. G. (1973). *Faraday Disc. Chem. Soc.*, **55**, 22.
Gordon, R. G. and Kim, Y. S. (1972). *J. Chem. Phys.*, **56**, 3122.
Gough, T. E., Miller, R. E., and Scoles, G. (1977). *Appl. Phys. Lett.*, **30**, 338.
Green, S. (1974). *J. Chem. Phys.*, **60**, 2654.
Green, S. (1975). *J. Chem. Phys.*, **62**, 3568.
Green, S. (1975a). *J. Chem. Phys.*, **62**, 2271.
Green, S., Garrison, B. J., and Lester, Jr., W. A. (1975). *J. Chem. Phys.*, **63**, 1154.
Green, S. and Thaddeus, P. (1974). *Astrophys. J.*, **191**, 653.
Green, S. and Thaddeus, P. (1976). *Astrophys. J.*, **205**, 762.
Green,S. and Monchick, L. (1975). *J. Chem. Phys.*, **63**, 4198.
Greene, E. F., Hall, R. B., and Sondergaard, N. A. (1977). *J. Chem. Phys.*, **66**, 3171.
Greene, E. F., Hoffman, L. F., Lee, M. W., Ross, J., and Young, C. E. (1969a). *J. Chem. Phys.*, **50**, 3450.
Greene, E. F., Lau, M. H., and Ross, J. (1969b). *J. Chem. Phys.*, **50**, 3122.
Greene, E. F., Moursund, A. L., and Ross, J. (1966). *Adv. Chem. Phys.*, **10**, 135.
Habets, A. H. M. (1977). *Thesis*, Technische Hogeschool, Eindhoven, Netherlands.
Hariharan, P. C. and Kutzelnigg, W. (1977). Private communication.
Harris, R. M. and Wilson, J. F. (1971). *J. Chem. Phys.*, **54**, 2088.
Harris, S. J., Janda, K. C., Novick, S. E., and Klemperer, W. (1975). *J. Chem. Phys.*, **63**, 881.
Helbing, R. K. B. (1968). *J. Chem. Phys.*, **48**, 472.
Helbing, R. K. B. (1969). *J. Chem. Phys.*, **51**, 3628.
Helbing, R., Gaide, W., and Pauly, H. (1968). *Z. Physik*, **208**, 215.
Helbing, R. K. and Rothe, E. W. (1968). *J. Chem. Phys.*, **48**, 3945.
Henrichs, J. M., DeBie, R. P. M., Simons, C. G. H., and Verster, N. F. (1977). *Abstracts of papers*, X ICEAC Paris, p. 778.
Hepburn, J. Scoles, G., and Penco, R. (1975). *Chem. Phys. Lett.*, **36**, 451.
Hassel, M. M. and Kusch, P. (1968). *J. Chem. Phys.*, **43**, 305.
Hinze, J. (1974). *Adv. Chem. Phys.*, **26**, 213.
Hirschy, V. L. and Aldridge, J. P. (1970). *Rev. Sci. Instr.*, **42**, 381.
Holmgren, S. L., Waldman, M. and Klemperer, W. (1978). *J. Chem. Phys.*, **69**, 1661.
Hostettler, H. U. and Bernstein, R. B. (1960). *Rev. Sci. Instr.*, **42**, 381.
Huisken, F. (1978). *Thesis*, Göttingen, Germany.
Hunter, L. W. (1975). *J. Chem. Phys.*, **62**, 2855.
Jacobs, M. and Reuss, J. (1977). *Chem. Phys.*, **25**, 425.
Jackob, M. and Wick, G. C. (1959). *Ann. Phys.*, **7**, 404.
Karny, Z., Estler, R. C., and Zare, R. N. (1978). *J. Chem. Phys.*, **69**, 5199.
Keil, M., Parker, G. A., and Kuppermann, A. (1978). *Chem. Phys. Lett.*, **59**, 443.
Keil, M., Slankas, J. T., and Kuppermann, A. (1979). *J. Chem. Phys.*, to be published.
Kessener, H. P. M. and Reuss, J. (1975). *Chem. Phys. Lett.*, **31**, 212.
Kim, Y. S. (1978). *J. Chem. Phys.*, **68**, 5001.
Kim, Y. S. and Gordon, R. G. (1974a). *J. Chem. Phys.*, **60**, 1842.

Kim, Y. S. and Gordon, R. G. (1974b). *J. Chem. Phys.*, **61**, 1.
King, D. L. (1974). *Thesis*, Harvard University.
King, D. L., Loesch, H. J., and Herschbach, D. R. (1973). *Faraday Disc. Chem. Soc.*, **55**, 222.
Kinsey, J. L. (1977). *Ann. Rev. Phys. Chem.*, **28**, 349.
Kinsey, J. L. (1977a). *J. Chem. Phys.*, **66**, 2560.
Klaassen, D., Thuis, H., Stolte, S., and Reuss, J. (1978). *Chem. Phys.*, **27**, 107.
Klar, H. (1969). *Z. Phys.*, **228**, 59.
Klar, H. (1971). *Nuovo Cimento*, **4A**, 529.
Klar, H. (1973). *J. Phys. B*, **6**, 2139.
Kolb, C. E., Baum, H. R., and Tsait, K. S. (1972). *J. Chem. Phys.*, **57**, 3409.
Kolb, C. E. and Elgin, J. B. (1977). *J. Chem. Phys.*, **66**, 119.
Kouri, D. J., Heil, T. G., and Shimoni, Y. (1976). *J. Chem. Phys.*, **65**, 1462.
Kouri, D. J. and McGuire, P. (1974). *Chem. Phys. Lett.*, **29**, 414.
Kramer, K. H. and Bernstein, R. B. (1964). *J. Chem. Phys.*, **40**, 200.
Kramer, H. L. and LeBreton, P. R. (1967). *J. Chem. Phys.*, **47**, 3367.
Kreek, H. and LeRoy, R. J. (1975). *J. Chem. Phys.*, **63**, 238.
Kuipers, J. W. and Reuss, J. (1973). *Chem. Phys.*, **1**, 64.
Kuipers, J. W. and Reuss, J. (1974). *Chem. Phys.*, **4**, 277.
Kukolich, S. G., Oates, D. E., and Wang, J. H. S. (1974). *J. Chem. Phys.*, **61**, 686.
Kuppermann, A., Gordon, R. J., and Coggiola, M. J. (1973). *Faraday Disc. Chem. Soc.*, **55**, 145.
Kusunoki, I. (1977). *J. Chem. Phys.*, **67**, 2224.
LaBudde, R. A. and Bernstein, R. B. (1973). *J. Chem. Phys.*, **59**, 3687.
Lang, N. C., Polanyi, J. C., and Wanner, J. (1977). *Chem. Phys.*, **24**, 219.
Langhoff, P. W., Gordon, R. G., and Karplus, M. (1971). *J. Chem. Phys.*, **55**, 2126.
LeBreton, R., and Kramer, K. H. (1969). *J. Chem. Phys.*, **51**, 3627.
Lee, Y. T. (1978). Private communication.
Lee, Y. T., McDonald, J. D., LeBreton, P. R., and Herschbach, D. R. (1969). *Rev. Sci. Instr.*, **40**, 1402.
Leone, S. R., Macdonald, R. G., and Moore, C. B. (1975). *J. Chem. Phys.*, **63**, 4735.
LeRoy, R. J., Carley, J. S., and Grabenstetter, J. E. (1977). *Faraday Disc. Chem. Soc.*, **62**, 169.
LeRoy, R. J. and Kranendonk, van, J. (1974). *J. Chem. Phys.*, **61**, 4750.
Lester, W. A. (1975). *Adv. Quantum Chem.*, **9**, 199.
Levine, R. D. (1968). *Chem. Phys. Lett.*, **2**, 76.
Loesch, H. J. (1976). *Chem. Phys.*, **18**, 431.
Loesch, H. J. and Beck, D. (1971). *Ber. Bunsen. Phys. Chem.*, **75**, 736.
Loesch, H. J. and Herschbach, D. R. (1972) *J. Chem. Phys.*, **57**, 2038.
Macdonald, R. G., Moore, C. B., Smitz, J. W. M., and Wodarczyk, F. J. (1975). *J. Chem. Phys.*, **62**, 2934.
Magnus, W. (1954). *Commun. Pure Appl. Math.*, **7**, 649.
Malthan, H. (1976). Max-Planck-Institut für Strömungsforschung, Göttingen, Bericht Nr. 16/1976.
Margenau, H. and Kestner, N. R. (1969). *Theory of Intermolecular Forces*, Pergamon Press, Oxford.
Mariella, Jr., R. P., Herschbach, D. R., and Klemperer, W. (1974). *J. Chem. Phys.*, **61**, 4575.
Massey, H. S. W. and Burhop, E. H. S. (1952). In *Electronic and Ionic Impact Phenomena*, Clarendon Press, Oxford.
Matzen, M. K. and Fisk, G. A. (1977). *J. Chem. Phys.*, **66**, 1514.

McDonald, J. D., LeBreton, P. R., Lee, Y. T., and Herschbach, D. R. (1972). *J. Chem. Phys.*, **56**, 769.
McGinnis, R. P. and Greene, E. F. (1978). To be published.
McGuire, P. (1973). *Chem. Phys. Lett.*, **23**, 575.
McGuire, P. (1974). *Chem. Phys.*, **4**, 249.
McGuire, P. (1975). *J. Chem. Phys.*, **62**, 525.
McGuire, P. (1976). *Chem. Phys.*, **13**, 81.
McGuire, P. (1977). *J. Chem. Phys.*, **66**, 1761.
McGuire, P. and Kouri, D. J. (1974). *J. Chem. Phys.*, **60**, 2488.
Meerts, W. L., Terhorst, G., Reinartz, J. M. L. J., and Dymanns, A. (1978). *Chem. Phys.*, **35**, 253.
Meyer, G. and Toennies, P. (1978). *Book of Abstracts of MOLEC II*, Branbjerg Hojstole, Danmark, p. 75.
Meyer, H. D. (1974). Max-Planck-Institut für Strömungsforschung, Göttingen, Bericht Nr. 113/1974.
Meyer, W. (1976). *Chem. Phys.*, **27**, 27.
Meyer, W. (1978). Private communication.
Micha, D. A. (1974). *Adv. Quantum Chem.*, **8**, 231.
Miller, W. H. (1969). *J. Chem. Phys.*, **50**, 3124.
Miller, W. H. (1970). *J. Chem. Phys.*, **53**, 1949.
Miller, W. H. (1971). *Chem. Phys. Lett.*, **11**, 535.
Miller, W. H. (1974). *Adv. Chem. Phys.*, **25**, 69.
Miller, W. H. (1975). *Adv. Chem. Phys.*, **30**, 77.
Miller, W. H. and Raczkowski, A. W. (1973). *Faraday Disc. Chem. Soc.*, **55**, 45.
Moerkerken, H., Prior, M. H., and Reuss, J. (1970). *Physica*, **50**, 499.
Moerkerken, H., Zandee, L., and Reuss, J. (1973). *Chem. Phys. Lett.*, **23**, 320.
Moerkerken, H., Zandee, L., and Reuss, J. (1975). *Chem. Phys.*, **11**, 87.
Monchick, L. (1977). *J. Chem. Phys.*, **67**, 4626.
Muntz, E. P. (1962). *Phys. Fluids*, **5**, 80.
Neilsen, W. B. and Gordon, R. G. (1973). *J. Chem. Phys.*, **58**, 4149.
Nielson, G. C., Parker, G. A., and Pack, R. T. (1976). *J. Chem. Phys.*, **64**, 2055.
Nielson, G. C., Parker, G. A. and Pack, R. T. (1977). *J. Chem. Phys.*, **66**, 1396.
Nikitin, E. E. (1975). In *The Physics of Electronic and Atomic Collisions* (Eds. J. S. Risley and R. Geballe), Invited Lectures of the IX ICPEAC Seattle, p. 275.
Nikulin, V. K. (1971). *Sov. Phys. Tech. Phys.*, **16**, 28.
Novick, S. E., Davies, P., Harris, S. J., and Klemperer, W. (1973). *J. Chem. Phys.*, **59**, 2273.
Odiorne, T. J., Brooks, P. R., and Kasper, J. V. V. (1971). *J. Chem.*, **55**, 1980.
Olson, R. E. and Bernstein, R. B. (1968). *J. Chem. Phys.*, **49**, 126.
Olson, R. E. and Bernstein, R. B. (1969). *J. Chem. Phys.*, **50**, 246.
Pack, R. T. (1972). *Chem. Phys. Lett.*, **14**, 393.
Pack, R. T. (1974). *J. Chem. Phys.*, **60**, 633.
Pack, R. T. (1975). *J. Chem. Phys.*, **62**, 3143.
Pack, R. T. (1976). *J. Chem. Phys.*, **64**, 1659.
Pack, R. T. (1977). *J. Chem. Phys.*, **66**, 1557.
Pack, R. T. (1978). *Chem. Phys. Lett.*, **55**, 197.
Parker, G. A. and Pack, R. T. (1976). *J. Chem. Phys.*, **64**, 2010.
Parker, G. A. and Pack, R. T. (1978). *J. Chem. Phys.*, **68**, 1585.
Parker, G. A. and Pack, R. T. (1978a). *J. Chem. Phys.*, **69**, 3268.
Parker, G. A., Snow, R. L., and Pack, R. T. (1976). *J. Chem. Phys.*, **64**, 1668.
Pauly, H. (1975). *Phys. Chem.*, Vol. VIB, Academic Press, New York, London, p. 553.

Pauly, H. and Toennies, J. P. (1965). *Adv. Atom. Mol. Phys.*, **1**, 195.
Pauly, H. and Toennies, J. P. (1968). *Methods of Experimental Physics*, Vol. 7A, Academic Press, New York, p. 227.
Pechukas, P. and Light, J. C. (1966). *J. Chem. Phys.*, **44**, 3897.
Pfeffer, G. and Secrest, D. (1977). *J. Chem. Phys.*, **67**, 1394.
Polanyi, J. C. and Sathyamurthy, N. (1978). *Chem. Phys.*, **29**, 9.
Polanyi, J. C., Sathyamurthy, N., and Schreiber, J. L. (1977). *Chem. Phys.*, **24**, 105.
Polanyi, J. C. and Woodall, K. B. (1972). *J. Chem. Phys.*, **56**, 1563.
Preston, R. K. and Pack, R. T. (1977). *J. Chem. Phys.*, **66**, 2480.
Pruett, J. G., Grabiner, F. R., and Brooks, P. R. (1974). *J. Chem. Phys.*, **60**, 3335.
Quigley, G. P. and Wolga, G. J. (1975). *J. Chem. Phys.*, **63**, 5263.
Rabitz, H. (1972). *J. Chem. Phys.*, **57**, 1718.
Rabitz, H. (1975). *J. Chem. Phys.*, **63**, 520.
Rabitz, H. (1976). In *Modern Theoretical Chemistry* (Ed. W. H. Miller), Vol. III, Plenum Press, New York.
Rabitz, H. A. and Gordon, R. G. (1970a). *J. Chem. Phys.*, **53**, 1815.
Rabitz, H. A. and Gordon, R. G. (1970b). *J. Chem. Phys.*, **53**, 1831.
Raczkowski, A. W. and Miller, W. H. (1974). *J. Chem. Phys.*, **61**, 5413.
Rae, A. J. M. (1973). *Chem. Phys. Lett.*, **18**, 574.
Raith, W. (1976). *Adv. Atom. Mol. Phys.*, **12**, 281.
Ree, van de J. (1971). *J. Chem. Phys.*, **54**, 3249.
Reed, K. A. and Wharton, L. (1977). *J. Chem. Phys.*, **66**, 3399.
Reuss, J. (1975). *Adv. Chem. Phys.*, **30**, 389.
Reuss, J. and Stolte, S. (1969). *Physica*, **42**, 111.
Richman, E. and Wharton, L. (1970). *J. Chem. Phys.*, **53**, 945.
Ross, J. and Greene, E. F. (1970). "Enrico Fermi", *Proceedings of the International School of Physics*, Course XLIV (Ed. Chr. Schlier), Academic Press, New York.
Rulis, A. M. and Scoles, G. (1977). *Chem. Phys.*, **25**, 183.
Rulis, A. M., Smith, K. M., and Scoles, G. (1977). *Chemical Physics Research Report*, University of Waterloo, Canada.
Saha, S. and Guka, E. (1975). *J. Phys. B: Atom. Mol. Phys.*, **8**, 2293.
Saha, S., Guka, E., and Barua, A. K. (1973). *J. Phys. B: Atom. Mol. Phys.*, **6**, 1824.
Saha, S., Guka, E., and Barua, A. K. (1974). *J. Phys. B: Atom. Mol. Phys.*, **7**, 2264.
Schapiro, M. and Tamir, M. (1976). *Chem. Phys.*, **13**, 215.
Schepper, W. (1978). Private communication.
Schepper, W., Ross, U., and Beck, D. (1978). *Book of Abstracts of MOLEC II*, Brandjerg, Hojskole, Denmark.
Schepper, W., Ross, U., and Beck, D. (1979). *Z. Phys.*, **A290**, 131.
Schlier, Ch. (1969). *Ann. Rev. Phys. Chem.*, **20**, 191.
Schlier, Ch. (1978). Private communication.
Schneider, B. (1973). *J. Chem. Phys.*, **58**, 4447.
Scholtheis, G. (1976). Unpublished work, Bielefeld.
Schwartz, H. L., Stolte, S., and Reuss, J. (1973). *Chem. Phys.*, **2**, 1.
Scoles, G. (1978). Private communication.
Scott, P. B., Mincer, T. R., and Muntz, E. P. (1973). *Chem. Phys. Lett.*, **22**, 71.
Scott, P. B., Mincer, T. R., and Muntz, E. P. (1974). *Rev. Sci. Instr.*, **45**, 207.
Secrest, D. (1973). *Ann. Rev. Phys. Chem.*, **24**, 379.
Secrest, D. (1975). *J. Chem. Phys.*, **62**, 710.
Shafer, R. and Gordon, R. G. (1973). *J. Chem. Phys.*, **58**, 5422.
Silvera, I. F. and Tommasini, F. (1976). *Phys. Rev. Lett.*, **37**, 136.
Sinha, M. P., Caldwell, C. D., and Zare, R. N. (1974). *J. Chem. Phys.*, **61**, 491.

Sköld, K. (1968). *Nucl. Instr. and Meth.*, **63**, 114.
Sloane, T. M., Tang, S. Y., and Ross, J. (1972). *J. Chem. Phys.*, **57**, 2745.
Smith, I. W. M. (1976). *Accts. Chem. Res.*, **9**, 161.
Stolte, S., Reuss, J., and Schwartz, H. L. (1972). *Physica*, **57**, 254.
Stolte, S., Reuss, J., and Schwartz, H. L. (1973). *Physica*, **66**, 211.
Suzukawa, H. H., Wolfsberg, M., and Thompson, D. L. (1978). *J. Chem. Phys.*, **68**, 455.
Takayanagi, K. (1963). *Progr. Theory. Phys. (Kyoto) Suppl.*, **25**, 40.
Takayanagi, K. (1965). *Adv. Atom. Mol. Phys.*, **1**, 149.
Tamir, M. and Schapiro, M. (1975). *Chem. Phys. Lett.*, **31**, 166.
Tamir, M. and Schapiro, M. (1976). *Chem. Phys. Lett.*, **39**, 79.
Tang, K. T. (1976). *Chem. Phys. Lett.*, **40**, 372.
Tang, K. T. and Toennies, J. P. (1977). *J. Chem. Phys.*, **66**, 1496.
Tang, K. T. and Toennies, J. P. (1978). *J. Chem. Phys.*, **68**, 5509.
Thomas, L. D. (1977). *Chem. Phys. Lett.*, **51**, 35.
Thomas, L. D., Kraemer, W. P., Diercksen, G. H. F., and McGuire, P. (1978). *Chem. Phys.*, **27**, 237.
Toennies, J. P. (1965). *Z. Phys.*, **182**, 257.
Toennies, J. P. (1966). *Z. Phys.*, **193**, 76.
Toennies, J. P. (1974). *Phys. Chem.*, Vol. VIA, Academic Press, New York/London, p. 227
Toennies, J. P. (1976). *Ann. Rev. Phys. Chem.*, **27**, 225.
Toennies, J. P. (1978). Private communication.
Truhlar, D. G. (1973). *J. Chem. Phys.*, **58**, 3109.
Tsait, K. S., Kolb, C. E., and Baum, H. R. (1973). *J. Chem. Phys.*, **59**, 3128.
Tsien, T. P. and Pack, R. T. (1970). *Chem. Phys. Lett.*, **6**, 54.
Tsien, T. P. and Pack, R. T. (1971). *Chem. Phys. Lett.*, **8**, 579.
Tsien, T. P., Parker, G. A., and Pack, R. T. (1973). *J. Chem. Phys.*, **59**, 5373.
Tsou, L. Y., Auerbach, D., and Wharton, L. (1977). *Phys. Rev. Lett.*, **38**, 20.
Tully, F. P. and Lee, Y. T. (1972). *J. Chem. Phys.*, **57**, 866.
Valbusa, U. (1978). Private communication.
Veberne, J., Ozier, I., Zandee, L., and Reuss, J. (1978). *Mol. Phys.*, **35**, 1649.
Verter, M. R. and Rabitz, H. A. (1973). *J. Chem. Phys.*, **59**, 3816.
Verter, M. R. and Rabitz, H. A. (1974). *J. Chem. Phys.*, **61**, 3707.
Warnock, T. T. and Bernstein, R. B. (1968). *J. Chem. Phys.*, **49**, 1878.
Weitz, E. and Flynn, G. (1974). *Ann. Rev. Phys. Chem.*, **25**, 275.
Wilcomb, B. E. and Dagdigian, P. J. (1977). *J. Chem. Phys.*, **67**, 3829.
Wilcomb, B. E., Habermann, J. A., Bickes, R. W., Mayer, T. M., and Bernstein, R. B. (1976). *J. Chem. Phys.*, **64**, 3501.
Wilkins, R. L. (1975). *J. Chem. Phys.*, **63**, 534.
Winicur, D. H., Moursand, A. L., Devereaux, W. R., Martin, L. R., and Kuppermann, A. (1970). *J. Chem. Phys.*, **52**, 3299.
Zandee, L. and Reuss, J. (1977a). *Chem. Phys.*, **26**, 327.
Zandee, L. and Reuss, J. (1977b). *Chem. Phys.*, **26**, 345.
Zandee, L., Verbene, J., and Reuss, J. (1976). *Chem. Phys. Lett.*, **37**, 1.
Zandee, L., Verbene, J., and Reuss, J. (1977). *Chem. Phys.*, **26**, 1.
Zare, R. N. and Dagdigian, P. J. (1974). *Science*, **185**, 739.
Zarur, G. and Rabitz, H. (1973). *J. Chem. Phys.*, **59**, 943.
Zarur, G. and Rabitz, H. (1974). *J. Chem. Phys.*, **60**, 2057.

Potential Energy Surfaces
Edited by K. P. Lawley

COLLISIONAL IONIZATION

K. LACMANN

Hahn–Meitner-Institut für Kernforschung Berlin GmbH, Bereich Strahlenchemie, Glienicker Straße 100, D-1000 Berlin 39, West Germany

CONTENTS

I. INTRODUCTION

In the past decade collisional ionization processes have gained more and more importance, since they occur in many fields, such as plasma chemistry and ionospheric chemistry. Investigations of these processes have also contributed to a better understanding of reaction dynamics, e.g. non-adiabatic transitions between potential energy hypersurfaces. Experiments coupled with trajectory calculations have led to a satisfactory description of collisional ionization. Furthermore, several potential parameters have been experimentally evaluated (e.g. electron affinity values, dissociation energies, and energy splittings between adiabatic potential surfaces) and have led to improvements in semiempirical calculations.

Collisional ionization processes include all reactions between neutral particles in which an electron transfer takes place. A survey of the different types of charge transfer processes is given in Table I. This review will concentrate mainly on ion-pair formation (reaction type 1 to 3).

In Section II a very introduction into the theoretical background is given together with literature references. Detailed discussions are available in many excellent books and review articles published during the last years and many aspects are reviewed in the chapter by Tully in this book. This section will concentrate on equations and definitions used in the following parts.

At the beginning diabatic and adiabatic potential surfaces of a two-particle system and their crossing behaviour are described. The crossing distance and probability for non-adiabatic transitions as given by the Landau–Zener model are presented. A semiempirical relation between the coupling matrix element and the crossing distances is given and finally the total cross section as a function of the collision velocity is obtained.

Then follows a description of the angular dependent polar cross section

TABLE I. Types of collisional ionization processes between atoms A and molecules BC.

$A + BC \longrightarrow A^+ + BC^-$	(1)	} Ion-pair formation by charge transfer (IPF)	
$\longrightarrow A^- + BC^+$	(2)		
$\longrightarrow A^+ + B^- + C$	(3)	Dissociative IPF	
$\longrightarrow A^+ + (B + C) + e$	(4)	Ionization	
$\longrightarrow AB^+ + C + e$	(5)	} Chemiionization with rearrangement	
$\longrightarrow AB^+ + C^-$	(6)		
$\longrightarrow AB + C^+ + e$	(7)		
$\longrightarrow ABC^+ + e$	(8)	Associative ionization	
$\longrightarrow A + B^+ + C^-$	(9)	Collision-induced polar dissociation	

as a function of the deflection function. The theory is then extended to potential hypersurfaces of many-particle systems. According to symmetry rules, the matrix element is given as a function of the orientation of the target molecule. The molecular motion is decoupled from the relative translational motion of the system. In this way the many-particle problem is reduced to a two-particle problem for which simple classical trajectory calculations can be performed, while the vibrational molecular motion is described in terms of bond stretching during the collision. Its influence on the probability of ion-pair formation is described.

Part III describes a typical crossed beam apparatus. References to the different variations often used are given, since there are many different types and combinations of fast beam sources, target gas arrangements, and mass and energy analysers.

The limiting factors of the accuracy of the potential parameters determined by the different experimental techniques are also discussed in connection with the problems of energy calibration and the final width of the relative velocity distribution of the reactants in crossed beam and scattering gas experiments.

The review of experimental results in Part IV contains several sections. The first reviews mainly experiments in which cross sections for total ion-pair formation are determined as a function of the energy of the alkali atoms and molecules. It is subdivided into different sections which reflect different characteristics of the potential surfaces. The ionization studies at very low collisional energies reveal some discrepancies between experiment and calculations on diabatic potential surfaces. The superiority of adiabatic potential surfaces is shown and the pre-stretching effect on adiabatic surfaces is described which explains the observation of transitions forming negative ions without excitation at the threshold energy which can be used to obtain adiabatic electron affinity values directly.

The second section reviews measurements over a very wide energy range from post-threshold to several hundred or even thousand eV. The post-threshold energy range is marked by bond stretching phenomena during the collision, while at the high energies, the vibrational motion during the collision is 'frozen' and the collision behaves similarly to atom–atom collisions. A modified Landau–Zener model is applicable and the total cross section versus collision velocity reveals a maximum. The velocity at the maximum of the cross section is a function of the energy splitting of the adiabatic surfaces i.e. H_{12}) and the crossing distance R_c. The ratio of atomic to molecular ion yield is, at least in halogen molecules at high energies, a sensitive measure of the crossing distance and, therefore, of the vertical electron affinity. Both experiments together yield accurate estimations of crossing distance, vertical electron affinity, and the coupling matrix element H_{12}. The observed increase in the atomic ion yield over the molecular one with decreasing relative velocity is shown to result from collisions via dissociative excited states.

Another section examines the influence of excitation energy on the total cross section. The vibrational excitation is varied with the temperature of the target gas. The projectile atoms are electronically excited by laser reaction. In both cases the influence on the threshold energy is demonstrated.

Numerous studies examining the threshold energy of negative ion formation yield values of electron affinity and bond energies of the molecules and their dissociation products. Groups of molecules receiving special interest during the last years are disccussed separately. These are the hexafluorides or derivatives of methane, such as freons. Newly determined potential parameters of many other molecules studied during the last five years are also included, often in form of tables.

Another section reviews studies of differential cross sections mostly of the angular distribution of the product ions, but also of the energy loss spectra, where available. From the double differential measurements not only the excitation energy of the product ions is deduced but also the detailed predictions of dynamical models can be tested. A relation between impact parameter and product ion excitation can be found and compared with trajectory calculations. These measurements yield a good test of the quality of semiempirically constructed potential surfaces and the method of calculations, e.g. the surface-hopping trajectory method (SHT).

In the final section are given several references to other types of collisional ionization reactions which lead mainly to ion-pair formation and which also involve endoergic ion-pair formation processes at hyperthermal energies, e.g. ionization cross sections in collisions with non-alkali projectile atoms or molecules or collisions with highly excited non-alkali atoms. Some associative ionization reactions and the dissociation of polar molecules in collisions with atoms are also reviewed.

To discuss the fast-growing field of positive ionization (reaction type 4 in Table I) from ground or highly excited states (e.g. Penning ionization) or the entire field of chemiionization with rearrangement (reaction type 5 in Table I) would exceed the bounds of this chapter.

II. GENERAL THEORETICAL CONSIDERATIONS

A. Introduction

In this section we shall confine ourselves to the explanation of those models and equations necessary to follow the discussion of the experimental results. For a comprehensive description of the theory of chemical elementary processes see Nikitin and Zülicke (1978) and for a detailed description of ion-pair formation (IPF) processes see the review article by Baede (1975).

So far, the exact calculation of ion-pair formation processes is limited to two-particle interactions between alkali atoms and halogen atoms (Faist and Levine, 1976). The following results are all developed in a two-state approximation. It will be shown that the solutions for two particles can be applied also to three- or more-particle collisions (atom–molecule interactions) by approximate models which separate the internal molecular motions from the trajectories of the colliding particles.

B. Curve Crossing and Total Cross Sections

The IPF processes are governed by non-adiabatic transitions between adiabatic states. If these two states have the same symmetry and multiplicity, they do not cross. From the Schrödinger equation for the two-state problem with Born–Oppenheimer approximation (separation of the slow nuclear motion from the fast electronic motion) one gets by diagonalization of the 2×2 electronic Hamiltonian matrix ($H_{12} = H_{21}$) the adiabatic potentials V_1 and V_2 as eigenvalues:

$$V_{1,2} = \tfrac{1}{2}\{H_{11} + H_{22} \pm [(H_{22} - H_{11})^2 + 4H_{12}^2]^{1/2}\} \tag{1}$$

with V_1 and V_2 the lower and upper adiabatic potentials and H_{11} and H_{22} the diabatic covalent and ionic potentials, schematically shown in Fig. 1. The coupling of the diabatic states results from the non-diagonal Hamiltonian term H_{12} which is non-zero if both states have the same symmetry. H_{12} is only of importance near the crossing point at a distance R_c where the diabatic states cross ($H_{11}(R_c) = H_{22}(R_c)$).

In collisional ionization processes between A and B the separation of the two states at large (infinite) distances is given by the energy of reaction

$$\Delta E = I - EA \tag{2}$$

where I is the ionization energy of A and EA the electron affinity of B. If these values are known, the crossing distance R_c can easily be calculated. In the simplest case, the ionic potential H_{22} is purely Coulombic and the covalent potential H_{11} is constant, leading to a crossing distance R_c of

$$R_c = \frac{e^2}{\Delta E} \quad \text{or} \quad = \frac{14.4}{\Delta E}\,[\text{Å}] \tag{3}$$

where ΔE is in eV units. This is a good approximation for reactions with endoergicities smaller than 3 eV or $R_c > 5$ Å, where induction forces (especially by polarization) and van der Waals forces are negligible.

During a collision the crossing distance is passed twice. For ion-pair formation at least one non-adiabatic transition between the adiabatic potentials must take place; therefore, these processes are called non-adiabatic processes. Such a process can occur if the off-diagonal energy H_{12} is small

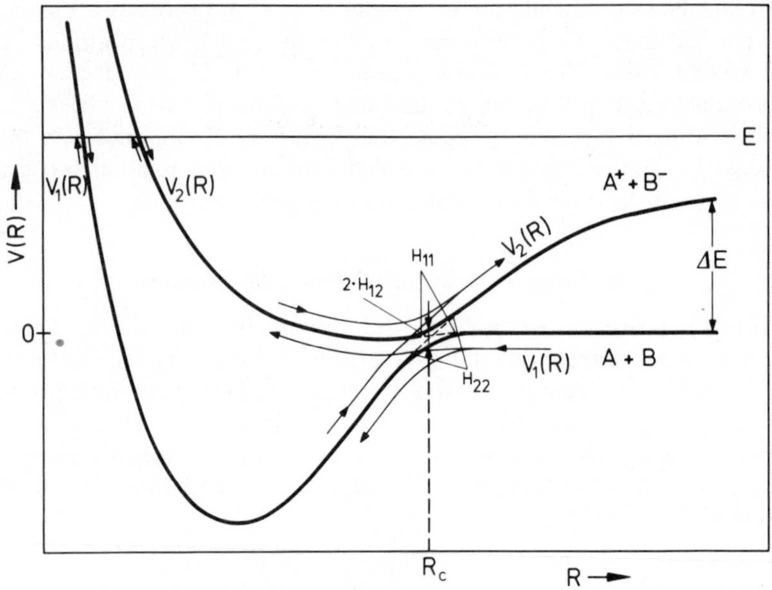

Fig. 1. Schematic potential curve of two adiabatic states $V_1(R)$ and $V_2(R)$ as function of internuclear distance R(A – B), and trajectories with adiabatic and non-adiabatic transitions (dashed lines show the diabatic states H_{11} (covalent) and H_{22} (ionic state)). The energy difference of the adiabatic states at R_c is $2H_{12}$.

and the relative velocity of the reactants is fast enough to prevent an electron rearrangement as the crossing region is traversed; i.e. if the time during which H_{12} is non-zero is too short, the system cannot change its electronic configuration and a non-adiabatic transition (surface hopping) occurs (Tully, 1976). Such processes are brought about by terms which couple the electronic wave functions to the motion of the nuclei and, which are neglected in the Born–Oppenheimer approximation.

The violation of the Born–Oppenheimer approximation is limited to internuclear distances around R_c where the wave function is strongly dependent on the nuclear distance R.

To describe such processes, the Landau–Zener model is often used because of its mathematical and physical simplicity. In the description of many experimental results it has proven its qualitative and even quantitative reliability. With some simplifying assumptions this model allows one to calculate the probability for non-adiabatic transitions. The assumptions demand that the transitions are restricted to a small area around the crossing distance where the radial velocity has to be constant. Then the transition probability p depends on the radial velocity v_{rad} at crossing distance R_c as:

$$p = \exp\left\{\frac{-2\pi H_{12}^2}{v_{rad}\left|\frac{d}{dR}(H_{22}-H_{11})\right|_{R=R_c}}\right\} \tag{4}$$

With a constant covalent potential H_{11} and a Coulombic potential $H_{22} = -e^2/R$, $v_{rad} = v(1 - b^2/R_c^2)^{1/2}$, where v is the relative velocity and b the impact parameter, p becomes:

$$p = \exp\left\{\frac{-2\pi H_{12}^2 R_c^2}{v}\left(1-\frac{b^2}{R_c^2}\right)^{-1/2}\right\} = \exp\left\{\frac{-v_0}{v}\left(1-\frac{b^2}{R_c^2}\right)^{-1/2}\right\} \tag{5}$$

p is zero (adiabatic transition probability $1-p=1$ at $v=0$ or $b=R_c$ (glancing collisions) and approaches unity (non-adiabatic transition) with increasing velocity v and $b<R_c$.

The coupling matrix element H_{12} depends exponentially on R_c. A semi-empirical relation is given by Olson *et al.* (1971) in the reduced form

$$\mathrm{H}_{12}^* = \mathrm{c}_1 \mathrm{R}_c^* \exp(-c_2 R_c^*) \tag{6}$$

with $$H_{12}^* = \frac{H_{12}}{(I\cdot EA)^{1/2}}, \qquad R_c^* = \frac{\sqrt{I}+\sqrt{EA}}{\sqrt{2}}R_c$$

with I and EA as defined for equation (2). With $c_1 = 1.0$ and $c_2 = 0.86$, 83% of the available H_{12}-values could be represented within a factor of 3 with this function.

As indicated by the fine lines with arrows in Fig. 1, there are four different paths possible in a collision and two times at which the crossing distance R_c is passed. At every crossing there is a probability p for a non-adiabatic transition (without change of electronic character) and $(1-p)$ to remain on the adiabatic potential (with change of the electronic character between ionic and covalent).

For a two-atom system the crossing probability p does not change between the first and second crossing, but for three- and more-particle collisions p also depends on the internal motions in the molecule during the collision, as will be shown later.

Two trajectories with either two adiabatic transitions or two non-adiabatic transitions lead to elastic scattering, while the other two paths consisting of one adiabatic and one non-adiabatic transition with the probability $p(1-p)$ or $(1-p)p$ lead to ion-pair formation. The total probability for ion-pair formation in a single collision is therefore

$$P = 2p(1-p) \tag{7}$$

From this it is clear that the maximum probability for ion-pair formation is 0.5 when $p = 0.5$. From equations (4) and (5) it follows that this occurs at only one radial velocity which in turn depends on the total velocity and the impact parameter. The total cross section Q results from integration over all impact parameters:

$$Q = 4\pi \int_0^{R_c} p(1-p)b\mathrm{d}b = 4\pi R_c^2 F(v_0/v) \qquad (8)$$

with v_0 as given in equation (5). $F(v_0/v)$ is shown later in Fig. 8 as F_1 as a function of the reduced velocity v/v_0. The maximum is reached at $v/v_0 = 2.36$ and amounts to $F = 0.113$ or $Q = 0.45\pi R_c^2$.

From total cross-section measurements at high velocities one can obtain the velocity v_{max}, at which the cross section reaches its maximum. Together with equation (5) the H_{12}-value can be determined or, if H_{12} is otherwise determined, the applicability of the Landau–Zener model can be tested. This has been done for the ion-pair formation in the systems Li + I and Na + I (Moutinho *et al.*, 1971). The values agree well with theoretical estimations by Grice and Herschbach (1974). Calculations over a wide velocity range covering the maximum of the cross section could be fitted well to the experimental curves (Faist and Levine, 1976). A detailed description of the atom–atom scattering results is given by Los and Kleyn (1979).

The determination of H_{12} for atom–molecule collisions from total cross-section measurements (Hubers *et al.*, 1976) will be reviewed in Part IV. B. 2.

C. Differential Cross Sections

Differential cross-section measurements (as a function of the scattering angle θ) give information about the dynamics of the process and about the potential surface of the molecule formed in ionizing collisions. The total cross section results from the integration of the differential cross section $\sigma(\theta)$ over all scattering angles θ:

$$Q = 2\pi \int_0^{\pi} \sin\theta\sigma(\theta)\mathrm{d}\theta \qquad (9)$$

The classical deflection function relates the scattering angle θ to the impact parameter b. A typical deflection function for ion-pair formation is shown in Fig. 2, calculated for Na + I (Delvigne and Los, 1973). The upper part (of the reduced deflection angle $\tau = \theta E$) results from scattering on the repulsive part of the potential surface. The lower part of the deflection function comes mainly from the attractive part of the potential: the small negative-angle part comes from covalent scattering, while the wide-angle part results from ionic scattering. Covalent scattering means the particles undergo a non-adiabatic transition at the first crossing, the change to ionic

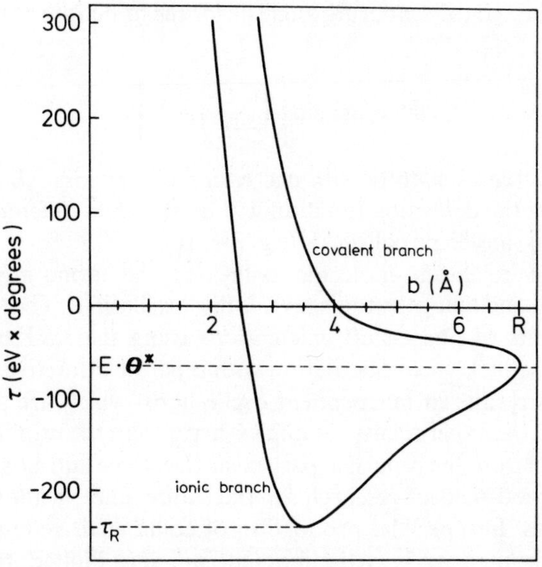

Fig. 2. Typical deflection function $\tau = E\theta$ versus impact parameter for ion-pair formation, calculated for Na + I collisions (Delvigne and Los, 1973). The covalent and ionic branches joint at $b = R_c$, τ_R is the rainbow angle (adjusted from Los and Kleyn, 1979).

character occurring adiabatically at the second crossing. Ionic scattered particles follow the lower adiabatic surface with strong Coulombic attraction and hop to the upper surface only at the second crossing (see Fig. 1). Both scattering parts of the deflection function meet at the maximum impact parameter $b_{max} = R_c$. The maximum attractive angle is called the rainbow angle τ_R. The differential cross section $\sigma(\theta)$ depends on the classical deflection function:

$$\sigma(\theta) = \frac{1}{\sin\theta} \sum_i p_i(1 - p_i) b \left| \frac{db}{d\theta} \right| \tag{10}$$

The experiment does not differ between positive and negative scattering angles. Therefore, the contributions from four impact parameters i have to be summed up for one scattering angle θ up to rainbow angle θ_R.

Above $|\tau_R|$ there are only two small contributions from repulsive collisions (upper part in Fig. 2). From a deflection function as given in Fig. 2 there result two maxima for the differential polar cross section $\sigma(\theta)\sin\theta$, one at small angles (covalent scattered particles), and one at wide angles (ionic scattering) with a maximum at the rainbow angle and a minimum at $b = b_{max}$. For a pure Coulombic scattering potential, for long-range crossing, Young *et al.* (1974) derived the scattering angle θ^* for the minimum of the differential cross section:

$$\theta^* = \text{arc sin}\left[\left(\frac{2E}{\Delta E}\right) - 1\right]^{-1} \tag{11}$$

θ^* (in deg) decreases slightly with decreasing energy loss ΔE. In Part IV.C these effects of the deflection function will be shown in differential measurements of atom–molecule collisions (e.g. Fig. 16).

In contrast to atom–molecule collisions the atom–atom scattering between alkali and halogen atoms reveals fine oscillations. They are explained with the results of phase-shift calculations using the JWKB method. An approximate semiclassical calculation with separate attractive and repulsive contributions results in independent oscillations, which are superposed to compare with the experiments. At angles larger than the rainbow angle only one oscillation from the repulsive part exists (Delvigne and Los, 1973).

This is a good field of research for the future since it allows a nice test of the theories. Because the production of collimated, state-selected atom beams of high intensity is at the moment still very limited, the differential measurements of the last years have been performed mainly between atoms and molecules. Therefore, the atom–atom scattering will not be reviewed in detail here. The readers are referred on this subject to reviews by Baede (1975), Los (1977), and Los and Kleyn (1979).

D. Atom – Molecule Collisions

While atom–atom collisions are described by an one-dimensional potential energy curve, atom–molecule collisions with N particles require $3N - 6$ degrees of freedom. Atom–diatomic collisions are described by a three-dimensional potential hypersurface. While for the atom–atom reactions accurate calculations of total and differential cross sections are available, the problem to be solved for three and more particles is much more difficult. Till now there have been only approximate calculations on hand, which do not allow an exact trajectory calculation of collisional ionization processes. For these, very accurate potential surfaces are needed. Up to now there exists only one *ab initio* calculation of the lightest system: $Li + F_2$ (Balint-Kurti, 1973), for which ion-pair formation has not yet been measured. Another non-empirical determination with several assumptions exists for $K + Cl_2$ (Nyeland and Ross, 1971) and recently the lowest diabatic surface of $Li + O_2$ and $Na + O_2$ (Alexander, 1978) has been determined by an exponential-rational approximant functional form, which naturally does not allow trajectory formulations but gives some potential depths for different electronic symmetries, which might be helpful in a qualitative discussion of differential cross sections.

For a trajectory calculation of ionization processes, at least two hypersurfaces and their coupling as a function of all dimensions are needed. A hopeful method to obtain semiempirical hypersurfaces of even more complicated systems in future is the diatomics-in-molecules (DIM) method (Kuntz, 1979). For details on potential surface calculation see the chapter of Tully in this book.

Several trajectory calculations exist which use semiempirically determined potential surfaces. The most crucial parameter in these calculations is the coupling matrix element, which in case of the crossing of two three-dimensional surfaces depends not only on the distance at the crossing but also on the orientation angle ϕ of the internuclear molecular axis with the distance vector between the atom and the centre-of-mass of the molecule. $H_{12}(\phi)$ is non-zero if the symmetry and multiplicity of the electron configuration in the collision complex of the reactants and products are equal. The multiplicity of the collision complex results from the multiplicities of the components. The symmetry depends on the alignment of the colliding particles described by the angle ϕ. C_{2v} is the isosceles or perpendicular geometry ($\phi = 90°$). $C_{\infty v}$ is the linear configuration ($\phi = 0$), and C_s the orientation for any angle in between $0 < \phi < 90°$.

Table II gives the different quasi-molecular states in the three geometries for collisions between alkali atoms M and halogen molecules X_2 in their ground states and the ionic products in the ground ($^2\Sigma_u^+$) and excited ($^2\Pi_g$) states and for alkali–O_2 reactions with O_2^- in the ground and two excited states. The table shows that adiabatic transitions in the isosceles geometry with the halogen ions are not possible: H_{12} is zero at $\phi = 90°$. In $C_{\infty v}$ and C_s geometry adiabatic transitions can take place ($H_{12} > 0$). This is well described by a coupling element H_{12} which varies continuously with the

TABLE II. Correlations of potential energy surfaces for covalent and ionic collision complexes in their ground and some excited states together with the angular dependent coupling matrix element $H_{12}(\phi)$ (M = alkali atom, X_2 = halogen molecule)

	$C_{\infty v}(\phi = 0)$	C_s	$C_{2v}(\phi = 90°)$	$H_{12}\phi$
$M(^2S_g) + X_2(^1\Sigma_g^+)$	$^2\Sigma^+$	$^2A'$	$^2A_1$	
$M^+(^1S_g) + X_2^-(^2\Sigma_u^+)$	$^2\Sigma^+$	$^2A'$	$^2B_2$	$H_{12}\cos\phi$
$M^+(^1S_g) + X_2^-(^2\Pi_g)$	$^2\Pi$	$^2A'$, $^2A''$	$^2B_2$, $^2A_2$	$H_{12}\sin 2\phi$
$M(^2S_g) + O_2(^3\Sigma_g^-)$	$^{4,2}\Sigma^-$	$^{4,2}A''$	$^{4,2}B_1$	
$M^+(^1S) + O_2^-(^2\Pi_g)$	$^2\Pi$	$^2A'$, $^2A''$	$^2A_2$, $^2B_2$	$H_{12}\sin 2\phi$
$M^+(^2S) + O_2^-(^4\Sigma_u^-)$	$^4\Sigma^-$	$^4A''$	$^4A_2$	$H_{12}\cos\phi$
$M^+(^2S) + O_2^-(^2\Pi_u)$	$^2\Pi$	$^2A'$, $^2A''$	$^2A_1$, $^2B_1$	$H_{12}\sin\phi$

angular orientation:

$$H_{12}(\phi) = H_{12} \cos \phi \tag{12}$$

This relation is confirmed by calculations for large R_c (Anderson and Herschbach, 1975). The formation of the excited state ($^2\Pi_g$) is only possible in the C_s geometry ($H_{12} = 0$ at 0 and 90°), H_{12} being described by the function (Gislason and Sachs, 1975)

$$H_{12}(\phi) = H_{12} \sin 2\phi \tag{13}$$

$O_2^-(^2\Pi_g)$ ground-state formation is possible only in C_s geometry and $H_{12}(\phi)$ is described as in equation (13). The excited $^4\Sigma_u^-$ state is not formed in the isosceles geometry ($H_{12}(\phi) \sim \cos \phi$). The excited $^2\Pi_u$ state is not formed in linear collisions, which is best described by:

$$H_{12}(\phi) = H_{12} \sin \phi \tag{14}$$

The transition probability p is an exponential function of $H_{12}^2(\phi)$ and is therefore strongly dependent on the orientation. $H_{12}(\phi)$ depends exponentially on the crossing distance R_c, as shown for H_{12} in equation (6).

E. Trajectory Calculations and Reactions Dynamics

Several trajectory calculations have been performed with semiempirical potentials which include the orientational effects of the molecule. The influence of several collision and potential parameters on the differential scattering in collisional ionization have been calculated by Düren (1973). He could show that calculations with an angle-averaged H_{12} agree well with the angle-dependent H_{12}. A Landau–Zener formula for atom–molecule collisions in an extended form has been developed, see Nikitin (1968). Trajectory calculation on one diabatic surface were extended to more surfaces by introduction of the surface-hopping trajectory (SHT) method (Tully and Preston, 1971) with calculation of the transition probabilities at the avoided crossing by an extended Landau–Zener model. This method has been successfully applied to ion-pair formation processes (Evers, 1978) and will be referred to in this review in Part IV.

Much simpler methods are, e.g. the multiple-crossing method by Bauer *et al.* (1969). It reduces the three-particle system to a two-particle multistate problem by a network of avoided crossing points, each representing a vibrational transition between the ionic–covalent states. The transition probability as a function of the coupling matrix element H_{12} and Franck–Condon factors has been calculated (Zembekov, 1975). It was applied successfully to the description of reactive collisions with an electron transfer at low energies (Gislason and Sachs, 1975; Yuan and Micha, 1976). Its application is limited to low energies and wide vibrational spacings to avoid interferences and overlaps by the energy uncertainty principle. Therefore, Kendall and Grice (1972) introduced an average electronic matrix element which determines the transition probability according to the Landau–Zener model, the Franck–Condon factors determining only the vibrational state distribution, as theoretically demonstrated by Child (1973).

During the last few years a 'bond stretching' model has been introduced by the group of Los. The main points of this model are: it decouples the internal motion in the molecule from the translational motion and so reduces the three-particle problem to a two-particle one. The model neglects momentum transfer between the nuclei. The rotational motion during the collision is frozen, which is reasonable since the collision time is only of the order of one vibrational period or even shorter.

This extended Landau–Zener model includes the angular dependence of the coupling matrix element $H_{12}(\phi)$ but, in addition, the dependence of the vertical electron affinity on the phase of the molecular vibration when the particles reach the crossing distance a second time. In other words, when the particles have approached the crossing distance following the adiabatic potential surface, the molecule changes vertically to the often repulsive part of the ionic potential; that means the molecule is now vibrationally excited and starts to increase its bond distance, and, as seen in Fig. 3a, its vertical electron affinity (or also called electron detachment energy) increases with the bond distance up to a maximum value near the turning point on the attractive wall (transition 3 in Fig. 3a). From equations (2) and (3) it follows that the crossing distance R_c increases with the electron affinity.

The crossing radii during a vibrational period are shown in Fig. 3(b) for the systems Na, K, and $Cs + Br_2$ (Aten *et al.*, 1976). The collision time could be in good approximation replaced by the Br–Br internuclear distance. The corresponding motion on the Br_2^- potential during the vibration are added to Fig. 3(a) (via transition (4)). The calculation corresponds to a transition at $r = 2.4$ Å with an EA-value of 2.0 eV.

The maximum EA is reached at about 3.5 Å. This corresponds to the maximum of R_c in Fig. 3(b), followed by a dip which corresponds to the classical turning point, in this example, at 3.9 Å. Then follows the equivalent inward vibration. The dash pointed line corresponds to a dissociative transition in the $K + Br_2$ system after a transition (2) in Fig. 3(a) at $r = 2.26$ Å with an electron affinity of about 1.3 eV.

The four straight lines correspond to trajectories with $b = 0$ and decreasing velocities (parameters a, b, c, d of 4; 2; 1; 0.5×10^4 m/s). The crossing of these lines with the R_c-function for $Na + Br_2$ gives the crossing distances at different collision energies (a′ to c′ for $K + Br_2$, respectively). A detailed discussion is given in Part IV.B.2 in connection with total cross-section measurements at high velocities (Hubers *et al.*, 1976).

In the system $Cs + Br_2$, R_c reaches its maximum at 228 Å. The crossing

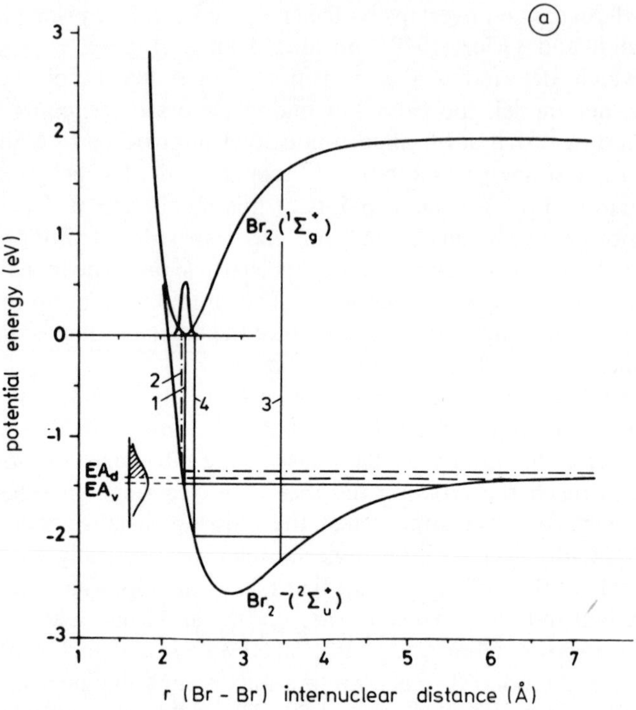

Fig. 3(a). Potential energy curves of the electronic ground states of Br_2 and Br_2^-. The Franck–Condon density distribution for a transition from $Br_2\,(v = 0)$ to Br_2^- is shown in the inset. EA_d is the electron affinity corresponding to a transition to Br_2^- at the dissociation limit, EA_v is the vertical electron affinity (transition 1).

points of the straight-line trajectories with the R_c-function show a strong increase during a collision. Replacing Na by K or Cs this increase is even more strongly increased. The coupling matrix element H_{12} decreases exponentially with increasing R_c (equation (6)) and according to the Landau–Zener model the transition probability p for a non-adiabatic transition depends exponentially on H_{12} (equation (4)) and will approach unity with a decreasing H_{12}. This means that particles which approach on an adiabatic surface will fly apart with a strongly increased non-adiabatic probability p' ($p < p' \approx 1$) as an ion pair. Equation (7) changes accordingly into

$$P = (1 - p)(p' + p) \tag{15}$$

For covalently scattered particles the crossing probability at the first and second crossings is assumed unchanged ($= p$).

This bond stretching model has been applied successfully to explain many experimental results in total and differential ionization cross-section

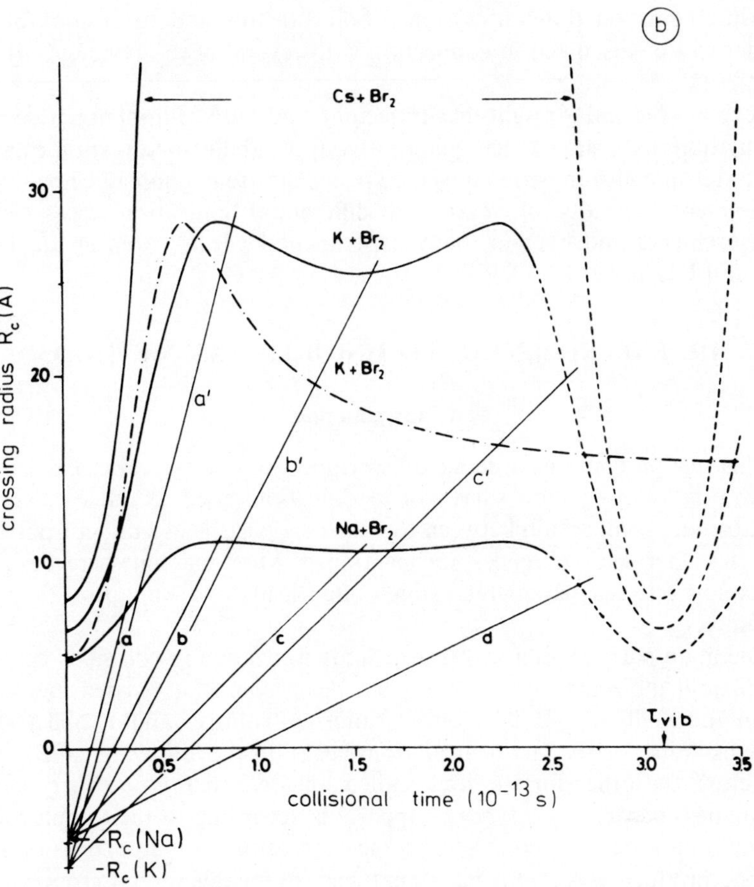

Fig. 3(b). The dependence of the crossing radius R_c for Na, K, and $Cs + Br_2$ on the collision time in the case of ionic scattering for a vibrational excitation as indicated in Fig. 3(a) by transition (4). Combination of the R_c resulting from the *EA*-values of (a) and equation (2) with a classical calculation of the Br_2^- motion (with $b = 0$) Fig. 3(a) can be mapped onto the time scale shown in the abscissa. For $K + Br_2$ two curves are shown. The oscillating curve corresponds to a transition to a bound state of the Br_2^- molecular ion at $r = 2.40$ Å (transition (4)). The minimum between the two maxima indicates the classical turning point in the Br_2^- vibrational motion. A dip is observed because the electron affinity of Br_2 is a maximum indicated as transition (3) in Fig. 3(a) at 3.5 Å. The other $K + Br_2$ curve (–.–.–) gives the R_c dependence after a transition (2) in part (a) to a continuum state of Br_2^- at $r = 2.26$ Å. The Na and $K + Br_2$ curves are accompanied by four straight-line trajectories with $b = 0$. The velocities are a) 4×10^4 m/s, b) 2×10^4 m/s, c) 1×10^4 m/s, and d) 5×10^3 m/s. The lines labeled a′, b′, c′ refer to $K + Br_2$. In the case of $Cs + Br_2$, R_c reaches a value of 228 Å at its maximum. For the transition (4) in par (a) at $r = 2.40$ Å the vibrational period τ_{vib} is indicated (adjusted from Aten *et al.*, 1976).

measurements on different systems. Some details and extensions of the model will be discussed in connection with several of the reviewed articles in Part IV.

Several classical straight-line trajectory and surface-hopping trajectory calculations have shown their quantitative applicability to ion-pair formation processes in collision with halogens. The results are in good agreement with experimental results of total and differential ionization cross-section measurements and explain many experimental effects (Aten *et al.*, 1976; Los and Kleyn, 1979).

III. EXPERIMENTAL TECHNIQUES AND METHODS

A. Introduction

This section contains a detailed description of one typical crossed beam apparatus followed by a survey of modifications used in other machines. An absolute comparison between the differing parts is usually not possible, but characteristic differences are mentioned. Most machines are designed to exploit selected parameters, which often leads to a limitation on other parameters.

The main parts of an apparatus to study ionization in collisions between atoms and molecules are: A fast atom beam source, a crossed molecular beam (normally under 90°) or a scattering volume, and two detection systems—one for the fast neutral atom beam (in the following called atom detector), the other for the ions (called ion detector). One can generally distinguish between two types of apparatus according to their application: (a) for total ionization cross-section measurements, with an extraction field for the product ions and a mass analyser; (b) for differential cross-section measurements with a rotatable mounting of either the beam sources or the detection system.

B. Crossed Beam Apparatus

In Fig. 4 is shown schematically a crossed beam apparatus for double differential collisional ionization studies (Kimura and Lacmann, 1978). The fast atom beam source consists of an ionizer which is commercially available for all alkali metals (Spectra Mat. Inc.). The source is easy to install because of its small dimensions of 18 mm diameter and 13 mm height. It is equipped with an indirect heating system and surrounded by a threefold heat shield. The ions are produced by thermal dissociation of an alkali-containing salt at more than 1500 K. The ions are extracted through an extraction lens system and accelerated to about 80 eV. They approach their final energy in a deceleration lens with a quadrupole lens for beam stirring

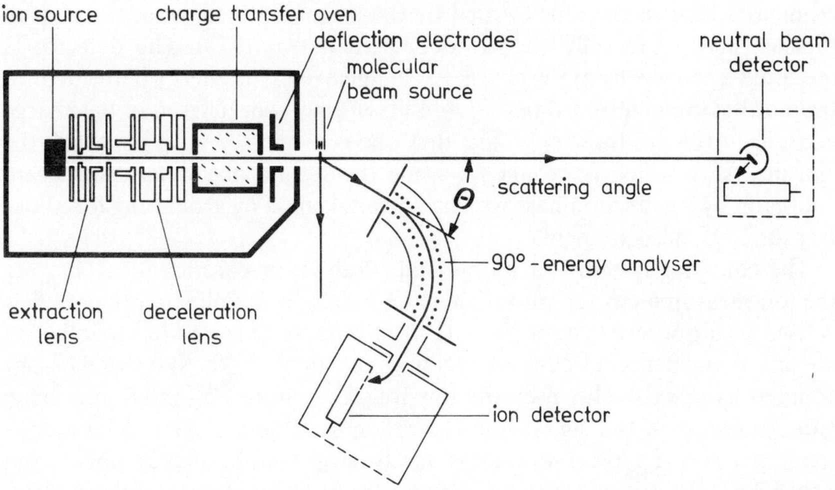

Fig. 4. Schematic drawing of a collisional ionization apparatus for the study of double differential ionization cross sections (Kimura and Lacmann, 1978).

focusing through a charge exchange oven (0.5 × 3 and 1 × 3 mm slits at entrance and exit, respectively) into the scattering volume. The charge exchange oven contains the alkali metal of the same species as the ions and is heated so as to give a vapour pressure of 0.1 to 1 P ($\sim 10^{-2}$–10^{-3} Torr) at 400–800 K depending on the alkali metal Cs to Li, respectively. The ions are neutralized in the oven over a path length of 30 mm.

Charged particles leaving the oven are removed by electrostatic deflection. The neutral beam has a divergence angle of 1.5° FWHM below 20 eV; at higher energies it is somewhat smaller. The total atom beam source is surrounded by copper walls, cooled by liquid nitrogen. The molecular beam effuses through a multichannel capillary array consisting of steel capillaries of 50 μm diameter and 3 mm length and 41% transparency. The array is confined by a slit of 2 × 3 mm at a distance of 1 cm from the atom beam axis. The two beams are crossed at right angles.

The energy of the product K^+ ions from the ionization process is measured in a 90° cylindrical electrostatic analyser with an entrance and exit slit width of 0.5 mm and a medium radius of 50 mm. The energy resolution is 1% FWHM. The distance from the scattering centre to the entrance slit amounts to 60 mm. The ions are detected through a magnetic multiplier (Bendix M306).

The fast neutral beam is monitored with a Langmuir–Taylor detector by surface ionization on either a lukewarm rhenium ribbon which does not ionize the thermal background alkali atoms, or on a hot ribbon with a slight retarding field between the ionizer and a magnetic multiplier which discriminates against the ions formed by the thermal atoms, while the ions of the fast atoms are only slightly depressed (Aten, 1972). The detector is mounted 2 m away from the charge exchange oven. The energy of the neutral beam is determined by a time-of-flight (TOF) method. In front of the charge exchange oven are three slits. The first one is connected with the third (the entrance slit), which is grounded, while the second one is used for beam chopping by a deflection field which can be taken off by grounding for about 1 μs for TOF measurements.

The counting systems for atoms and product ions are identical. During the ion measurement the atom detection system (multiplier + preamplifier) is used as a dummy system. Signals from both systems are fed together in an anti-coincidence circuit. Coinciding pulses in both systems are not counted as signals. This prevents low-frequency noise. In addition a delay gate generator in the signal line closes the scaler for a given time after a signal arrives. A typical value is 5 ms to stop high-frequency noise from power lines. In this way the noise rate is reduced to about 1 pulse in 10 s. The signal pulses are amplified and discriminated (from low-level background pulses created in the ion detector system) and stored in a multiscaling analyser. Each channel corresponds to a given product ion energy and scattering angle. The experiment is run semiautomatically. All parameters are pre-programmed and controlled by a Camac branch controller with a programmable read-only-memory (ROM).

C. On-line Computer

The trend in molecular research is to develop more sophisticated machines with finer selection of more and more parameters. But each additional parameter selected in an experiment usually reduces the intensity of the particles involved. Therefore, the experiments are often limited by the statistical error which is given by $\pm I^{1/2}$ where I is the number of particles collected for one measuring point, presuming the noise to background pulse ratio is small enough. This limitation of an experiment can be overcome by increasing the time for data collection; however, the time is limited in turn by the stability of the parameters of the experiment, such as potentials, fluxes, or human patience and abilities. The human element can be avoided by automation. This is nowadays done by an on-line computer which controls the experiment and data processing. The advantages of an on-line computer over a semiautomatic data acquisition system are a more flexible measuring programme with nearly unlimited control functions and an automatic work-up of the experimental data. In the apparatus described above, for example, the computer could calculate automatically a stepwise increase of the scattering angle in c.m. system and the energy analyser

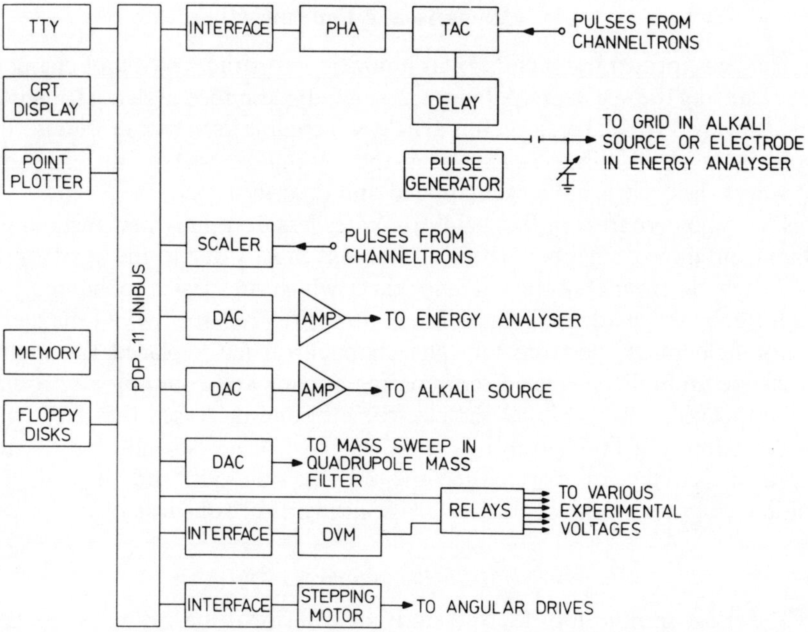

Fig. 5. Schematic data handling system of an on-line computer-controlled experiment (TTY—teletype, PHA—pulse height analyser, TAC—time to amplitude converter, DAC-digital to analog converter, DVM—digital voltmeter, CRT—cathode ray tube). (Reproduced by permission of Elsevier Scientific Publ. Co. from Warmack *et al.*, 1978a.)

setting (in Lab system) for particles of constant energy loss (in c.m. system). A quite complicated calculation, but very helpful for an optimal measuring programme. The other advantage is the ability of an immediate working-up of all data, as there are the transformation of the intensities from the laboratory into the centre-of-mass system and the calculation of the cross section from the intensities of product ions, and primary beam intensity as function of the collision energy.

A schematic drawing showing all the necessary parts for a computer-controlled experiment is given in Fig. 5. It is used by Warmack *et al.* (1978a) and similarly by Dispert and Lacmann (1978), the latter in connection with a Camac branch controller. A hardcopy unit is often used in addition or instead of the point plotter. It is faster but less precise in the scaled reproduction of the data. Computerized data processing with Camac interface system has been described by Sheen *et al.* (1977) for TOF measurements in a seeded beam apparatus for the investigation of ion-pair formation by polar dissociation.

D. Comparison of Equipment

The basic principles of collisional ionization apparatus have not changed much during the last years. All parts are already described in detail (Fluendy and Lawley, 1973). Therefore, this article will confine itself to a short description of the main differences between the alternative parts with references to where these parts have been applied and described.

The major progress in this field originates less from fully new techniques than from improvements of the old. This was mostly due to the availability of better electronic equipment and parts which are easier to handle e.g. high-frequency quadrupole mass filters now often replace the old magnetic sector field mass spectrometer. The channeltron has replaced the multi-electrode multiplier. Noise-discriminating circuits and counting electronics are improved, have replaced the electrometer and increased the sensitivity of detection to a limit given by statistical error or measuring time. Faster electronic circuits improved the time-of-flight methods, and through the on-line computer more sophisticated experiments became feasible.

1. *Fast Beam Sources*

The most specific tool for the study of collisional ionization is the fast atom beam source of variable energy in the eV range. Besides the resonance charge transfer source, there are several versions of charge transfer sources. The simplest version uses only one oven. Thermal atoms which leave the oven through a rear slit are ionized on a hot metal surface (with work function greater than the ionization energy of the atoms) and are accelerated by an electric field back into the oven where they are neutralized by resonance charge exchange without momentum transfer. This source was first developed for energies above 20 eV (Hollstein and Pauly, 1966) but by minor changes in the geometry around the ion path the space charge limitation could be reduced and the source could be applied to energies down to below 2 eV (Lacmann and Herschbach, 1970). Helbing and Rothe (1969) lowered this energy by mounting the ionizer and the accelerating electrode inside the oven. With the addition of a rotating multidisc velocity filter the absolute energy was determined. The energy spread of the source could be reduced by the velocity filter to the velocity spread of the filter of 3.4%. Above a beam energy of 7 eV the velocity filter is ineffective, the energy spread is given by the source itself to 0.5 eV.

A charge transfer source with an energy width of 1% FWHM was designed by Mochizuki and Lacmann (1976). It consists of an alkali oven with a hot porous tungsten disc as ionizer and a 127° cylindrical electrostatic energy filter which produces at 10 eV an ion beam of 0.1 eV FWHM. The beam is accelerated or decelerated to its final energy before it passes a second alkali oven for neutralization by resonance charge transfer.

The most frequently used beam source for low energies (a few eV) is the sputter source. A high-energy rare-gas (typically $\sim$ 9 kV Kr^+) ion beam hits a target. Atoms and ions—the latter are deflected—are produced by sputtering in a wide energy range up to over 100 eV with a maximum between 1 and 20 eV. The energy is selected by a high-speed rotating disc velocity selector (Politiek *et al.*, 1968; Young *et al.*, 1974; Kempter, 1975). At about 6 eV this source is with a velocity spread of typically $\Delta v/v \sim 4\%$ to 10% comparable to the charge transfer source. At higher energies its energy width becomes larger and its intensity is lower. The maximum speed of rotation of the mechanical velocity selector limits its application to low energies. The velocity selector has been most successfully applied to the study of chemical dynamics with thermal beams. Below $\sim$ 6 eV the sputter source produces higher beam intensities with lower energy widths than the charge transfer sources. The design is much more sophisticated compared to charge transfer sources because of the high-energy ion beam source and the high-speed mechanical velocity analyser.

Sputter sources are often used to produce non-alkali atom beams as of C, Ba, Ti, Ta, Fe, Al, U, and Ce. A review article by Wexler (1973) contains a survey of further publications in this field. The sputtering of chlorine-containing salts was applied to produce a fast chlorine atom beam (Können *et al.*, 1974).

The seeded supersonic nozzle beam source is used to produce an intense axial gas beam in the eV range. It was originally developed by Fenn (Abuaf *et al.*, 1967). Reactive and ion-pair formation processes are studied with a seeded UF_6 beam (Annis and Datz, 1978). Seeded beams are extensively applied in collisional polar dissociation studies (Wexler, 1973; Sheen *et al.*, 1977). By isentropic expansion of H_2 or He from high pressure in a nozzle, a strongly directed axial beam of high intensity with a translational energy of $\sim$ 0.1 eV is formed. By seeding the light gas with small amounts of a heavy gas (0.1 to some %), the heavy particles come into velocity equilibrium with the light ones by momentum transfer. The kinetic energy of the heavy particles is then increased at most by the ratio of their mass to the mass of the H_2 or He gas. In the case of xenon, beams of 6 to 18 eV are produced by seeding with H_2. The intensity of seeded beams amounts to 10^{18} to 10^{17} atoms sr^{-1} s^{-1} and exceeds by many orders of magnitude the intensities of charge exchange sources. The energy is varied by the nozzle temperature or by changing the composition and thereby the degree of velocity equilibrium or velocity slippage. The actual energy is measured by time-of-flight.

The seeded beam sources need a strong differential pumping between the nozzle and a skimmer ($\sim 10^4$ l/s). Through its orifice only the axial cone of the beam passes. A second stage of differential pumping may follow before the beam enters the reaction volume. Construction of these sources needs a lot of technical know-how and mechanical skill.

For details of seeded beams and other beam sources see Fluendy and Lawley (1973). The intensities of the different beam sources are compared in the review article by Kempter (1975).

Fast-energy homogeneous H-atom beams are produced by laser photodetachment of H^- ions (Aberle *et al.*, 1978; van Zyl *et al.*, 1976).

2. *Target Gas*

The target gas molecules are in most cases produced in a crossed beam by a multichannel array of glass or stainless steel capillaries ($\sim$ 20 to 50 μm dia). The angular divergence of the beam is smaller than the cosine distribution from the effusion sources but depends on the geometry of the array and the pressure in the gas reservoir. Alternatively, the beam can be produced by expansion from a high-pressure nozzle source. Wexler (1973) reported an angular distribution of only 7° FWHM. Simpler experiments involve the use of a scattering chamber for total cross-section measurements, but it has the disadvantage of a larger Doppler-broadening effect near the threshold because of the Maxwell–Boltzmann velocity distribution of the target gas molecules (Chantry, 1971). This distribution is well known as a function of the temperature and can be taken care of in unfolding procedures (Nalley *et al.*, 1973; Dispert and Lacmann, 1978). In experiments with a multichannel array the real beam divergence has not been determined. Any corrections are assumed to be negligible.

The functional relation of the Maxwell–Boltzmann distribution on the relative velocity distribution and its influence on the threshold behaviour of the ionization cross section will be discussed in Section III. E.

3. *Energy Analysers*

In connection with the above-described double differential crossed beam apparatus, several problems and methods of energy analysis have already been discussed or are described in detail elsewhere (Fluendy and Lawley, 1973).

In the following will be given only a comparative abstract with some recent publications describing energy-analysing techniques.

(a) The mechanical rotating multidisc velocity selector ($\Delta v/v > 3\%$) is superior for neutral beams below 5 eV. (b) Cylindrical electrostatic analysers and spherical electrostatic analysers, both have a typical ion energy resolution of about 1%, depending on the geometry. The spherical one consists of an inner and an outer hemispherical electrode. In its field the ions are focused in three dimensions, while the cylindrical one does not focus in the direction of the slit height (parallel to the electrodes). This may be a disadvantage in connection with channeltron detectors which often have a rather limited detection area. In connection with the magnetic multipliers

(Bendix Corp.) with a large detection area the differences are probably minor. (c) The time-of-flight method is most useful for chopped ion or neutral beams. Experiments with a continuous beam which is chopped only for TOF measurements are difficult because the beam width may be increased by chopping (Warmack *et al.*, 1978a) and the energy may even be displaced somewhat (Dispert and Lacmann, 1979).

The beam intensities of a continuous beam are decreased by chopping to about one tenth or less of the unchopped intensity. Therefore, the reactions are preferably studied in the continuous mode. If the product intensities are big enough, the TOF technique in connection with a gated detection system allows detection of only those reaction products that are formed in a short range of primary beam flight times. (Leffert *et al.*, 1972). This technique avoids the difficulties mentioned above and even allows an increase in the energy resolution by reducing the gate 'open' time to shorter intervals than the primary beam pulse length. The flight time between the scattering volume and the detector has to be short. Therefore, the product ions have to be accelerated immediately after the collision so that the flight time to the detector is short compared with the primary flight time and independent of it.

4. *Detection System*

The detection system is often differentially pumped or at least shielded from the other parts to reduce the background noise.

The fast neutral alkali beam is always detected by surface ionization. It applies the same effect as used in the charge transfer source to produce the ions. Particles with an ionization energy lower than the work function of the detector filament are ionized according to the Saha–Langmuir equation when they hit the surface and are detected on a negatively biased electrode or are amplified by a multiplier. Therefore, alkali metal atoms can be very efficiently detected on clean and therefore often-heated surfaces of rhenium, tungsten, platinum, or other metals of high work function. The ionization efficiency has been checked on a tungsten 110 surface and found to be unity for atoms of 1 to 10 eV of K or Na atoms. For thermal atoms it is 1 and 0.44 for K and Na, respectively. The work function is determined to 5.17 ± 0.01 eV very similar to the ionization energy of Na (5.12 eV) (Hurkmans *et al.*, 1976). The detection efficiency for sodium could be increased to near unity by an oxygen-covered tungsten surface. Its work function is higher than 6.0 eV (Hubers *et al.*, 1976).

Another method besides the above-mentioned ones to reduce the thermal background of alkali atoms which are ionized together with the fast atoms on a hot surface applies a Rabi-type magnet in front of the surface ionization detector. The magnet deflects the thermal alkali beam but the fast atoms are hardly influenced (Hubers *et al.*, 1976). These authors also studied the ionization efficiency on a hot surface at high energies. It is about unity up to 100 eV for K and Cs, for Na may be somewhat lower, at higher energies it drops for K to about 0.6.

Non-alkali neutral beams have to be ionized in electron impact ion sources. The ionization efficiency is only $\sim 1\%$. Intensity measurements of a molecular beam from a nozzle source are taken with an ionization pressure guage (Wexler, 1973).

For mass analysis the quadrupole mass filter or time-of-flight method has largely replaced the magnetic sector field. Mass analysis is mostly used in total ionization cross-section measurements to differentiate among the different negative product ion masses. For differential measurements this is normally not necessary because only the positive (alkali-)atomic ions of one species are studied. If in special cases of double differential measurements the negative molecular ions can be detected, it is helpful to analyse their masses. This has been done by the installation of an electronic chopper in front of a spherical energy analyser. In this way, at a given energy, the mass of the product ions can be analysed (Warmack *et al.*, 1978b).

For absolute total measurements of high intensities ($> 10^{-13}$ A) the ions are measured with an electrometer. But in most cases the intensity is much lower and an ion counting technique is applied. Most convenient is the small channeltron multiplier for intensities up to 10^5 particles per second. In comparison, the crossed field magnetic multiplier (Bendix Corp.) has a linear response up to 10^6 particles per second. The latter is very resistant to corrosive gases and has a larger lifetime. A small stray magnetic field can easily be shielded. The dimensions are somewhat larger than those of a channeltron.

For negative ion detection the multipliers have to be floated to ~ 2 kV cathode potential with the anode at ~ 4 kV.

Several descriptions of crossed beam apparatus for total ionization measurements with mass analysis are published, e.g. by Nalley *et al.* (1973), Tang *et al.* (1974c), and for differential measurements by Mochizuki and Lacmann (1976) and Warmack *et al.* (1978a).

E. Energy Calibration and Threshold Determination

In all collisional ionization experiments the energy calibration is very critical. The voltage applied between the ionizer and the charge exchange oven deviates by 1–3 eV from the energy determined by the TOF measurement. This was already observed in one of the first collisional ion-pair formation experiments with a mechanical rotating multidisc velocity selector (Helbing and Rothe, 1969). It originates mainly from surface potential differences between the hot surface of the ionizer (of rhenium, tungsten, or some other metal of high work function) and the alkali-metal-covered charge exchange oven. Between these two species arise the highest contact potential differences because of their extremely different work functions. This value is very sensitive to changes in the composition of the surfaces and is therefore often unstable with time. The energy correction is also dependent on the absolute energy. This is explained by space charge effects within the beam which lead to an energy separation within the beam cross section (Aten and Los, 1975). This effect is strongly dependent on the geometry of the ionizer and charge exchange oven and may vary with the beam alignment and leads also to instabilities of a few tenths of an eV during the experiment. Therefore, any time lapse between the calibration of the apparatus and the experimental measurements may easily lead to an error of 0.1–0.2 eV.

Even more difficult is the energy loss determination in double differential measurements, for, besides the problems with the primary beam energy, there are difficulties in determining the energy of the product ions. The ion energy distribution can easily be measured in an electrostatic cylindrical field analyser, but its absolute value is influenced by contact potential differences between the scattering volume and the energy analyser which are not accurately known and may add another error of 0.1–0.2 eV.

In some experiments the energy loss scale is calibrated from the limiting values for the reaction. The minimum energy loss ΔE is given by the equation (2) with EA the adiabatic electron affinity taken from the total cross-section measurements. This is reasonable only if the negative ions are formed in their ground state without excitation. The maximum energy loss is given by the collision energy itself: this means the particles are formed in the centre-of-mass without translational energy. This calibration point is available only at collision energies E near the threshold energy up to a maximum energy given by the maximum energy loss of the reaction which depends on the overlap of the distribution functions of the neutral and ionic molecule at the crossing distance R_c (Mochizuki and Lacmann, 1976).

The simplest way of energy calibration uses gases of well-known threshold energy, e.g. the accurately determined EA-values by photoelectron spectroscopy of Celotta *et al.* (1972) for O_2 of 0.44 eV, Celotta *et al.* (1974) for SO_2 of 1.097 eV, and Herbst *et al.* (1974) for NO_2 of 2.36 eV. The threshold energy is either determined by assuming a linearly increasing cross section, e.g. for O_3, SO_2, and SO_3 together with a small correction, according to the treatment of Chantry (1971) (Rothe *et al.*, 1975) or from the first rise as described by Compton *et al.* (1978a) in the study of hexafluoride molecules.

In the laboratory system in a crossed beam apparatus, the energy spread of the target gas is negligible and the spread of the fast beam is constant for each target molecule and, therefore, automatically included in the energy calibration by the test gases. This method has only a limited accuracy. The contact potentials may change somewhat between the test gases used and the gases to be investigated. The determination of the first rising may depend on the detection sensibility. The most accurate method of threshold determinations needs the exact velocity distributions of the fast beam determined by methods as mentioned in Part III.D.3 and of molecular beam as described in Part III.D.2.

The accuracy of the determination of the real ionization function from the experimentally determined curve involves a deconvolution of the relative velocity distribution of the colliding particles.

In practice, the threshold is more often determined in the reverse way. A hypothetical ionization function is convoluted with the experimentally determined relative velocity distribution. The threshold results from the best fitted curve. This method relies on an assumed cross-section function. Deviations of the fitted curve from the experimental one indicate an error either of the relative velocity distribution or the assumed trial function.

In the case of an exact crossed beam alignment of the fast atom and the thermal molecular beam the influence on the relative velocity distribution by the thermal beam is in the order of kT and negligible compared to the fast-beam spread. The velocity distribution is then determined only by the velocity distribution in the fast beam. It is typically of the order of several tenths of an eV, up to sometimes ~ 1 eV in the laboratory system.

In collisions between fast atoms and a scatter gas of thermal energy with a Maxwell–Boltzmann distribution, the relative velocity distribution depends on the mass ratios and the fast-beam energy and the effect of the thermal distribution amounting typically to some tenths of an eV at threshold energies of a few eV.

The influence of the thermal beam velocity distribution on the threshold energy is strongly dependent on the excitation function. Chantry (1971) calculated for a step function threshold behaviour the shift between the apparent threshold energy and the real threshold to be

$$\Delta E_{\text{thr}} \approx 0.6 W_{1/2} \tag{16}$$

$W_{1/2}(T)$ is the FWHM of the Doppler-broadened relative velocity distribution between a scattered gas (mass m_2) of Maxwellian distribution at the temperature T and a primary particle beam (mass m_1) at a relative collision energy E_0

$$W_{1/2}(T) \approx \left(11.1 \frac{m_1}{m_1 + m_2} E_0 kT \right)^{1/2} \tag{17}$$

The projectile atoms A of the fast primary beam may have an energy spread ΔE_{Lab} FWHM which is transformed into the centre-of-mass system $W_{1/2}(\text{A})$

$$W_{1/2}(\text{A}) = \frac{m_2}{m_1 + m_2} \Delta E_{\text{Lab}} \tag{18}$$

The total half width is

$$W_{1/2} = [W_{1/2}(T) + W_{1/2}(A)]^{1/2} \tag{19}$$

The relative contribution of beam and scattered gas to the total energy width depends strongly on the mass ratio. Assuming an energy half width of 0.5 eV for the fast alkali beam and a scatter gas at 293 K a collision energy near the threshold value, i.e. for the reaction $Li + Br_2 \rightarrow Li^+ + Br_2^-$ at 2.8 eV and for $Cs + Cl_2 \rightarrow Cs^+ + Cl_2^-$ at 1.4 eV, then the former reaction results in $W_{1/2}(T) = 0.2$ eV and $W_{1/2}(A) = 0.5$ while the latter yields: $W_{1/2}(T) = 0.5$ and $W_{1/2}(A) = 0.2$ eV. For both reactions $W_{1/2}$ will be ≈ 0.55 or $\Delta E_{thr} \approx$ 0.3 eV. In the first example with $W_{1/2}(A) > W_{1/2}(T)$ the application of equation (19) in equation (16) is only approximatively correct. If the fast-beam atoms are much lighter than the scatter gas, the scatter gas influences the threshold behaviour less than the beam width of 0.5 eV, while for atoms heavier than the scatter gas, the contributions to the energy spread are just the opposite. In these heavy atom–light molecule reactions, a finite divergence angle of a crossed molecular beam may become crucial. All methods which neglect the relative velocity distribution could easily result in an error of several tenths of an electron volt.

In crossed beam experiments the Doppler broadening is negligible only if the angular divergence is small enough, while the primary fast-beam distribution is not negligible.

Equation (16) is only correct for a step function—like excitation function. If the excitation function increases linearly—over an energy range larger than the $W_{1/2}$ value—the simplest way to determine the threshold energy is the extrapolation of the linear part to the zero line. Its error is only in the order of $1.5\gamma kT$ where $\gamma = m_1/(m_1 + m_2)$ (Chantry, 1971).

The assumption of a linearly increasing cross-section function is not always warranted (Rothe *et al.*, 1975). The only solution to the problem is to obtain the velocity (and angular distribution in crossed beam experiments) of the fast beam and the molecular gas. If these are available, then the quality of the fitting is a sensitive measure of the accuracy of the excitation function (Nalley *et al.*, 1973; Dispert and Lacmann, 1978).

A theoretical description of the threshold behaviour is given by Wigner (1948). For endoergic reactions with two oppositely charged product atoms the threshold behaviour is given by a step function. This was observed for instance in polar dissociation reactions with chemical rearrangement (Parks *et al.*, 1977). This theory, extended to atom–molecule collisions means that with increasing energy new threshold values of higher vibrationally excited product ions are matched. A stepwise increasing cross section could be expected which will be smoothed because the spread in relative velocities is generally larger than the vibrational spacing. The increase must not necessarily be in accord with the Franck–Condon transition probabilities, because the nuclear distances within the molecule may change at low velocities in the vicinity of two approaching potential surfaces. This will be discussed in the following chapter in connection with the pre-stretching effect in an extended bond stretching model.

Experimental and theoretical derived energy dependences of the ionization cross section often differ considerably (Baede, 1975).

IV. REVIEW OF EXPERIMENTAL RESULTS

A. Introduction

The last review in this field with the title 'Charge transfer between neutrals at hyperthermal energies' was published four years ago in this series by Baede (1975). This article will therefore concentrate on the results of the years 1975–1978; the older publications are only referred to in so far as their results or techniques are needed in connection with newer publications. As mentioned in Part II, the field of atom–atom collisions will be not reviewed again.

In this section total ionization and differential cross sections will be discussed separately. Ion-pair formation processes with non-alkali atoms will be cursorily reviewed and also some other endoergic reactions forming ion pairs as polar dissociation will be discussed.

To compact the results somewhat, especially the multitude of partial ionization experiments for energy threshold determination, the most essential potential parameters which could be deduced from these studies, such as electron affinities and bond energies, are listed in tables (Table III to VI).

B. Total Cross Sections

1. *Low-Energy Collisions, Pre-stretching Effect and Adiabatic Electron Affinity*

In the theoretical part, a model has been already described which decouples the molecular motion from the translational motion between the atom and the molecule. The vibrational motion of the molecule during the collision is included and led to the so-called 'bond stretching model'. This model has been used to explain several aspects of differential and total cross-section measurements. However, simple straight-line trajectory calculations of the total cross sections using diabatic potential surfaces differ considerably from this model at low collision energies near the threshold (Aten *et al.*, 1976).

The inferiority of diabatic potential surfaces has been pointed out also in theoretical calculations (Faist and Levine, 1976). This is especially obvious at low collision energies where the shape of the adiabatic surface in the crossing region becomes important. The diabatic potentials have to change their ionic–covalent character abruptly at R_c and the molecular ion is formed by a vertical transition at the equilibrium distance of the molecule. Besides the inappropriateness of having a cusp in the potential function, the vertical electron transition at the equilibrium distance of the molecule would lead to a threshold energy determined by the vertical electron affinity, while the experimental results are mostly in agreement with an adiabatic electron affinity. This discrepancy has been explained by vibrational relaxation during the collision as shown in theoretical calculations by Zembekov (1971) and calculations with the multiple jump model of the system $Cs + Br_2$ by Kendall and Grice (1972).

Diabatic potentials also fail to give the total cross section properly. According to the bond stretching model the maximum cross section is πR_c^2. At low velocities, during the approach, most particles follow the ionic scattering trajectories (equation (4) $p \approx 0$). By vibration of the negative ion the bond stretches during the collision and R_c increases, corresponding to the increase of the electron affinity, as shown in Fig. 3 and at the second crossing the particles leave the collision area in a non-adiabatic surface hopping process as ions ($p' \approx 1$ in equation (15)). All particles approaching with an impact parameter $b \leq R_c$ will undergo a collisional ionization reaction and the cross section is accordingly πR_c^2. But in some reactions bigger cross sections have been experimentally determined, which indicates that the maximum cross section is given by πR_{Cad}^2 with R_{Cad} an increased crossing radius. The experimental results can be explained by the dynamics on an adiabatic potential which continuously changes its character from ionic to covalent over a crossing seam a, which is limited by the Heisenberg uncertainty principle to

$$\frac{a}{v} \lesssim \frac{h}{2H_{12}} \tag{20}$$

In Fig. 1 it is seen that the lower adiabatic state $V_1(R)$ is already attractive at distances larger than R_c. Fig. 6 shows a cut through the potential surface along the Br–Br internuclear distance at a distance $K–Br_2$ somewhat larger than R_c. The dashed lines are the diabatic covalent and ionic potential curves of Br_2 and Br_2^- while the solid lines give the adiabatic potentials. At the equilibrium distance of Br_2 the adiabatic potential is already repulsive. That means, before the wavefunction has changed completely to ionic character, the bond distance increases. This effect is called pre-stretching (Aten and Los, 1977). The pre-stretching effect increases the distance of R_c up to a maximum value given by the equilibrium distance of the negative molecular ion. In that case the pre-stretching is complete and the electron affinity is equal to the adiabatic value.

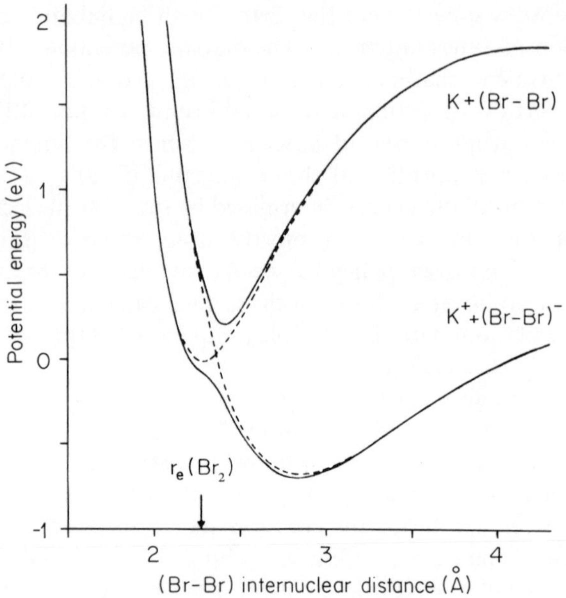

Fig. 6. Section along the Br-Br axis through the potential energy surfaces for the K–Br_2 interaction in the linear configuration. The K–Br distance is 5.8 Å, somewhat larger than R_c for the vertical electron affinity. Both the diabatic (dashed lines) and the adiabatic potential surfaces (solid lines) are shown. The equilibrium distance of Br_2 is indicated by an arrow. (Reproduced by permission of North-Holland Publ. Co. from Aten and Los, 1977.)

This model explains the increase of the total cross section near the threshold over πR_c^2 and also the observations that the threshold energy corresponds to the adiabatic and not vertical electron affinity. The model is in agreement with theoretical calculations. Replacing diabatic by adiabatic potential surfaces leads to an adiabatic crossing radius which is larger than the diabatic one (Faist and Levine, 1976). The model has proven its reliability in several cases. One test of the model is the comparison of some adiabatic electron affinity values which have been determined meanwhile very accurately by photodetachment with photoelectron energy spectroscopy and are given in the section 'energy calibration' for O_2, SO_2, and NO_2. In many other cases the adiabatic *EA*-values are well established by several other experimental methods or by a accurate *ab initio* calculations, e.g. for O_2 (Das *et al.*, 1978).

Energy loss measurements at different energies near the threshold indicate the same tendency. The energy loss near threshold increases with the collision energy up to a value given by vertical Franck–Condon transitions (Mochizuki and Lacmann, 1977). In differential cross-section measurements the minimum between covalently and ionically scattered particles is shifted with the collisional energy to higher scattering angles. According to equation (11) this indicates a shift of the electron affinity to smaller values and so proves the contribution of pre-stretching at lower energies (Aten and Los, 1977; Kimura and Lacmann, 1978).

Total cross sections for the reaction $K + Br_2$ on adiabatic surfaces over a wide energy range have been calculated with the trajectory surface-hopping method. Near the threshold the cross section exceeds the value πR_c^2 by about 40% (Evers, 1978). The experimental results by Baede *et al.* (1973) and Hubers *et al.* (1976) are in good agreement with calculated cross sections in the entire post-threshold region.

2. *Review of Total Cross-Section Measurements Extended up to High Collision Energies*

The study of ionization reactions over a wide energy range from post-threshold up to a few thousand eV reveals reaction dynamical effects (relations between bond stretching effects and vibrational phases during the collision) and yields potential parameters such as H_{12} values and accurate vertical EA values. The reactions of alkali metals (Na, K, Cs) with Cl_2, Br_2, I_2, ICl, and IBr have been studied up to 2×10^5 m/s (Hubers *et al.*, 1976). Two types of experiments are reported: total cross-section measurements and the ratios of atomic to molecular ion yields (X^-/XY^-).

The ratio X^-/XY^- versus velocity decreased up to velocities of $\sim 2 \times 10^4$ m/s; at higher velocities it is nearly constant. In alkali–molecule collisions this function is very sensitive to the position of the repulsive wall of X_2^- ion relative to the equilibrium distance of the X_2 molecule as indicated in Fig. 3(a). The total cross sections are obtained by integration over all EA-values of the product of the specific cross sections for XY^- formation at each EA-value with the normalized Franck–Condon overlap factors up to EA_d, the electron affinity leading to dissociation into $X + Y^-$. The specific cross section is analogous to the cross section for atom–atom interactions given in equation (8) for an effective impact parameter $b = b_{max}/2^{1/2}$.

The function $F(v_0/v)$ results from integration of equation (5) with $v_0(EA) \sim H_{12}^2(EA) \times R_c^2(EA)$. This relation is fundamental for the small trends of the fractions observed as a function of the relative velocity and will be described later.

The calculations are performed in a two-state approximation. At high velocities the vibrational motion is not considered. At low velocities it is introduced as described by the bond stretching model.

The function $F(v_0/v)$ is modified for the atom–molecule interactions by

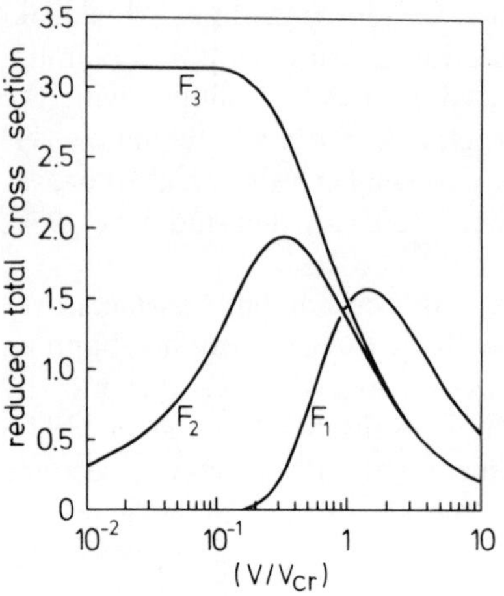

Fig. 7. Reduced total cross sections calculated using an effective trajectory. The curve F_1 represents the atom–atom case, curve F_2 the atom–molecule case without bond stretching, and curve F_3 the atom–molecule case with a completely stretched bond of the molecular ion. (Reproduced by permission of North-Holland Publ. Co. from Hubers *et al.*, 1976.)

introducing an orientation-dependent coupling matrix element according to equation (12) and averaging over all orientations. The result is plotted in Fig. 7 as F_2 in a reduced form of the total cross section versus the reduced velocity (v_{cr} in Fig. 7 corresponds to v_0). Because the averaged H_{12}-value is smaller, F_2 for atom–molecule collisions peaks at lower velocities ($v_{max} = 0.34v_0$) than in atom–atom interactions. This case is also seen in the figure as F_1.

The velocity v_{max} leading to a maximum cross section is determined experimentally. From F_2 follows then a relation between H_{12} and R_c according to:

$$H_{12} = 0.56 \times 10^{-2} v_{max}^2 / R_c (\text{eV}) \qquad (21)$$

where v_{max} is in m/s and R_c in Å.

F_1 and F_2 are calculated at high velocities without bond stretching. The function F_3 is included in Fig. 7 and represents the reduced total cross section for complete bond stretching; with p' in equation (15) equal to unity. At low velocity the total cross section is about twice as large as the maximum of F_1 *or* F_2.

The experimental results of the total cross-section measurements (Hubers *et al.*, 1976) are shown in Fig. 8 for alkali–bromine reactions as function of the relative velocity. The total cross section of collisions with sodium atoms shows a maximum at low velocities $\sim 8 \times 10^3$ m/s and decreases up to a velocity about 4×10^4 m/s. This behaviour is well described by the bond stretching model. In Fig. 3(b) are shown the crossing distances during a vibration of the negative ion for the different alkali–halogen systems. The internuclear distances of straight-line trajectories ($b = 0$) are included as function of the collision time for four different velocities. At the highest

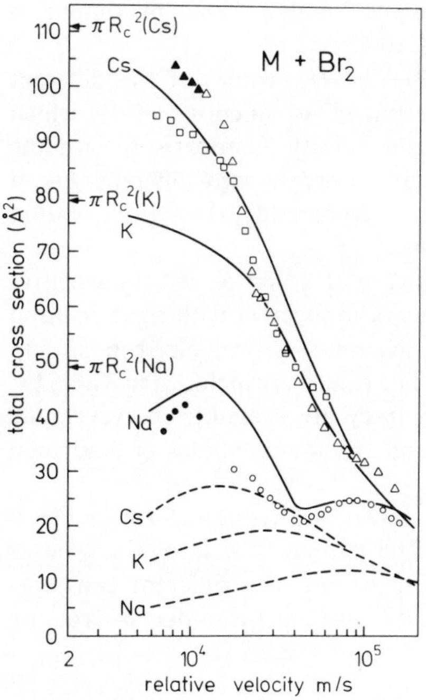

Fig. 8. The dependence of the total cross section for negative ion formation on the relative velocity in $M + Br_2$ ($M = Na$, K, Cs) collisions. The full lines represent on an absolute scale the calculated total (ionic + covalent) cross sections; the dotted lines give separately the covalent cross sections. The open symbols are the experimental results from Hubers *et al.* (1976), the solid symbols are the experimental points taken from Baede *et al.* (1973). The experimental total cross sections have been scaled to the calculated ones at the position of the (Landau–Zener) maximum of the corresponding covalent cross section. The various symbols indicate the alkali atom involved. (○ ●) $Na + Br_2$, (Δ▲) $K + Br_2$, (□) — $Cs + Br_2$, πR_c^2 corresponds to the vertical electron affinity EA_v (Table VI). (Reproduced by permission of North-Holland Publ. Co. from Aten *et al.*, 1976.)

velocity (a)$v = 4 \times 10^4$ m/s the influence of bond stretching on the $Na + Br_2$ reaction is negligible. The collision time $\tau \approx 0.3 \times 10^{-13}$ s amounts to about one tenth of the vibrational period (3.1×10^{-13} s). The line (a′) shows that in the case of $K + Br_2$ bond stretching is still effective and much larger. This is also reflected in Fig. 8 where in the case of $K + Br_2$ the function decreases at higher velocities than with sodium.

The second maximum for sodium reactions at higher velocities is attributed to the Landau–Zener maximum. At these velocities the vibrational motion is effectively frozen during the collision and the system can be treated as a two-particle system according to the Landau–Zener model and behaves analogously to F_2 in Fig. 7.

For the system $Na + Br_2$ the position of the Landau–Zener maximum is clearly visible in Fig. 7. In systems containing K and Cs the bond stretching affects the cross section at even higher velocities and the Landau–Zener maximum is shifted to smaller velocities. Therefore, the second maximum is in both cases concealed. The velocity yielding a maximum cross section has been estimated with an error of $\sim 50\%$.

The calculated cross sections are included in the figure. They are the results of straight-line trajectory calculations on adiabatic potential surfaces with an integration over all impact parameters (Aten *et al.*, 1976). The covalent cross section is included as a dashed line. It is not influenced by bond stretching and shows the shift of the Landau–Zener maximum to smaller velocities in going from Na to K to Cs.

From v_{max} together with equation (21) the H_{12}-values of the different alkali–halogen systems have been determined as functions of R_c which depend on the effective (vertical) electron affinity. Comparison with the relation (equation (6)) by Olson *et al.* (1971) reveals two different sets of straight lines with quite different slopes c_2. Apparently, the Olson relation is too simple for atom–molecule systems.

As mentioned before, the relation between H_{12} and R_c is very sensitive to the ratio X^-/X_2^-. By a careful fit of this ratio together with right position of the Landau–Zener maximum, very accurate vertical electron affinity values within 0.05 eV have been derived. The results are included in Table IV. Up to now no other values of this accuracy are available; however, the theoretical estimates of Person (1963) and the lowest values of Anderson and Herschbach (1975) agree within 25%.

The relation between H_{12} and R_c is shown in a reduced form in Fig. 9 together with the derived H_{12}-values. The straight line is of the type as given by Olson *et al.* (1971) in equation (6) but with different constants $c_1 = 1.73$ and $c_2 = 0.875$ and a different form of the reduced crossing

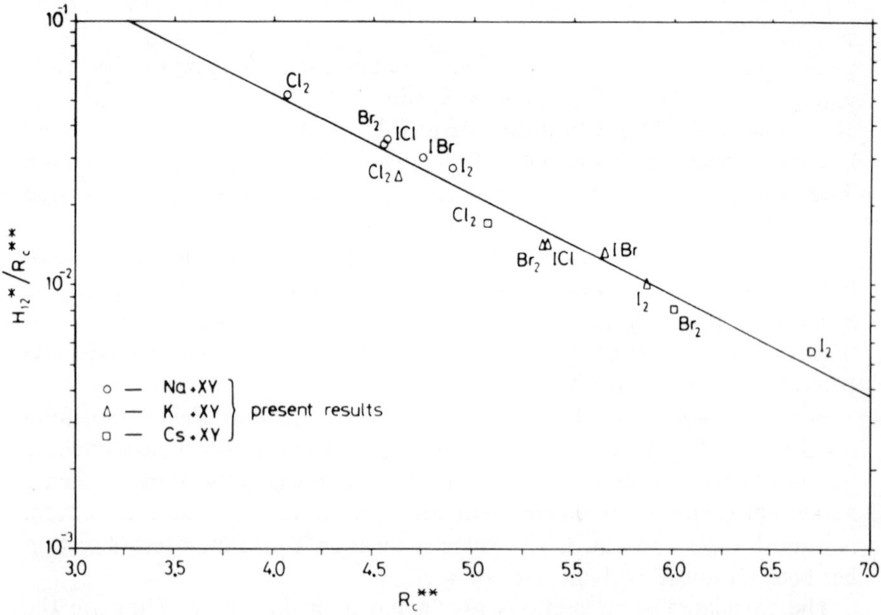

Fig. 9. Correlation of the reduced coupling matrix element H_{12}^*/R_c^{**} with the reduced crossing distance. For H_{12}^* the reduction proposed by Olson *et al.* (1971) is used (equation (6)). R_c is reduced according equation (22). The full line represents the best fit to the data of Hubers *et al.* (1976). (Reproduced by permission of North-Holland Publ. Co. from Hubers *et al.*, 1976.)

distance R_c^* which is now

$$R_c^{**} = \sqrt{2IR_c} \tag{22}$$

and is dependent on the electron affinity.

In the behaviour of the exponential part, this function is like the theoretically derived functions (see e.g. the review of Janev, 1976). The value for c_2 is in good agreement with results of Grice and Herschbach (1974).

The fractions X^-/XY^- are nearly constant at high velocities. At low velocities the ratios for the homonuclear systems decrease up to velocities of about 2×10^4 m/s^{-1}, while the Br^- or Cl^- ratios of the heteronuclear molecules are constant. However, the small I^- fractions show this trend even more pronouncedly as indicated in Fig. 10 (Hubers *et al.*, 1976).

Impact parameter calculations taking into account the bond stretching show no increase of X^- fraction at low velocities, in agreement with the bond stretching model, which does not change the vertical transition probability (Aten *et al.*, 1976).

This I^- formation at low velocities is an indication for the formation of electronically excited states. The $^2\Pi_{1/2}$ state with two antibonding σ electrons is not stable. In the case of I^- formation this is obvious. Auerbach *et al.* (1973) observed I^- formation and interpreted the I^- formation from heteronuclear molecules via an excited state ($^2\Pi_{1/2}$). The dissociative state of the heteronuclear halogen ion is not degenerate. In accord with the lower

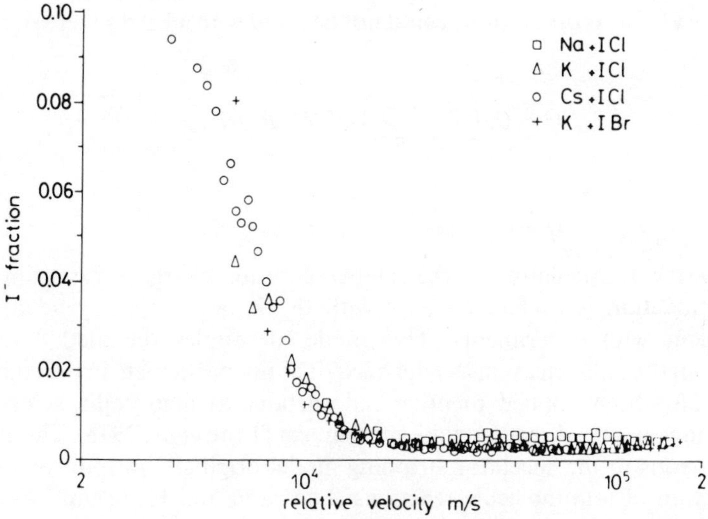

Fig. 10. The I^- fraction of the total negative ion yield in $M + IX$ collisions as a function of the relative velocity. (Reproduced by permission of North-Holland Publ. Co. from Hubers *et al.*, 1976.)

electron affinity of I^- compared with Br^- and Cl^-, the I^- state is above the Br^- or Cl^- state. Because of the smaller electronegativity of I, it will be preferably in antibonding states.

The electronically excited potentials have a less steep repulsive wall and smaller electron affinities. Therefore, the R_c-values and the bond stretching effect are much smaller and the cross sections of these excited states are also much smaller. Because of the smaller crossing distance at the first crossing, most particles are ionically scattered and without bond stretching would depart adiabatically on the covalent surface. A measurable cross section is therefore limited to low velocities and only if bond stretching is effective. From Cs to Li this limitation preponderates; accordingly, I^- is not observed in Na and Li collisions (Auerbach *et al.*, 1973).

The increase in the atomic yield at low velocities with homonuclear molecules is assumed to result also from electronically excited states. Double differential cross-section measurements as function of the energy loss support the formation of X^- via excited states up to maximum collision velocities (Kimura and Lacmann, 1977, 1978) (see Section IV. C).

Tang *et al.* (1975) have also measured the function X^-/X_2^- from 25 to 350 eV for Cl_2, Br_2, and I_2. The decrease in the ratio with increasing energy is explained according to calculations of Kendall and Grice (1972) by a distortion of the X_2^- potential.

Dissociative ionization processes have been studied in collisions between Cs and CF_3X with X = Cl, Br, and I up to several hundred eV (Rothe *et al.*, 1974). The ratios of F^-/X^- approach constant values at high energies (> 100 eV). The cross sections could not be fitted with a hard sphere equation of the type

$$Q = Q_0\left(1 - \frac{E_{\text{thr}}}{E}\right) \quad \text{for} \quad E \geq E_{\text{thr}} \tag{23}$$

and

$$Q = 0 \qquad\qquad \text{for} \quad E < E_{\text{thr}}.$$

However, a calculation of the relative collision energy E assuming that the momentum is exchanged only with the F or X atoms yielded good agreement with experiments. This model decouples the methyl radical fully from the halogen atom as if its mass does not participate in the collision. It has also been applied to other halocarbons at high collision energies indicating an 'ultradirect' impulse mechanism (Tang *et al.*, 1976). This model is analogous to the spectator stripping model originally introduced for the description of ion–molecule reactions (Lacmann and Henglein, 1965) and often applied to the description of many alkali–halogen reactive collisions, such as $K + Br_2$ (Herschbach, 1966).

Additional support for the bond stretching model results from total cross-section measurements of the system $M + O_2$ with $M = Na$, K, and Cs (Kleyn *et al.*, 1978). The system differs significantly from the halogen reactions because the molecular constants are quite different. The vertical electron affinity is much smaller (~ 0) and therefore the crossing radius is much smaller. Also, the vibrational period of O_2^- is about 10 times smaller than that of the halogens ($\tau \approx 3 \times 10^{-14}$ s). In this case the negative ion performs several vibrations during one collision even at collision energies well above the threshold. The periodic change of the crossing distances during a collision is indicated in Fig. 11. The lines a, b, c, and a′, b′, c′ correspond to the $Cs^+ - O_2^-$ and $Na^+ - O_2^-$ separations as function of the collision time for three different velocities. a. 1×10^4 m/s; b. 2×10^4 m/s, and c. 4×10^4 m/s.

The experimental results are shown in Fig. 12 together with a calculated curve. The differences between these calculations and the analogous one for the halogens are: the angular dependence of H_{12} is given by a $\sin(2\phi)$ dependence in accord with the symmetry rules (see equation (13) and Table II); a statistical weighting factor of 1/3 is included because two quartet

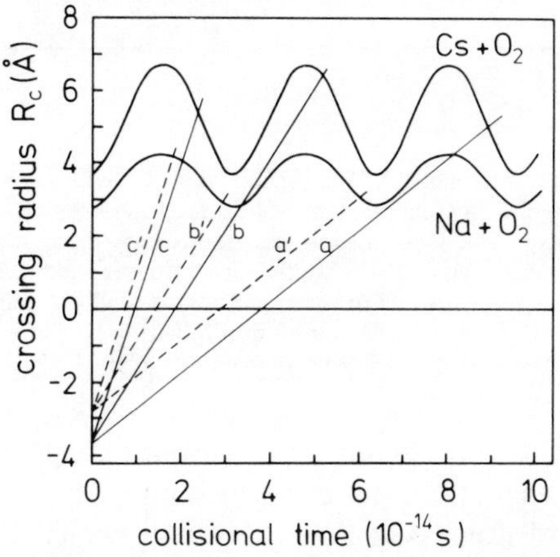

Fig. 11. The dependence of R_c on the collision time for Na and $Cs + O_2$ in the case of ionic scattering. At the first crossing, where electron transfer occurs, the O_2 is assumed to be at its equilibrium distance. To show the time scale of the process, straight lines have been drawn representing trajectories with impact parameter zero. All forces between M^+ and O_2^- are neglected. Three different velocities are shown: a) 1.0×10^4 m/s, b) 2.0×10^4 m/s, and c) 4.0×10^4 m/s. The full lines refer to Cs; the dotted lines labelled a′, b′, and c′ refer to Na. (Reproduced by permission of North-Holland Publ. Co. from Kleyn *et al.*, 1978.)

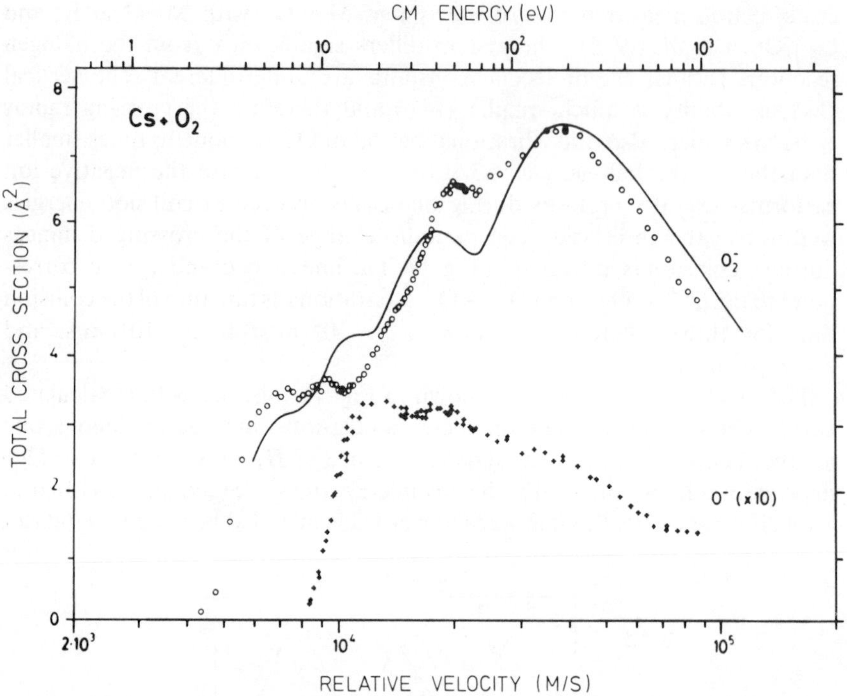

Fig. 12. Total cross section for negative ion formation in Cs + O_2 collisions as a function of the relative velocity and the c.m. energy. The solid line represents results of calculations on diabatic potential surfaces. The experimental O_2^- points have been scaled to the calculated cross section. According to Kleyn *et al.* (1978), the structure observed in both the experimental and the calculated cross sections is due to the rapid vibration of O_2^- during the collision. The experimental points for O^- formation have been multiplied by 10. (Reproduced by permission of North-Holland Publ. Co. from Kleyn *et al.*, 1978.)

states of the singlet–triplet reactant states do not couple with the M^+–O_2^- complex; and the calculations are performed on diabatic potential surfaces and are limited to Franck–Condon transitions which are stable against autodetachment.

The H_{12}–R_c relation of equation (6) could not be applied because the electron affinity is near zero. Instead, the following relation was used:

$$H_{12} = c_1 \exp(-c_2 R_c) \tag{24}$$

For the potential parameters of O_2 and O_2^-, well-established values which agree within 1% with the most accurate 76-configuration MCSCF calculations (Das *et al.*, 1978) were used. The only free parameters were c_1 and c_2. The parameter c_2 determines the amplitude of the oscillations and c_1 (with a fixed c_2) determines the position of the Landau–Zener maximum. They were fitted to the experimental results.

The bond stretching effect is much smaller than in the case of $Cs + Br_2$ (Fig. 3b). While with Br_2 R_c increases up to 228 Å, in the case of O_2 R_c only increases from 3.7 to 6.7 Å.

It is remarkable how well the experimental structure is reproduced by this simple calculation. The maximum at 3.8×10^4 m/s corresponds to a velocity at which the second crossing is just reached when the O_2^- has completed half a vibration and R_c reaches its maximum the first time, similar to trajectory c in Fig. 11. The minima correspond to velocities at which the collision time is just equal to a vibrational period. Four maxima are clearly resolved. Collisions with K show similar results. At low energies this structure can also be seen in earlier measurements of Moutinho *et al.* (1971). For Na almost no structure could be resolved. This is explained by the much smaller changes of the crossing distance during a vibration, also shown in Fig. 11.

The oscillatory structure is smeared out by averaging over the impact parameters and the O_2 internuclear distances. Nevertheless, the structure is still visible. This suggests that even in O_2 collisions with smaller crossing radii than in halogen collisions the O_2^- potential is not significantly deformed by the proximity of the Cs^+ ion.

3. *Internal Energy of Reactants*

a. Vibrational excitation. In a preceeding Section (III.E) several causes of errors in the determination of the threshold energy or adiabatic electron affinity determination have been pointed out. Another source is the internal energy of the molecule which may influence the cross section near the threshold energy. In several experiments the temperature dependence or the influence of the vibrational excitation on the total cross section has been studied at different translational energies. The results illustrate that near the threshold the total vibrational energy has at least the same or perhaps even a greater influence on the ionization cross section than the translational energy.

Moutinho *et al.* (1974) studied the total ionization cross section without mass analysis on the reaction between alkali metals Li, Na, K, and CH_3I and CH_3Br between 300 and 800 K. The true thresholds are determined by fitting a convoluted hard sphere cross-section function (equation (23)) to the experiments. The threshold behaviour leads to electron affinity values of 0.26 ± 0.1 eV for CH_3I and -0.46 ± 0.1 eV for CH_3Br at 300 K. The values are much smaller than the adiabatic electron affinity values of ~ 0.75 eV for I^- and 0.45 eV for Br^- formation. The values reflect the vertical or near-vertical electron affinity values in agreement with semiempirical potential curves of Wentworth *et al.* (1969) from the temperature dependence of electron attachment.

The CH_3I^- threshold was also determined by McNamee *et al.* (1973).

The result differs from the threshold values of other systems which yield the adiabatic electron affinity. The main difference from other potential surfaces is that the surface has at most only a shallow minimum while most other potentials have a minimum and a steeply repulsive potential wall at the equilibrium distance of the neutral molecule.

The temperature dependence on the cross section of CH_3I reactions for different collision energies is shown in Fig. 13. Even near the threshold energy, the influence of internal energy is more efficient than the translational energy. The strong increase in the cross section above 800 K is due to I-atom reactions from the thermal dissociation of CH_3I. Its threshold is lower.

The total cross section $Q(E, T)$ as function of the translational energy E and the temperature T is given by the sums of the specific cross sections $\sigma_v(E)$ for the molecules in each vibrational level multiplied with the Boltzmann factor f_v determining the fraction of molecules in the state v. Replacing the sum by an integral over the vibrational energy together with the semi-logarithmic dependence of the experimental results, as shown in Fig. 13, the total cross section is

$$Q(E, T) = C \exp(AT) = \int_0^\infty f_v(T)\sigma_v(E)\,dv. \tag{25}$$

C and A are only dependent on the collision energy. By an inverse Laplace transformation of the measured cross section, the total specific cross section as a function of the vibrational energy has been derived at fixed translational

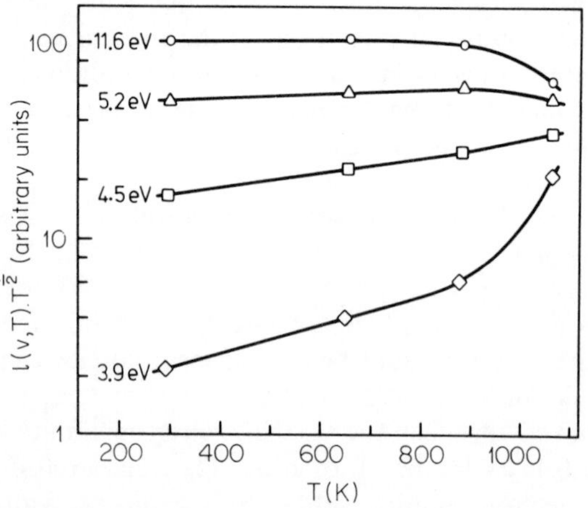

Fig. 13. Temperature dependence of the K^+ signal $i(v, T)$ multiplied by $T^{1/2}$, at several energies in the centre-of-mass system of K–CH_3I. (Reproduced by permission of North-Holland Publ. Co. from Moutinho *et al.*, 1974.)

energies and as a function of the translational energy for the different vibrational levels (Moutinho *et al.*, 1979). The experimental results are compared with simple calculations including bond stretching on diabatic surfaces. The experimentally determined specific cross sections show a stronger increase with the vibrational excitation than the calculated ones. This is qualitatively explained in term of pre-stretching effective only on adiabatic potential surfaces near the threshold, as discussed in the section 'pre-stretching'.

A detailed description of the mathematical procedures involved is given by Hubers and Los (1975) together with the temperature dependence measurements of SF_6. The temperature is varied between 300 and 850 K (a medium vibrational energy of 0.17–0.97 eV). The specific cross sections for SF_6^-, SF_5^-, and F^- have been evaluated in collisions with Li, Na, and K. One typical result for the SF_6^- cross section as function of the translational energy for different vibrational energies is shown in Fig. 14. Only near the threshold the vibrational energy and translational energy are equivalent, the ions show complete relaxation. At energies above threshold, the vibra-

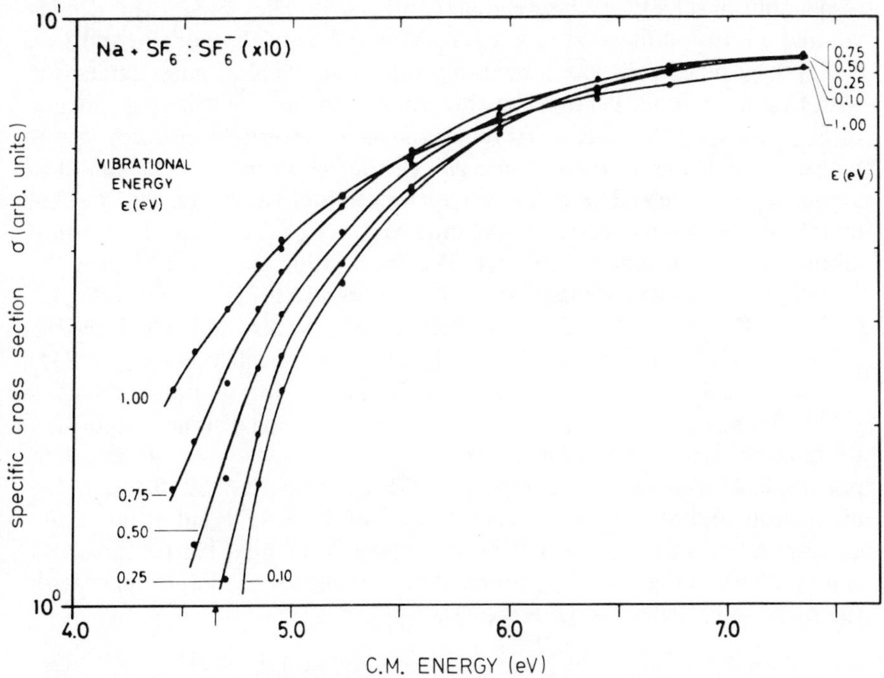

Fig. 14. Specific cross sections for SF_6^- formation from Na + SF_6 as a function of c.m. kinetic energy at various SF_6 vibrational energies. The arrow corresponds to the threshold for SF_6^- formation with the SF_6 source at room temperature. (Reproduced by permission of North-Holland Publ. Co. from Hubers *et al.*, 1975.)

tional energy dominates over the translational energy. This is interpreted as a relaxation mechanism based on the interaction of the alkali ion and the highly vibrationally excited SF_6^- ion. As a result SF_6^- is quenched leading to adiabatic threshold energies.

In the SF_5^- and F^- formation an intramolecular mechanism explains the equivalence of vibrational and translational energy up to 2 eV above the threshold similarly to the discussion of Tang *et al.* (1974c). In both cases the S—F bond is stretched. The discussion is based on the assumption that all vibrational modes, stretching and bending, contribute equally and no excited states are involved. The latter assumption is deduced from the fact that the ratios of F^-/SF_5^- are independent of the collision energy up to 200 eV in the c.m. system, indicating that no new pathways become accessible. The threshold values are derived by folding the hypothetical cross sections with the relative distributions of the alkali beam. From a best fit, the electron affinities and dissociation energies at room temperature and 0 K are derived (see Table III). The difference of 0.17 eV for SF_6 reflects the medium vibrational excitation energy at room temperature in excess of the zero-point vibrational energy of 0.58 eV. Leffert *et al.* (1974) report also a shift of 0.1–0.2 eV between 300 and 180 K. They obtained a deconvoluted electron affinity of 0.75 ± 0.1 eV with Cs at 0 K. Hubers and Los (1975) determined 0.32 ± 0.15 or 0.48 ± 0.1 eV at 300 K in good agreement with Compton *et al.* (1978a). They determined the temperature dependence of SF_5^- and SF_6^- by linear extrapolation of the threshold energies versus temperature. In centre-of-mass energies the difference between 300 and 0 K amounts to 0.12 eV. The different electron affinity values for SF_5^- (~ 2.8) and the dissociation energies of SF_6^- into $SF_5^- + F$ (~ 1.1 eV) of the different laboratories are given in Table III. The dissociation energy of SF_6 for the $EA(SF_5)$ calculation is taken from Hildenbrand (1973). Compton *et al.* (1978a) also examined SF_4; its EA-values and SF_3 are also added to Table III.

Hubers and Los (1975) found a slight increase in the difference $E_{thr} - I(M)$ (or the vertical electron transition energies) for SF_5^- formation in collisions with sodium and lithium. In these collisions the crossing distances according to equation (3) are 2.5 and 2.4 Å, respectively. At these distances the covalent potential is assumed to be repulsive. Momentum transfer between the alkali atom and SF_6 leads to a compression of the S—F bond distance and correspondingly an increase of the vertical electron affinity. The F^- threshold is found 3 eV above the SF_6 ground state which is in good agreement with the reaction enthalpy of the reaction

$$e + SF_6 \rightarrow SF_4 + F + F^- \qquad \Delta H = 3.1 \pm 0.4 \text{ eV} \tag{26}$$

This process needs an even higher degree of vibrational excitation of the SF_6^- than for the SF_5^- formation. This reaction mechanism indicates that more kinetic energy can be stored as potential energy in the molecular ion.

This energy has to be redistributed among the vibrational modes to allow reaction (26).

At the calculated threshold energy for the $SF_5 + F^-$ formation at about 0 eV, (well below SF_5^- formation), no F^- was observed, which is explained by the competitive reactive channels leading to $MF + SF_5$ formation.

b. Electronic excitation of the reactants. In view of the possibilities of laser excitation this will become an active field for future research. So far only preliminary results have been published.

TABLE III. Electron affinities (*EA*) and dissociation energies *D*(R—X) of hexafluoride molecules determined by collisional ionization.

Molecule R—X	*EA* (eV)	Temp. if not 300 K (K)	Bond R—X	*D*(R—X) (eV)	Reference
SF_6	0.75 ± 0.1	0	SF_5^-—F	1.14 ± 0.1	Leffert *et al.* (1974)
	0.46 ± 0.1	0		1.1 ± 0.1	Compton *et al.* (1978a)
	0.49 ± 0.1				Hubers and Los (1975)
	0.32 ± 0.15	0		1.0 ± 0.1	Hubers and Los (1975)
SF_5	$\geq 2.9 \pm 0.1$				Hubers and Los (1975)
	2.71 ± 0.2				Compton *et al.* (1978a)
SF_4	0.78 ± 0.2				Compton *et al.* (1978a)
SF_3	3.07 ± 0.2				Compton *et al.* (1978a)
TeF_6	3.3 ± 0.2	0	TeF_4—2F	6.9	Compton *et al.* (1978a)
	3.24 ± 0.2				Lacmann and Dispert (1975)
SeF_6	3.1 ± 0.2				Lacmann and Dispert (1975)
	2.9 ± 0.2	0			Compton *et al.* (1978a)
ReF_6	$> 5.1^{+0.2}_{-0.5}$				Compton *et al.* (1978a)
IrF_6	> 5.14				Compton *et al.* (1978a)
PtF_6	> 5.14				Compton *et al.* (1978a)
MoF_6	$> 5.1^{+0.2}_{-0.5}$				Compton *et al.* (1978a)
WF_6	> 4.9				Mathur *et al.* (1977)
WF_6 (exc. states)	3.7				Dispert and Lacmann (1977a)
WF_5	1.25		WF_5—F	5.1	Dispert and Lacmann (1977a)
			WF_5^-—F	7.6	Dispert and Lacmann (1977a)
			WF_5—F^-	5.4	Dispert and Lacmann (1977a)
UF_6	≥ 5.1		UF_5—F	3.0 ± 0.2	Compton (1977)
	> 4.3				Mathur *et al.* (1977)
UF_5	> 1.9				Mathur *et al.* (1977)
	4.0 ± 0.4				Compton (1977)

The threshold behaviour of $Na + SO_2$ and O_2 has been studied together with laser light of 5890 Å which excites small fractions of the $Na(^2S)$ into the excited ^2P state (Rothe *et al.*, 1977). The theoretical reaction threshold is reduced by 2 eV. The experimental curves show a definite reduction of the threshold energy but the energy threshold is not yet analysed.

4. *Reaction Threshold Measurements*

a. Hexafluorides. The hexafluoride molecules are of special interest in collisional ionization reactions because many of them have the highest electron affinities, so far observed. Studies of the oxidizing properties of hexafluorides yield electron affinity values for the third transition metal series (Barltett, 1968). The *EA*-values increase in the order $WF_6 < ReF_6 < OsF_6 < IrF_6 < PtF_6$ with $EA(PtF_6) > 6.75$ eV, and an increase of about 0.8 eV per unit atomic number.

The electron affinity values of these molecules exceed in some cases even the ionization energy of alkali atoms, i.e. the ion-pair formation processes are exothermic; collisions with thermal alkali atoms M should lead to ionization. At thermal or near-thermal energies reactive channels with even greater exothermicity may compete with ionization processes.

Since the asymptotic $M^+ + AF_6^-$ ground-state potential lies below the neutral state of the reactants, the potential hypersurfaces do not cross. Excited states are probably involved in this collisional ionization process. They may lead to increased threshold energies.

The competition among reactive channels at thermal energies has been demonstrated by Annis and Datz (1977) for several hexafluorides and at a relative enegy of 0.2–1.3 eV for $Cs + UF_6$ (Annis and Datz, 1978). With increasing energy, the ion-pair formation cross section rises to 32 Å^2 while the reactive CsF formation drops from 230 to 130 Å^2. They also show the dominance of chemical reactions of Cs and K with WF_6, MoF_6, and TeF_6. This may explain why Dispert and Lacmann (1977a) did not find ionization with K and $Na + WF_6$ at thermal energies but rather a steep rise at higher energies corresponding to an electron affinity of 3.7 eV, which may indicate collisions via an excited state.

The *EA*-values and dissociation energies determined by collisional ionization and chemiionization for various hexafluorides AF_6 and the dissociation products AF_5 are summarized in Table III. For a comparison with the results of other methods see Compton *et al.* (1978a).

The values for SeF_6 and TeF_6 are determined from the threshold energy for ion-pair formation in alkali–molecule collisions.

IrF_6^-, PtF_6^-, IrF_5^-, and PtF_5^- are formed even in collisions with thermal sodium atoms. Therefore, the electron affinities must be bigger than 5.1 eV, the ionization potential of sodium and the *EA*-values of IrF_5^- and PtF_5^-

must be even bigger than 5.1 plus the respective dissociation energies $D(AF_5—F)$.

For the molecules WF_6, MoF_6, ReF_6, and UF_6 the situation is somewhat different. WF_6 is not ionized in collisions with thermal potassium atoms but only in connection with the hot filament, normally used to ionize the alkali atoms and accelerate them into the charge exchange oven. But WF_6^- ions are already observed without an accelerating field. The intensity increases strongly above a minimum temperature. Therefore, Dispert and Lacmann (1977a) concluded that the thermal alkali atoms which hit the hot surface are reflected as neutrals with a Maxwell–Boltzmann velocity distribution according to the temperature of the hot filament. The high-energy tail of this distribution may reach several tenths of an eV, enough to reach the threshold energy. The electron affinity, therefore, should be smaller than the ionization energy of the potassium. Compton *et al.* (1978a) checked their results again and observed the same effect with sodium for WF_6, MoF_6, and UF_6 and interpreted it the same way. However, to get the electron affinity values, they assumed that the ionization potential of sodium is only a lower limit of the electron affinity value together with an increased error of – 0.5 eV. In the case of WF_6 this is not yet unequivocal, since these temperature effects are observed with potassium and sodium atoms. The difference in this ionization potential of 0.8 eV is higher than the thermal energies. In this case it seems to be more reasonable to assume a surface effect.

Another way to determine lower limits of the electron affinity follows from the reaction enthalpies of chemiionization reactions with alkali dimers of the type

$$M_2 + AF_6 \rightarrow M^+ + M + AF_6^- \tag{27}$$

Since some of these reactions have been observed at thermal energies (Mathur *et al.*, 1977), a lower bound on the *EA*-values follows:

$$EA(AF_6) \geq D(M_2) + I(M) \tag{28}$$

With caesium dimers this lower limit is 4.3 eV; for the potsssium dimer reaction it follows $EA > 4.85$ eV, e.g. as deduced from MoF_6 reactions. Reactions with alkali atoms can be excluded by applying an inhomogeneous magnetic field which deflects the atom beam but not the dimers.

In some cases the formation of $M_2F^+ + AF_5^-$ or $M^+ + MF + AF_5^-$ yields analogous lower limits for $EA(AF_5)$. The results are included in Table III.

b. Methyl derivatives. Methyl derivatives have been studied from different viewpoints. Firstly in environmental studies because of an increased interest in freons in connection with their possible reduction of the concentration of ozone in the upper atmosphere. Secondly from a reaction-dynamics viewpoint because the replacement of one to four hydrogen atoms by halogen atoms or radicals of high electron affinity forms new molecules which are very reactive and some of them form stable negative ions. The reaction dynamics of these molecules has often been found to depend strongly on the mass ratios over wide ranges between CH_3X and CX_4, with variation of the mass of X between fluorine and iodine (Herschbach, 1966; Kinsey *et al.*, 1976; and Bernstein and Wilcomb, 1977 and citations therein). Electron impact studies often yield only small amounts of parent ions and mainly X^- ions (Johnson *et al.*, 1977). The energy spectrum of the dissociative product ions often indicates the formation of electronically excited molecules or molecular ions (Franklin, 1976). Thermal electron attachment studies predict small bond energies for the parent ions of many halomethane ions which cannot be detected since the vertical transitions in electron impact experiments produce vibrationally excited ions above the dissociation limit (Wentworth *et al.*, 1969).

Collisional ionization experiments by atom impact have demonstrated that several parent negative ions are stable. The results are summarized in Table IV.

Collisional ionization studies in a crossed beam of Cs and CH_3NO_2 revealed 20% parent ion formation compared to electron impact measurements of a maximum of 2% (Tang *et al.*, 1974c). In experiments with CF_2Cl_2 no parent ions are observed. In a scattering chamber experiment performed by Dispert and Lacmann (1978), it was shown that in collisions with potassium at energies between 3.2 eV and 3.9 eV only $CF_2Cl_2^-$ ions are formed

TABLE IV. Electron affinities (*EA*) and dissociation energies *D*(R—X) of methyl derivative determined by collisional ionization

Molecule R—X	*EA*(eV) v = vertical	R—X	*D*(R—X⁻) (eV)	Reference
CH_3NO_2	$0.44^{+0.1}_{-0.2}$	$—NO_2^-$	0.56 ± 0.2	Compton *et al.* (1978b)
CF_3I	1.57 ± 0.2	$—I^-$	0.32 ± 0.2	Compton *et al.* (1978b)
	1.4 ± 0.2	$—I^-$	0.38 ± 0.1	Tang *et al.* (1976)
	2.2 ± 0.2			McNamee *et al.* (1973)
CF_3Br	0.91 ± 0.2	$—Br^-$	0.54 ± 0.2	Compton *et al.* (1978b)
CCl_4	2.0 ± 0.2	$—Cl^-$	1.4 ± 0.3	Dispert and Lacmann (1978)
$CFCl_3$	1.1 ± 0.3	$-Cl^-$	0.7 ± 0.3	Dispert and Lacmann (1978)
CF_2Cl_2	0.4 ± 0.3	$—Cl^-$	0.1 ± 0.3	Dispert and Lacmann (1978)
CH_3I	0.35 ± 0.2 v			McNamee *et al.* (1973)
CH_3I	0.26 ± 0.1 v			Moutinho *et al.* (1974)
CH_3Br	-0.46 ± 0.1 v			Moutinho *et al.* (1974)

which, above the threshold of Cl^- formation of 3.9 eV, decrease to about 10^{-3} of the Cl^- production.

This experiment is a typical example of where a scattering chamber experiment is superior to a crossed beam arrangement. In a scattering chamber the pathlength for collisions is usually much longer and the detection sensibility is accordingly much higher. The sensitivity is especially important to determine the threshold energies of parent ions of these type of molecules, which have only very small partial cross sections. The absolute cross section at a collision energy of 10 eV is estimated to be of the order of 10^{-17} Å^2.

The small relative intensity of 10^{-3} follows from the low bond energy. The threshold for Cl^- formation is only 0.1–0.4 eV above the $CF_2Cl_2^-$ threshold. The bond energies and *EA*-values of the dissociation products of CF_4, CF_2Cl_2, $CHCl_3$, and CCl_4 are also presented. The parent ions of CCl_4^- and $CFCl_3^-$ could be detected mass spectrometrically. All *EA*-values are enclosed in Table IV. Fig. 15 illustrates as one example the ionization cross section from $K + CFCl_3$ reactions of seven different product ions as a function of the collision energy. The following products have been determined. The appearance potentials *AP* are given in brackets in eV. A second *AP*-value indicates the opening of another channel via an excited state. $CFCl_3^-$ (3.2, 6.0), Cl^-(3.9), Cl_2^-(6.5), $CFCl_2^-$(6.4), F($\leq$ 6.5, 7.5), FCl^-(8.2), and CCl_2^-(8.2,

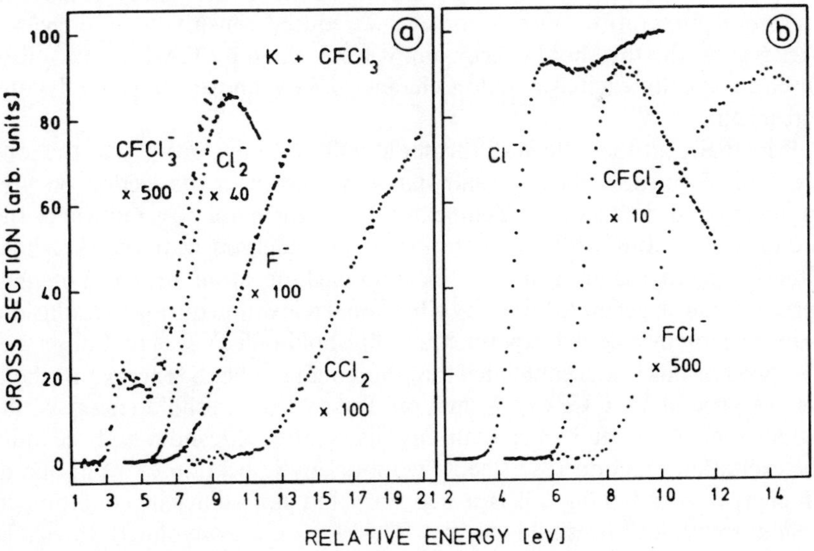

Fig. 15. Relative cross sections from the reaction $K + CFCl_3$ for the production of: (a) $CFCl_3^-$, CCl_2^-, Cl_2^-, and F^-; (b) $CFCl_2^-$, FCl^-, and Cl^-. (Reproduced by permission of Elsevier Scientific Publ. Co. from Dispert and Lacmann, 1978.)

12.0). Simple calculations of the reaction enthalpies of the ground states yield the following potential parameters. Together with the ionization energy of KI(K) = 4.34 eV follows for $CFCl_3$ an *EA*-value of 1.1 eV and together with the *AP* of Cl^- the dissociation energy results to $D(CFCl_2—Cl^-) = 0.7$ eV. With $EA(Cl) = 3.6$ follows for $EA(CFCl_2) = 1.1$ eV and $D(CFCl_2—Cl) = 3.2$ eV. This value is in good agreement with other thermodynamic values of the C—Cl bond energy of 3.0–3.2 eV. Cl^- is in all Cl-containing molecules the main ion product with 90–95% of the total ion yield. Its excitation function near the threshold could be best fitted by a linearly rising cross section which was convoluted with the Maxwell–Boltzmann distribution of the scatter gas and the energy spread of the fast K beam.

The contribution to the energy spread from both reactants is of the same order of magnitude, about 0.3–0.4 eV FWHM. The threshold values are determined to within ± 0.2 eV. Since the Cl bond energy is known to within 0.2 eV, the Cl^- threshold is in all cases a good calibration point of the energy scale and is always in good agreement with the energy determined with the TOF method. The parent ion $CFCl_3^-$ can be fitted by a step function cross section or a linear one. Both thresholds agree to within 0.2 eV but because of this difference and the extreme low intensities combined with an increased statistical error, the electron affinity of the parent ions is only determined to within ± 0.3 eV.

$CFCl_3$ may become a suitable gas for testing the sensitivity of a collisional ionization apparatus. With a somewhat reduced sensitivity or increased background the threshold energy jumps from 3.2 to 6 eV, where obviously an electronically excited reaction channel, 2.8 eV above the ground state, is reached.

Principally all these 'lowest' threshold values do not prove that the ions are really formed in their ground states. As shown in the section on pre-stretching and the theory of Zembekov (1971), there are possibilities for the production of ions without excitation at low energies near the threshold either through the influence of the approaching atom on the potential surface of the target molecule or by vibrational relaxation during the collision.

In the aforementioned experiment, a threshold of 6 eV was first observed. For two reasons it was concluded that this could not be the energy threshold for the ground state $CFCl_3^-$ formation. Firstly the threshold energy of the parent ion cannot be higher than any dissociative threshold and secondly the excitation function rose steeply without any indication of a tail due to an energy spread. This tail is a necessary but not sufficient condition to obtain the right threshold energy. Therefore, no convoluted threshold function could be fitted to the excitation function.

From the appearance potentials of the other dissociative product ions in Fig. 15 several *EA*-values and bond energies have been determined for the three- and four-atomic dissociation products. The dissociation energies derived are given in Table V. The *EA*-values are added to Table IV together with the results of Tang *et al.* (1976) for CF_3I^- formation and for the collision of alkali atoms with CH_3NO_2, CF_3Br, and CF_3I (Compton *et al.*, 1978b). In collisions with CH_3CN, no parent ions have been detected, although these ions had been observed in e.s.r. spectra and in charge exchange with highly excited Rydberg states. These seemingly contradictory observations are reconciled with the assumption that the electron is only weakly bound in a dipole field.

c. Miscellaneous molecules. Several other molecules have been studied during the last years. The potential parameters, electron affinities, and dissociation energies deduced from the measurements are summarized in Table VI. Only some of the experiments are mentioned in the following section.

TABLE V. Bond dissociation energies (D) calculated from appearance potentials of negative ion products in ion-pair formation processes between K and CF_4, $CFCl_3$, CF_2Cl_2, CCl_4, and $CHCl_3$. (Reproduced by permission of Elsevier Scientific Publ. Co. From Dispert and Lacmann, 1978.)

System	D(eV)
CF_3—F	$\lesssim 6.9$
$CFCl_2$—F	$\lesssim 6.5$
CCl_3—F	$\lesssim 5.6$
$CHCl_2$—Cl	3.4 ± 0.2
CF_2Cl—Cl	3.3 ± 0.2
$CFCl_2$—Cl	3.2 ± 0.2
CCl_3—Cl	3.0 ± 0.2
CF_2—F	$\gtrsim 3.3$,[a] 4.8[b]
CFCl—F	3.9 ± 0.3
CCl_2—F	4.7 ± 0.3
CCl_2—H	4.3 ± 0.3
CF_2—Cl	2.8 ± 0.3
CFCl—Cl	3.6 ± 0.3
CCl_2—Cl	3.7 ± 0.3
CHCl—Cl	$< 3.9 \pm 0.3$
CF_2Cl—Cl^-	$0.1 + 0.3$
$CFCl_2$—Cl^-	0.7 ± 0.3
CCl_3—Cl^-	1.4 ± 0.3

[a] with $D(CF_3—F) \lesssim 6.9$.
[b] with $D(CF_3—F) = 5.4$.

TABLE VI. Electron affinities (EA) and dissociation energies D(R—X) of miscellaneous molecules determined by collisional ionization

Molecule (R—X)	EA(eV) v = vertical	—X	D(R—X) (eV)	Reference
Cl_2,Br_2,I_2	2.50			Tang *et al.* (1975)
Cl_2	2.6 ± 0.2			Dispert and Lacmann (1977b)
Br_2	2.4 ± 0.2			Dispert and Lacmann (1977b)
	2.6 ± 0.2			Young *et al.* (1974)
Cl_2^a	$0.1\,^{+0.2}_{-0.4}$ v			Kimura and Lacmann (1978)
Br_2^a	0.65 ± 0.3 v			Kimura and Lacmann (1978)
Cl_2	1.02 ± 0.05 v			Hubers *et al.* (1976)
Br_2	1.47 ± 0.05 v			Hubers *et al.* (1976)
I_2	1.72 ± 0.05 v			Hubers *et al.* (1976)
ICl	1.48 ± 0.05 v			Hubers *et al.* (1976)
IBr	1.62 ± 0.05 v			Hubers *et al.* (1976)
PCl_3	0.82 ± 0.1	$—Cl^-$	0.49 ± 0.07	Mathur *et al.* (1976b)
$POCl_3$	$1.41\,^{+0.2}_{-0.1}$	$—Cl^-$	1.31 ± 0.1	Mathur *et al.* (1976b)
PBr_3	1.59 ± 0.15	$—Br^-$	0.78 ± 0.07	Mathur *et al* (1976b)
PCl_2Br	1.52 ± 0.2			Mathur *et al.* (1976b)
PBr_2Cl	1.63 ± 0.2			Mathur *et al.* (1976b)
$POCl_2$	$3.83\,^{+0.25}_{-0.2}$			Mathur *et al.* (1976b)
COS	0.46 ± 0.2			Compton *et al.* (1975)
CS_2	1.0 ± 0.2			Compton *et al.* (1975)
CO_2	-0.6 ± 0.2			Compton *et al.* (1975)
	-2.1 to -11 ± 0.25			Tang *et al.* (1974b)
	$-1.0\,^{+0.3}_{-0.2}$			Dispert and Lacmann (1979)
SO_3	$\geq 1.70 \pm 0.15$			Rothe *et al.* (1975)
SO_2	1.14 ± 0.15			Rothe *et al.* (1975)
O_3	2.14 ± 0.15			Rothe *et al.* (1975)
HNO_3	0.57 ± 0.15	$—NO_3^-$	1.28 ± 0.2	Mathur *et al.* (1976a)
NO_3	3.68 ± 0.2			Mathur *et al.* (1976a)
$C_6H_4O_2$	$1.89\,^{+0.2}_{-0.3}$			Cooper *et al.* (1975)
TCDM[b]	$2.8\,^{+0.05}_{-0.3}$	—CN	3.0 ± 0.4	Compton and Cooper (1977)

[a] Electronic. excited X_2^-.
[b] = Tetracyanoquinodimethane.

In the series CO_2, COS, and CS_2, the CO_2 is an especially interesting species since in collisions with alkali atoms CO_2^- ions are formed even though the electron affinity of CO_2 is negative i.e. CO_2^- is metastable with respect to autodetachment. The lifetime is determined to 4×10^{-5}s (Compton *et al.*, 1975). This is long enough to be detected. The stability results from the large distortion from the parent molecule. The bond length increases from 1.162 to ~ 1.25 Å for CO_2^- and the bond angle changes from 180 to $\sim 135°$.

The average electron affinity obtained from experiments with Cs and K amounts to -0.6 ± 0.2 eV (Compton *et al.*, 1975). In collisions with K atoms, Dispert and Lacmann (1979) determined -1.0 ± 0.3 eV. The difference may result from the fact that the crossing radius R_c amounts to 3.2 Å for Cs and to about 2.8 Å for K. At these small distances, equation (3) is probably no longer applicable, since at such small distances repulsive forces become important. This may lead to a compression of the C—O bond distance and result in higher vibrational excitation of the CO_2^- from collisions with K than Cs atoms. The results of Compton *et al.* (1975) also show a decrease in the average *EA*-value of 0.13 eV between Cs and K collisions. No CO_2^- is observed in collisions with sodium. In this case, R_c would amount to only 2.5 Å the crossing occurring at an even more repulsive part of the covalent potential surface, so that CO_2^- would only be formed above its dissociation limit.

An analogous effect is described in detail for the SF_5^- formation in collisions with Cs, K, Na, and Li. Similarly a decrease in the appearance potential is found for Na and Li (Hubers and Los, 1975). In both systems, for CO_2^- and SF_5^- formation, the reaction energies and therefore the crossing radii according to equation (3) are about equal.

All combinations of the single and mixed phosphorous halides of the type PX_2Y and Cl_3PO with X, Y either Cl or Br have been investigated and the results are included in Table VI (Mathur *et al.*, 1976b). The observed electron affinities are in the order $EA(PCl_3) < EA(Cl_3PO) < EA(PCl_2Br) \approx EA(PBr_2Cl) \approx EA(PBr_3)$ which is the order of increasing acceptor strength. PF_3^- should have no acceptor strength; accordingly, it was not found in Cs collisions (Tang *et al.*, 1974a).

The electron affinity of Cl_2PO^- is determined to be 3.8 eV, which is even higher than that of Cl (3.6 eV), and explains the higher yield of Cl_2PO^- ions than Cl^- ions.

Some new results from measurements of halogen molecules are also included in Table VI. The values are in agreement with other results already reviewed (Baede, 1975).

Another field of research which may attract even more interest in future is the study of complicated organic molecules, e.g. *p*-benzoquinone $C_6H_4O_2^-$, which is of special interest in biochemistry for studies of electron transport mechanism. In Cs collisions the dominant product ion is the parent ion with an electron affinity of 1.89 eV (Cooper *et al.*, 1975). Most of these studies deal with the formation of compound negative ion states by electron impact measurements which cannot be discussed in this article. The same arguments apply to the study of tetracyanoquinodimethane. This substance is often used as an acceptor molecule in many organic charge transfer species. An EA-value of 2.8 eV was determined which is in good agreement with EA-values determined by other experimental techniques and from calculations (Compton and Cooper, 1977).

Both articles show that collisional ionization can be successfully applied to the study of complicated organic molecules. In contrast to electron impact studies, collisional ionization yields in most cases negative parent ions of high intensity. This is right at least for molecules with positive electron affinities and bond energies. Measurements of total collisional ionization cross sections for analytic application are rather easy. Technically, the ion source of a standard mass spectrometer has to be replaced by a differentially pumped alkali oven with a scattering chamber. It could be either applied for mass analytic purposes or, by variation of the relative energy, to determine such thermodynamic data as electron affinities and bond energies.

C. Differential Cross Section

Differential and double differential measurements yield particular information about the dynamics of ion-pair formation processes.

In Part II.C the relation between the differential cross section $\sigma(\theta)$ and the deflection function is given in equation (10). In Fig. 2 was shown a typical deflection function calculated for the atom–atom system Na + I. In atom–molecule collisions the principal structure, i.e. the differentiation between covalent and ionic scattering, is still visible, but quantum effects are averaged out by the additional degrees of freedom.

A typical example of the differential cross section of the alkali–halogen system is shown in Fig. 16 for $K + Br_2$ at K-atom energies of 15 eV and 120 eV. The polar differential cross section $\sigma(\theta) \sin \theta$ is shown as a function of the reduced scattering angle $\tau = E\theta$ in the laboratory system. Besides the experimentally determined points, there are also included results of calculations on diabatic (dashed line) and adiabatic (solid line) surfaces (Aten and Los, 1977). All curves reflect the typical structure of differential ionization cross sections. The angular distribution peaks at small scattering angles (covalently scattered particles) near the forward direction of the K-atom beam, and then goes to a minimum followed by a second broad peak of ionically scattered particles.

All curves illustrate clearly the effect of bond stretching (see Section II.E), which leads to an increase of the ionically scattered particles over the covalently scattered ones. With increasing collision velocity this effect becomes

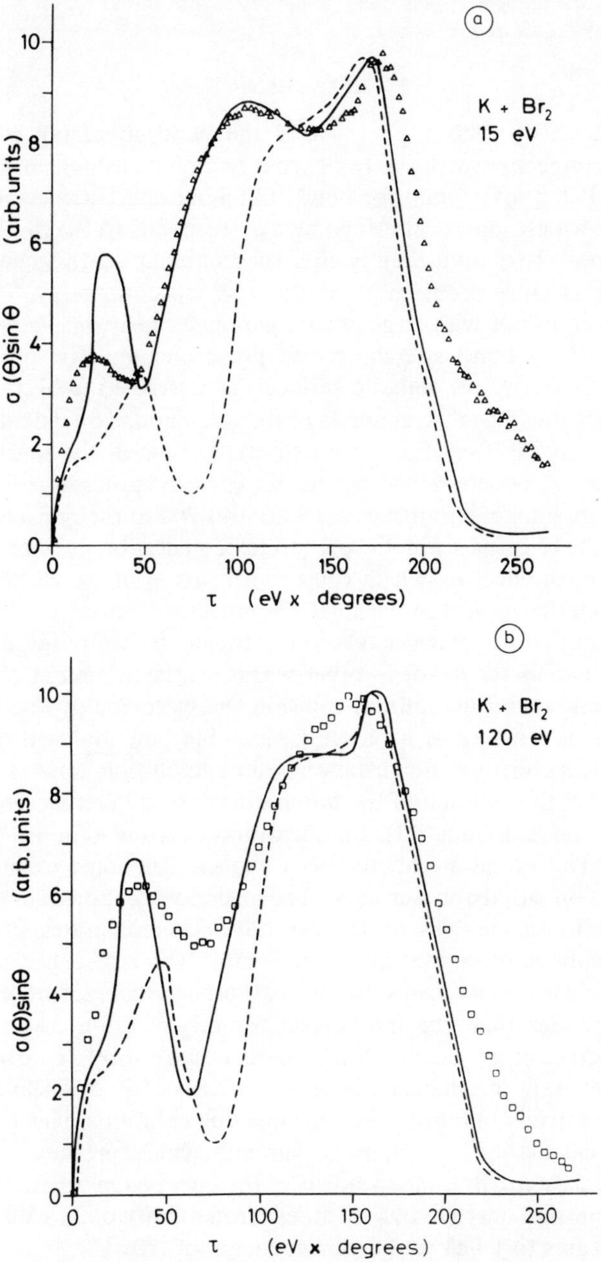

Fig. 16. Polar differential cross sections versus $\tau = E\theta$ for K^+ formation in $K + Br_2$ collisions (laboratory coordinates). The experimental points are represented by the symbols. The dashed lines represent the results of calculations on diabatic potential surfaces. The solid lines represent the results of similar calculations, done on adiabatic potential surfaces. (a) 15 eV = 8.6×10^3 m/s; (b) 120 eV = 2.43×10^4 m/s. (Reproduced by permission of North-Holland Publ. Co. from Aten and Los, 1977.)

smaller. At 120 eV with $\tau_{coll}/\tau_{vib} = 0.14$ the bond stretching effect is still visible. In connection with the total cross-section measurements, described in Part IV.B.2, a lower limit for bond stretching of 0.1 was estimated. The peak at the widest ionic scattering angle corresponds to the rainbow angle. The ionic peak rises immediately after the minimum and reaches a central peak which is more pronounced at the low collision energy. The peak is ascribed to collisions with large impact parameter near R_c. These collisions are increased by bond stretching and pre-stretching. The pre-stretching effect is not effective on diabatic surfaces (as described in Section IV.B.1). This explains the large discrepancies of the calculations on diabatic surfaces (dashed line in Fig. 16). The largest discrepancy is in the position of the minimum which is determined by the *EA* corresponding to collisions with the maximum impact parameter R_c. Since this is also the impact parameter tangent to the crossing seam, the pre-stretching effect on adiabatic potential surfaces is most effective, which causes an increase of the electron affinity in going from the vertical to the adiabatic values.

In equation (11) is given a relation between the scattering angle θ^* at the impact parameter R_c (or in other words, at the minimum of the cross-section function) and the ratio of collision energy to energy loss $E/\Delta E$. The assumption of crossing of a purely Coulombic potential with a covalent potential independent of the distance, will be fulfilled as long as R_c is large enough. From the position of the minimum of the differential cross section in Fig. 16 and equation (11), the apparent electron affinities have been calculated. The values are about 0.4 eV higher than those calculated from trajectories on adiabatic surfaces. The decrease of apparent *EA*-values between collision energies of 15 and 120 eV demonstrates that even at 15 eV (well above threshold) the electron affinity is larger than the vertical electron affinity. This indicates strongly that pre-stretching is more important at these energies than the mechanism for de-excitation involving three-particle interaction on diabatic surfaces, as described by Zembekov (1971). The de-excitation mechanism is most effective for collinear collisions, which are relatively infrequent in these ionizing collisions. This is confirmed by model calculations on diabatic surfaces which produce an electron affinity of 1.5 eV for Br_2, independent of the energy. On adiabatic surfaces, even at a collision energy of 15 eV an electron affinity of 2.1 eV is obtained, which decreases to 1.7 eV at a collision energy of 120 eV.

The results of the three-particle trajectory calculations are shown in Fig. 16. They are similar to the calculation described in connection with total cross-section measurements (Aten *et al.*, 1976), with the main difference that besides bond stretching of the molecular ion during the collision, and the initial distribution over internuclear distances of the molecule, the anisotropy of the interacting potential is included, as well as the orientation-dependent coupling matrix element.

These effects influence the differential cross section only in so far as they strongly reduce the rainbow angle peak, which is spread to lower scattering angles in better agreement with the experiments. For total cross-section calculations, this effect is negligible and was therefore not included in those calculations. A detailed description is given by Aten *et al.* (1977a) together with experimental results of the differential cross-section measurements in collisions of Na, K, and Cs with iodine. The experimental curves are compared with calculated cross sections on diabatic surfaces. The results for potassium and iodine are similar to the results shown in Fig. 16 for $K + Br_2$. For sodium, the central peak is not separated from the rainbow maximum, presumably because of the large ionization potential and correspondingly smaller R_c value and, consequently, reduced bond stretching (as demonstrated in Fig. 3b). Large impact parameter collisions contribute mainly to the central peak. The bond stretching effect in sodium collisions ceases at lower energies because of the lighter, and therefore faster sodium atoms at the same collision energy and the smaller R_c values. Accordingly, ionic and covalent scattering contribute about equally to the differential cross section already at 60 eV. Collisions with caesium exhibit just the opposite effect. Bond stretching is even greater than for potassium. In this case, the central peak is even higher than the maximum of the rainbow angle. The apparent EA-values reflect the same tendency: at 15 eV they increase from Na to Cs above the adiabatic values; with increasing energy the EA-values decrease. For sodium they are constant (1.35 eV) above 60 eV and somewhat smaller than the vertical value of (1.7 eV), while for caesium, even at 90 eV, the EA-value is still quite large (2.2. eV), indicating that pre-stretching is still important.

Three-dimensional surface-hopping trajectory (SHT) calculations by the method of Tully and Preston (1971) have been performed to calculate the differential and total cross section and fractions of atomic to molecular ion products (Evers, 1977). The calculations were performed with semi-empirical potential energy surfaces. A comparison with the experimental results of ion-pair formation of Na, K, and Cs with I_2 between 10 and 100 eV (Aten *et al.*, 1977a) indicates the usefulness of the approach especially since the SHT model involves no adjusted parameters. The SHT calculations are found to be quantitatively accurate to within the experimental uncertainties. The calculations are fairly insensitive to the precise value of the non-adiabatic coupling potential H_{12} (within a factor of 1.2). The numbers of trajectories reaching the avoided crossing seam is more important than the Landau–Zener probability. The SHT calculations confirm that trajectories scattered around the rainbow angle yield mainly dissociated atomic ions, while the trajectories with smaller scattering angle around the central peak yield preferably molecular ions.

The last effect was confirmed in double differential measurements between

Na, K, and Cs with I_2 at 8 eV (Aten *et al.*, 1977c), in which the polar cross section was measured as a function of the scattering angle for different product ion energies. The results were convoluted with the experimental energy resolution of $\Delta E = 0.1E$ and compared with the SHT theory. The experiments and calculations were in fair agreement and reflect the tendency of the ionically scattered particles to appear at smaller scattering angles with increasing product ion energy or with decreasing internal (vibrational) excitation.

The temperature dependence of the differential cross sections in collisions of alkali atoms with iodine has been studied by Aten *et al.* (1977b). With increasing temperature (300 to 700 K) an increase in the covalent scattering and a decrease in ionic scattering was observed. This effect is explained by the change in the shape of the Franck–Condon overlap function between I_2 and I_2^- at the first surface crossing. Simple trajectory calculations reflect the same tendency. The effect does not result from the increased internal energy but rather from the broader distribution over internuclear distances, influencing the vertical EA-values and therefore the crossing radii which depend exponentially on the coupling matrix element.

Differential cross sections of Li on several molecules (halogens X_2, HX, O_2, and SF_6) have been determined at collision energies only several eV above the threshold (Young *et al.*, 1974). With an iterative procedure they obtained a single function $\sigma(E\theta)$ which fitted the data at all energies. The results of long-range collisions (large electron affinities, e.g. $Li + X_2$) agree well with classical calculations. They applied the Landau–Zener model with modification for dissociations, which is similar to the bond stretching model, since it allows an increase in p to p' for the dissociative ion products, as in equation (15). The influence of variation of the potential parameters on the differential cross section was shown.

Double differential measurements, i.e. angular and energy loss determination with high-energy resolution (1–2%) in the centre-of-mass system have been carried out for reactions $Li + Cl_2$ and Br_2 (Kimura and Lacmann, 1977) and $K + Cl_2$ and Br_2 (Mochizuki and Lacmann, 1977; Kimura and Lacmann, 1978). The apparatus has been already described in the experimental Part (III.B).

For all systems the polar double differential cross section $I(\theta)\sin(\theta)$ $(d^2\sigma/dEd\Omega)$ was determined separately between 5 and 30 eV as a function of: a) the energy loss in forward direction ($\theta = 0°$) at different collision energies b) the energy loss at the different scattering angles at fixed collision energy, and c) the reduced scattering angle for different energy values at fixed collision energies.

The experimental results support in many points the bond stretching model with pre-stretching effects included. But, since the double differential measurements analyse the scattering angle in addition to the energy loss, they are able to test directly the results of the SHT calculation which also contained information on the energy loss and supported the above given interpretations. In addition, the energy loss distribution yields information on the formation of excited negative ions.

Fig. 17 shows the K^+ intensity from $K + Cl_2$ collisions determined in forward direction ($\theta = 0°$) versus the energy loss for different collision energies between 9.6 and 36 eV. The dashed lines mark the endoergicity for the formation of Br_2^- and Br^- in their ground states. The absolute energy loss was determined as described in Part III. B, i.e. the absolute energy scale may be erroneous by two tenths of an eV. The figure shows clearly that the Cl_2^- is formed at a collision energy of 9.6 eV with a most probable electron affinity ($= I(\mathrm{K}) - \Delta E$ or $4.34 - 2.95$) of 1.4 ± 0.2 eV. This indicates a slightly increased electron affinity over the vertical value of 1.0 eV, which is not influenced by prestretching or any other relaxation mechanism. This effect is much smaller than the pre-stretching effect reported above (Aten and Los, 1977), which was due to large impact parameter collisions. With increasing collision energy the medium *EA*-value shows a slight

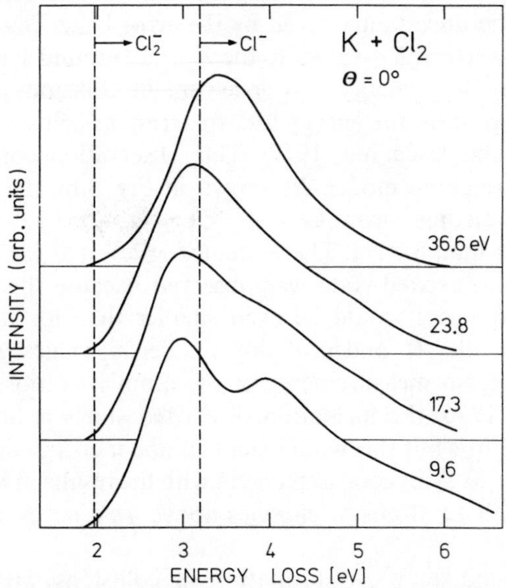

Fig. 17. Relative intensity of K^+ ions ($d^2\sigma/dEd\Omega$) in the direction of the incoming K beam as a function of energy loss for different colliding energies (all values in c.m. system). Minimum energy loss values for Cl_2^- and Cl^- formation are indicated by broken lines (Kimura and Lacmann, 1978).

shift to smaller values as expected if a small pre-stretching effect be still effective at lower energies in forward direction.

The second peak, clearly visible at about an energy loss of 4 eV at 9.6 eV collision energy, has nearly disappeared at higher collision energies. Similar effects have been also observed in $K + Br_2$ collisions. In collisions of Li with Br_2 or Cl_2 no second energy loss peak was observed at all. The second peak was assigned to the formation of a Br_2^- and Cl_2^- in the electronically excited $^2\Pi$ state, 0.8 ± 0.3 eV above the most probable transition to the Br_2^- ground state and $0.9 \pm ^{0,2}_{0,4}$ eV above the transition to the Cl_2^- ground state at a collision energy of about 9 eV for scattering into the forward direction.

Under neglect of any influence by pre-stretching because the impact parameter leading to forward scattering is much smaller than R_c, as seen schematically in Fig. 2, the low energy loss peak corresponds to the vertical *EA*. With the accurately determined vertical *EA*-values of the ground state (Table VI), the vertical *EA* of the excited $^2\Pi_g$ state of Br_2^- is determined to be 0.65 ± 0.3 eV and of Cl_2^- to be $0.1 \pm ^{0,2}_{0,4}$ eV. The curves developed from the molecular orbital considerations of Person (1963), which also show a clear trend to increasing vertical *EA*-values of the ($^2\Pi_g$) state in going from Cl_2^- to the corresponding Br_2^- state in agreement with above values within the uncertainty given by the error bars. This is analogous to the trend of the vertical *EA*-values to the $X_2^-(^2\Sigma_u)$ ground state.

The corresponding energy loss spectrum in collisions with Li did not show a second peak in the energy loss spectrum at collision energies above 9 eV (Kimura and Lacmann, 1977). This observation could be explained by the bond stretching model. As shown in Fig. 3(b), the bond stretching effect decreases strongly from Cs over K to Na. For Li it would be even a little bit smaller than for Na. Therefore, for Li, even at the lowest velocities $\approx 2 \times 10^4$ m/s, no excited state was observed because the bond stretching effect in excited states would be even smaller than for the ground state, owing to the smaller R_c and probably smaller ω_e-values. As shown in the Section IV.B.2 from measurements of the atomic to molecular ion yields (Hubers *et al.*, 1976) the formation of excited states is limited to smaller velocities $\approx 10^4$ m/s but this would yield an upper energy limit for collisions with K atoms of 20 eV in good agreement with the results in Fig. 17. However, in collisions with Li atoms at energies above 4 eV no excited states could be formed.

For the ground-state Cl_2^- formation in collisions with Li, the upper energy limit for bond stretching is calculated to be 20–30 eV or about 2.4–2.9×10^4 m/s. This corresponds to a collision time of about one tenth of the vibrational period of the Cl_2^- formed in a highly excited vibrational state by a vertical transition. The experimental results for the system $Li + Cl_2$ are shown in Fig. 18. The double differential cross section $I(\theta) \sin \theta$ is plotted

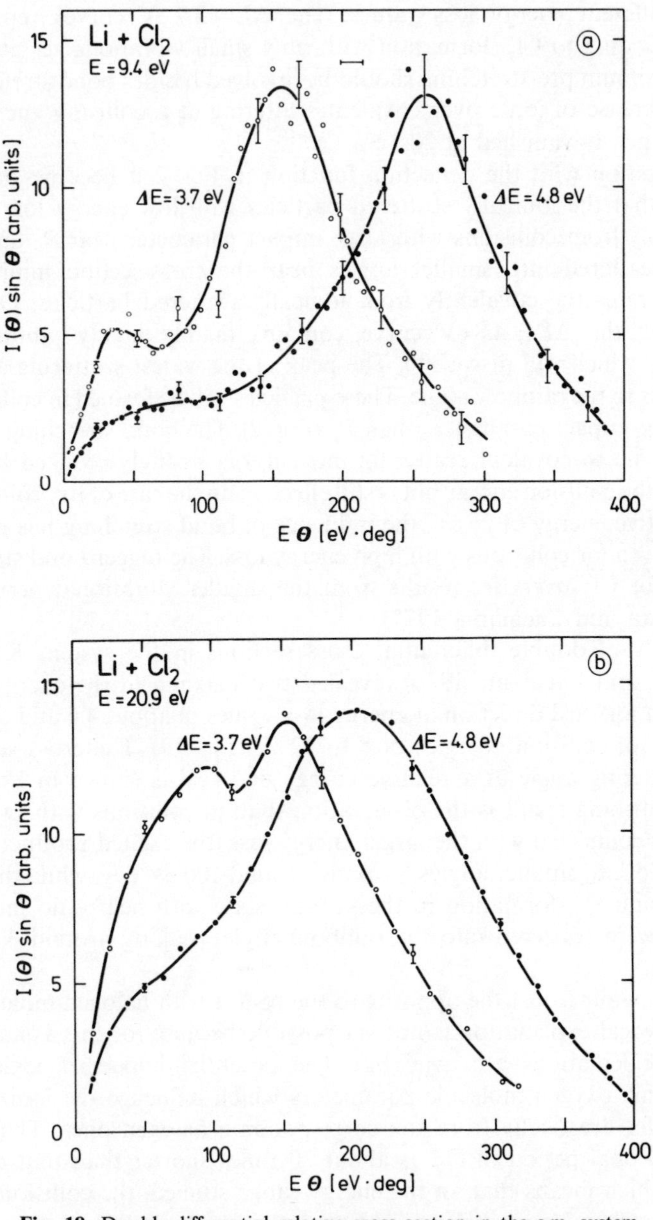

Fig. 18. Double differential relative cross section in the c.m. system for ion-pair formation in $Li + Cl_2$ collisions as a function of reduced scattering angle $E\theta$ at collision energies (a) $E = 9.4$ eV and (b) $E = 20.9$ eV. The energy loss $\Delta E = 3.7$ eV corresponds to Cl_2^- and $\Delta E = 4.8$ eV to Cl^- formation (Kimura and Lacmann, 1977).

against the reduced scattering angle at a collision energy of 9.4 and 20.9 eV for two different energy loss values. The $\Delta E = 3.7$ eV curves represent collisions leading to Cl_2^- formation with only small vibrational excitation, so that maximum pre-stretching should be involved besides bond stretching. The big increase of ionic over covalent scattering at a collision energy of 9.4 eV has nearly vanished at 20.9 eV.

In connection with the deflection function in Fig. 2 it becomes understandable that the ionically scattered particles with low energy loss must result mainly from collisions with large impact parameter near R_c and are therefore scattered into smaller angles near the cross-section minimum, which separates the covalently from ionically scattered particles. On the other hand, the $\Delta E = 4.8$ eV curve contains mainly highly vibrational excited Cl_2^- which will dissociate. The peak at the widest scattering angles corresponds to the rainbow angle. These particles will be formed in collisions with smaller impact parameters than R_c (Fig. 2). The bond stretching effect (ratio of ionic to covalent scattering intensity) is strongly reduced by the increase of the collision energy but is still effective. In the case of Br_2 collisions with a relative energy of 26 eV, the influence of bond stretching has nearly vanished, even for collisions with high energy loss. The bigger bond stretching effect for Cl_2 over Br_2 results from the smaller vibrational period of Cl_2^- (Kimura and Lacmann, 1977).

The study of double differential cross sections in the system $K + O_2$ (Mochizuki and Lacmann, 1976) revealed two maxima in the energy loss spectrum in forward direction at energy loss values of about 4.6 and 5.1 eV for a range of collision energies of 8 to 22 eV (Fig. 19). The cross section versus scattering angle at a relative energy of 14 eV is shown in Fig. 20. The most striking result is the observation that, in collisions with O_2, the trajectories connected with the larger energy loss (the excited product ions) are scattered into smaller angles, peaking around 100 eV deg, while the K^+ ions involving O_2^- formation in the ground state with nearly no internal excitation were scattered into the rainbow angle, peaking around 300 eV deg.

This behaviour is just the opposite to the results with halogen molecules. An unequivocal explanation is not yet possible because for this system no trajectory calculations are available. The potential hypersurface is not known. Some oxygen molecule parameters which influence the ionization process differ drastically from the corresponding halogen ones. The time for a vibrational period of O_2^- is about 10 times shorter than that of the halogens which means that, in the energy range studied, the collision time is of order or even longer than a vibrational period. This will have an averaging influence on the bond stretching effect, as demonstrated in Fig. 11. The other most important difference is the smaller crossing radius. At these

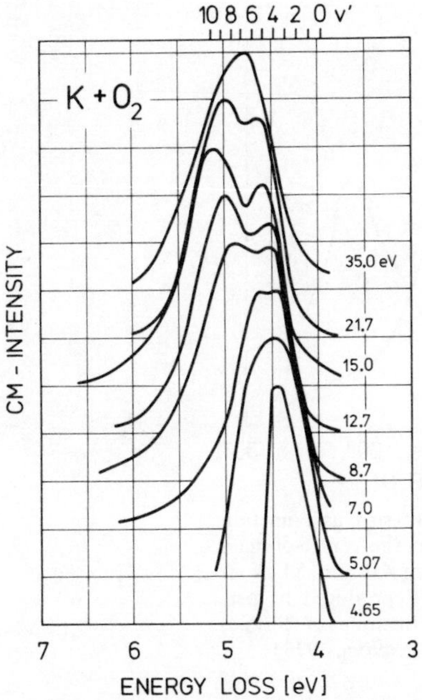

Fig. 19. Intensity of K^+ ($d^2\sigma/d\Omega dE$) ions, normalized to their maximum yield, as a function of the energy loss for different collision energies in the direction of the incoming K beam (all values in c.m. system). (Reproduced by permission of the American Institute of Physics from Mochizuki and Lacmann, 1976.)

smaller distances equation (3) is probably no longer accurate because repulsive forces will gain influence.

The maximum in Fig. 19 at 4.6 eV reflects, within an error of ± 0.2 eV, transition probabilities given by the Franck–Condon factors (on top are indicated the vibrational levels v' of the O_2^-). At collision energies less than 2 eV above the threshold energy of 3.9 eV, the energy loss shifts with decreasing collision energy to a value determined by the adiabatic *EA*-value, which, in this reaction, may result either from a mechanism like pre-stretching or a relaxation mechanism during the collision (Zembekov, 1971). The second peak, at an energy loss of 5.1 eV, was attributed to the formation of an electronically excited state like $^4\Sigma_u^-$ or $^2\Pi_u$ as included in Table II. This state must lie within 0.3 to 0.7 eV above the ground state. Meanwhile, published accurate *ab initio* calculations of the O_2^- ground and excited state by Das *et al.* (1978) revealed the lowest excited O_2^- as being 2.4 eV above the $^3\Sigma_g$ ground state. Therefore, a different process such as autodetachment may be strongly dependent on the vibrational excitation of the O_2^- and thereby explain the second peak in the energy loss spectrum.

Grosser and Meyer (1976) reported a hump in a broadened energy loss spectrum at K-atom energies between 12 and 23 eV at a fixed scattering angle of 8° in the laboratory system. A transformation into the c.m. system would result in a cross section at about 350–450 eV deg depending on the

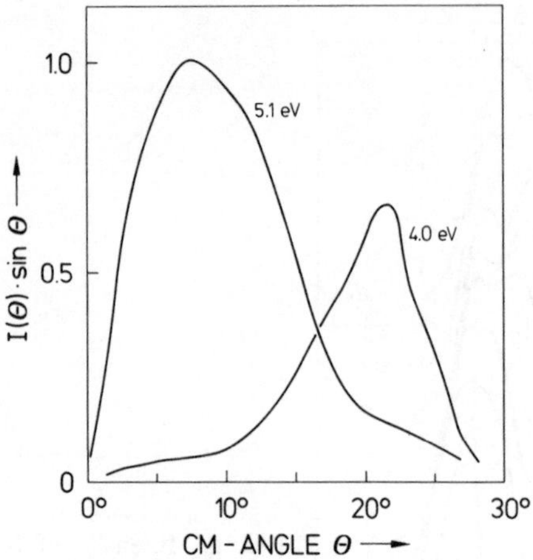

Fig. 20. Relative intensity $I(\theta)\sin\theta$ as a function of the scattering angle θ in the centre-of-mass system for an energy loss of 4.0 and 5.1 eV at 14 $eV_{c,m,}$ collision energy. (Reproduced by permission of the American Institute of Physics from Mochizuki and Lacmann, 1976.)

energy of the collision. This would be beyond the rainbow angle shown in Fig. 20 which was related to the ground-state O_2^- formation by Mochizuki and Lacmann (1976), and should therefore be explained by a different process.

Several excited states of O_2^- have so far been reported in different types of experiments, e.g. Durup *et al.* (1977). They observed excited molecular ion states in collisions with argon atoms.

Of all the systems considered in differential measurements up to now, only forward scattered positively charged alkali ions M^+ have been detected but not the negatively charged molecular ions AB^- or their dissociation products.

The negative ions would have been backward scattered in the c.m. system. From simple kinematic calculations it follows that, in the laboratory system, the energy would be very small and the particles would be scattered into wide angles, sometimes even backwards, when the mass of the projectile is lighter than the target gas molecule and the energy loss small compared to the collision energy. In the opposite case, if the projectile is much heavier than the target gas at collision energies near the threshold energy, even the backward scattered AB^- may have enough energy in the laboratory system to be detected.

An example is given in the reaction $Cs + D_2O \rightarrow Cs^+ + OD^- + D$. The energy threshold for the ground state $^2A_1$ of D_2O^-, which dissociates into $OD^-(^1\Sigma)$ and $H(^2S)$, amounts to 7.17 eV; which corresponds to a

Cs energy of 55 eV. The experimental results by Warmack *et al.* (1978b) are shown in Fig. 21.

Part (a) shows the Cs^+ intensity at different scattering angles as a function of the Cs^+ energy at a Cs impact energy of 96.3 eV (11.5 $eV_{c.m.}$), and part (b) shows analogously the OD^- energy distribution at an impact energy of 76.8 eV (10.0 $eV_{c.m.}$). All values are given, if not otherwise mentioned, in the laboratory system. The two peaks in both parts correspond to forward and backward scattered particles in the c.m. system. A central peak, only resolved in the OD^- intensity, corresponds to the velocity of the Cs–D_2O complex. All three peaks were identified by their mass spectra (by TOF, described in Part III. D. 4) as corresponding to OD^-. In part (c) the $-Q$-value is plotted, which is equivalent to the loss of relative energy in the reaction, determined from the energy distribution of K^+ ions scattered in

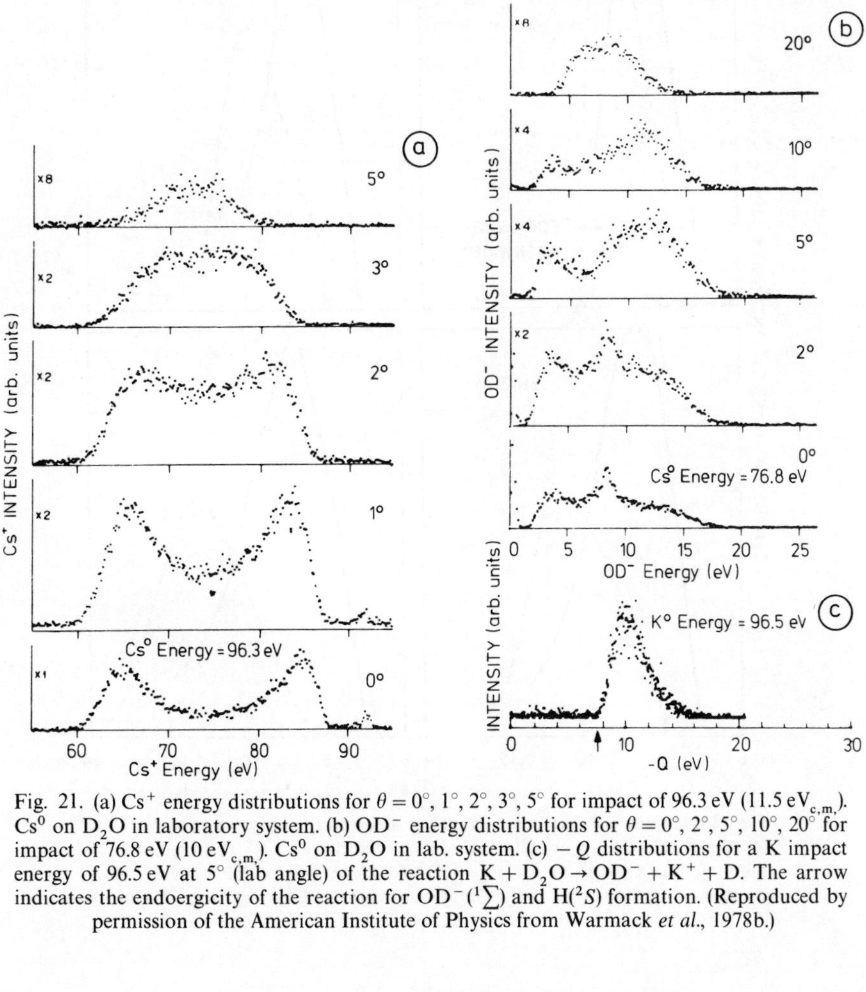

Fig. 21. (a) Cs^+ energy distributions for $\theta = 0°, 1°, 2°, 3°, 5°$ for impact of 96.3 eV (11.5 $eV_{c,m,}$). Cs^0 on D_2O in laboratory system. (b) OD^- energy distributions for $\theta = 0°, 2°, 5°, 10°, 20°$ for impact of 76.8 eV (10 $eV_{c,m,}$). Cs^0 on D_2O in lab. system. (c) $-Q$ distributions for a K impact energy of 96.5 eV at 5° (lab angle) of the reaction $K + D_2O \rightarrow OD^- + K^+ + D$. The arrow indicates the endoergicity of the reaction for $OD^-(^1\Sigma)$ and $H(^2S)$ formation. (Reproduced by permission of the American Institute of Physics from Warmack *et al.*, 1978b.)

the forward direction at a K impact energy of 96.5 eV (= 32.7 $eV_{c.m.}$). The arrow indicates the calculated Q-value of 7.6 eV for products formed in the ground state of the reaction without excitation. The product ions intensity increases steeply at the threshold values. The range of energy loss values suggests that the alkali collisions probe a variety of regions of the $^2A^1$ state ranging from near the asymptotic limit (corresponding to $EA = -3.2$ eV) to smaller values of the H—OH bond distance than those in the Franck–Condon region ($EA \approx -6.5$ eV).

Fig. 22 shows Cs^+ intensities versus energy loss (– Q-value) in the c.m.

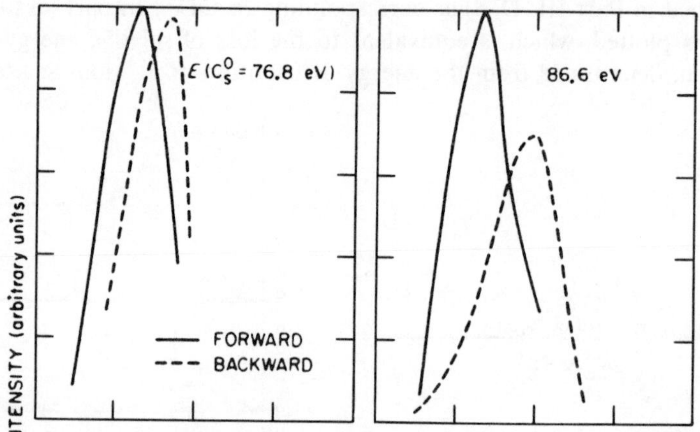

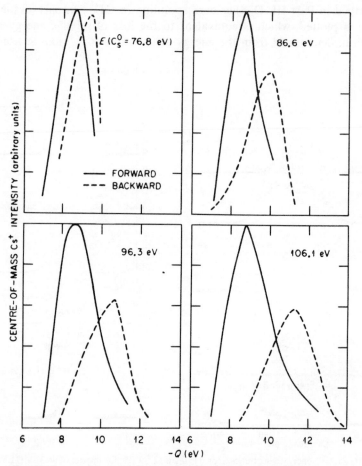

Fig. 22. Centre-of-mass Cs^+ intensity distribution plotted against $-Q$ for the forward (—) and backward (---) scattered Cs^+ distributions for the reaction $Cs + D_2O \rightarrow Cs^+ + OD^- + D$; parameter: Cs energy. (Reproduced by permission of the American Institute of Physics from Warmack *et al.*, 1978*b*.)

system at different collision energies between 77 and 106 eV (10 to 14 eV in c.m. system). The energy loss distribution of the forward scattered Cs^+ ions rises steeply at the energy threshold and peaks at about 2 eV above the threshold, independent of the collision energy, typical for collisions with large impact parameter ($b \approx R_c$) (grazing or 'fly-by' collisions). The backward scattered particles show a larger energy loss which increases with the collision energy. Therefore, they are related to 'head-on' encounters. The maximum energy loss observed is about equal to the collision energy corresponding to the central peak observed in Fig. 21, and is associated with a long-lived complex. The asymmetry between the peaks of the forward and backward scattered particles, illustrated in Fig. 22, suggests a collision mechanism different from complex formation. Because of the large negative electron affinity involved, energy loss values of this size (7 to ~ 12 eV), indicate much smaller crossing distances than in the systems described before. Smaller impact parameters, on the other hand, increase the relative probability of backward scattering. A comparative discussion of electron impact experiments and possible reaction channels is also given by Warmack *et al.* (1978*b*).

The double differential measurements give additional insight into the dynamics of ionizing collisions. Since these states have been studied intensively only for the last four years they still leave open several questions. While processes involving only ground electronic states are quite well understood, there are still many questions open, which are probably related to the involvement of excited states.

D. Collisional Ionization in Non-alkali Reactions

Ionizing collisions between fast halogen atoms produced by sputtering (compare Part III.D. 1) and several organic molecules (aniline, *n*-propylbenzene, carbon tetrachloride, benzene, propane, and cyclohexane) have been investigated (Können *et al.*, 1974). The total cross section versus collision energy for aniline–halogens are reported. The results show an analogy with electron impact and photoionization mass spectrometry. They are discussed in terms of potential curve crossing in the halogen-organic molecule system.

The following non-alkali reactions differ considerably from all other systems described so far, because the ionization energy for the non-alkali projectile atoms is generally much larger and therefore the threshold energies are much higher (around 10 eV). IPF processes therefore can no longer be described by the curve crossing of a constant covalent and a Coulombic potential because the crossing distances are so small (< 2Å) that chemical forces become important. In most examples, therefore, both the formation of $M^+ + e$ and of chemiionization processes (associative and rearrangement reactions) are studied together. One example is e.g. the reaction of $C + O_2$ which yields C^+, C^-, O_2^+, and O_2^- as well as free electrons and reactive products (with rearrangement) forming CO and CO^+ (Können *et al.*, (1975). This system reveals both ion pairs $C^+ + O_2^-$ and $C^- + O_2^+$, because the threshold energies differ by only 0.05 eV.

It can be shown that two different mechanisms are involved. The cross sections for C^+ or CO^+ and free electron formation are mugh bigger than for IPF. This is explained by the chemical forces between C and O_2 which play a dominant role at these small impact parameters in contrast to alkali–oxygen reactions. No associative ionization was detected although its threshold is much lower. Many other reactions of atoms M with O_2 yielded products of associative ionizations forming $MO_2^+ + e^-$ and $MO^+ + e$.

In chemiionization studies with accelerated (sputtered) beams of Al, Ba, Ta, and Ti with O_2 Cohen *et al.* (1973) observed associative ionization: $MO_2^+ + e$, with M = Ba and Ti, reactive ionization with IPF: $MO^+ + O^-$ for Ba and Ti, and electron transfer with IPF processes: $M^+ + O_2^-$ with Ba and Al. Angular distributions are presented for a non-velocity selected atom beam. From angular distributions the threshold energies for BaO_2^+ and TiO_2^+ are derived.

Chemiionization reactions between uranium and oxygen yielding UO_2^+, UO^+, and $U^+ + e$ were studied in detail as a function of the collision energy between 0.2 and 60 eV with threshold determination, angular distribution of the product ions and energy analysis of UO^+ and U^+ by the RPD method. UO^+ and U^+ are predominantly scattered in the forward direction, supporting the conclusion that the rearrangement ionization proceeds by a 'spectator stripping' mechanism (Young *et al.*, 1976a).

In a survey of chemiionization reactions in accelerated atom–O_2 crossed molecular beams, all thermodynamic threshold energies of M^+, MO^+, and MO_2^+ are given (Young *et al.*, 1976b). In general, they find that associative and rearrangement ionization reactions do not occur if the thermodynamic threshold for electron transfer is lower than for chemiionization.

Absolute cross section of $U + O_2$ and $U + O$ are determined at thermal energies for UO^+ and UO_2^+. Studies of uranium with oxygen (Fite *et al.*, 1974) show that, at thermal energies, if a neutral channel is open, mainly neutral products are formed. In the case of $U + O_2$, only 1% of the collisions yield $UO_2^+ + e(\sigma = 1.7 \times 10^{-17}\ cm^2)$, while for $U + O \rightarrow UO^+ + e$ reaction the cross section is $1.6 \times 10^{-15}\ cm^2$ about equal to the gas kinetic one.

Comparative studies of collisions between Ti, Zr, Gd, and Th and O and O_2 are also reported by Lo and Fite (1974).

Absolute cross sections are reported for several associative ionization reactions between uranium and thorium with ozone. The following ionic products could be detected: UO_2^+, UO^+, ThO_3^+, ThO_2^+, ThO^+, and small amounts of $O_2^- + ThO^+$ and $O^- + ThO_2^+$ (Patterson *et al.*, 1978).

Ion-pair formation together with rearrangement were observed in molecule–molecule collisions of $N_2 + CO$ by Utterback and Van Zyl (1978). The structure found in the total charge production cross section between 10 and 20 eV in the near-threshold region was mass spectrometrically identified as due to $NO^+ + CN^-$ formation. Its threshold is about 2 eV below the single CO^+ formation. In this energy range IPF is found to be of importance for the total charge production cross section.

Ionizing collision of H atoms, produced by photodetachment from negative ions (see Part III.D.1) with rare-gas atoms and O_2, have been studied between 20 and several hundred eV. H^+, Ar^+, H^-, and free electrons have been detected (Aberle *et al.*, 1978). Here also, IPF is the dominating process at low energies with strong oscillations of the Stückelberg type in the H^- cross sections. The cross section for free electron formation rises monotonically because of the crossing of a discrete potential curve with the continuum limit corresponding to the Rydberg states.

Another type of collisional ionization reaction with IPF results from collisions of noble-gas atoms (Xe, Kr) accelerated aerodynamically in H_2 with polar molecules such as thelium halides (TlCl, TlBr, TlI and their dimers) leading to collision-induced polar dissociation. The different partial cross sections as a function of the collision energies have been measured and have been reviewed by Wexler (1973).

Collisions of Xe and Kr with TlF and Tl_2F_2 have been investigated by Parks *et al.* (1977). The main difference to the other TlX molecules is a reduced cross section for TlF and TlF_2 and a high internal excitation of Tl_2F^+ at threshold. The authors developed a collision model in which dynamically constrained collisions result in the extraction of the light F^- ion from either TlF or Tl_2F_2.

Collision-induced IPF of CsCl and its dimers with rare gases (Ar, Kr, and Xe) has also been studied as a function of the collision energy. Partial cross sections for each product ion were determined mass spectrometrically by the TOF method yielding Cs^+, Cl^-, and CsM^+ in reactions with monomers, while dimer reactions produced $CsCl^+ + Cl^-$ and $Cs^+ + CsCl_2^-$. The threshold energies for all reaction channels were determined by deconvolution. The cross section near the threshold is described by a power law except for CsM^+ ions, which exhibit a step function behaviour. The threshold for collision-induced dissociation ($Cl^+ + Cl^-$) was determined most accurately from the peak of the $CsM^+ + Cl^-$ excitation function. The transition probability for non-adiabatic transition in CsCl is given by the small asymptotic energy difference between the ion and neutral states of only 0.28 eV, leading to a crossing distance of the order of 50 Å.

Numerous other experiments with dimers have been conducted and have provided insight into the reaction dynamics of IPF processes but most of the reactions are exothermic and have been studied at thermal energies without variation of the collision energy. Since it is not within the scope of this chapter to give a review of these results, the reader is referred to the following publications in addition to those mentioned already in connection with the hexafluoride experiments in Part IV. B. 4: IPF with alkali dimers is studied by Lin *et al.* (1973), Rothe *et al.* (1976), and Dispert and Lacmann (1977b).

A similar situation exists in the field of collisional ionization by highly excited atoms. Only some reactions which are directly related to the discussed IPF reactions are mentioned here.

Gillen *et al.* (1978) studied the differential cross section of $Ar(^3P) + I_2 \rightarrow Ar^+ + I_2^-$ at different collision energies. The experimental results are very similar to the alkali–halogen collisions described in Part IV.C. The ionization energy of the excited Ar* is about equal to that of K and the impact parameters are accordingly large. The results are not significantly altered by Penning ionization processes yielding $Ar + I_2^- + e$, because these occur at smaller impact parameters.

In the case of CH_3I, C_7F_{14}, and C_6F_6 the negative ions produced by collisional ionization of high Rydberg Xe atoms are consistent with those observed in free thermal electron attachment studied (Hildebrandt *et al.*, 1978).

Acknowledgements

I wish to express my gratitude to Dr. Philip Kuntz for reading the manuscript and for his many helpful suggestions, and to thank Mrs. G. Snoei for typing the manuscript.

References

Aberle, W., Brehm, B., and Grosser, J. (1978). *Chem. Phys. Lett.*, **55**, 71.
Abuaf, N., Anderson, J. B., Andres, R. P., Fenn, J. B., and Marsden, D. G. H. (1967). *Science*, **155**, 997.
Alexander, M. H. (1978). *J. Chem. Phys.*, **69**, 3502.
Anderson, R. W. and Herschbach, D. R. (1975). *J. Chem. Phys.*, **62**, 2666.
Annis, B. K. and Datz, S. (1977). *J. Chem. Phys.*, **66**, 4468.
Annis, B. K. and Datz, S. (1978). *J. Chem. Phys.*, **69**, 2553.
Aten, J. A. (1972). In 'Omladingsbron voor alkaliatomen van 5–250 ev.', FOM-Instituut, Amsterdam.
Aten, J. A. and Los, J. (1975). *J. Phys. E.: Sci. Instr.*, **8**, 408.
Aten, J. A. and Los, J. (1977). *Chem. Phys.*, **25**, 47.
Aten, J. A., Hubers, M. H., Kleyn, A. W., and Los, J. (1976). *Chem. Phys.*, **18**, 311.
Aten, J. A., Lanting, G. E. H., and Los, J. (1977a). *Chem. Phys.*, **19**, 241.
Aten, J. A., Lanting, G. E. H., and Los, J. (1977b). *Chem. Phys.*, **22**, 333.
Aten, J. A., Evers, C. W. A., deVries, A. E., and Los, J. (1977c). *Chem. Phys.*, **23**, 125.

Auerbach, D. J., Hubers, M. M., Baede, A. P. M., and Los, J. (1973). *Chem. Phys.*, **2**, 107.

Baede, A. P. M. (1975). *Adv. Chem. Phys.*, **30**, 463.

Baede, A. P. M., Auerbach, J., and Los, J. (1973). *Physica*, **64**, 134.

Balint-Kurti, G. G. (1973). *Mol. Phys.*, **25**, 393.

Bartlett, N. (1968). *Angew. Chem. Internat. Edit.*, **7**, 433.

Bauer, E., Fisher, E. R., and Gilmore, F. R. (1969). *J. Chem. Phys.*, **51**, 4173.

Bernstein, R. B. and Wilcomb, B. E. (1977). *J. Chem. Phys.*, **67**, 5809.

Celotta, R. J., Bennett, R. A., Hall, J. L., Siegel, M. W., and Levine, J. (1972). *Phys. Rev.* **A6**, 631.

Celotta, R. J., Bennett, R. A., and Hall, J. L. (1974). *J. Chem. Phys.*, **60**, 1740.

Chantry, P. J. (1971). *J. Chem. Phys.*, **55**, 2746.

Child, M. S. (1973). *Faraday Disc.*, *Soc.*, **55**, 30.

Cohen, R. B., Young, C. E., and Wexler, S. (1973). *Chem. Phys. Lett.*, **19**, 99.

Compton, R. N. (1977). *J. Chem. Phys.*, **66**, 4478.

Compton, R. N. and Cooper, C. D. (1977). *J. Chem. Phys.*, **66**, 4325.

Compton, R. N., Reinhardt, P. W., and Cooper, C. D. (1975). *J. Chem. Phys.*, **63**, 3821.

Compton, R. N., Reinhardt, P. W., and Cooper, C. D. (1978a). *J. Chem. Phys.*, **68**, 2023.

Compton, R. N., Reinhardt, P. W., and Cooper, C. D. (1978b). *J. Chem. Phys.*, **68**, 4360.

Cooper, C. D., Naff, W. T., and Compton, R. N. (1975). *J. Chem. Phys.*, **63**, 2752.

Das, G., Wahl, A. C., Zemke, W. T., and Stwalley, W. C. (1978). *J. Chem. Phys.*, **68**, 4252.

Delvigne, G. A. L. and Los, J. (1973). *Physica*, **67**, 166.

Dispert, H. and Lacmann, K. (1977a). *Chem. Phys. Lett.*, **45**, 311.

Dispert, H. and Lacmann, K. (1977b). *Chem. Phys. Lett.*, **47**, 533.

Dispert, H., and Lacmann, K. (1978). *Int. J. Mass Spectrom. Ion Phys.*, **28**, 49.

Dispert, H. and Lacmann, K. (1979). To be published.

Düren, R. (1973). *J. Phys. B*, **6**, 1801.

Durup, M., Parlant, G., Appell. J., Durup, J., and Ozenne, J. (1977).

Evers, C. (1977). *Chem. Phys.*, **21**, 355.

Evers, C. (1978). *Chem. Phys.*, **30**, 27.

Faist, M. B. and Levine, R. D. (1976). *J. Chem. Phys.*, **64**, 2953.

Fite, W. L., Lo, H. H. and Irving, P. (1974). *J. Chem. Phys.*, **60**, 1236.

Fluendy, M. A. D. and Lawley, K. P. (1973). *Chem. Applications of Molecular Beam Scattering*, Chapman and Hall.

Franklin, J. L. (1976). *Science*, **193**, 725.

Gillen, K. T., Gaily, T. D., and Lorents, D. C. (1978). *Chem. Phys. Lett.*, **57**, 192.

Gislason, E. A. and Sachs, J. G. (1975). *J. Chem. Phys.*, **62**, 2678.

Grice, R. and Herschbach, D. R. (1974). *Mol. Phys.*, **27**, 159.

Grosser, J. and Meyer, G. (1976). *Chem. Phys. Lett.*, **37**, 82.

Helbing, R. K. B. and Rothe, E. W. (1969). *J. Chem. Phys.*, **51**, 1607.

Herbst, E., Patterson, T. A., and Lineberger, W. C. (1974). *J. Chem. Phys.*, **61**, 1300.

Herschbach, D. R. (1966). *Adv. Chem. Phys.*, **10**, 319.

Hildebrandt, G. F., Kellert, F. G., Dunning, F. B., Smith, K. A., and Stebbings, R. F. (1978). *J. Chem. Phys.*, **68**, 1349.

Hildenbrand, D. L. (1973). *J. Phys. Chem.*, **77**, 897.

Hollstein, M. and Pauly, H. (1966). *Z. Physik*, **196**, 353.

Hubers, M. M. and Los, J. (1975). *Chem. Phys.*, **10**, 235.

Hubers, M. M., Kleyn, A. W., and Los, J. (1976). *Chem. Phys.*, **17**, 303.
Hurkmans, A., Overbosch, E. G., Kodera, K., and Los, J. (1976). *Nucl. Instr. and Meth.*, **132**, 453.
Janev, R. K. (1976). *Adv. Atom. Mol. Phys.*, **12**, 1.
Johnson, J. P., Christophorou, L. G., and Carger, J. G. (1977). *J. Chem. Phys.*, **67**, 2196.
Kempter, V. (1975). *Adv. Chem. Phys.*, **30**, 417.
Kendall, G. M. and Grice, R. (1972). *Mol. Phys.*, **24**, 1373.
Kimura, M. and Lacmann, K. (1977). *Chem. Phys. Lett.*, **51**, 585.
Kimura, M. and Lacmann, K. (1978). *J. Chem. Phys.*, **69**, 4938.
Kinsey, J. L., Kwei, G. H., and Herschbach, P. R. (1976). *J. Chem. Phys.*, **64**, 1914.
Kleyn, A. W., Hubers, M. M., and Los, J. (1978). *Chem. Phys.*, **34**, 55.
Können, G. P., Grosser, J., Haring, A. Eerkens, F., deVries, A. E., and Kistemaker, J. (1974). *Chem. Phys.*, **6**, 205.
Können, G. P., Haring, A., and deVries (1975). *Chem. Phys. Lett.*, **30**, 11.
Kuntz, P. J. (1979). In R. B. Bernstein (Ed.), *Atom–Molecule Collision Theory*, Chap. 3, Plenum Press, New York.
Lacmann, K. and Dispert, H. (1975). Private communication.
Lacmann, K. and Henglein, A. (1965). *Ber. Bunsen. Phys. Chem.*, **69**, 286.
Lacmann, K. and Herschbach, D. R. (1970). *Chem. Phys. Lett.*, **6**, 106.
Leffert, C. B., Jackson, W. M., Rothe, E. W., and Fenstermaker, R. W. (1972). *Rev. Sci. Instr.*, **43**, 917.
Leffert, C. B., Tang, S. Y., Rothe, E. W., and Cheny, T. C. (1974). *J. Chem. Phys.*, **61**, 4929.
Lin, S. M., Wharton, J. G., and Grice, R. (1973). *Mol. Phys.*, **26**, 317.
Lo, H. H. and Fite, W. L. (1974). *Chem. Phys. Lett.*, **29**, 39.
Los, J. (1977). In *Int. Conf. on the Physics of Electronic and Atomic Collisions*, X. ICPEAC, Inv. Lectures, eds.
Los, J. and Kleyn, A. W. (1979). In *The Alkali Halide Vapors* (Eds. P. Davidovits and D. McFadden Academic Press.
Mathur, B. P. Rothe, E. W., Tang, J. Y., Mahajan, K., and Reck, G. P. (1976a). *J. Chem. Phys.*, **64**, 1247.
Mathur, B. P., Rothe, E. W., and Tang, S. Y. (1976b). *J. Chem. Phys.*, **65**, 565.
Mathur, B. P., Rothe, E. W., and Reck, G. P. (1977). *J. Chem. Phys.*, **67**, 377.
McNamee, P. E., Lacmann, K., and Herschbach, D. R. (1973). *Faraday Disc. Chem. Soc.*, **55**, 318.
Mochizuki, T. and Lacmann, K. (1976). *J. Chem. Phys.*, **65**, 3257.
Mochizuki, T. and Lacmann, K. (1977). *Chem. Phys., Lett.*, **49**, 604.
Moutinho, A. M. C., Aten, J. A., and Los, J. (1971). *Physica*, **53**, 471.
Moutinho, A. M. C., Aten, J. A., and Los, J. (1974). *Chem. Phys.*, **5**, 84.
Moutinho, A. M. C., Kleyn, A. W., and Los, J. (1979). To be published.
Nalley, S. J., Compton, R. N., Schweinler, H. C., and Anderson, V. E. (1973). *J. Chem. Phys.*, **59**, 4125.
Nikitin, E. E. (1968). In *Chemische Elementarprozesse* (Ed. H. Hartmann), Springer, Berlin.
Nikitin, E. E. and Zülicke, L. (1978). *Theory of Chemical Elementary Processes in Lecture Notes in Chemistry*, Vol. 8, Springer-Verlag, Berlin.
Nyeland, C. and Ross, J. (1971). *J. Chem. Phys.*, **54**, 1665.
Olson, R. E., Smith, F. T., and Bauer, E. (1971). *Appl. Opt.*, **10**, 1848.
Parks, E. K., Kuhry, J. G., and Wexler, S. (1977). *J. Chem. Phys.*, **67**, 3014.
Patterson, T. A., Siegel, M. W., and Fite, W. L. (1978). *J. Chem. Phys.*, **69**, 2163.

Person, W. B. (1963). *J. Chem. Phys.*, **38**, 109.
Politiek, J., Rol, P. K., Los, J., and Ikelaar, P. G. (1968). *Rev. Sci. Instr.*, **39**, 1147.
Rothe, E. W., Tang, S. Y., and Reck, G. P. (1974). *Chem. Phys., Lett.*, **26**, 434.
Rothe, E. W., Tang, S. Y., and Reck, G. P. (1975). *J. Chem. Phys.*, **62**, 3829.
Rothe, E. W., Mathur, B. P., and Reck, G. P. (1976). *J. Chem. Phys.*, **65**, 2912.
Rothe, E. W., Mathur, B. P., and Reck, G. P. (1977). *Chem. Phys. Lett.*, **51**, 71.
Sheen, S. H., Dimoplon, G., Parks, E. K., and Wexler, S. (1977). *J. Chem. Phys.*, **68**, 4950.
Tang, S. Y., Rothe, E. W., and Reck, G. P. (1974a). *J. Chem. Phys.*, **60**, 4096.
Tang, S. Y., Rothe, E. W., and Reck, G. P. (1974b). *J. Chem. Phys.*, **61**, 2592.
Tang, S. Y., Rothe, E. W., and Reck, G. P. (1974c). *Int. J. Mass Spectrom. Ion Phys.*, **14**, 79.
Tang, S. Y., Leffert, C. B., and Rothe, E. W. (1975). *J. Chem. Phys.*, **62**, 132.
Tang, S. Y., Mathur, B. P., Rothe, E. W., and Reck, G. P. (1976). *J. Chem. Phys.*, **64**, 1270.
Tully, J. C. and Preston, R. K. (1971). *J. Chem. Phys.* **55**, 562.
Tully, J. C. (1976). In *Dynamics of Molecular Collisions* (Ed. W. H. Miller), Part B, Plenum Press, New York, pp. 217.
Utterback, N. G. and Van Zyl, B. (1978). *J. Chem. Phys.*, **68**, 2742.
Warmack, R. J., Stockdale, J. A. D., and Compton, R. N. (1978a). *Int. J. Mass Spectrom. Ion Phys.*, **27**, 239.
Warmack, R. J., Stockdale, J. A. D., and Compton, R. N. (1978b). *J. Chem. Phys.*, **68**, 916.
Wentworth, W. E., George, R., and Keith, H. (1969). *J. Chem. Phys.*, **51**, 1791.
Wexler, S. (1973). *Ber. Bunsen. Phys. Chem.*, **77**, 606.
Wigner, E. (1948). *Phys. Rev.*, **73**, 1002.
Young, C. E., Beuhler, R. J., and Wexler, S. (1974). *J. Chem. Phys.*, **61**, 174.
Young, C. E., Dehmer, P. M., Cohen, R. B., Pobo, L. G., and Wexler, S. (1976a), *J. Chem. Phys.*, **64**, 306.
Young, C. E., Cohen, R. B., Dehmer, P. M., Pobo, L. G., and Wexler, S. (1976b). *J. Chem. Phys.*, **65**, 2562.
Yuan, J.-M. and Micha, D. A. (1976). *J. Chem. Phys.*, **65**, 4876.
Zembekov, A. A. (1971). *Chem. Phys., Lett.*, **11**, 415.
Zembekov, A. A. (1975). *Theor. Exp. Chem. (USSR)*, **9**, 285.
Zyl, van B., Utterback, N. G., and Amme, R. C. (1976). *Rev. Sci. Instr.*, **47**, 814.

AUTHOR INDEX

SUBJECT INDEX